Grundlehren der
mathematischen Wissenschaften 318

A Series of Comprehensive Studies in Mathematics

Springer

Berlin
Heidelberg
New York
Barcelona
Budapest
Hong Kong
London
Milan
Paris
Santa Clara
Singapore
Tokyo

Shoshichi Kobayashi

Hyperbolic Complex Spaces

With 8 Figures

Springer

Shoshichi Kobayashi
Department of Mathematics
University of California
Berkeley, CA 94720
USA
e-mail: kobayash@math.berkeley.edu

Library of Congress Cataloging-in-Publication Data

Kobayashi, Shoshichi, 1932–
Hyperbolic complex spaces / Shoshichi Kobayashi.
p. cm. – (Grundlehren der mathematischen Wissenschaften, ISSN 0072-7830; 318)
Includes bibliographical references and index.
ISBN 3-540-63534-3 (hardcover: alk. paper)
1. Analytic spaces.   2. Holomorphic functions.   3. Distance geometry.
I. Title. II. Series. QA331.K716 1998 515'.94–dc21 98-16060 CIP

Mathematics Subject Classification (1991):
32H20, 32H15, 32H25 (primary)
32H35, 32H02, 32H04, 32H30 (secondary)

ISSN 0072-7830
ISBN 3-540-63534-3 Springer-Verlag Berlin Heidelberg New York

Cover design: MetaDesign plus GmbH, Berlin
Typesetting: Typeset in L^AT_EX by the author and reformatted by Kurt Mattes, Heidelberg, using a Springer T_EX macro-package
SPIN: 10573657   41/3143-5 4 3 2 1 0 Printed on acid-free paper

Table of Contents

Introduction

A Riemann surface is said to be elliptic, parabolic or hyperbolic according as its universal covering space is the Riemann sphere $P_1\mathbf{C}$, the finite plane $\mathbf{C}$ or the unit disc D. Most Riemann surfaces are hyperbolic. In particular, all compact Riemann surfaces of genus ≥ 2 are hyperbolic. From differential geometric view points, the type of a Riemann surface can be characterized by the sign of the curvature of the natural metric it carries. Namely, elliptic with curvature $+1$, parabolic with curvature 0, and hyperbolic with curvature -1. In the compact case, algebraic geometrically, these three types can be characterized by the Kodaira dimensions, -1, 0 and $+1$. We are interested in complex spaces of hyperbolic type since they represent the general case. There are several ways to extend the concept of hyperbolicity to higher dimensional complex spaces.

Our hyperbolicity is based on the existence of a certain intrinsic distance, and this intrinsic distance was originally introduced to generalize Schwarz' lemma to higher dimensional complex spaces. Schwarz' lemma, reformulated by Pick, says that every holomorphic map from a unit disc D of $\mathbf{C}$ into itself is distance-decreasing with respect to the Poincaré distance ρ, and is at the heart of geometric function theory. This lemma has been generalized to higher dimensional complex spaces in various ways.

In 1926 Carathéodory defined an intrinsic pseudo-distance c_X for a domain in $\mathbf{C}^n$ by setting $c_X(p,q) = \sup \rho(f(p), f(q))$, where the supremum is taken over all holomorphic maps f from X to D. If X is a bounded domain, c_X becomes a distance. The Carathéodory distance c_D of the unit disc D coincides with the Poincaré distance ρ, and any holomorphic map $f:(X, c_X) \to (Y, c_Y)$ between two complex spaces X and Y is distance-decreasing.

In 1938, Ahlfors generalized Schwarz' lemma as a comparison theorem between the Poincaré metric ds^2 of D (normalized so that the curvature is -1) and any Hermitian pseudo-metric $d\sigma^2$ of curvature ≤ -1 on D; Ahlfors's version states $d\sigma^2 \leq ds^2$. So if X is any Riemann surface with Hermitian pseudo-metric ds_X^2 of curvature ≤ -1, then every holomorphic map $f:(D, ds^2) \to (X, ds_X^2)$ is metric-decreasing, i.e., $f^*ds_X^2 \leq ds^2$.

In defining the Carathéodory distance of X, one considers the family $\mathrm{Hol}(X, D)$ of holomorphic maps *into* D. But Ahlfors's generalization suggests that it would be more natural to consider the family $\mathrm{Hol}(D, X)$ of holomorphic maps *from* D. Thus, in 1967 I introduced a new intrinsic pseudo-distance d_X by dualizing the Carathéodory's construction. It enjoys the same basic properties of the Carathéodory pseudo-distance. Namely, the intrinsic pseudo-distance d_D for the unit disc coincides with the Poincaré distance ρ, and any holomorphic map $f : (X, d_X) \to (Y, d_Y)$ is distance-decreasing. In fact, d_X can be characterized as the largest pseudo-distance such that all holomorphic maps $(D, \rho) \to (X, d_X)$ are distance-decreasing. In particular, d_X is larger than or equal to c_X, but is strictly larger in many cases. This makes d_X more useful than c_X. For example, if X is a compact complex space, $c_X \equiv 0$ since there are no nonconstant holomorphic functions on X. However, if X is a compact Riemann surface of genus ≥ 2, then d_X coincides with the distance function coming from the natural Hermitian metric of curvature -1. Therefore, as a generalization of the notion of Riemann surface of hyperbolic type, I named a complex space *hyperbolic* if this intrinsic pseudo-distance is actually a distance. (I prefer to put the adjective *hyperbolic* before "complex space" since a "complex hyperbolic space" would often mean a complex space form, i.e., a Kähler manifold of constant negative holomorphic sectional curvature, which is a very special example of hyperbolic complex space.)

This intrinsic pseudo-distance has found a number of applications, including generalizations of Picard's theorems. The little Picard theorem says that an entire function missing two values, say 0, 1, must be constant. This can be best understood in terms of hyperbolicity. Since $d_{\mathbf{C}} \equiv 0$ and $\mathbf{C} - \{0, 1\}$ is hyperbolic, the distance decreasing property of a holomorphic map $f : \mathbf{C} \to \mathbf{C} - \{0, 1\}$ immediately implies the little Picard theorem. The hard part in this argument is to show that $\mathbf{C} - \{0, 1\}$ is hyperbolic. In general, the most satisfactory way of showing that a complex space X admits no nonconstant entire map $f : \mathbf{C} \to X$ is by proving that X is hyperbolic. Since the hyperbolicity is defined only in terms of holomorphic maps from the disc, this is in accord with André Bloch's principle that a result on entire maps $f : \mathbf{C} \to X$ should be derived from results on holomorphic maps from the disc into X. This principle was announced as *"Nihil est in infinito quod non prius fuerit in finito"*, (see Bloch [4; p.2]).

I introduced the intrinsic pseudo-distance d_X in 1967 and published a little monograph *Hyperbolic Manifolds and Holomorphic Mappings* in 1970. This was followed by a long survey article in the Bulletin of the American Mathematical Society (1976). With increasing activities on hyperbolic complex analysis and geometry, in 1973 the Mathematical Reviews created two new subsections *"invariant metrics and pseudodistances"* and *"hyperbolic complex manifolds"* within the section *"analytic mappings"* (which is now called *"holomorphic mappings"*).

Since 1980 several books on intrinsic pseudo-distances and related topics have appeared, each emphasizing certain aspects of the theory:

- T. Franzoni and E. Vesentini, *Holomorphic Maps and Invariant Distances*, 1980.
- J. Noguchi and T. Ochiai, *Geometric Function Theory in Several Complex Variables*, 1984 (English translation in 1990).
- S. Lang, *Introduction to Complex Hyperbolic Spaces*, 1987.
- M. Abate, *Iteration Theory of Holomorphic Maps on Taut Manifolds*, 1989.

- S. Dinen, *The Schwarz Lemma*, 1989.
- M. Jarnicki and P. Pflug, *Invariant Distances and Metrics in Complex Analysis*, 1993.
- In addition, Encyclopaedia of Mathematical Sciences, vol. 9 (1989), *Several Complex Variables*, contains the following two chapters: .
  Chapter III "Invariant Metrics" by E. A. Poletskiĭ and B. V. Shabat,
  Chapter IV "Finiteness Theorems for Holomorphic Maps" by M. G. Zaidenberg and V. Ya. Lin.

A recent undergraduate level book by S. G. Krantz "*Complex Analysis: the Geometric Viewpoint*" (1990) in the Carus Mathematical Monographs series of the Mathematical Association of America is an elementary introduction to function theory from the viewpoint of hyperbolic analysis.

Our aim here is to give a systematic and comprehensive account of the theory of intrinsic pseudo-distances and applications to holomorphic mappings.

Since some of the basic results on holomorphic mappings make use of only metric properties of spaces together with the distance-decreasing property of maps, in Chapter 1 we assemble such results under the heading of Distance Geometry. The reader should skip Chapter 1 on the first reading. It would be best to return to Chapter 1 when it becomes necessary.

In Section 1 of Chapter 2, we prove Ahlfors' generalization of the Schwarz-Pick lemma. In the remainder of the chapter we present further generalizations as well as applications. The reader is again advised to read only up to Theorem (2.1.10) and then to proceed to Chapter 3. The rest of the chapter can be best understood if it is read when it becomes necessary.

Chapter 3, which forms a core of this book, is concerned with basic properties of hyperbolic and hyperbolically imbedded complex spaces and hyperbolicity criteria. The concept of hyperbolic imbedding, introduced in my 1970 monograph to generalize the big Picard theorem, is extensively studied. This concept is also essential in generalizing Montel's theory of normal families in Chapter 5.

Differential geometric criteria for hyperbolicity in terms of negative curvature are often diffcult to use. The most powerful criterion for hyperbolicity is the so-called Brody hyperbolicity. A complex space is said to be Brody-hyperbolic if it admits no nonconstant holomorphic maps from **C**. Every hyperbolic complex space is trivially Brody-hyperbolic. The theorem of Brody says that every compact Brody-hyperbolic complex space is hyperbolic. This may be regarded as the converse to the Bloch principle. Brody's criterion has been extended to a criterion for hyperbolic imbeddedness by Green, Zaidenberg and others. Brody's theorem, combined with Nevanlinna-Cartan theory of value distributions, seems to offer the best approach to the hyperbolicity question. The part of the Nevanlinna-Cartan theory we need for hyperbolicity criteria is summarized in a fairly self-contained manner in Appendix B.

A complex space X is said to be hyperbolic modulo a closed subset Δ if $d_X(p, q) > 0$ for distinct points p, q unless both p and q are in Δ. Although it is generally believed that a generic compact complex space is hyperbolic, concretely given compact complex spaces are often hyperbolic modulo a proper closed complex subspace Δ. The theorem of Brody does not extend to this situation. To find a usable criterion for hyperbolicity modulo a subspace is an important open problem.

Results in Sections 1 through 7 are basic and are used throughout the book. However, Sections 8 through 11 may be regarded as special topics.

Chapter 4 is more or less independent from the rest of the book. Using intrinsic distances, we compare various completeness concepts for domains in $\mathbf{C}^n$. We shall see that for strongly pseudoconvex domains with smooth boundary the Carathéodory and Kobayashi distances have the same boundary behavior. Sections 6 and 7 on extremal discs will be followed by Lempert's theorem that the two distances actually coincide for convex domains. Although some of the results are valid for domains in a complex Banach space, the reader interested in the infinite dimensional case should consult the books by Franzoni-Vesentini and by Dinen mentioned above. Geometry and analysis of bounded domains, particularly those of pseudoconvex domains are vast and rich areas. However, we have confined ourselves to the area directly touching the intrinsic distances. Our discussion on basic properties of the Bergman metric in Section 10 is more for the sake of comparison with the Carathéodory distance and the Kobayashi distance. It would take another book to give a good account of the Bergman metric.

Main applications of the intrinsic pseudo-distances are to holomorphic mappings, and in Chapter 5 we study first those results that depend largely on the distance-decreasing property of mappings. As in the work of Grauert-Reckziegel, Kaup and Wu, the emphasis in hyperbolic complex analysis was initially laid on normal families of holomorphic mappings. Section 1 may be regarded as a generalization of Montel's theory of normal families to holomorphic maps into higher dimensional spaces. For a satisfactory generalization of Montel's theory, the concept of hyperbolicity is not sufficient. The essence of Montel's theory is crystalized in the equivalence of hyperbolic imbeddedness and taut imbeddedness.

While Section 1 is concerned with topological properties (e.g. compactness or relative compactness) of spaces of holomorphic maps $\mathrm{Hol}(X, Y)$ into hyperbolic or hyperbolically imbedded spaces Y, Section 3 treats complex structures of $\mathrm{Hol}(X, Y)$.

Although we discuss in Section 5 Abate's results on iterates of holomorphic self-maps in hyperbolic complex spaces, we shall not go into a currently active area of complex dynamics in several variables. J. E. Fornaess' *"Dynamics in Several Complex Variables"* (1996) in the CBMS series of the American Mathematical Society presents applications of the intrinsic distance and hyperbolicity to complex dynamics.

The big Picard theorem is the pinnacle of the classical function theory. In Chapter 6, first we generalize the big Picard theorem to higher dimensional spaces as extension theorems for holomorphic maps. We combine these extension theorems with results of Chapter 5 to obtain finer structure theorems for $\mathrm{Hol}(X, Y)$. Various finiteness theorems for holomorphic maps and sections, largely due to

Noguchi, are proved in Sections 6 and 9. These finiteness theorems originate in the Mordell conjecture over function fields, proved by Grauert and Manin, and in Lang's conjectures in Diophantine geometry.

In Chapter 7 we consider the intrinsic measures or volume elements; they are useful in studying equidimensional holomorphic mappings. However, the condition of measure hyperbolicity is often not strong enough to yield interesting results. Most of the results in this chapter make use of the algebraic geometric condition of "general type", which is a little stronger than measure hyperbolicity. Generalizations of these results to the measure hyperbolic case are mostly open problems at this moment. In fact, it is not known if every measure hyperbolic compact complex space is of general type.

Chapter 8 is concerned with value distributions for holomorphic curves (mainly in P_nC). We follow here largely Chern [4], Cowen-Griffiths [1], Shabat [1] and Fujimoto [13]. We shall not go into value distrubutions for holomorphic maps from C^m with $m > 1$. In proving the theorem of Bloch-Ochiai in Section 9 of Chapter 3 we made use of part of the classical Nevanlinna theory. In Section 10 of Chapter 3 we used E. Borel's theorem which generalizes the little Picard theorem to a system of entire functions. Although the proof of Borel's theorem does not require Nevanlinna theory, we prefer to derive it as a consequence of the general defect relation.

An important application of the intrinsic distance we were not able to include in this book is the theorem of Royden on the Teichmüller metric. For this we refer the reader to the book of F. P. Gardiner *Teichmüller Theory and Quadratic Differentials*, (1987).

During the preparation of this manuscript, I was partially supported by the Japan Society for Promotion of Sciences (in 1990 at Hokkaido University) and the Alexander von Humbold Foundation (in 1992 and 1993 at Technische Universität in Berlin). I was a guest also at International Christian University in Tokyo, University of Tokyo, Seoul National University, Postech in Pohang, Academia Sinica in Taipei and Keio University. I would like to express my gratitude for hospitality to these institutions and my hosts at these institutions – Professors Haruo Suzuki, Udo Simon, Masakiti Kinukawa, Takushiro Ochiai, Hong-Jong Kim, Kang-Tae Kim, Shi-shyr Roan, and Yoshiaki Maeda. I would like to thank also the following mathematicians for their critical comments made on part of the manuscript at various stages of the preparation – Professors Marco Abate, Kazuo Azukawa, Akio Kodama, Myung He Kwack, and Junjiro Noguchi.

While working on the manuscript I could not help reminiscing good old days of scissors and paste. The computer has kept me revising the manuscript for the past two years.

Berkeley, January 1998   *S. Kobayashi*

Chapter 1. Distance Geometry

1 Pseudo-distances

Let X be a set. A **pseudo-distance** d on X is a function on $X \times X$ with values in the non-negative real numbers satisfying the following axioms:

D1	$d(p, q) = 0$ if $p = q$;	
D2	$d(p, q) = d(q, p)$,	(symmetry axiom);
D3	$d(p, r) \leq d(p, q) + d(q, r)$,	(triangular inequality).

A pseudo-distance d is called a **distance** if it satisfies, in addition to D2 and D3, the following

D1$'$ $d(p, q) = 0$ if and only if $p = q$.

A pseudo-distance d on X induces a topology on X in a natural manner. This topology is Hausdorff if and only if d is a distance.

In our applications, X will be a complex space. In such a case, X has two topologies, the complex space topology and the d-topology, i.e., the topology induced by d.

We consider only arcwise connected Hausdorff topological spaces and assume the following axiom.

D4.   If X is a topological space, then $d: X \times X \to \mathbf{R}$ is continuous.

In other words, the topology of X is at least as fine as the d-topology. For any of the pseudo-distances we construct in Chapter 3, D4 is satisfied.

Let X be a topological space with a pseudo-distance d. Given a curve $\gamma(t), a \leq t \leq b$, in X, the **length** $L(\gamma)$ of γ is defined by

$$(1.1.1) \qquad L(\gamma) = \sup \sum_{i=1}^{k} d(\gamma(t_{i-1}), \gamma(t_i)),$$

where the supremum is taken over all partitions $a = t_0 < t_1 < \ldots < t_k = b$ of the interval $[a, b]$. A curve γ is said to be **rectifiable** if its length $L(\gamma)$ is finite. The space (X, d) is said to be **finitely arcwise connected** if every pair of points p, q of X can be joined by a rectifiable curve. Then we define a new pseudo-distance d^i, called the **inner pseudo-distance** induced by d, by setting

(1.1.2)
$$d^i(p, q) = \inf L(\gamma),$$

where the infimum is taken over all d-rectifiable curves γ joining p and q. When X is a complex space, we modify the definition above by taking the infimum over all d-rectifiable, *piecewise differentiable* (C^1) curves γ joining p and q. This modification will remove technical difficulties which would arise when we discuss the Carathéodory pseudo-distance. The results in this section are valid with this modified definition of d^i as long as maps between complex spaces are assumed to be holomorphic.

From the definition of d^i it follows immediately that

(1.1.3)
$$d(p, q) \le d^i(p, q) \qquad \text{for} \quad p, q \in X.$$

The following proposition is in Rinow [1; p. 120].

(1.1.4) **Proposition.** *Let (X, d) be finitely arcwise connected. Then*

$$L^i(\gamma) = L(\gamma) \qquad \textit{for all rectifiable curves} \quad \gamma,$$

where L^i is the length defined by the induced inner pseudo-distance d^i.

Proof. Consider a partition $a = t_0 < t_1 < \ldots < t_k = b$ for the curve $\gamma(t)$, $a \le t \le b$. Let γ_j be the portion of γ corresponding to the interval $t_{j-1} \le t \le t_j$. Then

$$L(\gamma) = \sum_{j=1}^{k} L(\gamma_j).$$

From the definition of d^i, we obtain

$$d^i(\gamma(t_{j-1}), \gamma(t_j)) \le L(\gamma_j).$$

Hence,

$$\sum_{j=1}^{k} d^i(\gamma(t_{j-1}), \gamma(t_j)) \le L(\gamma).$$

Taking the supremum of the left hand side over all partitions of the interval $[a, b]$, we obtain

$$L^i(\gamma) \le L(\gamma).$$

The reverse inequality is obvious from $d \le d^i$. $\qquad\qquad\square$

(1.1.5) **Corollary.** *Let (X, d) be finitely arcwise connected. Then*

$$(d^i)^i = d^i.$$

We say that a pseudo-distance d is **inner** if $d^i = d$. The corollary above shows that the induced inner pseudo-distance d^i is inner, thus justifying our terminology.

Let (X, d) be finitely arcwise connected. It is said to be **without detour** (*ohne Umweg* in Rinow [1]) if for every point $p \in X$ and for every positive number ε, there exists a positive number δ such that every point $q \in X$ with $d(p, q) < \delta$ can be joined to p by a rectifiable curve γ of length $L(\gamma) < \varepsilon$.

Since $d \le d^i$, the topology induced by d^i is, in general, finer than the one induced by d. However, we have (Rinow [1; p. 119])

(1.1.6) Proposition. *Let (X, d) be finitely arcwise connected. Then d and d^i define the same topology on X if and only if (X, d) is without detour.*

Proof. It is clear from the definition above that (X, d) is without detour if and only if every ε-neighborhood of p with respect to d^i contains some δ-neighborhood of p with respect to d. $\qquad\qquad\square$

If d is inner, i.e., $d^i = d$, then it is without detour by (1.1.6). But in this case we can say a little more; if $d(p, q) < \varepsilon$, then q can be joined to p by a rectifiable curve γ of length $L(\gamma) < \varepsilon$. Hence,

(1.1.7) Proposition. *If d is inner, then for every $p \in X$ and for every positive real number ε, the open ε-ball $U(p; \varepsilon) = \{q \in X;\ d(p, q) < \varepsilon\}$ with center p is finitely arcwise connected. In fact, every $q \in U(p; \varepsilon)$ can be joined to p by a curve γ of length $L(\gamma) < \varepsilon$ lying in $U(p; \varepsilon)$.*

Let Δ be a closed subset of a topological space X with a pseudo-distance d. For $\delta > 0$, the δ-neighborhood of Δ is defined to be

$$U(\Delta; \delta) = \bigcup_{p \in \Delta} U(p; \delta).$$

Clearly, $U(\Delta; \delta)$ is an open neighborhood of Δ. However, given an open neighborhood V of Δ, there may or may not exist a δ-neighborhood $U(\Delta; \delta)$ contained in V.

Let $\{\Delta_i\}$ be a family of mutually disjoint closed subsets in X. We say that d is a **non-degenerate outside** $\{\Delta_i\}$ if $d(p, q) > 0$ unless $p = q$ or $p, q \in \Delta_i$ for some i. By collapsing each Δ_i into a single point and denoting it $[\Delta_i]$, we obtain a quotient space $X/\{\Delta_i\}$. The pseudo-distance d on X induces a distance, denoted also d, on $X/\{\Delta_i\}$. There are two topologies on $X/\{\Delta_i\}$, namely the quotient topology induced from the given topology of X and the metric topology defined by d. The former is finer. But we can say a little more.

(1.1.8) Proposition. *Let X be locally compact and d an inner pseudo-distance on X. Let $\{\Delta_i\}$ be a family of mutually disjoint closed subsets of X such that d is non-degenerate outside $\{\Delta_i\}$. Let $\Delta = \bigcup \Delta_i$. Then*

(1)   for every point $p \in X - \Delta$ and for every neighborhood $U \subset X - \Delta$ of p, there exists a δ-neighborhood V of p with respect to d such that $V \subset U$;

(2)   if each $\{\Delta_i\}$ is compact, the d-topology on $X/\{\Delta_i\}$ coincides with the quotient topology of $X/\{\Delta_i\}$;

(3)   if d is an inner distance (i.e. Δ is empty), it induces the given topology of X.

We note that (1) is stronger than the assertion that $d|_{X-\Delta}$ defines the given (relative) topology on $X - \Delta$. It says that both the quotient topology and the

d-topology give the same neighborhood systems for every point $[p] \in X/\{\Delta_i\}$ which is not of the type $[\Delta_i]$, i.e., for every point $[p]$ coming from $p \in X - \Delta$. However, the quotient topology gives a finer neighborhood system for the point $[\Delta_i]$ than the d-topology. In (2) we show that if Δ_i is compact, both topologies give the same neighborhood system for $[\Delta_i]$.

Proof. We follow the argument of Barth [3] who proved (3).

(1)   Since X is locally compact, there is a relatively compact open neighborhood W of p in X with $\bar{W} \subset U$. Let δ be the minimum value of the positive continuous function $d(p, \cdot)$ on the compact set $\partial W = \bar{W} - W$, and let $V = \{q \in X;\ d(p, q) < \delta\}$. Then $V \cap \partial W = \emptyset$. Since V is connected by (1.1.7) and $p \in W \cap V$, we have $V \subset W \subset U$.

(2)   If Δ_i is compact, we can find a relatively compact neighborhood W of Δ_i, and the rest of the argument is the same as in (1).

(3)   This is a special case of (2) where Δ is empty. $\square$

Let d be a distance function on X. We say that (X, d) is **Cauchy complete** or simply **complete** if every Cauchy sequence (with respect to d) converges. If every closed ball $B(o; r) = \{p \in X;\ d(o, p) \leq r\}$ with $o \in X$ and $r > 0$ is compact, then (X, d) is said to be **strongly complete** or **finitely compact**. Following Hristov [2] we say that (X, d) is **weakly complete** if for every point $o \in X$ there is an $r > 0$ (which depends on o) such that $B(o; r)$ is compact. It is clear that if X is locally compact and if d induces the given topology of X, then (X, d) is weakly complete. We shall later show the converse, see (1.1.10).

(1.1.9) **Proposition.** *Let d be a distance on a locally compact space X.*

(1)   *Then we have the following implications*:

$$\text{strongly complete} \Rightarrow \text{complete} \Rightarrow \text{weakly complete.}$$

(2)   *If d is inner, then completeness implies strong completeness.*

Proof. (1)   complete $\Rightarrow$ weakly complete.

Assume that (X, d) is not weakly complete. Then there is a point $o \in X$ such that for every $r > 0$ the ball $B(o; r)$ is not compact. For each natural number n, take a sequence of points $\{p_{nj}\}_{j=1}^{\infty}$ in $B(o; 1/n)$ without accumulation points.

We note that all sequences of of the type $\{q_n = p_{nj_n}\}_{n=1}^{\infty}$, where j_n are arbitrary natural numbers, are Cauchy and converges to o.

Fix a compact neighborhood U of o. For each fixed j, let N_j be the smallest integer such that $p_{nj} \in U$ for all $n > N_j$ (so $p_{N_j j} \notin U$). We put $A = \sup_j N_j \leq \infty$.

If $A = \infty$, then there is a subsequence $\{j(\lambda)\}_{\lambda=1}^{\infty}$ of $\{j\}_{j=1}^{\infty}$ such that $N_{j(\lambda)} \nearrow \infty$. Then $p_{N_{j(\lambda)} j(\lambda)} \notin U$ contradicts the fact that the sequence $\{p_{n_{j(\lambda)} j(\lambda)}\}_{\lambda=1}^{\infty}$ converges to o.

If $A < \infty$, take $n > A$. Then $\{p_{nj}\}_{j=1}^{\infty}$ are in a compact set $U \cap B(o; 1/n)$ and must have an accumulation point in $B(o; 1/n)$, which is also a contradiction.

(2)   complete $\Rightarrow$ strongly complete when d is inner.

Since d is an inner distance, it induces the given topology of X, see (1.1.8).

Lemma. *$B(o; r)$ is compact if there is a positive number b such that $B(p; b)$ is compact for every $p \in B(o; r)$.*

Proof of Lemma. It suffices to show that if such a positive number b exists and if $B(o; s)$ is compact, then $B(o; s + \frac{b}{2})$ is compact. Let $p_1, p_2, \ldots$ be points of $B(o; s + \frac{b}{2})$. Since d is inner, we can find points $q_i \in B(o; s)$ such that $d(p_i, q_i) < 3b/4$. Since $B(o; s)$ is compact, we may assume (by taking a subsequence) that $q_1, q_2, \ldots$ converges to some point $q \in B(o; s)$. Then $B(q; b)$ contains all p_i for large i. Since $B(q; b)$ is compact, a suitable subsequence of $p_1, p_2, \ldots$ converges to a point p of $B(q; b)$. Since $B(o; s + \frac{b}{2})$ is closed, p is in $B(o, s + \frac{b}{2})$. This completes the proof of Lemma.

In order to complete the proof of (2) we shall show that there is a positive number b such that $B(p; b)$ is compact for all $p \in X$. Assume the contrary. Then there is a point $p_1 \in X$ such that $B(p_1; \frac{1}{2})$ is noncompact. Applying lemma to $B(p_1; \frac{1}{2})$ we see that there is a point $p_2 \in B(p_1; \frac{1}{2})$ such that $B(p_2; \frac{1}{2^2})$ is noncompact. In this way we obtain a Cauchy sequence $p_1, p_2, \ldots$ such that $p_k \in B(p_{k-1}; \frac{1}{2^{k-1}})$ with noncompact $B(p_k; \frac{1}{2^k})$. Let p be the limit point of this Cauchy sequence. Let a be a positive number such that $B(p; a)$ is compact. For a sufficiently large k, $B(p_k; \frac{1}{2^k})$ is a closed set contained in $B(p; a)$ and hence must be compact. This is a contradiction. □

It should be pointed out that when (X, d) is not (Cauchy) complete, its completion X^* with respect to d need not be strongly complete. This is due to the fact that X^* need not be locally compact even if X is.

The following is due to Hristov [2].

(1.1.10) **Proposition**. *Let X be locally compact and d a distance function on X. If (X, d) is weakly complete, then d induces the given topology of X.*

Proof. Assume that the given topology is strictly finer than the d-topology. Then there exist a point $p \in X$ and an open neighborhood U of p in the given topology such that for every $\varepsilon > 0$ there is a point $q \in B(p; \varepsilon)$ with $q \notin U$. Since (X, d) is weakly complete, there is an $\varepsilon_0 > 0$ such that $B(p; \varepsilon_0)$ is compact. For $\varepsilon = \varepsilon_0/n$, we have

$$q_n \in B(p; \varepsilon_0/n) \subset B(p; \varepsilon_0), \quad q_n \notin U.$$

Since $B(p; \varepsilon_0)$ is compact, taking a subsequence we may assume that $\{q_n\}$ converges to a point q in $B(p; \varepsilon)$. Since we are assuming that d is continuous in the given topology, we have

$$d(p, q) = \lim d(p, q_n) = 0.$$

Hence, $p = q = \lim q_n$, in contradiction to $q_n \notin U$. □

The following proposition is obvious.

(1.1.11) **Proposition**. *Let X and X_i, $i \in I$, be subsets of a Hausdorff space Y such that $X = \bigcap_i X_i$. Let d and d_i be distances on X and X_i such that $d(p, q) \geq$*

$d_i(p, q)$ *for* $p, q \in X$. *If each* X_i *is finitely compact (resp. complete) with respect to* d_i, *then* X *is finitely compact (resp. complete) with respect to* d.

Proof. For $o \in X$ and $r > 0$, let $B(o; r) = \{p \in X; \ d(o, p) \leq r\}$ and $B_i(o; r) = \{p \in X_i; \ d_i(o, p) \leq r\}$. If each X_i is finitely compact with respect to d_i, then $K := \bigcap_i B_i(o; r)$ is a compact subset of $\bigcap_i X_i = X$. Since $B(o; r) \subset B_i(o; r)$ for all i, $B(o; r)$ is a closed subset of K, and hence is compact. This proves that X is finitely compact with respect to d. The proof for the second statement is even more trivial. $\qquad\square$

Let $f: X \to Y$ be a continuous map between topological spaces. Given a pseudo-distance d on Y, we can define the **induced pseudo-distance** $f^{-1}d$ on X by setting

$$(f^{-1}d)(p, q) = d(f(p), f(q)) \qquad \text{for} \quad p, q \in X.$$

Even if d is inner, $f^{-1}d$ needs not be inner. We define the **induced inner pseudo-distance** f^*d by

$$f^*d = (f^{-1}d)^i.$$

It is not hard to see that f^*d may be defined directly in the following manner. For two points $p, q \in X$, let γ be a curve from p to q. Let $L(f(\gamma))$ be the length of the curve $f(\gamma)$ with respect to d. Then

$$(1.1.12) \qquad\qquad (f^*d)(p, q) = \inf L(f(\gamma)),$$

where the infimum is taken over all curves γ connecting p to q.

From the construction of the induced inner pseudo-distance the following is evident.

$$d_Y(f(p), f(q)) \leq (f^*d_Y)(p, q) \qquad \text{for} \quad p, q \in X.$$

(1.1.13) Proposition. *Let* (X, d_X) *and* (Y, d_Y) *be topological spaces with inner pseudo-distances. If a continuous map* $f: X \to Y$ *has the property that every point* $p \in X$ *has a neighborhood* U *such that*

$$d_Y(f(q), f(r)) = d_X(q, r) \qquad \text{for} \quad q, r \in U,$$

then $d_X = f^*d_Y$.

Proof. Let $p, q \in X$, and $\gamma(t), a \leq t \leq b$, be a curve from p to q. Then

$$
\begin{aligned}
d_X(p, q) &= \inf_\gamma L(\gamma) = \inf_\gamma \sup \sum d_X(\gamma(t_{i-1}), \gamma(t_i)) \\
&= \inf_\gamma \sup \sum d_Y(f(\gamma(t_{i-1})), f(\gamma(t_i))) \\
&= \inf_\gamma L(f(\gamma)) = f^*d_Y(p, q),
\end{aligned}
$$

where the infimum is taken over all curves γ from p to q while the supremum is taken over all partitions $a = t_0 < t_1 < \ldots < t_k = b$. $\qquad\square$

2 Degeneracy of Inner Pseudo-distances

Let X be a topological space and d a pseudo-distance. We define an equivalence relation $\mathcal{R} \subset X \times X$ by

$$(1.2.1) \qquad \mathcal{R} = \{(p, q) \in X \times X; \ d(p, q) = 0\}$$

and obtain a quotient space $X^* = X/\mathcal{R}$ with a naturally induced distance d^*.

We define the **degeneracy set** for p by

$$(1.2.2) \qquad \Delta(p) = \{q \in X; \ d(p, q) = 0\}.$$

By collapsing $\Delta(p)$ into a single point p^*, we obtain X^*.

We can easily verify the following

(1.2.3) Proposition. *If d is an inner pseudo-distance on X, then d^* is an inner distance on X^*.*

The quotient topology on X^* is, in general, finer than the metric topology defined by d^*. If X^* is locally compact, the two topologies coincide by (1.1.8). However, local compactness of X does not, in general, guarantee that of X^*.

We say that a pseudo-distance d has **compact degeneracy** if $\Delta(p)$ is compact for every $p \in X$.

(1.2.4) Proposition. *If X is a locally compact space with a pseudo-distance d with compact degeneracy, then X^* is locally compact.*

(1.2.5) Corollary. *If X is a locally compact space with an inner pseudo-distance d with compact degeneracy, then the induced inner-distance d^* defines the quotient topology of X^*.*

(1.2.6) Proposition. *If X is a locally compact space with an inner pseudo-distance d with compact degeneracy, then the natural projection $j \colon X \to X^*$ is a proper map.*

Proof. Let $K \subset X^*$ be compact and let $p_n \in j^{-1} K$ be an infinite sequence. Then a subsequence of $\{j(p_n)\}$ converges to a point $q_0^* \in K$. Since $j^{-1}(q_0^*) = \Delta(q_0)$ is compact, it has a compact neighborhood $B(q_0; \delta) = j^{-1}(B(q_0^*; \delta))$. Then a subsequence of $\{p_n\}$ converges to a point of $B(q_0; \delta)$. $\qquad\square$

In spite of (1.1.7), $\Delta(p)$ may not be connected in general. However, compactness of $\Delta(p)$ again guarantees connectedness.

(1.2.7) Proposition. *Let d be an inner pseudo-distance on a locally compact space X. If $\Delta(p)$ is compact, then $\Delta(p)$ is connected.*

Proof. Suppose that $\Delta(p)$ is not connected, and let Δ_0 be the component containing p, and let U be a compact neighborhood of Δ_0 which meets no other components of $\Delta(p)$. Let ∂U be the boundary of U, and let δ be the distance between p and ∂U with respect to d. Since ∂U is compact and does not meet $\Delta(p)$, the distance δ is positive. Let q be any point of $\Delta(p)$ not in Δ_0. Any curve γ joining

p to q must go through the boundary ∂U, and so has length at least δ. This is a contradiction. $\qquad\square$

The proof of the following proposition is straightforward.

(1.2.8) Proposition. *Given two topological spaces X and Y with pseudo-distances d_X and d_Y, respectively, define a pseudo-distance $d_{X\times Y}$ on $X \times Y$ by*

$$d_{X\times Y}((x, y), (x', y')) = \max\{d_X(x, x'),\ d_Y(y, y')\}$$

$$for \quad (x, y), (x', y') \in X \times Y.$$

Then

$$(X \times Y)^* = X^* \times Y^*.$$

(1.2.9) Proposition. *Let (X, d_X) and (Y, d_Y) be two topological spaces with pseudo-distances. If $f: X \to Y$ is a distance-decreasing map, it induces a distance-decreasing map $f: X^* \to Y^*$ between the induced metric spaces.*

3 Mappings into Metric Spaces

Some of the results on holomorphic maps are direct consequences of purely topological results on maps into metric spaces. In this section we shall collect such topological results which will be used later.

Given two topological spaces X and Y, we denote by $\mathcal{C}(X, Y)$ the space of all continuous maps $f: X \to Y$ equipped with the compact-open topology. If Y is a metric space, then the compact-open topology coincides with the topology of uniform convergence on compact sets.

Let Y be a metric space with distance function d_Y. Let $\mathcal{F} \subset \mathcal{C}(X, Y)$. The family $\mathcal{F}$ is said to be **equicontinuous** at $x \in X$ if for every $\varepsilon > 0$ there exists a neighborhood U of x such that $d_Y(f(x), f(x')) < \varepsilon$ for all $x' \in U$ and all $f \in \mathcal{F}$.

(1.3.1) Arzela-Ascoli Theorem. *Let X be a locally compact, separable space and Y a locally compact metric space with distance function d_Y. Then a family $\mathcal{F} \subset \mathcal{C}(X, Y)$ is relatively compact in $\mathcal{C}(X, Y)$ (i.e., every sequence of maps $f_n \in \mathcal{F}$ contains a subsequence which converges to some map $f \in \mathcal{C}(X, Y)$ uniformly on every compact subset of X) if and only if*

   (a)   *$\mathcal{F}$ is equicontinuous at every point $x \in X$;*
   (b)   *for every $x \in X$, the set $\{f(x);\ f \in \mathcal{F}\}$ is relatively compact in Y.*

Proof. Assume that $\mathcal{F}$ is relatively compact. If (a) does not hold, there would exist an $\varepsilon > 0$, a sequence $x_n \to x$ and a sequence $f_n \in \mathcal{F}$ such that $d_Y(f_n(x), f_n(x_n)) \geq \varepsilon$. If $f \in \mathcal{C}(X, Y)$ is the limit to which a subsequence of $\{f_n\}$ converges, then we would have $d_Y(f(x), f(x)) \geq \varepsilon$, which is a contradiction. In order to prove (b), consider a sequence $\{f_n(x);\ f_n \in \mathcal{F}\}$ in Y. Choose a subsequence $\{f_{n_k}\}$ of $\{f_n\}$ that converges to some element f of $\mathcal{C}(X, Y)$. Then $\{f_{n_k}(x)\}$ converges to $f(x)$.

Assume (a) and (b). Let $\{x_k\}$ be a dense sequence of points in X. Given a sequence $\{f_n\}$ in $\mathcal{F}$, we are going to extract a subsequence which converges at all points x_k. By (b), we can find an array of indices

$$n_{11} < n_{12} < n_{13} < \dots$$
$$n_{21} < n_{22} < n_{23} < \dots$$
$$n_{31} < n_{32} < n_{33} < \dots$$

such that each row is a subequence of the preceding one and such that $\lim_j f_{n_{kj}}(x_k)$ exists for all k. Then the diagonal subsequence $\{f_{n_{jj}}\}$ converges at all points x_k.

We denote the subsequence $\{f_{n_{jj}}\}$ by $\{f_j\}$. For every $x \in X$, we have

$$d_Y(f_m(x), f_n(x))$$
$$\leq d_Y(f_m(x), f_m(x_k)) + d_Y(f_m(x_k), f_n(x_k)) + d_Y(f_n(x_k), f_n(x)).$$

Given $\varepsilon > 0$, because of (a) we can find x_k near x so that the first and third terms on the right become less than ε. Since f_n converges at all x_k, the middle term can be made also smaller than ε for large m and n. Hence, $\{f_n(x)\}$ is a Cauchy sequence. By (b) this Cauchy sequence has a limit. Thus we have a map $f: X \to Y$ to which $\{f_n\}$ converges pointwise. In order to show that f is continuous, let $q \in X$. Given $\varepsilon > 0$, let U_q be a neighborhood of q such that $d_Y(f_n(q), f_n(x)) < \varepsilon$ for all $x \in U_q$ and all n. (Such a neighborhood U_q exists because of (a)). Fix $x \in U_q$. Then there exists an integer n such that $d_Y(f_n(x), f(x)) < \varepsilon$. Taking n large, we may also assume that $d_Y(f(q), f_n(q)) < \varepsilon$. Then

$$\begin{aligned}
d_Y(f(q), f(x)) &\leq d_Y(f(q), f_n(q)) + d_Y(f_n(q), f_n(x)) + d_Y(f_n(x), f(x)) \\
&< 3\varepsilon.
\end{aligned}$$

This proves that f is continuous at q.

We shall complete the proof by showing that the convergence $f_n \to f$ is uniform on every compact set K. Let $\varepsilon > 0$. For each $q \in K$, choose an integer n_q such that $d_Y(f_n(q), f(q)) < \varepsilon$ for $n > n_q$. Let U_q be the neighborhood of q chosen above to prove the continuity of f. Then for any $x \in U_q$ and $n > n_q$, we have

$$\begin{aligned}
d_Y(f_n(x), f(x)) &\leq d_Y(f_n(x), f_n(q)) + d_Y(f_n(q), f(q)) + d_Y(f(q), f(x)) \\
&< \varepsilon + \varepsilon + 3\varepsilon = 5\varepsilon.
\end{aligned}$$

Now K can be covered by a finite number of U_q's, say U_{q_i}, $i = 1, \dots, s$. It follows that if $n > \max_i\{n_{q_i}\}$, then $d_Y(f_n(x), f(x)) < 5\varepsilon$ for all $x \in K$. $\qquad\square$

Let X and Y be locally compact, separable spaces with pseudo-distances d_X and d_Y, respectively. Let $\mathcal{D}(X, Y)$ denote the family of **distance-decreasing** maps $f \in \mathcal{C}(X, Y)$, i.e., maps f such that

$$d_Y(f(x), f(x')) \leq d_X(x, x') \qquad \text{for all} \quad x, x' \in X.$$

Then $\mathcal{D}(X, Y)$ is closed in $\mathcal{C}(X, Y)$.

(1.3.2) Corollary. *Let X be a locally compact, separable space with a pseudo-distance d_X. Let Y be a locally compact metric space with distance function d_Y. Then a subfamily $\mathcal{F} \subset \mathcal{D}(X, Y)$ is relatively compact in $\mathcal{C}(X, Y)$ if and only if, for every $x \in X$, the set $\{f(x); \ f \in \mathcal{F}\}$ is relatively compact in Y.*

Proof. Since $f \in \mathcal{F}$ is distance-decreasing, the equicontinuity condition (a) is automatically satisfied. $\qquad\square$

(1.3.3) Corollary. *Let X and Y be as in (1.3.2) and assume further that Y is strongly complete. Then a subfamily $\mathcal{F} \subset \mathcal{D}(X, Y)$ is relatively compact in $\mathcal{C}(X, Y)$ if and only if, for some $x_0 \in X$, the set $\{f(x_0); \ f \in \mathcal{F}\}$ is relatively compact in Y.*

Proof. Assume that for some $q \in X$ the set $\{f(q); \ f \in \mathcal{F}\}$ is contained in a compact subset S of Y. Let $x \in X$ and $a = d_X(q, x)$. Then the set $\{f(x); \ f \in \mathcal{F}\}$ is contained in $\{y \in Y; \ d_Y(y, S) \le a\}$, which is compact because Y is strongly complete. $\qquad\square$

(1.3.4) Corollary. *Let X and Y be as in (1.3.2) and assume further that Y is strongly complete. Then $\mathcal{D}(X, Y)$ is locally compact.*

Proof. Let K be a compact subset of X, and V a relatively compact open subset of Y. Let $\mathcal{F} = \{f \in \mathcal{D}(X, Y); \ f(K) \subset V\}$. By (1.3.3) $\mathcal{F}$ is relatively compact. $\qquad\square$

A sequence $f_n \in \mathcal{C}(X, Y)$ is said to be **compactly divergent** if given any compact subsets $K \subset X$ and $L \subset Y$ there exists an integer N such that $f_n(K) \cap L = \emptyset$ for all $n \ge N$.

The following proposition is essentially in Kaup [4].

(1.3.5) Proposition. *Let X be a locally compact, separable space and Y a locally compact metric space. For a closed family $\mathcal{F} \subset \mathcal{C}(X, Y)$ the following four conditions are mutually equivalent:*

(a) Every sequence in $\mathcal{F}$ is either compactly divergent or contains a convergent subsequence;

(b) The canonical map

$$\Phi: X \times \mathcal{F} \to X \times Y,$$

defined by $\Phi(x, f) = (x, f(x))$, is proper;

(c) For all compact subsets $K \subset X$ and $L \subset Y$, the subfamily $\mathcal{F}_{K,L} = \{f \in \mathcal{F}; \ f(K) \cap L \ne \emptyset\}$ is compact;

(d) If $Y^* = Y \cup \{\infty\}$ denotes the one-point compactification of Y and if ∞ denotes also the constant map which sends X to ∞, then $\mathcal{F} \cup \{\infty\}$ is a compact subset of $\mathcal{C}(X, Y^*)$.

Proof. $(a) \Rightarrow (b)$. Let $\{x_n\} \subset X$ and $\{f_n\} \subset \mathcal{F}$ such that $\{x_n\}$ and $\{f_n(x_n)\}$ are both convergent. Then $\{f_n\}$ cannot be compactly divergent. Hence, $\{f_n\}$ has a convergent subsequence. This proves that Φ is proper.

$(b) \Rightarrow (c)$. Let $\pi: X \times \mathcal{F} \to \mathcal{F}$ be the projection. Since

$$\Phi^{-1}(K \times L) = \{(x, f) \in K \times \mathcal{F}; \ f(x) \in L\},$$

we have $\mathcal{F}_{K,L} = \pi(\Phi^{-1}(K \times L))$. Since $\Phi^{-1}(K \times L)$ is assumed to be compact, $\mathcal{F}_{K,L}$ is also compact.

$(c) \Rightarrow (d)$. A sequence $\{f_n\} \subset \mathcal{F}$ converges to ∞ in $C(X, Y^*)$ if and only if for given compact subsets $K \subset X$ and $L \subset Y$ there exists an integer N such that $f_n(K) \cap L = \emptyset$ for $n > N$. Assume that $\{f_n\}$ does not converge to ∞. Then there exist compact subsets $K \subset X$ and $L \subset Y$ such that $f_n(K) \cap L \neq \emptyset$ for infinitely many n. Hence, $f_n \in \mathcal{F}_{K,L}$ for infinitely many n. Since $\mathcal{F}_{K,L}$ is compact, a subsequence of $\{f_n\}$ converges to an element of $\mathcal{F}_{K,L} \subset \mathcal{F}$.

$(d) \Rightarrow (a)$. Any sequence $\{f_n\} \subset \mathcal{F}$ converges either to an element of $\mathcal{F}$ or to the constant map ∞. In the latter case, it is compactly divergent. $\qquad \square$

A family $\mathcal{F} \subset C(X, Y)$ is said to be **normal** if its closure $\overline{\mathcal{F}}$ in $C(X, Y)$ satisfies one of the equivalent conditions in (1.3.5). We prove the following result due to Wu [1].

(1.3.6) Theorem. *Let X and Y be as in (1.3.2) and assume further that Y is strongly complete. Then $\mathcal{D}(X, Y)$ is a normal family.*

Proof. Let $\mathcal{F} = \{f_n\} \subset \mathcal{D}(X, Y)$ be a sequence which is not compactly divergent. Then there exist compact subsets $K \subset X$ and $L \subset Y$ such that for infinitely many f_n of $\mathcal{F}$, we have $f_n(K) \cap L \neq \emptyset$. By taking a subsequence we may assume that this holds for all $f_n \in \mathcal{F}$. Let δ be the diameter of K with respect to d_X. Then the diameter of $f_n(K)$ with respect to d_Y is at most δ. Hence, all $f_n(K)$ are contained in the closed δ-neighborhood of L. Choose $x \in K$ and apply (1.3.3). Then $\mathcal{F}$ is relatively compact and contains a subsequence which converges in $C(X, Y)$. $\qquad \square$

(1.3.7) Remark. We note that (1.3.6) is stronger than (1.3.4). In fact, as we see from (1.3.5), *if $\mathcal{F} \subset C(X, Y)$ is normal, then its closure $\overline{\mathcal{F}}$ is locally compact*, (see Wu [1]).

For later use we state an immediate consequence of (1.3.2) in the following form.

(1.3.8) Theorem. *Let X be a locally compact, separable space with a pseudo-distance d_X. Let $Y \subset Z$ be locally compact metric spaces with distance functions d_Y and d_Z such that $d_Z \leq d_Y$ on Y. Then $\mathcal{D}(X, Y)$ is relatively compact in $C(X, Z)$ if and only if, for each $x \in X$, the set $\{f(x); \ f \in \mathcal{D}(X, Y)\}$ is relatively compact in Z.*

Proof. Apply (1.3.2) to the subfamily $\mathcal{F} = \mathcal{D}(X, Y) \subset C(X, Y)$. $\qquad \square$

(1.3.9) Corollary. *Let X with d_X and $Y \subset Z$ with d_Y and d_Z be as in (1.3.8). If Y is relatively compact in Z, then $\mathcal{D}(X, Y)$ is relatively compact in $C(X, Z)$.*

Let Y be a locally compact metric space with distance d_Y, and $Y^* = Y \cup \infty$ the one-point compactification of Y. If Y is not strongly complete with respect to d_Y, then we can extend d_Y to a distance function d_{Y^*} on Y^* by setting

(1.3.10) $d_{Y^*}(y, \infty) = \sup\{r; \ \{y' \in Y; \ d_Y(y, y') \leq r\} \ \text{ is compact}\}.$

(1.3.11) **Theorem.** *Let Y be a locally compact metric space with distance d_Y, and $Y^* = Y \cup \infty$ the one-point compactification of Y. Let X be a locally compact space with pseudo-distance d_X. Then $\mathcal{D}(X, Y)$ is relatively compact in $\mathcal{C}(X, Y^*)$, i.e., every sequence $f_n \in \mathcal{D}(X, Y)$ has a subsequence which converges to a map $f \in \mathcal{C}(X, Y^*)$. If Y is strongly complete with respect to d_Y, then every sequence $f_n \in \mathcal{D}(X, Y)$ either has a subsequence which converges to a map $f \in \mathcal{C}(X, Y)$ or converges to the constant map ∞.*

Proof. (i). Consider first the case where Y is strongly complete. By (1.3.6) $\mathcal{D}(X, Y)$ is a normal family. By (1.3.5), every sequence $f_n \in \mathcal{D}(X, Y)$ either contains a subsequence which converges in $\mathcal{C}(X, Y)$ or is compactly divergent, i.e., converges to the constant map ∞.

(ii). Assume that Y is not strongly complete. Let Y^* be its one-point compactification and d_{Y^*} the extended distance function, see (1.3.10). Then the assertion follows from (1.3.9). $\square$

Given a pseudo-distance d_Y on a topological space Y and a closed subset Δ of Y, we say that d_Y is a **distance modulo** Δ if $d_Y(y, y') > 0$ unless $y = y'$ or $y, y' \in \Delta$.

(1.3.12) **Proposition.** *Let X and Y be topological spaces with inner pseudo-distances d_X and d_Y, respectively. Let Δ be a (possibly empty) closed subset of Y. Assume that d_Y is a distance modulo Δ. Let $f: X \to Y$ be a distance-decreasing mapping. If any one of the following three conditions is satisfied, then d_X is a distance modulo $f^{-1}(\Delta)$.*

(1)   Given $x, x' \in X, x \neq x'$ with $f(x) = f(x')$, there is a neighborhood V of $f(x)$ in Y such that x and x' are in different connected components of $f^{-1}(V)$;

(2)   Every $x \in X$ has a neighborhood U such that f is a homeomorphism from U onto an open set $f(U)$;

(3) X is locally compact, and for every $y \in Y$ the inverse image $f^{-1}(y)$ is finite.

Proof. Let $x, x' \in X$ with $x \neq x'$ and $x \notin f^{-1}(\Delta)$. If $f(x) \neq f(x')$, then $d_X(x, x') \geq d_Y(f(x), f(x')) > 0$. We consider the case where $f(x) = f(x')$, and set $y = f(x) = f(x')$. Assume $d_X(x, x') = 0$.

(1)   Let V be a neighborhood of $f(x)$ in Y described in condition (1). Since $f(x) \notin \Delta$, without loss of generality we may assume that V is an ε-neighborhood of $f(x)$ for some $\varepsilon > 0$. Let U be the ε-neighborhood of x in X. Then U contains x' and is finitely arcwise connected by (1.1.7). Let γ be a rectifiable curve from x to x' in U. Since f is distance-decreasing, f maps U into V, and the curve γ lies in $f^{-1}(V)$. This is a contradiction.

(2)   Let U be a neighborhood of x described in condition (2). Let $V = f(U)$, and apply (1).

(3)   Let U be an open neighborhood of x with compact closure $\bar{U}$ such that $\bar{U} \cap f^{-1}(\Delta) = \emptyset$ and $f^{-1}(y) \cap \bar{U} = \{x\}$. Then $d_X(x, \partial U) \geq d_Y(y, f(\partial U)) > 0$.

Since any curve from x to x' has length at least $d_X(x, \partial U)$, this is a contradiction.
$\square$

(1.3.13) Corollary. *Let X and Y be topological spaces with an inner pseudo-distances d_X and d_Y, respectively. If $f: X \to Y$ is a distance-decreasing covering projection and if d_Y is a complete distance, so is d_X.*

Proof. We have only to check the statement concerning the completeness. Assume that Y is complete with respect to d_Y. Let $\{x_n\}$ be a Cauchy sequence in X. Since f is distance-decreasing, $\{f(x_n)\}$ is also a Cauchy sequence in Y. Let $q \in Y$ be the limit point of $\{f(x_n)\}$. Let V be a 2ε-neighborhood of q such that V is homeomorphic to each connected component of $f^{-1}(V)$. Let V' be the ε-neighborhood of q. Since $f(x_n) \in V'$ for $n > N$, all $\{x_n\}_{n>N}$ are contained in one of the connected components, say U, of $f^{-1}(V)$. Let $p \in U$ be the point such that $f(p) = q$. Then $\{x_n\}$ converges to p.
$\square$

We say that a distance d_Y modulo Δ is **complete modulo** Δ if for each Cauchy sequence $\{y_n\}$ in Y with respect to d_Y, we have one of the following:

(a) $\{y_n\}$ converges to a point q in Y;

(b)   for every open neighborhood V of Δ in Y, there exists an integer n_0 such that $y_n \in V$ for $n > n_0$.

(1.3.14) Corollary. *Let X and Y be topological spaces with inner pseudo-distances d_X and d_Y, respectively. Let Δ be a compact subset of Y. If $f: X \to Y$ is a distance-decreasing proper finite-to-one map and if d_Y is a complete distance modulo Δ, then d_X is a complete distance modulo $f^{-1}(\Delta)$.*

Proof. We have only to check the statement concerning the completeness. Let $\{x_n\}$ be a Cauchy sequence in X with respect to d_X. Consider first the case the Cauchy sequence $\{f(x_n)\}$ is not convergent. Given a neighborhood U of $f^{-1}(\Delta)$ in X, we take a neighborhood V of Δ such that $f^{-1}(V) \subset U$. Let n_0 be an integer such that $f(x_n) \in V$ for $n > n_0$. Then $x_n \in U$ for $n > n_0$.

Next, we consider the case $\{f(x_n)\}$ converges to a point q in Y. If $q \in \Delta$, we argue as in case (b) above. So we assume that $q \notin \Delta$. We take a compact neighborhood V of q which is disjoint from Δ so that d_X is a distance on $f^{-1}(V)$. Let $f^{-1}(q) = \{p_1, \ldots, p_k\}$. Let V be a compact neighborhood of q. Since $f^{-1}(V)$ is compact, a subsequence of $\{x_n\}$ converges to one of $\{p_1, \ldots, p_k\}$, say p_1. Being a Cauchy sequence, $\{x_n\}$ must converge to p_1.
$\square$

4 Norms and Indicatrices

The results in this section will be used only in Section 5 of Chapter 3.

Let V be an n-dimensional complex vector space, and V^* its dual space. Let F be a real nonnegative function defined on a subset of V such that if F is defined at $v \in V$, it is defined at tv for all $t \in \mathbf{C}$ and

$$(1.4.1) \qquad\qquad F(tv) = |t| F(v).$$

We allow F to take the value ∞. We call such a function F a **quasi-norm**. We do not assume that F is defined on all of V. Nor do we assume that F is continuous.

Let

$$(1.4.2) \qquad \Gamma(F) = \{v \in V; \; F(v) \text{ is defined and } F(v) \le 1\}.$$

Then $\Gamma(F)$ is a **star-shaped circular subset** of V in the sense that if $v \in \Gamma(F)$ then $tv \in \Gamma(F)$ for $|t| \le 1$. We call $\Gamma(F)$ the **indicatrix** of F.

Conversely, given a star-shaped circular subset Γ of V, there is a unique quasi-norm F such that $\Gamma = \Gamma(F)$.

We note that $F(v_0) = 0$ if and only if the complex line $\mathbf{C}v_0 = \{tv_0; \; t \in \mathbf{C}\}$ is contained in $\Gamma(F)$, while $F(v_0) = \infty$ if and only if no points of the complex line $\mathbf{C}v_0$, except the origin 0, are in $\Gamma(F)$.

A quasi-norm F on V is called **pseudo-norm** if it satisfies the following convexity condition:

$$(1.4.3) \qquad F(u + v) \le F(u) + F(v) \qquad u, v \in V.$$

This convexity condition is equivalent to the convexity of the indicatrix Γ. A pseudo-norm F is always continuous on V. A pseudo-norm F is called a **norm** if $0 < F(v) < \infty$ for all nonzero $v \in V$.

Given a quasi-norm F on V and the corresponding star-shaped circular subset $\Gamma = \Gamma(F)$, we define the **dual quasi-norm** F^* on the dual space V^* by

$$(1.4.4) \qquad F^*(\lambda) = \sup_{v \in \Gamma} |\lambda(v)| \qquad \text{for} \quad \lambda \in V^*.$$

Clearly the dual quasi-norm F^* is defined everywhere on V^*. Whether F satisfies the convexity condition (1.4.3) or not, the dual quasi-norm F^* always satisfies the convexity condition:

$$(1.4.5) \qquad F^*(\lambda + \mu) \le F^*(\lambda) + F^*(\mu) \qquad \lambda, \mu \in V^*.$$

The indicatrix $\Gamma^* = \Gamma(F^*)$ of the norm F^* is not only star-shaped and circular but also convex. It is given also as an intersection of closed circular cylinders:

$$(1.4.6) \qquad \Gamma^* = \bigcap_{v \in \Gamma} \{\lambda \in V^*; \; |\lambda(v)| \le 1\}.$$

(1.4.7) Proposition. (1) *$F^*(\lambda) > 0$ for all nonzero $\lambda \in V^*$ if and only if $F(e_1) < \infty, \ldots, F(e_n) < \infty$ for some basis $e_1, \ldots, e_n$ of V;*

(2) $F^*(\lambda) < \infty$ for all $\lambda \in V^*$ if and only if Γ is bounded;

(3) If a quasi-norm F is positive (i.e., $0 < F(v) \le \infty$ for all nonzero $v \in V$) and satisfies the convexity condition (1.4.3), then $F^*(\lambda) < \infty$ for all $\lambda \in V^*$.

Proof. (1) Let U be the subspace of V spanned by Γ. Given $\lambda \in V^*$, $F^*(\lambda) = 0$ if and only if $\lambda(v) = 0$ for all $v \in U$. Hence, $F^*(\lambda) > 0$ for all nonzero $\lambda \in V^*$ if and only if $U = V$. On the other hand, $U = V$ if and only if $F(e_1) < \infty, \ldots, F(e_n) < \infty$ for some basis $e_1, \ldots, e_n$ of V.

(2) If Γ is bounded so that $\overline{\Gamma}$ is compact, then

$$\sup_{v \in \Gamma} |\lambda(v)| = \max_{v \in \overline{\Gamma}} |\lambda(v)| < \infty.$$

If Γ is not bounded, let $v_1, v_2, \ldots$ be a unbounded sequence of points in Γ. In terms of a basis $e_1, \ldots, e_n$ of V we write $v_\alpha = \sum_{i=1}^n a_\alpha^i e_i$. Then the sequence $a_1^i, a_2^i, \ldots$ is unbounded for some i. Define $\lambda \in V^*$ by $\lambda(\sum x^j e_j) = x^i$. Then $F^*(\lambda) = \infty$.

(3) It suffices to prove that Γ is bounded. Let

$$V' = \{v \in V; \ F(v) < \infty\}.$$

Since F satisfies the convexity condition, V' is a vector subspace of V. Clearly, $\Gamma \subset V'$. Since $F|_{V'}$ is a norm on V', the indicatrix Γ is a bounded subset of V'. $\square$

We identify the double dual V^{**} of V, i.e., the dual space of V^*, with V in a natural manner.

(1.4.8) Proposition. *If F is a quasi-norm on V, then $F^{**} = (F^*)^*$ is a pseudo-norm on $V = V^{**}$, and $F^{**} \leq F$.*

Proof. By (1.4.5), F^{**} satisfies the convexity condition and hence is a pseudo-norm. Let $v \in V$ and $\lambda \in V^*$ such that $F(v)$ is defined. From the definition of F^* we obtain $|\lambda(v)| \leq F^*(\lambda) F(v)$. Hence,

$$\sup_{\lambda \in \Gamma^*} |\lambda(v)| \leq F(v).$$

From the definition of F^{**} we obtain $F^{**}(v) \leq F(v)$. $\square$

Each $\lambda \in V^*$ defines a closed circular cylinder

$$C(\lambda) = \{v \in V; \ |\lambda(v)| \leq 1\},$$

which is obviously a convex set.

(1.4.9) Lemma. *If a closed subset $S \subset V$ is circular (i.e., invariant under scalar multiplication by any complex number of absolute value 1), then its convex hull $\hat{S}$ is given as an intersection of all closed circular cylinders that contain S.*

Proof. Let $\tilde{S}$ be the intersection of all closed circular cylinders that contain S. Since it is convex, we have $\hat{S} \subset \tilde{S}$. Let $v_0 \in V$ be a point not contained in $\hat{S}$. Then there is a real hyperplane separating v_0 from S. In terms of a coordinate system $v = (z^1, \ldots, z^n)$, $z^k = x^k + iy^k$, of V and the dual coordinate system in V^*, the real hyperplane may be given in the form:

$$\operatorname{Re}(\lambda(v)) = \sum a_k x^k - \sum b_k y^k = 1,$$

where $\lambda = (a_1 + ib_1, \ldots, a_n + ib_n) \in V^*$, and S is in the half-space defined by $\operatorname{Re}(\lambda(v)) \leq 1$ while $\operatorname{Re}(\lambda(v_0)) > 1$. Since S is circular, $\operatorname{Re}(\lambda(cv)) \leq 1$ for $v \in S$ and $c \in \mathbf{C}$, $|c| = 1$. Hence, $|\lambda(v)| \leq 1$ for $v \in S$. On the other hand,

$|\lambda(v_0)| \geq \mathrm{Re}(\lambda(v_0)) > 1$. This shows that the circular cylinder $C(\lambda)$ contains S but not v_0. $\qquad\square$

(1.4.10) Proposition. *Given a quasi-norm F on V, the indicatrix Γ^{**} of F^{**} is the intersection of all closed circular cylinders that contain Γ and hence is the convex hull $\hat{\Gamma}$ of Γ.*

Proof. From $F^{**} \leq F$ in (1.4.8), we obtain the inclusion $\Gamma \subset \Gamma^{**}$.

Let $v_0 \in V$, and suppose that there is a circular cylinder $C(\lambda)$ containing Γ but not containing v_0. Then $\lambda \in \Gamma^*$. Since $|\lambda(v_0)| > 1$, we conclude $v_0 \notin \Gamma^{**}$. $\qquad\square$

(1.4.11) Corollary. *If F is a pseudo-norm on V so that Γ is convex, then $\Gamma^{**} = \Gamma$ and $F^{**} = F$.*

(1.4.12) Proposition. *Let V be an n-dimensional complex vector space. Let F be a quasi-norm on V with indicatrix Γ, and F^{**} its double-dual with indicatrix Γ^{**}. Then*

(1)   Every element $v \in \Gamma^{**}$ is in the simplex determined by the origin of V and at most $2n$ elements of Γ, i.e.,

$$v = \sum_{i=1}^{m} t_i u_i \qquad \text{with} \quad t_i \geq 0, \quad \sum t_i \leq 1, \quad u_i \in \Gamma,$$

where $m \leq 2n$;

(2)   Given $v \in \Gamma^{**}$ and $\varepsilon > 0$, there exist $v_1, \ldots, v_m \in \Gamma$ with $m \leq 2n$ such that

$$v = v_1 + \ldots + v_m \quad \text{and} \quad F(v_1) + \ldots + F(v_m) \leq F^{**}(v) + \varepsilon.$$

(1.4.13) Remark. In (2), if $F^{**}(v) > 0$ we can find $v_1, \ldots, v_m \in \Gamma$, $(m \leq 2n)$, such that

$$v = v_1 + \ldots + v_m \quad \text{and} \quad F(v_1) + \ldots + F(v_m) \leq F^{**}(v).$$

Then by $F^{**}(v_i) \leq F(v_i)$ we actually have an equality:

$$F(v_1) + \ldots + F(v_m) = F^{**}(v).$$

We start the proof with the following lemma due to Carathéodory.

(1.4.14) Lemma. *Let V be a real vector space, and Γ a subset containing the origin $0 \in V$. Let $\hat{\Gamma}$ denote the convex hull of Γ. Then $v \in \hat{\Gamma}$ if and only if v is contained in a finite dimensional simplex having its vertices in Γ and having 0 as one of its vertices.*

Proof. We first prove Lemma in the special case when Γ is a finite set, say $\Gamma = \{0, v_1, \ldots, v_m\}$. The proof is by induction on m. Let $v \in \hat{\Gamma}$ so that

$$v = (1 - \sum_{i=1}^{m} t_i) \cdot 0 + \sum_{i=1}^{m} t_i v_i \quad \text{with} \quad t_i \geq 0, \quad \sum t_i \leq 1.$$

If $v_1, \ldots, v_m$ are linearly independent, there is nothing to prove. If not, we have a nontrivial relation

$$s_1 v_1 + \ldots + s_m v_m = 0.$$

We may assume that $\sum s_i \geq 0$ (by multiplying the equation by -1 if necessary). Looking at

$$t_1 - as_1, \ldots, t_m - as_m,$$

we let a increase from 0 until one of $t_i - as_i$ becomes 0 for the first time, and let a be that value. Without loss of generality we may assume that $t_m - as_m = 0$. Then

$$v = \sum_{i=1}^{m} t_i v_i - a \sum_{i=1}^{m} s_i v_i = \sum_{i=1}^{m-1} (t_i - as_i) v_i$$

and

$$t_i - as_i \geq 0 \quad \text{and} \quad \sum_{i=1}^{m-1} (t_i - as_i) = \sum_{i=1}^{m} t_i - a \sum_{i=1}^{m} s_i \leq 1.$$

This shows that v is in the convex hull of $\{0, v_1, \ldots, v_{m-1}\}$, thus completing the induction.

Now we prove Lemma in the general case. We consider r-dimensional simplices, each with its vertices in Γ and one of its vertices at 0. We denote the union of all such simplices by S_r and set $S = \bigcup_r S_r$. Then S is convex. To prove this, consider a point z on the line segment joining two points v and w of S. Since v is in a simplex with vertices, say $0, v_1, \ldots, v_k$, and w is in a simplex with vertices, say $0, v_{k+1}, \ldots, v_m$, the point z is in the convex hull of $\{0, v_1, \ldots, v_m\}$. (The sets $\{v_1, \ldots, v_k\}$ and $\{v_{k+1}, \ldots, v_m\}$ need not be disjoint). Using the finite case of Lemma just proved, we see that z is in S. Hence, $\hat{\Gamma} \subset S$. Since $S_r \subset \hat{\Gamma}$ for all r, we have $S \subset \hat{\Gamma}$. Thus, we have $S = \hat{\Gamma}$. $\qquad\square$

Proof of (1.4.12). Part (1) clearly follows from Lemma.

For any positive real number r, we write $r\Gamma$ for the set $\{rv; v \in \Gamma\}$. Similarly, we set $r\hat{\Gamma} = \{rv; v \in \hat{\Gamma}\}$. Then $r\hat{\Gamma}$ is the convex hull of $r\Gamma$.

We prove (2). Given $v \in \Gamma^{**} = \hat{\Gamma}$, let $r = F^{**}(v)$. Then $v \in (r + \varepsilon)\hat{\Gamma}$ for any $\varepsilon > 0$. By Lemma there exist $u_1, \ldots, u_m \in (r + \varepsilon)\Gamma$ with $m \leq 2n$ such that

$$v = \sum_{i=1}^{m} t_i u_i \quad \text{with} \quad t_i > 0 \quad \sum t_i \leq 1.$$

Then

$$\sum F(t_i u_i) = \sum t_i F(u_i) \leq (r + \varepsilon) \sum t_i \leq (r + \varepsilon).$$

By setting $v_i = t_i u_i$, we obtain the desired inequality. $\qquad\square$

We shall show that if $F^{**}(v) > 0$ then the stronger inequality indicated in Remark (1.4.13) holds.

Let $F^{**}(v) = r > 0$. Then $v \in r\hat{\Gamma}$. By Lemma, there exist $u_1, \ldots, u_m \in r\Gamma$ such that

$$v = \sum t_i u_i \quad \text{with} \quad t_i > 0, \quad \sum t_i \leq 1.$$

The remainder of the argument is the same as in the proof of (2) above.

Chapter 2. Schwarz Lemma and Negative Curvature

1 Schwarz Lemma

In this section we prove Ahlfors' generalization of the classical Schwarz lemma
in function theory of one complex variable and its variants. For the general theory
of intrinsic distances, we need only the classical Schwarz-Pick lemma in the form
(2.1.7).

Let X be a Riemann surface, i.e., a 1-dimensional complex manifold. Let

$$d\sigma^2 = 2\lambda dz d\bar{z}$$

be a Hermitian pseudo-metric on X expressed in terms of a local coordinate z,
and let

$$\omega = i\lambda dz \wedge d\bar{z}$$

be its **associated Kähler form**. The term **pseudo-metric** means that $d\sigma^2$ is only
positive semidefinite, i.e., $\lambda \geq 0$.

We recall the notation

$$d^c = i(d'' - d') \quad \text{so that} \quad dd^c = 2id'd''.$$

To ω we asscoiate the **Ricci form**

$$(2.1.1) \qquad\qquad \mathrm{Ric}(\omega) = -dd^c \log \lambda = 2K\omega,$$

where

$$(2.1.2) \qquad\qquad K = -\frac{1}{\lambda}\frac{\partial^2 \log \lambda}{\partial z \partial \bar{z}}$$

is the (Gaussian) **curvature** of $d\sigma^2$. Both $\mathrm{Ric}(\omega)$ and K are defined wherever λ
is positive.

Let D_a denote the open disc of radius a in the Gaussian plane $\mathbf{C}$,

$$D_a = \{z \in \mathbf{C};\ |z| < a\}.$$

Then the **Poincaré metric**

$$(2.1.3) \qquad\qquad ds_a^2 = \frac{4a^2 dz d\bar{z}}{A(a^2 - |z|^2)^2}, \qquad (A > 0)$$

on D_a is complete and has curvature $-A$. In the special case where $a = 1$, the unit disc D_1 will be denoted D and the Poincaré metric ds_1^2 with $A = 1$ will be denoted ds^2.

Let φ be the Kähler form of the Poincaré metric ds^2. Then

$$\mathrm{Ric}(\varphi) = -2\varphi,$$

since the curvature K is -1.

We prove a generalization of Schwarz-Pick Lemma by Ahlfors [1].

(2.1.4) Theorem. *Let ds^2 denote the Poincaré metric of curvature -1 on the unit disc D. Let $d\sigma^2$ be any Hermitian pseudo-metric on D whose curvature is bounded above by -1. Then*

$$d\sigma^2 \leq ds^2.$$

In terms of the Ricci forms $\mathrm{Ric}(\varphi)$ and $\mathrm{Ric}(\omega)$ of ds^2 and $d\sigma^2$ respectively, (2.1.4) may be stated as follows:

$$(2.1.4)' \qquad\qquad \mathrm{Ric}(\omega) \leq -2\omega \Rightarrow \omega \leq \varphi.$$

As we shall see in the proof, the theorem holds if $d\sigma^2$ is only continuous at zero points of $d\sigma^2$ and is twice differentiable at the points where it is positive (and hence the curvature is defined). This technical point is important in applications. Ahlfors proved the theorem for an upper semicontinuous $d\sigma^2$ with "supporting pseudo-metric". This will be explained later.

Proof. Let D_a be the disc of radius a with the Poincaré metric ds_a^2 of curvature -1 given by

$$ds_a^2 = 2\mu_a dz d\bar{z} \qquad \text{where} \quad \mu_a = 2a^2/(a^2 - |z|^2)^2.$$

We compare this metric with $d\sigma^2 = 2\lambda dz d\bar{z}$. Let u_a be the nonnegative function on D_a defined by

$$u_a = \lambda/\mu_a, \quad \text{i.e.,} \quad d\sigma^2 = u_a ds_a^2.$$

The problem is to show that $u_1 \leq 1$ on D. From the explicit expression for μ_a, it follows that $u_a(z_o) \to u_1(z_o)$ at every point $z_o \in D$ as $a \to 1$. Hence, it suffices to show that $u_a \leq 1$ on D_a for every $a < 1$. From the explicit expression for μ_a we see that $u_a(z) \to 0$ as z approaches the boundary of D_a. Therefore, u_a must attain its maximum in the interior of D_a, say at $z_o \in D_a$. If $u_a(z_o) = 0$, then $u_a \equiv 0$, and there is nothing to prove. So we assume $u_a(z_o) > 0$ and calculate the complex Hessian of $\log u_a$ at z_o. Since $u_a = \lambda/\mu_a$ and since the curvature of ds_a^2 is -1, we have

$$\frac{\partial^2 \log u_a}{\partial z \partial \bar{z}} = \frac{\partial^2 \log \lambda}{\partial z \partial \bar{z}} - \frac{\partial^2 \log \mu_a}{\partial z \partial \bar{z}}$$
$$= -\lambda K - \mu_a = \mu_a(-u_a K - 1).$$

Since the complex Hessian of $\log u_a$ must be nonpositive at the maximum point z_o, we obtain the inequality $-u_a(z_o)K(z_o) - 1 \le 0$. Hence, $u_a(z_o) \le -1/K(z_o) \le 1$. Since $u_a(z_o)$ is the maximum value of u_a, we have $u_a \le 1$ everywhere on D_a.

$\square$

(2.1.5) Corollary. *Let X be a Riemann surface with a Hermitian pseudo-metric ds_X^2 whose curvature (wherever defined) is bounded above by -1. Then every holomorphic map $f : D \to X$ is distance-decreasing, i.e.,*

$$f^* ds_X^2 \le ds^2.$$

Proof. Set $d\sigma^2 = f^* ds_X^2$. Then $d\sigma^2$ is a Hermitian pseudo-metric on D. If we denote the curvature of ds_X^2 by K_X, then the curvature of $d\sigma^2$ is given by $f^* K_X$; this is clear from the definition of the curvature (2.1.2). Now, (2.1.5) follows from (2.1.4).

$\square$

The following classical Schwarz-Pick lemma is immediate from (2.1.5).

(2.1.6) Corollary. *Let D be the unit disc with the Poincaré metric ds^2. Then every holomorphic map $f : D \to D$ is distance-decreasing, i.e.,*

$$f^* ds^2 \le ds^2.$$

Comparing the coefficients of $dz d\bar{z}$ in the inequality in (2.1.6) we can write

$$\frac{|f'(z)|}{1 - |f(z)|^2} \le \frac{1}{1 - |z|^2}, \qquad \text{for} \quad z \in D.$$

This is the usual form of Schwarz-Pick lemma.

If f is a holomorphic automorphism of D, then (2.1.6) applied to both f and f^{-1} implies that f is an isometry, i.e., $f^* ds^2 = ds^2$.

Let ρ denote the **Poincaré distance** on D, i.e., the distance function defined by the Poincaré metric ds^2. We shall find the explicit formula for ρ. Let $0 < a < 1$. If $z(t) = x(t) + iy(t), 0 \le t \le 1$, is a curve in D joining the origin $0 \in D$ to $a \in D$, its arc-length l with respect to ds^2 satisfies

$$
\begin{aligned}
l &= \int_0^1 \frac{2(x'(t)^2 + y'(t)^2)^{\frac{1}{2}}}{1 - x(t)^2 - y(t)^2} dt \\
&\ge \int_0^1 \frac{2|x'(t)|}{1 - x(t)^2} dt \ge \int_0^a \frac{2dx}{1 - x^2} = \log \frac{1+a}{1-a}.
\end{aligned}
$$

This shows that the ordinary line segment from 0 to a is the shortest path and

$$\rho(0, a) = \log \frac{1+a}{1-a}.$$

Since the Poincaré metric is invariant under the rotations, we have

$$\rho(0, a) = \log \frac{1 + |a|}{1 - |a|} = 2 \tanh^{-1} |a| \qquad \text{for} \quad a \in D.$$

Given two points a and b in D, the transformation

$$w = \frac{z - b}{1 - \bar{b}z}$$

is an automorphism of D that sends b to 0 and a to $(a - b)/(1 - a\bar{b})$. From the invariance of ρ we obtain

$$\rho(a, b) = \log \frac{|1 - a\bar{b}| + |a - b|}{|1 - a\bar{b}| - |a - b|} = 2 \tanh^{-1} \left| \frac{a - b}{1 - a\bar{b}} \right|.$$

However, we shall rarely use this explicit expression for ρ.

The integrated form of (2.1.6) reads as follows:

$$(2.1.7) \qquad\qquad \rho(f(a), f(b)) \leq \rho(a, b) \qquad \text{for} \quad a, b \in D.$$

In order to weaken the differentiability assumption on $d\sigma^2$ in (2.1.4), we first prove the following

(2.1.8) **Proposition.** *Let $d\sigma_0^2$ and $d\sigma_1^2$ be two smooth Hermitian metrics with curvature K_0 and K_1 defined in a neighborhood of a point $z_0 \in D$. If $d\sigma_0^2 \leq d\sigma_1^2$ in a neighborhood of z_0 with equality holding at z_0, then $K_1 \leq K_0$ at z_0.*

Proof. Let $d\sigma_i^2 = 2\lambda_i dz d\bar{z}$, and $u = \lambda_0/\lambda_1$. As in the proof of (2.1.4), we calculate $\partial^2 \log u / \partial z \partial \bar{z}$. Then using (2.1.2) we obtain

$$\frac{1}{\lambda_1} \frac{\partial^2 \log u}{\partial z \partial \bar{z}} = -u K_0 + K_1.$$

We evaluate this at z_0. Since u attains the maximum value 1 at z_0, we have $0 \geq -K_0 + K_1$ at z_0. $\qquad\square$

Let $d\sigma^2$ be an upper semicontinuous Hermitian pseudo-metric on D. A pseudo-Hermitian metric $d\sigma_0^2$ is called a **supporting pseudo-metric** for $d\sigma^2$ at $z_0 \in D$ if it is defined and of class C^2 in a neighborhood U of z_o and satisfies the following condition:

$$d\sigma^2 \geq d\sigma_o^2 \quad \text{in} \quad U, \quad \text{and with equality at} \quad z_0.$$

If $d\sigma^2$ is not smooth, we define its curvature $K_{d\sigma^2}$ by

$$(2.1.9) \qquad\qquad K_{d\sigma^2}(z_0) = \inf K_{d\sigma_o^2}(z_0),$$

where the infimum is taken over all supporting pseudo-metrics $d\sigma_o^2$ for $d\sigma^2$ at z_0.

Then (2.1.4) is generalized to non-smooth metrics, (Ahlfors [1]):

(2.1.10) **Theorem.** *Let ds^2 be the Poincaré metric on D, and $d\sigma^2$ an upper semicontinuous Hermitian pseudo-metric on D with its curvature ≤ -1. Then*

$$d\sigma^2 \leq ds^2.$$

The proof of (2.1.10) is exactly the same as that of (2.1.4). Since $d\sigma^2$ is upper semicontinuous, the existence of a maximum for the function u_a is assured. If z_o is a maximum point for $u_a = d\sigma^2/ds_a^2$, it is also a maximum point for the function $v_a = d\sigma_o^2/ds_a^2$ (where $d\sigma_o^2$ is a supporting pseudo-metric for $d\sigma^2$ at z_o). Then we have only to calculate the Hessian of $\log v_a$ instead of $\log u_a$. $\square$

(2.1.11) **Remark**. The classical Schwarz lemma asserts also that if an equality holds in (2.1.6) at one point of D, then the equality holds everywhere. This part of Schwarz lemma has been also generalized by Heins [1]. Namely, if an equality holds in (2.1.10) at one point of D, then $d\sigma^2 = ds^2$ everywhere on D.

For a finite family of of pseudo-metrics, (2.1.4) may be generalized (see $(2.1.4)'$) as follows (Cowen-Griffiths [1]):

(2.1.12) **Theorem**. *If a family of pseudo-metrics $d\sigma_k^2 = 2\lambda_k dz d\bar{z}$, $(k = 1, \ldots, N)$, on D with Kähler forms $\omega_k = i\lambda_k dz \wedge d\bar{z}$, has its curvatures collectively bounded by -1 in the sense that*

$$\sum \mathrm{Ric}(\omega_k) \leq -2 \sum \omega_k,$$

then the pseudo-metric

$$d\sigma^2 = 2\lambda dz d\bar{z}, \quad \text{where} \quad \lambda = (\lambda_1 \ldots \lambda_N)^{1/N}$$

has its curvature bounded by -1 and, consequently,

$$d\sigma^2 \leq ds^2.$$

Proof. Let $\omega = i\lambda dz \wedge d\bar{z}$ be the Kähler form of $d\sigma^2$. Then

$$
\begin{aligned}
-\mathrm{Ric}(\omega) &= dd^c \log \lambda = \frac{1}{N} \sum dd^c \log \lambda_k \\
&= \frac{1}{N} \sum \mathrm{Ric}(\omega_k) \geq \frac{2}{N} \sum \omega_k \\
&= \frac{2i}{N} \sum \lambda_k dz \wedge d\bar{z} \geq 2i(\lambda_1 \ldots \lambda_N)^{1/N} dz \wedge d\bar{z} = 2\omega,
\end{aligned}
$$

showing that the curvature of $d\sigma^2$ is bounded by -1. $\square$

The following is also a simple application of the maximum principle.

(2.1.13) **Theorem**. *Let $d\sigma^2$ be an upper semicontinuous Hermitian pseudo-metric on D. Let ξ be a holomorphic vector field on D and $|\xi|$ be the length of ξ measured by $d\sigma^2$. If $d\sigma^2$ has negative curvature in the sense of (2.1.9), then $|\xi|$ cannot attain a maximum in the interior of D unless $|\xi| \equiv 0$.*

Proof. Assume that $|\xi|$ attains a maximum at $z_o \in D$. Without loss of generality we may assume that $d\sigma^2$ is of class C^2 at z_o. (In fact, let $d\sigma_o^2$ be as in (2.1.9) and $|\xi|'$ the length of ξ measured by $d\sigma_o^2$. Then $|\xi|'$ attains a maximum at z_0.)

Let $d\sigma^2 = 2\lambda dz d\bar{z}$. If $\xi = f(\partial/\partial z)$, then $|\xi|^2 = 2\lambda f \bar{f}$. Assuming that $|\xi|^2 > 0$ at z_o, we calculate the Hessian of $\log |\xi|^2$. Thus

$$\frac{\partial^2 \log |\xi|^2}{\partial z \partial \bar{z}} = \frac{\partial^2 \log \lambda}{\partial z \partial \bar{z}} = -\lambda K,$$

where K is curvature of $d\sigma_o^2$. The left hand side is nonpositive at z_o while the right hand side is positive everywhere. This is a contradiction. $\qquad\square$

Now we state the Schwarz lemma for Hermitian metrics with a logarithmic singularity at the origin. This will be useful in logarithmic geometry.

(2.1.14) **Theorem** . *Let ds^2 be the Poincaré metric of curvature -1 on the unit disk. Let $d\sigma^2 = 2\lambda dz d\bar{z}$ be a Hermitian pseudo-metric on the punctured disc D^* such that $|z|^2 d\sigma^2$ becomes upper semicontinuous on D. If the curvature of $d\sigma^2$ is bounded above by $-|z|^2$, then $|z|^2 d\sigma^2 \leq ds^2$.*

Proof. Let $d\tilde{\sigma}^2 = |z|^2 d\sigma^2$. Then the curvature $\tilde{K}$ of $d\tilde{\sigma}^2$ is given by

$$\tilde{K} = -\frac{1}{\tilde{\lambda}} \frac{\partial^2 \log \tilde{\lambda}}{\partial z \partial \bar{z}},$$

where $\tilde{\lambda} = |z|^2 \lambda$. This shows that it is related to the curvature K of $d\sigma$ by $K = |z|^2 \tilde{K}$. Hence, $\tilde{K} \leq -1$, and (2.1.10) applied to $d\tilde{\sigma}^2$ yields $d\tilde{\sigma}^2 \leq ds^2$. $\qquad\square$

The following theorem of Sibony [4] is a version of Schwarz lemma and will be used to define the Sibony pseudo-distance.

(2.1.15) **Theorem**. *Let u be an upper semicontinuous function on D such that* (i) $0 \leq u < 1$, (ii) $\log u$ *is subharmonic,* (iii) $u(0) = 0$, *and* (iv) $u(z)/|z|^2$ *is bounded. Then*
 (1) $u(z) \leq |z|^2$ *for $z \in D$ with equality at some point $\neq 0$ if and only if* $u(z) \equiv |z|^2$;
 (2) *If, in addition, u is of class C^2 in a neighborhood of 0, then $\partial^2 u/\partial z \partial \bar{z} \leq 1$ at 0, with equality if and only if $u(z) \equiv |z|^2$.*

Proof. (1). Since u attains its minimum at 0, du vanishes at 0. Define in D^* the function $v(z) = u(z)/|z|^2$. Since $\log v$ is subharmonic in D^*, v is subharmonic in D^*. Since v is also bounded, there is a subharmonic extension $\tilde{v}$ of v in D. Since $\limsup_{z \to \partial D} v \leq 1$, it follows that $\tilde{v} \leq 1$. Hence $u(z) \leq |z|^2$.
 If $u(z_0) = |z_0|^2$ for some $z_0 \neq 0$, then $\tilde{v}(z_0) = 1$. This implies $\tilde{v}(z) \equiv 1$.
 (2). Let $z = x + iy$. Denoting partial differentiation by subscripts, we have

$$|z|^2 \geq u(z) = \frac{1}{2}(u_{xx}(0)x^2 + 2u_{xy}(0)xy + u_{yy}(0)y^2) + o(|z|^2).$$

Replacing y by $-y$ and adding the resulting ineqaulity to the above, we obtain

$$u_{xx}(0)x^2 + u_{yy}(0)y^2 + o(|z|^2) \leq 2|z|^2.$$

Hence,

$$u_{z\bar{z}}(0) = \frac{1}{4}(u_{xx}(0) + u_{yy}(0)) \leq 1.$$

If $u_{z\bar{z}}(0) = 1$, then $\tilde{v}(0) = u_{z\bar{z}}(0) = 1$. Hence, $\tilde{v} \equiv 1$. $\square$

We note that in (2.1.15) if u is of class C^2 in a neighborhood of 0, then $u(z)/|z|^2$ is automatically bounded in D.

For historical comments concerning Schwarz-Pick-Ahlfors lemma, see Royden [10].

2 Negatively Curved Riemann Surfaces

The results in this section will be used in Section 7 of Chapter 3.

In the preceding section we showed that the Poincaré metric on the unit disc D has curvature -1. It is sometimes more convenient to use the upper halfplane in place of the disc. Let

$$H = \{w = u + iv \in \mathbf{C}; \ v > 0\}.$$

We have a well known correspondence between the unit disc D and the upper halfplane H. It is given by

$$(2.2.1) \qquad z = \frac{i - w}{i + w} \in D \qquad \text{for} \quad w \in H.$$

Pulling back the Poincaré metric of D by the correspondence above, we obtain the Poincaré metric

$$(2.2.2) \qquad ds_H^2 = \frac{dw\,d\bar{w}}{v^2}$$

of curvature -1 on H. Since it corresponds to the Poincaré metric of D, ds_H^2 is also invariant by the automorphisms of H and is complete.

Let D^* be the **punctured disc**, i.e.,

$$D^* = \{z \in \mathbf{C}; \ 0 < |z| < 1\}.$$

Let $p: H \to D^*$ be the covering projection defined by

$$z = p(w) = e^{2\pi i w} \qquad \text{for} \quad w \in H.$$

Since the Poincaré metric ds_H^2 is invariant by the automorphisms of H, in particular, by the covering transformations $w \mapsto w + n$, $(n \in \mathbf{Z})$, there is a unique metric $ds_{D^*}^2$ on D^* such that $p^*(ds_{D^*}^2) = ds_H^2$. In order to find the explicit form of $ds_{D^*}^2$, we solve $z = p(w)$ for w locally in terms of z:

$$w = \frac{1}{2\pi i} \log z.$$

Substituting this into (2.2.2) we obtain the following complete metric

$$(2.2.3) \qquad ds^2_{D^*} = \frac{4dzd\bar{z}}{|z|^2(\log 1/|z|^2)^2}$$

of curvature -1 on D^*. Its area element, i.e., its Kähler form, is given by

$$\mu_{D^*} = \frac{idz \wedge d\bar{z}}{|z|^2(\log |z|^2)^2}.$$

Although the origin 0 of D is at infinity with respect to the metric $ds^2_{D^*}$, the area around 0 with respect to μ_{D^*} is bounded. More explicitly, let $D^*_a = \{0 < |z| < a\}$ with $a < 1$. Then

$$(2.2.4) \qquad \int_{D^*_a} \mu_{D^*} < \infty.$$

In fact, when we choose

$$F = \{w = u + iv \in H;\ 0 \le u < 1\}$$

as a fundamental domain for D^*, the subset corresponding to D^*_a is given by $F_b = \{w \in F;\ v > b\}$ with a suitable positive number b, and its area is given by

$$\int_{D^*_a} \mu_{D^*} = \int_{F_b} \frac{dudv}{v^2} = \frac{1}{b}.$$

We consider the twice-punctured plane $X = \mathbf{C} - \{0, 1\}$ and show that it carries a complete Hermitian metric with curvature $K \le -1$. This is obvious if we make use of the fact that the elliptic modular function, usually denoted λ, is a covering projection from H onto $\mathbf{C} - \{0, 1\}$, the latter being identified with $SL(2; \mathbf{Z})_2 \backslash H$ where $SL(2; \mathbf{Z})_2 = \{A \in SL(2; \mathbf{Z});\ A \equiv I \bmod 2\}$, (see, for example, Ahlfors [3; p. 269]). The metric induced from the Poincaré metric of H has constant curvature -1. Instead, we shall construct explicitly a complete Hermitian metric ds^2_X on $X = \mathbf{C} - \{0, 1\}$ whose curvature K, although not constant, remains bounded above by -1.

More generally, we consider the Riemann sphere minus k points, $k \ge 3$. Let $Z = (z^0, z^1)$ be a homogeneous coordinate in $P_1\mathbf{C}$, and let

$$z = z^1/z^0$$

be its inhomogeneous coordinate. For $W = (w^0, w^1)$, we set

$$\langle Z, W \rangle = z^0\bar{w}^0 + z^1\bar{w}^1.$$

We use the notaion

$$d^c = i(d'' - d') \quad \text{so that} \quad dd^c = 2id'd''.$$

Then the associated Kähler form of the Fubini-Study metric ds^2 of curvature 1 is given by

$$\Phi = dd^c \log\langle Z, Z\rangle = dd^c \log(1 + |z|^2) = \frac{2i\,dz \wedge d\bar{z}}{(1 + |z|^2)^2}.$$

Given k points $A_j = (a_j^0, a_j^1)$, $j = 1, \ldots, k$, in $P_1\mathbf{C}$, we shall construct explicitly a positive function ρ on $P_1\mathbf{C} - \{A_1, \ldots, A_k\}$ such that

(a) ρ goes to infinity at each of $A_1, \ldots, A_k$ (so that ρds^2 becomes a complete metric on $P_1\mathbf{C} - \{A_1, \ldots, A_k\}$;

(b) the curvature of ρds^2 is bounded above by a negative constant;

(c) the area of $P_1\mathbf{C} - \{A_1, \ldots, A_k\}$ with respect to $\rho\Phi$ is finite, i.e.,

$$\int_{P_1\mathbf{C}-\{A_1,\ldots,A_k\}} \rho\Phi < \infty.$$

To construct such a function, we may assume that none of A_j is the point at infinity, i.e., that $a_j^0 \neq 0$ for all j. We denote the polar of A_j by $A_j^{\perp}$. Thus,

$$A_j^{\perp} = (\bar{a}_j^1, -\bar{a}_j^0),$$

so that $\langle A_j, A_j^{\perp}\rangle = 0$.

Set

$$a_j = a_j^1/a_j^0,$$

and

(2.2.5)
$$\rho = \prod_{j=1}^{k} \frac{1}{\sigma_j(\log c/\sigma_j)^2},$$

where σ_j is the chordal distance between Z and A defined by

$$\sigma_j = \frac{|\langle Z, A_j^{\perp}\rangle|^2}{|Z|^2|A_j|^2} = \frac{|z - a_j|^2}{(1 + |z|^2)(1 + |a_j|^2)} \leq 1$$

and c is a large positive number yet to be determined.

Then (a) is clearly satisfied. To prove (b), write

$$\rho\Phi = 2\lambda i\,dz \wedge d\bar{z} \quad \text{with} \quad \lambda = (1 + |z|^2)^{k-2} \prod_{j=1}^{k} \frac{1 + |a_j|^2}{|z - a_j|^2(\log c/\sigma_j)^2}.$$

Since

$$\frac{\partial^2 \log(\log c/\sigma_j)^2}{\partial z \partial \bar{z}} = \frac{2}{\log c/\sigma_j} \frac{1}{(1 + |z|^2)^2} - \frac{2}{(\log c/\sigma_j)^2}\left|\frac{1}{z - a_j} - \frac{\bar{z}}{1 + |z|^2}\right|^2,$$

the curvature K of $\rho\Phi$ is given by

$$K = -\frac{1}{\lambda}(\mathrm{I} + \mathrm{II} + \mathrm{III}),$$

where

$$\mathrm{I} = \frac{k-2}{(1+|z|^2)^2},$$

$$\mathrm{II} = -\sum \frac{2}{(1+|z|^2)^2 \log c/\sigma_j},$$

$$\mathrm{III} = \sum \frac{2}{(\log c/\sigma_j)^2} \left| \frac{1}{z-a_j} - \frac{\bar{z}}{1+|z|^2} \right|^2.$$

Given $\varepsilon > 0$, we can choose c so large that

$$\left| \sum \frac{2}{\log c/\sigma_j} \right| < \varepsilon.$$

Then $\mathrm{I} + \mathrm{II} > 0$ for $k > 2$. Since (III) is non-negative, it follows that $K < 0$. However,

$$\lim_{z \to a_j} \frac{1}{\lambda} = C_j \lim_{z \to a_j} |z - a_j|^2 (\log c/\sigma_j)^2$$

with a suitable positive constant C_j. Hence,

$$\lim_{z \to a_j} -\frac{1}{\lambda}(\mathrm{I} + \mathrm{II}) = 0.$$

In order to show that K is bounded away from 0, we have to examine terms in (III). Since the j-th term in III is equal to

$$\frac{2}{(\log c/\sigma_j)^2} \left(\frac{1}{|z-a_j|^2} + \frac{|z|^2}{(1+|z|^2)^2} - \frac{1}{1+|z|^2} \left(\frac{z}{z-a_j} + \frac{\bar{z}}{\bar{z}-\bar{a}_j} \right) \right),$$

we have

$$\frac{1}{\lambda}(\mathrm{III}) = 2C_j + \ldots,$$

where the dots $\ldots$ indicate the terms which approach 0 as $z \to a_j$. It follows that K remains away from 0 as $z \to a_j$.

Finally, to prove (c), we consider a neighborhood $U_j = \{|z - a_j| < \varepsilon\}$ of a_j with $\varepsilon < 1$ and write

$$z - a_j = re^{i\theta}.$$

Then

$$\lambda \leq \frac{M}{r^2 (\log r^2)^2} \quad \text{in} \quad U_j.$$

Hence,

$$\int_{U_j} \rho\Phi \leq 4M \int_0^{2\pi} d\theta \int_0^\varepsilon \frac{dr}{r(\log r^2)^2} \leq \frac{2\pi M}{\log 1/\varepsilon} < \infty.$$

This proves (c).

In summary, we have

(2.2.6) Theorem. *Let X be P_1C minus k points, $k \geq 3$. Then X admits a complete Hermitian metric with curvature $K \leq -1$ and finite total area. Such a metric can be obtained by multiplying the Fubini-Study metric of P_1C by a function ρ of the form (2.2.5).*

The construction of the metric given here is a special case of the volume form constructed by Carlson-Griffiths [1]. Earlier constructions of similar merics for $k = 3$ are due to Ahlfors [1], Robinson [1] and Grauert-Reckziegel [1]. Such metrics were used to give "elementary" proofs (i.e., proofs without recourse to the elliptic modular function) of several important results in function theory including Picard's theorems; see Ahlfors's monograph [4] for further details. Though not smooth, the metric constructed by Ahlfors is more explicit and satisfies the conditions of (2.1.10). He used it to obtain a good lower bound for the Bloch constant.

Every compact Riemann surface X of genus ≥ 2 has the upper halfplane H as its universal covering space. Though well known, this is a nontrivial fact. It is obvious from this fact that X carries a Hermitian metric of constant curvature -1. We shall construct a Hermitian metric ds_X^2 with curvature $K \leq -1$ on X in an elementary fashion (i.e., without recourse to the uniformization theorem). We follow Grauert-Reckziegel [1].

Take two linearly independent holomorphic 1-forms ω_1, ω_2 on X and set

$$d\sigma^2 = 2(\omega_1\bar{\omega}_1 + \omega_2\bar{\omega}_2).$$

In terms of a local coordinate system z of X, we may write

$$d\sigma^2 = 2(f_1\bar{f}_1 + f_2\bar{f}_2)dzd\bar{z}, \qquad \text{where} \quad \omega_i = f_i dz.$$

Outside the set of common zeros $p_1, \ldots, p_k$ of ω_1, ω_2, the pseudo-metric $d\sigma^2$ is positive and its Gaussian curvature $K(d\sigma^2)$ is given by

$$K(d\sigma^2) = -\frac{|f_1' f_2 - f_1 f_2'|^2}{(f_1\bar{f}_1 + f_2\bar{f}_2)^3}.$$

Since ω_1 and ω_2 are linearly independent, f_1/f_2 is not constant. Since $K(d\sigma^2)$ vanishes exactly where $(f_1/f_2)'$ vanishes, $K(d\sigma^2)$ is negative except at finitely many points $p_{k+1}, \ldots, p_m$ of $X - \{p_1, \ldots, p_k\}$. Choose disjoint neighborhoods $V_1, \ldots, V_m$ of these points $p_1, \ldots, p_m$. We shall show how to modify $d\sigma^2$ in each V_i so that the curvature becomes negative everywhere on X. Using a local coordinate system z in V_i with $z(p_i) = 0$, let r be a positive number such that $B = \{z; \ |z| < r\} \subset V_i$. Let $B' = \{z; \ |z| < r/2\}$. Choose a C^∞ function $a(z, \bar{z})$ on V_i such that (i) $0 \leq a(z, \bar{z}) \leq 1$ on V_i, (ii) $a(z, \bar{z}) = 1$ on B', and (iii) $a(z, \bar{z}) = 0$ on $V_i - B$. Let c be a constant, $0 < c < 1$, and set

$$g = f + c^2 h,$$

where $f = f_1 \bar{f_1} + f_2 \bar{f_2}$ and $h = (1 + z\bar{z})a(z, \bar{z})$. Then the metric $ds_X^2 = 2g dz d\bar{z}$ coincides with $d\sigma^2$ on $V_i - B$. Since the curvature of $d\sigma^2$ is bounded above by a negative constant on the compact set $\bar{B} - B'$, the curvature of the metric ds_X^2 is strictly negative on $\bar{B} - B'$ if c is sufficiently small. In order to calculate the curvature of ds_X^2 on B', we set $f_3 = c$ and $f_4 = cz$ so that

$$ds_X^2 = 2 \sum_{j=1}^{4} |f_j|^2 dz d\bar{z} \qquad \text{on} \quad B'.$$

Its curvature K is given by

$$K = -\frac{(\sum |f_i'|^2)(\sum |f_k|^2) - (\sum f_i' \bar{f_i})(\sum f_k \bar{f_k'})}{(\sum |f_i|^2)^3}.$$

Since the vectors

$$(f_1, f_2, f_3, f_4) = (f_1, f_2, c, cz) \quad \text{and} \quad (f_1', f_2', f_3', f_4') = (f_1', f_2', 0, c)$$

are linearly independent at every point of B', it follows that K is negative everywhere on B'. By multiplying ds_X^2 with a suitable positive constant, we can make the curvature K bounded above by -1.

In summary,

(2.2.7) **Theorem**. *Every compact Riemann surface X of genus ≥ 2 carries a Hermitian metric ds_X^2 with curvature $K \leq -1$.*

(2.2.8) **Remark**. The curvature of the Hermitian metric $2(1 + |z|^2) dz d\bar{z}$ on $\mathbf{C}$ is given by

$$K = -\frac{1}{(1 + |z|^2)^3} < 0.$$

Clearly, the curvature approaches 0 as $|z|$ tends to infinity. As we shall see later, $\mathbf{C}$ cannot admit a Hermitian metric with curvature bounded above by a negative constant. If r denotes the geodesic distance from the origin to z with respect to the above metric, then $|z|^2 \sim \sqrt{2}r$ for $|z|$ large so that $K \sim -1/(\sqrt{2}r)^3$ as $r \to \infty$, see Remark (3.7.2).

3 Negatively Curved Complex Spaces

In this section we shall generalize results of Section 1 to higher dimensional complex spaces. The results will be used in Section 7 of Chapter 3.

Let X be a complex space. Let f be a holomorphic map sending a neighborhood U_f of the origin 0 in $\mathbf{C}$ into X. Let $g: U_g \to X$ be another such map. Then f and g are said to define the same 1-**jet** at 0 if $f(0) = g(0)$ and if they have the same first derivative at 0. Geometrically stated, this means that two holomorphic curves f and g have the same velocity at 0. The 1-jet represented by f will be denoted $f'(0)$. So by definition, $f'(0) = g'(0)$ if and only if f and g define the same 1-jet at 0.

For each $x \in X$, we call the set $\check{T}_x X$ of all 1-jets $f'(0)$ the **tangent cone** of X at x, and $\check{T} X = \bigcup_x \check{T}_x X$ the **tangent cone** of X. The tangent cone defined here is in general smaller than the usual tangent cone which consists of tangent vectors to real C^∞ curves in X. If x is a regular point of X, then $\check{T}_x X$ coincides with the tangent space $T_x X$.

Let $\mathcal{O}_x$ be the ring of germs of holomorphic functions at $x \in X$. Let $\mathbf{m}_x$ denote its maximal ideal consisting of functions vanishing at x. Then the **Zariski cotangent space** of X at x, is defined to be $\mathbf{m}_x / \mathbf{m}_x^2$, and its dual space, denoted $T_x X$, is called the **Zariski tangent space** of X at x. Although $T X = \bigcup_x T_x X$ may not be a fibre bundle if X is singular since the dimension of $T_x X$ may vary with x, it will be still called the Zariski tangent bundle of X.

Then we have a natural inclusion $\check{T} X \subset T X$. In fact, for a holomorphic function φ defined in a neighborhood of $x = f(0)$, set

$$(f'(0))\varphi = \frac{d}{dz}\varphi(f(z))|_{z=0}.$$

Then $f'(0)$ may be regarded as a derivation from the algebra of holomorphic functions at $x = f(0)$ into $\mathbf{C}$. If X is nonsingular, then $\check{T} X$ coincides with $T X$.

If (z, w) is the coordinate system for $\mathbf{C}^2$ and if X is the curve defined by $w^2 = z^3$, then the Zariski tangent space at the singular point $(0, 0)$ is the 2-dimensional vector space spanned by $\partial/\partial z$ and $\partial/\partial w$. The tangent cone at $(0, 0)$ defined here contains only the zero vector. In fact, if $f \in \mathrm{Hol}(D, X)$ with $f(0) = (0, 0)$, then expressing f by a pair of power series

$$(z, w) = \left(\sum_{j=1}^{\infty} a_j t^j, \sum_{k=1}^{\infty} b_k t^k \right)$$

and using the condition $w^2 = z^3$, we obtain $a_1 = b_1 = b_2 = 0$. Hence, the 1-jet of f at 0 vanishes.

For a systematic account of Zariski tangent spaces, tangent cones and other possible "tangent spaces", see Whitney [1].

A **pseudo-length function** on X is a real non-negative function F on $T X$ such that

$$F(av) = |a| F(v) \qquad \text{for} \quad a \in \mathbf{C}, \quad v, av \in T X.$$

We normally assume that F is smooth. However, for applications it is sometimes necessary to consider upper-semicontinuous pseudo-length functions. If $F(v) > 0$ for all nonzero $v \in T X$, then we call F a **length function** on X.

We say that a pseudo-length function F is **convex** if

$$F(v + v') \leq F(v) + F(v').$$

A convex (pseudo-)length function is also called a **Finsler (pseudo-)metric**.

Every pseudo-length function F on a complex space X gives rise to an inner pseudo-distance d by

$$(2.3.1) \qquad d(p, q) = \inf_{\gamma} \int_a^b F(\gamma'(t))dt,$$

where $\gamma(t)$, $a \leq t \leq b$, is a piecewise differentiable curve from p to q and $\gamma'(t)$ is the velocity vector of γ at $\gamma(t)$. If F is a length function, then d is a distance function.

A Hermitian metric on a complex manifold X defines a length function. However, a length function is much more general than a Hermitian metric or even a Finsler metric. On the other hand, restricted to a holomorphic curve a length function is a Hermitian metric. More precisely, if F is a pseudo-length function on X and if $f\colon D \to X$ is a holomorphic map, then f^*F defines a Hermitian pseudo-metric on D, i.e.,

$$(2.3.2) \qquad f^*F^2 = 2\lambda dz d\bar{z},$$

where λ is a non-negative function on D. If F is a length function and if f is everywhere non-degenerate, then λ is positive and f^*F^2 defines a Hermitian metric on D.

We consider the curvature K_{f^*F} of the pseudo-Hermitian metric f^*F^2 defined by (2.1.2) (by (2.1.9) if not smooth); it is defined where λ is positive. Let $v \in \check{T}_x X$, let $[v]$ denote the complex line spanned by v. Given an upper semicontinuous pseudo-length function F on X, we define the **holomorphic sectional curvature** $K_F([v])$ in the direction of $[v]$ by

$$(2.3.3) \qquad K_F([v]) = \sup K_{f^*F}(0),$$

where the supremum is taken over all $f \in \mathrm{Hol}(D, X)$ such that $f(0) = x$ and $[v]$ is tangent to $f(D)$.

Let $\varphi\colon X' \to X$ be a holomorphic map from a complex space X' into another complex space X. Given a pseudo-length function F on X, we have the induced pseudo-length function φ^*F on X'. From the definition of the curvature we obtain

$$(2.3.4) \qquad K_{\varphi^*F}(v) \leq K_F(\varphi_* v) \qquad \text{for} \quad v \in \check{T}X'$$

wherever the curvature is defined. In particular, if X' is a complex subspace of X, then the curvature of $(X', F|_{X'})$ is bounded above by the curvature of (X, F).

We apply (2.3.4) to a holomorphic map $f\colon D \to X$ and (2.1.4) to the Hermitian pseudo-metric $d\sigma^2 = f^*F^2$. Then we have the following generalization of the Schwarz lemma, (Grauert-Reckziegel [1], Kobayashi [5]).

(2.3.5) Theorem. *Let F be a pseudo-length function on a complex space X. If its holomorphic sectional curvature is bounded above by* -1, *then*

$$f^*F^2 \leq ds^2 \qquad for \quad f \in \mathrm{Hol}(D, X),$$

where ds^2 is the Poincaré metric of D.

We show that for a Hermitian manifold the definition of the holomorphic sectional curvature given here coincides with the usual one.

Given a Hermitian metric $ds^2 = 2 \sum_{i,j=1}^{n} g_{i\bar{j}} dz^i d\bar{z}^j$, on a complex manifold X, the components of its curvature tensor are expressed by

$$(2.3.6) \qquad R_{i\bar{j}k\bar{l}} = -\frac{\partial^2 g_{i\bar{j}}}{\partial z^k \partial \bar{z}^l} + \sum g^{p\bar{q}} \frac{\partial g_{i\bar{q}}}{\partial z^k} \frac{\partial g_{p\bar{j}}}{\partial \bar{z}^l}.$$

Then given a unit tangent vector $v = \sum v^i (\partial/\partial z^i)$, the holomorphic sectional curvature in the direction of v is defined to be

$$(2.3.7) \qquad H_{ds^2}(v) = \sum R_{i\bar{j}k\bar{l}} v^i \bar{v}^j v^k \bar{v}^l.$$

Let X' be a complex submanifold of X. We choose a local coordinate system $z^1, \ldots, z^n$ in such a way that X' is defined by

$$z^{m+1} = \ldots = z^n = 0.$$

so that we may use $z^1, \ldots, z^m$ as a local coodinate system for X'. Then the induced Hermitian metric on X' is given by $2 \sum_{i,j=1}^{m} g_{i\bar{j}} dz^i d\bar{z}^j$. We shall compute its curvature $R'_{i\bar{j}k\bar{l}}$. Fix a point $p \in X'$. By a linear change of coordinates, we may assume

$$g_{i\bar{j}} = \delta_{ij} \quad \text{at} \quad p.$$

Then at the point p we have the following equation of Gauss:

$$(2.3.8) \qquad R'_{i\bar{j}k\bar{l}} = R_{i\bar{j}k\bar{l}} - \sum_{p=m+1}^{n} \frac{\partial g_{i\bar{p}}}{\partial z^k} \frac{\partial g_{p\bar{j}}}{\partial \bar{z}^l}$$

for $i, j, k, l = 1, \ldots, m$.

The following proposition is immediate from (2.3.8)

(2.3.9) **Proposition.** *Let X' be a complex submanifold of a Hermitian manifold X. Then the holomorphic sectional curvature H'_{ds^2} of X' does not exceed the holomorphic sectional curvature H_{ds^2} of X, i.e.,*

$$H'_{ds^2}(v) \leq H_{ds^2}(v) \qquad \textit{for} \quad v \in TX'.$$

Fix a point $p \in X$, and let $v \in T_p X$ be a unit tangent vector. If X' is a holomorphic curve in X tangent to v at p, then its Gaussian curvature at p is bounded by the holomorphic sectional curvature $H_{ds^2}(v)$ of X in the direction v. We shall show that there is a holomorphic curve tangent to v whose Gaussian curvature at p equals $H_{ds^2}(v)$. We start with any holomorphic curve X' tangent to v at p. We may assume that a local coordinate system $z^1, \ldots, z^n$ with origin p was chosen in such a way that $g_{i\bar{j}} = \delta_{ij}$ and X' is given by $z^2 = \ldots = z^n = 0$. From

(2.3.8) we know that if $\partial g_{i\bar{q}}/\partial z^k = 0$ at p for $q = 2, \ldots, n$, then X' has already desired property. If not, we consider the following coordinate transformation:

$$z^1 = w^1, \quad z^q = w^q - \frac{1}{2}a^q(w^1)^2, \quad (q = 2, \ldots, n),$$

where $a^2, \ldots, a^n$ are constants to be chosen appropriately. Substitute the coordinate transformation above into $ds^2 = \sum g_{i\bar{j}}dz^i d\bar{z}^j$ to express it in terms of $w^1, \ldots, w^n$. Then we obtain $ds^2 = \sum h_{i\bar{j}}dw^i d\bar{w}^j$, with

$$h_{1\bar{q}} = g_{1\bar{q}} - \sum_{r=2}^{n} g_{r\bar{q}}a^r w^1.$$

It follows that if we set $a^q = (\partial g_{1\bar{q}}/\partial z^1)_p$, then $(\partial h_{1\bar{q}}/\partial w^1)_p = 0$, so that the holomorphic curve defined by $w^2 = \ldots = w^n = 0$ has the desired property. This proves the following (Wu [5])

(2.3.10) **Proposition.** *For a Hermitian manifold (X, ds^2) the holomorphic sectional curvature defined by (2.3.3) coincides with the classical holomorphic sectional curvature defined by (2.3.7).*

(2.3.11) **Remark.** We note that for K_F the assertion corresponding to (2.3.9) was immediate from the definition (2.3.3), (see (2.3.4)). However, (2.3.8) gives much more than (2.3.9). In fact, if

$$H'_{ds^2}(v) = H_{ds^2}(v) \qquad \text{for} \quad v \in T_p X',$$

then X' is totally geodesic at p, i.e., the second fundamental form of X' vanishes at p, provided either dim $X' = 1$ or X is Kähler.

From (2.3.5) we obtain

(2.3.12) **Corollary** *Let (X, ds^2_X) be a Hermitian manifold whose holomorphic sectional curvature is bounded above by -1. Then*

$$f^*ds^2_X \leq ds^2 \qquad for \quad f \in \mathrm{Hol}(D, X),$$

where ds^2 is the Poincaré metric of D.

We extend (2.1.13) to higher dimensional spaces X.

(2.3.13) **Theorem.** *Let X be a complex space with an upper semicontinuous pseudolength function F with the property that, at each $v \in \check{T}X$ with $F(v) > 0$, F has negative holomorphic curvature. Let ξ be a holomorphic vector field on X. Then $F(\xi)$ cannot attain a maximum in (the interior of X) unless $F(\xi) \equiv 0$.*

Proof. Assume that $F(\xi)$ attains a positive maximum at $x_o \in X$. Choose an imbedding $f: D \to X$ such that $f(0) = x_o$ and ξ is tangent to $f(D)$. Then apply (2.1.13) to the pseudo-Hermitian metric f^*F^2 on D. $\qquad\square$

(2.3.14) **Corollary.** *Let X and F be as in (2.3.13). If X is compact, it admits no holomorphic vector fields.*

The generalized Schwarz lemma (2.3.5) has been extended to holomorphic maps from a complex manifold M with a length function satisfying certain curvature conditions to a complex manifold X satisfying the condition of (2.3.5), see Yau [3], Chen-Cheng-Lu [1], Royden [7, 8], Yang-Chen [1], Q. H. Yu [1], Chen-Yang [1], Matsuura [2], Burns-Krantz [1].

4 Ricci Forms and Schwarz Lemma for Volume Elements

The results in this section, which generalize those of Sections 1 and 2 to volume elements, will not be used until Chapter 7.

Let L be a holomorphic line bundle over a complex manifold X with local coordinate system $z^1, \ldots, z^n$, and h a pseudo-metric on L, i.e., h is a non-negative smooth function on L such that

$$h(c\xi) = |c|^2 h(\xi) \qquad \text{for} \quad \xi \in L, \ c \in \mathbf{C}.$$

If $h(\xi) > 0$ for all nonzero $\xi \in L$, then h is called a metric on L.

To each pseudo-metric h we associate a closed $(1,1)$-form $\mathrm{Ric}(h)$ called the **Ricci form** as follows. Taking a local non-vanishing holomorphic section ξ, we set

$$(2.4.1) \qquad \mathrm{Ric}(h) = -dd^c \log h(\xi) = -2i\partial\bar{\partial} \log h(\xi) = 2i \sum R_{j\bar{k}} dz^j \wedge d\bar{z}^k,$$

where

$$(2.4.2) \qquad\qquad R_{j\bar{k}} = -\frac{\partial^2 \log h(\xi)}{\partial z^j \partial \bar{z}^k}.$$

If ξ is replaced by another non-vanishing holomorphic section $\eta = f\xi$ with f holomorphic, then $h(\eta) = |f|^2 h(\xi)$ and $\partial\bar{\partial} \log h(\eta) = \partial\bar{\partial} \log h(\xi)$, which shows that $\mathrm{Ric}(h)$ does not depend on the choice of ξ. The Ricci form $\mathrm{Ric}(h)$ is defined only at the points where h is strictly positive.

If h is a metric on L, then $\mathrm{Ric}(h)$ is globally defined on X and represents $4\pi c_1(L)$, where $c_1(L)$ is the first Chern class of L.

Sometimes, the Ricci form $\mathrm{Ric}(h)$ of a pseudo-metric can be globally defined (even where h vanishes). We say that h has a **holomorphic degeneracy** if h is locally of the form $|a|^{2q} g$, where g is strictly positive, a is holomorphic, and q is a rational number. Thus, if ξ is a non-vanishing local holomorphic section of L, then $h(\xi) = |a|^{2q} g$, and $h(\xi)$ vanishes only where a vanishes. In this case,

$$\partial\bar{\partial} \log h(\xi) = \partial\bar{\partial} \log g,$$

and $\mathrm{Ric}(h)$ is defined everywhere on X.

In order to generalize (2.1.4) to the case of higher dimension, we consider a pseudo-volume form and the associated Ricci form on a complex manifold X. In terms of a local coordinate system $z^1, \ldots, z^n$ of X, a **pseudo-volume form** v on X can be locally written as

$$(2.4.3) \qquad v = V \prod_{j=1}^{n} (i dz^j \wedge d\bar{z}^j),$$

where V is a (locally defined) nonnegative function. A pseudo-volume form v is a pseudo-metric on the anti-canonical line bundle $K^{-1} = \bigwedge^n TX$.

If V is positive everywhere, i.e., if v is a metric on the line bundle K^{-1}, then v is a **volume form** of X. To each v given by (2.4.3) we associate the Ricci form $\mathrm{Ric}(v)$ by

$$(2.4.4) \qquad \mathrm{Ric}(v) = -dd^c \log V = 2i \sum R_{j\bar{k}} dz^j \wedge d\bar{z}^k,$$

where $R_{j\bar{k}} = -\partial^2 \log V / \partial z^j \partial \bar{z}^k$. Of course, this is a special case, i.e., $L = K^{-1}$, of the construction (2.4.1).

If X is a Hermitian manifold with fundamental 2-form ω and the volume form $v = \omega^n$, then $(R_{j\bar{k}})$ represents the components of the **Ricci tensor** in the classical sense. Hence the name "Ricci form" for $\mathrm{Ric}(v)$. It is, however, important to associate the Ricci form directly to a volume form rather than to a Hermitian metric.

If $\dim X = 1$, then

$$(2.4.5) \qquad \mathrm{Ric}(v) = 2Kv,$$

where K is the Gaussian curvature of X. It is then natural to define K_v when $\dim X = n$ by the following formula.

$$(2.4.6) \qquad K_v = -\frac{(-\mathrm{Ric}(v))^n}{n!(n+1)^n v}.$$

We say that $\mathrm{Ric}(v)$ is **negative** if the matrix $(R_{j\bar{k}})$ is negative wherever defined (i.e., where $v > 0$). We say that it is **negatively bounded** if it is negative and if $K_v \leq c < 0$ wherever defined. If v is negatively bounded, we can always normalize it (by multiplying it with a positive constant) so that $K_v \leq -1$.

Let B_a^n be the ball of radius a in $\mathbf{C}^n$:

$$B_a^n = \{ \mathbf{z} = (z^1, \ldots, z^n) \in \mathbf{C}^n; \ \|\mathbf{z}\|^2 = |z^1|^2 + \ldots + |z^n|^2 < a^2 \}.$$

The unit ball B_1^n will be denoted B^n.

An invariant volume element μ_a of B_a^n is given by

$$(2.4.7) \qquad \mu_a = \frac{2^n a^2}{(a^2 - \|\mathbf{z}\|^2)^{n+1}} \prod_{j=1}^{n} (i dz^j \wedge d\bar{z}^j).$$

For $a = 1$, we write μ for μ_1. A simple calculation shows

$$(2.4.8) \qquad \mathrm{Ric}(\mu_a) = -2i(n+1) \sum_{j,k} \frac{\delta_{jk}(a^2 - \|\mathbf{z}\|^2) + \bar{z}^j z^k}{(a^2 - \|\mathbf{z}\|^2)^2} dz^j \wedge d\bar{z}^k,$$

and

$$(2.4.9) \qquad K_{\mu_a} = -\frac{(-\mathrm{Ric}(\mu_a))^n}{n!(n+1)^n \mu_a} = -1.$$

Let D_a^n be the polydisc of radius a in $\mathbf{C}^n$:

$$(2.4.10) \qquad D_a^n = \{(z^1, \ldots, z^n) \in \mathbf{C}^n; \; |z^1| < a, \ldots, |z^n| < a\}.$$

The unit polydisc D_1^n will be denoted D^n.

An invariant volume element ν_a of D_a^n is given by

$$(2.4.11) \qquad \nu_a = (2a)^{2n} \prod_{j=1}^{n} \frac{i\,dz^j \wedge d\bar{z}^j}{(a^2 - |z^j|^2)^2}.$$

For $a = 1$, we write ν for ν_1. By a simple calculation, we obtain

$$(2.4.12) \qquad \mathrm{Ric}(\nu_a) = -4i \sum_{j=1}^{n} \frac{a^2}{(a^2 - |z^j|^2)^2} dz^j \wedge d\bar{z}^j,$$

and

$$(2.4.13) \qquad K_{\nu_a} = -\frac{(-\mathrm{Ric}(\nu_a))^n}{n!(n+1)^n \nu_a} = -\frac{1}{(n+1)^n}.$$

The following theorem generalizes (2.1.4).

(2.4.14) Theorem. *Let v be any pseudo-volume form on the unit ball B^n with negatively bounded Ricci form $\mathrm{Ric}(v)$ and normalized in such a way that $K_v \leq -1$. Then $v \leq \mu$.*

Proof. For a, $0 < a < 1$, we use the volume form μ_a on the ball B_a^n defined by (2.4.7). Let u_a be the nonnegative function on B_a^n defined by

$$u_a = v/\mu_a.$$

As in the proof of (2.1.4), u_a attains its maximum at some interior point, say z_0 of B_a^n, and it suffices to prove $u_a(z_0) \leq 1$. Assuming that $u_a(z_0) > 0$, we calculate the complex Hessian of $\log u_a$ at z_0. From (2.4.2) we obtain

$$dd^c \log u_a = 2i \sum \frac{\partial^2 \log u_a}{\partial z^j \partial \bar{z}^k} dz^j \wedge d\bar{z}^k = \mathrm{Ric}(\mu_a) - \mathrm{Ric}(v).$$

Since the complex Hessian $dd^c \log u_a$ must be nonpositive at the maximum point z_0 and since $\mathrm{Ric}(v)$ is negative, we have

$$0 < -\mathrm{Ric}(v) \leq -\mathrm{Ric}(\mu_a) \qquad \text{at} \quad z_0.$$

Taking the nth exterior power of this inequality, we obtain

$$0 < -K_v v \leq -K_{\mu_a} \mu_a \qquad \text{at} \quad z_0.$$

From $K_v \leq -1 = K_{\mu_a}$, we obtain

$$v \le \mu_a \qquad \text{at} \quad z_0,$$

i.e., $u_a(z_0) \le 1$. $\qquad\qquad\qquad\qquad\qquad\qquad\qquad\qquad\qquad\qquad\qquad\qquad$ $\square$

In order to state a corollary to the theorem above, we need to explain the concept of meromorphic map. In general, a **meromorphic map** f from a complex space X into a complex space Y is a correspondence satisfying the following conditions:

(1) For each point x of X, $f(x)$ is a nonempty compact subset of Y;

(2) The graph $G_f = \{(x, y) \in X \times Y; \; y \in f(x)\}$ is a connected complex subspace of $X \times Y$ with $\dim G_f = \dim X$;

(3) There exists a dense subset X^* of X such that $f(x)$ is a single point for $x \in X^*$.

Let $\pi \colon G_f \to X$ be the projection defined by $\pi(x, y) = x$; it is a proper map. Then $\{x\} \times f(x) = \pi^{-1}(x)$, and $f(x)$ is a complex subspace of Y.

Let $E \subset G_f$ be the set of points where π is degenerate, i.e.,

$$E = \{(x, y) \in G_f; \; \dim f(x) > 0\},$$

and let

$$S = \pi(E) = \{x \in X; \; \dim f(x) > 0\}.$$

Then E is a closed complex subspace of codimension ≥ 1 of G_f, and S is a closed complex subspace of codimension ≥ 2 of X, (see Remmert [1]). It is then clear that $f \colon X - S \to Y$ is holomorphic. The subspace S is called the **singular locus** of f. The graph G_f of f is the topological closure of the graph of $f|_{X-S}$, i.e., the closure of $\{(x, f(x)); \; x \in X - S\}$.

Now, as a consequence of (2.4.14) we have the following generalization of (2.1.5).

(2.4.15) **Corollary.** *Let X be an n-dimensional complex manifold with pseudo-volume form v such that $\mathrm{Ric}(v)$ is negatively bounded and $K_v \le -1$. Then every meromorphic map $f \colon B^n \to X$ is volume-decreasing in the sense that $f^*v \le \mu$.*

Proof. If f is holomorphic, the proof is the same as that of (2.1.5); simply apply (2.4.14) to f^*v. If f is meromorphic, let $S \subset B^n$ be the singular locus of f. Since $\mathrm{Ric}(v)$ is negative, the coefficient V of v is plurisubharmonic, (in fact, $\log V$ is plurisubharmonic). Hence, the coefficient of f^*v is a plurisubharmonic function on $B^n - S$. Since the codimension of S is 2, f^*v extends across S. Now (2.4.14) can be applied to f^*v. $\qquad\qquad\qquad\qquad\qquad\qquad\qquad\qquad\qquad$ $\square$

(2.4.16) **Corollary.** *Every holomorphic map f of B^n into itself is volume-decreasing with respect to μ, i.e., $f^*\mu \le \mu$.*

If we want to use the polydisc D^n instead of the ball B^n as our "model" domain, then in view of (2.4.13) we have to use the "normalized" volume form $(n + 1)^n v$ rather than v.

(2.4.14) and its corollaries can be generalized to more general domains, in particular to symmetric bounded domains, see Kobayashi [6], [7; p. 33] and Hahn-Mitchell [1].

If the domain of a mapping f is a compact manifold rather than a noncompact space such as B^n, then the argument in (2.4.14) becomes simpler. For example, we have (Kobayashi [6], [7])

(2.4.17) **Theorem.** *Let X and Y be n-dimensional complex manifolds with volume form v_X and pseudo-volume form v_Y, respectively. Assume*
   (a)   *The Ricci forms $\mathrm{Ric}(v_X)$ and $\mathrm{Ric}(v_Y)$ are negatively bounded;*
   (b) $K_{v_Y}(y)/K_{v_X}(x) \geq 1$ *for $x \in X$ and $y \in Y$;*
   (c)   *X is compact.*
Then every meromorphic map $f: X \to Y$ is volume-decreasing in the sense that $f^* v_Y \leq v_X$.

Proof. Although f is not holomorphic, $f^* v_Y$ is a well-defined pseudo-volume form on X; see the proof of (2.4.15). Consider a non-negative function $u = f^* v_Y / v_X$ on X. Since X is compact, it attains its maximum at some point, say $x_0 \in X$. To complete the proof, consider $dd^c \log u$ at x_0 and follow the argument in the proof of (2.4.14). $\qquad\qquad\square$

(2.4.18) **Corollary.** *In (2.4.17), if Y is also compact, then*

$$\deg f \leq \mathrm{vol}(X)/\mathrm{vol}(Y),$$

where $\mathrm{vol}(X) = \int_X v_X$ and $\mathrm{vol}(Y) = \int_Y v_Y$.

Proof.

$$\deg f = \frac{\int_X f^* v_Y}{\int_Y v_Y} \leq \frac{\int_X v_X}{\int_Y v_Y}.$$

$\square$

If v is a volume element of a compact complex manifold, then its first Chern class $c_1(X)$ is represented by the Ricci form (modulo a constant factor):

(2.4.19) $$c_1(X) = \frac{1}{4\pi}[\mathrm{Ric}(v_X)].$$

By (2.4.6)

$$\int_X (-K_{v_X}) v_X = \frac{(-4\pi)^n}{n!(n+1)^n} c_1(X)^n.$$

Let

$$a_X = \min_X (-K_{v_X}) \quad \text{and} \quad b_X = \max_X (-K_{v_X}).$$

Then

(2.4.20) $$a_X \cdot \mathrm{vol}(X) \leq \frac{(-4\pi)^n}{n!(n+1)^n} c_1(X)^n \leq b_X \cdot \mathrm{vol}(X).$$

Similarly, we define a_Y and b_Y, and we obtain for Y inequalities similar to (2.4.20).

We note that condition (b) in (2.4.17) simply says $a_Y/b_X \geq 1$. If we do not make assumption (b), instead of the inequality $f^*(v_Y) \leq v_X$ we have

$$f^*(v_Y) \leq \frac{b_X}{a_Y} v_X.$$

From the proof of (2.4.18) and (2.4.20) we see that if X and Y are compact complex manifolds with volume forms v_X and v_Y whose associated Ricci forms are negative, then

$$(2.4.21) \qquad\qquad \deg f \leq \frac{b_X b_Y c_1(X)^n}{a_X a_Y c_1(Y)^n}.$$

As we will see soon, using not only volume forms but Kähler metrics we can get rid of these constants a_X, b_X, a_Y, b_Y from the formula above. But some remarks are in order.

The equidimensional Schwarz lemma (2.4.15) in higher dimension was first obtained by Dinghas [1] for holomorphic maps from D^n into Einstein-Kähler manifolds X of negative Ricci curvature and was generalized by Chern [2] to maps into Einstein-Hermitian manifolds X of negative Ricci curvature. By considering the Ricci form associated to a pseudo-volume form rather than to a Kähler metric, Kobayashi [6] got rid of the Einstein condition. As we shall see later, for algebraic geometric applications it is essential to get rid of the Einstein condition. The Ricci form associated to a volume form goes back to Koszul, who used it to study homogeneous complex manifolds.

However, on a Kähler manifold we can compare its Ricci tensor with the metric tensor. The simplest situation is when both X and Y are compact Einstein-Kähler manifolds with negative Ricci tensor. Let Φ_X be the Kähler form and $v_X = \frac{1}{n!}\Phi_X^n$ be the volume form of X. Set $\mathrm{Ric}_X = \mathrm{Ric}(v_X)$. Then our assumption is that $\mathrm{Ric}_X = -c\Phi_X$, where c is a positive constant. When Φ_X is multiplied by a positive constant, its Ricci form Ric_X remains unchanged. So we may assume that $c = 1$, i.e., $\mathrm{Ric}_X = -\Phi_X$. In this case we have $a_X = b_X$ and (2.4.20) becomes an equality.

We make a similar assumption on Y, so that $a_Y = b_Y = a_X = b_X$. Then (2.4.18) and (2.4.20) imply

$$\deg f \leq c_1(X)^n/c_1(Y)^n.$$

Now, if X is a compact complex manifold with negative first Chern class $c_1(X)$, then X admits an Einstein-Kähler metric by Aubin [1, 2], see also Yau [4]. Hence,

(2.4.22) **Theorem.** *Let X and Y be compact complex manifolds with negative first Chern class. Then every meromorphic map $f: X \to Y$ satisfies*

$$\deg f \leq c_1(X)^n/c_1(Y)^n.$$

5 Metrics on Jet Bundles

In this section, Schwarz' lemma (2.3.5) will be generalized to higher order jet metrics. Although the language of jets and jet bundles will be used in Section 9 of Chapter 3, the results of this section will not be used otherwise in the rest of the book.

First, we explain the notion of jet and jet bundle introduced by Ehresmann. Let X be an n-dimensional complex space. We consider holomorphic mappings f from domains in $\mathbf{C}$ into X, the domain of f depending on f. We say that two such maps f and g **osculate to order** k at $a \in \mathbf{C}$ or have the same k-**jet** at z if they have the same derivatives of order $0, 1, \ldots, k$ at a:

$$f(a) = g(a), \ f'(a) = g'(a), \ \ldots, f^{(k)}(a) = g^{(k)}(a).$$

The equivalence class thus defined is the k-**jet** of f at a and is denoted $j_a^k(f)$. We call a the **source** and $f(a)$ the **target** of the jet.

Choosing the origin $0 \in \mathbf{C}$ as a source and fixing $x \in X$, we denote the set of k-jets $j_0^k(f)$ with $f(0) = x$ by $J_x^k X$. If $x = f(0)$ is a nonsingular point of X, in a fixed local coordinate system around x, every k-jet $j_0^k(f)$ determines a point of $\mathbf{C}^{kn}$

$$(2.5.1) \qquad\qquad (f'(0), \ f''(0), \ldots, f^{(k)}(0)),$$

and this gives a one-to-one correspondence between $J_x^k X$ and $\mathbf{C}^{kn}$. The isomorphism $J_x^k X \cong \mathbf{C}^{kn}$ depends on the chosen local coordinate system of X.

Sometimes, we use k-jets with source at a point $a \neq 0$. In such cases, we identify $j_a^k(f)$ with $j_0^k(f_a)$, where f_a is defined by $f_a(z) = f(z + a)$. Then $j_a^k(f)$ determines the point

$$(f_a'(0), \ f_a''(0), \ldots, f_a^{(k)}(0)) = (f'(a), \ f''(a), \ldots, f^{(k)}(a)).$$

Clearly, $J_x^1 X$ is the tangent cone $\check{T}_x X$. Even when x is a nonsigular point of X, for $k \geq 2$, $J_x^k X$ does not have an intrinsic vector space structure. Nevertheless, $J_x^k X$ has the origin 0_k represented by the constant map to x, and it admits also scalar multiplication which is defined as follows:

$$(2.5.2) \qquad c \cdot j_0^k(f) = j_0^k(f \circ \mu_c), \qquad \text{where} \quad \mu_c(z) = cz.$$

In terms of coordinates, this may be expressed as follows:

$$(2.5.3) \qquad c \cdot j_0^k(f) = (cf'(0), c^2 f''(0), \ldots, c^k f^{(k)}(0)).$$

Put

$$J^k X = \bigcup_{x \in X} J_x^k X.$$

Then $J^k X$ is a fibre space over X; it is a fibre bundle over the regular part X_{reg} of X with fibre $\mathbf{C}^{kn}$. A local coordinate system $x = (x^1, \ldots, x^n)$ of X_{reg} induces a local coordinate system $(x, \xi_1, \ldots, \xi_k)$, where each ξ_j has n components $(\xi_j^1, \ldots, \xi_j^n)$, see (2.5.1). In particular, if X is nonsingular, J^1X is the tangent bundle TX. For $k \geq 2$, J^kX is not a vector bundle.

There is a sequence of natural projections:

$$J_x^k X \to J_x^{k-1} X \to \ldots \to J_x^1 X.$$

If x is a nonsingular point of X, each fibre of $J_x^k X \to J_x^{k-1} X$ has an intrinsic affine space structure, and the fibre over the origin 0_{k-1} has a natural vector space structure. This can be seen from the chain rule; if $x = (x^1, \ldots, x^n)$ and $y = (y^1, \ldots, y^n)$ are two local coordinate systems, then (in symbolic notation omitting indices)

$$\frac{d^k y}{dz^k} = \frac{\partial y}{\partial x} \frac{d^k x}{dz^k} + \frac{\partial^2 y}{\partial x \partial x} \frac{dx}{dz} \frac{d^{k-1} x}{dz^{k-1}} + \ldots.$$

Hence, the natural projection

$$J^k X_{\text{reg}} \to J^{k-1} X_{\text{reg}}$$

makes $J^k X_{\text{reg}}$ into an affine bundle over $J^{k-1} X_{\text{reg}}$ with fibre $\mathbf{C}^n$.

Following Green-Griffiths [1] and Grauert [7] we define a **jet pseudo-metric** F on $J^k X$ to be a continuous non-negative function on $J^k X$ which is smooth except when zero and is homogeneous of degree 1 in the following sense:

$$(2.5.4) \qquad F(c\xi) = |c| F(\xi) \qquad \xi \in J^k X, \; c \in \mathbf{C}.$$

For $k = 1$, it is a pseudo-length function or pseudo-metric.

In terms of a local coordinate system $(x, \xi_1, \ldots, \xi_k)$,

$$(2.5.5) \qquad F(x, c\xi_1, \ldots, c^k \xi_k) = |c| F(x, \xi_1, \ldots, \xi_k).$$

Every holomorphic map $f \colon D \to X$ lifts to a holomorphic map

$$j^k f \colon D \to J^k X, \qquad j^k f \colon a \mapsto j_a^k(f) = j_0^k(f_a), \quad a \in D.$$

By pulling back the jet pseudo-metric F by $j^k f$ we obtain a non-negative function $f^* F$ on D:

$$(f^* F)(a) = F(j_a^k f).$$

This gives rise to a pseudo-Hermitian metric

$$ds^2_{f^* F} := 2(f^* F)^2 dz d\bar{z}$$

on D. Its curvature $K_{f^* F}$ is given by (see (2.1.2))

$$(2.5.6) \qquad i K_{f^* F} dz \wedge d\bar{z} = -\frac{1}{f^* F^2} dd^c \log f^* F = -f^* \left(\frac{1}{F^2} dd^c \log F \right);$$

it is defined only where $f^* F$ is positive. Since f^* in (2.5.6) is the pull-back by the differential of $j^k f$, it follows that $K_{f^* F}(a)$, $a \in D$, depends only on $j_a^{k+1} f$, not on f. We define the **curvature** K_F of F to be a function on $J^k X - \{0_k\}$ given by

$$(2.5.7) \qquad K_F(\xi) = \sup_f K_{f^*F}(a), \qquad \xi \in J^k X, \ \xi \neq 0,$$

where the supremum is taken over all $f \in \mathrm{Hol}(D, X)$ such that $\xi = j_0^k f$. Then

$$(2.5.8) \qquad K_F(c\xi) = K_F(\xi), \qquad c \neq 0.$$

If $K_F \leq -1$, then the Ahlfors-Schwarz lemma (2.1.4) states

$$(2.5.9) \qquad ds_{f^*F}^2 \leq ds^2, \quad \text{or} \quad F(j_a^k(f)) = f^* F(a) \leq \frac{1}{(1 - |a|^2)^2},$$

where ds^2 is the Poincaré metric of D.

Now (2.3.5) generalizes as follows (see Green-Griffiths [1]):

(2.5.10) Theorem. *Given a jet pseudo-metric F on the jet bundle $J^k X$, we define a pseudo-length function $F_1 \colon J^1 X \to \mathbf{R}$ by setting*

$$F_1(v) = \inf_{\pi(\xi)=v} F(\xi),$$

where the infimum is taken over all $\xi \in J^k X$ which are mapped to $v \in TX = J^1 X$ by the projection $\pi \colon J^k X \to J^1 X$. Let $\|u\|$ denote the Poincaré length of $u \in TD$. If $K_F \leq -1$, then

$$F_1(f_* u) \leq \|u\| \qquad u \in TD, \ f \in \mathrm{Hol}(D, X).$$

Proof. Since D is homgeneous, we may assume that u is a vector at the origin. By mutiplying u by a suitable constant, we may assume that $u = (d/dz)_0$. Let $f \in \mathrm{Hol}(D, X)$, and $j^k f \colon D \to J^k X$ its lift. Then

$$f^* F(0) = F(j_0^k(f)) = F(f(0), f'(0), \ldots, f^{(k)}(0)).$$

Since $f_* u = (f(0), f'(0))$, we have $F_1(f_* u) \leq f^* F(0)$. Hence, by (2.5.9) we have $F_1(f_* u) \leq \|u\|$. $\qquad\square$

For simplicity we shall assume that X is nonsingular in the remainder of this ection. The k-jet bundle $J^k X$ is a generalization of the tangent bundle $TX = J^1 X$. We shall now generalize the cotangent bundle $T^* X$. Given a local coordinate system $(x^1, \ldots, x^n)$ in a coordinate neighborhood U, we consider the following symbols:

$$dx^1, \ldots, dx^n, \ d^2 x^1, \ldots, d^2 x^n, \ d^3 x^1, \ldots, d^3 x^n, \ \ldots$$

and assign weight p to each of $d^p x^1, \ldots, d^p x^n$. At each point $x \in U$ let $D_x^{(k,m)}$ be the set of polynomials of weight m in $dx^i, d^2 x^i, \ldots, d^k x^i$, $(i = 1, 2, \ldots, n)$, and set

$$(2.5.11) \qquad D_X^{(k,m)} = \bigcup_{x \in X} D_x^{(k,m)}.$$

Then $D_X^{(k,m)}$ is a vector bundle over X, defined independently of the coordinate system $(x^1, \ldots, x^n)$. A (local) holomorphic section of this bundle is called a k-**jet differential of weight** m.

For example, a 2-jet differential of weight 4 is locally of the form

$$(2.5.12) \qquad \omega = \sum a_{ijkl} dx^i dx^j dx^k dx^l + \sum b_{ijk} dx^i dx^j d^2 x^k + \sum c_{ij} d^2 x^i d^2 x^j,$$

where a_{ijkl} are holomorphic functions, symmetric in all indices while b_{ijk} and c_{ij} are holomorphic functions symmetric in i and j.

Clearly, $D_X^{(1,1)}$ is the cotangent bundle T^*X. More generally,

$$(2.5.13) \qquad D_X^{(1,m)} = S^m T^* X,$$

where $S^m T^* X$ denotes the m-th symmetric tensor power of $T^* X$.

Evidently, we have

$$D_X^{(1,m)} \subset D_X^{(2,m)} \subset \ldots \subset D_X^{(m,m)} = D_X^{(m+1,m)} = \ldots,$$

and the natural multiplication

$$D_X^{(k,m)} \times D_X^{(k,m')} \to D_X^{(k,m+m')}.$$

The differentiation operation

$$d: \mathcal{O}(D_X^{(k,m)}) \to \mathcal{O}(D_X^{(k+1,m+1)})$$

is defined by the Leibniz rule.

Every k-jet differential ω of weight m defines a holomorphic function on the jet bundle $J^k X$ such that

$$(2.5.14) \qquad \omega(c\xi) = c^m \omega(\xi), \qquad \xi \in J^k X, \ c \in \mathbf{C}.$$

In fact, if ω is of the form (2.5.12), then for $\xi = (x, \xi_1, \xi_2) \in J^2 X$ we have

$$\omega(\xi) = \sum a_{ijkl}(x)\xi_1^i \xi_1^j \xi_1^k \xi_1^l + \sum b_{ijk}(x)\xi_1^i \xi_1^j \xi_2^k + \sum c_{ij}(x)\xi_2^i \xi_2^j.$$

If $S^m T^* X$ is very ample for sufficiently large m, then $T^* X$ is said to be ample. In such a case X admits a Finsler metric of negative holomorphic curvature, (see Kobayashi [13]). Now we want to generalize this fact to $k \geq 2$.

The scalar multiplication defined by (2.5.2) leads us to the concept of weighted projective space. In general, given positive integers $m_0, m_1, \ldots, m_r$, define a scalar multiplication on $\mathbf{C}^{r+1}$ by

$$(2.5.15) \qquad c(z_0, z_1, \ldots, z_r) = (c^{m_0} z_0, c^{m_1} z_1, \ldots, c^{m_r} z_r), \qquad c \in \mathbf{C}.$$

This defines the action of the group $\mathbf{C}^*$ on $\mathbf{C}^{r+1} - \{0\}$. The resulting quotient variety

$$(2.5.16) \qquad P^{(m_0, m_1, \ldots, m_r)} = (\mathbf{C}^{r+1} - \{0\})/\mathbf{C}^*$$

is called the **weighted projective space** of weights $(m_0, m_1, \ldots, m_r)$. If $m_0 = m_1 = \ldots = m_r$, then we obtain an ordinary projective space $P_r\mathbf{C}$.

A weighted projective space is, in general, a singular variety since the action of $\mathbf{C}^*$ is not free. However, it can be described as a quotient of $P_r\mathbf{C}$ by a finite abelian group. Let ζ_j be a primitive m_j-th root of 1. Then the abelian group $\mathbf{Z}_{m_0} \times \mathbf{Z}_{m_1} \times \ldots \times \mathbf{Z}_{m_r}$ acts on $P_r\mathbf{C}$, with generators being given by

$$(z_0, \ldots, z_j, \ldots, z_r) \mapsto (z_0, \ldots, \zeta_j z_j, \ldots, z_r).$$

The quotient of $P_r\mathbf{C}$ by this group is naturally isomorphic to the weighted projective space $P^{(m_0, m_1, \ldots, m_r)}$.

Thus, if we divide $\mathbf{C}^{kn}$ by the action of $\mathbf{C}^*$ defined by (2.5.2), then we obtain $P^{(n\cdot 1, n\cdot 2, \ldots, n\cdot k)}$, where $n \cdot j$ denotes j repeated n times.

We put

$$(2.5.17) \qquad\qquad P(J^k X) = (J^k X - \{0_k\})/\mathbf{C}^*.$$

Then $P(J^k X)$ is a fibre bundle over X with fibre $P^{(n\cdot 1, n\cdot 2, \ldots, n\cdot k)}$.

Let $W \subset H^0(X, D_X^{(k,m)})$ be a linear space of holomorphic sections of $D_X^{(k,m)}$. Let $\omega_0, \omega_1, \ldots, \omega_N$ be a basis for W. Define $\varphi \colon J^k X \to \mathbf{C}^{N+1}$ by

$$(2.5.18) \qquad\qquad \xi \mapsto (\omega_0(\xi), \omega_1(\xi), \ldots, \omega_N(\xi)).$$

Since

$$\varphi(c\xi) = c^m \varphi(\xi),$$

φ induces a map $\tilde{\varphi} \colon P(J^k X) \to P_N\mathbf{C}$, provided that W has no base locus. We say that W defines an immersion of $P(J^k X)$ into $P_N\mathbf{C}$ if $\tilde{\varphi}$ immerses $P(J^k X)$ into $P_N\mathbf{C}$.

Let $w = (w^0, w^1, \ldots, w^N)$ be the natural coordinate system for $\mathbf{C}^{N+1}$ so that

$$\omega_\alpha = w^\alpha \circ \varphi = \varphi^* w^\alpha.$$

Let $ds^2_{P_N}$ be the Fubini-Study metric for $P_N\mathbf{C}$. Its associated Kähler form Φ_{P_N} is given by

$$\Phi_{P_N} = dd^c \log \|w\|^2.$$

We may consider $ds^2_{P_N}$ as a pseudo-metric which is degenerate in the fibre direction of the fibering $\mathbf{C}^{N+1} - \{0\} \to P_N\mathbf{C}$. If W immerses $P(J^k X)$ into $P_N\mathbf{C}$, then $\varphi^* ds^2_{P_N}$ is a pseudo-metric on $J^k X - \{0_k\}$, which is degenerate only in the fibre direction of the fibering $J^k X - \{0_k\} \to P(J^k X)$. Set

$$\Psi_{J^k X} = \varphi^* \Phi_{P_N}.$$

The following is a variant of a theorem in Green-Griffiths [1].

(2.5.19) Theorem. *Let X be a compact complex manifold such that a suitable subspace $W \subset H^0(X, D_X^{(k,m)})$ defines an immersion of $P(J^k X)$ into $P_N\mathbf{C}$. Let $\omega_0, \omega_1, \ldots, \omega_N$ be a basis for W, and define a jet metric F on $J^k X$ by*

$$F(\xi) = \left(\sum_\alpha |\omega_\alpha(\xi)|^2 \right)^{1/2m},$$

where the ω_α are considered as functions on $J^k X$. Then the holomorphic sectional curvature K_F of F is bounded above by a negative constant.

Proof. Since K_F is a function on $J^k X - \{0_k\}$ and satisfies the invariance condition (2.5.8), it may be considered as a function on the projective bundle $P(J^k X)$. Since $P(J^k X)$ is compact, it suffices to show that $K_F < 0$ on $J^k X - \{0_k\}$.

Let $f: D \to X$, and $j^k f: D \to J^k X$ be its lift. As before, for $(j^k f)^*$ we write f^*. Thus

$$f^* F = (j^k f)^* F \quad \text{and} \quad f^* \Psi_{J^k X} = (j^k f)^* \Psi_{J^k X}.$$

Since

$$dd^c \log f^* F = dd^c \log f^*(\varphi^* \|w\|^{2/m}) = \frac{1}{m} f^* \Psi_{J^k X},$$

from (2.5.6) we have

$$(2.5.20) \qquad i K_{f^* F} dz \wedge d\bar{z} = -\frac{1}{m} f^* \left(\frac{1}{F^2} \Psi_{J^k X} \right).$$

Let $\xi \in J^k X$ with $\xi \neq 0$, and take $f \in \mathrm{Hol}(D, X)$ such that $\xi = j_a^k f$, where $a \in D$. Since $F > 0$ on $J^k X - \{0_k\}$, $f^* F(a) > 0$ and $K_{f^* F}(a)$ is defined. We want to show that $K_{f^* F}(a) < 0$. From (2.5.20) it is clear that $K_{f^* F} \leq 0$. Suppose $K_{f^* F}(a) = 0$. Then $f^* \Psi_{J^k X} = 0$ at a, which means that the differential of the map

$$j^k: D \to J^k X, \qquad z \mapsto j_z^k f,$$

at $a \in D$ is vertical (i.e., in the fibre direction) in the fibering $J^k X - \{0_k\} \to P(J^k X)$. With respect to a local coordinate system of X, $j_z^k f$ is given by

$$j_z^k f = (f(z); f'(z), f''(z), \ldots, f^{(k)}(z)) \in X \times \mathbf{C}^{kn}, \qquad z \in D.$$

Its derivative at a is given by

$$(f'(a); f''(a), f'''(a), \ldots, f^{(k+1)}(a)) \in T_{f(a)} X \times \mathbf{C}^{kn}.$$

The condition that it is vertical amounts to the following:

$$f'(a) = 0, \ (f''(a), f'''(a), \ldots, f^{(k+1)}(a)) = c(f'(a), f''(a), \ldots, f^{(k)}(a)).$$

Since

$$c(f'(a), f''(a), \ldots, f^{(k)}(a)) = (cf'(a), c^2 f''(a), \ldots, c^k f^{(k)}(a)),$$

we obtain successively

$$0 = f'(a) = f''(a) = f'''(a) = \ldots = f^{(k+1)}(a),$$

which shows that $j_a^k f = 0$. This is in contradiction to the assumption $j_a^k f = \xi \neq 0$.
□

The following generalization is also a variant of a result in Green-Griffiths [1]. The special case where $D_X^{(1,m)} = S^k T^* X$ is due to Noguchi [1].

(2.5.21) **Corollary.** *Let X be a compact complex manifold with a very ample line bundle L. Assume that the vector bundle $D_X^{(k,m)} \otimes L^{-1}$ admits a section for some m. Let $\sigma_0, \ldots, \sigma_N$ be a basis of $H^0(X, L)$, and $\tau_0, \ldots, \tau_q$ be a basis for $H^0(X, D_X^{(k,m)} \otimes L^{-1})$. We define a jet pseudo-metric F on $J^k X$ by*

$$F(\xi) = \left(\sum_{i,\alpha} |(\tau_i \sigma_\alpha)(\xi)|^2 \right)^{1/2m}.$$

Then the curvature K_F of F is bounded above by a negative constant.

Proof. Set $\omega_{i,\alpha} = \tau_i \sigma_\alpha \in H^0(X, D_X^{(k,m)})$, and apply the proof of (2.5.19).   □

The pseudo-metric F constructed in (2.5.21) is degenerate on the base locus of $H^0(X, D_X^{(k,m)} \otimes L^{-1})$, i.e., the common zeros of $\tau_0, \ldots, \tau_q$.

Chapter 3. Intrinsic Distances

1 Two Intrinsic Pseudo-distances

Throughout this section we denote the unit disc by D, its Poincaré metric by ds^2, and the Poincaré distance by ρ, (see Chapter 2, Section 1).

In order to generalize the Schwarz-Pick lemma to holomorphic mappings between higher dimensional domains, Carathéodory [1], [2] introduced what is now known as the **Carathéodory pseudo-distance**. Given a complex space X, let $\mathrm{Hol}(X, D)$ denote the set of holomorphic mappings $f\colon X \to D$. Carathéodory defined a pseudo-distance c_X on X by setting

$$(3.1.1) \qquad c_X(p, q) = \sup_f \rho(f(p), f(q)) \qquad \text{for} \quad p, q \in X,$$

where the supremum is taken over all $f \in \mathrm{Hol}(X, D)$. Since D is homogeneous, it suffices to take the supremum over the subfamily $\mathcal{F} = \{f \in \mathrm{Hol}(X, D); \ f(p) = 0\}$. Applying the Arzela-Ascoli Theorem (1.3.1) and the following proposition (3.1.2) to this family $\mathcal{F}$ we see that $\mathcal{F}$ is compact. Hence the supremum in (3.1.1) is actually achieved by a mapping belonging to the family $\mathcal{F}$.

(3.1.2) **Proposition.** (1) *If X and Y are two complex spaces, then*

$$c_Y(f(p), f(q)) \le c_X(p, q) \qquad \text{for} \quad f \in \mathrm{Hol}(X, Y) \quad \text{and} \quad p, q \in X,$$

that is, $f\colon X \to Y$ is distance-decreasing;

(2) *For $X = D$, the Carathéodory pseudo-distance c_D coincides with the Poincaré distance ρ, i.e.,*

$$c_D = \rho.$$

Proof. (1) is immediate from the definition of the Carathéodory pseudo-distance. The Schwarz-Pick lemma (2.1.7) implies $c_D \le \rho$ while the consideration of the identity transformation of D implies $c_D \ge \rho$. $\qquad\square$

Clearly, (3.1.2) may be regarded as a generalization of the Schwarz-Pick lemma.

Dualizing the construction of the Carathéodory pseudo-distance, we define first a function d'_X on $X \times X$ by setting

$$(3.1.3) \qquad\qquad d'_X(p, q) = \inf \rho(a, b),$$

where the infimum is taken over all holomorphic maps $f: D \to X$ and all pairs of points $a, b \in D$ such that $f(a) = p$ and $f(b) = q$. By definition, $d'_X(p, q) = \infty$ if there is no $f \in \mathrm{Hol}(D, X)$ such that $p, q \in f(D)$. Although this function is symmetric, i.e., $d'_X(p, q) = d'_X(q, p)$, it may not satisfy the triangular inequality. We define the **Kobayashi pseudo-distance** d_X as the largest pseudo-distance bounded by d'_X. More explicitly, we construct d_X as follows, (Kobayashi [4]).

We call an element of $\mathrm{Hol}(D, X)$ a **holomorphic disc** in X. Given two points p, q of X, we consider a **chain of holomorphic discs** from p to q, that is, a chain of points $p = p_0, p_1, \ldots, p_k = q$ of X, pairs of points $a_1, b_1, \ldots, a_k, b_k$ of D, and holomorphic mappings $f_1, \ldots, f_k \in \mathrm{Hol}(D, X)$ such that

$$f_i(a_i) = p_{i-1} \quad \text{and} \quad f_i(b_i) = p_i \qquad \text{for} \quad i = 1, \ldots, k.$$

Denoting this chain by α, we define its **length** $l(\alpha)$ by

$$(3.1.4) \qquad\qquad l(\alpha) = \rho(a_1, b_1) + \ldots + \rho(a_k, b_k),$$

and the pseudo-distance d_X by

$$(3.1.5) \qquad\qquad d_X(p, q) = \inf_{\alpha} l(\alpha),$$

where the infimum is taken over all chains α of holomorphic discs from p to q. Then

(3.1.6) Proposition. (1)   *If X and Y are two complex spaces, then*

$$d_Y(f(p), f(q)) \le d_X(p, q) \qquad \text{for} \quad f \in \mathrm{Hol}(X, Y), \quad p, q \in X;$$

(2)   *For the unit disc D, the pseudo-distance d_D coincides with ρ, i.e.,*

$$d_D = \rho.$$

The proof of (3.1.6) is similar to that of (3.1.2).

Among the pseudo-distances which share the properties stated in (3.1.2) and (3.1.6), c_X and d_X represent the two extremes in the following sense.

(3.1.7) Proposition. *Let X be a complex space.*
(1)   *If δ_X is a pseudo-distance such that*

$$\rho(f(p), f(q)) \le \delta_X(p, q) \qquad \text{for} \quad f \in \mathrm{Hol}(X, D), \quad p, q \in X,$$

then

$$c_X(p, q) \le \delta_X(p, q) \qquad \text{for} \quad p, q \in X;$$

(2)   *If δ_X is a pseudo-distance such that*

$$\delta_X(f(a), f(b)) \le \rho(a, b) \qquad \text{for} \quad f \in \mathrm{Hol}(D, X), \quad a, b \in D,$$

then

$$\delta_X(p, q) \le d_X(p, q) \qquad \text{for} \quad p, q \in X.$$

Proof. We prove only (2) since (1) is even more trivial. As in the definition of d_X, let $p_0, \ldots, p_k, a_1, b_1, \ldots, a_k, b_k, f_1, \ldots, f_k$ be a chain of holomorphic discs from p to q. Then

$$\delta_X(p, q) \leq \sum \delta_X(p_{i-1}, p_i) = \sum \delta_X(f_i(a_i), f_i(b_i)) \leq \sum \rho(a_i, b_i).$$

Hence,

$$\delta_X(p, q) \leq \inf \sum \rho(a_i, b_i) = d_X(p, q).$$

$\square$

(3.1.8) Corollary. *For any complex space X, we have*

$$c_X(p, q) \leq d_X(p, q) \qquad for \quad p, q \in X.$$

(3.1.9) Theorem. *For any complex spaces X and Y, we have*

$$d_{X \times Y}((x, y), (x', y')) = \max\{d_X(x, x'), d_Y(y, y')\} \qquad x, x' \in X, \ y, y' \in Y.$$

Proof. Since the projection $\pi: X \times Y \to X$ is holomorphic and hence is distance-decreasing, we have

$$d_{X \times Y}((x, y), (x', y')) \geq d_X(x, x').$$

Similarly,

$$d_{X \times Y}((x, y), (x', y')) \geq d_Y(y, y').$$

Without loss of generality, we may assume that $d_X(x, x') \geq d_Y(y, y')$. Then it remains to prove

$$d_{X \times Y}((x, y), (x', y')) \leq d_X(x, x').$$

Given a chain α of holomorphic discs of length $l(\alpha)$ from x to x' in X, we want to construct a chain γ of holomorphic discs of length $l(\gamma)$ from (x, y) to (x', y') in $X \times Y$ such that $l(\gamma) \leq l(\alpha)$. By our assumption there is a chain β of holomorphic discs of length $l(\beta)$ from y to y' in Y such that $l(\beta) \leq l(\alpha)$. Let α and β consist of

$$\alpha : \begin{cases} x = x_0, \ldots, x_k = x' \in X; \\ a_1, a_1', \ldots, a_k, a_k' \in D; \\ f_1, \ldots, f_k \in \text{Hol}(D, X); \end{cases}$$

$$\beta : \begin{cases} y = y_0, \ldots, y_m = y' \in Y; \\ b_1, b_1', \ldots, b_m, b_m' \in D; \\ g_1, \ldots, g_m \in \text{Hol}(D, Y). \end{cases}$$

We may assume that $k = m$ and $\rho(a_i, a_i') \geq \rho(b_i, b_i')$ for all $i = 1, \ldots, k$ by refining these chains as follows.

Let a_i'' be a point on the geodesic from a_i to a_i' so that

$$\rho(a_i, a_i') = \rho(a_i, a_i'') + \rho(a_i'', a_i').$$

Let $x_i' = f_i(a_i'')$. Then we obtain a refined chain α' from x to x' consisting of $k + 1$ holomorphic discs:

$$\alpha' : \begin{cases} x = x_0, \ldots, x_{i-1}, x_i', x_i, \ldots, x_k = x' \in X; \\ a_1, a_1', \ldots, a_i, a_i'', a_i'', a_i', \ldots, a_k, a_k' \in D; \\ f_1, \ldots, f_i, f_i, \ldots, f_k \in \mathrm{Hol}(D, X). \end{cases}$$

We repeat this process a finite number of times and call the resulting chain α' a **refinement** of the chain α. This process does not really alter the geometric picture of the chain; in particular, $l(\alpha) = l(\alpha')$.

It is now clear that, after suitable refinements of α and β, we may assume that $k = m$ and $\rho(a_i, a_i') \geq \rho(b_i, b_i')$. By composing f_i and g_i with suitable automorphisms of D, we may also assume that $a_i = b_i = 0$ and that a_i', b_i' are all positive real numbers. Then $0 < b_i' \leq a_i' < 1$. Let $\varphi_i \colon D \to D$ be the multiplication by the positive number b_i'/a_i'. We define a holomorphic map $h_i \colon D \to X \times Y$ by

$$h_i(z) = (f_i(z), g_i(\varphi_i(z))) \qquad \text{for} \quad z \in D.$$

Let γ be the chain from (x, y) to (x', y') consisting of

$$(x, y) = (x_0, y_0), (x_1, y_1), \ldots, (x_k, y_k) = (x', y') \in X \times Y;$$

$$0, a_1', \ldots, 0, a_k' \in D; \quad h_1, \ldots, h_k \in \mathrm{Hol}(D, X \times Y).$$

Then γ possesses the required properties. $\qquad\qquad\square$

The theorem above, proven in a different manner in Royden [2], sharpens the result in Kobayashi [4; p. 47].

In particular,

(3.1.10) **Corollary.** *For a polydisc* $D^n = D \times \ldots \times D$, *we have*

$$d_{D^n}((x_1, \ldots, x_n), (y_1, \ldots, y_n)) = \max_i \{d_D(x_i, y_i)\}.$$

A similar product property for the Carathéodory pseudo-distance, due to Jarnicki and Pflug [4], will be proved in Section 9 of Chapter 4. Here we give a weaker result, which is easy to prove.

(3.1.11) **Proposition.** *Let* X *and* Y *be complex spaces. For* $(x, y), (x', y') \in X \times Y$, *we have*

$$\max\{c_X(x, x'), c_Y(y, y')\} \leq c_{X \times Y}((x, y), (x', y')) \leq c_X(x, x') + c_Y(y, y').$$

Proof. The first inequality is proven in the same way as in the proof of (3.1.9). In order to prove the second inequality, we observe first

$$c_{X \times Y}((x, y), (x', y)) = c_X(x, x').$$

This follows from the distance-decreasing property of the holomorphic maps $(x, y) \in X \times Y \to x \in X$ and $x \in X \to (x, y) \in X \times Y$. Now,

$$\begin{aligned} c_{X \times Y}((x, y), (x', y')) &\leq c_{X \times Y}((x, y), (x', y)) + c_{X \times Y}((x', y), (x', y')) \\ &= c_X(x, x') + c_Y(y, y'). \end{aligned}$$

$$\square$$

However, for a polydisc D^n, we have the result as precise as (3.1.10).

(3.1.12) **Corollary.** *For $D^n = D \times \ldots \times D$, we have*

$$c_{D^n}((x_1, \ldots, x_n), (y_1, \ldots, y_n)) = \max\{c_D(x_i, y_i)\}.$$

Proof. Applying an automorphism to each factor D, we may assume that $x_1 = \ldots = x_n = 0$ and $y_1, \ldots, y_n$ are all nonnegative real numbers. Without loss of generality we may assume that $c_D(x_i, y_i) \leq c_D(x_1, y_1)$ for all i, i.e., $y_i \leq y_1$. Let $c_i = y_i/y_1$ so that $0 \leq c_i \leq 1$. Consider the holomorphic map $f\colon D \to D^n$ defined by

$$f(z) = (c_1 z, \ldots, c_n z) \qquad \text{for} \quad z \in D.$$

Since f is distance-decreasing, we have

$$c_{D^n}((0, \ldots, 0), (y_1, \ldots, y_n)) = c_{D^n}(f(0), f(y_1)) \leq c_D(0, y_1).$$

This combined with the first inequality in (3.1.11) yields the desired equality.

$\square$

(3.1.13) **Proposition.** *For a complex space X, both c_X and d_X are continuous (as mappings from $X \times X$ to $\mathbf{R}$).*

Proof. Because of (3.1.8) it suffices to prove that d_X is continuous. In order to prove the continuity of d_X at $(p, q) \in X \times X$, let $\{(p_n, q_n)\}$ be a sequence in $X \times X$ which converges to (p, q) in the complex space topology of $X \times X$. From the triangular inequality we obtain

$$|d_X(p_n, q_n) - d_X(p, q)| \leq d_X(p_n, p) + d_X(q_n, q).$$

Hence the proof is reduced to showing that $d_X(p_n, p) \to 0$ as $p_n \to p$. Let U be a neighborhood of p. Since $d_X \leq d_U$ in U by (3.1.6), it suffices to show that $d_U(p_n, p) \to 0$ as $p_n \to p$ in U.

If p is a nonsingular point of X, then we may assume that U is a polydisc D^n, and our assertion follows from (3.1.10).

Consider the case where p is a singular point, and assume that there is a positive number δ such that $d_U(p_n, p) \geq \delta$ for all n. Let $\pi\colon \tilde{U} \to U$ be a resolution of the singularity, and $\{q_n\}$ a sequence in $\tilde{U}$ such that $\pi(q_n) = p_n$. Since π is proper, by taking a subsequence we may assume that $\{q_n\}$ converges to a point $q \in \tilde{U}$. By continuity, $\pi(q) = p$. Let V be a polydisc neighborhood of q in $\tilde{U}$. Since $\pi\colon V \to U$ is distance-decreasing and d_V is continuous by (3.1.10), we have

$$d_U(p_n, p) = d_U(\pi(q_n), \pi(q)) \leq d_V(q_n, q) \to 0$$

as $n \to \infty$, which is a contradiction.

$\square$

The proof of (3.1.13) given here is due to Barth [3]. It would be of some interest to find a proof that does not rely on resolutions of singularities.

Given $p, q \in X$, let α be a chain of holomorphic discs:

$$\alpha : \begin{cases} p = p_0, p_1, \ldots, p_k = q \in X; \\ a_1, b_1, \ldots, a_k, b_k \in D; \\ f_1, \ldots, f_k \in \mathrm{Hol}(D, X). \end{cases}$$

Let C_i be the (unique) geodesic from a_i to b_i in D. Joining $f_1(C_1), \ldots, f_k(C_k)$ consecutively, we obtain a curve from p to q in X. This curve, denoted by $|\alpha|$, will be called the **thread** of the chain α. It is rectifiable with respect to d_X and its length $L(|\alpha|)$ in the sense of (1.1.1) is clearly bounded by $l(\alpha)$, i.e.,

$$(3.1.14) \qquad\qquad L(|\alpha|) \le l(\alpha).$$

(3.1.15) **Theorem.** *For any complex space X, the pseudo-distance d_X is inner.*

Proof. Let d_X^i be the inner pseudo-distance induced by d_X, (see (1.1.2)):

$$d_X^i(p, q) = \inf_{\gamma} L(\gamma),$$

where the infimum is taken over all rectifiable curves γ from p to q. Since the inequality $d_X \le d_X^i$ holds always (see (1.1.3)), it suffices to prove $d_X^i \le d_X$. We have

$$d_X^i(p, q) \le \inf_{\alpha} L(|\alpha|) \le \inf_{\alpha} l(\alpha) = d_X(p, q),$$

where the infima are taken over all holomorphic chains α of discs from p to q. $\qquad\square$

In the course of the proof we obtained also the following equality:

$$(3.1.16) \qquad\qquad d_X(p, q) = \inf_{\alpha} L(|\alpha|),$$

where the infimum is taken over all chains α of holomorphic discs from p to q.

As we shall see soon, the Carathéodory pseudo-distance c_X is not always inner. Since d_X is inner, it is without detour by (1.1.6). But (1.1.7) says more:

(3.1.17) **Corollary.** *Let X be a complex space. Then for every $p \in X$ and for every positive real number ε, the open ε-ball*

$$U(p; \varepsilon) = \{q \in X; \; d_X(p, q) < \varepsilon\}$$

with center p is finitely arcwise connected. In fact, every $q \in U(p; \varepsilon)$ can be joined to p by a curve of length $< \varepsilon$ lying in $U(p; \varepsilon)$.

Following (1.2.2) we define the **degeneracy set** $\Delta(p)$ for $p \in X$ by

$$\Delta(p) = \{q \in X; \; d_X(p, q) = 0\}.$$

The following is an immediate consequence of (1.2.7).

(3.1.18) **Corollary.** *Let X be a complex space and $p \in X$. If $\Delta(p)$ is compact, so in particular if X is compact, then $\Delta(p)$ is connected.*

In connection with (3.1.17) we note that every $q \in U(p; \varepsilon)$ can be joined to p by a chain α of holomorphic discs of length $l(\alpha) < \varepsilon$ whose thread $|\alpha|$ lies in $U(p; \varepsilon)$. However, α itself may not stay within $U(p; \varepsilon)$. The following proposition rectifies this situation.

(3.1.19) Proposition. *Let X be a complex space. Given $o \in X$ and positive numbers ρ and ε there is a constant $C > 1$ such that, for any $\delta > 0$, every pair of points $p, q \in U(o; \rho)$ can be joined by a chain β of holomorphic discs of length $l(\beta) < C(d_X(p, q) + \delta)$ which lies in $U(o; 3\rho + \varepsilon)$. In particular,*

$$d_{U(o;3\rho+\varepsilon)}(p, q) \leq C \cdot d_X(p, q) \qquad for \quad p, q \in U(o, \rho).$$

Proof. Let r, $0 < r < 1$, be the positive number determined by $d_D(0, r) = \varepsilon$ so that the disc $D_r = \{z \in \mathbf{C}; \ |z| < r\}$ of radius r has radius ε with respect to the Poincaré distance d_D. We choose C in such a way that

$$d_{D_r}(0, a) \leq C \cdot d_D(0, a) \qquad for \quad a \in D_{r/2}.$$

In order to show that this C has the desired property, we join $p, q \in U(o; \rho)$ by a chain α of holomorphic discs in X of length $l(\alpha) < d_X(p, q) + \delta < 2\rho$, (where δ is a small number such that the second inequality holds). Since the length of thread $|\alpha|$ is less than 2ρ, it follows that $|\alpha|$ is contained in $U(o; 3\rho)$. Let $f_i \colon D \to X$ be the i-th holomorphic disc of the chain α sending $a_i, b_i \in D$ to $p_{i-1}, p_i \in X$. Without loss of generality, we may assume that $a_i = 0$ and $|b_i| < r/2$. Since $p_{i-1} \in U(o; 3\rho)$, we have $f_i(D_r) \subset U(o; 3\rho+\varepsilon)$. For if $z \in D_r$, then $d_D(0, z) < \varepsilon$ and $d_X(p_{i-1}, f_i(z)) = d_X(f_i(0), f_i(z)) < \varepsilon$. This shows that we obtain a desired chain β by restricting the holomorphic disc $f_i \colon D \to X$ to D_r; in order to normalize the smaller disc D_r we set

$$g_i(z) = f_i(rz) \qquad for \quad z \in D,$$

and let β be the chain consisting of $g_i \colon D \to X$. Then β is a chain from p to q and lies in $U(o; 3\rho + \varepsilon)$ and

$$l(\beta) \leq C \cdot l(\alpha) < C(d_X(p, q) + \delta).$$

Since δ is arbitrary, we obtain the desired inequality. $\qquad\qquad\square$

(3.1.20) Proposition. *Let $\{X_m\}$ be a monotone increasing sequence of subdomains in a complex space X such that $X = \bigcup X_m$. Then*

(a) $\qquad c_X(p, q) = \lim c_{X_m}(p, q).$

(b) $\qquad d_X(p, q) = \lim d_{X_m}(p, q).$

Proof. (a)   Clearly, we have

$$c_{X_m} \geq c_{X_{m+1}} \geq c_X \qquad on \quad X_m.$$

Given $p, q \in X$, choose $f_m \in \mathrm{Hol}(X_m, D)$ such that $f_m(p) = 0$ and $c_{X_m}(p, q) = \rho(0, f_m(q))$. Then the usual argument (using the Arzela-Ascoli theorem (1.3.1)) shows that a suitable subsequence of $\{f_m\}$ converges to a map $f \in \mathrm{Hol}(X, D)$. Then

$$\lim c_{X_m}(p, q) = \lim \rho(0, f_m(q)) = \rho(0, f(q)) \leq c_X(p, q).$$

(b)   Similarly, we have

$$d_{X_m} \geq d_{X_{m+1}} \geq d_X \qquad \text{on} \quad X_m.$$

Conversely, given $p, q \in X$ and $\varepsilon > 0$, let α be a chain of holomorphic disks from p to q with its length $l(\alpha) < d_X(p, q) + \varepsilon$. Let α consist of holomorphic disks $f_i \in \mathrm{Hol}(D, X)$. We shrink each holomorphic disk f_i by composing it with the multiplication by $r < 1$. We set $f_i^{(r)}(z) = f_i(rz)$. If r is sufficiently close to 1, then $\{f_i^{(r)}\}$ defines a chain $\alpha^{(r)}$ from p to q. Given any $\varepsilon > 0$, the length $l(\alpha^{(r)}) < l(\alpha) + \varepsilon$ for r sufficiently close to 1. Since the chain $\alpha^{(r)}$ is contained in a compact subset K of X, it is contained in X_m for $m > m_0$. Hence,

$$d_{X_m}(p, q) \leq l(\alpha^{(r)}) < l(\alpha) + \varepsilon < d_X(p, q) + 2\varepsilon.$$

$\square$

The proposition above is in Hristov [1, 3], where he discusses also the situation where X is the limit of a monotone decreasing sequence of complex spaces X_m.

(3.1.21) **Example.** For the Gaussian plane $\mathbf{C}$ and the punctured plane $\mathbf{C}^* = \mathbf{C} - \{0\}$, we have

$$d_{\mathbf{C}} = 0, \quad d_{\mathbf{C}^*} = 0, \quad c_{\mathbf{C}} = 0, \quad c_{\mathbf{C}^*} = 0.$$

In fact, given two points $p, q \in \mathbf{C}$ and an arbitrarily small positive number ε, there is a map $f \in \mathrm{Hol}(D, \mathbf{C})$ such that $f(0) = p$ and $f(\varepsilon) = q$. Hence, $d_{\mathbf{C}}(p, q) \leq \varepsilon$. In order to prove $d_{\mathbf{C}^*} = 0$, we consider the surjective holomorphic map $z \in \mathbf{C} \to e^z \in \mathbf{C}^*$. Since it is distance-decreasing, $d_{\mathbf{C}} = 0$ implies $d_{\mathbf{C}^*} = 0$. The remaining two statements follow from (3.1.8).

The statement $c_{\mathbf{C}} = 0$ is nothing but Liouville's theorem (that every bounded entire function is constant).

Both $\mathbf{C}$ and $\mathbf{C}^*$ are complex Lie groups. Generally we have

(3.1.22) **Example.** For any connected complex Lie group G, we have

$$d_G = 0 \quad \text{and} \quad c_G = 0.$$

In fact, given $p, q \in G$, there is a sequence of maps $f_1, \ldots, f_k \in \mathrm{Hol}(\mathbf{C}, G)$ such that $p \in f_1(\mathbf{C}), q \in f_k(\mathbf{C})$ and $f_i(\mathbf{C}) \cap f_{i+1}(\mathbf{C}) \neq \emptyset$ for $i = 1, \ldots, k - 1$. (These f_i's are suitable translates of complex 1-parameter subgroups of G.) Our assertion follows from (3.1.21) and the fact that the maps f_i are distance-decreasing.

Sitll more generally,

(3.1.23) **Example.** If a connected complex Lie group G acts on a complex space X, then $d_X(p, q) = 0$ for $p, q \in X$ belonging to the same G-orbit. In particular, if X is a complex space on which a complex Lie group G acts with a dense orbit, then

$$d_X = 0 \quad \text{and} \quad c_X = 0.$$

In fact, for a fixed $p_0 \in X$ the mapping $G \to X$ that sends $g \in G$ to $g(p_0) \in X$ is holomorphic. Hence, $d_X(g(p_0), g'(p_0)) \le d_G(g, g') = 0$.

(3.1.24) **Example.** Let $p: \mathbf{C}^n \to \mathbf{R}^+$ be a norm (not necessarily the Euclidean norm), and $B = \{z \in \mathbf{C}^n; \; p(z) < 1\}$ the unit ball for this norm. Then

$$c_B(0, z) = d_B(0, z) = \rho(0, p(z)) \qquad \text{for} \quad z \in B.$$

In fact, given $z \in B$, $z \ne 0$, define $f: D \to B$ by $f(t) = tz/p(z)$ so that $f(p(z)) = z$. Then

$$c_B(0, z) \le d_B(0, z) \le \rho(0, p(z)).$$

On the other hand, for every $z \in \mathbf{C}^n$, there exists a linear functional $\lambda_z: \mathbf{C}^n \to \mathbf{C}$ such that $\lambda_z(z) = p(z)$ and $|\lambda_z(w)| \le p(w)$ for all $w \in \mathbf{C}^n$. Then λ_z sends B into D and, if $z \in B$, then

$$\rho(0, p(z)) \le c_B(0, z).$$

If p is the Euclidean norm, then B is homogeneous. So in this case, c_B and d_B coincide not only at the origin but everywhere.

Although d_X is always an inner pseudo-distance, c_X is not necessarily inner. The following examples are due to Barth [6].

(3.1.25) **Example.** Fix $0 < r < 1$ and $0 < s < 1$, and consider the Hartogs figure

$$X = D^2 - \{(z, w); \; |z| \le r \quad \text{and} \quad |w| \ge s\}$$

in $\mathbf{C}^2$. Since every bounded holomorphic function $f: X \to D$ extends to a holomorphic function $\tilde{f}: D^2 \to D$ with the same bound, c_X is the restriction of c_{D^2}. Take $p = (z_0, w_0) \in X \subset D^2$ with $|z_0| > r$, $|w_0| > s$, and $\rho(|w_0|, s) > \rho(z_0, 0)$, and let $q = (-z_0, w_0) \in X$. A curve $\gamma = (\gamma_1, \gamma_2)$ joining p to q in X must go around or under the notch in the bidisc. That is, either $|\gamma_1(t)| \ge r$ for all t, or $|\gamma_1(t_0)| < r$ and $|\gamma_2(t_0)| < s$ for some t_0. In the first case, we have

$$L(\gamma) \ge K > c_X(p, q),$$

where K is the length of a curve (measured in the Poincaré metric) from $-z_0$ to z_0 which stays outside the disc of radius r while $c_X(p, q) = c_D(z_0, -z_0)$ is the geodesic distance from z_0 to $-z_0$ in D. In the second case, we have

$$\begin{aligned} L(\gamma) \; &\ge \; c_X(p, \gamma(t_0)) + c_X(\gamma(t_0), q) \ge \rho(w_0, \gamma_2(t_0)) + \rho(\gamma_2(t_0), w_0) \\ &> \; 2\rho(|w_0|, s) > 2\rho(z_0, 0) = c_X(p, q). \end{aligned}$$

Since $\rho(|w_0|, s) - \rho(z_0, 0)$ is independent of γ, we have

$$L(\gamma) > K' > c_X(p, q),$$

where K' is a constant independent of γ. Thus

$$c_X^i(p, q) = \inf_{\gamma} L(\gamma) > c_X(p, q).$$

The domain X in the example above is not a domain of holomorphy. Barth [6] pointed out that the Carathéodory distance of the following domain of holomorphy constructed by Sibony [1] is not inner.

(3.1.26) **Example**. In order to define a domain $X = M(D, V) \subset \mathbf{C}^2$, let $\{a_n\}$ be a discrete sequence of points in the unit disc D such that every point of the boundary circle ∂D is the non-tangential limit of a subsequence. Let $\{\lambda_n\}$ be a sequence of positive real numbers such that $\sum_n \lambda_n < \infty$. We set

$$\varphi(z) = \sum_n \lambda_n \log \frac{1}{2}|z - a_n|$$

and

$$V(z) = e^{\varphi(z)}.$$

The function φ is negative and subharmonic in D, and the function $V(z)$ is also subharmonic and $0 \leq V(z) < 1$ in D. Since $\{a_n\}$ is discrete in D, $V(z)$ is continuous and vanishes only at a_n. Define

$$X = M(D, V) = \{(z, w) \in D \times \mathbf{C};\ |w| < e^{-V(z)}\} \subset D \times D.$$

Since the function $|w|e^{V(z)}$ is plurisubharmonic in $D \times \mathbf{C}$, the domain X is pseudo-convex and, by Oka's theorem, it is a domain of holomorphy. Sibony has shown that X is a proper subdomain of D^2 and every bounded holomorphic function on X extends to a bounded holomorphic function on D^2. It follows that c_X is the restriction of c_{D^2}. Then Barth chooses the sequence $\{a_n\}$ in such a way that $a_1 = -a_2 = z_0$ and $a_n \neq 0$ for all n. Then X is contained in a Hartogs-figure (as in the example above) containing the discs $\{\pm z_0\} \times D$. Taking w_0, p and q as in the example above, we see that c_X is not inner.

Then Barth raised the following question: *If a bounded domain X is finitely compact with respect to c_X, is c_X inner ?* The following counter-example was found by Vigué [2].

(3.1.27) **Example**. Let $X = \{(z, w) \in \mathbf{C}^2;\ |z| + |w| < 1, |zw| < 1/16\}$. This is a generalized analytic polyhedron (defined in Chapter 4, Section 1). Every generalized analytic polyhedron is finitely compact with respect to its Carathéodory distance. Vigué [2] shows that

$$c_X((0, 0), (z, z)) < c_X^i((0, 0), (z, z)) \qquad \text{for}\ \ \frac{1}{8} < |z| < \frac{1}{4}.$$

We note that X is also a Reinhardt domain, i.e., it is invariant under the transformations $(z, w) \mapsto (e^{is}z, e^{it}w)$.

The following simple 1-dimensional example was given by Jarnicki-Pflug [5]:

(3.1.28) **Example**. For the annulus

$$A = \left\{ z \in \mathbf{C}; \ \frac{1}{R} < |z| < R \right\}, \quad R > 1,$$

the Carathéodory distance c_A is not inner.

(3.1.29) **Remark**. A system which assigns a pseudo-distance δ_X to each complex space X is called (by L. A. Harris [1] who considered domains in normed linear spaces) a **Schwarz-Pick system** if the following conditions are satisfied:
   (i)   it assigns to D the Poincaré distance ρ,
   (ii) every $f \in \mathrm{Hol}(X, Y)$ is distance-decreasing, i.e.,

$$\delta_Y(f(x), f(x')) \le \delta_X(x, x') \qquad x, x' \in X.$$

The Schwarz-Pick system of Carathéodory pseudo-distances and that of Kobayashi pseudo-distances are the two extremes in the sense of (3.1.7). An example of Schwarz-Pick system which lies between the Carathéodory pseudodistance and the Kobayashi pseudodistance will be discussed in Section 3 of Chapter 4.

(3.1.30) **Remark**. Demailly-Lempert-Shiffman [1] proved that for a quasi-projective variety X, the pseudo-distance d_X can be defined by means of chains of algebraic curves. More precisely,

$$d_X(p, q) = \inf \sum_{i=1}^{k} d_{C_i}(p_{i-1}, p_i),$$

where the infimum is taken over all chains of points $p = p_0, p_1, \ldots, p_k = q$ and irreducible algebraic curves $C_1, \ldots, C_k$ in X such that

$$p_0 \in C_1, \ p_1 \in C_1 \cap C_2, \ldots, p_{k-1} \in C_{k-1} \cap C_k, \ p_k \in C_k.$$

(3.1.31) **Remark**. Hahn [3] considers the pseudodistance obtained using only injective holomorphic discs in definition (3.1.5). Clearly, this pseudodistance dominates the Kobayashi pseudodistance. However, Overholt [1] shows that for a domain in $\mathbf{C}^n$, $n \ge 3$, it coincides with the Kobayashi pseudodistance. The same paper briefly summarizes earlier contributions by Minda [1], Vesentini [8], Vigué [10], and J. H. Zhang [2].

(3.1.32) **Remark**. Wu [7] introduced an invariant metric related to d_x but with more smoothness. See akso Cheung-Kim [1, 2].

(3.1.33) **Remark**. For systematic accounts of intrinsic pseudo-distances for domains in infinite dimensional normed spaces, see L. A. Harris [1], Franzoni-Vesentini [1], Dineen [1], and Barth [9].

(3.1.34) **Remark**. A construction similar to that of d_X defines a pseudonorm on the homology group of X, see Chern-Levine-Nirenberg [1], Królikowski-Tovar [1], and Zhang [1].

2 Hyperbolicity

We shall now study complex spaces X for which the Kobayashi pseudo-distance d_X is a distance. A complex space X is said to be **hyperbolic** if d_X is a distance, i.e., if $d_X(p, q) > 0$ for every pair $p, q \in X$ with $p \neq q$. A hyperbolic complex space X is said to be **complete** if it is Cauchy-complete with respect to d_X. Since d_X is inner by (3.1.15), a complete hyperbolic X is finitely compact with respect to d_X by (1.1.9).

Since d_X is inner (see (3.1.15)) and since every inner distance on a locally compact Hausdorff space X induces the given topology of X (see (1.1.8)), we have the following theorem of Barth [3], see also Barth [9].

(3.2.1) **Theorem.** *Let X be a hyperbolic complex space. Then d_X defines the topology of X.*

(3.2.2) **Proposition.** *Let X be a complex subspace of a complex space Y.*
    (1)   *If Y is hyperbolic, so is X;*
    (2)   *If Y is complete hyperbolic and X is closed, X is also complete hyperbolic.*

Proof. This follows from the fact that the injection $i\colon X \to Y$ is holomorphic and hence distance-decreasing. $\qquad\qquad\square$

From (3.1.9) we have

(3.2.3) **Proposition.** *For complex spaces X and Y, the product $X \times Y$ is (complete) hyperbolic if and only if both X and Y are (complete) hyperbolic.*

(3.2.4) **Proposition.** *Let X and Y be complex spaces, and $f\colon X \to Y$ a holomorphic map. Let Y' be a complex subspace of Y, and define $X' = f^{-1}Y'$. If both X and Y' are complete hyperbolic, so is X'.*

Proof. Let G_f denote the graph of $f\colon X \to Y$; it is a closed complex subspace of $X \times Y$. Let f' be the restriction of f to X', and $G_{f'}$ its graph. Then $G_{f'} = G_f \cap (X \times Y')$. Hence, $G_{f'}$ is closed in $X \times Y'$. By (3.2.3), $X \times Y'$ is complete hyperbolic. By (3.2.2) $G_{f'}$ is complete hyperbolic. Since the projection $X \times Y' \to X$ induces a holomorphic isomorphism from $G_{f'}$ onto X', it follows that X' is complete hyperbolic. $\qquad\qquad\square$

In the following proposition, what we have in mind for applications is the situation where X and X_i are all domains in a complex space Y. The proof is immediate from (1.1.11).

(3.2.5) **Proposition.** *Let X and X_i, $i \in I$, be complex subspaces of a complex space Y such that $X = \bigcap_i X_i$. If all X_i are complete hyperbolic, so is X.*

Although the hyperbolicity is a global concept, we can localize it as follows.

(3.2.6) **Proposition.** *Let X be a complex space. If there exist a family of points $p_\alpha \in X$ and positive numbers δ_α such that, for each α, the δ_α-neighborhood*

$$U_\alpha = \{q \in X; \ d_X(p_\alpha, q) < \delta_\alpha\}$$

is hyperbolic and that $\{U_\alpha\}$ is an open cover of X, then X is hyperbolic.

In particular, if for every $p \in X$ there is a positive number δ such that the δ-neighborhood $U(p; \delta) = \{q \in X; \; d_X(p, q) < \delta\}$ is hyperbolic, then X is hyperbolic.

Proof. Take positive numbers ρ_α and ε_α such that $3\rho_\alpha + \varepsilon_\alpha = \delta_\alpha$. By (3.1.19) there is a constant $C_\alpha > 1$ such that

$$d_{U_\alpha}(p, q) \leq C_\alpha \cdot d_X(p, q) \qquad \text{for} \quad p, q \in U_\alpha.$$

This shows that $d_X(p, q)$ is positive for $p, q \in U_\alpha$, $q \neq p$. If p, q are not in the same U_α, clearly $d_X(p, q) > 0$. $\qquad\square$

(3.2.7) **Proposition.** *Let X be a complex space. If there is a positive number δ such that for every $p \in X$ the δ-neighborhood $U(p; \delta)$ is complete hyperbolic, then X is complete hyperbolic.*

Proof. By (3.2.6) X is hyperbolic. Let $\{p_n\}$ be a Cauchy sequence in X. Let ρ and ε be positive number such that $\delta = 3\rho + \varepsilon$. We may assume, by omitting a finite number of points, that $d_X(p_m, p_n) < \rho$ for all m, n. Then by (3.1.19) we have

$$d_{U(p_1; \delta)}(p_m, p_n) \leq C \cdot d_X(p_m, p_n),$$

which shows that $\{p_n\}$ is a Cauchy sequence with respect to $d_{U(p_1; \delta)}$. Since $U(p_1; \delta)$ is complete with respect to $d_{U(p_1; \delta)}$, the sequence converges. $\qquad\square$

Let $f: X \to Y$ be a holomorphic mapping between complex spaces. Let $f^*d_Y = (f^{-1}d_Y)^i$ be the inner pseudo-distance on X induced from d_Y by f (see (1.1.12)). As we pointed out in Section 1 of Chapter 1, the inequality $f^*d_Y \leq d_X$ follows directly from the definition of f^*d_Y. It is reasonable to expect that the equality holds if f is a covering projection. In this connection we prove

(3.2.8) **Theorem.** *Let X be a complex space and $\pi: \tilde{X} \to X$ a covering space of X. Then*

(1) *If $p, q \in X$ and $\tilde{p}, \tilde{q} \in \tilde{X}$ with $\pi(\tilde{p}) = p$ and $\tilde{\pi}(\tilde{q}) = q$, then*

$$d_X(p, q) = \inf_{\tilde{q}} d_{\tilde{X}}(\tilde{p}, \tilde{q}),$$

where the infimum is taken over all $\tilde{q} \in \tilde{X}$ such that $\pi(\tilde{q}) = q$;

(2) *$\tilde{X}$ is (complete) hyperbolic if and only if X is (complete) hyperbolic;*

(3) *If X is hyperbolic, then $\pi: (\tilde{X}, d_{\tilde{X}}) \to (X, d_X)$ is a local isometry, and*

$$d_{\tilde{X}} = \pi^*d_X.$$

Proof. (1)   Since π is holomorphic, we have

$$d_X(p, q) \leq d_{\tilde{X}}(\tilde{p}, \tilde{q}).$$

Let α be a chain of holomorphic discs from p to q. We lift α to a chain $\tilde{\alpha}$ of holomorphic discs in $\tilde{X}$ starting from $\tilde{p}$. Then $\tilde{\alpha}$ ends at some point $\tilde{q} \in \tilde{X}$ with $\pi(\tilde{q}) = q$, and its length $l(\tilde{\alpha})$ is equal to the length $l(\alpha)$ of α. This proves (1).

(2)   Assume that $\tilde{X}$ is hyperbolic. Let $p, q \in X$ be such that $d_X(p, q) = 0$. Let $\tilde{p} \in \tilde{X}$ be such that $\pi(\tilde{p}) = p$. By (1) there exists a sequence $\{\tilde{q}_n\} \subset \tilde{X}$ such that $\pi(\tilde{q}_n) = q$ and $\lim d_{\tilde{X}}(\tilde{p}, \tilde{q}_n) = 0$. By (3.2.1) $\{\tilde{q}_n\}$ converges to $\tilde{p}$. Then $\{\pi(\tilde{q}_n)\}$ converges to p. But $\pi(\tilde{q}_n) = q$ and hence $p = q$.

Assume that $\tilde{X}$ is complete hyperbolic. If $\tilde{B}_{r+\delta}$ is the closed ball of radius $r + \delta$ around $\tilde{p} \in \tilde{X}$ and if B_r is the closed ball of radius r around $p \in X$, then (1) implies

$$B_r \subset \pi(\tilde{B}_{r+\delta}) \qquad \text{for} \quad \delta > 0.$$

Since $\tilde{B}_{r+\delta}$ is compact by assumption (see (1.1.9)) and B_r is closed, B_r is also compact. Hence X is complete.

If X is (complete) hyperbolic, so is $\tilde{X}$ by (1.3.13).

(3). We prove first that π is a local isometry. Let $\tilde{p} \in \tilde{X}$ and $p = \pi(\tilde{p}) \in X$. Let U be a 2ε-neighborhood of p such that U is homeomorphic to each component of $\pi^{-1}(U)$. We denote the component containing $\tilde{p}$ by $\tilde{U}$. Let V be the ϵ-neighborhood of p and $\tilde{V} = \tilde{U} \cap \pi^{-1}(V)$. We claim that π maps $\tilde{V}$ isometrically onto V. Let $\tilde{q}, \tilde{r} \in \tilde{V}$ and $q = \pi(\tilde{q}), r = \pi(\tilde{r}) \in V$. Since $d_X(q, r) \leq d_{\tilde{X}}(\tilde{q}, \tilde{r}) < \varepsilon$, there is a chain of holomorphic discs from q to r such that $l(\alpha) < \varepsilon$. The thread $|\alpha|$ of α remains within U since $L(|\alpha|) \leq l(\alpha)$, (see (3.1.14)). We lift α to a chain $\tilde{\alpha}$ starting from $\tilde{q}$ in $\tilde{X}$. Since $|\alpha|$ remains within U, $|\tilde{\alpha}|$ stays also within $\tilde{U}$. In particular, the end point of $\tilde{\alpha}$ must be $\pi^{-1}(r) \cap \tilde{U} = \tilde{r}$. This shows that, for every chain α from q to r with $l(\alpha) < \varepsilon$, there is a chain $\tilde{\alpha}$ from $\tilde{q}$ to $\tilde{r}$ with $l(\tilde{\alpha}) = l(\alpha)$. Hence, $d_X(q, r) \geq d_{\tilde{X}}(\tilde{q}, \tilde{r})$. The opposite inequality is a direct consequence of the fact that π is distance-decreasing.

The last statement follows from (1.1.13). $\square$

According to Zwonek [3], the infimum in (1) may not be attained in general.

Part of (3.2.8) is still valid when $\pi: \tilde{X} \to X$ is only a **spread**, i.e., when every point $\tilde{p} \in \tilde{X}$ has a neighborhood $\tilde{U}$ such that π maps $\tilde{U}$ biholomorphically onto the open subset $\pi(\tilde{U})$ of X. By (1.3.12) we have

(3.2.9) Proposition. *A spread $\tilde{X}$ over a hyperbolic complex space X is hyperbolic.*

From (3.2.2) and (3.2.9) we obtain

(3.2.10) Proposition. *If a complex space X is holomorphically immersed into a hyperbolic complex space Y, then X is also hyperbolic.*

The following proposition is of the same character as (3.2.9). From (1.3.12) and (1.3.14) we obtain

(3.2.11) Proposition. *Let $f: \tilde{X} \to X$ be a finite-to-one proper holomorphic map. If X is (complete) hyperbolic, so is $\tilde{X}$.*

A complex space X is said to be **normal** at x if the ring $\mathcal{O}_x$ of germs of holomorphic functions at x is integrally closed in its complete ring of quotients, and X is said to be normal if it is normal everywhere.

The following more geometric interpretation is useful for us. Let S be the singular locus of X, and $X_{\text{reg}} = X - S$. Given an open set $U \subset X$, a function holomorphic on U_{reg} and is bounded on $U_{\text{reg}} \cap K$ for every compact set $K \subset U$ is said to be **weakly holomorphic** on U. Then X is normal at x if and only if every weakly holomorphic function at x extends to a holomorphic function in a small neighborhood of x, see Narasimhan [1; p. 114]. The concept of weakly holomorphic mapping can be defined in terms of local coordinate systems of the target space.

A **normalization** of a complex space X is a pair $(\tilde{X}, \pi)$ consisting of a normal complex space $\tilde{X}$ and a surjective holomorphic mapping $\pi \colon \tilde{X} \to X$ such that

(a) π is proper and $\pi^{-1}(x)$ is finite for every $x \in X$,

(b)   If S is the singular locus of X, then $\tilde{X} - \pi^{-1}S$ is dense in $\tilde{X}$ and $\pi \colon \tilde{X} - \pi^{-1}S \to X - S$ is biholomorphic.

The normalization theorem of Oka (see Grauert-Remmert [3], Narasimhan [1; p. 118]) states that every complex space X has a unique (up to an isomorphism) normalization $\pi \colon \tilde{X} \to X$.

As an immediate consequence of (3.2.11) we have

(3.2.12) **Corollary**. *The normalization $\tilde{X}$ of a (complete) hyperbolic complex space X is (complete) hyperbolic.*

The following example by Kaliman-Zaidenberg [1] shows that the normalization $\tilde{X}$ of a non-hyperbolic complex space X can be hyperbolic.

(3.2.13) **Example**. Let (x, y, u, v) be a coordinate system in $\mathbf{C}^4$, and X be the affine algebraic surface given by the equations

$$y^4 = x^4 - 1$$
$$u^4 = y^4(v^4 - 1).$$

Its singular locus S is given by $y = 0$, which implies $x^4 = 1$ and $u = 0$ while v can be arbitrary. Since S consists of complex lines, X is not hyperbolic.

Let $\tilde{X}$ be the affine algebraic surface given by

$$y^4 = x^4 - 1$$
$$u^4 = v^4 - 1.$$

Then $\tilde{X}$ is nonsingular. Define the map $\pi \colon \tilde{X} \to X$ by $\pi(x, y, u, v) = (x, y, yu, v)$. Then $(\tilde{X}, \pi)$ is the normalization of X. Clearly, $\tilde{X}$ is a direct product of two copies of the affine algebraic curve in $\mathbf{C}^2$ defined by $y^4 = x^4 - 1$. This curve is (complete) hyperbolic. (In fact, its projective completion is a compact Riemann surface of genus 2 and hence hyperbolic by (3.2.8) or by (3.7.3)).

By suitably compactifying X and $\tilde{X}$, Kaliman and Zaidenberg obtain also a compact example.

We apply (3.2.6) to holomorphic maps other than covering projections.

(3.2.14) **Theorem**. *Let $\pi \colon X \to T$ be a holomorphic map of complex spaces. For $t \in T$ and $\delta > 0$, we set $U(t; \delta) = \{u \in T; \ d_T(t, u) < \delta\}$. If for every point $t \in T$ there is a positive number δ such that $\pi^{-1}(U(t; \delta))$ is hyperbolic, then X is hyperbolic.*

Proof. For every $p \in X$, its δ-neighborhood $\{q \in X; d_X(p, q) < \delta\}$ is contained in $\pi^{-1}(U(\pi(p); \delta))$ because π is distance-decreasing. By (3.2.6) X is hyperbolic. $\qquad\square$

The following result is due to Eastwood [1].

(3.2.15) **Theorem.** *Let $\pi: X \to T$ be a holomorphic map of complex spaces. If T is (complete) hyperbolic and if T has an open cover $\{U_i\}$ such that each $\pi^{-1}(U_i)$ is (complete) hyperbolic, then X is (complete) hyperbolic.*

Proof. For each $t \in T$, take $\delta > 0$ such that $U(t, \delta) \subset U_i$ for some U_i. Then $\pi^{-1}(U(t, \delta))$ is hyperbolic. By (3.2.14) X is hyperbolic.

We prove now completeness. Let $\{p_n\}$ be a Cauchy sequence in X. Then $\{\pi(p_n)\}$ is a Cauchy sequence in T and converges to a point $t_o \in T$. Take $\delta > 0$ such that $U(t_o, \delta) \subset U_i$ for some U_i. Take $\varepsilon > 0$ and $\rho > 0$ such that $3\rho + \varepsilon = \delta$. Omitting a finite number of p_n, we may assume that

$$d_T(t_o, \pi(p_1)) < \varepsilon \quad \text{and} \quad d_X(p_m, p_n) < \rho.$$

Let $V = \{x \in X; d_X(p_1, x) < \varepsilon\}$. Then by (3.1.19) there exists a constant $C > 0$ such that

$$d_V(p_m, p_n) \leq C \cdot d_X(p_m, p_n) \quad \text{for all} \quad m, n,$$

which shows that $\{p_n\}$ is Cauchy sequence in V with respect to d_V. Since $V \subset \pi^{-1}(U_i)$, $\{p_n\}$ is a Cauchy sequence in $\pi^{-1}(U_i)$ with respect to $d_{\pi^{-1}(U_i)}$. Since $\pi^{-1}(U_i)$ is complete hyperbolic, $\{p_n\}$ converges to a point in $\pi^{-1}(U_i)$. $\qquad\square$

(3.2.16) **Remark.** (i)   In general, even if T and all $\pi^{-1}(t)$, $t \in T$, are hyperbolic, X may not be hyperbolic. For example, the domain

$$X = \{(z, w) \in \mathbf{C}^2; \ |z| < 1, \ |zw| < 1\} - \{(0, w); \ |w| \geq 1\}$$

is not hyperbolic. Let $T = D = \{z; \ |z| < 1\}$ and $\pi(z, w) = z$. Then each $\pi^{-1}(t)$ is biholomorphic to a disc. To see that X is not hyperbolic, consider two points $p = (0, b)$ with $b \neq 0$ and $q = (0, 0)$. Set $p_n = (1/n, b)$. Then $d_X(p, q) = \lim d_X(p_n, q)$. Let $a_n = \min\{n, \sqrt{n/|b|}\}$. Then the mapping $t \in D \to (a_n t/n, a_n bt) \in X$ maps $1/a_n$ into p_n. Hence,

$$d_X(p, q) = \lim d_X(p_n, q) \leq \lim d_D(1/a_n, 0) = 0.$$

(ii)   However, as we shall see in (3.11.2), if $\pi: X \to T$ is a proper holomorphic map, hyperbolicity of T and $\pi^{-1}(t)$, $t \in T$, implies hyperbolicity of X.

(iii)   Let $\pi: X \to T$ be a holomorphic fibre bundle with fibre F in the sense that every point $t \in T$ has an open neighborhood U such that $\pi^{-1}(U)$ is biholomorphic to $U \times F$. If T and F are (complete) hyperbolic, then X is also (complete) hyperbolic by (3.2.3) and (3.2.15), (Kiernan [4]).

Conversely, if X is (complete) hyperbolic, so are T and F. In fact, each fibre is (complete) hyperbolic by (3.2.2). To prove (complete) hyperbolicity of T it suffices to show that the bundle is locally flat so that the pullback $\tilde{X}$ of X to the universal covering $\tilde{T}$ of T is holomorphically a product $\tilde{T} \times F$. (For if X is (complete) hyperbolic, so is $\tilde{X}$ by (3.2.8). Then T is (complete) hyperbolic by (3.2.3)). Thus the proof is reduced to showing the following (see Royden [5] and also (5.4.5) for details):

Let $f \in \mathrm{Hol}(D \times F, F)$, and write $f_t(y) = f(t, y)$. Then if f_0 is an automorphism of F, $f_t = f_0$ for all $t \in T$.

(iv)   If $\pi: X \to T$ is merely a fibre space, X can be hyperbolic without T being hyperbolic as the following example shows, (Kobayashi [7]). Let

$$X = \{(z, w) \in \mathbf{C}^2;\ 0 < |z|^2 + |w|^2 < 1\} \quad \text{and} \quad T = P_1\mathbf{C}$$

with the natural projection π which assigns to $(z, w) \in X$ the point of $P_1\mathbf{C}$ with homogeneous coordinates (z, w). Then $F = D^*$.

(3.2.17) Theorem. *Let X be a complete hyperbolic complex space and f a bounded holomorphic function on X. Then the open subspace*

$$X' = \{p \in X;\ f(p) \neq 0\} = X - \mathrm{Zero}(f)$$

is complete hyperbolic.

Proof. Without loss of generality we may assume that f maps X into the unit disc D. Then apply (3.2.4) with $Y = D$ and $Y' = D^*$. $\qquad\square$

Let Y be a complex space. We say that a complex subspace $X \subset Y$ is **locally complete hyperbolic in** Y if every point p of the closure $\overline{X}$ has neighborhood V_p in Y such that $V_p \cap X$ is complete hyperbolic. The condition is obviously satisfied by any point p of X. So this is a condition on the boundary points $p \in \partial X = \overline{X} - X$.

A **Cartier divisor** A in a complex space Y is a closed complex subspace that is locally defined as the zeros of a single holomorphic function. That is, each point $x \in A$ has a neighborhood V in Y such that $A \cap V$ is defined by one equation $f = 0$, where f is a holomorphic function on V.

(3.2.18) Corollary. *Let Y be a complex space and A a Cartier divisor of Y. Then*
   (1)   *$Y - A$ is locally complete hyperbolic in Y;*
   (2)   *If Y is (complete) hyperbolic, $Y - A$ is (complete) hyperbolic.*

Proof. Choose a complete hyperbolic V_p such that $A \cap V_p$ is given as the zeros of a bounded holomorphic function f in V_p and apply (3.2.17) with $X = V_p$ and $X' = V_p - A$. $\qquad\square$

While removing an analytic subset of codimension 1 from X can radically change the (pseudo-) distance d_X, removing a subset of large codimension does not, in general, change d_X. The following theorem is due to Campbell-Ogawa [1] and Campbell-Howard-Ochiai [1].

(3.2.19) Theorem. *Let X be a complex manifold and A a closed analytic subset of X of codimension at least 2. Then $\mathrm{Hol}(D, X - A)$ is dense in $\mathrm{Hol}(D, X)$ in the compact-open topology, and hence*

$$d_{X-A} = d_X \qquad \text{on} \quad X - A.$$

As pointed out by Campbell-Howard-Ochiai [1], the first part of (3.2.19) can be generalized as follows:

If A is a closed analytic subset of codimension $> k$ in an n-dimensional complex manifold X, then $\mathrm{Hol}(D^k, X - A)$ is dense in $\mathrm{Hol}(D^k, X)$.

In order to prove (3.2.19), we use the following lemma of Royden [4] which has other applications. (The proof of this lemma will be given in Appendix A of this Chapter.)

(3.2.20) Lemma. *If f is a holomorphic imbedding of a disc D_r of radius $r > 1$ into a complex manifold X of dimension n, there exists a holomorphic imbedding φ of the unit polydisc D^n into X such that*

$$f(z) = \varphi(z, 0, \ldots, 0,) \qquad \text{for} \quad z \in D.$$

Proof of (3.2.19). Let $f \in \mathrm{Hol}(D, X)$. We want to approximate f by elements of $\mathrm{Hol}(D, X - A)$. If we define $f_t \in \mathrm{Hol}(D, X)$ by $f_t(z) = f(tz)$, $0 < t < 1$, then each f_t extends past the boundary ∂D of D and $f_t \to f$ as $t \to 1$. Thus, if each f_t is in the closure of $\mathrm{Hol}(D, X - A)$, so is f. We may therefore assume that f extends to a slightly larger disc D_r, $r > 1$ and $\overline{f(D)} \subset X$. If we define $\tilde{f} \in \mathrm{Hol}(D, D \times X)$ by $\tilde{f}(z) = (z, f(z))$, then $\tilde{f}$ is an imbedding of D into $D \times X$. Let $\pi: D \times X \to X$ be the projection. If $g \in \mathrm{Hol}(D, D \times (X - A))$ approximates $\tilde{f}$, then $\pi \circ g \in \mathrm{Hol}(D, X - A)$ approximates f. We may therefore assume that f imbeds D_r, $r > 1$ into X.

Let $\varphi \in \mathrm{Hol}(D^n, X)$ be as in (3.2.20), and define $B = \varphi^{-1}(A)$. If the map $j \in \mathrm{Hol}(D, D^n)$ defined by $j(z) = (z, 0, \ldots, 0)$ can be approximated by $g \in \mathrm{Hol}(D, D^n - B)$, then the map $f = \varphi \circ j$ can be approximated by $\varphi \circ g \in \mathrm{Hol}(D, X - A)$.

The proof of the theorem is now reduced to the case where X is a domain in $\mathbf{C}^n$ and $\overline{f(D)} \subset X$. We consider the map $h : D \times A \to \mathbf{C}^n$ defined by $h(z, a) = f(z) - a$. Since $\dim(D \times A) < n$, $h(D \times A)$ is a meager set (i.e., a countable union of nowhere dense subsets) in $\mathbf{C}^n$ and, hence, there exists a sequence of points $c_1, c_2, \ldots$ of $\mathbf{C}^n - h(D \times A)$ converging to the origin $0 \in \mathbf{C}^n$. Define $f_m \in \mathrm{Hol}(D, \mathbf{C}^n)$ by $f_m(z) = f(z) - c_m$. Since $\overline{f(D)} \subset X$, there is an integer N such that $f_m(D) \subset X$ for $m > N$. The sequence $\{f_m, \ m > N\}$ converges to f in $\mathrm{Hol}(D, X)$, and, by construction, $f_m(D) \subset X - A$. $\qquad\square$

A few remarks are in order. First, as the following example shows, it is essential in (3.2.19) that X is nonsingular, (Campbell-Ogawa [1]).

(3.2.21) Example. Let $\pi: \mathbf{C}^{n+1} - \{0\} \to \mathbf{P}_n$ be the natural projection. Let $Y \subset \mathbf{P}_n$ be a hyperbolic algebraic manifold, e.g., a nonsingular curve of genus > 1. Let $X \subset \mathbf{C}^{n+1}$ be the cone over Y, i.e., $X = \pi^{-1}(Y) \cup \{0\}$. Then $d_X \equiv 0$ since X is a union of lines interesecting at the origin. Let $A = \{0\}$. Then d_{X-A} is nontrivial;

$$d_{X-A}(p, q) \geq d_Y(\pi(p), \pi(q)) > 0$$

if $p, q \in X - A$ do not lie on the same line through the origin. Note that A is the singular locus of X and that by choosing Y to be of large dimension, we can make the codimension of A in X as large as we wish.

We state, without proof, a generalization of (3.2.19) by Poletskiĭ-Shabat [1], see also Jarnicki-Pflug [10; p. 87]:

(3.2.22) **Theorem**. *Let X be an n-dimensional complex manifold, and A a closed subset with $(2n-2)$-dimensional Hausdorff measure equal to zero. Then* $\mathrm{Hol}(D, X - A)$ *is dense in* $\mathrm{Hol}(D, X)$ *in the compact-open topology, and hence*

$$d_{X-A} = d_X \qquad \text{on} \quad X - A.$$

A generalization of (3.2.19) to k-intrinsic measures (see Section 2 of Chapter 7 for intrinsic measures) was obtained by Kaliman-Zaidenberg [1].

Following Wu [1], we say that a complex space X with a distance function δ (which is assumed to induce the topology of X) is δ-**tight** if $\mathrm{Hol}(D, X)$ is equicontinuous with respect to δ and that X is **tight** if it is δ-tight for some δ. If X is hyperbolic, then it is d_X-tight (by (3.1.6)) and hence tight. Conversely (Kiernan [2]), we have

(3.2.23) **Theorem**. *A complex space X is hyperbolic if and only if it is tight.*

Proof. Assume that X is δ-tight. Let p and q be two distinct points of X. Let U be an open hyperbolic neighborhood of p such that $q \notin U$. Let W be a smaller neighborhood of p, relatively compact in U. Let $\varepsilon > 0$ be such that the ε-neighborhood $V = \{x \in X; \ \delta(W, x) < \varepsilon\}$ of W is relatively compact in U.

Since $\mathrm{Hol}(D, X)$ is an equicontinuous family, there exists a positive number $r < 1$ such that if $f \in \mathrm{Hol}(D, X)$ with $f(0) \in W$, then $f(D_r) \subset V$.

Let $c > 0$ be a constant such that $d_D(0, b) > c \cdot d_{D_r}(0, b)$ for all $b \in D_{r/2}$. Let

$$\alpha = \{p = p_0, p_1, \ldots, p_k = q; \ a_1, b_1, \ldots, a_k, b_k; \ f_1, \ldots, f_k\}$$

be a chain of holomorphic disks from p to q. We may assume that

$$p_0, p_1, \ldots, p_{j-1} \in W, \ p_j \notin W, \ a_1 = \ldots = a_k = 0, \ b_1, \ldots, b_k \in D_{r/2}.$$

Then $p_j \in V$, and

$$
\begin{aligned}
l(\alpha) \ &\geq \ \sum_{i=1}^{j} d_D(0, b_i) \geq c \sum_{i=1}^{j} d_{D_r}(0, b_i) \\
&\geq \ c \sum_{i=1}^{j} d_U(p_{i-1}, p_i) \geq c \cdot d_U(p_0, p_j).
\end{aligned}
$$

Take $c' > 0$ such that $d_U(p_0, V - W) \geq c'$. Then $d_X(p, q) \geq l(\alpha) \geq cc'$. $\qquad \square$

(3.2.24) Remark. If a complex space X is hyperbolic, then every holomorphic map f of $\mathbf{C}$ into X is necessarily constant since

$$d_X(f(a), f(b)) \leq d_{\mathbf{C}}(a, b) = 0$$

by (3.1.21). Hence, every holomorphic map f of $P_1\mathbf{C}$ or a complex torus into a hyperbolic complex space X is constant. As we shall see in (3.6.3) a compact complex space X is hyperbolic if there is no nonconstant holomorphic map of $\mathbf{C}$ into X. We say that a complex space X if **algebraically hyperbolic** if there is no nonconstant holomorphic map of $P_1\mathbf{C}$ or a complex torus into X. Ballico [1] proved that a generic hypersurface of large degree in $P_{n+1}\mathbf{C}$ is algebraically hyperbolic. It is not known if every algebraically hyperbolic algebraic manifold is hyperbolic.

It is important to generalize the concept of hyperbolicity to allow the Kobayashi distance to be partially degenerate. Let X be a complex space and Δ a closed subset of X. In applications, Δ is usually a closed complex subspace. We say that X is **hyperbolic modulo** Δ if for every pair of distinct points p, q of X we have $d_X(p, q) > 0$ unless both are contained in Δ. Then d_X induces a distance function $d_{X/\Delta}$ on the quotient space X/Δ in a natural way.

We say that X is **complete hyperbolic modulo** Δ if it is hyperbolic modulo Δ and if for each sequence $\{p_n\}$ in X which is Cauchy with respect to the pseudo-distance d_X, we have one of the following:

(a) $\{p_n\}$ converges to a point p in X;

(b)   for every open neighborhood U of Δ in X, there exists an integer N such that $p_n \in U$ for $n > N$.

Clearly, X is complete hyperbolic modulo Δ if and only if the quotient space X/Δ is complete with respect to the distance function $d_{X/\Delta}$.

Suppose that Δ and Δ' are two closed subsets of X and that X is hyperbolic modulo Δ as well as modulo Δ'. Then X is hyperbolic modulo $\Delta \cap \Delta'$. So we can speak of *the smallest* closed subset Δ such that X is hyperbolic modulo Δ. Clearly such a closed set is given by

$$\Delta_X = \text{the closure of } \{p \in X; \ d_X(p, q) = 0 \text{ for some } q \neq p\}.$$

From (1.1.8) we obtain the following

(3.2.25) Theorem. *If a complex space X is hyperbolic modulo a closed subset Δ, then*

(1)   for every point $p \in X - \Delta$ and for every neighborhood $U \subset X - \Delta$ of p, there exists a δ-neighborhood V of p with respect to d_X such that $V \subset U$;

(2)   if Δ is compact, on the quotient space X/Δ the pseudo-distance d_X induces the quotient topology.

The following proposition is a straightforward generalization of (3.2.2).

(3.2.26) Proposition. *Let X be a complex subspace of a complex space Y.*

(1)   *If Y is hyperbolic modulo Δ, then X is hyperbolic modulo $X \cap \Delta$;*

(2)   *If Y is complete hyperbolic modulo Δ and if X is closed, then X is complete hyperbolic modulo $X \cap \Delta$.*

From (3.1.9) we obtain

(3.2.27) Proposition. *If X and Y are (complete) hyperbolic modulo Δ and Δ' respectively, then $X \times Y$ is (complete) hyperbolic modulo $(\Delta \times Y) \cup (X \times \Delta')$.*

The following proposition generalizes (3.2.4).

(3.2.28) Proposition. *Let X and Y be complex spaces, and $f : X \to Y$ a holomorphic map. Let Y' be a complex subspace of Y, and set $X' = f^{-1}Y'$. If X is complete hyperbolic modulo Δ and if Y' is complete hyperbolic, then X' is complete hyperbolic modulo $X' \cap \Delta$.*

Proof. As in the proof of (3.2.4), let G_f be the graph of f, and $G_{f'}$ the graph of $f' = f|_{X'}$. By (3.2.27) $X \times Y'$ is complete hyperbolic modulo $\Delta \times Y'$. By (3.2.26) $G_{f'}$ is complete hyperbolic modulo $(\Delta \times Y') \cap G_{f'}$. Since $G_{f'}$ is biholomorphic to X' under the projection $X \times Y' \to X$, it follows that X' is complete hyperbolic modulo $\Delta \cap X'$. $\square$

The following two propostions can be proved in the same way as (3.2.6) and (3.2.7).

(3.2.29) Proposition. *Let X be a complex space and Δ a closed subset. If for every $p \in X$ there is a positive number δ such that the δ-neighborhood $U(p; \delta) = \{p \in X; \ d_X(p, q) < \delta\}$ is hyperbolic modulo $\Delta \cap U(p; \delta)$, then X is hyperbolic modulo Δ.*

(3.2.30) Proposition. *Let X be a complex space and Δ a closed subset. If there is a positive number δ such that for every $p \in X$ the δ-neighborhood $U(p; \delta)$ is complete hyperbolic modulo $\Delta \cap U(p; \delta)$, then X is complete hyperbolic modulo Δ.*

The following is a consequence of (1.3.12).

(3.2.31) Theorem. *If $\pi : \tilde{X} \to X$ is a spread and if X is hyperbolic modulo a closed subset Δ, then $\tilde{X}$ is hyperbolic modulo $\pi^{-1}(\Delta)$.*

(3.2.32) Theorem. *Let $\pi : \tilde{X} \to X$ be a covering space of a complex space X. Let Δ be a closed subset of X. Then $\tilde{X}$ is hyperbolic modulo $\pi^{-1}(\Delta)$ if and only if X is hyperbolic modulo Δ.*

Proof. Assume that $\tilde{X}$ is hyperbolic modulo $\tilde{\Delta} = \pi^{-1}(\Delta)$. Let $p, q \in X$ be such that $d_X(p, q) = 0$. Assuming that $p \notin \Delta$, we shall show that $p = q$. Let $\tilde{p} \in \tilde{X}$ be such that $\pi(\tilde{p}) = p$. Then $\tilde{p} \notin \tilde{\Delta}$. By (1) of (3.2.8) there exists a sequence $\{\tilde{q}_n\} \subset \tilde{X}$ such that $\pi(\tilde{q}_n) = q$ and $\lim d_{\tilde{X}}(\tilde{p}, \tilde{q}_n) = 0$. By (1) of (3.2.25) $\{\tilde{q}_n\}$ converges to $\tilde{p}$. Then $\{\pi(\tilde{q}_n)\}$ converges to $\tilde{p}$. Since $\pi(\tilde{q}_n) = q$, we have $p = q$. The opposite implication is already in (3.2.31). $\square$

The following is a direct consequence of (1.3.14).

(3.2.33) Theorem. *Let $\pi: \tilde{X} \to X$ be a finite-to-one proper holomorphic map. If X is (complete) hyperbolic modulo a compact subset Δ, then $\tilde{X}$ is (complete) hyperbolic modulo $\pi^{-1}(\Delta)$.*

(3.2.34) Remark. Given a complex space X, consider the equivalence relation R defined by the pseudodistance d_X, i.e., p is equivalent to q if and only if $d_X(p, q) = 0$. Then d_X induces a distance on X/R. However, X/R need not carry a complex structure which would make the projection $X \to X/R$ holomorphic. Even when X/R admits such a complex structure, the induced distance may not coincides with the intrinsic distance $d_{X/R}$ of the complex space X/R, see Horst [3]. For hyperbolic quotients of homogeneous complex manifolds, see Gilligan [1]. On the degeneracy set Δ for the pseudodistance d_X, see Hristov [4, 5, 6, 7] and Adachi-Suzuki [2].

3 Hyperbolic Imbeddings

Let Z be a complex space and Y a complex subspace with compact closure $\bar{Y}$. We call a point $p \in \bar{Y}$ a **hyperbolic point** if every neighborhood U of p contains a smaller neighborhood V of p, $\bar{V} \subset U$, such that

$$d_Y(\bar{V} \cap Y, Y - U) > 0.$$

We say that Y is **hyperbolically imbedded** in Z if every point of $\bar{Y}$ is a hyperbolic point. Clearly, Y is hyperbolically imbedded in Z if and only if, for every pair of distinct points p, q in $\bar{Y} \subset Z$, there exist neighborhoods U_p and U_q of p and q in Z such that $d_Y(U_p \cap Y, U_q \cap Y) > 0$.

In the definition above, there is no need to assume that Y is relatively compact. But in applications, Y is almost always a relatively compact open domain in Z. So, unless otherwise stated, we assume that Y is relatively compact in Z.

It is clear that a hyperbolically imbedded complex space Y is hyperbolic. The condition of hyperbolic imbedding says that the distance $d_Y(p_n, q_n)$ remains positive when two sequences $\{p_n\}$ and $\{q_n\}$ in Y approach two distinct points p and q of the boundary $\partial Y = \bar{Y} - Y$. The concept of hyperbolic imbedding was first introduced in Kobayashi [7] to obtain a generalization of the big Picard theorem. The term "hyperbolic imbedding" was first used by Kiernan [6].

We note that a compact hyperbolic complex space is hyperbolically imbedded in itself.

The proof of the following proposition is straightforward.

(3.3.1) Proposition. *If complex spaces Y and Y' are hyperbolically imbedded in Z and Z' respectively, then $Y \times Y'$ is hyperbolically imbedded in $Z \times Z'$.*

The following is obvious.

(3.3.2) Proposition. *If there is a distance function δ on $\bar{Y}$ such that*

$$d_Y(p, q) \geq \delta(p, q) \qquad for \quad p, q \in Y,$$

then Y is hyperbolically imbedded in Z.

Using the concept of length function and the induced distance function (see (2.3.1)) we state the converse, (see Kiernan [6] and Kiernan-Kobayashi [2]).

(3.3.3) **Theorem.** *Let Y be a relatively compact complex subspace of a complex space Z. Then the following are equivalent:*
(a) *Y is hyperbolically imbedded in Z;*
(b) *Given a length function F on Z there is a positive constant c such that*

$$f^*(c^2 F^2) \leq ds_D^2 \qquad for \quad f \in \mathrm{Hol}(D, Y).$$

Proof. Assume (a). If a constant c in (b) does not exist, then there exist a sequence $\{f_n\}$ in $\mathrm{Hol}(D, Y)$ and points $\{a_n\}$ in D such that

$$f_n^* F^2 > n \cdot ds_D^2 \qquad at \quad a_n.$$

Since D is homogeneous, we may assume that $a_n = 0$. Let e be a unit vector at $0 \in D$, measured by the Poincaré metric. Then the inequality above states

$$F(df_n(e))^2 > n.$$

Since $f_n(0) \in Y$ and $\bar{Y}$ is compact, by taking a subsequence we may assume that $\{f_n(0)\}$ converges to a point $p \in \bar{Y}$.

Let U be a complete hyperbolic neighborhood of p in Z, e.g., a neighborhood biholomorphic to a closed analytic subset of D^m. Assume that there exists a positive number $r < 1$ such that $f_n(D_r) \subset U$ for $n \geq n_0$. Since $f_n(0)$ belongs to a compact neighborhood of p in U and since U is complete hyperbolic, $\{f_n|_{D_r} \in \mathrm{Hol}(D_r, U)\}$ is relatively compact in $\mathrm{Hol}(D_r, U)$ by (1.3.3) and would have a subsequence which converges in $\mathrm{Hol}(D_r, U)$. But this is impossible since $F(df_n(e))^2 > n$. Thus no such r exists. This means that for each positive integer k, there exist a point $z_k \in D$ and an integer n_k such that $|z_k| < \frac{1}{k}$ and $f_{n_k}(z_k) \notin U$. Let $p_k = f_{n_k}(0)$ and $q_k = f_{n_k}(z_k)$. By taking a subsequence we may assume that $\{q_k\}$ converges to a point q not in U. Since

$$d_Y(p_k, q_k) \leq d_D(0, z_k) \to 0 \quad as \quad k \to \infty,$$

this contradicts the assumption that Y is hyperbolically imbedded in Z.

Assume (b). Let δ be the distance function on Z defined by the length function cF. Then
$$\delta(f(a), f(b)) \leq d_D(a, b) \qquad for \quad f \in \mathrm{Hol}(D, Y).$$
By (2) of (3.1.7), $\delta \leq d_Y$ on Y. Given two points $p, q \in \bar{Y}$, set $2\alpha = \delta(p, q)$ and let U_p and U_q be the open balls of δ-radius α around p and q. $\qquad \square$

In (3.2.18) we showed that if Y is the complement of a Cartier divisor in Z, then Y is locally complete hyperbolic in Z. In this connection we prove the following

(3.3.4) Theorem. *Let Y be hyperbolically imbedded in a complex space Z. If Y is locally complete hyperbolic in the sense that every point $p \in \bar{Y}$ has a neighborhood V_p in Z such that $V_p \cap Y$ is complete hyperbolic, then Y is complete hyperbolic.*

We start the proof with the following general lemma.

(3.3.5) Lemma. *Let Y be a complex subspace of a complex space Z. Let $p \in \bar{Y}$ and V_p a neighborhood of p in Z. Given a smaller neighborhood W_p of p such that*

$$\delta := d_Y(W_p \cap Y, Y - V_p) > 0$$

and a positive constant $\delta' < \delta/2$, there is a constant $c > 0$ such that

$$d_Y(q, q') \geq c \cdot d_{V_p \cap Y}(q, q') \quad \text{for} \quad q, q' \in W_p \cap Y \quad \text{with} \quad d_Y(q, q') < \delta'.$$

As a consequence, if a sequence of points in $W_p \cap Y$ is Cauchy with respect to d_Y, then it is Cauchy with respect to $d_{V_p \cap Y}$.

Proof. Let $q, q' \in W_p \cap Y$ such that $d_Y(q, q') < \delta'$. Consider a chain of holomorphic discs α from q to q' of length $l(\alpha) < \delta'$ consisting of points $q = y_0, y_1, \ldots, y_k = q'$ in Y, points $a_1, \ldots, a_k$ in D and maps $f_1, \ldots, f_k \in \mathrm{Hol}(D, Y)$ such that $f_i(0) = y_{i-1}$ and $f_i(a_i) = y_i$. Since $q \in W_p$ and $l(\alpha) < \delta'$, the d_Y-distance from W_p to y_{i-1} is less than δ'. Let r and r' be the positive real numbers defined by $d_D(0, r) = \delta/2$ and $d_D(0, r') = \delta'$ so that $r' < r$. Then $f_i(D_r)$ lies in the $(\delta/2)$-neighborhood of y_{i-1} and hence in V_p. Choose a constant c such that

$$d_D(0, z) \geq c \cdot d_{D_r}(0, z) \quad \text{for} \quad z \in D_{r'}.$$

Since $l(\alpha) < \delta'$, we have $a_i \in D_{r'}$ and

$$\sum d_D(0, a_i) \geq c \sum d_{D_r}(0, a_i) \geq c \sum d_{V_p \cap Y}(f_i(0), f_i(a_i)) \geq c \cdot d_{V_p \cap Y}(q, q').$$

Since this is valid for all holomorphic chains from q to q' of length less than δ', we have

$$d_Y(q, q') \geq c \cdot d_{V_p \cap Y}(q, q').$$

$\square$

Proof of (3.3.4). Suppose that Y is not complete hyperbolic. Then there would exist a sequence $\{p_n\}$ in Y which is Cauchy with respect to d_Y and such that $p_n \to p \notin Y$. Let V_p be a neighborhood of p in Z such that $V_p \cap Y$ is complete hyperbolic. By (3.3.5) $\{p_n\}$ is a Cauchy sequence with respect to $d_{V_p \cap Y}$ and must converges to a point in $V_p \cap Y$. This is a contradiction. $\square$

The case where Y is the complement of a Cartier divisor in Z is of particular interest. Combined with (3.2.18) the theorem yields the following

(3.3.6) Corollary. *Let Z be a compact complex space and A a Cartier divisor in Z. Let $Y = Z - A$. If Y is hyperbolically imbedded in Z, then Y is complete hyperbolic.*

In the corollary above, if Y is not hyperbolically imbedded in Z, then Y need not be complete even when it is hyperbolic. For example, let W be a compact complex manifold of dimension at least 2 with a divisor A such that the complement $W - A$ is hyperbolic. Let Z be the space obtained by blowing up W at a point p of $W - A$, and B the exceptional divisor obtained from p. Set $Y = W - (A \cup \{p\}) = Z - (A \cup B)$. Then Y is hyperbolic, but not complete by (3.2.19).

(3.3.7) **Theorem.** *Let Y be a complex space hyperbolically imbedded in a complex space Z. Let Z' be a complex space with a proper finite map $\pi: Z' \to Z$, and $Y' = \pi^{-1}(Y)$. Then Y' is hyperbolically imbedded in Z'.*

Proof. We know that Y' is hyperbolic by (3.2.11). Let $\bar{Y}'$ be the closure of Y' in Z'. We extend the distance function $d_{Y'}$ of Y' to the closure $\bar{Y}'$ by setting

$$d_{Y'}(z, z') = \inf d_{Y'}(y_m, y'_n), \qquad z, z' \in \bar{Y}',$$

where the infimum is taken over all sequences $\{y_m\}$ and $\{y'_n\}$ in Y' converging to z and z', respectively. In general, $d_{Y'}(z, z')$ can be infinite. Clearly, the extended $d_{Y'}$ is lower semicontinuous on $\bar{Y}' \times \bar{Y}'$. Since Y is hyperbolically imbedded in Z, $d_{Y'}(z, z') > 0$ if $\pi(z) \neq \pi(z')$. Given $z \in \bar{Y}'$, let

$$\Delta(z) = \{z' \in \bar{Y}'; \; d_{Y'}(z, z') = 0\}.$$

The problem is to show that $\Delta(z)$ is a singleton set $\{z\}$.

Since $\Delta(z) \subset \pi^{-1}(\pi(z))$ and since $\pi^{-1}(\pi(z))$ is a finite set, it suffices to prove that $\Delta(z)$ is connected. Let U be a compact neighborhood of z containing no other points of $\Delta(z)$. Let ∂U be the boundary of U. Since $d_{Y'}$ is lower semicontinuous, there is a positive δ such that $d_{Y'}(z, \partial U \cap \bar{Y}') \geq \delta$. Suppose $\Delta(z)$ contains another point z'. Any curve joining z to z' in $\bar{Y}'$ must go through ∂U and so has length at least δ. This is a contradiction. $\qquad\qquad\square$

We end this section with simple examples.

(3.3.8) **Example.** Let A be a finite subset of $P_1\mathbf{C}$ containing at least three points, say ∞, 0 and 1. Then $Y = P_1\mathbf{C} - A$ is hyperbolically imbedded in $P_1\mathbf{C}$. It is well known that the disc is the universal covering space of Y. By (3.2.8) Y is complete hyperbolic. (There is also a differential geometric argument to show that Y is complete hyperbolic, see Section 7). Since Y has only isolated boundary points, it is hyperbolically imbedded in $P_1\mathbf{C}$.

(3.3.9) **Example.** Let Q be a complete quadrilateral in $P_2\mathbf{C}$. That is, Q is the union of six projective lines which pass through a set of four points in general position. We shall show that $Y = P_2\mathbf{C} - Q$ is complete hyperbolic and hyperbolically imbedded in $P_2\mathbf{C}$. Let $\{p_1, p_2, p_3, p_4\}$ be four points in general position. For $i < j$, let l_{ij} be the (projective) line passing through p_i and p_j, (see Fig. i). Let p be any point of $\bar{Y} = P_2\mathbf{C}$. Without loss of generality we may assume that $p \notin l_{12}$. If we remove l_{12} as the line at infinity, we are left with an affine plane $\mathbf{C}^2$ with five lines, where l_{13} and l_{14} are parallel (i.e., meet at infinity) and l_{23} and l_{24} are also parallel, (see Fig. ii), so that

$$M := \mathbf{C}^2 - (l_{13} \cup l_{14} \cup l_{23} \cup l_{24}) \cong (\mathbf{C} - \{0, 1\}) \times (\mathbf{C} - \{0, 1\}).$$

Since $p \notin l_{12}$, p is in $\mathbf{C}^2$. It is clear from (3.3.8) and (3.3.1) that p has neighborhoods V and U with $\bar{V} \subset U$ such that $d_M(V \cap M, (\mathbf{C}^2 - U) \cap M) > 0$. Since $Y \subset M$ and $(\mathbf{C}^2 - U) \cap Y = (P_2\mathbf{C} - U) \cap Y$, we have

$$d_Y(V \cap Y, (P_2\mathbf{C} - U) \cap Y) > 0.$$

This proves that Y is hyperbolically imbedded in $P_2\mathbf{C}$. The fact that Y is complete hyperbolic follows from (3.3.4).

(3.3.10) Example. From projective plane $P_2\mathbf{C}$ we remove four lines, say l_0, l_1, l_0', l_1', in general position. The resulting space is not hyperbolic since it contains $\mathbf{C}^*$, namely the (projective) line l_∞ through points $l_0 \cap l_1$ and $l_0' \cap l_1'$ with these two points removed, (see Fig. iii). If we remove also l_∞ as the line at infinity, then we obtain $\mathbf{C}^2$ with two pairs of parallel lines removed, (see Fig. iv).

Set $Y = P_2\mathbf{C} - (l_0 \cup l_1 \cup l_0' \cup l_1' \cup l_\infty)$. Then $Y \cong (\mathbf{C} - \{0, 1\}) \times (\mathbf{C} - \{0, 1\})$ is complete hyperbolic. We shall show that Y is *not* hyperbolically imbedded in $P_2\mathbf{C}$. If we remove l_0' as the line at infinity, we obtain Figure v; l_1' and l_∞ are parallel, and l_0, l_1 and l_∞ are concurrent, i.e., meet at one point, say o. Consider another line l passing through o and take two points p and q on it, for instance, diametrically opposite to each other with respect to o. We shall show that as we rotate l around o toward l_∞, the distance $d_Y(p, q)$ approaches zero. Let $a = l \cap l_1'$ and let r be the distance between o and a with respect to the Euclidean metric in $\mathbf{C}^2$. As l rotates toward l_∞, the point a moves away toward infinity and r tends to infinity. On the line l we take the punctured disc around o with radius r and denote it D_r^*. Then $d_Y(p, q) \leq d_{D_r^*}(p, q)$. Since $d_{D_r^*}(p, q) \to 0$ as $r \to \infty$, we see that $d_Y(p, q)$ approaches zero as l rotates toward l_∞. This shows that Y is not hyperbolically imbedded in $P_2\mathbf{C}$. (It is not difficult to verify that the condition for hyperbolic imbedding is violated only by the points of l_∞).

On the other hand, the same space $Y \cong (\mathbf{C} - \{0, 1\}) \times (\mathbf{C} - \{0, 1\})$ can be hyperbolically imbedded in $P_1\mathbf{C} \times P_1\mathbf{C}$ in a natural manner.

We know that the space $\hat{P}_2$ obtained by blowing up two points of $P_2\mathbf{C}$ is biholomorphic to the space $\hat{Q}_2$ obtained by blowing up the quadric $Q_2 = P_1\mathbf{C} \times P_1\mathbf{C}$ once. More precisely, by setting one of the coordinates equal to 0, 1 or ∞ in Q_2, we obtain two sets of three parallel lines $\{l_0, l_1, m\}$ and $\{l_0', l_1', m'\}$ in Q_2 as in Figure vi. By blowing up the point $(\infty, \infty) = m \cap m'$ of Q_2 we obtain $\hat{Q}_2$ with an exceptional curve l_∞. By blowing down m and m' in $\hat{Q}_2$ we obtain $P_2\mathbf{C}$ with l_∞ as the line at infinity, (see Fig. vii). In terms of homogeneous coordinate systems (z^0, z^1, z^2) for $P_2\mathbf{C}$ and $((u^0, u^1), (v^0, v^1))$ for $P_1\mathbf{C} \times P_1\mathbf{C}$, the birational correspondence

$$\varphi: P_2\mathbf{C} \to P_1\mathbf{C} \times P_1\mathbf{C}$$

is given by

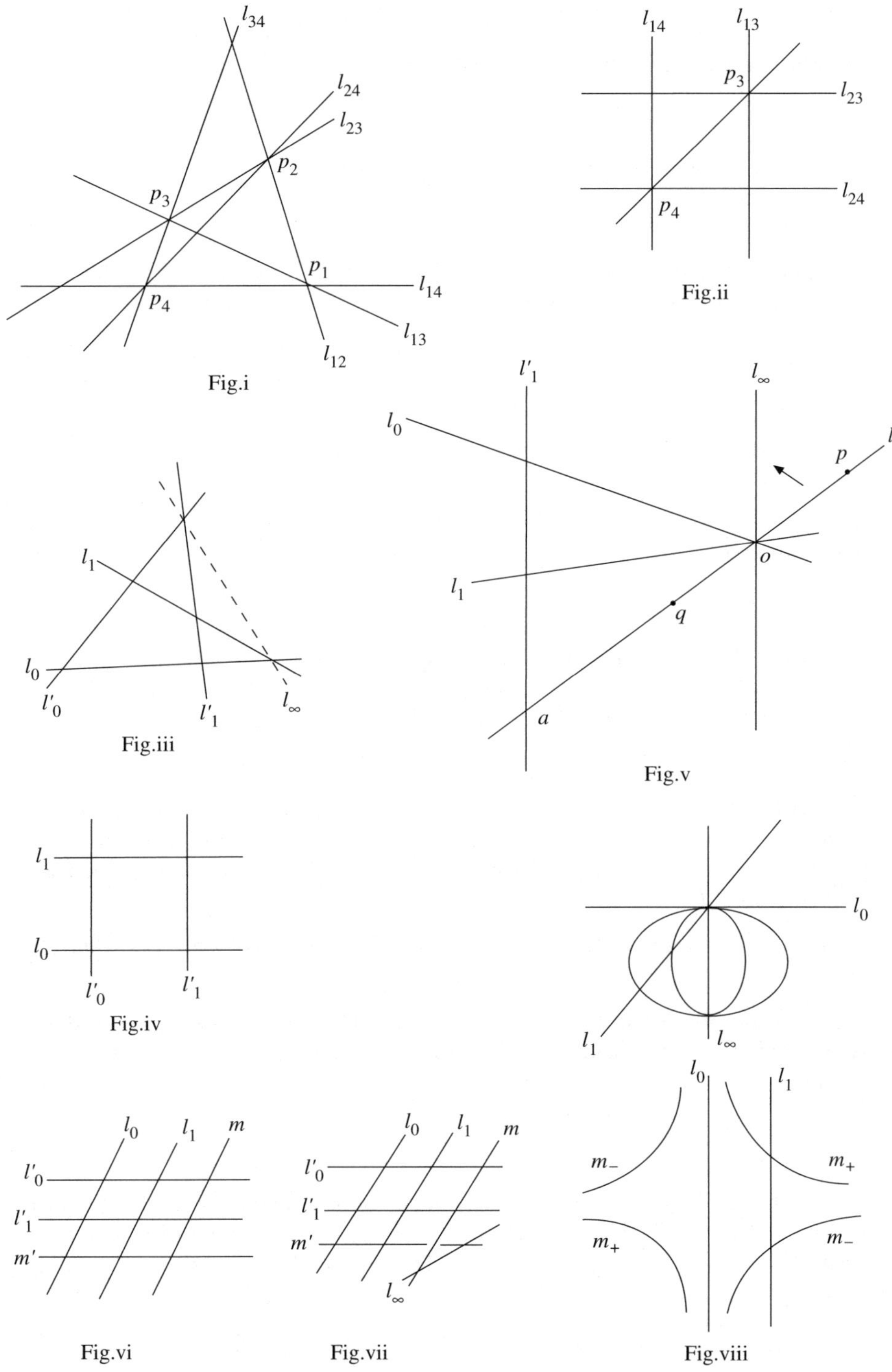

Fig.i

Fig.ii

Fig.iii

Fig.iv

Fig.v

Fig.vi

Fig.vii

Fig.viii

$$u^0 = z^0, \quad u^1 = z^1, \quad v^0 = (z^0)^2, \quad v^1 = z^1 z^2.$$

Then, in Figure vii, the lines l_∞, m and m' are given by

$$l_\infty = \{z^0 = 0\}, \quad m = \{u^0 = 0\}, \quad m' = \{v^0 = 0\}.$$

Finally, we shall construct a holomorphic map

$$f: D^* \times D^* \to Y$$

which does not extend to a map $f: D \times D \to P_2\mathbf{C}$. We may assume without loss of generality that the five lines in $P_2\mathbf{C}$ in Figure iii are given by

$$l_\infty = \{z^0 = 0\}, \quad l_0 = \{z^1 = 0\}, \quad l_1 = \{z^0 = z^1\},$$

$$l_0' = \{z^2 = 0\}, \quad l_1' = \{z^1 = z^2\}.$$

If we set $x = z^1/z^0$, $y = z^2/z^0$, then in Figure v we have

$$l_0 = \{x = 0\}, \quad l_1 = \{x = 1\}, \quad l_0' = \{y = 0\}, \quad l_1' = \{x = y\}.$$

We define f by

$$x = tu, \quad y = u/t \qquad \text{for} \quad (t, u) \in D^* \times D^*.$$

In terms of the homogeneous coordinate system of $P_2\mathbf{C}$, f is given by

$$f(t, u) = (1, tu, u/t) = (t, t^2 u, u) \qquad \text{for} \quad (t, u) \in D^* \times D^*.$$

Then f extends to all points of $D \times D$ except the origin $(0, 0)$.

The five lines we deleted from $P_2\mathbf{C}$ in the example above are not in general position since l_0, l_1 and l_∞ are concurrent. As we shall see later, when five lines in general position are removed from $P_2\mathbf{C}$, the resulting space is hyperbolically imbedded in $P_2\mathbf{C}$.

The following example due to Kiernan [5] involves not only lines but also curves.

(3.3.11) **Example.** From the projective plane $P_2\mathbf{C}$ with homogeneous coordinate system (z^0, z^1, z^2) we delete the following three lines l_∞, l_0, l_1 and two curves m_+, m_-:

$$l_\infty = \{z^0 = 0\}, \quad l_0 = \{z^1 = 0\}, \quad l_1 = \{z^1 = z^0\},$$

$$m_+ = \{z^1 z^2 = (z^0)^2\}, \quad m_- = \{z^1 z^2 = -(z^0)^2\}.$$

If we consider l_∞ as the line at infinity and use the inhomogeneous coordinate system $x = z^1/z^0$, $y = z^2/z^0$, then the resulting space Y looks like Figure viii, where $l_0 = \{x = 0\}, l_1 = \{x = 1\}, m_+ = \{xy = 1\}$ and $m_- = \{xy = -1\}$. Under the mapping

$$\varphi: (x, y) \in Y \to (x, xy) \in (\mathbf{C} - \{0, 1\}) \times (\mathbf{C} - \{-1, 1\}),$$

Y is biholomorphic to $(\mathbf{C}-\{0, 1\}) \times (\mathbf{C}-\{-1, 1\})$ and hence is complete hyperbolic. We shall show that Y is *not* hyperbolically imbedded in $P_2\mathbf{C}$. Take any two points on the line l_0, say $(0,0)$ and $(0,1)$ in terms of the inhomogeneous coordinate system (x, y). For each positive integer k,

$$f_k: t \mapsto (1/k, t)$$

sends the disc $D_k = \{t \in \mathbf{C}; |t| < k\}$ into Y. If we set

$$p_k = f_k(0) = (1/k, 0), \quad q_k = f_k(1) = (1/k, 1),$$

then $p_k \to (0, 0)$, $q_k \to (0, 1)$ and $d_Y(p_k, q_k) \le d_{D_k}(0, 1) \to 0$ as $k \to \infty$. This shows that Y is not hyperbolically imbedded in $P_2\mathbf{C}$.

In this case, we note that the holomorphic map

$$g: D^* \times D \to Y \subset P_2\mathbf{C}$$

defined by

$$g(t, u) = (1, t, u/t) = (t, t^2, u) \qquad \text{for} \quad (t, u) \in D^* \times D$$

does not extend to a map $g: D \times D \to P_2\mathbf{C}$. (It extends to all points of $D \times D$ except the origin $(0, 0)$).

The following example, also due to Kiernan [5], involves transcendental curves.

(3.3.12) **Example**. From the projective plane $P_2\mathbf{C}$ with homogeneous coordinate system (z^0, z^1, z^2) we delete the three lines l_∞, l_0, l_1 of the example above as well as the following two curves n_+, n_-:

$$n_+ = \{z^2 = z^0 e^{z_0/z_1}\} \quad n_- = \{z^2 = -z^0 e^{z_0/z_1}\}.$$

In terms of the inhomogeneous coordinate system (x, y) used in the example above, these curves are defined by $n_+ = \{y = e^{1/x}\}$ and $n_- = \{y = -e^{1/x}\}$. Then the resulting space Y is biholomorphic to $(\mathbf{C} - \{0, 1\}) \times (\mathbf{C} - \{-1, 1\})$ under the map

$$\psi: (x, y) \in Y \to (x, ye^{-1/x}) \in (\mathbf{C} - \{0, 1\}) \times (\mathbf{C} - \{-1, 1\}).$$

So, Y is complete hyperbolic. Using the same sequence of maps $f_k: D_k \to Y$ as in the preceding example, we can show that Y is *not* hyperbolically imbedded in $P_2\mathbf{C}$. We note that the holomorphic map $h: D^* \to Y \subset P_2\mathbf{C}$ defined by

$$h(t) = (1, t, 2e^{1/t}) \qquad \text{for} \quad t \in D^*$$

does not extend to a map $h: D \to P_2\mathbf{C}$.

In the preceding section we considered hyperbolicity modulo Δ. Similarly, let Δ be a closed subset of a complex space Z. Then we say that a relatively compact complex subspace $Y \subset Z$ is **hyperbolically imbedded modulo Δ in** Z if every point p of $\bar{Y} - \Delta$ is a hyperbolic point. This is equivalent to saying that for every pair of distinct points p, q of $\bar{Y}$ not both contained in Δ, there exist neighborhoods

U_p and U_q of p and q in Z such that $d_Y(U_p \cap Y, U_q \cap Y) > 0$. It is clear that if Y is hyperbolically imbedded modulo Δ in Z, then it is hyperbolic modulo $\Delta \cap Y$.

Generalizing (3.3.3) we have the following (Kiernan-Kobayashi [2])

(3.3.13) Theorem. *A relatively compact complex subspace Y of a complex space Z is hyperbolically imbedded modulo Δ if and only if given a compact neighborhood K of Y in Z and a length function F on K there is a continuous non-negative function φ on K such that*

(a)   *φ is strictly positive on $K - \Delta$,*

(b)   *$f^*(\varphi^2 F^2) \leq ds_D^2$ for all $f \in \mathrm{Hol}(D, Y)$.*

Proof. Assume that Y is hyperbolically imbedded modulo Δ in Z. The proof of (3.3.3) shows (see also Kiernan-Kobayashi [2]) that given a compact subset L of $K - \Delta$, we can choose a constant $c > 0$ such that $f^*(c^2 F) \leq ds_D^2$ in $f^{-1}(L)$ for all $f \in \mathrm{Hol}(D, Y)$. Let $L_1 \subset L_2 \subset \ldots$ be an increasing sequence of compact subsets of $K - \Delta$ such that $K - \Delta = \bigcup_{i=1}^{\infty} L_i$. Take a corresponding sequence of positive constants $c_1 \geq c_2 \geq \ldots$. Let φ be any non-negative continuous function on K such that $0 < \varphi \leq c_i$ on L_i.

The proof of the converse is the same as in (3.3.3). $\square$

Generalizing (3.3.4) we have the following (Kiernan-Kobayashi [2])

(3.3.14) Theorem. *Let Y be hyperbolically imbedded modulo Δ in Z. If Y is locally complete hyperbolic in the sense that every $p \in \bar{Y}$ has a neighborhood V_p in Z such that $Y \cap V_p$ is complete hyperbolic, then Y is complete hyperbolic modulo $\Delta \cap Y$.*

Proof. We note that every Cauchy sequence with respect to d_Y is a Cauchy sequence with respect to the pseudo-distance defined by the pseudo-length function φF of (3.3.13). Hence, if Y is not complete hyperbolic modulo $\Delta \cap Y$, then there exists a sequence $\{p_n\}$ in Y which is Cauchy with respect to d_Y such that $p_n \to p \notin Y \cup \Delta$. Let V_p be a neighborhood of p in Z such that $V_p \cap Y$ is complete hyperbolic. By (3.3.5) $\{p_n\}$ is Cauchy with respect to $d_{V_p \cap Y}$ and must converges to a point of $V_p \cap Y$. This is a contradiction. $\square$

Combined with (3.2.18) the theorem yields the following

(3.3.15) Corollary. *Let Z be a compact complex space, A a Cartier divisor in Z, and Δ a closed subset of Z. Let $Y = Z - A$. If Y is hyperbolically imbedded modulo Δ in Z, then Y is complete hyperbolic modulo $\Delta \cap Y$.*

(3.3.16) Remarks. We have discussed so far only very elementary examples of hyperbolically imbedded spaces. Nontrivial examples require techniques which will be developed in subsequent sections. At this point we mention the following basic conjecture (cf. Kobayashi [7]).

Conjecture 1. (i) *A generic hypersurface of degree $\geq 2n + 1$ in $P_n\mathbf{C}$ is hyperbolic, and* (ii) *the complement of such a hypersurface is complete hyperbolic and is hyperbolically imbedded in $P_n\mathbf{C}$.*

As we shall see from (3.6.12), parts (i) and (ii) are related.

We list several supporting evidence for the conjecture. The classical result in support of (ii) is that the complement of $2n + 1$ hyperplanes in general position in $P_n\mathbf{C}$ is hyperbolically imbedded in $P_n\mathbf{C}$. This will be proved in Section 10.

The first result in support of (i) is the following examples of nonsingular hyperbolic surfaces in $P_3\mathbf{C}$ by Brody-Green [1]:

$$(z^0)^d + \ldots + (z^3)^d + s(z^0z^1)^{d/2} + t(z^0z^2)^{d/2} = 0$$

with an even degree $d \geq 50$ and generic $s, t \in \mathbf{C}^*$. This was also the first example of a simply connected compact hyperbolic manifold.

Based on the example above, Azukawa-Suzuki [1] constructed the first examples of smooth curves in $P_2\mathbf{C}$ with hyperbolically imbedded complements. They proved that the complement of the curve

$$(z^0)^d + (z^1)^d + (z^2)^d + \varepsilon(z^0z^1)^{d/2} + \varepsilon(z^0z^2)^{d/2} = 0$$

in $P_2\mathbf{C}$ is hyperbolically imbedded in $P_2\mathbf{C}$ if either $d \geq 30$ and ε is a complex number with $\varepsilon^2 \neq 0, 4$ or $d \geq 14$ and $\varepsilon^2 = 2$. (The curve above is nonsingular if $\varepsilon^2 \neq 2$.)

Nadel [1] also obtained similar examples of hyperbolic surfaces in $P_3\mathbf{C}$ and curves in $P_2\mathbf{C}$ with hyperbolically imbedded complements.

Zaidenberg [7] has shown that, for each $d \geq 5$, the set of smooth curves of degree d in $P_2\mathbf{C}$ with hyperbolically imbedded complements forms a nonempty open set in the set of all curves of degree d (in the usual topology, not in the Zariski topology).

In order to state more systematic results, let us say that a curve C in $P_2\mathbf{C}$ has degree $(d_1, \ldots, d_k)$ if it consists of irreducible components $C_1, \ldots, C_k$ of degrees $d_1, \ldots, d_k$. Then the complement of a generic curve C of degrees $(d_1, \ldots, d_k)$ in $P_2\mathbf{C}$ is known to be hyperbolically imbedded in $P_2\mathbf{C}$ in the following cases:

(i) $k \geq 5$ with any degree (Babets [3]);

(ii) $k = 4$ with degree $(2, 1, 1, 1)$ (Green [7]);

(iii) $k = 4$ with any degree as long as $\sum d_i \geq 5$ (Green [2], Dethloff-Schumacher-Wong [1]);

(iv) $k = 3$ with $d_1, d_2, d_3 \geq 2$ (Dethloff-Schumacher-Wong [1, 2]).

Masuda and Noguchi [1] have found an algorithm to produce nonsingular hyperbolic hypersurfaces of every degree $d \geq d(n)$ in $P_n\mathbf{C}$ with hyperbolically imbedded complements. However, $d(n)$ is still very high compared with $2n + 1$.

A generic surface of degree $d \geq 5$ in $P_3\mathbf{C}$ contains no rational or elliptic curves, which makes such a surface very close to being hyperbolic, see Zaidenberg [12] and Remark (3.2.24).

We mention a different kind of hyperbolicity criterion for plane curves. According to Grauert-Peternell [1], the complement of a smooth curve C of genus $g \geq 2$ in $P_2\mathbf{C}$ is complete hyperbolic and hyerpbolically imbedded in $P_2\mathbf{C}$ if the dual curve C^* of C has only ordinary double points and i cusps with $i < 2g - 2$.

We cannot expect to lower the degree $2n+1$ in Conjecture 1. In fact, Zaidenberg [6] made the following conjecture with strong supporting evidence.

Conjecture 2. *For every hypersurface A of degree $\leq 2n$ in $P_n\mathbf{C}$, there is a line which meets A only in at most two points. In particular, the complement of A is not hyperbolic.*

He verified this for a generic hypersurface A of degree $\leq 2n$. As we shall see later in (3.10.13) a theorem of Snurnitsyn implies that the complement of $2n$ hyperplanes in $P_n\mathbf{C}$ is never hyperbolic. As another supporting evidence we mention Green [6], where it is shown that the complement of a curve of degree 4 in $P_2\mathbf{C}$ is not hyperbolic.

As we shall see in (3.10.25), the complement of $n+2$ hyperplanes in general position in $P_n\mathbf{C}$ is hyperbolic modulo the diagonal hyperplanes. Now we state another conjecture, which has much less evidence than the preceding ones.

Conjecture 3. (i) *A generic hypersurface of degree $\geq n+2$ in $P_n\mathbf{C}$ is hyperbolic modulo a proper algebraic subset, and* (ii) *the complement of such a hypersurface in $P_n\mathbf{C}$ is complete hyperbolic and hyperbolically imbedded in $P_n\mathbf{C}$ modulo a proper algebraic subset of $P_n\mathbf{C}$.*

There are some results in this direction in terms of the number of irreducible components of the hypersurface rather than in terms of the degree, see Nishino [1], Adachi-Suzuki [1].

We note that Part (i) is a special case of the following general conjecture.

Conjecture 4. *A projective algebraic manifold of general type is hyperbolic modulo a proper algebraic subset.*

4 Relative Intrinsic Pseudo-distance

Let Z be a complex space and Y a complex subspace with closure $\bar{Y}$. Let d_Y and d_Z denote the intrinsic pseudo-distances of Y and Z, respectively.

We shall now introduce a **relative pseudo-distance** $d_{Y,Z}$ on $\bar{Y}$ so that Y is hyperbolically imbedded in Z if and only if $\bar{Y}$ is compact and $d_{Y,Z}$ is a distance. (Although we call $d_{Y,Z}$ a pseudo-distance, $d_{Y,Z}(p, q)$ can be ∞ when p and/or q are on the boundary ∂Y.)

Let $\mathcal{F}_{Y,Z} \subset \mathrm{Hol}(D, Z)$ be the family of holomorphic maps $f: D \to Z$ such that $f^{-1}(Z - Y)$ is either empty or a singleton. Thus, $f \in \mathcal{F}_{Y,Z}$ maps all of D, with the exception of possibly one point, into Y. The exceptional point is of course mapped into $\bar{Y}$.

We define a pseudo-distance $d_{Y,Z}$ on $\bar{Y}$ in the same way as d_Z, but using only chains of holomorphic discs belonging to $\mathcal{F}_{Y,Z}$. Namely, writing $l(\alpha)$ for the length of a chain α of holomorphic discs as in (3.1.4), we set

$$(3.4.1) \qquad\qquad d_{Y,Z}(p, q) = \inf_{\alpha} l(\alpha), \qquad p, q \in \bar{Y},$$

where the infimum is taken over all chains α of holomorphic discs from p to q which belong to $\mathcal{F}_{Y,Z}$. (We want to emphasize that chains of points $p = p_0, p_1, \ldots, p_k = q$ we use for chains of holomorphic discs α are points of $\bar{Y}$. If we take $p_1, \ldots, p_{k-1}$ only from Y, then we would not get the triangular inequality for $d_{Y,Z}$).

If p or q is in the boundary of Y, such a chain may not exist. In such a case, $d_{Y,Z}(p, q)$ is defined to be ∞. For example, if Y is a convex bounded domain in $\mathbf{C}^n$, any holomorphic disc passing through a boundary point of Y goes outside the closure $\bar{Y}$ or is contained in the boundary $\partial Y = \bar{Y} - Y$, so that $d_{Y,\mathbf{C}^n}(p, q) = \infty$ if p is a boundary point of Y. On the other hand, if Y is Zariski-open in Z, any pair of points p, q in $\bar{Y} = Z$ can be joined by a chain of holomorphic discs beloning to $\mathcal{F}_{Y,Z}$, so that $d_{Y,Z}(p, q) < \infty$ for $p, q \in Z = \bar{Y}$.

The relative pseudo-distance $d_{Y,Z}$ is a generalization of the pseudo-distance d_Y in the sense that

$$d_Y = d_{Y,Y}.$$

Since

$$\mathrm{Hol}(D, Y) \subset \mathcal{F}_{Y,Z} \subset \mathrm{Hol}(D, Z),$$

we have

$$(3.4.2) \qquad d_Z \leq d_{Y,Z} \leq d_Y,$$

where the second inequality holds on Y while the first is valid on $\bar{Y}$.

For the punctured disc $D^* = D - \{0\}$, we have

$$(3.4.3) \qquad d_{D^*,D} = d_D.$$

The inequality $d_{D^*,D} \geq d_D$ is a special case of (3.4.2). Using the identity map $\mathrm{id}_D \in \mathcal{F}_{D^*,D}$ as a holomorphic disc joining two points of D yields the opposite inequality.

Let $Y' \subset Z'$ be another pair of complex spaces with $\bar{Y}'$ compact. If $f: Z \to Z'$ is a holomorphic map such that $f(Y) \subset Y'$, then

$$(3.4.4) \qquad d_{Y',Z'}(f(p), f(q)) \leq d_{Y,Z}(p, q), \qquad p, q \in \bar{Y}.$$

The proof of the following proposition is the same as that of (3.1.7).

(3.4.5) **Proposition.** *If $\delta_{\bar{Y}}$ is a pseudo-distance on $\bar{Y}$ such that*

$$\delta_{\bar{Y}}(f(a), f(b)) \leq d_D(a, b) \qquad for \quad a, b \in D \quad and \quad f \in \mathcal{F}_{Y,Z},$$

then

$$\delta_{\bar{Y}}(p, q) \leq d_{Y,Z}(p, q) \qquad for \quad p, q \in \bar{Y}.$$

The proof of the following theorem is the same as that of (3.1.9).

(3.4.6) **Theorem.** *Let $Y \subset Z$ and $Y' \subset Z'$. Then for $p, q \in \bar{Y}$ and $p', q' \in \bar{Y}'$ we have*

$$d_{Y \times Y', Z \times Z'}((p, p'), (q, q')) = \max\{d_{Y,Z}(p, q), d_{Y',Z'}(p', q')\}.$$

From (3.1.10), (3.4.3) and (3.4.6) we obtain

(3.4.7) **Corollary**. *For $(D^*)^k \times D^{n-k} \subset D^n$, we have*

$$d_{(D^*)^k \times D^{n-k}, D^n} = d_{D^n}.$$

(3.4.8) **Proposition**. *Let $Y \subset Z$. Then*

(a) $d_{Y,Z}$ *is continuous on $Y \times Y$, and is lower semicontinuous on $\bar{Y} \times \bar{Y}$;*

(b)   *If Y is the complement of a closed analytic subset A of Z so that $\bar{Y} = Z$, then $d_{Y,Z}$ is continuous on $Z \times Z$.*

Proof. The first assertion in (a) follows from (3.1.13) and (3.4.2). The second follows from the definition of $d_{Y,Z}$.

In order to prove (b), as in the proof of (3.1.13) it suffices to show that, for every $p \in Z$ and for a small neighborhood U of p, $d_{Y \cap U, U}(p_j, p) \to 0$ as $p_j \to p$. Resolving the singularity of (Z, A) we may assume that $U = D^n$ and $Y \cap U = (D^*)^k \times D^{n-k}$. Now use (3.4.7), (see the proof of (3.1.13) for details).

(3.4.9) **Remark**. Since $d_{Y,Z}(p, q)$ can be infinite when p is in ∂Y, $d_{Y,Z}$ may not be continuous on $\bar{Y} \times \bar{Y}$.

Making use of (1.1.8), (3.4.8) and the proof of (3.1.15), we can generalize (3.1.15) and (3.2.1) as follows:

(3.4.10) **Proposition**. (a)   *Let $Y \subset Z$ such that $d_{Y,Z}(p, q) > 0$ for all $p, q \in \bar{Y}$, $p \neq q$. Then $d_{Y,Z}$ is an inner distance on Y and defines the topology of Y.*

(b)   *If, moreover, Y is the complement of a closed analytic subset A of Z, then $d_{Y,Z}$ is an inner distance on Z and defines the topology of Z.*

(3.4.11) **Theorem**. *Let $Y \subset Z$ with $\bar{Y}$ compact. Among the following conditions, (a) and (b) are equivalent, and they imply (c). Thus*

$$\text{(a)} \Leftrightarrow \text{(b)} \Rightarrow \text{(c)}.$$

(a)   *given a length function F on Z, there is a constant $c > 0$ such that*

$$f^*(c^2 F^2) \leq ds_D^2 \qquad \text{for all} \quad f \in \mathcal{F}_{Y,Z};$$

(b)   *for $p, q \in \bar{Y}$, $p \neq q$,*

$$d_{Y,Z}(p, q) > 0;$$

(c)   *Y is hyperbolically imbedded in Z.*

In (3.6.20) we shall show that (c) is also equivalent to (b)

Proof. (a) $\Rightarrow$ (b). Let δ_Z be the distance function defined by the length function cF in (a). Then the restriction of δ_Z to $\bar{Y}$ satisfies the condition for $\delta_{\bar{Y}}$ in (3.4.5). Hence,

(3.4.12) $\delta_Z \leq d_{Y,Z}$ on $\bar{Y}$.

(b) $\Rightarrow$ (a). This implication, due to Joseph-Kwack [1], can be proved in the same way as (3.3.3). If such a constant c does not exists, then there exist a sequence $\{f_n\}$ in $\mathcal{F}_{Y,Z}$ and tangent vectors $v_n \in TD$ such that

$$F(df_n(v_n)) > n \cdot |v_n|,$$

where $|v_n|$ is the length of v_n with respect to ds_D^2. Then, by Arzela-Ascoli theorem (1.3.1) $\{f_n\}$ is not equicontinuous at some point, say $a \in D$. By taking a subsequence of $\{f_n\}$ we can find a sequence $\{a_n\}$ in D converging to a such that

$$\lim f_n(a_n) \neq \lim f_n(a).$$

On the other hand, we have

$$d_{Y,Z}(f_n(a_n), f_n(a)) \leq d_D(a, a_n) \to 0,$$

which contradicts (b).

(b) $\Rightarrow$ (c). Let δ_Z be the distance function defined by cF in (a). Since $\delta_Z \leq d_Y$ on Y by (3.4.2) and (3.4.12), Y is hyperbolically imbedded in Z by (3.3.2). $\square$

(3.4.13) **Remark**. The inequality (3.4.12) established in the course of the proof is important. Since δ_Z does not become infinite and defines the topology of Z, it is often better to use δ_Z in place of $d_{Y,Z}$.

The pseudo-distance $d_{Y,Z}$ was introduced in Kobayashi [24], [25] where the equivalence of (b) and (c) was proved.

Using chains of holomorphic punctured discs, we may define a slightly different pseudo-distance d_Y^*. More precisely, We use $\mathrm{Hol}(D^*, Y)$ with the punctured disk D^* equipped with the restriction of the Poincaré distance $d_D = \rho$, (not d_{D^*}). Then

(3.4.14) $$d_Y^* \leq d_Y.$$

In fact, in the definition (3.1.4) of $l(\alpha)$ we may assume that points $a_1, b_1, \ldots, a_k, b_k$ are not the origin of D. Then by restricting $f_1, \ldots, f_k$ to D^*, we obtain from each chain of holomorphic discs a chain of holomorphic punctured discs, hence the desired inequality.

The following proposition is immediate from the definition of d_X^* and d_Y^*.

(3.4.15) **Proposition**. *Every holomorphic map $f: X \to Y$ between complex spaces is distance-decreasing with respect to d_X^* and d_Y^*.*

(3.4.16) **Proposition**. *If Y is a complex subspace of Z with the property that every map $f \in \mathrm{Hol}(D^*, Y)$ extends to a map $\bar{f} \in \mathrm{Hol}(D, Z)$, then*

$$d_Z \leq d_{Y,Z} \leq d_Y^* \leq d_Y \quad \text{on} \quad Y.$$

Proof. If $f \in \mathrm{Hol}(D^*, Y)$, then $\bar{f} \in \mathcal{F}_{Y,Z}$. Hence, every chain of holomorphic punctured discs in Y gives rise to a chain of holomorphic discs used in the construction of $d_{Y,Z}$. Hence, $d_{Y,Z} \leq d_Y^*$. The remaining inequalities are from (3.4.2) and (3.4.14) $\square$

We cannot claim the equality $d_{Y,Z} = d_Y^*$ in (3.4.16) since a chain of holomorphic discs used in the construction of $d_{Y,Z}$ may not induce a chain of holomorphic punctured discs if some of the connecting points $p_1, \ldots, p_{k-1}$ are in ∂Y.

However, using the natural injection $D^* \to D$ and the the identity map $D^* \to D^*$, we obtain

$$(3.4.17) \qquad d_D^* = d_D \qquad \text{and} \qquad d_{D^*}^* = d_D \quad \text{on} \quad D^*.$$

For $(D^*)^k \times D^{n-k}$, our knowledge is less precise:

$$(3.4.18) \qquad d_{D^n} \leq d_{(D^*)^k \times D^{n-k}}^* \leq n \cdot d_{D^n}.$$

The first inequality is a special case of (3.4.16). To prove the second inequality, given $p = (a_1, \ldots, a_n)$ and $q = (b_1, \ldots, b_n)$ in $(D^*)^k \times D^{n-k}$, we consider a chain of points $p = p_0, p_1, \ldots, p_{n-1}, p_n = q$ given by

$$p_i = (b_1, \ldots, b_i, a_{i+1}, \ldots, a_n).$$

Then using an obvious chain of holomorphic punctured discs and (3.4.17) we obtain

$$d_{(D^*)^k \times D^{n-k}}^*(p, q) \leq \sum d_D(a_i, b_i) \leq n \cdot d_{D^n}(p, q).$$

(3.4.19) Proposition. *Let Y be a relatively compact complex subspace of Z satisfying conditions* (a) *and* (b) *of (3.4.11). Assume that every map $f \in \mathrm{Hol}(D^*, Y)$ extends to a map $\bar{f} \in \mathrm{Hol}(D, Z)$. Then every map $h \in \mathrm{Hol}((D^*)^k \times D^{n-k}, Y)$ extends to a map $\bar{h} \in \mathrm{Hol}(D^n, Z)$.*

As we shall see later in (6.3.7), the assumption on the extendability of f is superfluous; it will be shown there that when Y is hyperbolically imbedded in Z, every $f \in \mathrm{Hol}(D^*, Y)$ extends to $\bar{f} \in \mathrm{Hol}(D, Z)$.

Proof. Set $X = D^n$ and $X - A = (D^*)^k \times D^{n-k}$. Let δ_Z be the distance function on Z defined by the length function cF of (3.4.11) so that $\delta_Z \leq d_{Y,Z}$. Given any point $p_0 \in A$, take a sequence of points $p_i \in X - A$ converging to p_0. Taking a subsequence we may assume that $h(p_i)$ converges to a point $q_0 \in \bar{Y}$. As we remarked in (3.4.13), we use δ_Z which gives the topology of Z. Given a neighborhood V of q_0 in Z, take an ε-neighborhood

$$W = \{q \in Z; \ \delta_Z(q, q_0) < \varepsilon\}$$

such that $W \subset V$. Let U be the $(\varepsilon/4n)$-neighborhood of p_0 in $X = D^n$ with respect to d_{D^n}. We have only to show that $h(U \cap (X - A)) \subset V$. Let $p \in U \cap (X - A)$. Choose m such that $p_m \in U$ and $\delta_Z(q_0, h(p_m)) < \varepsilon/2$. Then by (3.4.18)

$$d_{X-A}^*(p_m, p) \leq n \cdot d_X(p_m, p) \leq n(d_X(p_m, p_0) + d_X(p_0, p)) < \varepsilon/2.$$

By (3.4.12), (3.4.16) and (3.4.15)

$$\begin{aligned} \delta_Z(h(p_m), h(p)) &\leq d_{Y,Z}(h(p_m), h(p)) \leq d_Y^*(h(p_m), h(p)) \\ &\leq d_{X-A}^*(p_m, p) < \varepsilon/2. \end{aligned}$$

Hence, $\delta_Z(q_0, h(p)) < \varepsilon$, which proves that $h(p) \in W \subset V$. Hence, h extends. $\qquad\square$

In the proof of (3.4.19), consider more generally a complex space X and a closed complex subspace A and try to extend a map $f \in \mathrm{Hol}(X - A, Y)$ to a map $\bar{f} \in \mathrm{Hol}(X, Z)$. In the proof above we made an essential use of the fact that there is a constant n such that $d^*_{X-A} \le n \cdot d_X$. The existence of such a constant seems to depend on the nature of the singular loci of X and A. Consider the following examples.

(3.4.20) Example. Let

$$
\begin{aligned}
X &= D \times D = \{(z, w);\ |z| < 1,\ |w| < 1\}, \\
A &= (0 \times D) \cup (D \times 0) \cup \mathrm{diag}(D \times D) \\
  &= \{(z, w) \in D \times D;\ zw(z - w) = 0\}, \\
Z &= P_1 \mathbf{C} \quad \text{and} \quad Y = P_1 \mathbf{C} - \{\infty, 0, 1\} = \mathbf{C} - \{0, 1\}.
\end{aligned}
$$

and

$$
f: X - A \to Y \qquad f(z, w) = w/z.
$$

Let $a, b \in \mathbf{C} - \{0, 1\}$, $a \neq b$, and t a nonzero complex number. Since $f: (X - A, d^*_{X-A}) \to (Y, d^*_Y)$ is distance-decreasing, we have

$$
d^*_{X-A}((t, at), (t, bt)) \ge d^*_Y(f(t, at), f(t, bt)) = d^*_Y(a, b) \ge d_{Y,Z}(a, b) > 0,
$$

showing that $d^*_{X-A}((t, at), (t, bt))$ is bounded below by a positive constant independent of t. On the other hand, $d_X((t, at), (t, bt)) \to 0$ as $t \to 0$. Hence, there is no constant c such that $d^*_{X-A} \le c \cdot d_X$.

(3.4.21) Example. Let P be a homogeneous polynomial on $\mathbf{C}^3$, and S be the affine surface defined by $P = 0$; it is the cone over the projective plane curve C defined by $P = 0$. Let B be the open unit ball around the origin in $\mathbf{C}^3$. We set $X = S \cap B$ and let A be the origin of $\mathbf{C}^3$, i.e., the vertex of the cone S. Let $\pi: X - A \to C$ be the natural projection. Assume that C is non-singular and that the degree of P is greater than 3 so that the genus of C is greater than 1. As we shall see in (3.7.3), C is then hyperbolic. We shall show that there is no constant c such that $d^*_{X-A} \le c \cdot d_X$.

Let p and q be linearly independent unit vectors in $\mathbf{C}^3$ that lie in S. Let $g, h \in \mathrm{Hol}(D, X)$ be defined by

$$
g(z) = zp \quad \text{and} \quad h(z) = zq, \qquad z \in D.
$$

Let $\{a_n\}$ be a sequence of points of D converging to 0. Using the chain of holomorphic disks consisting of g and h, we see that $\lim_n d_X(a_n p, a_n q) = 0$.

On the other hand, since the projection $\pi: (X - A, d^*_{X-A}) \to (C, d^*_C)$ is distance-decreasing, we have

$$
d^*_{X-A}(a_n p, a_n q) \ge d^*_C(\pi(a_n p), \pi(a_n q)) = d^*_C(\pi(p), \pi(q)).
$$

Since C is hyperbolic, every map $f \in \mathrm{Hol}(D^*, C)$ extends to a map $\bar{f} \in \mathrm{Hol}(D, C)$ as we shall show in (6.3.5). Hence, $d_C^* = d_C$ by (3.4.16). Thus, $d_{X-A}^*(a_n p, a_n q)$ is bounded below by a positive number $d_C(\pi(p), \pi(q))$ which is independent of n.

5 Infinitesimal Pseudo-metric F_X

Given a complex space X, we shall define an **intrinsic pseudo-metric**, i.e., the **infinitesimal form** $\hat{F}_X$ **of the Kobayashi pseudo-distance** d_X. All necessary algebraic results on norms are summarized in Section 4 of Chapter 1.

Let $T_x X$ and $T_x^* X$ denote the Zariski tangent space and cotangent space of X at x. Since there are some unresolved technical problems when X has singularities, we often have to assume that X is a complex manifold. We first define a quasi-norm F_X^* on the Zariski cotangent space $T_x^* X$ by setting

$$(3.5.1) \qquad F_X^*(\lambda) = \sup \| f^* \lambda \| \qquad \text{for} \quad \lambda \in T_x^* X,$$

where $\| f^* \lambda \|$ is the length of the cotangent vector $f^* \lambda \in T^* D$ measured by the Poincaré metric ds^2 of the unit disc D, and the supremum is taken over all $f \in \mathrm{Hol}(D, X)$ with $x \in f(D)$. Because of the homogeneity of D, it suffices to take the supremum over all $f \in \mathrm{Hol}(D, X)$ with $f(0) = x$.

For some λ we may have $F_X^*(\lambda) = +\infty$. On the other hand, if X is a complex manifold, then we have

$$(3.5.2) \qquad F_X^*(\lambda) > 0 \qquad \text{for} \quad 0 \neq \lambda \in T_x^* X.$$

In fact, given $\lambda \neq 0$, choose an element $f_*(u)$ of $T_x X$ such that $\lambda(f_*(u)) \neq 0$. Then $f^* \lambda \neq 0$ and $F_X^*(\lambda) \geq \| f^* \lambda \| > 0$.

From the defintion (3.5.1) we can easily verify the convexity property for F_X^*:

$$(3.5.3) \qquad F_X^*(\lambda + \mu) \leq F_X^*(\lambda) + F_X^*(\mu) \qquad \text{for} \quad \lambda, \mu \in T_x^* X.$$

Let

$$(3.5.4) \qquad \tilde{T}_x^* X = \{\lambda \in T_x^* X; \ F_X^*(\lambda) < \infty\}.$$

From (3.5.3) it follows that $\tilde{T}_x^* X$ is a vector subspace of $T_x^* X$.

We recall (see (1.4.2)) that the indicatrix of F_X^* at x is the subset Γ_x^* of $T_x^* X$ defined by

$$(3.5.5) \qquad \Gamma_x^* = \{\lambda \in T_x^* X; \ F_X^*(\lambda) \leq 1\}.$$

It is convex and circular and is contained in $\tilde{T}_x^* X$.

Let $\lambda_1, \ldots, \lambda_r$ be a basis for $\tilde{T}_x^* X$, and let $F^*(\lambda_1), \ldots, F^*(\lambda_r) \leq M$. Then by (3.5.3) for any $\lambda = \sum a_i \lambda_i$ with $|a_i| \leq 1$, we have $F_X^*(\lambda) \leq r M$. It follows that F_X^* is bounded on any bounded subset of $\tilde{T}_x^* X$. It follows also that given $\varepsilon > 0$, there is a neighborhood of 0 in $\tilde{T}_x^* X$ such that $F_X^*(\mu) < \varepsilon$ for μ belonging to that neighborhood. Using again (3.5.3) we see that F_X^* is continuous in $\tilde{T}_x^* X$. Fix a norm in $\tilde{T}_x^* X$ and consider the unit sphere $S_x \subset \tilde{T}_x^* X$ with respect to the norm. Since F_X^* is continuous on S_x and positive (see (3.5.2)), it follows that Γ_x^* is a closed, bounded subset of $\tilde{T}_x^* X$. Summarizing, we have

(3.5.6) **Proposition.** *The indicatrix Γ_x^* of F_X^* is a compact, convex, circular subset of $\tilde{T}_x^* X$.*

Let $\hat{F}_X$ be the quasi-norm on $T_x X$ dual to F_X^*, i.e.,

$$(3.5.7) \qquad \hat{F}_X(v) = F_X^{**}(v) = \sup_{\lambda \in \Gamma_x^*} |\lambda(v)| \qquad \text{for} \quad v \in T_x X.$$

Since Γ_x^* is compact, the supremum in the definition above is actually attained, i.e.,

$$(3.5.8) \qquad \hat{F}_X(v) = \max_{\lambda \in \Gamma_x^*} |\lambda(v)| \qquad \text{for} \quad v \in T_x X.$$

Hence, if (3.5.2) holds, in particular if X is a complex manifold, then

$$(3.5.9) \qquad \hat{F}_X(v) < \infty \qquad \text{for} \quad v \in T_x X.$$

We also have (see (1.4.5))

$$(3.5.10) \qquad \hat{F}_X(v + v') \leq \hat{F}_X(v) + \hat{F}_X(v') \qquad \text{for} \quad v, v' \in T_x X.$$

Thus, $\hat{F}_X$ defines a pseudo-norm on $T_x X$.

The following proposition corresponds to (3.1.6).

(3.5.11) **Proposition.** (1) *Let $f : X \to Y$ be a holomorphic mapping between complex spaces. Then*

$$F_X^*(f^* \lambda) \leq F_Y^*(\lambda) \qquad \textit{for} \quad \lambda \in T^* Y,$$

$$\hat{F}_Y(f_* v) \leq \hat{F}_X(v) \qquad \textit{for} \quad v \in T X;$$

(2) *For the unit disc D, both F_D^* and $\hat{F}_D$ coincide with the norms defined by the Poincaré metric, i.e.,*

$$\hat{F}_D^2 = ds^2.$$

(3.5.12) **Proposition.** *Let X be a complex space. Then*

$$\| f^* \lambda \| \leq F_X^*(\lambda) \qquad \textit{for} \quad f \in \mathrm{Hol}(D, X) \quad \textit{and} \quad \lambda \in T^* X.$$

Conversely, if E^* is any quasi-norm on $T_x^* X$ such that

$$\| f^* \lambda \| \leq E^*(\lambda) \qquad \lambda \in T_x^* X$$

for all $f \in \mathrm{Hol}(D, X)$ with $f(0) = x$, then

$$F_X^*(\lambda) \le E^*(\lambda) \qquad \text{for} \quad \lambda \in T_x^* X.$$

Proof. Taking the supremum over f of the following inequality

$$\|f^*\lambda\| \le E^*(\lambda),$$

we obtain

$$F_X^*(\lambda) = \sup \|f^*(\lambda)\| \le E^*(\lambda). \qquad \square$$

Dualizing (3.5.12) we have the following infinitesimal version of (3.1.7).

(3.5.13) Proposition. *Let X be a complex space. Then*

$$\hat{F}_X(f_*u) \le \|u\| \qquad \text{for} \quad f \in \mathrm{Hol}(D, X) \quad \text{and} \quad u \in TD.$$

Conversely, if E is any quasi-norm on $T_x X$ such that

$$E(f_*u) \le \|u\| \qquad u \in T_0 D$$

for all $f \in \mathrm{Hol}(D, X)$ with $f(0) = x$, then

$$E^{**}(v) \le \hat{F}_X(v) \qquad \text{for} \quad v \in T_x X.$$

It should be noted that we are not claiming a stronger inequality $E \le \hat{F}_X$.

Proof. Fix $f \in \mathrm{Hol}(D, X)$ with $f(0) = x$ and also $\lambda \in T_x^* X$. When u varies in $T_0 D$ and v in $T_x X$, we have

$$\begin{aligned}
\|f^*\lambda\| &= \sup_{\|u\| \le 1} |(f^*\lambda)(u)| = \sup_{\|u\| \le 1} |\lambda(f_*u)| \le \sup_{E(f_*u) \le 1} |\lambda(f_*u)| \\
&\le \sup_{E(v) \le 1} |\lambda(v)| = E^*(\lambda).
\end{aligned}$$

Thus,

$$\|f^*\lambda\| \le E^*(\lambda) \qquad \text{for} \quad \lambda \in T_x^* X.$$

Using this, we obtain

$$\hat{F}_X(v) = \sup_{F_X^*(\lambda) \le 1} |\lambda(v)| \ge \sup_{E^*(\lambda) \le 1} |\lambda(v)| = E^{**}(v). \qquad \square$$

At a point x of a complex space X, we consider the tangent cone $\check{T}_x X$. We recall that $\check{T}_x X$ is a subset of the Zariski tangent space consisting of vectors of the form $f_*(u)$, where $f \in \mathrm{Hol}(D, X)$ and $u \in TD$. As we remarked in Section 3 of Chapter 2, $\check{T}_x X$ is in general smaller than what is usually called the tangent cone. If x is a regular point of X, then $\check{T}_x X = T_x X$.

The quasi-norm F_X^* on $T_x^* X$ defined by (3.5.1) can be obtained as the dual quasi-norm of a more geometric quasi-norm F_X on the tangent cone $\check{T}_x X$. For $v \in \check{T}_x X$, we set

$$(3.5.14) \qquad F_X(v) = \inf\{\|u\|;\ u \in TD \quad \text{and} \quad f_*(u) = v\},$$

where $\|u\|$ is the length of the tangent vector u measured by the Poincaré metric ds^2 of the unit disc D, and the infimum is taken over all $f \in \mathrm{Hol}(D, X)$ and $u \in TD$ such that $f_*(u) = v$. Because of the homogeneity of D, it suffices to take the infimum over tangent vectors u at the origin 0 of D which are pointing in the positive direction of the real axis. If x is a regular point, then given $v \in T_x X$, there is a vector $u \in TD$ such that $f_*(u) = v$, so that $F_X(v) < \infty$. However, if x is a singular point and if no such u exists, then we set $F_X(v) = \infty$.

The quasi-norm F_X was defined in Kobayashi [4] as a direct infinitesimal analogue of d_X, and $\hat{F}_X$ was introduced in Kobayashi [22] as an improved infinitesimal form of d_X, see (3.5.23) below.

(3.5.15) **Remark.** If X is complete hyperbolic and if $F_X(v_0) < \infty$, then the infimum in the definition (3.5.14) of $F_X(v_0)$ is actually attained, i.e., $F_X(v_0) = \|u_0\|$ for some $u_0 \in T_0 D$ and some $f_0 \in \mathrm{Hol}(D, X)$ with $f_{0*}(u_0) = v_0$. To find such an f_0, we consider a minimizing sequence

$$\{(u_k, f_k); \; u_k \in T_0 D, \; f_k \in \mathrm{Hol}(D, X)\}$$

such that $f_{k*}(u_k) = v_0$ and $F_X(v_0) = \lim \|u_k\|$. Then by passing to a subsequence if necessary, we may assume that $\{u_k\}$ converges to some vector $u_0 \in T_0 D$. By applying (1.3.3) to the family $\{f_k\}$, we see that a subsequence of $\{f_k\}$ converges to a map $f_0 \in \mathrm{Hol}(D, X)$. This proof is valid under the weaker condition that X is taut. The concept of taut complex space will be introduced in Section 1 of Chapter 5.

The following alternate definition of F_X is sometimes useful. Let D_R be the disc $\{z \in \mathbf{C}; \; |z| < R\}$ of radius R with Poincaré metric

$$ds_R^2 = 4R^2 dz d\bar{z}/(R^2 - |z|^2)^2 = j_R^* ds^2,$$

where $j_R : D_R \to D$ is the isomorphism sending z to z/R and ds^2 is the Poincaré metric of the unit disc D defined in Section 1 of Chapter 2. Let e denote the tangent vector $(\partial/\partial z)_0$ of D_R at the origin 0. The vector $f_*(e) \in T_{f(0)} X$ is traditionally denoted $f'(0)$. From the fact that e has a length $2/R$ with respect to ds_R^2, it follows that F_X may be defined as follows:

$$(3.5.16) \qquad\qquad F_X(v) = \inf \frac{2}{R},$$

where the infimum is taken over all positive real numbers R for which there is a holomorphic map $f : D_R \to X$ such that $f'(0) = v$.

We easily see that F_X defines a quasi-norm on $T_x X$ (in the sense of Section 4 of Chapter 1). This quasi-norm is defined only on the tangent cone $\check{T}_x X \subset T_x X$. If x is a regular point, it satisfies the condition

$$(3.5.17) \qquad\qquad F_X(v) < \infty \qquad \text{for} \quad v \in T_x X.$$

The proof of the following proposition is similar to that of (3.1.6) and is straightforward.

(3.5.18) **Proposition.** (1) *If X and Y are two complex spaces, then*

$$F_Y(f_*v) \le F_X(v) \quad for \quad f \in \mathrm{Hol}(X, Y) \quad and \quad v \in \check{T}X;$$

(2) *For the unit disc D, F_D coincides with the Poincaré metric, i.e.,*

$$F_D^2 = ds^2.$$

Corresponding to (2) of (3.1.7), we have

(3.5.19) **Proposition.** *Let X be a complex space. Then*

$$F_X(f_*u) \le \|u\| \quad for \quad f \in \mathrm{Hol}(D, X) \quad and \quad u \in TD.$$

Conversely, if E is a quasi-norm function (continuous or otherwise) defined on the tangent cone bundle $\check{T}X = \bigcup \check{T}_xX$ such that

$$E(f_*u) \le \|u\| \quad for \quad f \in \mathrm{Hol}(D, X) \quad and \quad u \in TD,$$

then

$$E(v) \le F_X(v) \quad for \quad v \in \check{T}X.$$

(3.5.20) **Proposition.** *Let X be a complex space and $x \in X$. Then F_X^* defined by (3.5.1) is dual to F_X defined by (3.5.14), i.e.,*

$$F_X^*(\lambda) = \sup_{F_X(v) \le 1} |\lambda(v)| \quad for \quad \lambda \in T_x^*X,$$

where the supremum is taken over all $v \in \check{T}_xX$ with $F_X(v) \le 1$.

Proof. Let $\mathrm{Hol}_x(D, X)$ denote the subset of $\mathrm{Hol}(D, X)$ consisting of maps f with $f(0) = x$. From the definition (3.5.14) of F_X we obtain

$$\{v \in \check{T}_xX; \ F_X(v) < 1\} = \{f_*u; \ u \in T_0D, \ \|u\| < 1, \ f \in \mathrm{Hol}_x(D, X)\}.$$

Hence, for any $\lambda \in T_x^*X$ we have

$$F_X^*(\lambda) = \sup_f \|f^*\lambda\| = \sup_{f, \|u\|<1} |(f^*\lambda)(u)| = \sup_{f, \|u\|<1} |\lambda(f_*u)| = \sup_{F_X(v)<1} |\lambda(v)|,$$

where the suprema are taken over $f \in \mathrm{Hol}_x(D, X)$, $u \in T_0D$ with $\|u\| < 1$, and $v \in \check{T}_xX$ with $F_X(v) < 1$. $\square$

(3.5.21) **Corollary.** *Let F_X^{**} denote the double dual of F_X. Then*

$$\hat{F}_X(v) = F_X^{**}(v) \quad for \quad v \in \check{T}X.$$

This may be restated in terms of the indicatrices Γ_x, $\hat{\Gamma}_x$ and Γ_x^{**} of F_X, $\hat{F}_X$ and F_X^{**} at $x \in X$:

(3.5.22) **Corollary.** *In T_xX, $\hat{\Gamma}_x = \Gamma_x^{**}$ and $\hat{\Gamma}_x$ is the convex hull of Γ_x.*

Since $\hat{F}_X$ is the double dual of F_X, $\hat{F}_X$ may be characterized as follows:

(3.5.23) Corollary. *$\hat{F}_X$ is the pseudo-length function on TX determined by the following conditions:*

(a) $\hat{F}_X \leq F_X$ *on* $\check{T}X$;

(b) $\hat{F}_X(v + v') \leq \hat{F}_X(v) + \hat{F}_X(v')$ *for* $v, v' \in T_xX$;

(c)   *$\hat{F}_X$ is the largest pseudo-length function satisfying conditions (a) and (b).*

As a consequence, $\hat{F}_X$ may be defined more directly in terms of F_X.

(3.5.24) $$\hat{F}_X(v) = \inf_{v=\sum v_i} F_X(v_i) \qquad \text{for} \quad v \in T_xX,$$

where the infimum is taken over all possible ways to write v as a finite sum of vectors v_i in $\check{T}_xX$.

The proof of the following proposition is simpler than that of the corresponding proposition (3.1.9).

(3.5.25) Proposition. *For any complex spaces X and Y we have*

(1) $F_{X \times Y}(u, v) = \max\{F_X(u),\ F_Y(v)\}$ *for* $u \in \check{T}X,\ v \in \check{T}Y$;

(2) $\hat{F}_{X \times Y}(u, v) = \max\{\hat{F}_X(u),\ \hat{F}_Y(v)\}$ *for* $u \in TX,\ v \in TY$.

Also, corresponding to (1) of (3.2.8) we have

(3.5.26) Proposition. *Let X be a complex space and $\pi : \tilde{X} \to X$ a covering space of X. Then*
$$F_{\tilde{X}} = \pi^* F_X \quad \text{and} \quad \hat{F}_{\tilde{X}} = \pi^* \hat{F}_X.$$

(3.5.27) Theorem. *If X is a nonsingular complex manifold, then F_X and $\hat{F}_X$ are upper semicontinuous as a function on TX while F_X^* is lower semicontinuous on T^*X.*

Royden [2] has shown that F_X is upper semicontinuous. From his result it follows that F_X^* is lower semicontinuous and that $\hat{F}_X$ is upper semicontinuous. It is possible to prove directly that F_X^* is lower semicontinuous. In any case, the proof makes use of the following extension lemma (Royden [4]); see Appendix A of this chapter.

(3.5.28) Lemma. *If f is a holomorphic map of D_R, where $R > 1$, into a complex manifold X of dimension n such that its differential f_* is nonzero at the origin 0, then there is a holomorphic map φ of the polydisc $D^n = D \times D^{n-1}$ into X such that φ is biholomorphic in some neighborhood of the origin and*
$$f(z) = \varphi(z, 0, \ldots, 0) \quad \text{for} \quad z \in D.$$

Proof of Theorem. We shall first prove that F_X is upper semicontinuous. Let $v \in TX$. From the definition of F_X it follows that, given $\varepsilon > 0$, there is a pair, $f \in \mathrm{Hol}(D, X)$ and $u \in T_0 D$, such that
$$f_* u = v \qquad \text{and} \qquad F_D(u) < F_X(v) + \varepsilon.$$

Let $R > 1$ and $j_R : D_R \to D$ be the isomorphism defined by $j_R(z) = z/R$. By applying (3.5.28) to $f \circ j_R \in \mathrm{Hol}(D_R, X)$, we obtain a holomorphic map $\varphi \in \mathrm{Hol}(D^n, X)$ which maps a neighborhood U of 0 in D^n biholomorphically onto a neighborhood V of $f(0)$ in X and satisfies

$$f \circ j_R(z) = \varphi(z, 0, \ldots, 0) \qquad \text{for} \quad z \in D.$$

We choose $R > 1$ sufficiently close to 1 so that

$$F_D(Ru) = R \cdot F_D(u) < F_X(v) + \varepsilon.$$

By imbedding D into D^n in an obvious manner (i.e., by sending z to $(z, 0, \ldots, 0)$), we consider the tangent vector $Ru \in TD$ as a tangent vector of TD^n. By (3.5.25) we have

$$F_{D^n}(Ru) = F_D(Ru).$$

Since F_{D^n} is continuous by (3.5.25), there exists a neighborhood U' of Ru in TD^n such that

$$F_{D^n}(u') < F_{D^n}(Ru) + \varepsilon \qquad \text{for} \quad u' \in U'.$$

We take U' sufficiently small so that the elements of U' are vectors at points of U. Let $v' \in \varphi_*(U')$. Combining all these relations, we obtain

$$F_X(v') = F_X(\varphi_*(u')) \leq F_{D^n}(u') < F_{D^n}(Ru) + \varepsilon = F_D(Ru) + \varepsilon < F_X(v) + 2\varepsilon.$$

This proves that F_X is upper semicontinuous.

In order to prove that F_X^* is lower semicontinuous, let $\lambda_k \in T^*X$ be a sequence converging to $\lambda \in T^*X$.

Assume that $F_X^*(\lambda) < \infty$. Given any $\varepsilon > 0$, there is a map $f \in \mathrm{Hol}(D, X)$ such that $F_X^*(\lambda) - \varepsilon < \| f^*\lambda \|$. By restricting f to a slightly smaller disc, we may assume that f can be extended to a holomorphic map $\varphi : D \times D^{n-1} \to X$ so that $f(z) = \varphi(z, 0)$, $z \in D$, and the differential φ_* is non-degenerate at $(0, 0)$, (see (3.5.28)). Then

$$F_X^*(\lambda) - \varepsilon < \| f^*\lambda \| = \lim \| \varphi^*\lambda_k \| \leq \liminf F_X^*(\lambda_k).$$

This shows that F_X^* is lower semicontinuous at λ.

Assume that $F_X^*(\lambda) = \infty$. Given an arbitrarily large number M, there is a map $f \in \mathrm{Hol}(D, X)$ with an extension φ as above such that

$$M < \| f^*\lambda \| = \lim \| \varphi^*\lambda \| \leq \liminf F_X^*(\lambda_k),$$

which shows that F_X^* is lower semicontinuous at λ.

In order to show that $\hat{F}_X$ is upper semicontinuous, we make the following observation. We know that, for each $x \in X$, Γ_x^* is compact, (see (3.5.6)). Since F_X^* is lower semicontinuous, it follows that for any compact set $K \subset X$ the union $\bigcup_{x \in K} \Gamma_x^* \subset T^*X$ is also compact.

Now, let $v_k \in T_{x_k} X$ be a sequence converging to $v_0 \in T_{x_0} X$. We want to prove the inequality

$$\hat{F}_X(v_0) \geq \limsup \hat{F}_X(v_k).$$

Assume the contrary. Taking a subsequence, we may assume that $\hat{F}_X(v_0) < \lim \hat{F}_X(v_k)$. By (3.5.8), $\hat{F}_X(v_k) = |\lambda_k(v_k)|$ for some $\lambda_k \in \Gamma^*_{x_k}$. Taking a subsequence, we may assume that λ_k converges to some $\lambda_0 \in T^*_{x_0}X$. Since F^*_X is lower semicontinuous, we have $\lambda_0 \in \Gamma^*_{x_0}$. Then

$$\hat{F}_X(v_0) \geq |\lambda_0(v_0)| = \lim |\lambda_k(v_k)| = \lim \hat{F}_X(v_k).$$

This is a contradiction. Hence, $\hat{F}_X$ must be upper semicontinuous. $\qquad\square$

If X is singular, we consider a desingularization $\pi: \tilde{X} \to X$. Let $v \in \check{T}X$. As in the nonsingular case, given $\varepsilon > 0$ there is a pair $f \in \mathrm{Hol}(D, X)$ and $u \in T_0 D$ such that $f_*(u) = v$ and $F_D(u) < F_X(v) + \varepsilon$.

Assume that we can lift f to $\tilde{f} \in \mathrm{Hol}(D, \tilde{X})$ such that $f = \pi \circ \tilde{f}$ and set $\tilde{v} = \tilde{f}_*(u)$ so that $\pi_*(\tilde{v}) = v$. Then

$$F_X(v) \leq F_{\tilde{X}}(\tilde{v}) \leq F_D(u) < F_X(v) + \varepsilon.$$

Since this holds for any $\varepsilon > 0$, we obtain $F_{\tilde{X}}(\tilde{v}) = F_X(v)$.

Since $F_{\tilde{X}}$ is upper semicontinuous, given $\varepsilon > 0$ there is a neighborhood $\tilde{V}$ of $\tilde{v}$ in $T\tilde{X}$ such that $F_{\tilde{X}}(\tilde{v}') < F_{\tilde{X}}(\tilde{v}) + \varepsilon$ for all $\tilde{v}' \in \tilde{V}$. Let $V = \pi_*(\tilde{V})$. For any $v' \in V$ choose $\tilde{v}' \in \tilde{V}$ such that $v' = \pi_*\tilde{v}'$. Then

$$F_X(v') \leq F_{\tilde{X}}(\tilde{v}') < F_{\tilde{X}}(\tilde{v}) + \varepsilon = F_X(v) + \varepsilon.$$

This shows that F_X is upper semicontinuous at v (under the assumption that f can be lifted to $\tilde{f}$).

We consider now the question of lifting $f \in \mathrm{Hol}(D, X)$ to $\tilde{f} \in \mathrm{Hol}(D, \tilde{X})$. Let S be the singular locus of X.

(3.5.29) Lemma. *A holomorphic map $f \in \mathrm{Hol}(D, X)$ can be lifted to a holomorphic map $\tilde{f} \in \mathrm{Hol}(D, \tilde{X})$ if $f(D) \not\subset S$.*

Proof. The map $\pi^{-1}: X \to \tilde{X}$ is meromorphic with singularity set S. Hence, $\pi^{-1} \circ f: D \to \tilde{X}$ is meromorphic with singularity set $f^{-1}(S)$. Since $\dim D = 1$, the singularities of $\pi^{-1} \circ f$ are all removable and $\pi^{-1} \circ f$ is actually holomorphic. $\qquad\square$

(3.5.30) Theorem. *Let X be a complex space with singular locus S. Then F_X and $\hat{F}_X$ are upper semicontinuous on $X - S$, and F^*_X is lower semicontinuous on $X - S$. In fact, F_X is upper semicontinuous at $v \in \tilde{T}X$ unless $v \in \tilde{T}S$.*

Proof. Given $\varepsilon > 0$ there is a pair $f \in \mathrm{Hol}(D, X)$ and $u \in T_0 D$ such that $f_*(u) = v$ and $F_D(u) < F_X(v) + \varepsilon$. If $v \notin \tilde{T}S$, by (3.5.29) f can be lifted to a map $\tilde{f} \in \mathrm{Hol}(D, \tilde{X})$, and F_X is upper semicontinuous at v by the argument above.

Let $\lambda \in T^*_x X$ with $x \notin S$. Let $f \in \mathrm{Hol}(D, X)$ be a map used in the direct proof of lower semicontinuity for F^*_X, (see the proof of (3.5.27)). Since $f(D) \not\subset S$, by (3.5.29) f can be lifted to a map $\tilde{f} \in \mathrm{Hol}(D, \tilde{X})$. By (3.5.28) we may assume that $\tilde{f}$ extends to $\tilde{\varphi} \in \mathrm{Hol}(D^n, \tilde{X})$. (If necessary, we first shrink $\tilde{f}$ to a slightly smaller disc). Set $\varphi = \pi \circ \tilde{\varphi} \in \mathrm{Hol}(D^n, X)$. Now by the same argument as in the proof of (3.5.27) F_X^* is lower semicontinuous at λ.

From lower semicontinuity of F_X^* we obtain upper semicontinuity of $\hat{F}_X$ as in the proof of (3.5.27). $\qquad\qquad\square$

Do Duc Thai [1] also discusses upper semicontinuity of F_X for complex spaces X with singularities. See also Venturini [6] and Remark (3.5.46).

Let X be a nonsingular complex manifold. Since $\hat{F}_X$ is upper semicontinuous, for any piecewise smooth curve γ joining two points, say p and q of X we can define the integral $\int_\gamma \hat{F}_X$.

(3.5.31) Theorem. *Let X be a complex manifold. Then*

$$d_X(p, q) = \inf_\gamma \int_\gamma F_X = \inf_\gamma \int_\gamma \hat{F}_X,$$

where the infimum is taken over all piecewise smooth curves γ joining p to q.

The first equality is due to Royden [2].

Proof. Set

$$\hat{d}_X(p, q) = \inf_\gamma \int_\gamma \hat{F}_X, \qquad \bar{d}_X(p, q) = \inf_\gamma \int_\gamma F_X.$$

Since both $\hat{F}_D^2$ and F_D^2 agree with the Poincaré metric of D (see (3.5.11) and (3.5.18)), we know that

$$\hat{d}_D(a, b) = \bar{d}_D(a, b) = d_D(a, b).$$

By integrating (3.5.19) we obtain

$$\bar{d}_X(f(a), f(b)) \leq d_D(a, b) \qquad \text{for} \quad f \in \mathrm{Hol}(D, X) \quad \text{and} \quad a, b \in D.$$

By (3.1.7), we have $\bar{d}_X \leq d_X$. From (3.5.23) we have $\hat{d}_X \leq \bar{d}_X$. Thus

(3.5.32). $$\hat{d}_X \leq \bar{d}_X \leq d_X$$

We shall first prove the equality $\bar{d}_X = d_X$, which is due to Royden. Let $\gamma = \gamma(t), 0 \leq t \leq 1$, be a curve from p to q such that

$$\int_\gamma F_X < \bar{d}_X(p, q) + \varepsilon.$$

Let $\gamma'(t)$ denote the velocity vector of γ at time t. Since $F_X(\gamma'(t))$ is upper semicontinuous on $[0, 1]$, it is the limit of a monotone decreasing sequence of continuous functions $h_n(t)$ on the interval $[0, 1]$. By the Lebesgue convergence theorem, we have

$$\int_\gamma F_X = \lim \int_0^1 h_n(t)\,dt.$$

Let $h = h_n$ for a fixed large n so that

$$\int_\gamma F_X < \int_0^1 h(t)\,dt < \bar{d}_X(p, q) + \varepsilon.$$

Since h is continuous, it is Riemann integrable, and there is a positive number δ such that if $0 = t_0 < t_1 < \ldots < t_k = 1$ is any subdivision of $[0, 1]$ with $t_i - t_{i-1} < \delta$ and $s_1, \ldots, s_k$ are points of $[0, 1]$ with $|s_i - t_i| < \delta$, then

$$\sum_{i=1}^k h(s_i)(t_i - t_{i-1}) < \bar{d}_X(p, q) + \varepsilon.$$

Now we shall prove the following "mean value theorem".

(3.5.33) **Lemma.** *Let $\gamma = \gamma(t)$, $a \le t \le b$, be a C^1 curve in X. Then given $\varepsilon > 0$ and s, $a < s < b$, there exists $\delta > 0$ such that for all t in the interval $|t - s| < \delta$ we have*

$$d_X(\gamma(t), \gamma(s)) < (F_X(\gamma'(s)) + \varepsilon)|t - s|.$$

Proof of Lemma. From the definition of F_X (see (3.5.16)), there exist a disc D_R of radius R and a holomorphic map $f: D_R \to X$ such that

(i) $\qquad f(0) = \gamma(s), \quad f'(0) = \gamma'(s) \quad$ and $\quad \dfrac{2}{R} < F_X(\gamma'(s)) + \varepsilon.$

Let W be a polydisc neighborhood of $\gamma(s)$. We choose $\delta_1 > 0$ such that $\gamma(t) \in W$ for $|t - s| < \delta_1$. Since f restricted to the real axis and γ are two parametrized curves with the same tangent vector at $f(0) = \gamma(s)$, there exists $\delta_2 > 0$ such that for $|t - s| < \delta_2$ we have

$$d_W(\gamma(t), f(t - s)) < \varepsilon|t - s|.$$

Since the injection $W \to X$ is distance-decreasing, we have

(ii) $\qquad d_X(\gamma(t), f(t - s)) < \varepsilon|t - s|.$

On the other hand, from the distance-decreasing property of f and from the expression for the Poincaré metric of D_R with curvature -1 (see (2.1.3)) it follows that there exists $\delta_3 > 0$ such that

$$d_X(f(t - s), f(0)) \le d_{D_R}(t - s, 0) \le \left(\frac{2}{R} + \varepsilon\right)|t - s|$$

for $|t - s| < \delta_3$. Combining this with (i) we have

(iii) $\qquad d_X(f(t - s), f(0)) < (F_X(\gamma'(s)) + 2\varepsilon)|t - s|$

for $|t - s| < \delta_3$. By (ii), (iii) and the triangular inequality we have

$$d_X(\gamma(t), \gamma(s)) < (F_X(\gamma'(s)) + 3\varepsilon)|t - s|,$$

thereby proving Lemma (3.5.33).

For each $s \in [0, 1]$, let I_s denote the interval $|t - s| < \delta$ obtained in Lemma (3.5.33). We apply the Lebesgues covering lemma (see, for example, Kelley [1; p. 154]) to the open cover $\{I_s; \ s \in [0, 1]\}$. (The Lebesgues covering lemma states that, given an open cover $\mathcal{U}$ of a compact metric space A, there is a positive number η such that the open η-ball about each point of A is contained in some member of $\mathcal{U}$). Thus there is $\eta > 0$ such that if $t, t' \in [0, 1]$ and $|t - t'| < \eta$, then $t, t' \in I_s$ for some s. Let $0 = t_0 < t_1 < \ldots < t_k = 1$ be a subdivision of $[0, 1]$ with $t_i - t_{i-1} < \eta$, and choose s_i so that $t_{i-1}, t_i \in I_{s_i}$. Then

$$d_X(p, q) \leq \sum \overline{d_X(\gamma(t_{i-1}), \gamma(t_i))} < \sum (h(s_i) + \varepsilon)|t_i - t_{i-1}| < \bar{d}_X(p, q) + 2\varepsilon.$$

This completes the proof of the equality $d_X = \bar{d}_X$.

The equality $\bar{d}_X = \hat{d}_X$ has little to do with complex analysis and is a direct consequence of the following theorem in Finsler geometry.

(3.5.34) **Theorem.** *Let X be a (real) manifold. Let $F: TX \to \mathbf{R}$ be an upper semicontinuous pseudo-length function on X, and let $\hat{F}$ be the convex pseudo-length function defined by the property that its indicatrix at each $x \in X$ is the convex hull of the indicatrix of F at x. Then the pseudo-distance d defined by F coincides with the pseudo-distance $\hat{d}$ defined by $\hat{F}$.*

This theorem has been proved by Busemann and Mayer [1] under the assumption that F is continuous and strictly positive. For the proof of (3.5.34), see Kobayashi [23]. $\qquad\square$

If X is a singular complex space, F_X and $\hat{F}_X$ may not be upper semicontinuous. However, using the upper integral we can still define

$$\bar{d}_X(p, q) = \inf_\gamma \overline{\int_\gamma} F_X,$$

$$\hat{d}_X(p, q) = \inf_\gamma \overline{\int_\gamma} \hat{F}_X.$$

Obviously, we have

$$\hat{d}_X(p, q) \leq \bar{d}_X(p, q) \leq d_X(p, q).$$

But the question remains whether $\hat{d}_X$, $\bar{d}_X$ and d_X are all equal when X is singular.

The proof of (3.2.19) gives also the following infinitesimal analogue of (3.2.19).

(3.5.35) **Proposition.** *If X is a complex manifold and A is a closed analytic subset of codimension at least 2, then*

$$F_{X-A} = F_X|_{X-A} \quad \text{and} \quad \hat{F}_{X-A} = \hat{F}_X|_{X-A}.$$

The second equality is a direct result of the first.

It is not easy to determine F_X even for relatively simple domains. The following example is due to Graham-Wu [1].

(3.5.36) **Example**. Let

$$X = D \times \mathbf{C} - \left\{ (z, w); \ |z| \geq \frac{1}{2}, \ |w| \geq 1 \right\},$$

and let $p: X \to D$ and $q: X \to \mathbf{C}$ be the projections defined by $p(z, w) = z$ and $q(z, w) = w$. Let $x_0 = (z_0, w_0)$ be a point of X and v_0 a nonzero tangent vector at x_0.

If $|z_0| < 1/2$, then

$$F_D(p_*(v_0)) \leq F_X(v_0) \leq F_{D_{1/2}}(p_* v_0).$$

If $|z_0| > 1/2$ (and hence $|w_0| < 1$), then $F_X(v_0) > 0$. More precisely, let $\delta = d_D(z_0, z_0/2|z_0|) > 0$ and choose r, $0 < r < 1$, such that $d_D(0, r) < \delta/2$. We shall show

$$\max\{F_D(p_* v_0), r F_D(q_* v_0)\} \leq F_X(v_0) \leq \max\{F_D(p_* v_0), F_D(q_* v_0)\}.$$

(Since $q(x_0) = w_0 \in D$, we are considering here $q_* v_0$ as a tangent vector of D). The second inequality giving an upper bound for $F_X(v_0)$ is trivial since $D \times D \subset X$. The inequality $F_D(p_* v_0) \leq F_X(v_0)$ is also trivial since p is distance-decreasing. Assuming $F_X(v_0) < r F_D(q_* v_0)$ we shall derive a contradiction. From the definition of $F_X(v_0)$ there exist $u \in T_0 D$ and $f \in \mathrm{Hol}(D, X)$ such that $f_* u = v_0$ and $F_X(v_0) \leq F_D(u) < r F_D(q_* v_0)$. If $q(f(D_r)) \subset D$, then $F_D(q_* f_*(u)) \leq F_{D_r}(u)$, and

$$r F_D(q_* v_0) = r F_D(q_* f_*(u)) \leq r F_{D_r}(u) = F_D(u),$$

which is a contradiction. Hence, $q(f(D_r)) \not\subset D$, and there exists a point $a \in D_r$ such that $q(f(a)) \notin D$. Hence, $p(f(a)) \in D_{1/2}$. Then $\delta \leq d_D(p(f(a)), p(x_0))$. On the other hand,

$$d_D(p(f(a)), p(x_0)) = d_D(p(f(a)), p(f(0)) \leq d_D(a, 0) < \delta/2,$$

which is also a contradiction.

Finally, let $|z_0| = 1/2$. Trivially,

$$F_D(p_*(v_0)) \leq F_X(v_0).$$

It is not clear, however, whether $F_X(v_0) = 0$ when $p_*(v_0) = 0$.

As the following example shows, F_X or $\hat{F}_X$ is, in general, not continuous.

(3.5.37) **Example**. Let $X = \{(z, w) \in \mathbf{C}^2; \ |zw| < 1\}$; it is a domain of holomorphy. Consider a vector field $\xi = \frac{\partial}{\partial z} - \frac{\partial}{\partial w}$. We claim that

(i) $F_X(\xi) = 0$ at $(a, a) \in X, a \neq 0,$

(ii) $F_X(\xi) \neq 0$ at $(0, 0) \in X.$

To prove (i), we use the **C**-action on X defined by

$$t : (z, w) \mapsto (e^t z, e^{-t} w), \quad t \in \mathbf{C}.$$

The tangent vector to its orbit at (z, w) is given by

$$\eta = z \frac{\partial}{\partial z} - w \frac{\partial}{\partial w}.$$

Obviously, $F_X(\eta) = 0$. Since $\eta = a\xi$ at (a, a), we have $F_X(\xi) = 0$ at (a, a), $a \neq 0$.

To see (ii), let $f \colon D \to X$ be a holomorphic map with $f(0) = (0, 0)$. The map $f = (f_1, f_2)$ is of the form

$$f_1(t) = at + O(2), \quad f_2(t) = bt + O(2).$$

Then

$$f_1(t) f_2(t) = abt^2 + O(3), \quad t \in D.$$

Using $|f_1(t) f_2(t)| < 1$ and Cauchy's integral formula, we estimate the second derivative of $f_1 f_2$ at $t = 0$ and obtain $|ab| < 1$. In order for the derivative of f at 0 to be in the direction of ξ at $(0, 0)$, the map f must satisfy the condition $b = -a$. Hence $|a|^2 < 1$. Since

$$f_* \left(\left(\frac{d}{dt} \right)_0 \right) = a \left(\frac{\partial}{\partial z} - \frac{\partial}{\partial w} \right)_{(0,0)},$$

we have

$$F_X(\xi) \geq \frac{2}{|a|} > 2.$$

In certain cases, F_X is not only upper semicontinuous but continuous. The following is due to Royden [2]:

(3.5.38) **Proposition.** *If a complex space X is complete hyperbolic, then F_X, F_X^* and $\hat{F}_X$ are continuous.*

Proof. In order to prove that F_X is continuous at $v_0 \in \check{T}X$, let $v_k \in \check{T}X$ be a sequence of vectors converging to v_0. Since X is complete hyperbolic, by (3.5.15) there exist vectors $u_k \in T_0 D$ and maps $f_k \in \mathrm{Hol}(D, X)\}$ such that $f_{k*}(u_k) = v_k$ and $F_X(v_k) = \|u_k\|$. Since F_X is upper semicontinuous, $\{F_X(v_k)\}$ is bounded. Hence, by passing to a subsequence if necessary, we may assume that $\{u_k\}$ converges to some vector $u_0 \in T_0 D$. By applying (1.3.3) to the family $\{f_k\}$, we see that a subsequence of $\{f_k\}$ converges to a map $f_0 \in \mathrm{Hol}(D, X)$. Then $f_{0*}(u_0) = v_0$ and

$$\lim F_X(v_k) = \lim \|u_k\| = \|u_0\| \geq F_X(v_0),$$

showing that F_X is lower semicontinuous at v_0.

The continuity of F_X^* and $\hat{F}_X$ follows from (3.5.20) and (3.5.21). $\qquad \square$

(3.5.39) **Remark**. As Royden states and as we remarked in (3.5.15), the proof above is valid under the weaker assumption that X is taut. The concept of taut complex space will be introduced in Section 1 of Chapter 5.

Wright [1] proved that if X is a projective algebraic manifold of general type, then F_X is continuous even when X is not hyperbolic.

Although F_X is not very smooth, in the hyperbolic case we can find a smooth length function which can often take place of F_X.

(3.5.40) **Proposition**. *If a complex space X is hyperbolic modulo a closed set Δ and if E is a length function on X, then there exists a nonnegative continuous function φ on X which is positive outside Δ such that*

$$f^*(\varphi^2 E^2) \leq ds_D^2 \quad for \quad f \in \mathrm{Hol}(D, X).$$

Proof. Let $\{U_m\}$ be an increasing sequence of relatively compact domains in $X - \Delta$ which exhaust $X - \Delta$, that is, $\bar{U}_m \subset U_{m+1}$ and $\bigcup U_m = X - \Delta$. As in the proof of (3.3.13) we can find a sequence of positive constants $c_1 \geq c_2 \geq \ldots$ such that

$$f^*(\varphi_m^2 E^2) \leq ds_D^2 \quad \text{on} \quad f^{-1}(U_m) \quad \text{for} \quad f \in \mathrm{Hol}(D, X).$$

Take a nonnegative continuous function φ on X such that $0 < \varphi \leq c_m$ on U_m. Then $f^*(\varphi^2 E^2) \leq ds_D^2$ for $f \in \mathrm{Hol}(D, X)$. $\qquad\square$

(3.5.41) **Corollary**. *If X is hyperbolic modulo Δ, then given a length function E on X, there exists a nonnegative continuous function φ on X such that $\varphi E \leq F_X$ and $\varphi > 0$ on $X - \Delta$.*

We can define also an **intrinsic relative pseudo-metric**, i.e., the **infinitesimal form of the relative pseudo-distance** $d_{Y,Z}$. Let Y be a complex subspace of a complex space Z. As in Section 4, let $\mathcal{F}_{Y,Z}$ be the family of holomorphic maps $f: D \to Z$ such that $f^{-1}(Z - Y)$ is either empty or a singleton. Then using the family $\mathcal{F}_{Y,Z}$ instead of $\mathrm{Hol}(D, Y)$, we define $F_{Y,Z}^*$, $\hat{F}_{Y,Z}$ and $F_{Y,Z}$ exactly as in (3.5.1), (3.5.7) and (3.5.14).

We do not state obvious basic properties of these pseudo-length functions. However, we note that the proof of (3.5.27) shows that if Z is non-singular and Y is the complement of a divisor with no worse than normal crossing singularities, then $\hat{F}_{Y,Z}$ and $F_{Y,Z}$ are upper semi-continuous and $F_{Y,Z}^*$ is lower semi-continuous.

We state the following converse to (3.4.11); it strengthens (3.3.3). The proof will be given in the course of the proof of (3.6.20)

(3.5.42) **Theorem**. *Let a complex space Y be hyperbolically imbedded in a complex space Z. Given a length function E on Z, there is a constant $c > 0$ such that*

$$f^*(c^2 E^2) \leq ds_D^2 \quad for \quad f \in \mathcal{F}_{Y,Z}.$$

(3.5.43) Corollary. *Let Y be hyperbolically imbedded in Z. Given a length function E on Z, there is a constant $c > 0$ such that $cE \leq \hat{F}_{Y,Z}$ on $\overline{Y}$.*

(3.5.44) Remark. For $Z = P_1 \mathbf{C}$ and $Y = P_1 \mathbf{C} - \{\infty, 0, 1\}$, there is an estimate for $F_{Y,Z}$ by Landau, (see Carathéodory [4, vol. 2, p. 198]). In fact, Landau proved that if $f(z) = a_1 z + a_2 z^2 + \ldots$ is holomorphic in $|z| < R$ and if f does not take values 0 or 1 in $0 < |z| < R$, then $R \leq 16/|a_1|$ and that this bound is the best possible. This may be restated as $F_{Y,Z}((d/dz)_0) = 1/8$.

(3.5.45) Remark. According to the definition of the curvature given by (2.3.3), F_X has holomorphic sectional curvature ≥ -1, provided X is complete hyperbolic (or taut), see B. Wong [1], Masaaki Suzuki [1] and Royden [9]. Given a nonzero $v \in T_x X$, we want to show that $g^* F_X$ has curvature ≥ -1 for some $g \in \mathrm{Hol}(D, X)$ such that $g(0) = x$ and v is tangent to $g(D)$. Since X is complete hyperbolic, there is a map $g \in \mathrm{Hol}(D, X)$ such that $F_X(v) = \|u\|$ for some $u \in T_0 D$ with $g_* u = v$, (see the proof of (3.5.38)). Then in the inequality $g^* F_X^2 \leq ds^2$, the equality holds at 0. Take a supporting metric $d\sigma^2$ for $g^* F_X^2$ at 0. Then it is a supporting metric for ds^2 at 0. By (2.1.8) it has curvature ≥ -1. Hence, $g^* F_X$ has curvature ≥ -1.

(3.5.46). Remark. In order to generalize Royden's result (3.5.31) to singular complex spaces Venturini [6] extended the definition of $F_X(\xi)$ to vectors ξ of higher order osculation.

6 Brody's Criteria for Hyperbolicity and Applications

Theorem (3.6.3) of Brody is the simplest and most useful criterion for hyperbolicity. We give several technical improvements of Brody's criterion and their applications.

The following is immediate from (3.1.6) and (3.1.21).

(3.6.1) Proposition. *If X is a hyperbolic complex space, then every holomorphic map $f : \mathbf{C} \to X$ is constant.*

Proof. For $a, b \in \mathbf{C}$, we have

$$d_X(f(a), f(b)) \leq d_{\mathbf{C}}(a, b) = 0.$$

Hence, $f(a) = f(b)$. $\qquad\qquad\square$

We shall now prove the converse when X is compact. The proof is based on the following **Reparametrization Lemma** (3.6.2) of Brody [1]. Wu [6] points out that this lemma was first proved by Landau [1; pp. 618–619] in the context of holomorphic functions on the unit disc and then by Zalcman [1] for meromorphic functions on the unit disc.

The Poincaré metric ds_R^2 of curvature -1 on the disc D_R of radius R is given by (see (2.1.3))

$$ds_R^2 = \frac{4R^2 dz d\bar{z}}{(R^2 - |z|^2)^2}.$$

In the following we use the metric $R^2 ds_R^2$ of curvature $-1/R^2$, which agrees with the Euclidean metric $4dzd\bar{z}$ at the origin 0.

(3.6.2) Lemma. *Let X be a complex space with a length function F. Given $f \in \mathrm{Hol}(D_R, X)$, define a function*

$$u = f^* F^2 / R^2 ds_R^2$$

on D_R. If $u(0) > c > 0$, then there is a map $g \in \mathrm{Hol}(D_R, X)$ such that

(a)   the function $g^* F^2 / R^2 ds_R^2$ is bounded by c on D_R and attains the maximum value c at the origin;

(b) $g = f \circ \mu_r \circ \varphi$, where φ is a holomorphic automorphism of D_R and μ_r is the multiplication by suitable r, $0 < r < 1$, (i.e., $\mu_r(z) = rz$ for $z \in D_R$).

Proof. For $t \in [0, 1)$, define $f_t \in \mathrm{Hol}(D_R, X)$ by

$$f_t(z) = f \circ \mu_t(z) = f(tz) \qquad \text{for} \quad z \in D_R.$$

Set $u_t = f_t^* F^2 / R^2 ds_R^2 = (f \circ \mu_t)^* F^2 / R^2 ds_R^2$. Then

$$u_t = \frac{\mu_t^* f^* F^2}{\mu_t^* (R^2 ds_R^2)} \cdot \frac{\mu_t^* ds_R^2}{ds_R^2} = \mu_t^*(u) \frac{t^2 (R^2 - |z|^2)^2}{(R^2 - |tz|^2)^2}.$$

Set

$$U(t) = \sup_{z \in D_R} u_t(z) = \sup_{z \in D_R} u(tz) \frac{t^2 (R^2 - |z|^2)^2}{(R^2 - |tz|^2)^2}.$$

From the explicit expression for $u_t(z)$ given above, we see that, for each $t \in [0, 1)$, $u_t(z)$ approaches zero at the boundary of D_R and hence $\sup_{z \in D_R} u_t(z)$ is attained in the interior of D_R. It is easy to see that $U(t)$ is continuous in the interval $[0,1)$. Since $u_t(0) = u(0)t^2 > ct^2$, we have $U(t) > c$ for t sufficiently close to 1. On the other hand, $U(0) = 0$. Thus, $c = U(r)$ for some $r \in (0, 1)$. Let $z_0 \in D_R$ be a point where $c = \sup_{z \in D_R} u_r(z)$ is attained. Let φ be a holomorphic automorphism of D_R which sends 0 to z_0. Then $g = f \circ \mu_r \circ \varphi$ possesses all the desired properties.
$\square$

In the preceding section (see (3.5.14) and (3.5.16)) we defined the pseudo-length function F_X as an infinitesimal form of d_X. Since we need here only its definition and its most basic property (3.5.18), we shall quickly review F_X. Given a point x in a complex space X, the tangent cone $\check{T}_x X$ consists of vectors of the form $f_*(u)$, where $u \in TD$ and $f \in \mathrm{Hol}(D, X)$. Then $F_X : \check{T}_x X \to \mathbf{R}$ is defined by

$$F_X(v) = \inf\{\|u\|; \ u \in TD \quad \text{and} \quad f_*(u) = v\}, \qquad v \in \check{T}_x X,$$

where $\|u\|$ is the length of u measured by the Poincaré metric of D, and the infimum is taken over all $u \in TD$ and $f \in \mathrm{Hol}(D, X)$ such that $f_*(u) = v$. Alternatively, F_X may be defined in terms of a fixed vector $e = (\partial/\partial z)_0 \in T_0\mathbf{C}$ and discs D_R of varying radius R:

$$F_X(v) = \inf \frac{2}{R},$$

where the infimum is taken over all positive real numbers R for which there is a holomorphic map $f: D_R \to X$ such that $f_*(e) = v$.

Let E be any (continuous) pseudo-length function on X such that $E \le F_X$. Then

$$E(f_*u) \le F_X(f_*u) \le \|u\| \quad \text{for} \quad u \in TD, \ f \in \text{Hol}(D, X).$$

It follows that if δ denotes the pseudo-distance defined by E, then every holomorphic map $f: (D, \rho) \to (X, \delta)$ is distance-decreasing so that (by (3.1.7))

$$\delta \le d_X.$$

In particular, if E is a length function so that δ is a distance, then X is hyperbolic.

The same argument shows that if X is a relatively compact complex subspace of a complex space Y and if there is a length function E on Y such that $E \le F_X$ on X, then X is hyperbolically imbedded in Y.

It is useful to introduce the concept of complex line following Zaidenberg [2]. Let z denote the natural coordinate system on $\mathbf{C}$. Let X be a complex space, and E a length function defined on X. A nonconstant holomorphic map $h: \mathbf{C} \to X$ such that

$$h^*E^2 \le C dz d\bar{z}$$

for some constant $C > 0$ is called a **complex line**. If $h(\mathbf{C})$ is contained in a compact subset of X, then this condition is independent of E. Let S be a subset (often a domain) in X. We say that a complex line $h: \mathbf{C} \to X$ is a **limit complex line coming from** S if on each disc $D_R \subset \mathbf{C}$ of radius R the mapping $h|_{D_R}$ is a limit of holomorphic mappings of D_R into S. In this case, we have $h(\mathbf{C}) \subset \bar{S}$. Every complex line in X is a limit complex line coming from X.

We are now in a position to prove the following theorem of Brody [1].

(3.6.3) **Theorem.** *Let X be a compact complex space. If X is not hyperbolic, then there is a complex line $h: \mathbf{C} \to X$.*

Proof. Let E be a length function on X. Assume that X is not hyperbolic. Let F_X be the pseudo-length function defined above. If there is a positive number a such that $a \cdot E \le F_X$, then X would be hyperbolic as explained above. Hence, there is a sequence of tangent vectors $v_n \in TX$ such that $E(v_n) = 1$ and $F_X(v_n) < 1/n$. By applying the second definition of F_X above, we find an increasing sequence of concentric discs D_{R_n} of radius R_n with $\lim R_n = \infty$ and a sequence of maps $f_n \in \text{Hol}(D_{R_n}, X)$ such that $f_n'(0) = v_n$. (By $f_n'(0)$ we mean $df_n(e)$, where $e = (\partial/\partial z)_0$ is the tangent vector of $\mathbf{C}$ at the origin 0.) Since the length $\|e\|_n$ of e measured by the Poincaré metric $ds_{R_n}^2 = 4R_n^2 dz d\bar{z}/(R_n^2 - |z|^2)^2$ of D_{R_n} is equal to $2/R_n$, the function $u_n = f_n^*E^2/R_n^2 ds_{R_n}^2$ on D_{R_n}, evaluated at the origin 0, gives

$$u_n(0) = E(f_n'(0))^2/R_n^2 \|e\|_n^2 = E(v_n)^2/2^2 = 1/4.$$

By applying (3.6.2) to each f_n and a constant $0 < c < 1/4$, we obtain a sequence of maps $g_n \in \mathrm{Hol}(D_{R_n}, X)$ such that

(a) $g_n^* E^2 \le c R_n^2 ds_{R_n}^2$ on D_{R_n} and the equality holds at the origin 0;

(b) $\overline{g_n(D_{R_n})} \subset f_n(D_{R_n})$.

By (a), the family of maps $g_n \in \mathrm{Hol}(D_{R_n}, X)$ is equicontinuous. To be more precise, since

$$g_n^* E^2 \le c R_n^2 ds_{R_n}^2 \le c R_m^2 ds_{R_m}^2 \qquad \text{for} \quad n \ge m,$$

the family $\mathcal{F}_m = \{g_n | D_{R_m}, n \ge m\}$ is equicontinuous for each fixed m.

Since the family $\mathcal{F}_1 = \{g_n | D_{R_1}\}$ is equicontinuous, the Arzela-Ascoli theorem (1.3.1) implies that we can extract a subsequence which converges to a map $h_1 \in \mathrm{Hol}(D_{R_1}, X)$. (We note that this is where we use the compactness of X). Applying the same theorem to the corresponding sequence in $\mathcal{F}_2$, we extract a subsequence which converges to a map $h_2 \in \mathrm{Hol}(D_{R_2}, X)$. In this way we obtain maps $h_k \in \mathrm{Hol}(D_{R_k}, X)$, $k = 1, 2, \ldots$, such that each h_k is an extension of h_{k-1}. Hence, we have a map $h \in \mathrm{Hol}(\mathbf{C}, X)$ which extends all h_k.

Since $g_n^* E^2$ at the origin 0 is equal to $(c R_n^2 ds_{R_n}^2)_{z=0} = 4cd z d\bar{z}$, it follows that

$$(h^* E^2)_{z=0} = \lim_{n \to \infty} (g_n^* E^2)_{z=0} = 4cd z d\bar{z} \ne 0,$$

which shows that h is nonconstant.

Since $g_n^* E^2 \le c R_n^2 ds_{R_n}^2$, in the limit we have

$$h^* E^2 \le 4cd z d\bar{z}.$$

By suitably normalizing h we obtain $h^* E^2 \le dz d\bar{z}$.   □

(3.6.4) **Corollary.** *Let X be a compact complex space. Given a nonconstant holomorphic map $f: \mathbf{C} \to X$, there is a complex line $h: \mathbf{C} \to X$ such that $h(\mathbf{C}) \subset \overline{f(\mathbf{C})}$.*

Proof. Let D_{R_n} be an increasing sequence of concentric discs with $\lim R_n = \infty$. Let f_n be the restriction of f to D_{R_n}. Moving the origin of $\mathbf{C}$ if necessary, we may assume that f is non-degenerate at 0. As in the proof of (3.6.3), let $e = (\partial/\partial z)_0$ be the tangent vector of $\mathbf{C}$ at 0. Set $v = df(e) = f'(0)$. Multiplying E by a suitable constant, we may assume that $E(v) = 1$. We set $v_n = v$ for all n. Since $F_X(v) \le F_{\mathbf{C}}(e) = 0$, we are now in a position to apply the proof of (3.6.3). Given a constant c, $0 < c < 1/4$, by (3.6.2) we obtain a sequence of maps $g_n \in \mathrm{Hol}(D_{R_n}, X)$ satisfying conditions (a) and (b) in the proof of (3.6.3). According to (b) of (3.6.2), g_n is of the form

$$g_n = f \circ \mu_{r_n} \circ \varphi_n,$$

where μ_{r_n} is the multiplication by a suitable r_n, $0 < r_n < 1$, while φ_n is an automorphism of D_{R_n}. Repeating the argument in the proof of (3.6.3) we obtain a

complex line $h: \mathbf{C} \to X$ as a limit of a subsequence of $\{g_n\}$. Clearly, $h(\mathbf{C}) \subset \overline{f(\mathbf{C})}$.
$\square$

With very little change in the proof we can extend Brody's criterion (3.6.3) to certain noncompact complex spaces, (Urata [5]).

(3.6.5) Theorem. *Let Z be a complex space, and Y a relatively compact complex subspace of Z. If Y is not hyperbolically imbedded in Z, then there is a limit complex line $h: \mathbf{C} \to Z$ coming from Y so that $h(\mathbf{C}) \subset \bar{Y}$.*

Conversely, if Y is hyperbolically imbedded in Z, then Z contains no limit complex lines coming from Y.

Proof. Let E be a length function on Z. Assume that Y is not hyperbolically imbedded in Z. As we explained earlier (see the paragraph preceding (3.6.3)), there is no length function F on Z such that $F \le F_Y$ on Y. Hence, there is no positive constant a such that $a \cdot E \le F_Y$ on Y. The remainder of the proof is essentially the same as that of (3.6.3).

If there is a limit complex line $h: \mathbf{C} \to Z$ coming from Y, then any pair of points $p, q \in h(\mathbf{C}) \subset \bar{Y}$ would violate the condition for Y to be hyperbolic imbedded in Z.
$\square$

The following example by D. Eisenman and L. Taylor shows that (3.6.3) does not hold for some noncompact manifolds.

(3.6.6) Example. The domain

$$X = \{(z, w) \in \mathbf{C}^2;\ |z| < 1,\ |zw| < 1\} - \{(0, w);\ |w| \ge 1\}$$

is not hyperbolic, but there is no nonconstant holomorphic map : $\mathbf{C} \to X$.

Proof. The mapping $h: (z, w): \mapsto (z, zw)$ sends X into the unit bidisc and is one-to-one except on the set $z = 0$. If $f: \mathbf{C} \to X$ is holomorphic, then $h \circ f: \mathbf{C} \to D^2$ is holomorphic and hence constant by Liouville's theorem. It follows that either f is constant or f maps $\mathbf{C}$ into the set $\{(0, w) \in X\}$. But this set is equivalent to the unit disc $\{w \in \mathbf{C};\ |w| < 1\}$. Hence, f is constant in either case. Since h is distance-decreasing, we see that $d_X(p, q) > 0$ for $p \ne q$ unless both p and q are in the subset $\{(0, w) \in X\}$. We shall show that if p and q are in this subset, then $d_X(p, q) = 0$. Let $p = (0, b)$ with $b \ne 0$ and $q = (0, 0)$. Set $p_n = (1/n, b)$. Then $d_X(p, q) = \lim d_X(p_n, q)$. Let $a_n = \min\{n, \sqrt{n/|b|}\}$. Then the mapping $t \in D \to (a_n t/n, a_n bt) \in X$ maps $1/a_n$ into p_n. Hence,

$$\lim d_X(p_n, q) \le \lim d_D(1/a_n, 0) = 0,$$

which shows that X is not hyperbolic.
$\square$

In this example, let $Z = P_1\mathbf{C} \times P_1\mathbf{C}$ be a natural compactification of $\mathbf{C}^2$. Then the holomorphic mapping $h: \mathbf{C} \to \bar{X} \subset Z$ given by $h(z) = (0, z)$ satisfies the inequality $h^* E^2 \le dz d\bar{z}$ with equality at $z = 0$ (with respect to the product metric E^2 coming from the Fubini-Study metric of $P_1\mathbf{C}$).

Following Lang [3] we can strengthen (3.6.5) also in the following form.

(3.6.7) **Theorem.** *Let Z be a complex space, and Y a relatively compact subset of Z. Let $\{U_n\}$ be a decreasing sequence of relatively compact open subsets of Z such that $\bigcap U_n = Y$. Assume that none of these U_n is hyperbolically imbedded in Z. Then for each n we can find a limit complex line $h_n \colon \mathbf{C} \to Z$ coming from U_n such that the sequence $\{h_n\}$ converges to a complex line $h \colon \mathbf{C} \to Z$ with its image $f(\mathbf{C})$ in $\bar{Y}$.*

Proof. Let E be a length function Z. If U_n is not hyperbolically imbedded in Z, then by (3.6.5) there is a limit complex line $h_n \colon \mathbf{C} \to Z$ coming from U_n so that $h_n(\mathbf{C}) \subset \bar{U}_n$. Then $h_n^* E^2 \le C_n dz d\bar{z}$ with $C_n > 0$. By composing h_n with a suitable affine transformation $z \mapsto az + b$, we may assume that $h_n^* E^2 \le dz d\bar{z}$ with equality holding at $z = 0$. Applying Arzela-Ascoli Theorem (1.3.1) to the family $\{h_n\}$ we obtain the desired result. $\qquad\square$

Letting Y to be a compact complex subspace of Z in the theorem above, we obtain:

(3.6.8) **Corollary.** *Let Z be a complex space, and Y a compact complex subspace of Z. If Y is hyperbolic, there is a relatively compact neighborhood U of Y which is hyperbolically imbedded in Z.*

As in (1.2.2), for $x \in X$ consider the degeneracy set

$$\Delta(x) = \{y \in X;\ d_X(x, y) = 0\}.$$

We know (see (1.2.7) and (3.1.18)) that $\Delta(x)$ is connected if X is compact. As an application of (3.6.7) we obtain

(3.6.9) **Corollary.** *Let X be a compact complex space, and $x \in X$. If $\Delta(x)$ is non-trivial, i.e., if it contains more than one point, then there is a complex line $h \colon \mathbf{C} \to X$ such that $h(\mathbf{C}) \subset \Delta(x)$.*

Proof. Let
$$U_n = \{y \in X;\ d_X(x, y) < 1/n\}.$$
Then $\bigcap U_n = \Delta(x)$. By (3.1.19), for all n we have

$$d_{U_n}(x, y) = 0 \qquad \text{for}\quad y \in \Delta(x).$$

In particular, U_n is not hyperbolic. By (3.6.7), we have a holomorphic map $h \colon \mathbf{C} \to X$ with the stated property. $\qquad\square$

In order to consider the case where Y is the complement of a hypersurface in Z, we need a generalization of **Hurwitz theorem**. The classical theorem of Hurwitz states:

(3.6.10) **Theorem.** *The limit of a convergent sequence of nowhere vanishing holomorphic functions on a domain vanishes either nowhere or everywhere.*

This may be extended as follows.

(3.6.11) Theorem. *Let Z be a complex space and $S = \bigcup_{i=1}^{m} S_i$ a Cartier divisor in Z where each S_i is irreducible. Assume that a sequence $\{h_m\} \subset \mathrm{Hol}(D, Z - S)$ converges to a map $h \in \mathrm{Hol}(D, Z)$. Then $h(D)$ is either in $Z - S$ or in S. More precisely, $h(D)$ lies either in $Z - S$ or in $\bigcap_{i \in I} S_i - \bigcup_{j \in J} S_j$, where $I = \{i;\ h(0) \in S_i\}$ and $J = \{j;\ h(0) \notin S_j\}$.*

Proof. Suppose that $h(0) \in S$. Let V be a neighborhood of $h(0)$ in Z such that $V \cap S$ is defined by a holomorphic function $f = \prod_i f_i$, where $f_i = 0$ defines $V \cap S_i$. Take i such that $f_i(h(0)) = 0$. Apply the classical theorem of Hurwitz to a sequence of holomorphic functions $\{f_i \circ h_m\}$ which are nowhere zero. Its limit $f_i \circ h$ must be identically zero since $f_i(h(0)) = 0$. Hence h maps D into S_i. $\qquad\square$

We are now in a position to state the theorem of Green [7] and Howard.

(3.6.12) Theorem. *Let Z be a compact complex space with a length function E. Let S be a Cartier divisor in Z, and $Y = Z - S$. Then Y is complete hyperbolic and hyperbolically imbedded in Z if the following two conditions are satisfied:*
    (a)   *There are no complex lines in Y;*
    (b)   *There are no complex lines in S.*

Proof. Suppose that Y is not hyperbolically imbedded in Z. By (3.6.5) there is a limit complex line $h \in \mathrm{Hol}(\mathbf{C}, Z)$ coming from Y. Hence, either $h(\mathbf{C}) \subset Y$ or $h(\mathbf{C}) \subset S$ by the generalized Hurwitz theorem (3.6.11). This is a contradiction. From (3.3.6) we see that Y is complete hyperbolic. $\qquad\square$

More precisely, we have (Green [7])

(3.6.13) Theorem. *Let Z be a compact complex space with a length function E. Let S be a union of Cartier divisors $S_1, \ldots, S_m$. Then $Y = Z - S$ is complete hyperbolic and hyperbolically imbedded in Z if the following two conditions are satisfied:*
    (a)   *There are no complex lines in Y;*
    (b)   *For any partition of indices $I \cup J = \{1, 2, \ldots, m\}$, there are no complex lines in $\bigcap_{i \in I} S_i - \bigcup_{j \in J} S_j$.*

As we shall see later, the corollary above combined with Borel's Lemma imply that the complement of $2n + 1$ hyperplanes in general position in $P_n\mathbf{C}$ is complete hyperbolic and hyperbolically imbedded in $P_n\mathbf{C}$.

The following result of Zaidenberg [4, 5] follows from (3.6.13).

(3.6.14) Corollary. *Let Z be an n-dimensional compact complex manifold and $S = \bigcup_{i=1}^{m} S_i$ a divisor with only normal crossing singularities. Let $S^{(k)}$, $1 \le k \le n$, denote the stratum of S consisting of the points of S of multiplicity k, i.e.,*

$$S^{(k)} = S^k - S^{k+1}, \quad \text{where} \quad S^k = \bigcup_{1 \le i_1 < \ldots < i_k \le m} S_{i_1} \cap \ldots \cap S_{i_k}.$$

Then the domain $Y = Z - S$ is hyperbolically imbedded in Z if neither Y nor any of the strata $S^{(k)}$, $1 \le k \le n - 1$, contains complex lines.

Proof. In view of (3.6.10) it suffices to prove that each limit complex line $h \colon \mathbf{C} \to Z$ coming from Y is contained either in Y or in one of the strata $S^{(k)}$. Suppose that $h(\mathbf{C})$ is not contained in Y. By the generalized Hurwitz theorem (3.6.11), for each i either $h(\mathbf{C}) \subset S_i$ or $h(\mathbf{C}) \cap S_i = \emptyset$. Without loss of generality we may assume that $h(\mathbf{C}) \subset S_i$ for $i = 1, \ldots, k$ and $h(\mathbf{C}) \cap S_j = \emptyset$ for $j = k+1, \ldots, m$. Then $h(\mathbf{C})$ is contained in the stratum $S^{(k)}$. $\qquad\square$

In order to prove a partial converse to (3.6.14) we start with the following variation of Royden's extension lemma (3.A.1) by Zaidenberg [4].

(3.6.15) **Lemma.** *Let Z be a complex manifold and $S \subset Z$ a hypersurface. Let S_{reg} denote the set of regular points of S. Given a holomorphic mapping $f \colon D_R \to S_{\mathrm{reg}}$ with $R > 1$, there is a holomorphic map $\varphi \colon D \times D \to Z$ such that*

$$\varphi(z, 0) = f(z) \quad \text{for} \quad z \in D \quad \text{and} \quad \varphi(D \times D^*) \subset Z - S.$$

Proof. As in the proof of (3.A.1), by considering the graph of f we reduced the problem to the case where f is an imbedding. Thus, we have only to show that if $f \colon D_R \to S_{\mathrm{reg}}$ is a holomorphic imbedding, there is a holomorphic imbedding $\varphi \colon D \times D \to Z$ with the property above.

By (3.A.3) the vector bundle $TX|_{f(D_R)}$ over $f(D_R)$ splits:

$$TX|_{f(D_R)} = TS|_{f(D_R)} \oplus L,$$

where L is a line bundle over $f(D_R)$ and normal to S. By (3.A.4), there is a holomorphic affine connection in a neighborhood of $f(D_R)$. Now we repeat the the last of step of the proof of (3.A.2). Namely, we find a small neighborhood B of D in $L|_{f(D)} \cong D \times \mathbf{C}$ such that $B = D \times D$, and we set $\varphi = \exp|_B$. $\qquad\square$

(3.6.16) **Corollary.** *Let Z, S and S_{reg} be as in (3.6.15). Let $r < R$. Then every holomorphic mapping $f \colon D_R \to S_{\mathrm{reg}}$ can be approximated on D_r by holomorphic mappings of D_r into $Z - S$.*

Proof. Let $\mu_r \colon D_{R/r} \to D_R$ be the multiplication by r. Given $f \colon D_R \to S_{\mathrm{reg}}$, apply (3.6.15) to $f \circ \mu_r \colon D_{R/r} \to S_{\mathrm{reg}}$ to obtain a map $\varphi \colon D \times D \to Z$ such that (i) $\varphi(z, 0) = f(rz)$ for $z \in D$ and (ii) $\varphi(D \times D^*) \subset Z - S$. Set $\psi(z, w) = \varphi(z/r, w)$ for $(z, w) \in D_r \times D$. Then $\psi(z, 0) = f(z)$ for $z \in D_r$. Let $h_\varepsilon(z) = \psi(z, \varepsilon)$. Then $\{h_\varepsilon\}$ is an approximation of f on D. $\qquad\square$

(3.6.17) **Lemma** *Let Z be an n-dimensional complex manifold with smooth hypersurfaces $S_1, \ldots, S_k$, $k < n$, whose union $S = \bigcup_i S_i$ is a divisor with normal crossings. Then every holomorphic mapping $f \colon D_R \to \bigcap_i S_i$ can be approximated on each disc D_r, $r < R$, by holomorphic mappings $h \colon D_r \to Z - S$.*

Proof. The proof is by induction on k. The case $k = 1$ is a special case of (3.6.16) where S smooth. Assume that (3.6.17) holds for $k - 1$. Set $S^{[m]} = \bigcap_{i=1}^{m} S_i$. Then $S^{[k]}$ is a smooth hypersurface in a smooth manifold $S^{[k-1]}$ of dimension $n - k + 1$. Let $r < \rho < R$. Then by (3.6.16) the mapping $f|_{D_\rho}$ is approximated by mappings $g \colon D_\rho \to S^{[k-1]} - S^{[k]} \subset Z - S_k$. By the induction hypothesis applied to $Z' = Z - S_k, S_1' = S_1 - S_k, \ldots, S_{k-1}' = S_{k-1} - S_k$, the mapping $g|_{D_\rho}$ is approximated by mappings $h \colon D_r \to Z' - \bigcup_{i=1}^{k-1} S_i' = Z - \bigcup_{i=1}^k S_i$. $\square$

Now we are in a position to prove the following partial converse to (3.6.13).

(3.6.18) Theorem. *Let Z be a compact complex manifold with smooth hypersurfaces $S_1, \ldots, S_m$, whose union $S = \bigcup_i S_i$ is a divisor with normal crossings. If $Y = Z - S$ is hyperbolically imbedded in Z, then*

(a) *there are no complex lines in Y;*

(b) *for any partition of indices $I \cup J = \{1, 2, \ldots, m\}$, there are no complex lines in $\bigcap_{i \in I} S_i - \bigcup_{j \in J} S_j$.*

Proof. If (a) is violated, then Y is not hyperbolic. Assume that (b) is violated. Then there is $h \in \mathrm{Hol}(\mathbf{C}, Z)$ such that $h(\mathbf{C}) \subset \bigcap_{i=1}^k S_i - \bigcup_{j=k+1}^m S_j$. In view of (3.6.5) it suffices to show that this complex line is a limit line coming from Y. For any $R > 0$ we can find a neighborhood U of $h(\bar{D}_R)$ in Z such that $U \cap S_j = \emptyset$ for $j = k+1, \ldots, m$. Now apply (3.6.17) to the manifold U with hypersurfaces $S_1 \cap U, \ldots, S_k \cap U$. Then $h|_{D_R}$ can be approximated on each disk $D_r, r < R$, by holomorphic mappings of D_r into $U - \bigcup_{i=1}^k S_i \subset Z - \bigcup_{i=1}^m S_i$. $\square$

Similarly, we have the following partial converse to (3.6.14), (due to Zaidenberg [4, 5]).

(3.6.19) Theorem. *Let Z, $S = \bigcup_{i=1}^m S_i$, and $Y = Z - S$ be as in (3.6.18). If Y is hyperbolically imbedded in Z, then*

(a) *there are no complex lines in Y;*

(b) *there are no complex lines in any of the strata $S^{(k)}$, $1 \le k \le n - 1$.*

Proof. If (a) is violated, then Y is not hyperbolic. Assuming (b) is violated, let $h \in \mathrm{Hol}(\mathbf{C}, Z)$ be such that $h(\mathbf{C}) \subset S^{(k)} = S^k - S^{k+1}$. Since $h(\mathbf{C}) \subset S^k$, we may assume that $h(\mathbf{C}) \subset S_1 \cap \ldots \cap S_k$. Since $h(\mathbf{C}) \cap S^{k+1} = \emptyset$, $h(\mathbf{C})$ does not interesect any of $S_1 \cap \ldots \cap S_k \cap S_j$ for $j = k+1, \ldots, m$. Hence, $h(\mathbf{C}) \cap S_j = \emptyset$ for $j = k+1, \ldots, m$. This implies $h(\mathbf{C}) \subset \bigcap_{i=1}^k S_i - \bigcup_{j=k+1}^m S_j$. This violates (b) of (3.6.18). Hence, Y is not hyperbolically imbedded in Z. $\square$

Given a complex subspace $Y \subset Z$, we defined the pseudo-distance $d_{Y,Z}$ on $\bar{Y}$, see (3.4.1). Let $F_{Y,Z}$ be the infinitesimal form of $d_{Y,Z}$ defined in the same way as the infinitesimal form F_Z of d_Z, using the subfamily $\mathcal{F}_{Y,Z} \subset \mathrm{Hol}(D, Z)$, see the end of Section 5.

Now, using the argument in the proof of (3.6.3) we shall prove the implication (c) $\Rightarrow$ (b) in (3.4.11), thus making all three conditions in (3.4.11) mutually equivalent.

(3.6.20) Theorem. *If a complex space Y is hyperbolically imbedded in Z, then $d_{Y,Z}(p, q) > 0$ for all pairs $p, q \in \bar{Y}$, $p \ne q$.*

Proof. Let E be any length function on Z. In order to prove the theorem, it suffices to show that there is a positive constant c such that $cE \le F_{Y,Z}$ on $\bar{Y}$. Suppose that there is no such constant. Then there exist a sequence of tangent vectors v_n of $\bar{Y}$

with $E(v_n) = 1$, a sequence of holomorphic maps $f_n \in \mathcal{F}_{Y,Z}$ and a sequence of tangent vectors e_n of D with Poincaré length $\|e_n\| \searrow 0$ such that $df_n(e_n) = v_n$. Since D is homogeneous, we may assume that e_n is a vector at the origin of D.

As in the proof of (3.6.3), we replace the above $f_n \in \mathrm{Hol}(D, Z)$ by a new $f_n \in \mathrm{Hol}(D_{R_n}, Z)$ with $R_n \nearrow \infty$; instead of using the fixed disc D and varying vectors e_n, we use varying discs D_{R_n} and the fixed tangent vector $e = (d/dz)_0$ at the origin. Let $\mathcal{F}_{Y,Z}^{R_n}$ be the family of holomorphic maps $f: D_{R_n} \to Z$ such that $f^{-1}(Z - Y)$ is either empty or a singleton. Having replaced (D, e_n) by (D_{R_n}, e), we may assume that $f_n \in \mathcal{F}_{Y,Z}^{R_n}$ and $df_n(e) = v_n$.

By applying Brody's lemma (3.6.2) to each f_n and a constant $0 < c < 1/4$ we obtain holomorphic maps $g_n \in \mathrm{Hol}(D_{R_n}, Z)$ such that

(a) $g_n^* E^2 \leq c R_n^2 ds_{R_n}^2$ on D_{R_n} and the equality holds at the origin 0;

(b) $\mathrm{Image}(g_n) \subset \mathrm{Image}(f_n)$.

Since g_n is of the form $g_n = f_n \circ \mu_{r_n} \circ h_n$, where h_n is an automorphism of D_{R_n} and μ_{r_n}, $(0 < \mu_{r_n} < 1)$, is the multiplication by r_n, each g_n maps all of D_{R_n}, except possibly one point, into Y.

Now, as in the proof of (3.6.3) we shall construct a nonconstant holomorphic map $h: \mathbf{C} \to Z$ to which a suitable subsequence of $\{g_n\}$ converges. In fact, since

$$g_n^* E^2 \leq c R_n^2 ds_{R_n}^2 \leq c R_m^2 ds_{R_m}^2 \qquad \text{for} \quad n \geq m,$$

the family $\mathcal{F}_m = \{g_n | D_{R_m}; \ n \geq m\}$ is equicontinuous for each fixed m. Since the family $\mathcal{F}_1 = \{g_n | D_{R_1}\}$ is equicontinuous, the Arzela-Ascoli theorem implies that we can extract a subsequence which converges to a map $h_1 \in \mathrm{Hol}(D_{R_1}, Z)$. (We note that this is where we use the compactness of $\bar{Y}$.) Applying the same theorem to the corresponding sequence in $\mathcal{F}_2$, we extract a subsequence which converges to a map $h_2 \in \mathrm{Hol}(D_{R_2}, Z)$. In this way we obtain maps $h_k \in \mathrm{Hol}(D_{R_k}, Z)$, $k = 1, 2, \ldots$, such that each h_k is an extension of h_{k-1}. Hence, we have a map $h \in \mathrm{Hol}(\mathbf{C}, Z)$ which extends all h_k.

Since $g_n^* E^2$ at the origin 0 is equal to $(c R_n^2 ds_{R_n}^2)_{z=0} = 4c \, dz d\bar{z}$, it follows that

$$(h^* E^2)_{z=0} = \lim_{n \to \infty} (g_n^* E^2)_{z=0} = 4c \, dz d\bar{z} \neq 0,$$

which shows that h is nonconstant.

Since $g_n^* E^2 \leq c R_n^2 ds_{R_n}^2$, in the limit we have

$$h^* E^2 \leq 4c \, dz d\bar{z}.$$

By suitably normalizing h we obtain

$$h^* E^2 \leq dz d\bar{z} \quad \text{with the equality holding at} \quad z = 0.$$

We may assume that $\{g_n\}$ itself converges to h. Since h is the limit of $\{g_n\}$, clearly $h(\mathbf{C}) \subset \bar{Y}$. Let p, q be two points of $h(\mathbf{C})$, say $p = h(a)$ and $q =$

$h(b)$. Taking a subsequence and suitable points $a, b \in \mathbf{C}$ we may assume that $g_n(a), g_n(b) \in Y$. Then $\lim g_n(a) = p$ and $\lim g_n(b) = q$.

If there is a subsequence of $\{g_n\}$, still denoted $\{g_n\}$, such that $g_n(D_{R_n}) \subset Y$, then

$$d_Y(g_n(a), g_n(b)) \leq d_{D_{R_n}}(a, b) \to 0 \quad \text{as} \quad n \to \infty,$$

contradicting the assumption that Y is hyperbolically imbedded in Z.

If there is no such subsequence, there is a subsequence, again denoted $\{g_n\}$, such that each g_n maps exactly one point, say $c_n \in D_{R_n}$ into the boundary ∂Y. If the set $\{c_n\}$ is unbounded in $\mathbf{C}$, by taking subsequence we may assume that $|c_n| \nearrow \infty$. Then we can find $R'_n < |c_n| < R_n$ such that $R'_n \nearrow \infty$. If we denote the restriction of g_n to $D_{R'_n}$ by g'_n so that $g'_n(D_{R'_n}) \subset Y$, then

$$d_Y(g'_n(a), g'_n(b)) \leq d_{D_{R'_n}}(a, b) \to 0 \quad \text{as} \quad n \to \infty,$$

again contradicting the assumption that Y is hyperbolically imbedded in Z.

Therefore we assume that $\{c_n\}$ is bounded. Taking a subsequence, we may assume that $\{c_n\}$ converges to, say $c \in \mathbf{C}$. We may assume that c is the origin of $\mathbf{C}$. (For each n we consider a disk $D(c, R'_n) \subset D_{R_n}$ of center c and radius R'_n such that $R'_n \nearrow \infty$, and we restrict g_n to $D(c, R'_n)$.)

Now we assume that $\lim c_n = 0$. Set

$$a_n = a - c_n, \quad b_n = b - c_n, \quad R'_n = R_n - |c_n|, \quad g'_n(z) = g_n(z + c_n).$$

Then

$$g'_n(a_n) = g_n(a), \quad g'_n(b_n) = g_n(b), \quad g'_n(0) \in \partial Y.$$

Thus, g'_n maps the punctured disk $D^*_{R'_n} = \{0 < |z| < R'_n\}$ into Y. Hence,

$$d_Y(g'_n(a_n), g'_n(b_n)) \leq d_{D^*_{R'_n}}(a_n, b_n) \to 0 \quad \text{as} \quad n \to \infty,$$

contradicting the assumption that Y is hyperbolically imbedded in Z. We note that since $a_n \to a$ and $b_n \to b$, in order to conclude $d_{D^*_{R'_n}}(a_n, b_n) \to 0$ it suffices to show

$$d_{D^*_R}(a, b) \to 0 \quad \text{as} \quad R \to \infty.$$

But this can be seen from the expression for the infinitesimal metric $ds^2_{D^*_R}$ corresponding to $d_{D^*_R}$:

$$ds^2_{D^*_R} = \frac{4 dz d\bar{z}}{|z|^2 (\log R^2 - \log |z|^2)^2},$$

which can be obtained from (2.2.3) by replacing z by z/R. $\square$

We say that the cotangent bundle T^*X of a compact complex space X is **ample** (or the tangent bundle TX is **negative** in the sense of Grauert) if the zero section $0(X)$ of TX can be blown down to a point, namely there exists a holomorphic map $\pi: TX \to Y$ onto a complex space $Y = TX/0(X)$ which maps $0(X)$ to a single point, say $y_0 \in Y$, and $TX - 0(X)$ biholomorphically onto $Y - \{y_0\}$. This negativity of TX implies the existence of a length function F with negative curvature. As we shall see in (3.7.1), such a space is hyperbolic. Instead of the differential geometric argument just described (see Kobayashi [13]) we present the proof by Urata [5] which uses Brody's criterion.

(3.6.21) **Theorem**. *Let X be a compact complex space with ample cotangent bundle. Then X is hyperbolic.*

Proof. With the notation above, let V be a small (hence hyperbolic) neighborhood of y_0 in $Y = TX/0(X)$. We choose a length function E on X in such a way that every tangent vector of E-length ≤ 1 belongs to $\pi^{-1}(V)$. Assume that X is not hyperbolic. By (3.6.3) there exists a nonconstant holomorphic map $f: \mathbf{C} \to X$ such that its differential $f': \mathbf{C} \to TX$ has the property that $E(f'(z)) \leq 1$ for all $z \in \mathbf{C}$. Since V is hyperbolic, the map $\pi \circ f': \mathbf{C} \to V$ must be constant and $\pi(f'(\mathbf{C})) = \{y_0\}$. Hence, $f' \equiv 0$, which implies that f is constant. This is a contradiction. $\qquad\square$

As another application of Brody's lemma, Urata [5] proved the following

(3.6.22) **Theorem**. *Let X be a complex space with a length function E, and $G = \mathrm{Aut}(X, E)$ the group of holomorphic isometries. Assume that X/G is compact. Then X is complete hyperbolic if there is no complex line $h: \mathbf{C} \to X$.*

Proof. Let K be a compact subset of X such that $G(K) = X$. Suppose that X is not hyperbolic. Let e denote the tangent vector $(d/dz)_0$ of $\mathbf{C}$ at 0. Let D_n denote the disk of radius n with its Poincaré metric $ds_n^2 = 4n^2 dz d\bar{z}/(n^2 - |z|^2)^2$. Then for each n, there exists a holomorphic mapping $f_n: D_n \to X$ such that $|df_n(e)|_E > 1$. We repeat the argument in the proof of (3.6.3). By Brody's lemma (3.6.2) and from $G(K) = X$ it follows that for each n there exists a holomorphic map $g_n: D_n \to X$ such that

(i) $g_n(0) \in K$,

(ii) $g_n^* E^2 \leq n^2 ds_n^2$,

with equality holding at $0 \in D_n$.

Let δ_X be the distance function on X defined by E. From (ii) above we have

$$\delta_X(g_n(0), g_n(z)) \leq \int_0^{|z|} \frac{2n^2 dt}{n^2 - t^2} = n \log \frac{n + |z|}{n - |z|} \leq 4|z|$$

for $|z| \leq n/2$.

Since X/G is compact, X is complete with respect to δ_X. The estimate above shows that for each fixed $z \in \mathbf{C}$, the set $\{g_n(z); n \geq n_0\}$, where $n_0 \geq 2|z|$, is relatively compact in X. By (ii) the family $\{g_n\}$ is also equicontinuous. By Arzela-Ascoli theorem (1.3.1), a subsequence of $\{g_n\}$ converges to a holomorphic map $g: \mathbf{C} \to X$. By (ii), $g^* E^2 \leq 4 dz d\bar{z}$ with equality holding at 0. This is a contradiction, showing that X is hyperbolic. Since d_X is invariant by G and since $G(K) = X$ with K compact, d_X is a complete distance. $\qquad\square$

(3.6.23) **Corollary**. *A homogeneous Hermitian manifold X with an invariant Hermitian metric ds^2 is complete hyperbolic if there is no complex line $h: \mathbf{C} \to X$.*

7 Differential Geometric Criteria for Hyperbolicity

The results in Chapter 2 yield differential geometric criteria for hyperbolicity. We shall combine them with Brody's criteria explained in the preceding section.

(3.7.1) Theorem. *Let X be a complex space. If there is a length function F with (holomorphic sectional) curvature K_F bounded above by a negative constant, then X is hyperbolic. If, moreover, F defines a complete distance on X, then X is complete hyperbolic.*

Proof. By normalizing F we may assume that $K_F \leq -1$. Let δ be the distance function on X defined by F. By (2.3.5) every holomorphic map $f \colon (D, \rho) \to (X, \delta)$ is distance-decreasing. By (2) of (3.1.7), we have $\delta \leq d_X$. Our assertion now follows. $\qquad\qquad\square$

(3.7.2) Remark. Milnor [1] observed that for a Riemann surface the curvature condition of (3.7.1) can be relaxed as follows. Given a Riemann surface X with a Hermitian metric, let r denote the geodesic distance from a fixed point of X. If the curvature K satisfies $K(r) \leq -1/(r^2 \log r)$ asymptotically, then X is hyperbolic. Remark (2.2.8) shows that the asymptotic condition $K(r) \leq -1/r^3$ is too weak to imply hyperbolicity. For a higher dimensional analogue of Milnor's result, see Greene-Wu [2; p. 113, Theorem G'].

We know from (2.2.6) that $P_1\mathbf{C}$ minus at least three points carries a complete Hermitian metric with curvature $K \leq -1$ and from (2.2.7) that every compact Riemann surface of genus ≥ 2 admits a Hermitian metric with curvature $K \leq -1$. Hence,

(3.7.3) Corollary. (1)   *The Riemann sphere $P_1\mathbf{C}$ minus at least three points is complete hyperbolic.*

(2)   *Every compact Riemann surface of genus ≥ 2 is complete hyperbolic.*

We can generalize (3.7.1) to a pseudo-length function F. Considering F as a nonnegative function on TX, we assume that it is continuous everwhere and twice differentiable wherever it is positive so that if $F(v) > 0$ at $v \in TX$ then the curvature $K_F(v)$ is defined. We say that (X, F) is **negatively curved** if there is a negative constant c such that $K_F(v) \leq c$ for all $v \in TX$ for which $F(v) > 0$.

In accordance with (2.1.10) we can weaken the conditions on F as follows.

(a) F is upper semicontinuous;

(b)   For each $v \in TX$ with $F(v) > 0$ there is a length function F_v, defined and of class C^2 in a neighborhood $U \subset TX$ of v, such that (i) $F \geq F_v$ in U, (ii) $F(v) = F_v(v)$, and (iii) the curvature of F_v is bounded above by a negative constant c independent of v.

Then we say that (X, F) is **negatively curved**.

A point $x \in X$ is called a **degeneracy point** of F if $F(v) = 0$ for some nonzero $v \in T_xX$. The set of such degeneracy points is called the **degeneracy set** of F. Then the argument in the proof of (3.7.1) yields the following result. (The second statement in the theorem will not be used and can be skipped).

(3.7.4) Theorem. *Let X be a complex space. If there is a pseudo-length function F such that (X, F) is negatively curved in the sense defined above, then X is hyperbolic modulo the degeneracy set of F.*

More generally, if F is a jet pseudo-metric on the jet bundle $J^k X$ such that $K_F \leq -1$, then X is hyperbolic modulo the degeneracy set of F_1, where F_1 is the pseudo-length function defined in (2.5.10).

As we see from (2.2.8), for X to be hyperbolic it is not sufficient that X admits a Hermitian metric with negative holomorphic sectional curvature. It is important that the curvature is not only negative but also bounded away from zero. See Remark (3.7.15) for more comments on this point.

On the other hand, Brody's criterion can be strengthened if the holomorphic sectional curvature is only non-positive, (Kobayashi [15]).

(3.7.5) Lemma. *Let $2\lambda dz d\bar{z}$ with $0 \leq \lambda \leq 1$ be a pseudo-metric on $\mathbf{C}$ expressed in terms of the natural coordinate function z of $\mathbf{C}$. If its Gaussian curvature K is nonpositive at every point where $\lambda > 0$, then λ is constant.*

Proof. The Gaussian curvature K is given by

$$K = -\frac{1}{\lambda} \frac{\partial^2 \log \lambda}{\partial z \partial \bar{z}}.$$

Since $K \leq 0$, we have

$$\frac{\partial^2 \log \lambda}{\partial z \partial \bar{z}} \geq 0$$

wherever $\lambda > 0$. This means that $\log \lambda$ is a subharmonic function on $\mathbf{C}$. On the other hand, $\log \lambda \leq 0$ since $\lambda \leq 1$. But a bounded subharmonic function on $\mathbf{C}$ is constant. Hence, λ is constant. $\qquad\square$

(3.7.6) Lemma. *Let X be a complex space with a length function E whose holomorphic sectional curvature K_E is nonpositive. If $f: \mathbf{C} \to X$ is a holomorphic map such that $f^*E^2 \leq dz d\bar{z}$ with equality holding at some point, then $f^*E^2 = dz d\bar{z}$, i.e., f is an isometric holomorphic immersion with respect to the metric of X defined by E and the Euclidean metric $dz d\bar{z}$ of $\mathbf{C}$.*

Proof. Set $f^*E^2 = \lambda dz d\bar{z}$, and apply (3.7.5). $\qquad\square$

If X is a Hermitian manifold, we can say a little more about $f(\mathbf{C})$.

(3.7.7) Lemma. *Let M be a Hermitian manifold with metric ds_M^2 whose holomorphic sectional curvature is nonpositive. If $f: \mathbf{C} \to M$ is a holomorphic map such that $f^*ds_M^2 \leq dz d\bar{z}$ with equality holding at some point, then f is a totally geodesic, isometric holomorphic immersion.*

Proof. By (3.7.6) f is an isometric holomorphic immersion. Since M has nonpositive holomorphic sectional curvature and $f(\mathbf{C})$ is flat, M has vanishing holomorphic sectional curvature in the direction of $f(\mathbf{C})$ by (2.3.9). By Remark (2.3.11), $f(\mathbf{C})$ is totally geodesic in M. $\qquad\square$

We apply Lemmas (3.7.6) and (3.7.7) to (3.6.3), (3.6.5), (3.6.9), (3.6.11) and (3.6.12) to obtain the following results, (3.7.8) through (3.7.12).

(3.7.8) Theorem. *Let X be a compact complex space with a length function E whose holomorphic sectional curvature is nonpositive. If X is not hyperbolic, there is an isometric holomorphic immersion $h\colon \mathbf{C} \to X$. In fact, if $x \in X$ and if the degeneracy set $\Delta(x)$ is non-trivial, then we can find such a map h with its image $h(\mathbf{C})$ in $\Delta(x)$.*

If, moreover, X is a compact complex subspace of a Hermitian manifold Y with nonpositive holomorphic sectional curvature, such a map $h\colon \mathbf{C} \to Y$ is totally geodesic.

(3.7.9) Theorem. *Let Z be a complex space with a length function E whose holomorphic sectional curvature is nonpositive, and Y a relatively compact open subset of Z. If Y is not hyperbolically imbedded in Z, then there is an isometric holomorphic immersion $h\colon \mathbf{C} \to Z$ such that $h(\mathbf{C}) \subset \bar{Y}$.*

If, moreover, Z is a Hermitian manifold with nonpositive holomorphic sectional curvature, such a map h is totally geodesic.

(3.7.10) Theorem. *Let Z be a compact complex space with a length function E whose holomorphic sectional curvature is nonpositive. Let S be a union of Cartier divisors $S_1, \ldots, S_m$. Then $Y = Z - S$ is complete hyperbolic and hyperbolically imbedded in Z if the following two conditions are satisfied:*

(a)   There are no isometric holomorphic immersions $h\colon \mathbf{C} \to Y$;

(b)   For any partition of indices $I \cup J = \{1, 2, \ldots, m\}$, there are no isometric holomorphic immersions

$$h\colon \mathbf{C} \to \bigcap_{i \in I} S_i - \bigcup_{j \in J} S_j \subset Z.$$

(3.7.11) Theorem. *Let Z be a complex space with a length function E whose holomorphic sectional curvature is nonpositive, and Y a relatively compact subset of Z. Let $\{U_n\}$ be a decreasing sequence of relatively compact open neighborhoods of $\bar{Y}$ such that $\bigcap U_n = \bar{Y}$. If none of these U_n is hyperbolically imbedded in Z, then there is an isometric holomorphic immersion $h\colon \mathbf{C} \to Z$ such that $h(\mathbf{C}) \subset \bar{Y}$.*

If, moreover, Z is a Hermitian manifold with nonpositive holomorphic sectional curvature, such a map h is totally geodesic.

From (3.7.8) we obtain the following result, conjectured by Lang [1] and proved by Green [8].

(3.7.12) Theorem. *A closed complex subspace X of a complex torus T is hyperbolic if there is no nonconstant affine map $h\colon \mathbf{C} \to T$ such that $h(\mathbf{C}) \subset X$, or equivalently, if X contains no translate of a complex subtorus of T.*

Proof. Suppose that X is not hyperbolic. In (3.7.8), let $Y = T$. Since T admits a flat Hermitian metric, there is a totally geodesic, isometric holomorphic map $h: \mathbf{C} \to T$ such that $h(\mathbf{C}) \subset X$. By translating $h(\mathbf{C})$ in T, we may assume that $H = h(\mathbf{C})$ is a subgroup of T. We claim that the Zariski closure $\bar{H}$ of H in T is a subgroup of T, i.e., $(\bar{H})^{-1} \subset \bar{H}$ and $\bar{H} \cdot \bar{H} \subset \bar{H}$, and hence is a complex torus. This follows from the fact that although the operation $T \times T \to T$, $(x, y) \mapsto x^{-1}y$, is not Zariski continuous, the operations $x \mapsto x^{-1}$ and $x \mapsto xa$ (with a fixed) are homeomorphisms of T onto itself in the Zariski topology. (See Lang [3; p. 84] for details. We note that the closure in the usual topology would yield merely a real subtorus). $\qquad\square$

Another result, also conjectured by Lang and proved by Green, follows from (3.7.10):

(3.7.13) **Theorem**. *Let T be a complex torus and S a complex hypersurface. If S contains no complex subtorus, then $Y = T - S$ is complete hyperbolic and hyperbolically imbedded in T.*

Proof. In view of (3.7.10) it suffices to show that there is no totally geodesic, isometric holomorphic immersion $h: \mathbf{C} \to Y$. Assuming that such a map h exists, we shall obtain a contradiction. Let A be the topological closure of $h(\mathbf{C})$ in T; it is a real subtorus of T.

First we consider the case $A \cap S = \emptyset$. Let $\{t_n\}$ be a convergent sequence of translations in T such that $t_n(A) \cap S = \emptyset$ and $(\lim t_n)(A) \cap S \neq \emptyset$. Choose points $a_n \in \mathbf{C}$ such that $\lim t_n(h(a_n)) \in S$. Let U_n be the unit disc neighborhood of a_n in $\mathbf{C}$. Then we have a sequence of holomorphic maps $t_n \circ h: U_n \to Y$ such that $\lim t_n(h(U_n)) \cap S \neq \emptyset$. By identifying U_n with the unit disc D around the origin 0 in $\mathbf{C}$, we obtain a sequence of convergent holomorphic maps $f_n: D \to Y$ with the limit map $f = \lim f_n$ such that $f(D) \cap S \neq \emptyset$. By the generalized Hurwitz theorem (3.6.11), we have $f(D) \subset S$. From the construction of f_n and f it is clear that f extends to an affine map $\mathbf{C} \to T$. Then $f(\mathbf{C})$ must be also contained in S. This contradicts the assumption of the theorem.

Next we consider the case $A \cap S \neq \emptyset$. Choose points $a_n \in \mathbf{C}$ such that $\lim h(a_n) \in S$. Let U_n be the unit disc neighborhood of a_n in $\mathbf{C}$. The remainder of the proof is the same as in the first case; simply let t_n be the identity transformation in the proof of the first case. $\qquad\square$

(3.7.14) **Corollary**. *Let T be a simple torus, i.e., a complex torus containing no proper complex subtorus. Then*
    (1)   *Every closed complex subspace of T is hyperbolic;*
    (2)   *For every complex hypersurface S, its complement $T - S$ is complete hyperbolic and hyperbolically imbedded in T.*

(3.7.15) **Remark**. In general, a hyperbolic complex manifold may not admit a negatively curved Hermitian metric, see Demailly [3]. Cheung [1] discusses a special class of hyperbolic manifolds which admit negatively curved Hermitian metrics.

8 Subvarieties of Quasi Tori

In this section we shall study subvarieties of complex tori by purely differential geometric means. For an algebraic geometric approach, we refer the reader to Ueno [1]. First we quickly review basic facts on the second fundamental form and the equations of Gauss-Codazzi for a Kähler submanifold. We follow here Nagano-Smyth [1], who studied minimal submanifolds in real tori. Since a complex submanifold of a Kähler manifold is a minimal submanifold, we can apply their results. Because of complex analyticity in our case, singularities of complex subspaces present no essential difficulty.

We start with the Riemannian case. For (3.8.1) through (3.8.6) we refer the reader to Kobayashi-Nomizu [1; vol.1]. Let M be an $(n + p)$-dimensional Riemannian manifold with metric g and with covariant differentiation $\tilde{\nabla}$, and X an n-dimensional submanifold with covariant differentiation ∇. Given vector fields u, v on X, we have the following formula of Gauss:

$$(3.8.1) \qquad\qquad \tilde{\nabla}_u v = \nabla_u v + \alpha(u, v),$$

where $\alpha: T_x X \times T_x X \to T_x^\perp X$ is a symmetric bilinear map called the **second fundamental form** of $X \subset M$.

If ξ is a section of the normal bundle $T^\perp X$, then we have the following formula of Weingarten:

$$(3.8.2) \qquad\qquad \tilde{\nabla}_u \xi = -A_\xi(u) + \nabla_u^\perp \xi,$$

where A_ξ defines a symmetric linear transformation of TX and $\nabla^\perp$ defines a connection in the normal bundle $T^\perp X$. Then α and A are related by

$$(3.8.3) \qquad\qquad g(A_\xi(u), v) = g(\alpha(u, v), \xi).$$

The connection ∇ of TX combined with the connection $\nabla^\perp$ of $T^\perp X$ defines a connection in $T^* X \otimes T^* X \otimes T^\perp X$, which we shall denote by ∇^*. In particular, for α, we have

$$(3.8.4) \qquad (\nabla_u^* \alpha)(v, w) = \nabla_u^\perp(\alpha(v, w)) - \alpha(\nabla_u v, w) - \alpha(v, \nabla_u w).$$

The curvature $\tilde{R}$ of a Riemannian manifold M is a 2-form with values in the endomorphism bundle $\mathrm{End}(TM)$. We associate to $\tilde{R}$ a quadrilinear map, also denoted by $\tilde{R}$, by

$$\tilde{R}(v_1, v_2, v_3, v_4) = g(\tilde{R}(v_3, v_4)v_2, v_1).$$

Now we state the **equation of Gauss**:

(3.8.5) **Theorem.** *Let $\tilde{R}$ and R be the Riemannian curvature tensors of M and X, respectively. Then for any vector fields v_1, v_2, v_3, v_4 on X, we have*

$$\tilde{R}(v_1, v_2, v_3, v_4) = R(v_1, v_2, v_3, v_4)$$
$$+ g(\alpha(v_1, v_4), \alpha(v_2, v_3)) - g(\alpha(v_1, v_3), \alpha(v_2, v_4)).$$

The equation of Gauss expresses the tangential component of $\tilde{R}(v_3, v_4)v_2$ in terms of R and α. Its normal component is described by the following **equation of Codazzi**:

(3.8.6) **Theorem**. *For any vector fields u, v, w of X, the normal component of $\tilde{R}(u, v)w$ is given by*

$$(\tilde{R}(u, v)w)^{\perp} = (\nabla_u^*\alpha)(v, w) - (\nabla_v^*\alpha)(u, w).$$

In particular, if M is a space of constant curvature, then

$$(\nabla_u^*\alpha)(v, w) = (\nabla_v^*\alpha)(u, w).$$

We define the **relative nullity space** at x to be

(3.8.7) $N_x = \{u \in T_xX;\ \alpha(u, v) = 0 \quad \text{for all} \quad v \in T_xX\}.$

Let

$$v = \min_{x \in X} \dim N_x,$$

$$U = \{x \in X;\ \dim N_x = v\}.$$

Then U is an open subset of X, and $N = \bigcup_{x \in U} N_x$ is a subbundle of $T(U)$ of rank v, i.e., a v-dimensional distribution on U.

(3.8.8) **Lemma**. *Assume that M is a space of constant curvature and that $v > 0$. Then the distribution N is integrable and each maximal integral submanifold of N is totally geodesic not only in U but also in M.*

Proof. Let u, w be sections of the bundle N, and v any vector field on U. Then by (3.8.4), we have

$$(\nabla_u^*\alpha)(v, w) = -\alpha(v, \nabla_u w), \qquad (\nabla_v^*\alpha)(u, w) = 0.$$

Hence, (3.8.6) implies $\alpha(v, \nabla_u w) = 0$. Since this holds for all vector fields v on U, $\nabla_u w$ is a section of N. This shows that N is integrable and totally geodesic in U. Since

$$\tilde{\nabla}_u w = \nabla_u w + \alpha(u, w) = \nabla_u w$$

and since $\nabla_u w$ is in N, it follows that N is totally geodesic in M as well. $\square$

Now we consider the case where M is a Kähler manifold and X is a complex submanifold. Then the second fundamental form α satisfies the following (see Kobayashi-Nomizu [1; vol.2]):

(3.8.9) $\alpha(Ju, v) = \alpha(u, Jv) = J(\alpha(u, v)), \qquad u, v \in T_xX,$

where J is the complex structure of M. This implies that the relative nullity space N_x is a complex subspace, i.e., invariant by J.

From (3.8.5) and (3.8.9) we obtain

$$(3.8.10) \qquad \tilde{R}(u, Ju, v, Jv) = R(u, Ju, v, Jv) + 2g(\alpha(u, v), \alpha(u, v)).$$

Let $\tilde{S}$ and S be the Ricci tensors of M and X, respectively; they define symmetric bilinear forms on each tangent space. If we choose a local orthonormal basis of TM in such a way that $e_1, \ldots, e_n, Je_1, \ldots, Je_n$ are tangent to X and $e_{n+1}, \ldots, e_{n+p}, Je_{n+1}, \ldots, Je_{n+p}$ are normal to X, then

$$\tilde{S}(u, u) = \sum_{i=1}^{n+p} \tilde{R}(e_i, Je_i, u, Ju),$$

$$S(u, u) = \sum_{i=1}^{n} R(e_i, Je_i, u, Ju).$$

Using (3.8.10) we obtain

$$S(u, u) = \tilde{S}(u, u) - 2\sum_{i=1}^{n} g(\alpha(e_i, u), \alpha(e_i, u)) - \sum_{k=n+1}^{n+p} \tilde{R}(e_k, Je_k, u, Ju).$$

From now on, we assume that M is flat. Then

$$(3.8.11) \qquad S(u, u) = -2\sum_{i=1}^{n} g(\alpha(e_i, u), \alpha(e_i, u)) \leq 0.$$

If $u \in N_x$, then $S(u, u) = 0$ by (3.8.11). Conversely, if $u \in T_x X$ and $S(u, u) = 0$, then $\alpha(e_i, u) = 0$ for $i = 1, \ldots, n$ by (3.8.11) and $\alpha(Je_i, u) = 0$ for $i = 1, \ldots, n$ by (3.8.9), and hence $u \in N_x$. Thus,

$$(3.8.12) \qquad N_x = \{u \in T_x X; \ S(u, u) = 0\}.$$

Since S is symmetric and negative semi-definite, we have

$$(3.8.13) \qquad N_x = \{u \in T_x X; \ S(u, v) = 0 \quad \text{for all} \quad v \in T_x X\}.$$

This shows that the relative nullity space N_x, defined originally in terms of the second fundamental form α, can be defined in terms of an intrinsic invariant of X, namely the Ricci tensor S.

Now, let M be a locally flat Kähler manifold and X a closed complex subspace. We can apply the results above to the regular locus X_{reg} of X. Let N be the relatively nullity distribution defined on an open set U of X_{reg}. Since the relative nullity spaces N_x are all complex vector spaces, we denote by ν the minimum of the complex dimensions of N_x, $x \in U$. For $x \in U$, let $N(x)$ be the maximal integral submanifold through x defined by the distribution N.

Let M be a commutative complex Lie group with an invariant flat Kähler metric; M is of the form $\mathbf{C}^{n+p}/\Gamma$, where Γ is a discrete subgroup of $\mathbf{C}^{n+p}$. By parallel translation we can compare tangent spaces at different points. This fact makes it possible to define the Gauss map. Let X be an n-dimensional closed complex subspace of M. Let $G(n, n + p)$ be the complex Grassmann manifold of n-dimensional subspaces in $\mathbf{C}^{n+p}$. The Gauss map

$$G: X_{\mathrm{reg}} \to G(n, n + p)$$

assigns to each regular point x the n-dimensional subspace of the tangent space $T_0 M$ parallel to $T_x X$. We shall compute the differential $dG: T_x X \to T_{G(x)} G(n, n + p)$. We take a local orthonormal frame field $e_1, \ldots, e_{n+p}$ around x such that $e_1, \ldots, e_n$ are tangent to X_{reg} (and $e_{n+1}, \ldots, e_{n+p}$ normal to X_{reg}). Then $T_{G(x)} G(n, n + p)$ is identified with the space of complex $(n \times p)$-matrices. Set

$$(3.8.14) \qquad de_i = \sum_{j=1}^{n} \omega_i^j e_j + \sum_{k=n+1}^{n+p} \alpha_i^k e_k, \qquad 1 \le i \le n.$$

The differential dG is identified with (α_i^k). Comparing (3.8.14) with the definition (3.8.1) of the second fundamental form α, we see that

$$(3.8.15) \qquad \alpha(u, e_i) = \sum_{k=n+1}^{n+p} \alpha_i^k(u) e_k,$$

which says that dG can be identified with the second fundamental form. It follows from (3.8.15) that the relative nullity space N_x agrees with the kernel of dG at x:

$$(3.8.16) \qquad N_x = \mathrm{Ker}(dG_x), \qquad x \in X_{\mathrm{reg}}.$$

From this we immediately obtain

(3.8.17) **Lemma.** *let M be a commutative complex Lie group with an invariant flat Kähler metric and X a closed complex subspace. Then each maximal intergal submanifold $N(x)$, $x \in U$ is a connected component of a level set of the Gauss map G, and hence it is closed in U.*

For a different proof on closedness of $N(x)$, see Nagano-Smyth [1]. The use of the Gauss map was suggested by Wu, see Fischer-Wu [1].

Since, by (3.8.8), $N(x)$ is totally geodesic in $M = \mathbf{C}^{n+p}/\Gamma$, it extends to X. This extension is a translate of a connected ν-dimensional complex subgroup, say M'_x, of M and is the closure $\overline{N(x)}$ of $N(x)$ in M. Hence, this subgroup M'_x is closed in M. In general, M'_x may vary with x. However, it is independent of x if M is a quasi torus in the following sense.

A commutative complex Lie group $M = \mathbf{C}^N/\Gamma$ is said to be a **quasi torus** if it is an extension of a complex torus T by $(\mathbf{C}^*)^k$. Thus we have an exact sequence of commutative complex Lie groups:

$$0 \to (\mathbf{C}^*)^k \to M \to T \to 0.$$

Actually, a connected complex Lie group is automatically commutative if it is an extension of T by $(\mathbf{C}^*)^k$, see Iitaka [3]. Let $e_1, \ldots, e_N$ be the natural basis for $V = \mathbf{C}^N$. After a linear coordinate change for $\mathbf{C}^N$, the quasi lattice Γ has a $\mathbf{Z}$-basis $e_1, \ldots, e_k, b_1, \ldots, b_{2m}$, where $k + m = N$ with the following property. Let $V_1 = \mathbf{C}^k$ and Γ_1 be, respectively, the subspace of V and the subgroup of Γ spanned by $e_1, \ldots, e_k$ so that $(\mathbf{C}^*)^k \cong V/\Gamma_1$. Let $V_2 = V/V_1$, and $\bar{b}_1, \ldots, \bar{b}_{2m}$ be the images of $b_1, \ldots, b_{2m}$ in V_2 so that the group Γ_2 generated by $\bar{b}_1, \ldots, \bar{b}_{2m}$ is the lattice for the complex torus T, i.e., $T = V_2/\Gamma_2$.

(3.8.18) Lemma. (1)   *Every connected closed complex subgroup M' of a quasi torus M is a quasi torus, and the quotient group M/M' is also a quasi torus;*

(2)   *A quasi torus contains only countably many quasi subtori.*

Proof. (1). Set $G = (\mathbf{C}^*)^k$. Each $\mathbf{m} = (m_1, \ldots, m_k) \in \mathbf{Z}^k$ defines a character $\chi_{\mathbf{m}}$ of G, i.e., a homomorphism of G into $\mathbf{C}^*$ by

$$\chi_{\mathbf{m}}(t_1, \ldots, t_k) = (t_1^{m_1}, \ldots, t_k^{m_k}), \qquad (t_1, \ldots, t_k) \in (\mathbf{C}^*)^k.$$

Conversely, every character of G is of the form $\chi_{\mathbf{m}}$. Thus, the set G^* of characters of G is a commutative group isomorphic to $\mathbf{Z}^k$. By duality, there is a one-to-one correspondence between the subgroups of G^* and the closed subgroups of G, one being the annihilator of the other. In particular, every connected closed subgroup of G is isomorphic to $(\mathbf{C}^*)^l$.

Let M' be a connected closed complex subgroup of M. Set $H = M' \cap G$, where $G = (\mathbf{C}^*)^k$ as above. Then M'/H is a complex subgroup of $M/G = T$. To see that M'/H is compact, let K be the maximal compact subgroup of M. (Since M is commutative, there is only one maximal compact subgroup). Since $M = GK$, it follows that $M' = H(K \cap M')$. Let $p: M \to T = M/G$ be the projection. Since $p(K) = T$, we have $p(K \cap M') = M'/H$, which shows that M'/H is compact. This proves that M' is a quasi torus.

It is now obvious that M/M' is also a quasi torus.

(2). If $M = V/\Gamma$ is a quasi torus and $M' = V'/\Gamma'$ is a quasi subtorus, then $V' \subset V$ and $\Gamma' \subset \Gamma$. Since there are only countably many subgroups Γ' of Γ and since V' is determined by Γ' (i.e., spanned by Γ'), there are at most countably many quasi subtori of M. $\qquad\square$

(3.8.19) Lemma. *Let M be a quasi torus and X a closed complex subspace. Then the subspaces $\overline{N(x)}$, $x \in U$, are all parallel translates of one quasi subtorus, say M', of M.*

Proof. By (1) of (3.8.18) $M'_x = \overline{N(x)}$ is a quasi subtorus of M, and by (2) of (3.8.18) M'_x is independent of x since $\overline{N(x)}$ depends continuously on x. $\qquad\square$

In order to explain the subgroup M', we need the following result of Bochner.

(3.8.20) Lemma. *An infinitesimal isometry of constant length on a Riemannian manifold with negative semi-definite Ricci tensor S is parallel and satisfies $S(v, v) = 0$.*

Proof. This follows from the formula:

$$\Delta(g(v, v)) = g(\nabla v, \nabla v) - S(v, v),$$

which is proved, for example, in Yano-Bochner [1] and Kobayashi [8].   □

(3.8.21) **Lemma.** *Let $X \subset M$ and M' be as in (3.8.19). Then M' is the largest connected subgroup of M leaving X invariant.*

Proof. Since every one-parameter subgroup of M' induces a vector field, say v, tangent to $N(x)$, it leaves X invariant.

Conversely, if v is any vector field on M generating a one-parameter subgroup of M leaving X invariant, then it defines an infinitesimal holomorphic isometry of constant length on X_{reg}. By (3.8.20), v must satisfy $S(v, v) = 0$ on X_{reg}. By (3.8.12), the one-parameter group generated by v is in M'.   □

By (3.8.19) the relative nullity distribution N on U extends to a parallel distribution N on X_{reg}; in fact, the vector fields coming from the action of M' are all parallel vector fields on X_{reg}. Hence, the distribution $N^{\perp}$ orthogonal to N is also a parallel distribution on X_{reg}. By the theorem of de Rham on the holonomy decomposition of a Riemannian manifold, a neighborhood of each point $x \in X_{\mathrm{reg}}$ is locally a Riemannian direct product of the foliation defined by N and the orthogonal foliation.

In summary, we have

(3.8.22) **Theorem.** *Let X be an n-dimensional closed complex subspace of a quasi torus $M = \mathbf{C}^{n+p}/\Gamma$ with an invariant flat Kähler metric g, and M' be the largest quasi subtorus of M leaving X invariant. Let T'_x denote the subspace of the tangent space $T_x X$ spanned by all vector fields induced by the action of M'. Let S be the Ricci tensor of X_{reg}, and let $N_x = \{v \in T_x X;\ S(v, v) = 0\}$. Let $M'' = M/M'$ with projection $\pi: M \to M''$. Then*
   (1) $N_x = T'_x$ *on a dense open subset U of X_{reg};*
   (2) X *is a principal M'-bundle over $\pi(X) = X/M'$;*
   (3)   *The M'-orbits, i.e., the fibers in the above fibering, are the flat factor in the local holonomy decomposition of X_{reg};*
   (4).   *With respect to the induced metric, the Ricci tensor of $\pi(X)$ is negative definite on a dense open subset.*

(3.8.23) **Corollary.** *Let $X \subset M$, M' and S be as above. Then $M' = 0$ if and only if the Ricci tensor S is negative definite on a dense open subset of X_{reg}.*

Let $M = \mathbf{C}^{n+p}/\Gamma$, and $w^1, \ldots, w^{n+p}$ the natural coordinate system of $\mathbf{C}^{n+p}$. Then using family of holomorphic n-forms $dw^{i_1} \wedge \ldots \wedge dw^{i_n}$, $i_1 < \ldots < i_n$, on X, we obtain a meromorphic map

$$(3.8.24) \qquad \Phi: X \to P_r \mathbf{C}, \qquad r = \binom{n + p}{n} - 1$$

which sends $x \in U$ to a point with homogeneous coordinates

$$(\langle dw^{i_1} \wedge \ldots \wedge dw^{i_n}, e_1(x) \wedge \ldots \wedge e_n(x)\rangle)_{i_1 < \ldots < i_n}.$$

We recall the definition of the **Plücker imbedding** $P: G(n, n + p) \to P_r\mathbf{C}$. Given a point of $G(n, n + p)$, i.e., an n-dimensional vector subspace V of $\mathbf{C}^{n+1}$, choose a basis $v_1, \ldots, v_n$ for V and assign the line in $\bigwedge^n \mathbf{C}^{n+p}$ spanned by $v_1 \wedge \ldots \wedge v_n$, which is an element of the projective space $P(\bigwedge^n \mathbf{C}^{n+p})$. More explicitly, express each v_j by its components $v_j = (v_j^1, \ldots, v_j^{n+p})$ and consider $\binom{n+p}{n}$ numbers

$$v^{i_1\ldots i_n} = \begin{vmatrix} v_1^{i_1} & \cdots & v_n^{i_1} \\ \cdot & \cdots & \cdot \\ v_1^{i_n} & \cdots & v_n^{i_n} \end{vmatrix}, \qquad i_1 < \ldots < i_n,$$

which are called the Plücker coordinates of the point $V \in G(n, n + p)$. Thus, P maps V to a point of $P_r\mathbf{C}$ with homogeneous coordinates $(v^{i_1\ldots i_n})$.

From the definitions of Φ and P we obtain

(3.8.25) Lemma. *In terms of the Gauss map $G: X_{\text{reg}} \to G(n, n + p)$ and the Plücker imbedding $P: G(n, n + p) \to P_r\mathbf{C}$, the map Φ can be written as follows:*

$$\Phi = P \circ G.$$

(3.8.26) Corollary. *Let $X \subset M$ and M' be as in (3.8.22). If $M' = 0$, then $\Phi: X \to P_r\mathbf{C}$ is a meromorphic immersion.*

Proof. Since P is an imbedding and since by (3.8.16) and (3.8.19) G is a meromorphic immersion, $\Phi = P \circ G$ is a meromorphic immersion. $\qquad\square$

(3.8.27) Corollary. *Let $X \subset M = \mathbf{C}^{n+p}/\Gamma$ and M' be as in (3.8.22). Let $\iota^*: X \to M$ denote the imbedding. If $M' = 0$, then X has $n + 1$ holomorphic 1-forms $\omega_1, \ldots, \omega_{n+1}$ that are linear combinations of $\iota^*(dw^1), \ldots, \iota^*(dw^{n+p})$ such that $n + 1$ holomorphic n-forms*

$$\{\omega_1 \wedge \ldots \wedge \widehat{\omega_j} \wedge \ldots \wedge \omega_{n+1}\}_{1 \leq j \leq n+1}$$

are linearly independent. Without loss of generality, we may assume that the coordinate system $w^1, \ldots, w^{n+p}$ for $\mathbf{C}^{n+p}$ is such that

$$\omega_1 = \iota^*(dw^1), \quad \ldots, \quad \omega_{n+1} = \iota^*(dw^{n+1}).$$

This fact (in the torus case) is exactly what Ochiai [2] needed to complete his proof of Bloch conjecture; and it was proved by Kawamata [1].

Proof. The meromorphic mapping $\Phi: X \to P_r\mathbf{C}$ defined in (3.8.24) is a meromorphic immersion by (3.8.26). By taking a suitable projection $p: P_r\mathbf{C} \to P_n\mathbf{C}$ (which is a meromorphic map) we obtain an equidimensional meromorphic immersion

$$p \circ \Phi: X \to P_n\mathbf{C}.$$

Now, our assertion is obvious from the definition of the map Φ. $\qquad\square$

From (3.8.26) we recover the following algebraic geometric result (see Ueno [1; p. 74]). For the concept of Kodaira dimension and that of general type, (for details, see Section 4 of Chapter 7).

(3.8.28) Corollary. *Let T be a complex torus, and X a closed complex subspace. Let T' be the largest connected subgroup of T leaving X invariant. Then X/T' is of general type, i.e., the Kodaira dimension $\kappa(X/T')$ is equal to $\dim X/T'$.*

Proof. On account of (3.8.22), replacing T by T/T' and X by X/T' we may assume that $T' = 0$. By (3.8.26), Φ is a meromorphic immersion. Since Φ is defined using only sections of K_X, X is of general type. $\qquad\square$

(3.8.29) Remarks. (1) Our proof shows that in order to prove $\kappa(X/T') = \dim X/T'$ it is sufficient to consider $K_{X/T'}$; there is no need to take higher tensor powers of $K_{X/T'}$.

(2) In (3.8.28) we have

$$\kappa(X) = \kappa(X/T') = \dim X/T'.$$

The second equality is nothing but (3.8.28). By (3.8.16), $\Phi = P \circ G$ has rank equal to $\dim X/T'$ at $x \in U$. Hence, $\kappa(X) \geq \dim X/T'$. The opposite inequality follows from Iitaka's theorem (Iitaka [1], see also Ueno [1; p. 74]).

(3) From the fact that X/T' is of general type, hence Moishezon, it follows that X/T' is projective algebraic and the smallest complex subtorus of T/T' containing X/T' is an Abelian variety, (see Ueno [1; p. 120]). Moreover, if T is an Abelian variety, then there exist finite unramified coverings $\tilde{T}$ and $\tilde{T}''$ of T and $T'' = T/T'$, respectively, such that $\tilde{T} \cong T' \times \tilde{T}''$, and accordingly a finite unramified covering $\tilde{X}$ of X splits as a direct product $\tilde{T}' \times \tilde{X}/\tilde{T}'$. If a Hodge metric is used for T, then this splitting is compatible with the holonomy decomposition of de Rham explained above.

(4) If X is a closed complex subspace of a quasi torus M, we can derive from (3.8.25) a statement similar to (3.8.28) on the logarithmic Kodaira dimension of X/M'.

As an application of (3.8.25) we have a simple proof of the following theorem (see Ueno [1; p. 117]).

(3.8.30) Corollary. *A closed complex subspace X of a complex torus T is a translate of a complex subtorus if and only if its geometric genus is 1, i.e., $\dim \Gamma(K_X) = 1$.*

Proof. If the geometric genus of X is 1, then the map Φ defined by (3.8.24) is a constant map. Then by (3.8.25) the Gauss map is also constant, and $N_x = T_x X$ at all $x \in U$, showing that X is a translate of T'. The converse is trivial. $\qquad\square$

(3.8.31) Remarks. Again, the logarithmic version of (3.8.30) can be similarly derived if X is closed complex subspace of a quasi torus M.

For closed nonsingular complex submanifolds X of a complex torus T, the fibering $X \to X/T'$ has been studied by Matsushima-Stoll [1], Matsushima [1], Howard-Matsushima [1], and Smyth [1] as well as by Nagano-Smyth [1].

For structures of complex submanifolds of complex parallelizable manifolds, see Kodama-Sakane [1] and Huckleberry-Winkelmann [1].

9 Theorem of Bloch-Ochiai

Using results of the preceding section we shall prove the theorem of Bloch and Ochiai for holomorphic curves in an abelian variety. What we need from Nevanlinna theory is summarized in Appendix B of this chapter.

A quasi torus M, which is an extension of a complex torus T by $(\mathbf{C}^*)^k$, is called a **quasi-abelian variety** or a **semi-abelian variety** if T is an abelian variety. Consider M as a bundle over T with fiber $(\mathbf{C}^*)^k$. Compactify M by compactifying each fiber to either $(P_1\mathbf{C})^k$ or $P_k\mathbf{C}$. Then the compactification of M is projective if M is quasi-abelian.

The following theorem corresponds to Theorem A in Ochiai [2].

(3.9.1) **Theorem.** *Let X be an n-dimensional closed algebraic subspace of a quasi-abelian variety $M = \mathbf{C}^{n+p}/\Gamma$ with imbedding $\iota\colon X \to M$ and with natural coordinate system $w^1, \ldots, w^{n+p}$ for $\mathbf{C}^{n+p}$. Assume that the largest connected subgroup M' of M leaving X invariant is trivial. Let*

$$\omega_1 = \iota^*(dw^1), \ \ldots, \ \omega_{n+1} = \iota^*(dw^{n+1})$$

be holomorphic 1-forms obtained in (3.8.27). Let $f\colon D_R \to X \subset \mathbf{C}^{n+p}/\Gamma$ be a holomorphic map with a lift $\tilde{f}\colon D_R \to \mathbf{C}^{n+p}$ given by

$$\tilde{f}(t) = (f_1(t), \ldots, f_{n+p}(t)).$$

Assume that $f(D_R)$ is not contained in the singular locus of X and that f is nondegenerate with respect to $\omega_1, \ldots, \omega_{n+1}$ in the sense that $f(D_R)$ does not lie in a divisor of the form

$$\sum_{j=1}^{n+1} a_j \omega_1 \wedge \ldots \wedge \widehat{\omega}_j \wedge \ldots \wedge \omega_{n+1} = 0, \qquad (a_1, \ldots, a_{n+1}) \neq (0, \ldots, 0).$$

Let φ be a rational function on X such that $\varphi \circ f$ is defined as a meromorphic function on D_R. Then $\varphi \circ f$ is algebraic over the field

$$K = \mathbf{C}(f_1', \ldots, f_{n+1}', f_1'', \ldots, f_{n+1}'', \ldots, f_1^{(n)}, \ldots, f_{n+1}^{(n)})$$

generated by meromorphic functions

$$f_1', \ldots, f_{n+1}', f_1'', \ldots, f_{n+1}'', \ldots, f_1^{(n)}, \ldots, f_{n+1}^{(n)}.$$

Proof. Only after (3.9.14) we shall use the assumption that M is quasi-abelian and X is a closed algebraic subspace of M. Until then, M will be just a quasi-torus and X a closed complex subspace of M.

We consider the k-jet bundle $J^k M$ over a quasi torus $M = \mathbf{C}^{n+p}/\Gamma$. Using the natural coordinate system $(w^1, \ldots, w^{n+p})$ for $\mathbf{C}^{n+p}$ (or more exactly, using the 1-forms $dw^1, \ldots, dw^{n+p}$), we identify $J^k M$ with the product bundle $M \times (\mathbf{C}^{n+p})^k$, (for jet bundles, see Section 5 of Chapter 2). Let

$$p^{(k)} \colon J^k M \to (\mathbf{C}^{n+p})^k$$

be the natural projection. The natural coordinate system in $(\mathbf{C}^{n+p})^k$ is denoted by

$$(3.9.2) \qquad \begin{pmatrix} W^{1(1)} & \cdots & W^{1(k)} \\ \vdots & \cdots & \vdots \\ W^{n+p(1)} & \cdots & W^{n+p(k)} \end{pmatrix};$$

thus, if $j_0^k f \in J^k M$ with f given by $w^i = w^i(t)$, then the coordinates for $p^{(k)}(j_0^k f)$ are given by

$$W^{i(\alpha)}(p^{(k)}(j_0^k f)) = \left(\frac{d^\alpha w^i}{dt^\alpha} \right)_{t=0}, \qquad 1 \le i \le n+p, \ 1 \le \alpha \le k;$$

thus the second superscript (α) in $W^{i(\alpha)}$ indicates the number of times differentiation is taken.

Let X be an n-dimensional closed complex subspace of M with imbedding map $\iota \colon X \to M$. Let $J^k X$ be the k-jet bundle of X; since X is imbedded in a nonsingular manifold M, it is defined as a complex subspace of $J^k M$. When X is singular, $J^k X$ is only a fibre space over X. The imbedding $\iota \colon X \to M$ induces an imbedding

$$\iota^{(k)} \colon J^k X \to J^k M;$$

Let

$$q^{(k)} = p^{(k)} \circ \iota^{(k)} \colon J^k X \to (\mathbf{C}^{n+p})^k.$$

In a Zariski neighborhood U of some regular point of X, holomorphic 1-forms $\omega_1, \ldots, \omega_n$ are linearly independent. For otherwise, $\omega_1 \wedge \ldots \wedge \omega_n$ would vanish identically on X, contradicting the fact that $n+1$ holomorphic n-forms

$$(*) \qquad \{\omega_1 \wedge \ldots \wedge \widehat{\omega_j} \wedge \ldots \wedge \omega_{n+1}\}_{1 \le j \le n+1}$$

are linearly independent, see (3.8.27). (In the following we shall replace the Zariski open set U by a smaller Zariski open set whenever necessary without explicitly saying so).

Since $(\iota^* w^1, \ldots, \iota^* w^n)$ can be taken as a local coordinate system in U, we set $z^i = \iota^* w^i$, $i = 1, \ldots, n$. The imbedding $\iota \colon U \to M$ is given by

$$(3.9.3) \qquad w^i = \begin{cases} z^i & i = 1, \ldots, n, \\ F^i(z^1, \ldots, z^n) & i = n+1, \ldots, n+p. \end{cases}$$

Using the coordinate system $(z^1, \ldots, z^n)$ (or more exactly, using the 1-forms $dz^1, \ldots, dz^n$), we identify $J^k U$ with $U \times (\mathbf{C}^n)^k$. We denote the coordinate system in $(\mathbf{C}^n)^k$ by

$$
(3.9.4) \qquad
\begin{pmatrix}
Z^{1(1)} & \cdots & Z^{1(k)} \\
\vdots & \cdots & \vdots \\
Z^{n(1)} & \cdots & Z^{n(k)}
\end{pmatrix}.
$$

Then $q^{(k)}: U \times (\mathbf{C}^n)^k \to (\mathbf{C}^{n+p})^k$ is given by

$$
(3.9.5) \quad q^{(k)}:
\begin{pmatrix}
z^1 & Z^{1(1)} & \cdots & Z^{1(k)} \\
\vdots & \vdots & \cdots & \vdots \\
z^n & Z^{n(1)} & \cdots & Z^{n(k)}
\end{pmatrix}
\mapsto
\begin{pmatrix}
Z^{1(1)} & \cdots & Z^{1(k)} \\
\vdots & \cdots & \vdots \\
Z^{n(1)} & \cdots & Z^{n(k)} \\
Q^{n+1(1)} & \cdots & Q^{n+1(k)} \\
\vdots & \cdots & \vdots \\
Q^{n+p(1)} & \cdots & Q^{n+p(k)}
\end{pmatrix},
$$

where $Q^{i(\alpha)}$ is a polynomial in $Z^{j(\beta)}$ ($1 \le j \le n$, $1 \le \beta \le \alpha$), whose coefficients are partial derivatives of F^i of order $\le \alpha$ in $z^1, \ldots, z^n$. In order to find $Q^{i(\alpha)}$ explicitly, we have only to calculate derivatives

$$
\frac{dw^i}{dt} = \sum \frac{\partial w^i}{\partial z^j} \frac{dz^j}{dt},
$$

$$
\frac{d^2 w^i}{dt^2} = \sum \frac{\partial w^i}{\partial z^j} \frac{d^2 z^j}{dt^2}, + \sum \frac{\partial^2 w^i}{\partial z^j \partial z^k} \frac{dz^j}{dt} \frac{dz^k}{dt}
$$

$$
\cdots\cdots\cdots\cdots\cdots\cdots\cdots\cdots\cdots,
$$

and then to replace $d^\alpha w^i / dt^\alpha$ and $d^\beta z^j / dt^\beta$ by $Q^{i(\alpha)}$ and $Z^{j(\beta)}$, respectively. Thus

(3.9.6) **Lemma.** *If we set $F^i_j = \partial F^i / \partial z^j$, $F^i_{jk} = \partial^2 F^i / \partial z^j \partial z^k$, etc., then we can write $Q^{i(\alpha)}$ as follows:*

$$
Q^{i(1)} = \sum F^i_j Z^{j(1)},
$$

$$
Q^{i(2)} = \sum F^i_j Z^{j(2)} + \sum F^i_{jk} Z^{j(1)} Z^{k(1)}
$$

$$
\cdots\cdots\cdots\cdots\cdots\cdots\cdots\cdots
$$

Given a holomorphic map $f: D_R \to X$, its k-jet $j^k f: D_R \to J^k X$, sending t to $j^k_t f$, is given locally by

$$
(3.9.7) \qquad Z^{i(\alpha)} = \frac{d^\alpha z^i}{dt^\alpha}, \qquad i = 1, \ldots, n, \ \alpha = 1, \ldots, k.
$$

The map $q^{(k)} \circ j^k f: D_r \to (\mathbf{C}^{n+p})^k$ is given by

$$
(3.9.8) \quad W^{i(\alpha)} =
\begin{cases}
d^\alpha z^i / dt^\alpha & i = 1, \ldots, n, \\
d^\alpha F^i (z^1(t), \ldots, z^n(t)) / dt^\alpha & i = n+1, \ldots, n+p.
\end{cases}
$$

We shall now ignore $w^{n+2}, \ldots, w^{n+p}$, and consider only $F(z^1, \ldots, z^n) = F^{n+1}(z^1, \ldots, z^n)$ and $Q^{(\alpha)} = Q^{n+1(\alpha)}$ in (3.9.3) and (3.9.5). Thus

$$(3.9.9) \qquad w^i = \begin{cases} z^i & i = 1, \dots, n, \\ F(z^1, \dots, z^n) & i = n+1, \end{cases}$$

and

$$(3.9.10) \qquad \omega_i = \begin{cases} dz^i & i = 1, \dots, n, \\ \sum_j F_j dz^j & i = n+1, \end{cases}$$

where $F_j = \partial F / \partial z^j$.

In stead of $q^{(k)}$ in (3.9.5), we consider $q \colon U \times (\mathbf{C}^n)^n \to (\mathbf{C}^{n+1})^n$ given by

$$(3.9.11) \qquad q \colon \begin{pmatrix} z^1 & Z^{1(1)} & \cdots & Z^{1(n)} \\ \vdots & \vdots & \cdots & \vdots \\ z^n & Z^{n(1)} & \cdots & Z^{n(n)} \end{pmatrix} \mapsto \begin{pmatrix} Z^{1(1)} & \cdots & Z^{1(n)} \\ \vdots & \cdots & \vdots \\ Z^{n(1)} & \cdots & Z^{n(n)} \\ Q^{(1)} & \cdots & Q^{(n)} \end{pmatrix},$$

and now (3.9.6) reads as follows:

$$\begin{aligned} Q^{(1)} &= \sum F_i Z^{i(1)} \\ Q^{(2)} &= \sum F_i Z^{i(2)} + \sum F_{ij} Z^{i(1)} Z^{j(1)} \end{aligned}$$

$$\cdots\cdots\cdots\cdots\cdots\cdots$$

Let $f \in \mathrm{Hol}(D_R, X)$. We calculate the Jacobian $\det(dq)$ of the equidimensional map q defined by (3.9.11), and we evaluate it at a point $j_t^n f \in J^n U$. Clearly,

$$\det(dq) = \pm \det(\partial Q^{(\alpha)} / \partial z^k).$$

Evaluating $\partial Q^{(\alpha)} / \partial z^k$ at $j_t^n f$, and writing $z(t)$ for $(z^1(t), \dots, z^n(t))$, we have

$$(3.9.12) \qquad \left(\frac{\partial Q^{(\alpha)}}{\partial z^k} \right)_{j_t^n f} = \frac{d^\alpha F_k(z(t))}{dt^\alpha}, \qquad k = 1, \dots, n.$$

To see this, we consider for example the case $\alpha = 2$.

$$\begin{aligned} \left(\frac{\partial Q^{(2)}}{\partial z^k} \right)_{j_t^n f} &= \sum F_{ik} \frac{d^2 z^i}{dt^2} + \sum F_{ijk} \frac{dz^i}{dt} \frac{dz^j}{dt} \\ &= \sum F_{ki} \frac{d^2 z^i}{dt^2} + \sum F_{kij} \frac{dz^i}{dt} \frac{dz^j}{dt} \\ &= \frac{d^2 F_k}{dt^2}. \end{aligned}$$

Hence, $\det(dq)$ at $j_t^n(f)$ is given by the Wronskian

$$(3.9.13) \qquad \det(dq)_{j_t^n f} = \pm \begin{pmatrix} F_1' & \cdots & F_n' \\ F_1'' & \cdots & F_n'' \\ \vdots & \cdots & \vdots \\ F_1^{(n)} & \cdots & F_n^{(n)} \end{pmatrix},$$

where $F_i^{(\alpha)} = d^\alpha F_i(z(t))/dt^\alpha$. If this Wronskian is identically equal to zero, then there is a nontrivial linear relation

$$c_1 F_1'(z(t)) + \ldots + c_n F_n'(z(t)) \equiv 0,$$

or equivalently,

$$(3.9.14) \qquad c_1 F_1(z(t)) + \ldots + c_n F_n(z(t)) + c_{n+1} \equiv 0.$$

On the other hand, from (3.9.10) we have

$$\sum_{i=1}^{n+1} (-1)^{n-i} c_i \omega_1 \wedge \ldots \wedge \widehat{\omega_i} \wedge \ldots \wedge \omega_{n+1} = \left(\sum_{i=1}^{n} c_i F_i + c_{n+1} \right) \omega_1 \wedge \ldots \wedge \omega_n.$$

Hence, (3.9.14) is equivalent to saying that $\sum (-1)^{n-i} c_i \omega_1 \wedge \ldots \wedge \widehat{\omega_i} \wedge \ldots \wedge \omega_{n+1}$ vanishes at every point of $f(D_R)$. Since we are assuming that f is non-degenerate with respect to $\omega_1, \ldots, \omega_{n+1}$, $\det(q) \neq 0$ at $j_t^n f$ for some $t \in D_R$.

Now, we impose the condition that M is quasi-abelian and X is algebraic. Then $\omega_1, \ldots, \omega_{n+1}$ are regular rational 1-forms on X. From (3.9.10) it follows that

$$F_i = (-1)^{n-i} (\omega_1 \wedge \ldots \wedge \widehat{\omega_i} \wedge \ldots \wedge \omega_{n+1})/(\omega_1 \wedge \ldots \wedge \omega_n), \qquad i = 1, \ldots, n,$$

which shows that $F_1, \ldots, F_n$ are rational functions on X. Moreover, with the notation in the proof of (3.8.27) the meromorphic immersion $p \circ \Phi \colon X \to P_n \mathbf{C}$ is given by $(F_1, \ldots, F_n, 1)$. In particular, $F_1, \ldots, F_n$ form a transcendental basis for the field of rational functions on X. Thus, every rational function φ on X is algebraic over the field $\mathbf{C}(F_1, \ldots, F_n)$.

In a suitable Zariski open set U, the rational functions $F_1, \ldots, F_n$ can be taken as a local coordinate system. (whereas $z^1, \ldots, z^n$ are transcendental functions). Coordinate functions $W^{i(\alpha)}$ given in (3.9.2) are rational functions on $J^k M$ since they are defined by means of rational 1-forms $dw^1, \ldots, dw^{n+p}$ of M. Similarly, coordinate functions $Z^{i(\alpha)}$ given in (3.9.4) are rational functions on $J^k X$ since they are defined by means of rational 1-forms $dz^1, \ldots, dz^n$ of X.

Since $dF_i = \sum F_{ij} dz^j$ and since F_i and dz^j are rational, each F_{ij} is a rational function on X. Similarly, from $dF_{ij} = \sum F_{ijk} dz^k$ we see that each F_{ijk} is a rational function on X, and so on.

Now, we look at the map q given by (3.9.11). We use rational functions $F_1, \ldots, F_n$ instead of $z^1, \ldots, z^n$ as a local coordinate system in U. Since $Q^{(\alpha)}$ in (3.9.11) is a polynomial in rational functions $Z^{i(\beta)}$, F_i, $F_{ij}, \ldots$ with constant coefficients, it follows that q is a rational map. From (3.9.13) and (3.9.14) we see that if the Wronskian (3.9.13) vanishes identically, then $f(D_R)$ would be contained in a proper algebraic subspace of X defined by $\sum c_i F_i = 0$. Thus, $\det(q) \neq 0$ at $j_t^n f$ for some $t \in D_R$. It follows that q is a dominant rational map of a finite degree. Thus, the field

$$\mathbf{C}(F_1, \ldots, F_n, Z^{1(1)}, \ldots, Z^{1(n)}, \ldots, Z^{n(1)}, \ldots, Z^{n(n)})$$

is a finite extension of the field

$$\mathbf{C}(Q^{(1)}, \ldots, Q^{(n)}, Z^{1(1)}, \ldots, Z^{1(n)}, \ldots, Z^{n(1)}, \ldots, Z^{n(n)}).$$

Pull back these fields by f (i.e., by $j^n f$). Since

$$Z^{i(\alpha)}(j_t^n f) = f_i^{(\alpha)}(t)$$

by (3.9.7) and since

$$Q^{(\alpha)}(j_t^n f) = Q^{n+1(\alpha)}(j_t^n f) = f_{n+1}^{(\alpha)}(t)$$

from the definition of $Q^{(\alpha)}$, we see that the field

$$\mathbf{C}(F_1 \circ f, \ldots, F_n \circ f, f_1', \ldots, f_1^{(n)}, \ldots, f_n', \ldots, f_n^{(n)})$$

is a finite extension of the field

$$K = \mathbf{C}(f_1', \ldots, f_1^{(n)}, \ldots, f_n', \ldots, f_n^{(n)}, f_{n+1}^{(1)}, \ldots, f_{n+1}^{(n)}).$$

This shows that $F_1 \circ f, \ldots, F_n \circ f$ are algebraic over K.

Let φ be a rational function on X such that $\varphi \circ f$ is defined as a meromorphic function on D_R. Since φ is algebraic over $\mathbf{C}(F_1, \ldots, F_n)$, $\varphi \circ f$ is algebraic over $\mathbf{C}(F_1 \circ f, \ldots, f_n \circ f)$ and hence algebraic over K. $\square$

Now we state the theorem of Bloch [2] and Ochiai [2] in the form generalized by Noguchi [4]. However, we prove it only in the case M is an abelian variety. For the general case of M quasi-abelian, see Noguchi [4]. Our proof follows Ochiai [2] and Noguchi-Ochiai [1].

(3.9.15) **Theorem**. *Let X be an n-dimensional closed algebraic subspace of a quasi-abelian variety $M = \mathbf{C}^{n+p}/\Gamma$. Assume that it is not a translate of a quasi-abelian subvariety of M. Let regular rational 1-forms $\omega_1, \ldots, \omega_{n+1}$ be as in (3.9.1). Then every holomorphic map $f: \mathbf{C} \to X$ has its image $f(\mathbf{C})$ in a proper closed algebraic subspace of X. More precisely, $f(\mathbf{C})$ lies either in the singular locus of X or in a divisor of the form*

$$\sum_{j=1}^{n+1} a_j \omega_1 \wedge \ldots \wedge \widehat{\omega}_j \wedge \ldots \wedge \omega_{n+1} = 0, \qquad (a_1, \ldots, a_{n+1}) \neq (0, \ldots, 0).$$

Proof. We shall use the notation and results from Nevanlinna theory summarized in Appendix B of this chapter.

Let M' be the largest connected subgroup of M leaving X invariant. Considering $X/M' \subset M/M'$, we assume that $M' = 0$. Assume that $f(\mathbf{C})$ is contained neither in the singular locus of X nor in any divisor of the form above. Then we are in a position to apply (3.9.1) to f.

Take an imbedding $j: M \to P_N \mathbf{C}$, and let $\zeta^0, \ldots \zeta^N$ be a homogeneous coordinate system for $P_N \mathbf{C}$. Let

$$\varphi: \mathbf{C} \xrightarrow{f} X \xrightarrow{\iota} M \xrightarrow{j} P_N \mathbf{C}.$$

Restricted to X, ζ^i/ζ^0, $(i = 1, \ldots, N)$, are rational functions on X. We may assume that $f(\mathbf{C})$ is not contained in the hyperplane $\zeta^0 = 0$, and we define meromorphic functions $\varphi_i \circ f$, $(i = 1, \ldots, N)$, on $\mathbf{C}$ by

$$\varphi_i(t) = \frac{\zeta^i}{\zeta^0}((j \circ \iota \circ f)(t)), \qquad i = 1, \ldots, N.$$

Then φ is given by $(1, \varphi_1, \ldots, \varphi_N)$. By (3.B.21)

$$T(r, \varphi) \leq \sum_{i=1}^{N} T(r, \varphi_i) + O(1).$$

On the other hand, from the fact (see (3.9.1)) that φ_i are algebraic over the field K generated by $f_j^{(\alpha)}$, $\alpha = 1, \ldots, n$; $j = 1, \ldots, n+1$, we obtain

$$(*) \qquad\qquad T(r, \varphi_i) \leq O(\max T(r, f_j^{(\alpha)})),$$

where the maximum is taken over $\alpha = 1, \ldots, n$ and $j = 1, \ldots, n+1$. Postponing the proof of $(*)$ to the end we shall continue the proof. Applying (i) of (3.B.25) to f_j', we have $T(r, f_j^{(\alpha)}) \leq O(T(r, f_j')) + O(\log r)$ outside an exceptional set E. Hence,

$$T(r, \varphi) \leq O(\max_j T(r, f_j')) + O(\log r) \qquad \|.$$

The Kähler form Ψ of the invariant flat metric on $M = \mathbf{C}^{n+p}/\Gamma$ is given by

$$\Psi = \sum_{l=1}^{n+p} dd^c |w^j|^2.$$

We consider the order function of $\iota \circ f : \mathbf{C} \to M$ with respect to Ψ:

$$T_\Psi(r, \iota \circ f) = \int_0^r \frac{d\rho}{\rho} \int_{D_\rho} f^* \Psi.$$

Since $f^* \iota^* \Psi = \sum_{l=1}^{n+p} dd^c |f_j|^2$, (3.B.26) implies

$$m(r, f_j') \leq \log(O(T_\Psi(r, \iota \circ f))) + O(\log r) \qquad \|.$$

Combining all these, we have

$$T(r, \varphi) \leq \log(O(T_\Psi(r, \iota \circ f))) + O(\log r) \qquad \|.$$

The order function $T(r, \varphi)$ was defined by means of the Kähler form Φ of the Fubini-Study metric of $P_N \mathbf{C}$. Now, we make use of compactness of M. Since Ψ and $j^* \Phi$ are comparable, $T(r, \varphi)$ and $T_\Psi(r, \iota \circ f)$ are of the same order. Hence,

$$T(r, \varphi) \leq \log(O(T(r, \varphi))) + O(\log r) \qquad \|.$$

But this contradicts the following fact:

$$(3.9.16) \qquad O(T(r, \varphi)) = O(T_\Psi(r, \iota \circ f)) \geq O(r^2).$$

In order to prove (3.9.16), choose i, $1 \leq i \leq n+1$, such that f_i is nonconstant, and consider its power series expansion:

$$f_i(t) = a_0 + a_1 t + a_2 t^2 + \ldots.$$

Then

$$\begin{aligned} T_\Psi(r, \iota \circ f) &\geq \int_0^r \frac{d\rho}{\rho} \int_{D_\rho} dd^c |f_i|^2 \\ &= \int_0^{2\pi} |f_i(re^{i\theta})|^2 d\theta + C \qquad \text{by (3.B.9)} \\ &= 2\pi(a_0 + a_1 r^2 + a_2 r^4 + \ldots) + C, \end{aligned}$$

establishing (3.9.16). The contradiction comes from the assumption that $f(\mathbf{C})$ is not contained in any divisor of the form stated in the theorem.

In order to complete the proof we shall now prove $(*)$. The proof for the following lemma of Valiron [1] is taken from Noguchi-Ochiai [1, p. 222].

(3.9.17). **Lemma.** *If $a_0, \ldots, a_k$ are entire functions with $a_0 \neq 0$ and if φ is a meromorphic function on $\mathbf{C}$ satisfying the polynomial equation*

$$a_0(t)\varphi(t)^k + a_1(t)\varphi(t)^{k-1} + \ldots + a_{k-1}(t)\varphi(t) + a_k(t) \equiv 0,$$

then

$$T(r, \varphi) \leq \sum_{j=1}^k T(r, a_j) + \text{constant}.$$

Proof. Take $t \in \mathbf{C}$ such that $a_0(t) \neq 0$, fixing it temporarily. Setting $A_i = a_i(t)$, we define a polynomial in w by

$$P(w) = A_0 w^k + A_1 w^{k-1} + \ldots + A_k.$$

Let $\beta_1, \beta_2, \ldots, \beta_k$ be the roots of the equation $P(w) = 0$:

$$P(w) = A_0(w - \beta_1)(w - \beta_2) \ldots (w - \beta_k).$$

Since $\varphi(t)$ is one of the roots, we assume that $\beta_1 = \varphi(t)$. We have

$$\frac{1}{2\pi} \int_0^{2\pi} \log|P(e^{i\theta})| d\theta = \log|A_0| + \sum_{j=1}^k \frac{1}{2\pi} \int_0^{2\pi} \log|e^{i\theta} - \beta_j| d\theta.$$

Making use of the following formula (which follows from Cauchy's formula):

$$\frac{1}{2\pi} \int_0^{2\pi} \log|e^{i\theta} - \beta| d\theta = \log^+ |\beta|,$$

we obtain

$$\frac{1}{2\pi} \int_0^{2\pi} \log |P(e^{i\theta})| d\theta = \log |A_0| + \sum_{j=1}^{k} \log^+ |\beta_j|$$

$$\geq \log |A_0| + \log \sqrt{1 + |\beta_1|^2} - \log 2.$$

On the other hand, since

$$\log |P(e^{i\theta})| = \log \left| \sum_{j=0}^{k} A_j e^{i(k-j)\theta} \right| \leq \sum_{j=0}^{k} \log^+ |A_j| + \log(k+1),$$

we have

$$\frac{1}{2\pi} \int_0^{2\pi} \log |P(e^{i\theta})| d\theta \leq \sum_{j=0}^{k} \log^+ |A_j| + \log(k+1).$$

Hence,

$$\log |A_0| + \log^+ |\beta_1| \leq \sum_{j=0}^{k} \log^+ |A_j| + \log 2 + \log(k+1).$$

Substituting $A_j = a_j(t)$ and $\beta_1 = \varphi(t)$ back into this, we obtain

$$\log |a_0(t)| + \log^+ |\varphi(t)| \leq \sum_{j=0}^{k} \log \sqrt{1 + |a_j(t)|^2} + \text{const.}$$

Integrating this inequality over the circle D_r yields

$$\frac{1}{2\pi} \int_0^{2\pi} \log |a_0(re^{i\theta})| d\theta + m(r, \varphi) \leq \sum_{j=0}^{k} T(r, a_j) + \text{const.}$$

But (3.B.13) applied to the map $f(t) = (a_0(t), 1) = (1, 1/a_0(t)): \mathbf{C} \to P_1\mathbf{C}$ gives

$$\frac{1}{2\pi} \int_0^{2\pi} \log |a_0(re^{i\theta})| d\theta = N(r, a_0, 0) + C.$$

In order for φ to satisfy the given polynomial equation, every pole of φ must be a zero of a_0 of at least the same order. Thus,

$$N(r, \varphi) \leq N(r, a_0, 0).$$

Hence,

$$N(r, \varphi) + m(r, \varphi) \leq \sum_{j=0}^{k} T(r, a_j) + \text{const.}$$

$\square$

(3.9.18). **Corollary**. *If each a_i is a polynomial of entire functions $\psi_1, \ldots, \psi_m$ and if φ is a meromorphic function on $\mathbf{C}$ satisfying the polynomial equation in (3.9.17), then*

$$T(r, \varphi) \le C \cdot \max_j T(r, \psi_j).$$

Proof. For an entire function ψ_j, we have $N(r, \psi_j) = 0$ since $N(r, \psi_j)$ in our notation counts poles of ψ_j. Hence, $T(r, \psi_j) = m(r, \psi_j)+$ constant by the first main theorem. By (3.B.22) and (3.B.23) we have

$$T(r, a_j) \le c_1 T(r, \psi_1) + \ldots + c_m T(r, \psi_m) + c_0,$$

where $c_0, c_1, \ldots, c_m$ are positive constants. Substitute this into the inequality in (3.9.17). $\qquad\square$

We can reformulate (3.9.15) as follows:

(3.9.19) Theorem. *Let f be a holomorphic map of $\mathbf{C}$ into a quasi-abelian variety M. Then the Zariski closure of $f(\mathbf{C})$ is a translate of a quasi-abelian subvariety of M.*

The original theorem of Bloch-Ochiai is stated as follows:

(3.9.20) Theorem. *Let X be an n-dimensional projective algebraic manifold with irregulalrity, i.e., $\dim H^0(X, \Omega^1)$, greater than n. Then every holomorphic map $f: \mathbf{C} \to X$ has its image in a propoer closed algebraic subset of X.*

Proof. To derive this from (3.9.15), let $\alpha: X \to A_X$ be the Albanese map. Now apply (3.9.15) to its image $\alpha(X) \subset A_X$. $\qquad\square$

(3.9.21) Corollary. *A nonsingular algebraic surface with irregularity > 2 is hyperbolic if and only if it has no curves of genus 0 or 1.*

It is desirable to strengthen the conclusion of (3.9.15) as follows:
Conjecture. *Let X and $\omega_1, \ldots, \omega_{n+1}$ be as in (3.9.15). Then X is hyperbolic modulo its singularity locus and a divisor of the form*

$$\sum_{j=1}^{n+1} a_j \omega_1 \wedge \ldots \wedge \widehat{\omega_j} \wedge \ldots \wedge \omega_{n+1} = 0, \qquad (a_1, \ldots, a_{n+1}) \ne (0, \ldots, 0).$$

(3.9.22) Remarks. R. Kobayashi [1] gave a proof of (3.9.15) by establishing the second main theorem for holomorphic curves in abelian varieties. A proof of (3.9.19), which is more arithmetic in spirit, was given by McQuillan [1].

For a smooth algebraic surface X of irregularity 2, a theorem similar to (3.9.20) has been obtained by Grant [1] under the assumption that the Albanese variety of X is simple.

10 Projective Spaces with Hyperplanes Deleted

E. Borel [1] observed that the little Picard theorem may be restated in the following form.

If two holomorphic functions f and g on $\mathbf{C}$ vanish nowhere and satisfy the identity

$$(3.10.1) \qquad\qquad f + g \equiv 1,$$

then they are constant.

In fact, they omit two values 0 and 1 and hence must be constant by the little Picard theorem. Conversely, if f is an entire function omitting two values 0 and 1, then f and $g = 1 - f$ satisfy the identity above.

Now we state Borel's generalization of the little Picard theorem in the following three equivalent forms.

(3.10.2) Theorem. (1) *Assume that entire functions $g_0, g_1, \ldots, g_N$ vanish nowhere on $\mathbf{C}$ and satisfy the identity*

$$g_0 + g_1 + \ldots + g_N \equiv 0.$$

Partition the index set $I = \{0, 1, \ldots, N\}$ into subsets I_α, $I = \bigcup_{\alpha=0}^{p} I_\alpha$, putting two indices i and j in the same subset I_α if and only if g_i/g_j is constant. Then, for each α we have

$$\sum_{i \in I_\alpha} g_i \equiv 0;$$

(2) *Assume that entire functions $f_1, \ldots, f_N$ vanish nowhere on $\mathbf{C}$ and satisfy the identity*

$$f_1 + \ldots + f_N \equiv 1.$$

Then at least one of the f_i's is constant;

(3) *Under the same assumption as in (2), $f_1, \ldots, f_N$ are linearly dependent (over $\mathbf{C}$).*

We shall first show that these three statements are equivalent and then give applications of the theorem. The theorem will be proved in Appendix B of this chapter. For a more direct and shorter proof, see for example Noguchi-Ochiai [1].

Proof of equivalence. In order to derive (1) from (2), we set

$$h_\alpha = \sum_{i \in I_\alpha} g_i$$

so that $h_0 + \ldots + h_p \equiv 0$. Since each h_α is a constant multiple of a function g_i with $i \in I_\alpha$, it follows that either $h_\alpha \equiv 0$ or h_α vanishes nowehere. Assume that (1) does not hold. Without loss of generality, we may assume that $h_0, \ldots, h_m$, $m \geq 1$, vanish nowhere and $h_{m+1} = \ldots = h_p \equiv 0$. Set $f_\alpha = -h_\alpha/h_0$, $1 \leq \alpha \leq m$, so that $f_1 \ldots + f_m \equiv 1$. Then one of the f_α's, i.e., say $f_1 = -h_1/h_0$, is constant by (2). Since h_0 is a constant multiple of a function g_i with $i \in I_0$ and, similarly, h_1 is a constant multiple of a function g_j with $j \in I_1$. This means that g_j/g_i is constant, contradicting the definition of I_α.

We shall derive (2) from (3). Since $f_1, \ldots, f_N$ are linearly dependent, without loss of generality we may assume the following linear relation:

$$c_1 f_1 + \ldots + c_{N-1} f_{N-1} + f_N \equiv 0.$$

By subtracting this identity from

$$f_1 + \ldots + f_{N-1} + f_N \equiv 1,$$

we obtain

$$(1 - c_1) f_1 + \ldots + (1 - c_{N-1}) f_{N-1} \equiv 1.$$

By applying (3) and the same argument to this identity, we obtain a shorter identity. Finally, we end up with the identity $c f_1 \equiv 1$.

Finally, we derive (3) from (1). We set $f_0 = -1$ so that $f_0 + f_1 \ldots + f_N \equiv 0$, and apply (1) to this identity. Let I_0 be the index set that contains 0. If $I = I_0$, then the functions $f_1, \ldots f_N$ are all constant and hence linearly dependent. If $I_0 \neq I$, then $\sum_{i \in I_\alpha} f_i \equiv 0$ for every α such that $I_\alpha \neq I_0$, thus yielding a nontrivial linear relation. $\qquad\square$

Now we shall give various applications of Borel's theorem. In treating $n + 2$ hyperplanes $H_0, H_1, \ldots, H_{n+1}$ in general position in P_n, it is most convenient to consider $P_n \mathbf{C}$ as a hyperplane in $P_{n+1} \mathbf{C}$ given by

$$(3.10.3) \qquad\qquad w^0 + w^1 + \ldots + w^{n+1} = 0$$

in terms of the homogeneous coordinate system $w^0, w^0 \ldots, w^{n+1}$ of $P_{n+1} \mathbf{C}$. By a linear change of coordinates, the defining equations for $H_0, H_1, \ldots, H_{n+1}$ may be reduced to the following simple forms:

$$(3.10.4) \qquad H_0 = \{w^0 = 0\}, \quad \ldots \quad, \quad H_n = \{w^n = 0\}, \quad H_{n+1} = \{w^{n+1} = 0\}.$$

This gives an equal status to all $n + 2$ hyperplanes.

We partition the index set $I = \{0, 1, \ldots, n + 1\}$ into two disjoint subsets $J = \{j_0, j_1, \ldots, j_p\}$ and $K = \{k_0, k_1, \ldots, k_q\}$, where $p + q = n$. The intersection $L_J = H_{j_0} \cap \ldots \cap H_{j_p}$ of $p + 1$ hyperplanes defines a linear subspace of dimension $q - 1$ in $P_n \mathbf{C}$, and the intersection $L_K = H_{k_0} \cap \ldots \cap H_{k_q}$ of the remaining $q + 1$ hyperplanes defines a linear subspace of dimension $p - 1$ in $P_n \mathbf{C}$. Then L_J and L_K span a unique hyperplane H_{JK} of $P_n \mathbf{C}$. If we set

$$(3.10.5) \qquad\qquad F_{JK} = \sum_{j \in J} w^j = -\sum_{k \in K} w^k,$$

then the defining equation for H_{JK} is given by

$$F_{JK} = 0.$$

If J contains only one index, say j, then $L_J = H_j$ and L_K is empty. So we consider nontrivial partitions $I = J \cup K$ such that both J and K contain at least two indices and call H_{JK} a **diagonal hyperplane** of the configuration $H_0, H_1, \ldots, H_{n+1}$. For a quadrangle, (i.e., four lines in general positions), H_0, H_1, H_2, H_3 in the projective plane $P_2\mathbf{C}$, there are three diagonal lines, (see Figure i in Section 3).

Every holomorphic map $\tilde{f}: \mathbf{C} \to \mathbf{C}^{n+2} - \{0\}$, composed with the projection $\pi: \mathbf{C}^{n+2} - \{0\} \to P_{n+1}\mathbf{C}$ induces a holomorphic map $\pi \circ \tilde{f}: \mathbf{C} \to P_{n+1}\mathbf{C}$. Conversely, every holomorphic map $f: \mathbf{C} \to P_{n+1}\mathbf{C}$ is thus obtained. In fact, using an open cover $\{U_\alpha\}$ of $\mathbf{C}$, we lift f locally to a holomorphic map $\tilde{f}_\alpha: U_\alpha \to \mathbf{C}^{n+2}-\{0\}$. Then $\tilde{f}_\alpha = h_{\alpha\beta}\tilde{f}_\beta$ on $U_\alpha \cap U_\beta$, where $h_{\alpha\beta}: U_\alpha \cap U_\beta \to \mathbf{C}^*$ is holomorphic. Then $\{h_{\alpha\beta}\}$ defines an element of $H^1(\mathbf{C}, \mathcal{O}^*) = 0$ so that $h_{\alpha\beta} = \lambda_\alpha^{-1}\lambda_\beta$ with $\lambda_\alpha: U_\alpha \to \mathbf{C}^*$ holomorphic. Set $\tilde{f} = \tilde{f}_\alpha\lambda_\alpha = \tilde{f}_\beta\lambda_\beta$. We sometimes denote a lift $\tilde{f}$ of f by the same symbol f.

Let

$$f: \mathbf{C} \to P_n\mathbf{C} - \bigcup_{i=0}^{n+1} H_i \subset P_{n+1}\mathbf{C}$$

be a holomorphic map and $\tilde{f}: \mathbf{C} \to \mathbf{C}^{n+2} - \{0\}$ a lift of f. Then $\tilde{f}$ is given by $n + 2$ entire functions $(f^0(z), f^1(z), \ldots, f^{n+1}(z))$ satisfying

$$f^0 + f^1 + \ldots + f^{n+1} \equiv 0.$$

Since f avoids the $n + 2$ hyperplanes given by (3.10.4), $f^0, f^1, \ldots, f^{n+1}$ vanish nowhere.

Partition the index set $I = \{0, 1, \ldots, n, n + 1\}$ as in (1) of (3.10.2). If all f^i/f^j are constant, then f is a constant map. Assume that f is not a constant map. Then the partition $I = \cup_{\alpha=0}^p I_\alpha$ is nontrivial. If I_α contains only one index, say i, then the identity $f^i = \sum_{i \in I_\alpha} f^i \equiv 0$ obtained in (3.10.2) contradicts the assumption that f misses the hyperplane H_i. Hence, each I_α contains at least two indices. Set $J = I_0$ and $K = \bigcup_{\alpha \neq 0} I_\alpha$. Then

$$\sum_{j \in J} f^j \equiv -\sum_{k \in K} f^k \equiv 0,$$

and the diagonal hyperplane H_{JK} defined by $\sum_{j \in J} w^j = 0$ contains the image $f(\mathbf{C})$. Thus, from the theorem of E. Borel we have derived the following theorem of A. Bloch [1] and H. Cartan [1].

(3.10.6) **Theorem** *Let $H_0, H_1, \ldots, H_{n+1}$ be $n + 2$ hyperplanes in general position in $P_n\mathbf{C}, n \geq 2$. If $f: \mathbf{C} \to P_n\mathbf{C} - \bigcup_\alpha H_\alpha$ is a nonconstant holomorphic map, then its image lies in one of the diagonal hyperplanes.*

The following theorem which strengthens (3.10.6) is due to Dufresnoy [1, Théorème XVI], see also Fujimoto [4, 5], Green [1], Lang [3]. For generalizations to mappings from $\mathbf{C}^k$, $k > 1$, see Fujimoto [4, 5].

(3.10.7) **Theorem**. *If a holomorphic map $f: \mathbf{C} \to P_n\mathbf{C}$ has its image in the complement of $n + p$ hyperplanes $H_1, \ldots, H_{n+p}$ in general position, then this image is contained in a linear subspace of dimension $\leq [n/p]$.*

Proof. Let $F_i: \mathbf{C}^{n+1} \to \mathbf{C}$ be a linear form defining H_i. Let $\tilde{f}: \mathbf{C} \to \mathbf{C}^{n+1} - \{0\}$ be a lift of f. Set $h_i = F_i \circ \tilde{f}$. Since f misses H_i, the entire function h_i vanishes nowhere in $\mathbf{C}$. We partition the index set $I = \{1, 2, \ldots, n + p\}$ into a disjoint union $I = \bigcup_{\alpha=1}^{q} I_\alpha$ by putting two indices i and j into the same I_α if and only if h_i / h_j is constant.

We claim that the complement of any I_{α_0}, i.e., $\bigcup_{\beta \neq \alpha_0} I_\beta$, contains at most n elements. If our claim is false, we would obtain a set J with $n + 2$ indices by taking $n + 1$ of them from the complement of I_{α_0} and one from I_{α_0}. Then we partition J in the same way, i.e., $J = \bigcup J_\alpha$ with $J_\alpha = J \cap I_\alpha$, (dropping those J_α that are empty). As we have seen in the proof of (3.10.6) above, each (nonempty) J_α must contain at least two elements. But we know from the construction of J that J_{α_0} contains exactly one index. This contradiction proves our claim that the complement of each I_{α_0} contains at most n indices.

Hence, each I_α, $\alpha = 1, \ldots, q$, contains at least p elements so that $pq \leq n + p$.

Let I' be any subset of $I = \{1, \ldots, n + p\}$ consisting of exactly $n + 1$ elements. Since $H_1, \ldots, H_{n+p}$ are in general position, the linear forms F_i, $i \in I'$, are linearly independent. Write $I' = \bigcup_\alpha I'_\alpha$, where $I'_\alpha = I' \cap I_\alpha$. (Some I'_α may be empty). Let k_α be the cardinality of I'_α. Each I'_α, if nonempty, gives rise to a $k_\alpha - 1$ linearly independent equations:

$$F_i - c_i F_{i_0} = 0, \quad (i_0, i \in I'_\alpha, \ i \neq i_0),$$

where $c_i = h_i / h_{i_0}$. Hence, $h_1, \ldots, h_{n+p}$ satisfy at least $k_1 - 1 + \ldots + k_q - 1$ linearly independent equations. But

$$k_1 - 1 + \ldots + k_q - 1 = n + 1 - q \geq n + 1 - \frac{n + p}{p} = n - \frac{n}{p}.$$

Hence, $f(\mathbf{C})$ lies in a linear subspace of dimension $\leq n/p$. $\qquad\qquad\square$

In W. W. Chen [1], (3.10.7) is derived from Borel's theorem by a matrix method.

(3.10.8) **Corollary**. *If a holomorphic map $f: \mathbf{C} \to P_n\mathbf{C}$ misses $2n + 1$ or more hyperplanes in general position, then it is a constant map.*

Using (3.10.8) and (3.6.13) we can strengthen (3.10.8) as follows.

(3.10.9) **Corollary**. *The complement of $2n + 1$ or more hyperplanes in general position in $P_n\mathbf{C}$ is complete hyperbolic and hyperbolically imbedded in $P_n\mathbf{C}$.*

Proof. It suffices to consider the case of $2n + 1$ hyperplanes in general position. Let $H_1, \ldots, H_{2n+1}$ be hyperplanes in general position in $P_n\mathbf{C}$, and let $X = P_n\mathbf{C} - \bigcup_{i=1}^{2n+1} H_i$. By (3.10.8), Condition (a) of (3.6.13) is satisfied. Consider a partition of indices $\{1, 2, \ldots, 2n + 1\} = I \cup J$, and let $L_I = \bigcap_{i \in I} H_i$. If I contains k

elements, L_I is an $n - k$ dimensional linear subspace. The intersections $H_j \cap L_I$, $(j \in J)$, are $2n + 1 - k$ hyperplanes in general position in $L_I \cong P_{n-k}\mathbf{C}$. Since $2n+1-k > 2(n-k)+1$, by (3.10.8) all holomorphic maps of $\mathbf{C}$ into $L_I - \bigcup_{j \in J} H_j$ are constant, which shows that Condition (b) of (3.6.13) is also satisfied. $\square$

(3.10.10) **Remark.** Dufresnoy [1] has shown that, for $X = P_n\mathbf{C} - \bigcup_{i=1}^{2n+1} H_i$ as in (3.10.9), $\mathrm{Hol}(D, X)$ is relatively compact in $\mathrm{Hol}(D, P_n\mathbf{C})$. We shall show later that, in general, X is hyperbolically imbedded in Y if and only if $\mathrm{Hol}(D, X)$ is relatively compact in $\mathrm{Hol}(D, Y)$. So we may say that Dufresnoy had a result equivalent to (3.10.9).

In order to consider more general arrangements of hyperplanes, following Zaidenberg [2] we say that a set of hyperplanes $H_1, \dots, H_N$ in $P_n\mathbf{C}$ is in **hyperbolic configuration** or satisfies **condition (h)** for short if each projective line l in $P_n\mathbf{C}$ intersects $\bigcup_i H_i$ in at least three points while it is in **hyperbolic-imbedding configuration** or satisfies **condition (hi)** for short if each projective line l intersects $\bigcup_{H_i \not\supset l} H_i$, (i.e.,the union of those H_i that do not contain l), in at least three points. (These conditions (hi) and (h) are called conditions (a) and (b) in Zaidenberg's paper). Clearly, condition (hi) is stronger than condition (h).

Condition (h) is violated if and only if there is a pair of points $p, q \in P_n\mathbf{C}$ such that each H_i passes through either p or q, but not both. Condition (hi) is violated if and only if there is a pair of points $p, q \in P_n\mathbf{C}$ such that each H_i passes through p or q, possibly both.

The following theorem is due to Snurnitsyn [1].

(3.10.11) **Theorem.** *If hyperplanes $H_1, \dots, H_N$ in $P_n\mathbf{C}$ are in hyperbolic configuration, then $N \geq 2n + 1$.*

We start with the proof of

(3.10.12) **Lemma.** *Let $H_1, \dots, H_{k+1}$, $(2 \leq k \leq n)$, be hyperplanes in $P_n\mathbf{C}$ such that* $\dim \bigcap_{i=1}^{k} H_i = n - k$ *and* $\bigcap_{i=1}^{k} H_i \subset H_{k+1}$. *Then there is an index m, $1 \leq m \leq k$, such that* $\bigcap_{i=1, i \neq m}^{k} H_i \not\subset H_{k+1}$.

Proof. Set $E = \bigcap_{i=1}^{k} H_i$ and $E_j = \bigcap_{i=1, i \neq j}^{k} H_i$. Assume that $E_j \subset H_{k+1}$ for all $j = 1, \dots, k$. Then $H_{k+1} \cap (\bigcap_{i=1}^{l} H_i)$ contains $E_{l+1}, \dots, E_k$, $(1 \leq l \leq k - 1)$. The first condition in Lemma implies $E_{l+1} \not\subset H_{l+1}$ for all $l = 0, \dots, k - 1$, and hence $H_{k+1} \cap (\bigcap_{i=1}^{l} H_i) \not\subset H_{l+1}$. So

$$H_{l+1} \cap H_{k+1} \cap \left(\bigcap_{i=1}^{l} H_i\right) \neq H_{k+1} \cap \left(\bigcap_{i=1}^{l} H_i\right)$$

and

$$\dim\left(H_{k+1} \cap \left(\bigcap_{i=1}^{l+1} H_i\right)\right) = \dim\left(H_{k+1} \cap \left(\bigcap_{i=1}^{l} H_i\right)\right) - 1, \qquad l = 0, \dots, k - 1.$$

Hence,

$$
\begin{aligned}
n - 1 \;&=\; \dim H_{k+1} = \dim(H_{k+1} \cap H_1) + 1 \\
&=\; \dim\Big(H_{k+1} \cap \big(\bigcap_{i=1}^{2} H_i\big)\Big) + 2 = \ldots \\
&=\; \dim\Big(H_{k+1} \cap \big(\bigcap_{i=1}^{k} H_i\big)\Big) + k = \dim(H_{k+1} \cap E) + k \\
&=\; \dim E + k = n - k + k = n.
\end{aligned}
$$

This contradiction proves Lemma.

Proof of (3.10.11). Assuming that $N \leq 2n$ we shall show that $H_1, \ldots, H_N$ are not in hyperbolic configuration, that is, we shall produce a pair of points $p, q \in P_n\mathbf{C}$ such that each H_i passes through exactly one of these two points. Obviously, it suffices to consider the case $N = 2n$.

If $\bigcap H_i$ is nonempty, it suffices to take p in $\bigcap H_i$ and q in $P_n\mathbf{C} - \bigcap H_i$.

Hence we shall assume that $\bigcap H_i$ is empty. We renumber $H_1, \ldots, H_{2n}$ in such a way that

$$
\dim \bigcap_{i=1}^{k} H_i = n - k \qquad \text{for} \quad k = 1, \ldots, n.
$$

We set $Q_0 = \bigcap_{i=1}^{n} H_i$; Q_0 is a point. Renumbering the remaining hyperplanes $H_{n+1}, \ldots, H_{2n}$, we assume that

$$
Q_0 \in \bigcap_{i=1}^{r_0} H_i \quad \text{and} \quad Q_0 \notin \bigcup_{j=r_0+1}^{2n} H_j, \qquad (n \leq r_0 < 2n).
$$

We set $P_0 = \bigcap_{j=r_0+1}^{2n} H_j$. Clearly, $Q_0 \notin P_0$ and $\dim P_0 \geq r_0 - n \geq 0$. If $P_0 \not\subset \bigcup_{i=1}^{r_0} H_i$, then it suffices to take $p \in P_0$ and $q = Q_0$.

Hence assume that $P_0 \subset \bigcup_{i=1}^{r_0} H_i$. Then $P_0 \subset H_{j_0}$ for some j_0, $1 \leq j_0 \leq r_0$. We consider two cases:

(a) $1 \leq j_0 \leq n$,
(b) $n + 1 \leq j_0 \leq r_0$.

In case (a) we renumber $H_1, \ldots, H_n$ so that $j_0 = n$, i.e., $P_0 \subset H_n$. We set $Q_1 = \bigcap_{i=1}^{n-1} H_i$; $\dim Q_1 = 1$. Then we renumber $H_n, \ldots, H_{2n}$ (including H_n) so that $Q_1 \subset \bigcap_{i=1}^{r_1} H_i$, $(r_1 \geq n-1)$, and $Q_1 \not\subset \bigcup_{i=r_1+1}^{2n} H_i$, (after this renumbering H_n is now one of $H_{r_1+1}, \ldots, H_{2n}$). We set $P_1 = \bigcap_{i=r_1+1}^{2n} H_i$. The subspace P_0 is, by definition, the intersection of the $2n - r_0$ hyperplanes (called $H_{r_0+1}, \ldots, H_{2n}$ earlier) that do not contain Q_0. Since $Q_0 \subset Q_1 \subset H_i$ for $i = 1, \ldots, r_1$, these $2n - r_0$ hyperplanes are now among $2n - r_1$ hyperplanes $H_{r_1+1}, \ldots, H_{2n}$. In addition, P_0 was contained also in another hyperplane (called H_n before the last renumbering) in this group. Hence, P_0 is given as an intersection of $2n - r_0 + 1$ hyperplanes among $H_{r_1+1}, \ldots, H_{2n}$. Since P_1 can be written as the intersection of P_0 with hyperplanes among $H_{r_1+1}, \ldots, H_{2n}$ that do not contain P_0, we have

$$\begin{aligned}
\dim P_1 &\geq \dim P_0 - [(2n - r_1) - (2n - r_0 + 1)] \\
&\geq r_0 - n - r_0 + r_1 + 1 = r_1 - (n - 1) \geq 0.
\end{aligned}$$

Hence, $P_1 \neq \emptyset$.

In case (b) we construct P_1 and Q_1 as follows. Since $Q_0 = \bigcap_{i=1}^{n} H_i = \bigcap_{i=1}^{r_0} H_i \subset H_{j_0}$ and since $\dim Q_0 = 0$, Lemma implies that there is an index m, $1 \leq m \leq n$, such that $\bigcap_{i=1, i \neq m}^{n} H_i \not\subset H_{j_0}$. After renumbering of $H_1, \ldots, H_n$ we may assume that $m = n$. We set $Q_1 = \bigcap_{i=1}^{n-1} H_i$ and renumber $H_n, \ldots, H_{2n}$ so that $Q_1 = \bigcap_{i=1}^{r_1} H_i$ and $Q_1 \not\subset \bigcup_{i=r_1+1}^{2n} H_i$. (After this renumbering, H_n is now among $H_{r_1+1}, \ldots, H_{2n}$). Set $P_1 = \bigcap_{i=r_1+1}^{2n} H_i$. Since the hyperplane which had index j_0 before the last renumbering does not contain Q_1, it is among $H_{r_1+1}, \ldots, H_{2n}$. By its very definition, H_{j_0} contains P_0. Hence, as in case (a) we obtain

$$\dim P_1 \geq r_1 - (n - 1) \geq 0.$$

If $P_1 \not\subset \bigcup_{i=1}^{r_1} H_i$, then it suffices to take $p \in P_1 - \bigcup_{i=1}^{r_1} H_i$ and $q \in Q_1 - \bigcup_{i=r_1+1}^{2n} H_i$.

Hence, we assume that $P_1 \subset \bigcup_{i=1}^{r_1} H_i$. Then $P_1 \subset H_{j_1}$ for some j_1, $1 \leq j_1 \leq r_1$. Proceeding as above, we construct subspaces P_2 and Q_2, etc. If this process continues to the k-th step, then we obtain numbers $r_k \leq r_{k-1} \leq \ldots \leq r_0$ and subspaces P_k and Q_k such that

(1) $Q_k = \bigcap_{i=1}^{n-k} H_i$, $\dim H_k = k$;

(2) $Q_k = \bigcap_{i=1}^{r_k} H_i$, $Q_k \not\subset \bigcup_{i=r_k+1}^{2n} H_i$;

(3) $P_k = \bigcap_{i=r_k+1}^{2n} H_i$, $\dim P_k \geq r_k - (n - k) \geq 0$.

If $P_k \not\subset \bigcup_{i=1}^{r_k} H_i$, then it suffices to take $p \in P_k - \bigcup_{i=1}^{r_k} H_i$ and $q \in Q_k - \bigcup_{i=r_k+1}^{2n} H_i$.

If not, the process continues. If it continues to the $(n - 1)$-st step, we have

$$Q_{n-1} = H_1; \qquad P_{n-1} = \bigcap_{i=2}^{2n} H_i, \qquad \dim P_{n-1} \geq 0.$$

(Since $\dim Q_{n-1} = n - 1$, Q_1 cannot be contained in any other H_i than H_1 and $r_{n-1} = 2$.) Since by assumption $\bigcap_{i=1}^{2n} H_i$ is empty, we have $P_{n-1} \not\subset H_1$, and it suffices to take $p \in P_{n-1} - H_1$ and $q \in Q_{n-1} - \bigcup_{i=2}^{2n} H_i$. $\square$

Kiernan [1] proved that *the complement of $2n$ hyperplanes in general position in $P_n \mathbf{C}$ is not hyperbolic* and conjectured that *the complement of $2n$ hyperplanes in any position in $P_n \mathbf{C}$ is never hyperbolic*; he verified the conjecture for $n \leq 5$. His conjecture follows from (3.10.11), Snurnitsyn [1]. More strongly, we have

(3.10.13) **Theorem.** *A set of hyperplanes $H_1, \ldots, H_N$ in $P_n \mathbf{C}$ are in hyperbolic configuration if the complement $X = P_n \mathbf{C} - \bigcap_{i=1}^{N} H_i$ is hyperbolic. In particular, the complement of $2n$ hyperplanes in $P_n \mathbf{C}$ is never hyperbolic.*

Proof. If $H_1, \ldots, H_N$ are not in hyperbolic configuration, there is a pair of points p, q in $P_n\mathbf{C}$ such that each H_i passes through p or q, but not both. Let l be the projective line passing through p and q. Then X would contain $l - \{p, q\} \cong \mathbf{C}^*$ and would not be hyperbolic. $\qquad\square$

The following question seems to be open. *If $H_1, \ldots, H_N$ are in hyperbolic configuration, is X hyperbolic ?*

(3.10.14) **Remark**. We shall give a special configuration of $2n + 1$ hyperplanes in $P_n\mathbf{C}$ such that its complement is complete hyperbolic. Using a homogeneous coordinate system $(w^0, \ldots, w^n)$, define $2n + 1$ hyperplanes by

$$w^0 w^1 \ldots w^n (w^0 - w^1)(w^1 - w^2) \ldots (w^{n-1} - w^n) = 0.$$

We shall show, by induction on n, that the complement X of these hyperplanes is biholomorphic to $(\mathbf{C} - \{0, 1\})^n$. Using $w^{n-1} = 0$ as the hyperplane at infinity, we introduce the inhomogeneous coordinate system $z^0 = w^0/w^{n-1}, \ldots, z^{n-2} = w^{n-2}/w^{n-1}$, $z^n = w^n/w^{n-1}$. Then the equation above is written as

$$z^0 z^1 \ldots z^{n-2} z^n (z^0 - z^1)(z^1 - z^2) \ldots (z^{n-2} - 1)(1 - z^n) = 0.$$

This equation defines $2n$ hyperplanes in the affine space $\mathbf{C}^n$. Separating the variable z^n from the others, we rewrite the above equation as equations defining hyperplanes in $\mathbf{C}^n = \mathbf{C}^{n-1} \times \mathbf{C}$:

$$z^0 z^1 \ldots z^{n-2}(z^0 - z^1)(z^1 - z^2) \ldots (z^{n-2} - 1) = 0 \quad \text{or} \quad z^n(1 - z^n) = 0.$$

The second equation defines two points $\{0, 1\}$ in the last factor $\mathbf{C}$ of $\mathbf{C}^n$. The first equation defines $2(n - 1)$ hyperplanes in the factor $\mathbf{C}^{n-1}$. The complement of these $2(n - 1)$ hyperplanes in $\mathbf{C}^{n-1}$ is biholomorphic to the complement of the following $2n - 1$ hyperplanes in $P_{n-1}\mathbf{C}$:

$$w^0 w^1 \ldots w^{n-1}(w^0 - w^1)(w^1 - w^2) \ldots (w^{n-2} - w^{n-1}) = 0,$$

which, by induction, is biholomorphic to $(\mathbf{C} - \{0, 1\})^{n-1}$.

We shall construct a holomorphic map $f: (D^*)^n \to X$ which does not extend to a map $f: D^n \to P_n\mathbf{C}$. For $(z^1, \ldots, z^n) \in (D^*)^n$ we set

$$w^0 = z^1, \ w^1 = z^1 z^2, \ w^2 = z^1 z^2 z^3, \ldots, \ w^{n-1} = z^1 z^2 \ldots z^n, \ w^n = z^2 z^3 \ldots z^n.$$

This map extends to $D^n - \{0\}$ but not through the origin. As we shall see later in (6.3.9), this implies that X is not hyperbolically imbedded in $P_n\mathbf{C}$. This example generalizes the configuration of Example (3.3.10).

Theorem (3.10.7) and its Corollary (3.10.8) are concerned with hyperplanes in general position and may be regarded as a generalization of the little Picard theorem. For hyperplanes which are not in general position, we have the following result.

(3.10.15) **Theorem.** *Let $H_1, \ldots, H_N$, $(N \geq 3)$, be distinct hyperplanes which are not in general position in $P_n\mathbf{C}$. Then the image of any holomorphic map $f: \mathbf{C} \to P_n\mathbf{C} - \bigcup H_i$ lies in a hyperplane.*

Proof. Let F_i be a linear form on $\mathbf{C}^{n+1}$ which defines H_i. Then there is a nontrivial linear relation:
$$\sum c_i F_i = 0.$$

Without loss of generality, we may assume that $c_1, \ldots, c_k$ are nonzero and $c_{k+1} = \ldots = c_N = 0$. Replacing F_i by $c_i F_i$ for $1 \leq i \leq k$ we may assume that $c_1 = \ldots = c_k = 1$ so that
$$F_1 + \ldots + F_k = 0.$$

We may further assume that this is the shortest linear relation. Since $H_1, \ldots H_k$ are distinct, we have $k \geq 3$. Let $\tilde{f}: \mathbf{C} \to \mathbf{C}^{n+1} - \{0\}$ be a lift of f, and set $h_i = F_i \circ \tilde{f}$, $(i = 1, \ldots, k)$. Then the entire functions $h_1, \ldots, h_k$ vanish nowhere and satisfy the identity
$$h_1 + \ldots + h_k \equiv 0.$$

As in (1) of (3.10.2) we partition $I = \{1, 2, \ldots k\}$ into subsets I_α, $(\alpha = 1, \ldots, p)$ so that $\sum_{i \in I_\alpha} h_i \equiv 0$. Since $h_i \not\equiv 0$, each I_α contains more than one element. If $I_1 = I$, then f is a constant map. Otherwise, by (1) of (3.10.2) we have a linear relation $\sum_{i \in I_1} h_i \equiv 0$, where I_1 is a proper subset of I and contains more than one element. This implies that the image $f(\mathbf{C})$ is contained in the hyperplane defined by $\sum_{i \in I_1} F_i = 0$. $\square$

Following Zaidenberg [2] we strengthen (3.10.8) as follows:

(3.10.16) **Theorem.** *If a set of hyperplanes $H_1, \ldots, H_N$ in $P_n\mathbf{C}$ are in hyperbolic configuration, then every holomorphic map $f: \mathbf{C} \to P_n\mathbf{C} - \bigcup_{i=1}^{N} H_i$ is constant.*

Proof. By (3.10.11), $N \geq 2n + 1$. Applying (3.10.6) (if the given hyperplanes are in general position) or (3.10.15) (if not), we see that the image $f(\mathbf{C})$ lies in a hyperplane, say H. Set $\tilde{H}_i = H \cap H_i$. Then the set of hyperplanes $\tilde{H}_1, \ldots, \tilde{H}_N$ in $H \cong P_{n-1}(\mathbf{C})$ are in hyperbolic configuration. Repeating this process, we find after n steps that $f(\mathbf{C})$ is 0-dimensional. $\square$

If a set of hyperplanes $H_1, \ldots, H_N$ are in hyperbolic configuration, then $N \geq 2n + 1$. Now, we have

(3.10.17) **Theorem.** (1) *If a set of hyperplanes $H_1, \ldots, H_N$, $N \geq 2n + 1$, in $P_n\mathbf{C}$ are in general position, then they are in hyperbolic-imbedding configuration.*

(2) *If a set of hyperplanes $H_1, \ldots, H_{2n+1}$ in $P_n\mathbf{C}$ are in hyperbolic-imbedding configuration, then they are in general position.*

Proof. (1) Assume that $H_1, \ldots, H_N$ are in general position. If condition (hi) is not satisfied, then there is a pair of points p, q such that each H_i passes through at least one of these points. Then at least $n + 1$ of $H_1, \ldots, H_N$ must have p or q in common. This is a contradiction.

(2)   Assume that $H_1, \ldots, H_{2n+1}$ are not in general position. By renumbering these hyperplanes, we may assume that the last $n+1$ hyperplanes $H_{n+1}, \ldots, H_{2n+1}$ have a point, say p, in common. Since $\bigcap_{i=1}^{n} H_i$ is nonempty, it contains a point, say q. Then the pair p, q violates condition (hi). $\square$

(3.10.18) **Theorem.** *Given a set of hyperplanes $H_1, \ldots, H_N$ in $P_n\mathbf{C}$, set $X = P_n\mathbf{C} - \bigcup_{i=1}^{N} H_i$. If the given set of hyperplanes is in hyperbolic-imbedding configuration, then X is complete hyperbolic and is hyperbolically imbedded in $P_n\mathbf{C}$. Conversely, if X is hyperbolically imbedded in $P_n\mathbf{C}$, then the given set of hyperplanes is in hyperbolic-imbedding configuration.*

Proof. Assume that the given set of hyperplanes satisfies condition (hi). We shall apply (3.6.13). By (3.10.16) condition (a) of (3.6.13) is satisfied, i.e., there are no complex lines in X. For each subset $I \subset \{1, \ldots, N\}$, put $P_I = \bigcap_{i \in I} H_i$. Then the set of hyperplanes $\{\tilde{H}_j = H_j \cap P_I\}_{j \notin I}$ in the projective space P_I satisfies condition (hi). By (3.10.16) condition (b) of (3.6.13) is also satisfied, i.e., there are no complex lines in $\bigcap_{i \in I} H_i - \bigcup_{j \notin I} H_j$. (If $N = 2n + 1$, this follows also from (3.10.17) and (3.10.9).)

Assume that the given set of hyperplanes does not satisfy condition (hi). Then there is a pair of points p, q such that each H_i passes through at least one of these points. Let I be the set of i such that both p and q are in H_i. Let l be the projective line passing through p and q. Then $l - \{p, q\} \subset \bigcap_{i \in I} H_i - \bigcup_{j \notin I} H_j$. (If $I = \emptyset$, then by $\bigcap_{i \in I} H_i$ we mean $P_n\mathbf{C}$.) Then the map exp: $\mathbf{C} \to \mathbf{C}^* \cong l - \{p, q\}$ is a complex line in $\bigcap_{i \in I} H_i - \bigcup_{j \notin I} H_j$ (with respect to the Fubini-Study metric of $P_n\mathbf{C}$ since $|e^z|^2 dz d\bar{z}/(1+|e^z|^2)^2 \le dz d\bar{z}$). By (3.6.19), X cannot be hyperbolically imbedded in $P_n\mathbf{C}$. $\square$

From (3.10.17) and (3.10.18) we obtain the following converse to (3.10.9) as well as (3.10.9). (The "only if" part is due to Zaidenberg [2]).

(3.10.19) **Corollary.** *The complement of a set of $2n + 1$ hyperplanes in $P_n\mathbf{C}$ is hyperbolically imbedded in $P_n\mathbf{C}$ if and only if they are in general position.*

Given a set of hyperplanes $H_1, \ldots, H_N$ in $P_n\mathbf{C}$, set $X = P_n\mathbf{C} - \bigcup_{i=1}^{N} H_i$. Then there are three cases: (i) condition (hi) is satisfied, or equivalently (by (3.10.18)) X is hyperbolically imbedded in $P_n\mathbf{C}$; (ii) condition (h) is not satisfied, and hence (by (3.10.13)) X is not hyperbolic; and (iii) condition (h) is satsified but not condition (hi). This last case is difficult to understand in general. In case (iii) there is a pair of points p, q such that each H_i passes through at least one of these points, and moreover, some H_i passes through both points.

However, for $n = 2$ case (iii) will be completely analyzed in the following example (see Zaidenberg [4]).

(3.10.20) **Example.** Let $l_1, \ldots, l_N$ be a set of lines in $P_2\mathbf{C}$ satisfying condition (h) but not condition (hi). Then there is a pair of points p, q such that one of the lines, say l_N passes through both points and each of the remaining lines $l_1, \ldots, l_{N-1}$ passes through exactly one of the points. Put $X = P_2\mathbf{C} - \bigcup_{i=1}^{N} l_i$. Suppose that the first j lines pass through p and the next k lines pass through q. Then $N = j+k+1$.

If we consider l_N as the line at infinity and identify $P_2\mathbf{C} - l_N$ with $\mathbf{C}^2$, then the first j lines are mutually parallel in $\mathbf{C}^2$, and similarly, the next k lines are also mutually parallel. Hence,

$$X \cong (\mathbf{C} - \{j \ \text{points}\}) \times (\mathbf{C} - \{k \ \text{points}\}).$$

It follows that X is hyperbolic if and only if $j, k \geq 2$. Together with (3.10.13) and (3.10.17) this answers a question raised by Iitaka [4].

We consider a set of hyperplanes $H_1, \ldots, H_N$ in $P_n\mathbf{C}$ as a point in $(P_n^*\mathbf{C})^N$, where $P_n^*\mathbf{C}$ denotes the dual projective space. The symmetric group S_N acts on $(P_n^*\mathbf{C})^N$ in an obvious manner. In order to define the moduli space $\mathcal{M}(n, N)$ of all (unordered) sets of distinct N hyperplanes in $P_n\mathbf{C}$, we have to first remove all points of $(P_n^*\mathbf{C})^N$ which are fixed by some elements (other than the identity) of S_N and then divide by S_N. This yields a nonsingular complex manifold of dimension nN. We then have to divide it by the natural action of the projective linear group $PGL(n; \mathbf{C})$.

Clearly, the set of points in $\mathcal{M}(n, N)$ not satisfying condition (hi) is closed. Thus, if $X - \bigcup_{i=1}^N H_i$ is hyperbolically imbedded in $P_n\mathbf{C}$, then under a small perturbation of these N hyperplanes, X remains to be hyperbolically imbedded.

On the other hand, the set of points in $\mathcal{M}(n, N)$ not satisfying condition (h) is not closed. In fact, the set of points of $\mathcal{M}(n, N)$ satisfying (h) but not satisfying (hi) is closed.

The truncated defect relation of Cartan (3.B.42), or rather one of its consequences (3.B.46), can be applied to prove the following results of Green [2] on Fermat hypersurfaces.

(3.10.21) **Example.** Let $w^0, w^1, \ldots, w^{n+1}$ be the homogeneous coordinate system for $P_{n+1}\mathbf{C}$, and let

$$F(n, d): \quad (w^0)^d + (w^1)^d + \ldots + (w^{n+1})^d = 0$$

be the **Fermat hypersurface** of degree d in $P_{n+1}\mathbf{C}$. If $d > n(n + 2)$, then every holomorphic map $f: \mathbf{C} \to F(n, d)$ has its image in a hyperplane section. In fact, its image lies in a linear subspace of dimension $\leq [n/2]$.

To see this, let $P_n\mathbf{C}$ be the hyperplane in $P_{n+1}\mathbf{C}$ defined by

$$w^0 + w^1 + \ldots + w^{n+1} = 0.$$

Then under the projection

$$\pi: (w^0, w^1, \ldots, w^{n+1}) \to ((w^0)^d, (w^1)^d, \ldots, (w^{n+1})^d),$$

the Fermat hypersurface $F(n, d)$ is a covering space of $P_n\mathbf{C}$ ramified over the hyperplanes $H_j = \{w^j = 0\}$, $j = 0, \ldots, n+1$. Assume that the image of $\pi \circ f$ does not lie in a lower-dimensional linear subspace. If $\pi \circ f$ does not interesect H_j, the truncated defect $\delta^{[n]}(\pi \circ f, H_j) = 1$. If it intersects H_j, it intersects with multiplicity at least d. By the argument in the proof of (3.B.46), $\delta^{[n]}(\pi \circ f, H_j) \geq 1 - \frac{n}{d}$. Hence, we have

$$(n + 2)\left(1 - \frac{n}{d}\right) \leq n + 1,$$

which would imply $d \leq n(n + 2)$. Hence, $\pi \circ f$ satisfies a linear equation $\sum_{j=0}^{n+1} a_j w^j = 0$ in addition to the linear equation $\sum_{j=0}^{n+1} w^j = 0$, so that $f = (f^0, \ldots, f^{n+1})$ satisfies the two homogeneous equations of degree d:

$$\sum_{j=0}^{n+1}(f^j)^d = 0, \qquad \sum_{j=0}^{n+1} a_j (f^j)^d = 0.$$

In order to prove the second assertion, we may assume that $f^j \equiv 0$ does not hold for any j. (For, if $f^j \equiv 0$ for some j, the problem is reduced to a lower dimensional case.) We claim that f^i/f^j is constant for some pair (i, j) with $i \neq j$. Without loss of generality, we may assume that $a_{n+1} = 1$. Then taking the difference of the two equations above, we have

$$(a_0 - 1)(f^0)^d + \ldots + (a_n - 1)(f^n)^d = 0.$$

Replacing each f^j by $(a_j - 1)^{1/d} f^j$ yields a Fermat hypersurface of lower dimension. Inductively, we obtain an equation of the form $a(f^0)^d + b(f^1)^d = 0$, proving our claim.

We partition the index set $\{0, 1, \ldots, n + 1\}$ into $I_1, \ldots, I_m$ under the equivalence $i \sim j$ if and only if f^i/f^j is constant. From each I_r we pick i_r and set $f^j = b_j f^{i_r}$ for $j \in I_r$. Then

$$\sum_{j \in I_r}(f^j)^d = c_r(f^{i_r})^d, \quad \text{where} \quad c_r = \sum_{j \in I_r} b_j^d,$$

and

$$\sum_{r=1}^{m} c_r(f^{i_r})^d = 0.$$

Unless all the c_r are zero, the equation above defines a Fermat hypersurface of dimension $\leq m - 1$. (Set $g^r = c_r^{1/d} f^{i_r}$ so that $\sum_{r=1}^{m}(g^r)^d = 0$). Then, by the claim above, f^{i_p}/f^{i_q} is constant for some $p \neq q$, which is impossible since $i_p \in I_p$ and $i_q \in I_q$. Hence, all $c_r = 0$. Thus,

$$\sum_{j \in I_r}(f^j)^d = 0.$$

In particular, every I_r contains at least two indices. The image of f lies in the linear subspace given by the family of hyperplanes

$$w^j - b_j w^{i_r} = 0, \quad j \in I_r, \ j \neq i_r, \ r = 1, \ldots, m.$$

(3.10.22) **Example.** We consider the complement of the Fermat hypersurface $F(n - 1, d)$ in $P_n\mathbf{C}$. If $d > n(n + 1)$, then every holomorphic map $f : \mathbf{C} \to P_n\mathbf{C} - F(n - 1, d)$ has its image in a hyperplane. In fact, its image lies in a linear subspace of dimension $\leq [n/2]$.

Let $w^0, \ldots, w^n$ be the homogeneous coordinate system for $P_n\mathbf{C}$. We consider the $n+2$ hyperplanes $H_j = \{w^j = 0\}$, $j = 0, \ldots, n$, and $P_{n-1}\mathbf{C} = \{w^0 + \ldots + w^n = 0\}$. Under the projection

$$\pi : (w^0, \ldots, w^n) \to ((w^0)^d, \ldots, (w^n)^d),$$

$P_n\mathbf{C}$ is a covering space of $P_n\mathbf{C}$ ramified over these hyperplanes. Assuming that the image of $\pi \circ f$ does not lie in a lower-dimensional linear subspace, apply (3.B.42) to the map $\pi \circ f$. Then we have

$$(n+1)\left(1 - \frac{n}{d}\right) + 1 \le n + 1,$$

which implies $d \le n(n+1)$. We omit the remainder of the proof, which is similar to that of (3.10.21), see Green [2] for details.

In order to strengthen (3.10.6), we consider sequences of holomorphic maps from D into $P_n\mathbf{C}$ missing $n+2$ hyperplanes $H_0, H_1, \ldots, H_{n+1}$ in general position. We represent $P_n\mathbf{C}$ by a hyperplane (3.10.3) in $P_{n+1}\mathbf{C}$ and $H_0, H_1, \ldots, H_{n+1}$ by (3.10.4). Following Kiernan-Kobayashi [2], we shall draw a geometric consequence (3.10.27) from the following theorem of Cartan [1; p. 58].

(3.10.23) **Theorem**. *Given an infinite sequence $f_\lambda = (f_\lambda^0, f_\lambda^1, \ldots, f_\lambda^{n+1})$ of systems of $n+2$ nowhere-vanishing holomorphic functions on the unit disc D satisfying the identity*

$$f_\lambda^0 + f_\lambda^1 + \ldots + f_\lambda^{n+1} = 0,$$

there is a subsequence, still denoted by f_λ for which one of the following (a) or (b) holds.

(a)   *The index set $I = \{0, 1, \ldots, n+1\}$ is partitioned into two disjoint subsets J and K, J containing at least two indices and K possibly empty, such that*
    (1)   *for $i, j \in J$, the sequence $\{f_\lambda^i / f_\lambda^j\}_{\lambda = 1, 2, \ldots}$ converges to a nowhere-vanishing holomorphic function;*
    (2)   *for $j \in J$ and $k \in K$, the sequence $\{f_\lambda^k / f_\lambda^j\}_{\lambda = 1, 2, \ldots}$ converges to zero;*
    (3)   *for $j \in J$, the sequence $\{(\sum_{i \in J} f_\lambda^i)/f_\lambda^j\}_{\lambda = 1, 2, \ldots}$ converges to zero.*

(b)   *There are two disjoint subsets I' and I'' of $I = \{0, 1, \ldots, n+1\}$, each containing at least two indices and having partitions $I' = J' \cup K'$ and $I'' = J'' \cup K''$ with Properties (1), (2) and (3) of case (a).*

We note that in case (a) Property (3) is a consequence of (2). However, in case (b), it is independent of (1) and (2).
    Set

$$Z = P_n\mathbf{C}, \qquad Y = P_n\mathbf{C} - \bigcup_{i=0}^{n+1} H_i.$$

If K is empty so that $J = I$ in case (a), then the subsequence $\{f_\lambda\}$ converges to a map in $\mathrm{Hol}(D, Y)$. If K is singleton, say $\{n+1\}$, in case (a), then the subsequence $\{f_\lambda\}$ converges to a map in $\mathrm{Hol}(D, H_{n+1}) \subset \mathrm{Hol}(D, Z)$. In the remaining cases of

(a) and in case (b), we have a subset J of I containing at least two but no more than $n-1$ indices such that, for each $i \in J$, the subsequence $\{(\sum_{j \in J} f_\lambda^j)/f_\lambda^i\}_{\lambda=1,2,\ldots}$ converges to zero. Hence,

(3.10.24) **Corollary.** *Given a sequence $\{f_\lambda\}$ of maps from D into Y as in the theorem above, there is a subsequence, also denoted by $\{f_\lambda\}$, for which one of the following holds:*

(a)   *The subsequence $\{f_\lambda\}$ converges to a map in $\mathrm{Hol}(D, Z)$;*

(b)   *There exist a subset J of $\{0, 1, \ldots, n+1\}$ containing at least two but no more than $n - 1$ indices such that, for each $i \in J$, the subsequence $\{(\sum_{j \in J} f_\lambda^j)/f_\lambda^i\}_{\lambda=1,2,\ldots}$ converges to zero.*

In case (b) of the corollary above, we have the following convergence for the subsequence:

$$(3.10.25) \qquad \frac{\sum_{j \in J} f_\lambda^j}{\sqrt{\sum_{i=0}^{n+1} |f_\lambda^i|^2}} \to 0.$$

(3.10.26) **Corollary.** *Let Y and Z be as above, and let Δ be the union of diagonal hyperplanes. Given a sequence of maps $f_\lambda \in \mathrm{Hol}(D, Y)$, there is a subsequence, also denoted by $\{f_\lambda\}$, for which one of the following holds:*

(a)   *The subsequence converges in $\mathrm{Hol}(D, Z)$;*

(b)   *Given a positive $r < 1$ and a neighborhood U of Δ in Z, there is an integer λ_0 such that $f_\lambda(D_r) \subset U$ for $\lambda \geq \lambda_0$.*

In the terminology of Section 1 of Chapter 5, this simply says that Y is tautly imbedded modulo Δ in Z. As we shall show in (5.1.13), this implies the following geometric theorem.

(3.10.27) **Theorem.** *The complement of $n + 2$ hyperplanes in general position in $P_n\mathbf{C}$ is hyperbolically imbedded in $P_n\mathbf{C}$ modulo the diagonal hyperplanes.*

Cartan [1] developed the idea in an earlier paper of Bloch [1] and strengthened the result of Bloch. (There were also some gaps in Bloch's argument). The main difference between their results is that Bloch had to restrict himself to sequences of holomorphic maps f with fixed $f(0)$ in the complement of $\Delta = \bigcup_{i=0}^{n+1} H_i$ (or at least $f(0)$ staying in a compact subset of $P_n\mathbf{C} - \Delta$) whereas Cartan imposed no such condition. Thus, Bloch's result seems to give hyperbolicity modulo Δ. As we observed in Kiernan-Kobayashi [2], the full strength of (3.10.23) is yet to be geometrically explained. In Lang [3] the proof of Cartan's theorem (3.10.23) is reproduced. Cartan conjectured something stronger than (3.10.23). However, it has been disproved by Erëmenko [2].

For $n = 2$, Cowen [3] obtained by a direct differential geometric method an explicit lower bound for the infinitesimal intrinsic metric F_Y for the complement Y of five lines in general position in $P_2\mathbf{C}$, thus establishing hyperbolicity of Y. A similar result was obtained, independently, by Hall [1], who used a Bloch-Cartan type method.

11 Deformations and Hyperbolicity

In this section we shall prove stability results for hyperbolicity and hyperbolic imbeddedness as applications of (3.6.7) and (3.6.8), see Brody [1], Kalka [1], Wright [1] and Zaidenberg [7].

A **complex fiber space** (X, π, R) consists of complex spaces X, R and a surjective holomorphic map $\pi \colon X \to R$. We set

$$X_r = \pi^{-1}(r), \quad X_U = \pi^{-1}(U) \quad \text{for} \quad r \in R, \quad U \subset R.$$

The following theorem which is immediate from (3.6.8) shows stability of hyperbolicity under deformations.

(3.11.1) Theorem. *Let (X, π, R) be a complex fiber space with compact fibers. If there is a point $r_0 \in R$ such that (every connected component of) the fiber X_{r_0} is hyperbolic, then there is a neighborhood U of r_0 in R such that (each connected component of) X_U is hyperbolic and hyperbolically imbedded in X. In particular, each X_r, $r \in U$ is hyperbolic.*

From (3.2.15) and (3.11.1) we have

(3.11.2) Corollary. *Let (X, π, R) be a complex fiber space with compact hyperbolic fibers. If R is (complete) hyperbolic and each connected component of of X_r is (complete) hyperbolic, then X is (complete) hyperbolic.*

We consider the following example.

(3.11.3) Example. Let $Z = P_1\mathbf{C} \times D$ be the product bundle over the disc D with fibre $P_1\mathbf{C}$. Let $(z : w)$ be the homogeneous coordinate system for $P_1\mathbf{C}$. Taking w/z as an inhomogeneous coordinate for the finite part $\mathbf{C}$ and $(0 : w)$ as the point ∞, we identify $P_1\mathbf{C}$ with $\mathbf{C} \cup \infty$. Consider the following three divisors

$$B_0 = \{(z : 0)\} \times D, \quad B_\infty = \{(0 : w)\} \times D, \quad B_1 = \{((tw : w), t) \in P_1\mathbf{C} \times D\}$$

and a point $z_1 = ((1 : 1), 0)$ in Z. We may consider B_0 as the zero-section and B_∞ as the section at ∞.

Define a subfibre space $Y = Z - (B_0 \cup B_\infty \cup B_1 \cup \{z_1\})$. Each fibre Y_t, $t \in D$, is the complement of 3 points in the fibre $Z_t = P_1\mathbf{C}$. Hence Y_t is complete hyperbolic and hyperbolically imbedded in Z_t.

However, for any neighborhood U of $0 \in D$, Y_U is not hyperbolic. In fact, take any two points $p = ((1 : a), 0)$ and $q = ((1 : b), 0)$ in the fibre Y_0. Consider the sequences $p_n = ((1 : a), 1/n)$ and $q_n = ((1 : b), 1/n)$. Then

$$d_{Y_U}(p, q) \le d_{Y_U}(p, p_n) + d_{Y_U}(q, q_n) + d_{Y_U}(p_n, q_n).$$

Since $d_{Y_U}(p, p_n) \to 0$, $d_{Y_U}(q, q_n) \to 0$, and $d_{Y_U}(p_n, q_n) \le d_{Y_{1/n}}(p_n, q_n) \to 0$ as $n \to \infty$, we have $d_{Y_U}(p, q) = 0$.

We shall now prove an analogue of (3.11.1) for a hyperbolically imbedded spaces, namely, stability under deformations of hyperbolical imbeddedness. In view of the example above, we cannot expect stability for a general hyperbolically imbedded space.

Given a complex fibre space $\pi\colon Z \to R$ and a point $r_0 \in R$, a Cartier divisor B in Z is said to be **transversal** to a fiber Z_{r_0} if the intersection of each component of B with Z_{r_0} is an irreducible Cartier divisor of Z_{r_0}.

(3.11.4) Theorem. *Let (Z, π, R) be a complex fibre space with compact fibers and a Cartier divisor B. Put $Y = Z - B$, and let (Y, π, R) be a subfiber space. Fix a point $r_0 \in R$, and assume that* (i) *the fiber Z_{r_0} is nonsingular,* (ii) *B is transveral to Z_{r_0} and* (iii) *$S = B \cap Z_{r_0}$ is a divisor of Z_{r_0} with no worse than normal crossing singularities. If Y_{r_0} is hyperbolically imbedded in Z_{r_0}, then there is a neighborhood U of r_0 such that Y_U is hyperbolically imbedded in Z. In particular, for each $r \in U$, Y_r is hyperbolically imbedded in Z_r.*

Proof. Let $B = \bigcup_{i=1}^{m} B_i$ be the decomposition into its irreducible components. Set $S_i = B_i \cap Z_{r_0}$ for $i = 1, \ldots, m$. Then $S_i, \ldots, S_m$ are all nonsingular hypersurfaces of Z_{r_0} and have no worse than normal crossing singularities.

Let U_n be a decreasing sequence of neighborhoods of r_0 such that $\bigcap U_n = \{r_0\}$. Then $\{Y_{U_n}\}$ is a sequence of open subsets of Z such that $\bigcap Y_{U_n} = Y_{r_0}$. Assuming that none of Y_{U_n} is hyperbolically imbedded in Z, we apply (3.6.7). Then for each n we obtain a limit complex line $h_n \in \mathrm{Hol}(\mathbf{C}, Z)$ in $Z_{\bar{U}_n}$ coming from Y_{U_n} and such that the sequence $\{h_n\}$ converges to a complex line $h \in \mathrm{Hol}(\mathbf{C}, Z)$. Since there is no nonconstant holomorphic map of $\mathbf{C}$ into U_n (for large n), $\pi \circ h_n$ is constant, say $r_n \in \bar{U}_n$. Thus, $h_n(\mathbf{C})$ lies in the fibre Z_{r_n}. By Hurwitz' theorem (3.6.11), each $h_n(\mathbf{C})$ is either in Y or in $(\bigcap_{i \in I} B_i - \bigcup_{j \in J} B_j)$ for some partition $I \cup J = \{1, \ldots, m\}$. Taking a subsequence, we may assume that all $h_n(\mathbf{C})$ are either in Y or in $\bigcap_{i \in I} B_i - \bigcup_{j \in J} B_j$ for some fixed partition $I \cup J = \{1, \ldots, m\}$. Again by Hurwitz' theorem, the limit complex line $h(\mathbf{C})$ is either in Y or in $\bigcap_{i \in I} B_i - \bigcup_{j \in J} B_j$ for some partition $I \cup J = \{1, \ldots, m\}$. On the other hand, since $\pi(h_n(\mathbf{C})) = r_n \to r_0$, $h(\mathbf{C})$ is in Z_{r_0}. Hence, $h(\mathbf{C})$ is either in Y_{r_0} or in $\bigcap_{i \in I} S_i - \bigcup_{j \in J} S_j$. This contradicts (3.6.18). Hence, for some n, Y_{U_n} is hyperbolically imbedded in Z. $\qquad\square$

We shall now investigate the behavior of the infinitesimal pseudo-metrics F_X and $\hat{F}_X$ under deformations. Let X be a complex manifold of dimension n. We shall first show (Wright [1], Kalka [1], Zaidenberg [3, 4]) that F_X and $\hat{F}_X$ are upper semicontinuous under deformations of the complex structure of X.

Let X be a complex manifold of dimension $m + n$, R a complex manifold of dimension m, and $\pi\colon X \to R$ a holomorphic map of maximal rank m everywhere. Then each fibre X_r is a complex submanifold of dimension n in X. We consider $\{X_r\}$ as a family of complex manifolds parametrized by $r \in R$.

Let $T^v X$ denote the subbundle of the tangent bundle TX consisting of vertical vectors, i.e., vectors which are annihilated by π; it is a vector bundle of rank n over X.

(3.11.5) Theorem. *Let $\pi\colon X \to R$ be a surjective holomorphic map of maximal rank between complex manifolds. Then the families of infinitesimal pseudo-metrics*

$\{F_{X_r}\}_{r \in R}$ and $\{\hat{F}_{X_r}\}_{r \in R}$, where $X_r = \pi^{-1}(r)$, are upper semicontinuous functions on $T^v X$.

Proof. We shall prove only upper semicontinuity of F_X. (From upper semicontinuity of F_X it follows by a general principle that its double dual $\hat{F}_X$ is also upper semicontinuous). We need the following parametrized version of (3.5.28).

(3.11.6) Lemma. *Given $f \in \mathrm{Hol}(D_R, X_{r_0})$ with $R > 1$ such that its differential f_* is nonzero at the origin $0 \in D_R$, there exist a neighborhood U of r_0 and a map $\varphi \in \mathrm{Hol}(D \times D^{n-1} \times U, X)$ which is biholomorphic in a neighborhood of $(0, 0, r_0)$ and which satisfies*

$$\pi(\varphi(z, w, r)) = r \quad \text{and} \quad \varphi(z, 0, r_0) = f(z) \quad \text{for} \quad z \in D, \ w \in D^{n-1}, \ r \in U.$$

With this lemma, the proof of (3.11.5) is similar to that of (3.5.27). As in the proof of (3.5.28) (see (3.A.1)), by considering the graph of f we can reduce the proof of (3.11.6) to the following

(3.11.7) Lemma. *If $f \in \mathrm{Hol}(D_R, X_{r_0})$ is a holomorphic imbedding, then there exist a neighborhood U of r_0 and a holomorphic imbedding $\varphi \in \mathrm{Hol}(D \times D^{n-1} \times U, X)$ such that*

$$\pi(\varphi(z, w, r)) = r \quad \text{and} \quad \varphi(z, 0, r_0) = f(z) \quad \text{for} \quad z \in D, \ w \in D^{n-1}, \ r \in U.$$

Proof. Let $1 < a < R$. By (3.A.5), given a neighborhood V of $f(D_a)$ in X there is a Stein neighborhood $W \subset V$ of $f(D_a)$. By (3.A.2), there is a holomorphic imbedding ψ of D^n into X_{r_0} such that

$$f(z) = \psi(z, 0) \qquad \text{for} \quad (z, 0) \in D \times D^{n-1}.$$

Replacing D^{n-1} by a smaller polydisc and rescaling it, we may assume that $\psi(D^n) \subset W$.

We consider the following exact sequence of vector bundles over X:

$$0 \to T^v X \to TX \to \pi^* TR \to 0.$$

Restricted to a Stein manifold W, this exact sequence splits (see (3.A.3)). Thus, $TX|_W$ has a horizontal subbundle $T^h X|_W \cong (\pi^* TR)|_W$.

Let U be a coordinate neighborhood of r_0 with local coordinate system $u^1, \ldots, u^m$. Let $\xi_1, \ldots, \xi_m$ be the horizontal lifts of the vector fields $\partial/\partial u^1, \ldots, \partial/\partial u^m$; they are holomorphic vector field on X_U belonging to $T^h X$.

For fixed $a^1, \ldots, a^n$, a vector field $\sum a^i \xi_i$ generates a *local* one-parameter group $\exp(t \sum a^i \xi)$ acting on X_U, i.e., the action of $\exp(t \sum a^i \xi_i)$ is defined for $|t| < \varepsilon$. For a compact subset of X_U a uniform ε can be found.

Since $\psi(\overline{D}^n) \subset W$ is compact, it follows that for a sufficiently small neighborhood U of r_0 the action of $\exp(t \sum u^i \xi_i)$ is defined on $\psi(\overline{D}^n)$ for $|t| \leq 1$, and for $(u^1, \ldots, u^m) \in U$. We define a map $\varphi \in \mathrm{Hol}(D^n \times U, X)$ by

$$\varphi(z, w, (u^1, \ldots, u^m)) = (\exp(\sum u^i \xi_i))\psi(z, w) \quad \text{for} \quad (z, w) \in D \times D^{n-1}.$$

Then φ has the required property. $\square$

The proof of the following theorem is the same as that of (3.5.38).

(3.11.8) **Theorem.** *Let* $\pi: X \to R$ *be a surjective holomorphic map of maximal rank between complex manifolds. If* $U \subset R$ *is an open set such that* $X_U = \pi^{-1}(U)$ *is complete hyperbolic, then the families of infinitesimal pseudo-metrics* $\{F_{X_r}\}_{r \in U}$ *and* $\{\hat{F}_{X_r}\}_{r \in U}$ *are continuous on* $T^v X_U$.

From (3.11.1) and (3.11.8) we obtain the following result of Wright [1].

(3.11.9) **Corollary.** *Let* $\pi: X \to R$ *be a proper surjective holomorphic map of maximal rank between complex manifolds. If* X_{r_0} *is hyperbolic, then there is an open neighborhood* U *of* r_0 *such that the families* $\{F_{X_r}\}_{r \in U}$ *and* $\{\hat{F}_{X_r}\}_{r \in U}$ *are continuous on* $T^v X_U$.

See also Kalka [1] and Zaidenberg [2, 3].

As pointed out by Wright [1], once (3.11.9) is established, the proof in Narasimhan-Simha [1] that the coarse moduli space of a compact complex manifold with ample canonical bundle is Hausdorff shows also that the coarse moduli space of a compact hyperbolic manifold is Hausdorff. Thus,

(3.11.10) **Theorem.** *Let* $V_{\mathbf{R}}$ *be a compact real analytic manifold and* M *the set of isomorphism classes of hyperbolic complex structures on* $V_{\mathbf{R}}$. *Then* M *has a natural structure of a complex space such that if* $\pi: X \to R$ *is any holomorphic family of hyperbolic complex structures on* V, *parametrized by a complex space* R, *then the map which sends* $r \in R$ *to the isomorphism class of* $X_r = \pi^{-1}(r)$ *is a holomorphic map from* R *into* M.

Proof. Following Narasimhan-Simha, we shall construct local coordinate charts for M. Since we have the intrinsic distance for a hyperbolic complex space, our case is simpler. However, we shall leave it the reader to check that these coordinate charts patch up to make M a complex space. For this last step, the argument in Narasimhan-Simha [1] is valid with no change at all. First, we recall the existence theorem for the Kuranishi family, Kuranishi [1, 2].

(3.11.11) **Theorem.** *Let* V_0 *be a compact complex manifold with the underlying real manifold* $V_{\mathbf{R}}$. *Then there is a holomorphic family* $p: V \to S$ *of complex structures on* $V_{\mathbf{R}}$ *parametrized by a complex space* S *together with an isomorphism (i.e., a biholomorphic map)* $V_0 \to V_{s_0}$ *for some* $s_0 \in S_0$ *having the following local completeness property:*

Given A holomorphic family $\pi: X \to R$ *of complex structures on* $V_{\mathbf{R}}$ *and an isomorphism* $f_0: X_{r_0} \to V_{s_0}$ *for some* $r_0 \in R$, f_0 *extends to a local isomorphism f of the family X to the family V in the following sense. Namely, there exist a neighborhood* R' *of* r_0 *in* R *and holomorphic maps* $\bar{f}: R' \to S$ *and* $f: X_{R'} \to V$ *such that*

(i) $p \circ f = \bar{f} \circ \pi$ *(f is fiber-preserving),*

(ii)   *$f: X_r \to V_{\bar{f}(r)}$ is an isomorphism for every $r \in R'$,*

(iii)   *$\bar{f}(r_0) = s_0$, and $f: X_{r_0} \to V_{s_0}$ coincides with f_0.*

Moreover, there exists a neighborhood N of the identity element in the group of diffeomorphisms of $V_{\mathbf{R}}$ (in the C^1-topology) such that if $s_1, s_2 \in S$, $\varphi \in N$ and $\varphi: V_{s_1} \to V_{s_2}$ is holomorphic, then $s_1 = s_2$ and $\varphi = \mathrm{id}_{V_{\mathbf{R}}}$.

From the last statement it follows that f satisfying (i), (ii) and (iii) is unique. The family V is called the **Kuranishi family** of V_0.

Now assume that the group $\mathrm{Aut}(V_0)$ is finite, which is the case if V_0 is hyperbolic by (5.4.4). By letting $X = V$ and f_0 an automorphism of V_0 in (3.11.11), we can find a neighborhood S' of s_0 and an automorphism f of $V|_{S'}$ which extends f_0 and induces a biholomorphic transformation $\bar{f}$ of S' fixing s_0. Thus, $\mathrm{Aut}(V_0)$ acts on S' fixing s_0. By shrinking S to S' we assume that $\mathrm{Aut}(V_0)$ acts on S.

We claim that if V_0 is hyperbolic, there is a neighborhood S' of s_0 in S such that for $s, s' \in S'$, the two complex manifolds V_s and $V_{s'}$ are isomorphic if and only if s and s' are in the same orbit of $\mathrm{Aut}(V_0)$. If such a neighborhood S' does not exist, we would have sequences $\{s_\nu\}$ and $\{s'_\nu\}$ in S converging to s_0 and isomorphisms $\varphi_\nu: V_{s_\nu} \to V_{s'_\nu}$ not coming from $\mathrm{Aut}(V_0)$. Let d_{s_ν} and $d_{s'_\nu}$ denote the intrinsic distances for V_{s_ν} and $V_{s'_\nu}$, respectively. Then φ_ν is an isometry with respect to these intrinsic distances. Fixing a local C^∞ product structure $V = S \times V_{\mathbf{R}}$, we consider $\{\varphi_\nu\}$ as a sequence of diffeomorphisms of $V_{\mathbf{R}}$ onto itself. Since the intrinsic distance d_{V_s} depends continuously on s (see (3.11.9)), the distances d_{s_ν} and $d_{s'_\nu}$ are uniformly equivalent to d_{V_0} as distance on V, namely there is a constant $C \geq 1$ such that

$$\frac{1}{C} d_{V_0} \leq d_{s_\nu} \leq C d_{V_0} \quad \text{and} \quad \frac{1}{C} d_{V_0} \leq d_{s'_\nu} \leq C d_{V_0}$$

for all ν. Thus, $\{\varphi_\nu\}$ is an equicontinuous family of diffeomorphisms of $V_{\mathbf{R}}$. Since $V_{\mathbf{R}}$ is compact, by the Ascoli-Arzela theorem, it has a subsequence which converges to a map f_0 from $V_{\mathbf{R}}$ into itself. By considering the sequence $\{\varphi_\nu^{-1}\}$, we see that f_0 is a homeomorphism of $V_{\mathbf{R}}$ onto itself. Since each φ_ν is holomorphic, the limit map f_0 is an automorphism of V_0. By (3.11.11) f_0 extends to a local isomorphism $f: V \to V$ which induces a holomorphic transformation $\bar{f}$ of S fixing s_0 (after S is suitably shrunk). If we put

$$s''_\nu = \bar{f}^{-1}(s'_\nu) \quad \text{and} \quad \psi_\nu = f^{-1} \circ \varphi_\nu: V_{s_\nu} \to V_{s''_\nu},$$

then $s''_\nu \to s_0$ and $\psi_\nu \to \mathrm{id}_{V_0}$. By the last statement in (3.11.11), $\psi_\nu = \mathrm{id}_{V_0}$ for large ν. Hence, $\varphi_\nu = f$ on V_{s_ν} for large ν, showing that, for large ν, φ_ν comes from f_0. This contradiction proves our assertion.

The quotient space $S/\mathrm{Aut}(V_0)$ of the complex space S by the finite group $\mathrm{Aut}(V_0)$ is a complex space. All points of S with isomorphic fibers being identified, $S/\mathrm{Aut}(V_0)$ has no redundancy. Given a point of M represented by V_0, we have a natural injective map $S/\mathrm{Aut}(V_0) \to M$, which will be taken as a local coordinate chart at the isomorphism class of V_0. As I said at the beginning it is left to the reader to verify that these coordinate charts patch up to give the structure of a complex space on M. $\square$

A Royden's Extension Lemma

Throughout this section, let X be a complex manifold of dimension n and D_R a disk of radius $R > 1$ with $D = D_1$.

In (3.A.1) below, we prove Royden's Lemma (3.2.20) which was used in the proof of (3.2.19). As we shall see, (3.A.2) implies (3.A.1), which was used in the proof of (3.5.27). The main step in Royden's proof of (3.A.2) is to show that if D_R is an imbedded disc in X, then $\overline{D}$ has a Stein neighborhood in X. This has been generalized by Seabury [1], Schneider [1], and Siu [3]. Here we follow Seabury [1], but making the presentation more differential geometric in terms of holomorphic affine connections.

(3.A.1) **Royden's Lemma.** *If f is a holomorphic map of D_R, where $R > 1$, into an n-dimensional complex manifold X such that its differential f_* is nonzero at the origin 0, then there is a holomorphic map $\tilde{f}$ of the polydisc $D^n = D \times D^{n-1}$ into X such that $\tilde{f}$ is biholomorphic in some neighborhood of the origin and*

$$f(z) = \tilde{f}(z, 0, \ldots, 0) \quad for \quad z \in D.$$

As shown in Royden [4], the proof can be reduced to the following

(3.A.2) **Proposition.** *If f is a holomorphic imbedding of D_R into a complex manifold X, then there is a holomorphic imbedding $\tilde{f}$ of D^n into X such that*

$$f(z) = \tilde{f}(z, 0, \ldots, 0) \quad for \quad z \in D.$$

To see that (3.A.2) implies (3.A.1), given f in (3.A.1), we consider its graph $g: D_R \to D_R \times X$ defined by

$$g(z) = (z, f(z)),$$

which is an imbedding of D_R into $D_R \times X$. Applying (3.A.2) to g we have an imbedding $\tilde{g}$ of D^{n+1} into $D_R \times X$ such that

$$g(z) = \tilde{g}(z, 0, \ldots, 0) \quad \text{for} \quad z \in D,$$

so that

$$(z, f(z)) = \tilde{g}(z, 0, \ldots, 0) \quad \text{for} \quad z \in D.$$

Let $p: D_R \times X \to X$ be the projection. Then

$$f(z) = p(\tilde{g}(z, 0, \ldots, 0)) \quad \text{for} \quad z \in D,$$

and it suffices to take $\tilde{f} = p \circ \tilde{g}$.

(3.A.3) **Lemma.** *Every exact sequence of holomorphic vector bundles*

$$0 \to E' \to E \to E'' \to 0$$

on a Stein manifold S splits.

Proof. The obstruction to a splitting lies in $H^1(S, \mathrm{Hom}(E'', E'))$, which is zero if S is Stein. $\square$

We need also the following characterization of a Stein manifold, see, for example, Hörmander [1; p. 116].

(3.A.4) Theorem. *A complex manifold S is a Stein manifold if and only if there exists a smooth strictly plurisubharmonic function φ such that $S_c = \{z \in S;\ \varphi(z) \leq c\}$ is compact for every real number c.*

The main part of the proof of (3.A.2) lies in establishing the following theorem in the special case where S is an imbedded disc D.

(3.A.5) Theorem. *Let S be a Stein submanifold of a complex manifold X. Then every compact subset $K \subset S$ has a fundamental system of open Stein neighborhoods.*

Proof. The first step is to construct a local coordinate system adapted to S. Let $m = \dim S$ and $m + r = n = \dim X$. We cover S by polydisc neighborhoods $\{U_\alpha\}$ in X with coordinate system $x_\alpha^1, \ldots, x_\alpha^m, y_\alpha^1, \ldots, y_\alpha^r$ such that $S \cap U_\alpha$ is given by $y_\alpha^1 = \ldots = y_\alpha^r = 0$. For simplicity we write $x_\alpha = (x_\alpha^1, \ldots, x_\alpha^m)$ and $y_\alpha = (y_\alpha^1, \ldots, y_\alpha^r)$. Then $\partial y_\alpha / \partial x_\beta = 0$ on S. Thus, when restricted to S, the Jacobian matrix of the coordinate change from (x_α, y_α) to (x_β, y_β) is of the following form:

$$(3.A.6) \qquad J_{\alpha\beta} = \begin{pmatrix} \dfrac{\partial x_\alpha}{\partial x_\beta} & \dfrac{\partial x_\alpha}{\partial y_\beta} \\[2mm] 0 & \dfrac{\partial y_\alpha}{\partial y_\beta} \end{pmatrix}.$$

In order to make $\partial x_\alpha / \partial y_\beta = 0$ on S by a suitable coordinate change, we use (3.A.3). Thus, we have a holomorphic decomposition

$$(3.A.7) \qquad TX|_S \cong TS \oplus N,$$

where N is the normal bundle $(TX|_S)/TS$.

This means that on each $S \cap U_\alpha$ there is a holomorphic $(m \times r)$-matrix $B_\alpha(x_\alpha)$ such that

$$\begin{pmatrix} I_n & -B_\beta(x_\beta) \\ 0 & I_r \end{pmatrix} J_{\alpha\beta} \begin{pmatrix} I_n & B_\alpha(x_\alpha) \\ 0 & I_r \end{pmatrix} = \begin{pmatrix} \dfrac{\partial x_\alpha}{\partial x_\beta} & 0 \\[2mm] 0 & \dfrac{\partial y_\alpha}{\partial y_\beta} \end{pmatrix}.$$

Then with respect to the new local coordinate system $\{(\xi_\alpha, \eta_\alpha)\}$ given by

$$\begin{aligned} \xi_\alpha &= x_\alpha - B_\alpha(x_\alpha) y_\alpha \\ \eta_\alpha &= y_\alpha, \end{aligned}$$

the Jacobian, restricted to S, takes a simpler form:

$$J'_{\alpha\beta} = \begin{pmatrix} \dfrac{\partial \xi_\alpha}{\partial \xi_\beta} & 0 \\ 0 & \dfrac{\partial \eta_\alpha}{\partial \eta_\beta} \end{pmatrix}.$$

We may therefore assume that

$$(3.A.8) \qquad J_{\alpha\beta} = \begin{pmatrix} \dfrac{\partial x_\alpha}{\partial x_\beta} & 0 \\ 0 & \dfrac{\partial y_\alpha}{\partial y_\beta} \end{pmatrix} \qquad \text{on} \quad S \cap U_\alpha \cap U_\beta.$$

The next step is to show that, by a suitable coordinate change, we may assume

$$(3.A.9) \qquad \frac{\partial^2 x_\alpha}{\partial y_\beta \partial y_\beta} = \left(\frac{\partial^2 x_\alpha^i}{\partial y_\beta^j \partial y_\beta^k} \right) = 0 \qquad \text{on} \quad S \cap U_\alpha \cap U_\beta.$$

In order to see the geometric idea behind, we shall explain holomorphic affine connections in cohomological terms. Given a complex manifold X in general, we cover it by a locally finite family of Stein open sets U_α with local coordinate system $(z_\alpha^1, \ldots, z_\alpha^n)$. Set

$$c_{\alpha\beta} = \sum \frac{\partial^2 z_\alpha^i}{\partial z_\beta^j \partial z_\beta^k} dz_\beta^j \otimes dz_\beta^k \otimes \frac{\partial}{\partial z_\alpha^i} \in \Gamma(U_\alpha \cap U_\beta, T^*X \otimes T^*X \otimes TX).$$

Then the chain rule for second partial derivatives means precisely that $\{c_{\alpha\beta}\}$ is a 1-cocyle with coefficients in $T^*X \otimes T^*X \otimes TX$ of the open cover $\{U_\alpha\}$. This 1-cocyle is cohomologous to zero if and only if there exists a 0-cochain $\{b_\alpha\}$,

$$b_\alpha = \sum \gamma_{jk}^i(z_\alpha) dz_\alpha^j \otimes dz_\alpha^k \otimes \frac{\partial}{\partial z_\alpha^i} \in \Gamma(U_\alpha, T^*X \otimes T^*X \otimes TX)$$

such that $c_{\alpha\beta} = b_\beta - b_\alpha$, which is nothing but the classical transformation law for the Christoffel symbols $\gamma_{jk}^i(z_\alpha)$. This shows that the 1-cocyle $\{c_{\alpha\beta}\}$ is cohomologous to zero if and only if there exists a holomorphic affine connection. Hence, if $H^1(X, T^*X \otimes T^*X \otimes TX) = 0$ (in particular, if X is a Stein manifold), then X admits a holomorphic affine connection. In the present situation, X is not Stein, but S is.

Let $\{c_{\alpha\beta}\}$ be as above. Consider the restriction of $T^*X \otimes T^*X \otimes TX$ to S. Then $\{c_{\alpha\beta}|_S\}$ is a 1-cocyle with coefficients in $(T^*X \otimes T^*X \otimes TX)|_S$. Since $H^1(S, (T^*X \otimes T^*X \otimes TX)|_S) = 0$, there exists a 0-cochain $\{b_\alpha\}$ with $b_\alpha \in \Gamma(U_\alpha \cap S, (T^*X \otimes T^*X \otimes TX)|_S)$ such that $c_{\alpha\beta}|_S = b_\beta - b_\alpha$ on $U_\alpha \cap U_\beta \cap S$.

Dualizing the splitting in (3.A.7), we have

$$T^*X|_S = T^*S \oplus N^*.$$

We consider only the $(N^* \otimes N^* \otimes TS)$-component $c'_{\alpha\beta}$ and the $(N^* \otimes N^* \otimes N)$-component $c''_{\alpha\beta}$ of $c_{\alpha\beta}$:

$$c'_{\alpha\beta} \;=\; \sum \frac{\partial^2 x^i_\alpha}{\partial y^j_\beta \partial y^k_\beta} dy^j_\beta \otimes dy^k_\beta \otimes \frac{\partial}{\partial x^i_\alpha}$$

$$c''_{\alpha\beta} \;=\; \sum \frac{\partial^2 y^i_\alpha}{\partial y^j_\beta \partial y^k_\beta} dy^j_\beta \otimes dy^k_\beta \otimes \frac{\partial}{\partial y^i_\alpha}$$

Then there exist

$$b'_\alpha \in \Gamma(U_\alpha \cap S, N^* \otimes N^* \otimes TS), \qquad b''_\alpha \in \Gamma(U_\alpha \cap S, N^* \otimes N^* \otimes N)$$

such that

$$c'_{\alpha\beta} = b'_\beta - b'_\alpha, \qquad c''_{\alpha\beta} = b''_\beta - b''_\alpha.$$

Write

$$b'_\alpha = \sum \gamma'^i_{\alpha jk}(x_\alpha, 0) dy^j_\alpha \otimes dy^k_\alpha \otimes \frac{\partial}{\partial x^a_\alpha}, \qquad b''_\alpha = \sum \gamma''^i_{\alpha jk}(x_\alpha, 0) dy^j_\alpha \otimes dy^k_\alpha \otimes \frac{\partial}{\partial y^i_\alpha}.$$

We make the following change of local coordinate system:

$$\xi^i_\alpha \;=\; x^i_\alpha - \sum \gamma'^i_{\alpha jk}(x_\alpha, 0) y^j_\alpha y^k_\alpha,$$

$$\eta^i_\alpha \;=\; y^i_\alpha - \sum \gamma''^i_{\alpha jk}(x_\alpha, 0) y^j_\alpha y^k_\alpha.$$

Then

$$\frac{\partial^2 \xi^i_\alpha}{\partial \eta^j_\beta \partial \eta^k_\beta} = 0,$$

which verifies (3.A.9).

More generally, for any integer $p > 0$ there is a local coordinate system such that

$$(3.A.10) \qquad \frac{\partial^t x^i_\alpha}{\partial y^{j_1}_\beta \ldots \partial y^{j_t}_\beta} = 0 \qquad \text{on} \quad S \cap U_\alpha \cap U_\beta$$

for all $t \le p$.

We verified (3.A.10) for $p = 1, 2$. We shall indicate the proof for $p = 3$, which is all we need in the proof of (3.A.2). Starting with a local coordinate system $\{U_\alpha, (x^i_\alpha, y^j_\alpha)\}$ which satisfies (3.A.10) for $p = 2$, we set

$$c'_{\alpha\beta} \;=\; \sum \frac{\partial^2 x^i_\alpha}{\partial y^j_\beta \partial y^k_\beta \partial y^l_\beta} dy^j_\beta \otimes dy^k_\beta \otimes dy^l_\beta \otimes \frac{\partial}{\partial x^i_\alpha}$$

$$c''_{\alpha\beta} \;=\; \sum \frac{\partial^2 y^i_\alpha}{\partial y^j_\beta \partial y^k_\beta \partial y^l_\beta} dy^j_\beta \otimes dy^k_\beta \otimes dy^l_\beta \otimes \frac{\partial}{\partial y^i_\alpha}$$

Then there exist

$$b'_\alpha \in \Gamma(U_\alpha \cap S, N^* \otimes N^* \otimes N^* \otimes TS), \quad b''_\alpha \in \Gamma(U_\alpha \cap S, N^* \otimes N^* \otimes N^* \otimes N)$$

such that

$$c'_{\alpha\beta} = b'_\beta - b'_\alpha, \qquad c''_{\alpha\beta} = b''_\beta - b''_\alpha.$$

Write

$$b'_\alpha = \sum \gamma'^i_{\alpha jkl}(x_\alpha, 0) dy^j_\alpha \otimes dy^k_\alpha \otimes dy^l_\alpha \otimes \frac{\partial}{\partial x^i_\alpha},$$

$$b''_\alpha = \sum \gamma''^i_{\alpha jk}(x_\alpha, 0) dy^j_\alpha \otimes dy^k_\alpha \otimes dy^l_\alpha \otimes \frac{\partial}{\partial y^i_\alpha}.$$

The following change of local coordinate system yields a coordinate system with the desired property:

$$\xi^i_\alpha = x^i_\alpha - \sum \gamma'^i_{\alpha jkl}(x_\alpha, 0) y^j_\alpha y^k_\alpha y^l_\alpha,$$

$$\eta^i_\alpha = y^i_\alpha - \sum \gamma''^i_{\alpha jkl}(x_\alpha, 0) y^j_\alpha y^k_\alpha y^l_\alpha.$$

From now on we use only a local coordinate system $\{U_\alpha, (x^i_\alpha, y^j_\alpha)\}$ satisfying the condition of (3.A.10) for $p = 3$.

In order to complete the proof of (3.A.5) we want to construct a strictly plurisubharmonic function on a neighborhood of K in X. Since S is a Stein manifold, there is a strictly plurisubharmonic function φ on S satisfying the condition of (3.A.4). We may assume that $K = \{z \in S;\ \varphi(z) \le c\}$.

Since S is a Stein manifold, there are global sections $\sigma_1, \ldots, \sigma_p$ of N^* which generate all fibers of N^* over K. Each σ_j defines a function on N, which is linear on each fiber of N. Hence, $\sum_{j=1}^p |\sigma_j|^2$ is a plurisubharmonic function on N, strictly plurisubharmonic on each fiber. Let $\pi\colon N \to S$ be the projection. Then the function $\pi^*\varphi$ is plurisubharmonic on N, strictly plurisubharmonic in the horizontal direction. We want to transfer the functions $\pi^*\varphi$ and $\sum_{j=1}^p |\sigma_j|^2$ to a neighborhood of K in X.

The local coordinate (x^i_α, y^j_α) on U_α induces a local coordinate $(\xi^i_\alpha, \eta^j_\alpha)$ in $N|_{U_\alpha \cap S} = \pi^{-1}(U_\alpha \cap S)$. Namely, we consider $\partial/\partial y^1_\alpha, \ldots, \partial/\partial y^r_\alpha$ as a local basis for N. Then for $v \in \pi^{-1}(U_\alpha \cap S)$, we define $(\xi^i_\alpha(v), \eta^j_\alpha(v))$ by

$$\xi^i_\alpha(v) = x^i_\alpha(\pi(v)), \qquad v = \sum_j \eta^j_\alpha(v) \frac{\partial}{\partial y^j_\alpha}.$$

By identifying $(\xi^i_\alpha, \eta^j_\alpha)$ with (x^i_α, y^j_α), we define a map

$$h_\alpha\colon U_\alpha \to \pi^{-1}(U_\alpha \cap S) \subset N,$$

i.e., $x^i_\alpha = \xi^i_\alpha \circ h_\alpha$, $y^j_\alpha = \eta^j_\alpha \circ h_\alpha$.

For each α, we extend $\varphi|_{U_\alpha \cap S}$ to a function φ_α on U_α by setting

$$\varphi_\alpha = \varphi \circ \pi \circ h_\alpha.$$

If the coordinates (x_α, y_α) denote the point of U_α they represent, (in particular, $(x_\alpha, 0)$ denoting a point of S), then

$$\varphi_\alpha(x_\alpha, y_\alpha) = \varphi(x_\alpha, 0).$$

Clearly, $\varphi_\alpha - \varphi_\beta = 0$ on $S \cap U_\alpha \cap U_\beta$. Hence, the following functions all vanish on $S \cap U_\alpha \cap U_\beta$:

$$\frac{\partial \varphi_\alpha}{\partial x_\alpha^i} - \frac{\partial \varphi_\beta}{\partial x_\alpha^i}, \quad \frac{\partial^2 \varphi_\alpha}{\partial x_\alpha^i \partial \bar{x}_\alpha^j} - \frac{\partial^2 \varphi_\beta}{\partial x_\alpha^i \partial \bar{x}_\alpha^j}.$$

By (3.A.10), the following functions also vanish on $S \cap U_\alpha \cap U_\beta$:

$$\frac{\partial \varphi_\alpha}{\partial y_\alpha^i} - \frac{\partial \varphi_\beta}{\partial y_\alpha^i}, \quad \frac{\partial^2 \varphi_\alpha}{\partial y_\alpha^i \partial \bar{y}_\alpha^j} - \frac{\partial^2 \varphi_\beta}{\partial y_\alpha^i \partial \bar{y}_\alpha^j}, \quad \frac{\partial^2 \varphi_\alpha}{\partial x_\alpha^i \partial \bar{y}_\alpha^j} - \frac{\partial^2 \varphi_\beta}{\partial x_\alpha^i \partial \bar{y}_\alpha^j}.$$

Let $\{\rho_\alpha\}$ be a partition of unity subordinate to the open cover $\{U_\alpha\}$ of X. If we set

$$\tilde{\varphi} = \sum_\beta \rho_\beta \varphi_\beta,$$

then $\tilde{\varphi} = \varphi$ on $S \cap (\bigcup U_\alpha)$. It follows that $\partial \bar{\partial} \tilde{\varphi}$ is strictly positive on S in the tangential directions.

Similarly, for each α, the section σ_j of N^* induces a function $\sigma_{j\alpha}$ on U_α by

$$\sigma_{j\alpha} = \sigma_j \circ h_\alpha.$$

Again, in terms of the coordinates (x_α, y_α), write

$$\sigma_j = \sum_{k=1}^{r} s_{jk\alpha}(x_\alpha) dy_\alpha^k.$$

Then

$$\sigma_{j\alpha}(x_\alpha, y_\alpha) = \sum_{k=1}^{r} s_{jk\alpha}(x_\alpha) y_\alpha^k.$$

Define

$$\psi = \sum_\alpha \rho_\alpha \sum_{j=1}^{p} |\sigma_{j\alpha}|^2.$$

Since $\sigma_{j\alpha}$ vanishes on $S \cap U_\alpha$, $\partial \bar{\partial} \psi$ is positive semi-definite at every point of K and is strictly positive in the normal directions.

It follows that the function

$$\Phi = \tilde{\varphi} + \lambda \psi$$

is strictly plurisubharmonic at every point of K if λ is sufficiently large. It is then strictly plurisubharmonic in a neighborhood V of K.

Since $K = \{z \in S; \; \varphi(z) \le c\}$ and ψ vanishes on $S \cap (\bigcup U_\alpha)$, it follows that, for $c' > c$ sufficiently close to c, the set

$$W = \{z \in X; \; \Phi(z) < c'\}$$

is a relatively compact neighborhood of K contained in V. Then $\Phi/(c' - \Phi)$ is strictly plurisubharmonic in W and tends to infinity at the boundary of W. By (3.A.4) W is a Stein manifold, thus completing the proof of (3.A.5).

We are now in a position to complete the proof of (3.A.2). Let $K \subset S \subset X$ be as in (3.A.5), and let W be a Stein neighborhood of K. Let $TX|_S = TS \oplus N$ be a holomorphic decomposition, where N is the normal bundle of S. Let N' be a small neighborhood of the zero section of $N|_{W \cap S}$. As we saw in the course of the proof of (3.A.5) W admits a holomorphic affine connection. Then its exponential map $\exp\colon N' \to W$ is holomorphic. Taking both N' and W sufficiently small, we may assume that $\exp\colon V \to W$ is biholomorphic.

Together with the decomposition $TX|_S = TS \oplus N$, the holomorphic affine connection in W induces a holomorphic connection in the normal bundle $N_{W \cap S}$. Its curvature, being a holomorphic 2-form, must vanish if $\dim S = 1$. Hence, $N_{W \cap S}$ is holomorphically a product (by parallel displacement) if S is simply connected and 1-dimensional.

Applying this to $S = D_R$, we have $N|_D \cong D \times \mathbf{C}^{n-1}$. Then $D^n = D \times D^{n-1}$ may be identified with a neighborhood of the zero section of $N|_D$, and the holomorphic exponential map $D^n \to W$ gives a desired holomorphic imbedding. This completes the proof of (3.A.2).

B Nevanlinna-Cartan Theory

We quickly summarize Nevanlinna-Cartan theory for holomorphic curves in $P_n \mathbf{C}$ following Cartan [5]. Cartan's method gives the first and second main theorems and the defect relation more quickly than that of Weyl and Ahlfors. Assuming only Nevanlinna's lemma on logarithmic derivatives without proof, we prove everything else completely. In Chapter 8, we shall give a systematic exposition of the theory based on the idea of Ahlfors. We point out that although Cartan did not consider associated (or derived) curves as in Ahlfors [2], Fujimoto [11] obtained results for associated curves by extending Cartan's method.

Let $P_n \mathbf{C}$ be the n-dimensional projective space with homogeneous coordinate system $\mathbf{z} = (z^0, z^1, \ldots z^n)$. Each nonzero vector $\mathbf{a} = (a^0, a^1, \ldots, a^n) \in \mathbf{C}^{n+1}$ defines a hyperplane

$$H_{\mathbf{a}} : \langle \mathbf{z}, \mathbf{a} \rangle = \sum \bar{a}^i z^i = 0.$$

Since we are interested in the hyperplane $H_{\mathbf{a}}$ rather than the vector $\mathbf{a}$, we usually assume that $\mathbf{a}$ has unit length. We often call the hyperplane defined by the vector $\mathbf{e}_0 = (1, 0, \ldots, 0)$:

$$H_{\mathbf{e}_0} : \quad z^0 = 0$$

the **hyperplane at infinity**. The inhomogeneous coordinate system

$$z^1/z^0, \ldots, z^n/z^0$$

is valid outside the hyperplane at infinity $H_{\mathbf{e}_0}$.

The Kähler form of the Fubini-Study metric is given by

$$(3.B.1) \qquad \Phi = \frac{1}{2\pi} dd^c \log \|\mathbf{z}\|, \qquad \text{where} \quad \|\mathbf{z}\|^2 = \sum_{i=0}^{n} |z^i|^2.$$

Let $D_R = \{t \in \mathbf{C}; |t| < R\}$, $0 < R \le \infty$, and $f \in \mathrm{Hol}(D_R, P_n\mathbf{C})$. Assume that $f(D_R)$ does not lie in a hyperplane of $P_n\mathbf{C}$. In terms of the homogeneous coordinate system, we express f as follows:

$$f(t) = (f_0(t), \ldots, f_n(t)),$$

where $f_i(t)$, $i = 0, \ldots, n$, are all holomorphic functions and reduced in the sense that at each t some $f_i(t) \ne 0$. In terms of the inhomogeneous coordinate system $z^1/z^0, \ldots, z^n/z^0$, f is given by

$$f(t) = (1, \varphi_1(t), \ldots, \varphi_n(t)), \qquad \text{where} \quad \varphi_i(t) = f_i(t)/f_0(t).$$

Let $t = re^{i\theta}$. The **order function** $T(r, f)$, $0 < r < \infty$, of f is defined by

$$(3.B.2) \qquad T(r, f) = \int_0^r \tau(\rho) \frac{d\rho}{\rho}, \qquad \text{where} \quad \tau(\rho) = \int_{D_\rho} f^*\Phi.$$

The **proximity function** $m(r, f, \mathbf{a})$ of f at the hyperplane $H_\mathbf{a}$ is defined to be

$$(3.B.3) \qquad m(r, f, \mathbf{a}) = \frac{1}{2\pi} \int_0^{2\pi} \log \frac{\|f(re^{i\theta})\|}{|\langle f(re^{i\theta}), \mathbf{a}\rangle|} d\theta.$$

For each $\alpha \in D_R$, let

$$\nu(\alpha, f, \mathbf{a}) = \text{the order of zero of } \langle f(t), \mathbf{a}\rangle \text{ at } \alpha;$$

it is the multiplicity with which f maps α into the hyperplane $H_\mathbf{a}$. Let $r < R$, and set

$$(3.B.4) \qquad n(r, f, \mathbf{a}) = \sum_{\alpha \in \bar{D}_r} \nu(\alpha, f, \mathbf{a}).$$

Thus, $n(r, f, \mathbf{a})$ is the number (with multiplicity counted) of points of $\bar{D}_r$ that are mapped into the hyperplane $H_\mathbf{a}$. Taking the possibility of $\nu(0, f, \mathbf{a})$ being nonzero into account, we define the **counting function** by

$$(3.B.5) \qquad N(r, f, \mathbf{a}) = \int_0^r (n(\rho, f, \mathbf{a}) - \nu(0, f, \mathbf{a})) \frac{d\rho}{\rho} + \nu(0, f, \mathbf{a}) \log r.$$

The integral in the definition of the counting function $N(r, f, \mathbf{a})$ can be written as a finite sum. In fact, let $\alpha_1, \ldots, \alpha_k$ be the zeros of $\langle f(t), \mathbf{a}\rangle$ in $\bar{D}_r - \{0\}$ arranged in the increasing order of their absolute values: $|\alpha_1| \le |\alpha_2| \le \ldots \le |\alpha_k|$. Then

$$n(\rho, f, \mathbf{a}) - \nu(0, f, \mathbf{a}) = \sum_{0 < |\alpha_j| \le \rho} \nu(\alpha_j, f, \mathbf{a}).$$

Integrating this with respect to $d \log \rho$ yields

$$\int_0^r (n(\rho, f, \mathbf{a}) - v(0, f, \mathbf{a})) \frac{d\rho}{\rho} = \sum_{0 < |\alpha_j| \le \rho} \int_0^r v(\alpha_j, f, \mathbf{a}) \frac{d\rho}{\rho},$$

where $v(\alpha_j, f, \mathbf{a})$ on the right hand side should be considered as a step function of ρ which is zero for $|\rho| < |\alpha_j|$ and is equal to $v(\alpha_j, f, \mathbf{a})$ for $|\rho| \ge |\alpha_j|$. Then

$$\int_0^r v(\alpha_j, f, \mathbf{a}) \frac{d\rho}{\rho} = \int_{|\alpha_j|}^r v(\alpha_j, f, \mathbf{a}) \frac{d\rho}{\rho} = v(\alpha_j, f, \mathbf{a}) \log \frac{r}{|\alpha_j|}.$$

Hence,

$$\int_0^r (n(\rho, f, \mathbf{a}) - v(0, f, \mathbf{a})) \frac{d\rho}{\rho} = \sum_{0 < |\alpha_j| \le r} v(\alpha_j, f, \mathbf{a}) \log \frac{r}{|\alpha_j|}.$$

Substituting this into (3.B.5) we obtain

$$(3.B.6) \qquad N(r, f, \mathbf{a}) = \sum_{0 < |\alpha_j| \le r} v(\alpha_j, f, \mathbf{a}) \log \frac{r}{|\alpha_j|} + v(0, f, \mathbf{a}) \log r.$$

We shall later use truncated counting functions. Given an integer $m > 0$, we truncate $v(\alpha, f, \mathbf{a})$ at m by setting

$$v^{[m]}(\alpha, f, \mathbf{a}) = \min\{v(\alpha, f, \mathbf{a}), m\}.$$

We define $n^{[m]}(r, f, \mathbf{a})$ and the **truncated counting function** $N^{[m]}(r, f, \mathbf{a})$ by replacing $v(\alpha, f, \mathbf{a})$ with $v^{[m]}(\alpha, f, \mathbf{a})$ in the definition of $n(r, f, \mathbf{a})$ and $N(r, f, \mathbf{a})$,

For $\mathbf{a} = \mathbf{e}_0$, i.e., for the hyperplane at infinity, we drop $\mathbf{a}$ from $m(r, f, \mathbf{a})$, $n(r, f, \mathbf{a})$ and $N(r, f, \mathbf{a})$ and simply write $m(r, f)$, $n(r, f)$ and $N(r, f)$. Then

$$(3.B.7) \qquad m(r, f) = \frac{1}{2\pi} \int_0^{2\pi} \log \sqrt{1 + \sum |\varphi_i(re^{i\theta})|^2} d\theta,$$

$$(3.B.8) \qquad N(r, f) = \int_0^r (n(\rho, f) - n(0, f)) \frac{d\rho}{\rho} + n(0, f) \log r.$$

For any real valued function $v(t)$, $t = re^{i\theta}$, we have

$$d^c v = \frac{\partial v}{\partial r} r d\theta - \frac{\partial v}{\partial \theta} \frac{dr}{r},$$

and

$$\int_0^r \frac{d\rho}{\rho} \int_{\partial D_\rho} d^c v = \int_0^r d\rho \int_0^{2\pi} \frac{\partial v(\rho e^{i\theta})}{\partial \rho} d\theta = \int_0^{2\pi} (v(re^{i\theta}) - v(0)) d\theta.$$

Hence,

$$(3.B.9) \qquad \frac{1}{2\pi} \int_0^r \frac{d\rho}{\rho} \int_{D_\rho} dd^c v = \frac{1}{2\pi} \int_0^{2\pi} v(re^{i\theta}) d\theta - v(0).$$

We apply (3.B.9) to $v(t) = \log \|f(t)\|$. Since $|f(t)|^2 = \sum_{i=0}^n |f_i(t)|^2 > 0$, from (3.B.1) and (3.B.2) we obtain

$$(3.B.10) \qquad T(r, f) = \frac{1}{2\pi} \int_0^{2\pi} \log \|f(re^{i\theta})\| d\theta - \log \|f(0)\|.$$

Integrating

$$\log \|f(re^{i\theta})\| = \log \sqrt{1 + \sum |\varphi_i(re^{i\theta})|^2} + \log |f_0(re^{i\theta})|$$

with respect to $d\theta$ yields

$$(3.B.11) \qquad T(r, f) + \log \|f(0)\| = m(r, f) + \frac{1}{2\pi} \int_0^{2\pi} \log |f_0(re^{i\theta})| d\theta.$$

In order to calculate the integral on the right, we recall **Jensen's formula**:

(3.B.12) **Theorem.** *If h is a meromorphic function on D_R with Laurent expansion $h(t) = ct^m + \ldots$, $c \neq 0$, at 0, then for $r < R$ we have*

$$\log |c| = \frac{1}{2\pi} \int_0^{2\pi} \log |h(re^{i\theta})| d\theta - \sum m_j \log \frac{r}{|\alpha_j|} + \sum n_k \log \frac{r}{|\beta_k|} - m \log r,$$

where the $\alpha_j \neq 0$'s are zeros of h with multiplicity m_j in D_r while the $\beta_k \neq 0$'s are the poles of h with multiplicity n_k in D_r. In particular, if 0 is not a zero or pole of $h(t)$, then

$$\log |h(0)| = \frac{1}{2\pi} \int_0^{2\pi} \log |h(re^{i\theta})| d\theta - \sum m_j \log \frac{r}{|\alpha_j|} + \sum n_k \log \frac{r}{|\beta_k|}.$$

We apply this formula to the holomorphic function $f_0(t)$. If c is the leading coefficient of the Taylor expansion of $f_0(t) = ct^{\nu(0,f)} + \ldots$ at 0, then

$$\log |c| = \frac{1}{2\pi} \int_0^{2\pi} \log |f_0(re^{i\theta})| d\theta - \sum \nu(\alpha_j, f) \log \frac{r}{|\alpha_j|} - \nu(0, f) \log r.$$

With (3.B.6) this can be rewritten as

$$(3.B.13) \qquad \frac{1}{2\pi} \int_0^{2\pi} \log |f_0(re^{i\theta})| d\theta = N(r, f) + \log |c|.$$

Now, (3.B.11) reads as follows:

$$(3.B.14) \qquad T(r, f) = m(r, f) + N(r, f) + \log |c| - \log \|f(0)\|.$$

Since any hyperplane $H_{\mathbf{a}}$ can be considered as the hyperplane at infinity, we have

$$(3.B.15) \qquad T(r, f) = m(r, f, \mathbf{a}) + N(r, f, \mathbf{a}) + C.$$

This is the **first main theorem** of Nevanlinna theory. In particular, we have

$$(3.B.16) \qquad T(r, f) \geq N(r, f, \mathbf{a}) + C.$$

Suppose that f is defined on all of $\mathbf{C}$. Since $\tau(\rho)$ is monotone increasing, it follows that $T(r) \to \infty$ as $r \to \infty$. We define the **defect** $\delta(f, \mathbf{a})$ and the **truncated defect** $\delta^{[m]}(f, \mathbf{a})$ of $\mathbf{a}$ (or rather $H_\mathbf{a}$) by setting

$$\delta(f, \mathbf{a}) = \liminf_{r \to \infty}\left(1 - \frac{N(r, f, \mathbf{a})}{T(r, f)}\right), \qquad \delta^{[m]}(f, \mathbf{a}) = \liminf_{r \to \infty}\left(1 - \frac{N^{[m]}(r, f, \mathbf{a})}{T(r, f)}\right).$$

Then $0 \leq \delta(f, \mathbf{a}) \leq \delta^{[m]}(f, \mathbf{a}) \leq 1$. We note that $\delta(f, \mathbf{a}) = 1$ if $f(\mathbf{C}) \cap H_\mathbf{a} = \emptyset$.

Consider the case $n = 1$. A meromorphic function φ on D_R can be written as a quotient of two holomorphic functions with no common zeros:

$$\varphi(t) = f_1(t)/f_0(t).$$

Thus, it is considered as a holomorphic map $f \in \mathrm{Hol}(D_R, P_1\mathbf{C})$ with $f(t) = (f_0(t), f_1(t))$. In this case, we often write $T(r, \varphi), m(r, \varphi)$ and $N(r, \varphi)$ for $T(r, f)$, $m(r, f)$ and $N(r, f)$. On the other hand, we write $m(r, \varphi, 0)$ and $N(r, \varphi, 0)$ for $m(r, f, \mathbf{e}_1)$ and $N(r, f, \mathbf{e}_1)$. We note that $N(r, \varphi)$ counts zeros of $f_0(t)$, i.e., poles of φ while $N(r, \varphi, 0)$ counts zeros of $f_1(t)$, i.e., zeros of φ.

As a special case of (3.B.7), we have

$$(3.B.17) \qquad m(r, \varphi) = \frac{1}{2\pi}\int_0^{2\pi} \log\sqrt{1 + |\varphi(re^{i\theta})|^2}d\theta.$$

Using the notation $\log^+ x = \max\{0, \log x\}$, $m(r, \varphi)$ is often given by

$$\frac{1}{2\pi}\int_0^{2\pi} \log^+ |\varphi(re^{i\theta})|d\theta,$$

which is asymptotically the same as (3.B.17) since $\log^+ |x| \leq \log\sqrt{1 + |x|^2} \leq \log^+ |x| + \log 2$. We prefer to use (3.B.17).

If φ is holomorphic, then $f(t) = (1, \varphi(t))$ is in a reduced form, and (3.B.10) becomes

$$(3.B.18) \qquad T(r, \varphi) = \frac{1}{2\pi}\int_0^{2\pi} \log\sqrt{1 + |\varphi(re^{i\theta})|^2}d\theta - \log\sqrt{1 + |\varphi(0)|^2}.$$

For each i, $1 \leq i \leq n$, we consider the meromorphic function $\varphi_i(t) = f_i(t)/f_0(t)$ as a holomorphic map into $P_1\mathbf{C}$. Since

$$1 + |\varphi_i(t)|^2 \leq 1 + \sum_{i=1}^{n} |\varphi_i(t)|^2 \leq \prod_{i=1}^{n}(1 + |\varphi_i(t)|^2),$$

we have

$$(3.B.19) \qquad m(r, \varphi_i) \le m(r, f) \le \sum_{i=1}^{n} m(r, \varphi_i).$$

In order to count $n(\rho, \varphi_i)$ we reduce $(f_0(t), f_i(t))$ by factoring out the common zeros of $f_0(t)$ and $f_i(t)$, and then count the zeros of $f_0(t)$. Therefore we have

$$(3.B.20) \qquad N(r, \varphi_i) \le N(r, f) \le \sum_{i=1}^{n} N(r, \varphi_i).$$

The first main theorem together with (3.B.19) and (3.B.20) yields

$$(3.B.21) \qquad T(r, \varphi_i) \le T(r, f) + C \le \sum_{i=1}^{n} T(r, \varphi_i) + C'.$$

For two meromorphic functions φ and ψ on D_R, the inequality

$$1 + |\varphi\psi|^2 \le (1 + |\varphi|^2)(1 + |\psi|^2)$$

implies

$$(3.B.22) \qquad m(r, \varphi\psi) \le m(r, \varphi) + m(r, \psi),$$

while the inequality

$$1 + |\varphi + \psi|^2 \le 2(1 + |\varphi|^2)(1 + |\psi|^2)$$

implies

$$(3.B.23) \qquad m(r, \varphi + \psi) \le m(r, \varphi) + m(r, \psi) + \text{constant}.$$

Now we state Nevanlinna's **lemma on logarithmic derivative**. For its proof, see Nevanlinna [1; pp. 63–64], Noguchi-Ochiai [1; pp. 225–227], or Lang [3; p. 172].

(3.B.24) **Lemma.** *Let φ be a meromorphic function on* **C**. *Then*

$$m(r, \varphi'/\varphi) = O(\log^{+} T(r, \varphi) + \log r) \qquad \|.$$

Here, $\|$ indicates that the inequality holds outside an exceptional set E of finite Lebesgue measure, i.e., $\int_E dr < \infty$.

(3.B.25) **Corollary.** *Let φ be a meromorphic function on* **C**. *Then, for the p-th derivative $\varphi^{(p)}$ we have*

$$\text{(i)} \qquad T(r, \varphi^{(p)}) \le (p + 1)T(r, \varphi) + O(\log^{+} T(r, \varphi) + \log r) \qquad \|, \quad p \ge 0;$$

$$\text{(ii)} \qquad m(r, \varphi^{(p)}/\varphi) \le O(\log^{+} T(r, \varphi) + \log r), \qquad \|, \quad p \ge 1.$$

Proof. The following proof by induction on p is from Ochiai-Noguchi [1]. While (i) is trivial for $p = 0$, (ii) for $p = 1$ is the lemma above. Assume that (i) holds for $p - 1$ and (ii) for p. Then

$$m(r, \varphi^{(p)}) \;=\; m\!\left(r, \varphi \cdot \frac{\varphi^{(p)}}{\varphi}\right) \le m(r, \varphi) + m\!\left(r, \frac{\varphi^{(p)}}{\varphi}\right)$$
$$\le\; m(r, \varphi) + O(\log^+ T(r, \varphi) + \log r) \qquad \|.$$

Since the order of each pole of φ increases by 1 everytime φ is differentiated, we have (recalling that $N(r, \varphi)$ in our notation counts poles of φ)

$$N(r, \varphi^{(p)}) \le (p+1) N(r, \varphi).$$

Hence

$$
\begin{aligned}
T(r, \varphi^{(p)}) \;\le\;& N(r, \varphi^{(p)}) + m(r, \varphi^{(p)}) + C \\
\le\;& (p+1) N(r, \varphi) + m(r, \varphi) + O(\log^+ T(r, \varphi) + \log r) \qquad \| \\
\le\;& (p+1) T(r, \varphi) + O(\log^+ T(r, \varphi) + \log r) \qquad \|,
\end{aligned}
$$

which proves (i) for p.

By (3.B.24) and (i) for p just proved,

$$
\begin{aligned}
m\!\left(r, \frac{\varphi^{(p+1)}}{\varphi}\right) \;=\;& m\!\left(r, \frac{\varphi^{(p+1)}}{\varphi^{(p)}}\,\frac{\varphi^{(p)}}{\varphi}\right) \le m\!\left(r, \frac{\varphi^{(p+1)}}{\varphi^{(p)}}\right) + m\!\left(r, \frac{\varphi^{(p)}}{\varphi}\right) \\
\le\;& O(\log^+ T(r, \varphi^{(p)}) + \log r) + O(\log^+ T(r, \varphi) + \log r) \qquad \| \\
\le\;& O(\log^+ T(r, \varphi) + \log r) \qquad \|,
\end{aligned}
$$

which proves (ii) for $p + 1$. $\qquad\qquad\square$

The following corollary corresponds to Lemma (6.1.29) in Noguchi-Ochiai [1, p. 230]. We derive it from (3.B.24).

(3.B.26) **Corollary**. *Let φ be a holomorphic function on* **C**. *Then*

$$m(r, \varphi') \le \log \int_0^r \frac{d\rho}{\rho} \int_{D_\rho} dd^c |\varphi|^2 + O(\log r) \qquad \|.$$

Proof. Since $m(r, \varphi') \le m(r, \varphi'/\varphi) + m(r, \varphi)$ by (3.B.22), we first estimate $m(r, \varphi)$.

$$
\begin{aligned}
m(r, \varphi) \;=\;& \frac{1}{4\pi} \int_0^{2\pi} \log(1 + |\varphi(re^{i\theta})|^2)\,d\theta \\
\le\;& \frac{1}{2} \log\!\left(1 + \frac{1}{2\pi} \int_0^{2\pi} |\varphi(re^{i\theta})|^2 d\theta\right) && \text{(by concavity of } \log\text{)} \\
\le\;& \frac{1}{2} \log\!\left(1 + \frac{1}{2\pi} \int_0^r \frac{d\rho}{\rho} \int_{D_\rho} dd^c |\varphi|^2 + C\right) && \text{(by (3.B.9))} \\
\le\;& \frac{1}{2} \log^+ \int_0^r \frac{d\rho}{\rho} \int_{D_\rho} dd^c |\varphi|^2 + C'.
\end{aligned}
$$

On the other hand, an upper bound for $m(r, \varphi'/\varphi)$ is given by (3.B.24). Hence,

$$m(r, \varphi') \leq \frac{1}{2} \log^+ \int_0^r \frac{d\rho}{\rho} \int_{D_\rho} dd^c |\varphi|^2 + O(\log^+ T(r, \varphi) + \log r) \qquad \|.$$

We obtain the desired inequality by replacing $T(r, \varphi)$ by the following bound

$$T(r, \varphi) \leq \frac{1}{4\pi} \int_0^r \frac{d\rho}{\rho} \int_{D_\rho} dd^c |\varphi|^2 + C''.$$

This latter inequality follows from

$$dd^c \log(1 + |\varphi|^2) \leq \frac{dd^c |\varphi|^2}{1 + |\varphi|^2} \leq dd^c |\varphi|^2.$$

$\square$

Let $q \geq n$, and $\mathbf{a}_0, \ldots, \mathbf{a}_q \in \mathbf{C}^{n+1}$ be $q + 1$ unit vectors in general position with components

$$\mathbf{a}_\lambda = (a_\lambda^0, \ldots, a_\lambda^n), \qquad \lambda = 0, \ldots, q$$

and with the corresponding hyperplanes

$$H_\lambda: \quad \sum_{i=0}^n \bar{a}_\lambda^i z^i = 0, \qquad \lambda = 0, 1, \ldots, q.$$

Let

$$f(t) = (f_0(t), f_1(t), \ldots, f_n(t)): D_R \to \mathbf{C}^{n+1} - \{0\}$$

be as before. Set

(3.B.27) $$g_\lambda(t) = \sum_{i=0}^n \bar{a}_\lambda^i f_i(t), \qquad \lambda = 0, \ldots, q.$$

Take $n + 1$ integers $\{\lambda_0, \ldots, \lambda_n\}$ from $\{0, 1, \ldots, q\}$, and let $\{\lambda_{n+1}, \ldots, \lambda_q\}$ be the complementary set of integers. The Wronskians $W(g_{\lambda_0}(t), \ldots, g_{\lambda_n}(t))$ and $W(f_0(t), \ldots, f_n(t))$ are related by

$$W(g_{\lambda_0}, \ldots, g_{\lambda_n}) = c(\lambda_0, \ldots, \lambda_n) W(f_0, \ldots, f_n),$$

where $c(\lambda_0, \ldots, \lambda_n) = \det(\bar{a}_{\lambda_j}^i)_{0 \leq i, j \leq n}$ is a nonzero constant since $\mathbf{a}_0, \ldots, \mathbf{a}_q$ are in general position. Set

(3.B.28) $$W^*(g_{\lambda_0}, \ldots, g_{\lambda_n}) = \begin{vmatrix} 1 & 1 & \cdots & 1 \\ \dfrac{g'_{\lambda_0}}{g_{\lambda_0}} & \dfrac{g'_{\lambda_1}}{g_{\lambda_1}} & \cdots & \dfrac{g'_{\lambda_n}}{g_{\lambda_n}} \\ \dfrac{g''_{\lambda_0}}{g_{\lambda_0}} & \dfrac{g''_{\lambda_1}}{g_{\lambda_1}} & \cdots & \dfrac{g''_{\lambda_n}}{g_{\lambda_n}} \\ \cdots\cdots\cdots\cdots\cdots \\ \dfrac{g^{(n)}_{\lambda_0}}{g_{\lambda_0}} & \dfrac{g^{(n)}_{\lambda_1}}{g_{\lambda_1}} & \cdots & \dfrac{g^{(n)}_{\lambda_n}}{g_{\lambda_n}} \end{vmatrix}.$$

Then

$$(3.B.29) \qquad \frac{g_{\lambda_{n+1}} \cdots g_{\lambda_q}}{c(\lambda_0, \ldots, \lambda_n) W^*(g_{\lambda_0}, \ldots, g_{\lambda_n})} = \frac{g_0 g_1 \cdots g_q}{W(f_0, \ldots, f_n)}.$$

Since the right hand side is independent of the choice $\{\lambda_0, \ldots, \lambda_n\}$, so is the left hand side. We set

$$(3.B.30) \qquad h = \frac{g_0 g_1 \cdots g_q}{W(f_0, \ldots, f_n)}$$

so that

$$g_{\lambda_{n+1}} \cdots g_{\lambda_q} = c(\lambda_0, \ldots, \lambda_n) W^*(g_{\lambda_0}, \ldots, g_{\lambda_n}) h.$$

Then

$$\sum |g_{\lambda_{n+1}} \cdots g_{\lambda_q}|^2 = |h|^2 \sum |c(\lambda_0, \ldots, \lambda_n) W^*(g_{\lambda_0}, \ldots, g_{\lambda_n})|^2,$$

where the sum is taken over all choices of $\{\lambda_0, \ldots, \lambda_n\} \subset \{0, 1, \ldots, q\}$. For simplicity, we set

$$u = \left(\sum |c(\lambda_0, \ldots, \lambda_n) W^*(g_{\lambda_0}, \ldots, g_{\lambda_n})|^2 \right)^{1/2}$$

so that

$$(3.B.31) \qquad \sum |g_{\lambda_{n+1}} \cdots g_{\lambda_q}|^2 = |h|^2 u^2.$$

(3.B.32) Lemma. *Let $f(t) = (f_0(t), f_1(t), \ldots, f_n(t))$ and $g_0, g_1, \ldots, g_q$ be as in (3.B.27). Then*

$$\|f(t)\|^{2(q-n)} = \left(\sum_i |f_i(t)|^2 \right)^{q-n} \leq K \sum |g_{\lambda_{n+1}}(t) \cdots g_{\lambda_q}(t)|^2 \qquad on \quad D_R,$$

where K is a sufficiently large constant and the last sum is taken over all subsets $\{\lambda_0, \ldots, \lambda_n\}$ of $\{0, 1, \ldots, q\}$.

Proof. For each fixed $t_0 \in D_R$, we arrange $g_0(t_0), \ldots, g_q(t_0)$ in the increasing order of their absolute values:

$$|g_{\lambda_0}(t_0)| \leq |g_{\lambda_1}(t_0)| \leq \ldots \leq |g_{\lambda_q}(t_0)|.$$

Since $n+1$ vectors $\mathbf{a}_{\lambda_0}, \ldots, \mathbf{a}_{\lambda_n}$, are linearly independent, we can express all $f_i(t)$ as linear combinations of $g_{\lambda_0}(t), \ldots, g_{\lambda_n}(t),$. Let C be the maximum absolute value of the coefficients of these linear combinations. Since

$$|g_{\lambda_0}(t_0)| \leq \ldots \leq |g_{\lambda_n}(t_0)| \leq |g_{\lambda_j}(t_0)| \quad \text{for} \quad n+1 \leq j \leq q,$$

we have

$$(*) \qquad |f_i(t_0)| \leq (n+1) C |g_{\lambda_j}(t_0)| \quad \text{for} \quad 0 \leq i \leq n, \ n+1 \leq j \leq q,$$

$\square$

Combining (3.B.31) and (3.B.32) yields

(3.B.33) $$(q - n) \log \|f\| \leq \log |h| + \log u + \text{const.}$$

We integrate each term of (3.B.33) along the circle ∂D_r. By (3.B.10)

(3.B.34) $$\frac{1}{2\pi} \int_0^{2\pi} \log \|f(re^{i\theta})\| d\theta = T(r, f) + \log \|f(0)\|.$$

Jensen's formula (3.B.13) implies

$$\frac{1}{2\pi} \int_0^{2\pi} \log |h(re^{i\theta})| d\theta \leq \sum m_j \log \frac{r}{|\alpha_j|} + \log |c| + m \log r,$$

where c and m are given by the Laurent expansion $h(t) = ct^m + \dots$ while $\alpha_j \neq 0$ are zeros of h in D_R with multiplicity m_j. By (3.B.6),

$$N(r, h, 0) = \sum m_j \log \frac{r}{|\alpha_j|} + \log |c| + m \log r.$$

This combined with (3.B.34) yields

(3.B.35) $$\frac{1}{2\pi} \int_0^{2\pi} \log |h(re^{i\theta})| d\theta \leq N(r, h, 0) + \log |c|.$$

We shall bound $N(r, h, 0)$ in terms of truncated counting functions

$$N^{[n]}(r, g_\lambda, 0) = \int_0^r (n^{[n]}(\rho, g_\lambda, 0) - \nu^{[n]}(0, g_\lambda, 0)) \frac{d\rho}{\rho} + \nu^{[n]}(0, g_\lambda, 0) \log r.$$

Suppose that t_0 is a zero of the meromorphic function h in D_R. As in the proof of (3.B.32), let

$$|g_{\lambda_0}(t_0)| \leq |g_{\lambda_1}(t_0)| \leq \dots \leq |g_{\lambda_q}(t_0)|.$$

Then from $(*)$ in the proof of (3.B.32) we know that

$$0 < |g_{\lambda_{n+1}}(t_0)| \leq \dots \leq |g_{\lambda_q}(t_0)|.$$

Since h is equal to the left hand side of (3.B.29) and since the numerator $g_{\lambda_{n+1}} \dots g_{\lambda_q}$ does not vanish at t_0, the order of zero of h at t_0 is equal to the order of pole of $W^*(g_{\lambda_0}, \dots, g_{\lambda_n})$ at t_0. We estimate the order of pole of $W^*(g_{\lambda_0}, \dots, g_{\lambda_n})$ at t_0. Let μ_λ be the order of zero of g_λ at t_0. Then the order of pole of $g_\lambda^{(k)}/g_\lambda$ at t_0 is equal to k if $\mu_\lambda \geq k$ and to μ_λ if $\mu_\lambda < k$; in particular, it is bounded by $\min(\mu_\lambda, k)$. Hence, the order of pole of $W^*(g_{\lambda_0}, \dots, g_{\lambda_n})$ at t_0 is bounded by $\sum_{i=0}^n \min(\mu_{\lambda_i}, n)$, i.e., by $\sum_{i=0}^n \nu^{[n]}(t_0, g_{\lambda_i})$. Thus,

$$\nu(t_0, h, 0) \leq \sum_{i=0}^n \nu^{[n]}(t_0, g_{\lambda_i}, 0) \leq \sum_{\lambda=0}^q \nu^{[n]}(t_0, g_\lambda, 0).$$

Since this holds at every zero t_0 of h, we have

(3.B.36) $$N(r, h, 0) \leq \sum_{\lambda=0}^q N^{[n]}(r, g_\lambda, 0).$$

By the very definition of g_λ, we have

$$v(t_0, g_\lambda, 0) = v(t_0, f, \mathbf{a}_\lambda) \qquad v^{[n]}(t_0, g_\lambda, 0) = v^{[n]}(t_0, f, \mathbf{a}_\lambda).$$

Hence,

$$N(t_0, g_\lambda, 0) = N(t_0, f, \mathbf{a}_\lambda), \qquad N^{[n]}(t_0, g_\lambda, 0) = N^{[n]}(t_0, f, \mathbf{a}_\lambda),$$

and we can rewrite (3.B.36) as

$$(3.B.37) \qquad N(r, h, 0) \le \sum_{\lambda=0}^{q} N^{[n]}(r, f, \mathbf{a}_\lambda).$$

Substituting this into (3.B.35) yields

$$(3.B.38) \qquad \frac{1}{2\pi} \int_0^{2\pi} \log|h(re^{i\theta})|d\theta \le \sum_{\lambda=0}^{q} N^{[n]}(r, f, \mathbf{a}_\lambda) + \log|h(0)|.$$

Finally, we consider the integral of $\log|u|$ along the circle ∂D_r. Since each $W^*(g_{\lambda_0}, \ldots, g_{\lambda_n})$ is a polynomial of $g_\lambda^{(p)}/g_\lambda$, $(\lambda = 0, \ldots, q; \quad p = 1, \ldots, n)$, we have

$$\log u^2 \le K + K \sum_{\lambda=0}^{q} \sum_{p=1}^{n} \log^+ |g_\lambda^{(p)}/g_\lambda|,$$

where K is a constant. Integrating this over the circle ∂D_r, we obtain

$$(3.B.39) \qquad \frac{1}{2\pi} \int_0^{2\pi} \log u(re^{i\theta}) < K + K \sum_{\lambda=0}^{q} \sum_{p=1}^{n} m(r, g_\lambda^{(p)}/g_\lambda).$$

Substituting (3.B.34), (3.B.38) and (3.B.39) into the integral of (3.B.33) yields the following fundamental **inequality of H. Cartan**:

$$(3.B.40) \qquad (q-n)T(r, f) \le \sum_{\lambda=0}^{q} N^{[n]}(r, f, \mathbf{a}_\lambda) + S(r),$$

where

$$S(r) = K + K \sum_{\lambda=0}^{q} \sum_{p=1}^{n} m(r, g_\lambda^{(p)}/g_\lambda).$$

Assume that f is defined on all of $\mathbf{C}$. Then by (3.B.25), we have

$$m(r, g_\lambda^{(p)}/g_\lambda) \le O(\log^+ T(r, g_\lambda) + \log r) \qquad \|.$$

Since $|g_\lambda(t)|^2 \le \sum_i |a_\lambda^i|^2 \cdot \sum_j |f_j(t)|^2 = \|f(t)\|^2$, (3.B.10) and (3.B.18) imply

$$T(r, g_\lambda) \le T(r, f).$$

Hence,

$$(3.B.41) \qquad S(r) = O(\log^+ T(r, f) + \log r) \qquad \|.$$

We rewrite Cartan's inequality (3.B.40) as follows:

$$\sum_{\lambda=0}^{q} (T(r, f) - N^{[n]}(r, f, \mathbf{a}_\lambda)) \leq (n+1)T(r, f) + S(r).$$

Divide this by $T(r, f)$ and let $r \to \infty$. Since $\liminf S(r)/T(r, f) = 0$ by (3.B.41), we have the **truncated defect relation** of Cartan:

(3.B.42) **Theorem.** *Let* $\mathbf{a}_0, \ldots, \mathbf{a}_q \in \mathbf{C}^{n+1}$ *be* $q+1$ *vectors in general position. Then for any holomorphic map* $f: \mathbf{C} \to P_n\mathbf{C}$ *that is non-degenerate in the sense that its image is not contained in any linear subsapce, we have*

$$\sum_{\lambda=0}^{q} \delta^{[n]}(f, \mathbf{a}_\lambda) \leq n+1.$$

This is of course sharper than the usual **defect relation**

(3.B.43) $$\sum_{\lambda=0}^{q} \delta(f, \mathbf{a}_\lambda) \leq n+1.$$

We have now a generalization of the little Picard Theorem:

(3.B.44) **Corollary.** *If a holomorphic map* $f: \mathbf{C} \to P_n\mathbf{C}$ *misses* $n+2$ *hyperplanes in general position, then its image is contained in a hyperplane.*

This implies Borel's theorem (3.10.2), which was stated in three equivalent forms. We prove it in the form (3) of (3.10.2).

(3.B.45) **Corollary.** *If entire functions* $f_1, \ldots, f_n$ *vanishing nowhere on* $\mathbf{C}$ *satisfy the identity*

$$f_1 + \ldots + f_n \equiv 1,$$

then they are linearly dependent.

Proof. Consider the map $f: \mathbf{C} \to P_n\mathbf{C}$ given by

$$z^0 = 1, \quad z^1 = f_1(t), \quad \ldots, \quad z^n = f_n(t).$$

Then f misses the following $n+2$ hyperplanes which are in general position:

$$z^0 = 0, \quad z^1 = 0, \quad \ldots, \quad z^n = 0, \quad z^0 + z^1 + \ldots + z^n = 0.$$

Now Borel's theorem follows from (3.B.44). $\square$

(3.B.46) **Corollary.** *Let* $f: \mathbf{C} \to P_n\mathbf{C}$ *and* $\mathbf{a}_0, \ldots, \mathbf{a}_q$ *be as in Theorem* (3.B.42). *For each* λ, *let* H_λ *be the hyperplane perpendicular to* $\mathbf{a}_\lambda$, *and*

$$m_\lambda = \min\{\nu(\alpha, f, \mathbf{a}_\lambda); \ \alpha \in f^{-1}(H_\lambda)\}.$$

We set $m_\lambda = \infty$ *if* $f(\mathbf{C}) \cap H_\lambda = \emptyset$. *Then*

$$\sum_{\lambda=0}^{q} \max\left\{\left(1 - \frac{n}{m_\lambda}\right), 0\right\} \leq n + 1.$$

Proof. Since

$$\max\left\{\left(1 - \frac{n}{m_\lambda}\right), 0\right\} = 0$$

if $m_\lambda \leq n$, it suffices to show

$(*)$
$$1 - \frac{n}{m_\lambda} \leq \delta^{[n]}(f, \mathbf{a}_\lambda)$$

when $m_\lambda > n$. In this case,

$$v^{[n]}(\alpha, f, \mathbf{a}_\lambda) \leq \frac{n}{m_\lambda} v(\alpha, f, \mathbf{a}_\lambda),$$

and

$$N^{[n]}(r, f.\mathbf{a}_\lambda) \leq \frac{n}{m_\lambda} N(r, f, \mathbf{a}_\lambda).$$

Therefore,

$$1 - \frac{N^{[n]}(r, f, \mathbf{a}_\lambda)}{T(r, f)} \geq 1 - \frac{n}{m_\lambda} \frac{N(r, f, \mathbf{a}_\lambda)}{T(r, f)}.$$

Now the inequality $(*)$ follows from (3.B.16). $\qquad\qquad\square$

(3.B.47) **Remark**. In (3.B.42) and (3.B.46) we assumed that f is non-degenerate. These results have been generalized by Nochka [1] to the situation where $f(\mathbf{C})$ lies in a lower-dimensional linear subspaces, see also Fujimoto [14].

Chapter 4. Intrinsic Distances for Domains

1 Carathéodory Distance and Its Associated Inner Distance

Let X be a complex space. We denote its Carathéodory pseudo-distance by c_X, (see (3.1.1)), and the induced inner pseudo-distance by c_X^i, (see (1.1.2)). While the Kobayashi pseudo-distance d_X is always inner (see (3.1.15)), the Carathéodory pseudo-distance c_X need not be (see Examples (3.1.25), (3.1.26), (3.1.27) and (3.1.28).

Since every inner distance on a locally compact Hausdorff space X induces the given topology of X (see (1.1.8)), we obtain

(4.1.1) Theorem. *If c_X^i is a distance, it induces the complex space topology on X.*

In general, even if c_X is a distance, it may not induce the complex space topology of X. The following proposition is a special case of (1.1.10) (due to Hirstov [2]) and strengthens Sibony [2], in which c_X is assumed to be strongly complete.

(4.1.2) Proposition. *If c_X is a weakly complete distance, it induces the complex space topology on X.*

The following result is due to Sibony [2].

(4.1.3) Proposition. *If X is a relatively compact domain in a Stein space M, then c_X induces the complex space topology on X.*

Proof. Let $H(M)$ be the Fréchet algebra of holomorphic functions on M equipped with the topology of uniform convergence on compact sets. It is known that M is isomorphic to the spectrum of $H(M)$ with weak topology (see, for example, Gunning-Rossi [1]). Therefore, the topology of X is the weakest one that makes $f|_X$ continuous for every $f \in H(M)$. But $H(M)|_X$ is contained in the algebra $H^\infty(X)$ of bounded holomorphic functions, and every $f \in H^\infty(X)$ is continuous with respect to c_X. Hence, the c_X-topology is at least as fine as the complex space topology of X. $\qquad\square$

The following general observation is due to Barth [6], see also Barth [9].

(4.1.4) Proposition. *If c_Y induces the complex space topology on Y and if there is a holomorphic map $f: X \to Y$ that is a homeomorphism onto its image, then c_X induces the complex space topology on X.*

Proof. This follows from the fact that f decreases the Carathéodory distance.

$\square$

The first example of a complex space X whose Carathéodory distance c_X does not define the complex space topology was given by Vigué [5].

(4.1.5) Example. In the tri-disc $D^3 \subset \mathbf{C}^3$ we consider the following 2-dimensional complex space $X = X_1 \bigcup X_2$, where

$$
\begin{aligned}
X_1 &= \{(x, y, 0) \in D^3\} = D \times D \times \{0\}, \\
X_2 &= \{(0, y, z) \in D^3\} - \{(0, y, 0);\ |y| \le \tfrac{1}{2}\} \\
&= \{0\} \times D \times D - \{0\} \times \overline{D}_{1/2} \times \{0\}.
\end{aligned}
$$

Using the extension of c_X to $(D \times D \times \{0\}) \bigcup (\{0\} \times D \times D)$, we see that the sequence $p_n = (0, 0, 1/n)$ converges to the origin $(0,0,0)$ with respect to c_X. But it does not converges in the complex space topology of X since $D_{1/m} \times D_{1/m} \times \{0\}$, $(m = 1, 2, \ldots)$, form a neighborhood basis for the origin $(0,0,0)$ in X.

Later, Hayashi [1] gave an example of an open Riemann surface X such that c_X is a distance but does not induce the topology of X. Making use of Hayashi's construction, Jarnicki-Pflug-Vigué [1], (see also Jarnicki-Pflug [10]), proved the existence of a domain X of holomorphy in $\mathbf{C}^3$ such that c_X is a distance but does not induce the topology of X.

Because of (4.1.1), these provide also examples of complex spaces X with non-inner Carathéodory distance c_X.

We must be very careful with the topology defined by the Carathéodory distance. For $a \in X$ and $r > 0$, we set

$$B(a; r) = \{x \in X;\ c_X(a, x) < r\} \quad \text{and} \quad K(a; r) = \{x \in X;\ c_X(a, x) \le r\}.$$

Since c_X is continuous on $X \times X$, we have always $\overline{B(a; r)} \subset K(a; x)$. However, in general, $\overline{B(a; r)} \ne K(a; r)$ even when X is a strongly pseudoconvex bounded domain in $\mathbf{C}^n$, as shown by Jarnicki-Pflug-Vigué [2], (see also Jarnicki-Pflug [10]). There is a survey by Barth [9] on topologies defined by Carathéodory and other distances including the infinite dimensional case.

A complex space X is said to be **Carathéodory-hyperbolic** or C-**hyperbolic** if c_X is a distance and induces the complex space topology of X. A C-hyperbolic space X is said to be **complete** (resp. **strongly complete**) if X is Cauchy complete with respect c_X (resp. if all closed balls with respect to c_X are compact), (see Section 1 of Chapter 1).

(4.1.6) Proposition. *If a complex space X has a (complete) C-hyperbolic covering space $\tilde{X}$, then it is (complete) hyperbolc.*

Proof. Since $c_{\tilde{X}} \le d_{\tilde{X}}$, $\tilde{X}$ is (complete) hyperbolic. By (3.2.8) X is (complete) hyperbolic.

$\square$

Every bounded domain in $\mathbf{C}^n$ is C-hyperbolic and hence hyperbolic. According to Masaaki Suzuki [3], a complete pseudoconvex circular domain must be bounded if it is hyperbolic.

A useful criterion for strong completeness can be stated in terms of peak functions. Given a bounded domain $X \subset \mathbf{C}^n$ and a boundary point $x_0 \in \partial X$, a holomorphic function f defined in a neighborhood of the closure $\bar{X}$ is said to be a **peak function** (resp. **weak peak function**) for X at x_0 if $|f(x)| < |f(x_0)|$ for all $x \in \bar{X} - \{x_0\}$ (resp. $x \in X$). We may always assume that $|f(x_0)| = 1$. A **local (weak) peak function** for X at x_0 is by definition a (weak) peak function for $X \cap B(x_0, r)$ at x_0, where $B(x_0, r)$ is an open ball of small radius r about x_0.

(4.1.7) Theorem. *If $X \subset \mathbf{C}^n$ is a bounded domain such that there is a weak peak function for X at each point of ∂X, then it is C-hyperbolic and strongly complete with respect to c_X.*

Proof. For each boundary point $x \in X$, choose a weak peak function f_x such that $|f_x(x)| = 1$ (and $|f_x(x')| < 1$ for all $x' \in X$). Let $\mathcal{F}$ be the family of these peak functions. Then $\mathcal{F} \subset \mathrm{Hol}(X, D)$.

Given $o \in X$ and $a > 0$, we shall show that the closed ball $K(o, a) = \{x \in X; \ c_X(o, x) \leq a\}$ is compact. Choose a positive number $b < 1$ such that

$$\{z \in D; \ \rho(f(o), z) \leq a, \ f \in \mathcal{F}\} \subset \{z \in D; \ |z| \leq b\}.$$

Then

$$
\begin{aligned}
K(o, a) \ &= \ \{x \in X; \ \rho(f(o), f(x)) \leq a, \ f \in \mathrm{Hol}(X, D)\} \\
&\subset \ \{x \in X; \ \rho(f(o), f(x)) \leq a, \ f \in \mathcal{F}\} \\
&\subset \ \{x \in X; \ |f(x)| \leq b, \ f \in \mathcal{F}\}.
\end{aligned}
$$

The last set is compact since each boundary point x_0 of X has a neighborhood which does not meet the set $\{x \in X; \ |f_{x_0}(x)| \leq b\}$. Hence, $K(o, a)$ is compact.
$\square$

Let G be a domain in $\mathbf{C}^n$ and $f_1, \ldots, f_k$ holomorphic functions defined in G. Let P be a connected component of the open subset of G defined by

$$|f_1(z)| < 1, \ldots, |f_k(z)| < 1.$$

If the closure of P in $\mathbf{C}^n$ is compact and is contained in G, then P is called an **analytic polyhedron**. It is clear that at each boundary point of P, one of the functions $f_1, \ldots, f_k$ is a weak peak function for P.

(4.1.8) Corollary. *Every analytic polyhedron P is C-hyperbolic and is strongly complete with respect to c_P.*

We can apply (4.1.7) to slightly more general domains than analytic polyhedron. Let h_j be real analytic functions on G of the form

$$h_j = \sum_{m=1}^{\infty} |f_{jm}|^2, \qquad j = 1, \ldots, k,$$

where each f_{jm} is holomorphic in G. Let P be a connected component of the open subset of G defined by

$$h_1(z) < 1, \ldots, h_k(z) < 1.$$

If the closure of P in $\mathbf{C}^n$ is compact and is contained in G, we call P a **generalized analytic polyhedron**.

Let S be the set of sequences $\{a_m\}$ of complex numbers such that $\sum_{m=1}^{\infty} |a_m|^2 = 1$. Then, for each j, we have

$$(*) \qquad \{z \in G; \ h_j(z) < 1\} = \bigcap_{\{a_m\} \in S} \{z \in G; \ |\sum_{m=1}^{\infty} a_m f_{jm}(z)| < 1\}.$$

In fact, the left hand side of $(*)$ is contained in the right hand side since

$$|\sum_m a_m f_{jm}|^2 \le \sum_m |a_m|^2 \sum_m |f_{jm}|^2 = |h_j|^2.$$

If z belongs to the right hand side, let $a_m = \overline{f_{jm}(z)}/(\sum_m |f_{jm}(z)|^2)^{1/2}$. Then

$$1 > |\sum_m a_m f_{jm}(z)| = (\sum_m |f_{jm}(z)|^2)^{1/2} = h_j(z)^{1/2},$$

showing that z is contained in the left hand side. From $(*)$ we see that at each boundary point of P a function of the form $\sum_m a_m f_{jm}$ is a weak peak function for P. Hence

(4.1.9) Corollary. *Every generalized analytic polyhedron P is C-hyperbolic and is strongly complete with respect to c_P.*

(4.1.10) Corollary. *Every bounded convex domain $X \subset \mathbf{C}^n$ is C-hyperbolic and is strongly complete with respect to c_X.*

Proof. For each $x_0 \in \partial X$ there is a complex linear functional $\varphi \colon \mathbf{C}^n \to \mathbf{C}$ such that $\mathrm{Re}(\varphi(x)) < \mathrm{Re}(\varphi(x_0))$ for all $x \in X$. Then $f = e^{\varphi}$ is a weak peak function for X at x_0. $\qquad \square$

Pflug [3] has shown that every bounded complete pseudoconvex Reinhart domain X is complete with respect to c_X.

Since $c_X \le d_X$, if a complex space X is C-hyperbolic and is complete with respect to c_X, then it is complete hyperbolic. In particular, all domains described in (4.1.7) through (4.1.10) are complete hyperbolic. But we can say a little more.

(4.1.11) Corollary. *If $X \subset \mathbf{C}^n$ is a bounded domain such that at every boundary point of X there is a local weak peak function, then X is complete hyperbolic.*

Proof. Clearly, X is hyperbolically imbedded in $\mathbf{C}^n$. By (4.1.7) every boundary point $x \in \partial X$ has a neighborhood U in $\mathbf{C}^n$ such that $U \cap X$ is strongly complete with respect to $c_{U \cap X}$. Then $U \cap X$ is complete hyperbolic. By (3.3.4) X is complete hyperbolic. $\qquad\square$

(4.1.12) **Corollary.** *Every bounded strongly pseudoconvex domain X with C^2 boundary is complete hyperbolic.*

Proof. For every boundary point $x \in \partial X$ there is a neighborhood U and a biholomorphic map $f : U \to f(U)$ such that $f(U \cap X)$ is strongly convex. Hence, there is a local peak function for X at x. $\qquad\square$

(4.1.13) **Example.** As in (3.1.26), let $X = M(D, V) \subset \mathbf{C}^2$ be the domain constructed by Sibony. It is a proper subdomain of D^2, and as we explained in (3.1.26), c_X is the restriction of c_{D^2}. In particular, c_X is not complete. Following Eastwood [1] we shall show that X is complete hyperbolic. Let $\{x_n\} \subset X$ be a Cauchy sequence with respect to d_X. Set $x_n = (z_n, w_n) \in D \times D$. Then both $\{z_n\}$ and $\{w_n\}$ are Cauchy sequences in D with respect to d_D. Let $z_0 = \lim z_n$ and $w_0 = \lim w_n$. We have to show that $(z_0, w_0) \in X$. Let $\{a_p\}$ be the discrete sequence of points in D used in the construction of the domain X, see (3.1.26).

Case (i) $\quad z_0 = a_p$ for some p. In this case, $V(z_0) = 0$ and $|w| < 1 = e^{-V(z_0)}$ for every $w \in D$. Hence, $(z_0, w) \in X$ for all $w \in D$.

Case (ii) $\quad z_0 \neq a_p$ for all p. Let ρ be a small positive number such that the neighborhood $N = \{z \in D; \ d_D(z_0, z) < 4\rho\}$ contains none of the a_p's. Given a point $o \in X$ and a positive number δ, we set $U(o; \delta) = \{x \in X; \ d_X(o, x) < \delta\}$. Then by discarding a finite number of x_n, we may assume that all x_n are in the $U(o; \rho)$ for some $o \in X$; By (3.1.19) there is a positive constant C such that

$$d_{U(o;4\rho)}(p, q) < C \cdot d_X(p, q) \qquad \text{for} \quad p, q \in U(o; \rho).$$

This shows that $\{x_n\}$ is a Cauchy sequence in $U(o; 4\rho)$ with respect to $d_{U(o;4\rho)}$. Let π be the projection from X to D sending (z, w) to z. Since $U(o; 4\rho) \subset \pi^{-1}(N)$, $\{x_n\}$ is a Cauchy sequence in $\pi^{-1}(N)$ with respect to $d_{\pi^{-1}(N)}$. It suffices therefore to show that $\pi^{-1}(N)$ is complete hyperbolic.

We may assume that $z_0 = 0$ so that $N = \{z \in D; |z| < r\}$ for some r. Take a suitable branch of $\log(z - a_p)$ on N so that $\prod((z - a_p)/2)^{\lambda_p}$ converges to a holomorphic function γ on N. Then $V = |\gamma|$. Now we can write $\pi^{-1}(N)$ as follows:

$$\pi^{-1}(N) = \bigcap_u \{(z, w); \ |z| < r, \ |we^{u\gamma(z)}| < 1\},$$

where the intersection is taken over all complex numbers u of absolute value 1. This shows that $\pi^{-1}(N)$ is a generalized analytic polyhedron. By (4.1.9) and (4.1.6) $\pi^{-1}(N)$ is complete hyperbolic.

(4.1.14) **Example.** Following Pyatezkii-Shapiro [2] we define Siegel domains of the second kind. Let V be a convex cone in $\mathbf{R}^n$ containing no entire straight lines. A mapping $F : \mathbf{C}^m \times \mathbf{C}^m \to \mathbf{C}^n$ is said to be V-Hermitian if

(a) $F(u, v) = \overline{F(v, u)}$, $u, v \in \mathbf{C}^m$;

(b) $F(u, v)$ is complex linear in u;

(c) $F(u, u) \in \bar{V}$ (the closure of V), $u \in \mathbf{C}^m$;

(d) $F(u, u) = 0$ only when $u = 0$.

The subset $S = S(V, F)$ of $\mathbf{C}^{n+m}$ defined by

$$S = \{(z, u) \in \mathbf{C}^n \times \mathbf{C}^n;\ \mathrm{Im}(z) - F(u, u) \in V\}$$

is called the **Siegel domain of the second kind** defined by V and F.

If V is a special convex cone given by $y^1 > 0, \ldots, y^m > 0$ for a suitably chosen linearly independent set of linear functionals $y^1, \ldots, y^m$ on $\mathbf{R}^m$, (i.e., for a suitable coordinate system $y^1, \ldots, y^m$), then $S(V, F)$ is biholomorphic to a product of open balls (see Pyatezkii-Shapiro [2] or Kobayashi [7; pp. 29–31]), and hence is finitely compact with respect to its Carathéodory distance. Since a general convex cone containing no lines is an intersection of (in general, infinitely many) such special convex cones, a general Siegel domain of the second kind $S(V, F)$ is an intersection of domains, each of which is finitely compact with respect to its Carathéodory distance. It follows from (1.1.11) that $S(V, F)$ is finitely compact with respect to its Carathéodory distance. In particular, it is complete hyperbolic.

(4.1.15) **Remark**. In section 5, we shall give boundary estimates for both Kobayashi and Carathéodory distances of *strongly* pseudoconvex domains with smooth boundary and reprove, in particular, (4.1.12). However, it is not known if every bounded pseudoconvex domain with smooth boundary is complete hyperbolic.

Let F be a continuous Minkowski function on $\mathbf{C}^n$, i.e., a continuous nonnegative function on $\mathbf{C}^n$ such that $F(z) > 0$ for nonzero $z \in \mathbf{C}^n$ and $F(tz) = |t| F(z)$ for $z \in \mathbf{C}^n$ and $t \in \mathbf{C}$. The bounded domain $X_F = \{z \in \mathbf{C}^n;\ F(z) < 1\}$ is star-shaped and circular. Jarnicki-Pflug [9] has an example of F such that X_F is pseudoconvex but not complete hyperbolic.

According to Sibony (see Jarnicki-Pflug [10]), there is a bounded pseudoconvex domain, not complete hyperbolic, whose boundary is smooth (of class C^∞) except at one point.

2 Infinitesimal Carathéodory Metric

We define the infinitesimal form of the Carathéodory pseudo-distance. Given a complex space X, we define a nonnegative function E_X on the tangent bundle TX by setting

$$(4.2.1) \qquad\qquad E_X(v) = \sup_f \| f_* v \| \qquad \text{for} \quad v \in TX,$$

where $\| f_* v \|$ is the length of the tangent vector $f_* v$ of D measured by the Poincaré metric ds^2 of D, and the supremum is taken over all $f \in \mathrm{Hol}(X, D)$. This supremum is actually achieved by some $f \in \mathrm{Hol}(X, D)$ which sends the base point $\pi(v) \in X$ to $0 \in D$, (see the comment following the definition of c_X in (3.1.1).)

Clearly, E_X satisfies the following convexity condition:

$$(4.2.2) \qquad E_X(v + v') \leq E_X(v) + E_X(v') \qquad \text{for} \quad v, v' \in T_p X.$$

We call E_X the **Carathéodory pseudo-metric** or **Carathéodory pseudo-length**; it is a pseudo-length function in the sense of Section 3 of Chapter 2.

Corresponding to (3.1.2) we have

(4.2.3) Proposition. (1) *If X and Y are complex spaces, then*

$$E_Y(f_* v) \leq E_X(v) \qquad \text{for} \quad f \in \mathrm{Hol}(X, Y) \quad \text{and} \quad v \in TX;$$

(2) *For the unit disc D with Poincaré metric ds^2, we have*

$$E_D^2 = ds^2.$$

Corresponding to (3.1.7) we have

(4.2.4) Proposition. *If F is a pseudo-length function* (*continuous or otherwise*) *on X such that*

$$\| f_* v \| \leq F(v) \qquad \text{for} \quad f \in \mathrm{Hol}(X, D) \quad \text{and} \quad v \in TX,$$

then $E_X \leq F$.

In particular, if $\hat{F}_X$ and F_X denote the infinitesimal forms of d_X defined in (3.5.7), (3.5.14) and (3.5.16), then

$$(4.2.5) \qquad\qquad E_X \leq \hat{F}_X \leq F_X.$$

(4.2.6) Theorem. *The infinitesimal Carathédory pseudo-metric $E_X : TX \to \mathbf{R}$ is continuous.*

Proof. Let $v_k \in T_{p_k} X$ be a sequence of vectors converging to $v \in T_p X$. As we remarked above, there is a map $f \in \mathrm{Hol}(X, D)$ such that $E_X(v) = \| f_* v \|$ and $f(p) = 0$. Then

$$E_X(v) = \| f_* v \| = \lim \| f_* v_k \| \leq \lim E_X(v_k).$$

Also for each v_k there is a map $f_k \in \mathrm{Hol}(X, D)$ such that $E_X(v_k) = \| f_{k*} v_k \|$ and $f_k(p_k) = 0$.

Taking a convergent subsequence, we set $g = \lim f_k$. Then

$$E_X(v) \geq \| g_* v \| = \lim \| f_{k*} v_k \| = \lim E_X(v_k). \qquad \square$$

As we remarked after (1.1.2), in constructing the inner pseudo-distance c_X^i from c_X, we use only piecewise differentiable, c_X-rectifiable curves. Then we have the following result of Reiffen [1, 2]. (As pointed out by Jarnicki-Pflug [10], if all c_X-rectifiable curves are used, it is not clear whether (4.2.7) is valid or not).

(4.2.7) Theorem. *Let X be a complex space. The inner pseudo-distance c_X^i induced by c_X is obtained by integration of the Carathéodory pseudo-metric E_X.*

Proof. Let δ be the pseudo-distance obtained by integration of E_X; it is inner (see (2.3.1)). By (4.2.1) we have

$$E_X(v) \geq \|f_* v\| \qquad \text{for} \quad f \in \mathrm{Hol}(X, D), \quad v \in TX.$$

Integrating this inequality yields

$$\delta(p, q) \geq \rho(f(p), f(q)) \qquad \text{for} \quad f \in \mathrm{Hol}(X, D), \quad p, q \in X.$$

By the extremal property of c_X (see (3.1.7)), we have

$$\delta(p, q) \geq c_X(p, q) \qquad \text{for} \quad p, q \in X.$$

Since δ is inner, we have

$$\delta(p, q) = \delta^i(p, q) \geq c_X^i(p, q) \qquad \text{for} \quad p, q \in X.$$

In order to prove the opposite inequality, we consider a curve $\gamma(t)$ in X. Let v be the velocity vector of $\gamma(t)$ at $t = t_0$. Let $f \in \mathrm{Hol}(X, D)$ be such that $\|f_* v\| = E_X(v)$. From

$$\begin{aligned}
\delta(\gamma(t_0), \gamma(t)) \;&\geq\; c_X^i(\gamma(t_0), \gamma(t)) \geq c_X(\gamma(t_0), \gamma(t)) \\
&\geq\; \rho(f(\gamma(t_0)), f(\gamma(t))),
\end{aligned}$$

we obtain

$$\begin{aligned}
E_X(v) \;&=\; \lim_{t \to t_0} \frac{\delta(\gamma(t_0), \gamma(t))}{|t - t_0|} \geq \lim_{t \to t_0} \frac{c_X^i(\gamma(t_0), \gamma(t))}{|t - t_0|} \\
&\geq\; \lim_{t \to t_0} \frac{c_X(\gamma(t_0), \gamma(t))}{|t - t_0|} \geq \lim_{t \to t_0} \frac{\rho(f(\gamma(t_0)), f(\gamma(t)))}{|t - t_0|} \\
&=\; \|f_* v\| = E_X(v).
\end{aligned}$$

Fixing $t = a$, let $L_E(s)$ (resp. $L_c(s)$) be the arc-length of γ from $t = a$ to $t = s$ measured by E_X or, equivalently, by δ (resp. by c_X or, equivalently, by c_X^i). Then

$$L_E'(t_0) = E_X(v).$$

On the other hand, the length of γ from $t = t_0$ to $t = s$ measured by δ is at least as long as its length measured by c_X^i. Hence,

$$\frac{L_E(s) - L_E(t_0)}{s - t_0} \geq \frac{L_c(s) - L_c(t_0)}{s - t_0} \geq \frac{c_X(\gamma(t_0), \gamma(s))}{s - t_0}.$$

Letting $s \to t_0$, we obtain

$$E_X(v) = L_E'(t_0) \geq L_c'(t_0) \geq E_X(v).$$

This establishes the equality $L'_E(t_0) = L'_c(t_0)$ for all t_0. Hence, $L_E(t) = L_c(t)$ for all t, and $\delta = c^i_X$. $\qquad\square$

For a nonsingular X it can be shown (Franzoni-Vesentini [1]) that E_X is not only continuous but locally Lipschitz.

(4.2.8) Theorem. *For a complex manifold X, the infinitesimal Carathéodory pseudo-metric E_X is locally Lipschitz.*

Proof. Let $o \in X$ and fix a local coordinate system in a neighborhood U of o. Using the coordinate system we identify the tangent space $T_p X$ at $p \in U$ with $\mathbf{C}^n$. Thus we write $(p, \xi) \in U \times \mathbf{C}^n$ for $v \in T_p U$. Let B_r be the ball of radius r around o with respect to the coordinate. Let $p, q \in B_{r/2}$ and $\xi, \eta \in \mathbf{C}^n$. Suppose $E_X(p, \xi) \geq E_X(q, \xi)$. Then (the suprema being taken over all $f \in \mathrm{Hol}(X, D)$ such that $f(p) = 0$)

$$
\begin{aligned}
E_X(p, \xi) - E_X(q, \xi) \;&=\; \sup 2|f_*(p, \xi)| - \sup \frac{2|f_*(q, \xi)|}{1 - |f(q)|^2} \\[2mm]
&\leq\; 2 \sup\left\{ |f_*(p, \xi)| - \frac{|f_*(q, \xi)|}{1 - |f(q)|^2} \right\} \\[2mm]
&\leq\; 2 \sup\{ |f_*(p, \xi)| - |f_*(q, \xi)| \} \\[2mm]
&\leq\; 2 \sup |f_*(p, \xi) - f_*(q, \xi)| \\[2mm]
&\leq\; \frac{C}{r^2} \|p - q\| \cdot \|\xi\|.
\end{aligned}
$$

The last inequality follows from Cauchy's integral formula for the second derivative of f. Similarly, using Cauchy's formula for the first derivative of f we obtain

$$
|E_X(q, \xi) - E_X(q, \eta)| \leq E_X(q, \xi - \eta) = \sup |f_*(q, \xi - \eta)| \leq \frac{C'}{r} \|\xi - \eta\|,
$$

where the supremum is taken over all $f \in \mathrm{Hol}(X, D)$ such that $f(q) = 0$. Now the theorem follows from the above two estimates. $\qquad\square$

(4.2.9) Remark. We shall show that according to the definition of the curvature given by (2.3.3), E_X has holomorphic sectional curvature ≤ -1, see B. Wong [1] and Masaaki Suzuki [1]. Given a nonzero $v \in T_x X$, consider maps $g \in \mathrm{Hol}(D, X)$ such that $g(0) = x$ and v is tangent to $g(D)$. We want to show that $g^* E_X$ has curvature ≤ -1 for all g. Take a map $f \in \mathrm{Hol}(X, D)$ such that $f(x) = 0$ and $\|f_* v\| = E_X(v)$, where $\| \ \|$ denotes the length defined by the Poincaré metric ds^2 of D. Then $f^* ds^2 \leq E_X^2$ with equality at x. Then $g^* f^* ds^2 \leq g^* E_X^2$ with equality at 0. Thus, $g^* f^* ds^2$ is a supporting metric for $g^* E_X^2$ at 0. Since $g^* f^* ds^2$ is isometric to ds^2 by $f \circ g$, it has curvature -1. By (2.1.9), $g^* E_X$ has curvature ≤ -1.

This is in contrast to the fact (see Remark (3.5.45)) that F_X has holomorphic sectional curvature ≥ -1 provided X is complete hyperbolic.

Given a point $p \in X$, the **indicatrix** $\Gamma(X, p)$ of E_X at p is a domain in the tangent space $T_p X$ defined by

$$(4.2.10) \qquad \Gamma(X, p) = \{v \in T_p X; \; E_X(v) < 2\}.$$

If we denote

$$\mathrm{Hol}_p(X, D) = \{f \in \mathrm{Hol}(X, D); \; f(p) = 0\},$$

then the indicatrix may be described more directly:

$$(4.2.11) \qquad \Gamma(X, p) = \bigcap_{f \in \mathrm{Hol}_p(X, D)} \{v \in T_p X; \; |f_* v| < 1\}.$$

We remark that $|f_* v|$ is the ordinary absolute value and hence is a half of the Poincaré length $\|f_* v\|$ because of the normalization we adopted for the Poincaré metric ds^2 in (2.1.3). That is why we defined $\Gamma(X, p)$ by $E_X < 2$ in (4.2.10) rather than by $E_X < 1$; this is, of course, a matter of preference.

We state simple consequences of the definition of $\Gamma(X, p)$, (Carathéodory [2]).

(4.2.12) Proposition. *If X is a complex subspace of a complex space Y, then*

$$\Gamma(X, p) \subset \Gamma(Y, p) \qquad \text{for} \quad p \in X.$$

Proof. By (4.2.2), we have $E_Y \leq E_X$ on X. $\qquad\qquad\square$

(4.2.13) Proposition. *The indicatrix $\Gamma(X, p)$ is always a convex circular domain in $T_p X$.*

Proof. Either by (4.2.2) or by (4.2.11) $\Gamma(X, p)$ is convex. Since

$$E_X(\alpha v) = |\alpha| E_X(v) \qquad \text{for} \quad v \in TX, \; \alpha \in \mathbf{C},$$

the indicatrix $\Gamma(X, p)$ is invariant under multiplication by $e^{it}, t \in \mathbf{R}$, and hence is circular. $\qquad\qquad\square$

Conversely, every convex circular domain in $\mathbf{C}^n$ is the indicatrix $\Gamma(X, p)$ for some domain $X \subset \mathbf{C}^n$ and some point $p \in X$. In fact, we have

(4.2.14) Theorem. *Let X be a star-shaped circular domain in $\mathbf{C}^n$. Then under the natural identification $T_0 X \cong \mathbf{C}^n$, the indicatrix $\Gamma(X, 0)$ is the smallest convex circular domain containing X.*

In particular, if X is a convex circular domain, then $\Gamma(X, 0) = X$.

We recall that a domain $X \subset \mathbf{C}^n$ is said to be **star-shaped** if $p \in X$ and $|\alpha| \leq 1$ imply $\alpha p \in X$.

Proof. Let $p \in X$ and $f \in \mathrm{Hol}_0(X, D)$. Define $\varphi \in \mathrm{Hol}(\bar{D}, D)$ by

$$\varphi(t) = f(tp), \qquad t \in \bar{D}.$$

Then $\varphi(0) = 0$. Considering $p \in X \subset \mathbf{C}^n$ as a point in $T_0 X$ under the identification of $T_0 X$ with $\mathbf{C}^n$, we have $f_*(p) = \varphi'(0)$. Since $|\varphi'(0)| < 1$ by Schwarz lemma, it follows from (4.2.11) that $p \in \Gamma(X, 0)$. Hence, $X \subset \Gamma(X, 0)$.

In order to prove the opposite inclusion when X is convex, let q be a boundary point of X, and H the supporting real hyperplane at q. By applying a complex linear transformation to the coordinate system $z^1, \ldots, z^n$ of $\mathbf{C}^n$, we may assume that q has coordinates $(1, 0, \ldots, 0)$, H is the hyperplane $x^1 = 1$ (where $z^1 = x^1 + iy^1$) and X lies in the half-space $x^1 < 1$. Since X is circular, it follows that, for any real θ, $(e^{i\theta}, 0, \ldots, 0)$ is a boundary point of X with the support hyperplane

$$x^1 \cos\theta + y^1 \sin\theta = 1$$

and that X lies in the half-space

$$x^1 \cos\theta + y^1 \sin\theta < 1.$$

Since this holds for all $\theta \in \mathbf{R}$, the domain X is contained in the circular cylinder $C_q = \{(z^1, \ldots, z^n); \ |z^1| < 1\}$. Clearly,

$$X = \bigcap_{q \in \partial X} C_q.$$

Hence, it suffices to show that $\Gamma(X, 0)$ is contained in C_q.

Let $f \in \mathrm{Hol}(X, D)$ be the restriction to X of the projection $(z^1, \ldots, z^n) \mapsto z^1$. Let

$$v = \sum a^i \left(\frac{\partial}{\partial z^i} \right)_0 \in \Gamma(X, 0).$$

By (4.2.11), $|a^1| = |f_* v| < 1$. Hence, $(a^1, \ldots, a^n) \in C_q$, showing that $\Gamma(X, 0) \subset C_q$.

If X is not convex, let K denote the smallest convex circular domain containing X. From $X \subset K$, we obtain

$$X \subset \Gamma(X, 0) \subset \Gamma(K, 0) = K.$$

Since $\Gamma(X, 0)$ is a convex circular domain containing X and since K is the smallest such domain, we conclude $\Gamma(X, 0) = K$. $\qquad\square$

(4.2.15) Corollary. *Let X be a convex circular domain in $\mathbf{C}^n$. Then E_X^2 is a Hermitian metric if and only if X is affinely isomorphic to a unit ball $\{\mathbf{z} \in \mathbf{C}^n; \ \|z\| < 1\}$.*

Proof. Assume that E_X^2 is Hermitian, i.e.,

$$E_X^2(\mathbf{v}) = \sum h_{j\bar{k}}(p) v^j \bar{v}^k \qquad \text{for} \quad \mathbf{v} = (v^j) \in T_p X.$$

Then

$$\Gamma(X, 0) = \{(z^1, \ldots, z^n) \in \mathbf{C}^n; \ \sum h_{j\bar{k}}(0) z^j \bar{z}^k < 2\}.$$

By (4.2.14), $X = \Gamma(X, 0)$. $\qquad\square$

If X and Y are complex spaces and $f \in \mathrm{Hol}(X, Y)$ with $f(p) = q$, then the differential $f_*: T_p X \to T_q Y$ maps $\Gamma(X, p)$ into $\Gamma(Y, q)$. In particular, if f is biholomorphic, then $f_*: \Gamma(X, p) \to \Gamma(Y, q)$ is a linear isomorphism.

On the indicatrix of the Carathéodory metric, see Tishabaev [1].

3 Pseudo-distance Defined by Plurisubharmonic Functions

Following Sibony [2], Klimek [1, 2] and Azukawa [2, 3], we shall use plurisubharmonic functions to define a pseudo-distance p_X on each complex manifold X.

Let X be a complex manifold. Given a point $x_0 \in X$, let $\mathcal{P}_X(x_0)$ denote the set of upper semicontinuous functions φ on X such that (i) $0 \leq \varphi < 1$, (ii) $\varphi(x_0) = 0$, (iii) $\log \varphi$ is plurisubharmonic, and (iv) with respect to a local coordinate system $z = (z^1, \ldots, z^n)$ with origin at x_0, $\varphi/\|z\|$ is bounded in a neighborhood of x_0.

Following Klimek [1] we consider the following extremal function:

$$(4.3.1) \qquad \lambda_X(x, x_0) = \sup\{\varphi(x);\ \varphi \in \mathcal{P}_X(x_0)\}.$$

Then

(4.3.2) Lemma. *If $f: X \to Y$ is a holomorphic map between complex manifolds and if $\psi \in \mathcal{P}_Y(f(x_0))$, then $\psi \circ f \in \mathcal{P}_X(x_0)$ and*

$$\lambda_Y(f(x), f(x_0)) \leq \lambda_X(x, x_0).$$

Proof. All we have to show is that $\psi \circ f$ satisfies condition (iv). With respect to a local coordinate system $z = (z^1, \ldots, z^n)$ around x_0 and a local coordinate system $w = (w^1, \ldots, w^m)$ around $f(x_0)$, we have

$$\log \psi(f(x)) - \log \|z(x)\| = \log \psi(f(x)) - \log \|w(f(x))\| + \log \frac{\|w(f(x))\|}{\|z(x)\|},$$

which is bounded in a neighborhood of x_0. $\qquad\square$

For further properties of the extremal function, see Demailly [1] as well as Klimek [1].

Let ρ denote the Poincaré distance on D, and d'_X be the function on $X \times X$ defined in (3.1.3). If $\lambda_X(x, x_0) < 1$, then $\rho(\lambda_X(x, x_0), 0)$ is defined. If $\lambda_X(x, x_0) = 1$, then $\rho(\lambda_X(x, x_0), 0)$ is defined to be ∞. With this understanding, we have

$$(4.3.3) \qquad \rho(\lambda_X(x, x_0), 0) \leq d'_X(x, x_0).$$

To prove this, let $f \in \mathrm{Hol}(D, X)$ be a map passing through both x_0 and x. If no such map exists, then $d'_X(x, x_0) = \infty$ and the claimed inequality holds trivially. If such a map f exists, by composing it with an automorphism of D we may assume that $f(0) = x_0$ and $f(a) = x$ with $0 \leq a < 1$. Let $\varphi \in \mathcal{P}_X(x_0)$. Then by (4.3.2), $\varphi \circ f \in \mathcal{P}_D(0)$, and by (2.1.15) $\varphi(f(z)) \leq |z|$ for $z \in D$. In particular, $\varphi(f(a)) \leq a$. Since this holds for all $\varphi \in \mathcal{P}_X(x_0)$, we have

$$\lambda_X(x, x_0) = \lambda_X(f(a), x_0) = \sup_{\varphi} \varphi(f(a)) \leq a < 1.$$

Hence, $\rho(\lambda_X(x, x_0), 0) \leq \rho(a, 0)$. Taking the infimum over all pairs (f, a) such that $f(0) = x_0$ and $f(a) = x$, we obtain (4.3.3).

Define

$$(4.3.4) \qquad p'_X(x, x') = \max\{\rho(\lambda_X(x, x'), 0), \ \rho(\lambda_X(x', x), 0)\}, \qquad x, x' \in X.$$

Although $p'_X(x, x')$ is nonnegative and symmetric, it may not satisfy the triangular inequality. Let $p_X(x, x')$ be the largest pseudo-distance bounded by $p'_X(x, x')$. More explicitly, taking a chain of points $x = x_0, x_1, \ldots, x_k = x'$, we define (Klimek [1])

$$(4.3.5) \qquad p_X(x, x') = \inf \sum p'_X(x_{i-1}, x_i),$$

where the infimum is taken over all chains of points from x to x'. Then p_X is a pseudo-distance on X. Since $d'_X(x, x')$ is symmetric in x, x', (4.3.3) implies $p'_X(x, x') \leq d'_X(x, x')$. Since d_X is the largest pseudo-distance bounded by d'_X, we have $p_X \leq d_X$.

(4.3.6) Proposition. (1) *If $f: X \to Y$ is a holomorphic map between two complex manifolds, then*

$$p_Y(f(x), f(x')) \leq p_X(x, x'), \qquad x, x' \in X;$$

(2) *For the unit disc D, p_D coincides with the Poincaré distance ρ;*

(3) $$c_X \leq p_X \leq d_X.$$

Proof. (1) This is immediate from (4.3.2).

(2) By (1) p_X is invariant under biholomorphic automorphisms of X. Since $D = \{|z| < 1\}$ is homogeneous, it suffices to verify $p_D(z, 0) = \rho(z, 0)$. The function $|z|$ belongs to $\mathcal{P}_D(0)$. Hence, $\lambda_D(z, 0) \geq |z|$.

Conversely, let $\varphi \in \mathcal{P}_D(0)$. Then by (2.1.15) (applied to φ^2), we obtain $\varphi(z) \leq |z|$. Hence, $\lambda_D(z, 0) = |z|$, which implies

$$p_D(z, 0) = \rho(\lambda_D(z, 0), 0) = \rho(|z|, 0) = \rho(z, 0).$$

(3) We already proved the inequality $p_X \leq d_X$. The inequality $c_X \leq p_X$ follows from (1) and (2) above and (3.1.7). $\square$

For a unit ball $B = \{z \in \mathbf{C}^n; \ \|z\| < 1\}$, we have (see (3.1.24))

$$c_B(z, 0) = p_B(z, 0) = d_B(z, 0).$$

Since any two points of B can be interchanged by an automorphism of B, $\lambda_B(z, z')$ is symmetric in z and z'. From (4.3.3) and (4.3.4) we obtain $p'_B(z, 0) \leq d'_B(z, 0)$. Since $d'_B(z, 0) = d_B(z, 0)$, we have

$$c_B(z, 0) = p_B(z, 0) \leq p'_B(z, 0) \leq d_B(z, 0).$$

Hence,

$$\rho(\|z\|, 0) = c_B(z, 0) = p'_B(z, 0) = \rho(\lambda_B(z, 0), 0),$$

which implies

$$(4.3.7) \qquad\qquad \lambda_B(z, 0) = \|z\|.$$

Using this, we prove the following result of Klimek [1].

(4.3.8) Theorem. *The extremal function $\lambda_X(x, x_0)$ belongs to $\mathcal{P}_X(x_0)$.*

Proof. Clearly, $\log \lambda_X(x, x_0)$ is plurisubharmonic and nonpositive. By letting $X = D$ and $Y = X$ in (4.3.2), we see that it is strictly negative at x which can be joined to x_0 by a holomorphic disc. Since a plurisubharmonic function cannot attain its maximum unless it is a constant function, $\lambda_X(x, x_0)$ cannot take the value 1. To see that $\lambda_X(x, x_0)$ satisfies condition (iv), consider an open ball $U = \{\|z\| < r\}$ defined by a local coordinate system $z = (z^1, \ldots, z^n)$ with origin at x_0. By (4.3.2), $\lambda_X(x, x_0) \le \lambda_U(x, x_0)$ for $x \in U$. By (4.3.7), $\lambda_U(x, x_0) = \|z(x)\|/r$, and

$$\log \lambda_X(x, x_0) - \log \|z(x)\| \le \log \lambda_U(x, x_0) - \log \|z(x)\| = -\log r.$$

$\square$

Following Azukawa [2, 3], we shall now define also an infinitesimal pseudo-metric P_X. Let $v \in T_{x_0} X$. Essentially we want to set

$$P_X(v) = \sup\{|v(\varphi)|;\ \varphi \in \mathcal{P}_X(x_0)\}.$$

Since φ is not necessarily differentiable, taking a map $f \in \mathrm{Hol}(D, X)$ such that $f(0) = x_0$ and $f'(0) = v$, we set

$$(4.3.9) \qquad\qquad P_X(v) = \limsup_{t \to 0, t > 0} \frac{\lambda_X(f(t), x_0)}{t}.$$

As in Sibony [2], we may consider the subfamily $\mathcal{Q}(x_0) \subset \mathcal{P}(x_0)$ consisting of those functions $\varphi \in \mathcal{P}_X(x_0)$ which are of class C^2 in a neighborhood of x_0 and define an infinitesimal pseudo-metric Q_X by

$$(4.3.10) \qquad\qquad Q_X(v) = \sup\{|v(\varphi)|;\ \varphi \in \mathcal{Q}_X(x_0)\}.$$

It is a straightforward matter to verify the following

(4.3.11) Proposition. (1) *Let X and Y be a complex manifolds and $f \in \mathrm{Hol}(X, Y)$. Then*

$$f^* P_Y \le P_X, \qquad f^* Q_Y \le Q_X;$$

(2) *For the unit disc D, both P_D^2 and Q_D^2 coincide with the Poincaré metric ρ.*

Hence,

$$(4.3.12) \qquad\qquad E_X \le Q_X \le P_X \le F_X.$$

Intuitively, integrating P_X should yield the inner pseudo-distance p_X^i generated by p_X. However, no such precise relation is known. In fact, P_X may not even be upper semicontinuous in general. If X is a Stein manifold, P_X is known to be upper semicontinuous (see Demailly [1]). For more on this topic, see Jarnicki-Pflug [10].

Masaaki Suzuki [2] has shown that an unbounded strictly pseudoconvex domain X is p_X-hyperbolic. For P_X, see also Azukawa [3, 5] and Klimek [4].

4 Holomorphic Completeness

We shall first study the relationship between holomorphic completeness of a complex space X and strong completeness (or finite-compactness) of the Carathéodory distance c_X.

The following result (Kobayashi [14], Vesentini [1]) is stronger than just claiming that $c_X(p, x)$ is a plurisubharmonic function of x.

(4.4.1) Theorem. *Let X be a complex space, and fix a point $p \in X$. Then $\log c_X(p, x)$ is a plurisubharmonic function of $x \in X$.*

Proof. We make use of the following elementary fact.

(4.4.2) Lemma. *Fix a point a in the unit disc D. Then $\log \rho(a, z)$ is a subharmonic function of $z \in D$.*

In order to prove Lemma, we may assume that $a = 0$ because of the homogeneity of D. Then the assertion follows from the formula (see Section 1 of Chapter 2):

$$\rho(0, z) = \log \frac{1 + |z|}{1 - |z|}.$$

Now, by Lemma, for every fixed mapping $f \in \mathrm{Hol}(X, D)$ the function $\log \rho(f(p), f(x))$ is plurisubharmonic in $x \in X$. Being the upper-envelop of a family of plurisubharmonic functions $\{\log \rho(f(p), f(x)); \; f(x) \in \mathrm{Hol}(X, D)\}$, the function $\log c_X(p, x)$ is plurisubharmonic. $\qquad\qquad\square$

We denote the set of holomorphic functions on a complex space X by $H(X)$. Given a subset $K \subset X$, its **holomorphic hull** is defined to be

$$(4.4.3) \qquad \hat{K}_{H(X)} = \{x \in X; \; |f(x)| \le \sup_{y \in K} |f(y)| \quad \text{for all} \quad f \in H(X)\}.$$

A connected open subset R of a Stein space X is called a **Runge domain** in X if, for every compact subset $K \subset R$, its holomorphic hull $\hat{K}_{H(X)}$ in X is compact and is contained in R.

According to the classical theorem of Oka-Weil, R is a Runge domain in X if and only if R is a Stein space and every holomorphic function on R can be uniformly approximated on compact subsets of R by holomorphic functions on X, i.e., $H(X)|_R$ is dense in $H(R)$. In particular, a domain in $\mathbf{C}^n$ is a Runge domain if and only if it is a domain of holomorphy and the polynomials are dense in $H(R)$.

Now, as an immediate consequence of (4.4.1) we obtain

(4.4.4) Corollary. *Let X be a Stein space such that c_X is a strongly complete distance. Then for every $p \in X$ and every positive number r, the set*

$$X(p; r) = \{x \in X; \; c_X(p, x) < r\}$$

is a Runge domain in X.

Proof. By (4.4.1) $\log c_X(p, x)$ is plurisubharmonic on X, but not continuous at p. So, fix $r > 0$ and consider the following function

$$\varphi(x) = \max\left\{\log c_X(p, x), \ \log\frac{r}{2}\right\},$$

which is plurisubharmonic and continuous on X. Now,

$$X(p, r) = \{x \in X; \ \varphi(x) < \log r\}$$

is Runge by the following result (see, for example, Hörmander [1]).

(4.4.5) Lemma. *If φ is a continuous plurisubharmonic function on a Stein space X, then for every real number r the set $X_{\varphi,r} = \{x \in X; \varphi(x) < r\}$ is a Runge domain in X.*

The proof of (4.4.5) will not be given here. $\square$

(4.4.6) Theorem. *Let X be a complex space with the following properties*:
   (a) c_X *is a distance*;
   (b)   *There is a point $p \in X$ such that $X(p; r) = \{x \in X; \ c_X(p, x) \le r\}$ is compact for every $r > 0$.*
   Then X is a Stein space.

Proof. It is known that a complex space X is Stein if the following two conditions are met:
   (i)   For every point $p \in X$ there is a holomorphic map $f: X \to \mathbf{C}^N$, $N = N(p)$, such that p is an isolated point of $f^{-1}(f(p))$;
   (ii)   There exists a continuous plurisubharmonic function φ such that $X_{\varphi,r} = \{x \in X; \varphi(x) \le r\}$ is compact for every r, (Lelong pseudo-convexity, see Appendix A of this Chapter).
   Condition (i) is clearly satisfied if $H(X)$ separates the points of X, in particular, if c_X is a distance.
   Let $\varphi(x) = \log c_X(p, x)$. Then by (4.4.1) φ is plurisubharmonic, and Assumption (b) guarantees Condition (ii). $\square$

If c_X is strongly complete, by definition Condition (b) of (4.4.6) is satisfied by every point p of X. Hence,

(4.4.7) Corollary. *If X is a complex space such that c_X is a strongly complete distance, then X is a Stein space.*

This generalizes Horstmann's [1] result that a domain X in $\mathbf{C}^n$ with strongly complete c_X is a domain of holomorphy.
   We review now (4.4.7) from a slightly different angle. Let A be a set of holomorphic functions on X, i.e., $A \subset H(X)$. We say that X is A-**separable** if the elements of A separate the points of X. For every subset $K \subset X$, its A-**convex hull** $\hat{K}_A$ is defined by

$$\hat{K}_A = \{x \in X; \ |f(x)| \le \sup_{y \in K}|f(y)| \quad \text{for all} \quad f \in A\}.$$

We say that X is A-**convex** if for every compact subset $K \subset X$ its A-convex hull $\hat{K}_A$ is compact.

If X is $H(X)$-separable and $H(X)$-convex, then it is Stein space. If $A \subset B \subset H(X)$ and if X is A-separable (resp. A-convex), then X is B-separable (resp. B-convex). In particular, if X is A-separable and A-convex, then it is a Stein space.

Let $H^\infty(X)$ denote the algebra of bounded holomorphic functions on X. For each $p \in X$, the subalgebra of $H^\infty(X)$ consisting of those functions which vanish at p will be denoted by $H_p^\infty(X)$. Then the following theorem strengthens (4.4.7), (Kobayashi [7]).

(4.4.8) **Theorem**. *Let X be a complex space with strongly complete c_X. Then X is $H_p^\infty(X)$-separable and $H_p^\infty(X)$-convex for every $p \in X$, and hence it is $H^\infty(X)$-separable and H^∞-convex.*

Proof. Fix a point $p \in X$ and let a be a positive number. Let A denote the family of holomorphic maps $f: X \to D$ such that $f(p) = 0$. Since $A \subset H_p^\infty(X)$ and since every element of $H_p^\infty(X)$ is of the form cf for some $f \in A$ and $c \in \mathbf{C}$, it follows that the A-convex hull of a set $K \subset X$ coincides with the $H_p^\infty(X)$-convex hull of K.

Let $K = K(p, a) = \{x \in X; \ c_X(p, x) \leq a\}$. Since X is strongly complete with respect to c_X, the ball K is compact. Since every compact subset of X is contained in $K = K(p, a)$ for a sufficiently large a, it suffices to prove that A-convex hull $\hat{K}_A$ is compact. We shall actually show that $\hat{K}_A = K$.

$$
\begin{aligned}
\hat{K}_A \ &= \ \{x \in X; \ |f(x)| \leq \sup_{y \in K} |f(y)| \quad \text{for} \quad f \in A\} \\
&= \ \{x \in X; \ \rho(0, f(x)) \leq \sup_{y \in K} \rho(0, f(y)) \quad \text{for} \quad f \in A\} \\
&\subset \ \{x \in X; \ \rho(0, f(x)) \leq \sup_{y \in K} c_X(p, y) \quad \text{for} \quad f \in A\} \\
&= \ \{x \in X; \ \rho(0, f(x)) \leq a \quad \text{for} \quad f \in A\} = K.
\end{aligned}
$$

Since $\hat{K}_A$ contains K, we conclude $\hat{K}_A = K$. $\square$

The converse to (4.4.8) does not hold in general. See Sibony [1] and Ahern-Schneider [1] for counter-examples.

In order to strengthen (4.4.7), we shall now explain the concept of Banach-Stein space introduced by Fischer [1], (see also Ancona-Speder [1]). With the topology of uniform convergence on compact sets, the space $H(X)$ of holomorphic functions on X is a Fréchet space. We say that a complex space X is **Banach-Stein** if there exists a Banach space A with a continuous linear injection $i: A \to H(X)$ satisfying the following conditions:

(a) X is $i(A)$-separable;

(b)   For every infinite sequence of points $\{x_n\}$ in X without accumulation points, there exists a holomorphic function $f \in i(A) \subset H(X)$ which is not bounded on $\{x_n\}$;

(c)   For every automorphism φ of X, the induced automorphism φ^* of $H(X)$ maps $i(A)$ onto itself and induces an endomorphism of A;

(d)   For each complex space S and each S-automorphism Φ of $S \times X$ (i.e., Φ is of the form $\Phi(s, x) = (s, \varphi_s(x))$, where φ_s is an automorphism of X), the map $\tilde{\Phi} : S \to \mathrm{End}(A)$ defined by $\tilde{\Phi}(s) = \varphi_s^*$ is holomorphic.

(4.4.9) **Remark**. If X is $H(X)$-convex, then for every infinite sequence of points $\{x_n\}$ in X without accumulation points, there exists a holomorphic function $f \in H(X)$ which is not bounded on $\{x_n\}$, and vice-versa. But, Condition (b) is, in general, stronger than $i(A)$-convexity. For example, if $A = H^\infty(X)$, then (b) is never satisfied whether X is $H^\infty(X)$-convex or not.

The following result is due to Hirschowitz [1].

(4.4.10) **Theorem**. *If a complex space X is strongly complete with respect to its Carathéodory distance c_X, then it is Banach-Stein.*

Proof. Throughout the proof we write $\exp t$ for e^t. Fix a point $p \in X$ and let A be the family of holomorphic functions f such that

$$\| f \| := \sup_{x \in X} |f(x)| \exp(-c_X(p, x)) < \infty.$$

Then A is a Banach subspace of $H(X)$ with the norm $\| f \|$ defined above. We shall verify the four conditions (a), (b), (c) and (d).

(a)   Since A contains $H^\infty(X)$ and since X is $H^\infty(X)$-separable, X is A-separable.

(c)   Let $\varphi \in \mathrm{Aut}(X)$, and $f \in A$ with $\| f \| \le M$. Then

$$\begin{aligned}
|f(\varphi(x))| &\exp(-c_X(p, x)) \\
&= |f(\varphi(x))| \exp(-c_X(p, \varphi(x)) - c_X(p, x) + c_X(p, \varphi(x))) \\
&\le M \exp(-c_X(p, x) + c_X(\varphi^{-1}(p), x)) \\
&\le M \exp(c_X(p, \varphi^{-1}(p))),
\end{aligned}$$

showing that $\varphi^* f \in A$ and

$$\| \varphi^* f \| \le M \exp(c_X(p, \varphi^{-1}(p))).$$

This also shows that

$$\| \varphi^* \| \le \exp(c_X(p, \varphi^{-1}(p))).$$

(d)   We need the following

(4.4.11) **Lemma**. *Let S be a complex space and X a hyperbolic complex space. Given a map $\varphi \in \mathrm{Hol}(S \times X, X)$, set $\varphi_s(x) = \varphi(s, x)$. If φ_{s_0} is an automorphism of X for some $s_0 \in S$, then $\varphi_s = \varphi_{s_0}$ for all $s \in S$.*

This lemma, essentially due to H. Cartan [8] (see also Konrad Peters [1]), will be proved in (5.4.5); see also paragraphs following (5.4.4).

In verifying (d), we note that by this lemma the map $\tilde{\Phi}\colon S \to \mathrm{End}(A)$ defined by $\tilde{\Phi}(s) = \varphi_s^*$ is constant and hence, in particular, holomorphic.

(b) Let $\{x_n\}$ be an infinite sequence of points in X without accumulation points. By taking a subsequence, we may assume that $c_X(p, x_n)$ goes to infinity as $n \to \infty$. We can then choose $f_n \in \mathrm{Hol}(X, D)$ such that $f_n(p) = 0$ and $\lim |f_n(x_n)| = 1$. Taking a subsequence, we may assume that $\{f_n\}$ converges to a map $f \in \mathrm{Hol}(X, D)$.

For each n, we set

$$\Gamma_n = \{g \in A;\ |g(x_m)| \leq n \quad \text{for all} \quad m \geq n\}.$$

Clearly, Γ_n is closed in A. We shall show that it has an empty interior. That will show (by the theorem of Baire) that $A - \bigcup_n \Gamma_n$ is nonempty, and every element of $A - \bigcup_n \Gamma_n$ is unbounded on $\{x_n\}$.

Given $g \in \Gamma_n$ and $\varepsilon > 0$, we shall show that there is an $h \in A$ such that $\|h\|_A \leq \varepsilon$ and $g + h \notin \Gamma_n$. It suffices to find an $h \in A$ and an integer $m > n$ such that $\|h\|_A \leq \varepsilon$ and $|h(x_m)| > 2n$. We shall find such an h in the form $h_{k,m} = 3nf_m^k$. We set

$$K = \left\{x \in X;\ c_X(p, x) \leq -\log \frac{\varepsilon}{3n}\right\}.$$

Since X is strongly complete with respect to c_X, K is compact. Hence, $\sup_{x \in K} |f(x)| < 1$. Since $f_m \to f$, it follows that there exists $L < 1$ such that $\sup_{x \in K} |f_m(x)| \leq L$ for all m. Hence, there is an integer k such that $\sup_{x \in K} |f_m^k(x)| \leq \varepsilon/3n$ for all m, so that $\sup_{x \in K} |h_{k,m}(x)| \leq \varepsilon$ for all m. If $x \notin K$, then

$$|h_{k,m}(x)e^{-c_X(p,x)}| \leq \varepsilon |f_m^k(x)| < \varepsilon.$$

Hence,

$$\|h_{k,m}\|_A = \sup_{x \in X} |h_{k,m}(x)e^{-c_X(p,x)}| \leq \varepsilon.$$

Since

$$h_{k,m}(x_m) = 3n|f_m^k(x_m)| \to 3n \quad \text{as} \quad m \to \infty,$$

it follows that $|h_{k,m}(x_m)| > 2n$ for m large. $\square$

The theorem above was proved by Hirschowitz to show that a holomorphic fibre bundle over a Stein manifold with a fibre that is strongly complete in its Carathéodory distance is itself a Stein manifold.

Following Stehlé [1] we say that a Stein space X is **hyperconvex** if there exists a continuous plurisubharmonic negative function ψ on X such that

$$X_c = \{x \in X;\ \psi(x) \leq c\}$$

is compact for every $c < 0$. Intuitively, this means that ψ approaches 0 at the "boundary" of X.

(4.4.12) **Theorem.** *If a complex space X is strongly complete with respect to its Carathéodory distance c_X, then it is hyperconvex.*

Proof. By (4.4.7) X is Stein. Fix $a \in X$ and consider the family $H_a^\infty(X)$ of bounded holomorphic functions on X which vanish at a. By (4.4.8), X is $H_a^\infty(X)$-convex. The proof is now reduced to that of the following (Stehlé [1]):

(4.4.13) Lemma. *If a Stein space X is $H_a^\infty(X)$-convex, it is hyperconvex.*

Proof of Lemma. Let $\gamma(x) = \sup |f(x)|^2$, where the supremum is taken over all $f \in H_a^\infty(X) \cap \mathrm{Hol}(X, D)$. Being the upper-envelop of a family of plurisubharmonic functions $|f|^2$, γ is plurisubharmonic. Since

$$c_X(a, x) = \rho(0, \gamma(x)) = \log \frac{1 + \gamma(x)}{1 - \gamma(x)},$$

γ is continuous. Set $\psi(x) = \gamma(x)^2 - 1$. Then ψ satisfies the required conditions. $\qquad\Box$

(4.4.14) Remarks. According to Stehlé [1], a real positive function f on X is said to be m-**plurisubharmonic** if the function g defined by

$$g = \frac{1}{1 - m} f^{1-m} \qquad \text{for} \quad m \neq 1,$$

$$g = \log f \qquad \text{for} \quad m = 1,$$

is plurisubharmonic. A Stein space X is hyperconvex if and only if it admits an m-plurisubharmonic function f with $m > 1$ which goes to infinity at the "boundary" (i.e., such that $\{x \in X; \ f(x) \leq c\}$ is compact for every $c \in \mathbf{R}$). This can be seen by verifying that the function $g = \frac{1}{1-m} f^{1-m}$ is a negative plurisubharmonic function approaching 0 at the "boundary" if and only if f satisfies the condition above. Theorem (4.4.1) says that $c_X(a, x)$ is a 1-plurisubharmonic function. Hence, if X is strongly complete with respect to c_X, then $c_X(a, x)$ is a 1-plurisubharmonic function going to infinity at the "boundary". But this is a little less than saying that X is hyperconvex. For more results on hyperconvexity, see Section 2 of Chapter 5.

The concept of Banach-Stein complex space and that of hyperconvex Stein space were introduced in an attempt to prove an old conjecture of Serre (1953) that the total space of a holomorphic fibre bundle over a Stein manifold with a Stein fibre is also Stein. The conjecture has been verified under various additional assumptions. A brief summary of positive results can be found in Skoda [1], which gives a counter-example to the conjecture. For recent results on the subject, see Vâjâitu [1].

5 Strongly Pseudoconvex Domains

Let X be a bounded domain in $\mathbf{C}^n$ with coordinate system $(z^1, \ldots, z^n)$. It is said to be **strongly pseudoconvex** with C^2 boundary if there exists a real C^2 strongly plurisubharmonic function ϕ defined on a neighborhood U of the boundary ∂X such that

(i) $X \cap U = \{x \in X;\ \phi(x) < 0\}$;

(ii) $d\phi \neq 0$ in U.

The **Levi form** of ϕ at $x_0 \in \partial X$ is a Hermitian form given by

$$(4.5.1) \qquad L_{\phi, x_0}(\zeta) = \sum \frac{\partial^2 \phi}{\partial z^i \partial \bar{z}^j}(x_0) \zeta^i \bar{\zeta}^j, \qquad \zeta = (\zeta^1, \dots, \zeta^n).$$

The expression

$$(4.5.2) \quad p_x(z) = \sum \frac{\partial \phi}{\partial z^j}(x)(z^j - x^j) + \frac{1}{2} \sum \frac{\partial^2 \phi}{\partial z^j \partial z^k}(x)(z^j - x^j)(z^k - x^k)$$

is the **Levi polynomial** of ϕ at $x \in \partial X$. Expanding ϕ about $x_0 \in \partial X$, we obtain

$$(4.5.3) \qquad \phi(z) = 2\mathrm{Re}(p_{x_0}(z)) + L_{\phi, x_0}(z - x_0) + o(\|z - x_0\|^2),$$

where Re denotes the real part.

Assume that X is a strongly pseudoconvex domain with C^2 boundary. Since L_{ϕ, x_0} is positive definite and since ∂X is compact, there are positive constants c_1, c_2 such that

$$c_1 \|\zeta\|^2 \leq L_{\phi, x_0}(\zeta) \leq c_2 \|\zeta\|^2, \qquad x_0 \in \partial X, \quad \zeta \in \mathbf{C}^n.$$

Since $\phi(z) < 0$ in $X \cap U$, there is a neighborhood V_{x_0} of x_0 for which $\mathrm{Re}(p_{x_0}(z)) < 0$ in $V_{x_0} \cap X$. Since ∂X is compact, we can assume that V_{x_0} is of uniform size, i.e., there exists a fixed neighborhood V of the origin such that $V_{x_0} = x_0 + V$ for all $x_0 \in \partial X$.

Given a strongly pseudoconvex domain X, for each boundary point $x_0 \in \partial X$ there is a peak function $\Psi(x_0, z)$ that depends continuously on x_0. More precisely, we have the following result due to Henkin [1], Øvrelid [1] and Graham [1].

(4.5.4) **Theorem.** *Let $X \subset \mathbf{C}^n$ be a strongly pseudoconvex bounded domain with C^2 boundary. Then there exist a neighborhood X' of $\bar{X}$ and a continuous function $\Psi: \partial X \times X' \to \mathbf{C}$ such that for each fixed $x_0 \in \partial X$, $\Psi(x_0, \cdot)$ is holomorphic in X' and is a peak function for X at x_0, normalized so that $\Psi(x_0, x_0) = 1$ and $|\Psi(x_0, z)| < 1$ for $z \in \bar{X} - \{x_0\}$.*

The proof requires the following result of Kohn, (see Krantz [1] and Graham [1]). We denote by $L^2_{(0,1)}(X)$ (resp. $L^\infty_{(0,1)}(X)$) the space of $(0, 1)$-forms on X with square-integrable (resp. bounded) coefficients. Since X is bounded, we have $L^\infty_{(0,1)}(X) \subset L^2_{(0,1)}(X)$.

(4.5.5) **Lemma.** *Let X be as in (4.5.4). Given a d''-closed smooth $(0, 1)$-form ω in $L^2_{(0,1)}(X)$, there is a unique smooth solution $u \in L^2(X)$ of $d''u = \omega$ orthogonal to the space $H^2(X)$ of square-integrable holomorphic functions on X.*

Moreover, the linear operator $S: L^2_{(0,1)}(X) \to L^2(X)$ *defined by* $u = S\omega$ *is bounded.*

(4.5.6) Lemma. *Let X be as in (4.5.4). Let $M \subset \mathbf{R}^m$ be a compact set and $\omega: M \to L^\infty_{(0,1)}(X)$ be a continuous map such that $\omega_t = \omega(t)$ is smooth and d''-closed for every $t \in M$. Let $u_t = S\omega_t$. Then the function $u: M \times X \to \mathbf{C}$ given by $u(t, z) = u_t(z)$ is continuous on $M \times X$.*

Proof of (4.5.6) *assuming* (4.5.5). By (4.5.5) and the linearity of S, u_t converges to u_{t_0} uniformly on compact subsets of X as $t \to t_0$. This implies the joint continuity of $u(t, z)$. $\qquad\qquad\qquad\square$

Proof of (4.5.4). Let V be a neighborhood of the origin such that $\mathrm{Re}(p_{x_0}) < 0$ in $X \cap V_{x_0}$ (where $V_{x_0} = x_0 + V$) for every $x_0 \in \partial X$. Let

$$A(x_0) = \{z \in \mathbf{C}^n; \ p_{x_0}(z) = 0\}.$$

We claim there exist two Euclidean balls $B_2 \subset\subset B_1 \subset\subset V$ centered at the origin and a strongly pseudoconvex neighborhood X_1 of $\bar{X}$ such that

$$(*) \qquad\qquad (B_1(x_0) - B_2(x_0)) \cap X_1 \cap A(x_0) = \emptyset,$$

where $B_j(x_0) = x_0 + B_j$, $j = 1, 2$. In fact, let $\varepsilon > 0$ be smaller than the eigenvalues of L_{ϕ, x_0} for all $x_0 \in \partial X$. Using (4.5.3) we can choose a ball B_1 such that

$$A(x_0) \cap B_1(x_0) \cap \{z \in U; \ \phi(z) - \varepsilon\|z - x_0\|^2 = 0\} = \{x_0\}.$$

Then take any ball $B_2 \subset\subset B_1$ of radius r say, and put

$$X_1 = \{z \in U; \ \phi(z) < \varepsilon r^2\} \cup (X - U).$$

Finally, let X_2 be a strongly pseudoconvex domain with C^∞ boundary such that $X \subset\subset X_2 \subset\subset X_1$.

Let $\chi: \mathbf{C}^n \to [0, 1]$ be a smooth function with support contained in B_1 and identically 1 on B_2, and put $\chi_{x_0}(z) = \chi(z - x_0)$. Then for every $x_0 \in \partial X$, χ_{x_0}/p_{x_0} is well defined on X and $\mathrm{Re}(\chi_{x_0}/p_{x_0}) \leq 0$ on X. The d''-closed $(0, 1)$-form $\omega_{x_0} = d''(\chi_{x_0}/p_{x_0})$ has bounded smooth coefficients on X_2 and is jointly continuous in $z \in X_2$ and $x_0 \in \partial X$. Applying (4.5.6) with $M = \partial X$, the solution $u_{x_0} = S\omega_{x_0}$ yields a continuous function $u: \partial X \times X_2 \to \mathbf{C}$. Replacing X_2 by a slightly smaller domain we may assume that $|u| < k$ on $\partial X \times X_2$ for some constant k. Since

$$d''(\chi_{x_0}/p_{x_0}) - d''u_{x_0} = \omega_{x_0} - \omega_{x_0} = 0,$$

the functions $\chi_{x_0}/p_{x_0} - u_{x_0} - k$ are meromorphic on X_2 and have negative real part on $X \cup (X_2 - B_2(x_0))$ (for sufficiently large k).

The linear fractional transformation $h(w) = (w + 1)/(w - 1)$ maps the left half-plane to the unit disk, sending ∞ to 1. We set

$$\Psi_{x_0} = h(\chi_{x_0}/p_{x_0} - u_{x_0} - k).$$

Then by $(*)$ each $\Psi_{x_0} = \Psi(x_0, \cdot)$ is holomorphic on $X \cup (X_2 - B_2(x_0))$. On $B_2(x_0)$, we have

$$\Psi_{x_0} = \frac{1 - (u_{x_0} + k - 1)p_{x_0}}{1 - (u_{x_0} + k + 1)p_{x_0}}.$$

This is holomorphic when the denominator is not zero. Since $|u_{x_0}| < k$ on X_2, we have $|u_{x_0} + k + 1| < 2k + 1$ and $\operatorname{Re}(u_{x_0} + k + 1) > 1$ on X_2. In general, if two complex numbers α and β satisfy the relations $\operatorname{Re}(\beta) < 1/|\alpha|^2$ and $\operatorname{Re}(\alpha) > 1$, then $1 - \alpha\beta \neq 0$. Hence, if $\operatorname{Re}(p_{x_0}) < 1/(2k+1)^2$, then the denominator is not zero. Since the Levi polynomials are equicontinuous on X_2 and since $\operatorname{Re}(p_{x_0}) < 0$ on $X \cap B_1(x_0)$, there is a neighborhood $X' \subset X_2$ of $\bar{X}$ such that $\operatorname{Re}(p_{x_0}) < 1/(2k+1)^2$ on $X' \cap B_2(x_0)$ for all $x_0 \in \partial X$. Then the functions Ψ_{x_0} are all holomorphic on $X \cup (X_2 - B_2(x_0)) \cup (X' \cap B_2(x_0))$, which contains X'.

Set

$$\Psi(x_0, z) = \Psi_{x_0}(z).$$

The continuity of $\Psi(x_0, z)$ on $\partial X \times X'$ follows from that of $u(x_0, z)$, $p_{x_0}(z)$, and $\chi_{x_0}(z)$. From the construction of Ψ_{x_0}, it is clear that $\Psi_{x_0}(x_0) = 1$ and $|\Psi_{x_0}(z)| < 1$ on $\bar{X} - \{x_0\}$. $\qquad \square$

We shall now study the boundary behavior of c_X and d_X following Abate [1, 2]. See also Janicki-Pflug [10]. Some of the results go back to Vormoor [1]. For the boundary behavior of the corresponding infinitesimal metrics, see Graham [1], Aladro [1] and Ma [1, 2].

For $z \in X \subset \mathbb{C}^n$, its Euclidean distance from the boundary ∂X is denoted by $\delta(z, \partial X)$.

(4.5.7) Lemma. *Let B_r be the Euclidean ball of radius r centered at 0. Then for every $z \in B_r$,*

$$\log r - \log \delta(z, \partial B_r) \leq c_{B_r}(0, z) = d_{B_r}(0, z) \leq \log 2r - \log \delta(z, \partial B_r).$$

Proof. We have

$$d_{B_r}(0, z) = c_{B_r}(0, z) = \rho\left(0, \frac{\|z\|}{r}\right),$$

and

$$\delta(z, \partial B_r) = r - \|z\|.$$

Then, setting $t = \|z\|/r$, we have

$$\begin{aligned}
\log r - \log \delta(z, \partial B_r) &= \log \frac{1}{1 - t} \leq \log \frac{1 + t}{1 - t} = \rho(0, t) \\
&\leq \log \frac{2}{1 - t} = \log 2r - \log \delta(z, \partial B_r).
\end{aligned}$$

$\qquad \square$

A similar upper estimate holds for a general domain, pseudoconvex or not.

(4.5.8) Theorem. *Let $X \subset \mathbf{C}^n$ be a bounded domain with C^2 boundary, and K a compact subset of X. Then there is a constant $c_1 \in \mathbf{R}$ depending only on X and K such that*

$$d_X(z_0, z) \le c_1 - \log \delta(z, \partial X) \qquad \text{for} \quad z \in X, \; z_0 \in K.$$

Proof. For $x \in \partial X$, let $\mathbf{n}_x$ denote unit inward normal to ∂X at x. Since ∂X is of class C^2, we can take $\varepsilon > 0$ so small that $B(x + \varepsilon \mathbf{n}_x, \varepsilon) \subset X$ for all $x \in \partial X$. Let U_ε be the ε-neighborhood of ∂X. Put

$$c_1 = \sup\{d_X(z_0, z); \; z \in X - U_\varepsilon, \; z_0 \in K\} + \max\{0, \; \log \mathrm{diam}(X)\},$$

where $\mathrm{diam}(X)$ is the Euclidean diameter of X.

There are two cases:

(i) $z \in U_\varepsilon \cap X$. Let $x \in \partial X$ be such that $\|x - z\| = \delta(z, \partial X)$. Set $w = x + \varepsilon \mathbf{n}_x$ and $B = B(w, \varepsilon)$. Now (4.5.7) yields

$$
\begin{aligned}
d_X(z_0, z) \; &\le \; d_X(z_0, w) + d_X(w, z) \le d_X(z_0, w) + d_B(w, z) \\
&\le \; d_X(z_0, w) + \log 2\varepsilon - \log \delta(z, \partial B) \\
&\le \; c_1 - \log \delta(z, \partial X).
\end{aligned}
$$

(ii) $z \in X - U_\varepsilon$. Then

$$d_X(z_0, z) \le c_1 - \log \mathrm{diam}(X) \le c_1 - \log \delta(z, \partial X),$$

because $\delta(z, \partial X) \le \mathrm{diam}(X)$. $\qquad\square$

For a lower estimate, we use the strong pseudoconvexity of the domain.

(4.5.9) Theorem. *Let $X \subset \mathbf{C}^n$ be a strongly pseudoconvex bounded domain with C^2 boundary, and K a compact subset of X. Then there is a constant $c_2 \in \mathbf{R}$ depending only on X and K such that*

$$c_2 - \log \delta(z, \partial X) \le c_X(z_0, z) \qquad \text{for} \quad z \in X, \; z_0 \in K.$$

Proof. Let X' be a small neighborhood of $\bar{X}$, and $\Psi: \partial X \times X' \to \mathbf{C}$ be given by (4.5.4), and define $\varphi: \partial X \times X \times D \to D$ by

$$\varphi(x, z_0, \zeta) = \frac{1 - \overline{\Psi(x, z_0)}}{1 - \Psi(x, z_0)} \cdot \frac{\zeta - \Psi(x, z_0)}{1 - \overline{\Psi(x, z_0)}\zeta}.$$

Since there is r_0, $0 < r_0 < 1$, such that $|\Psi(x, z_0)| \le r_0 < 1$ for all $x \in \partial X$ and $z_0 \in K$, $\varphi(x, z_0, \zeta)$ is actually defined on $\partial X \times K \times D_{1/r_0}$. Then the map

$$\Phi(x, z_0, z) = \Phi_{x, z_0}(z) = \varphi(x, z_0, \Psi(x, z))$$

is defined and continuous on $\partial X \times K \times X'$ if X' is a sufficiently small neighborhood of $\bar{X}$, and each Φ_{x, z_0} is a holomorphic peak function for X at $x \in \partial X$ and satisfies $\Phi_{x, z_0}(z_0) = 0$.

Let $P(x, \varepsilon)$ be the polydisc of radius ε centered at x. For $x \in \partial X$, $z_0 \in K$ and $z \in P(x, \varepsilon)$ the Cauchy's integral formula for derivatives gives

$$
\begin{aligned}
|1 - \Phi_{x,z_0}(z)| &= |\Phi_{x,z_0}(x) - \Phi_{x,z_0}(z)| \leq \left\| \frac{\partial \Phi_{x,z_0}}{\partial z} \right\|_{P(x,\varepsilon)} \|z - x\| \\
&\leq \frac{c}{\varepsilon^2} \|\Phi\|_{\partial X \times K \times \partial P(x,2\varepsilon)} \|z - x\| = M\|z - x\|,
\end{aligned}
$$

where constant M is independent of z and x. Put

$$
c_2 = \min\{-\log M, \ \log \varepsilon\}.
$$

Noting that $B(x, \varepsilon) \subset P(x, \varepsilon)$, we set $U_\varepsilon = \bigcup_{x \in \partial X} B(x, \varepsilon)$, where Then there is an $\varepsilon > 0$ such that $U_\varepsilon \subset\subset X_0$ and U_ε is contained in a regular tubular neighborhood of ∂X. Then we have two cases:

(i) $z \in X \cap U_\varepsilon$. Choose $x \in \partial X$ so that $\delta(z, \partial X) = \|z - x\| < \varepsilon$. Since $\Phi_{x,z_0}(X) \subset D$ and $\Phi_{x,z_0}(z_0) = 0$, we have

$$
c_X(z_0, z) \geq \rho(\Phi_{x,z_0}(z_0), \Phi_{x,z_0}(z)) \geq \log \frac{1}{1 - |\Phi_{x,z_0}(z)|}.
$$

Now,

$$
1 - |\Phi_{x,z_0}(z)| \leq |1 - \Phi_{x,z_0}(z)| \leq M\|z - x\| = M\delta(z, \partial X);
$$

therefore

$$
c_X(z_0, z) \geq -\log M - \log \delta(z, \partial X) \geq c_2 - \log \delta(z, \partial X).
$$

(ii) $z \in X - U_\varepsilon$. Then $\delta(z, \partial X) \geq \varepsilon$. Hence,

$$
c_X(z_0, z) \geq 0 \geq \log \varepsilon - \log \delta(z, \partial X) \geq c_2 - \log \delta(z, \partial X).
$$

$\qquad\qquad\qquad\qquad\qquad\qquad\qquad\qquad\qquad\qquad\qquad\qquad\qquad\qquad\qquad\square$

Combining the two preceding theorems yields (Abate [1, 2])

(4.5.10) Corollary. *If $X \subset \mathbf{C}^n$ is a strongly pseudoconvex domain with C^2 boundary and $z_0 \in X$, then*

$$
\lim_{z \to \partial X} \frac{c_X(z_0, z)}{-\log \delta(z, \partial X)} = \lim_{z \to \partial X} \frac{d_X(z_0, z)}{-\log \delta(z, \partial X)} = 1,
$$

where the limits are locally uniform in the variable z_0.

Now we study the behavior of $d_X(z_1, z_2)$ when both z_1 and z_2 approach the boundary of X.

(4.5.11) Proposition. *Let $X \subset \mathbf{C}^n$ be a strongly pseudoconvex bounded domain with C^2 boundary, and $x_0 \in \partial X$. Then there exist $\varepsilon > 0$ and $c \in \mathbf{R}$ depending only on X and x_0 such that*

$$
c - \log \delta(z, \partial X) \leq c_X(z, z_0) \quad \text{for} \quad z \in X \cap B(x_0, \varepsilon), \ z_0 \in X - B(x_0, 2\varepsilon).
$$

Proof. Choose ε as in the proof of (4.5.9). Define $\varphi\colon \partial X \times X \times D \to D$ and $\Phi(x, z_0, z)$ as in the proof of (4.5.9). Since $z \in B(x_0, \varepsilon)$ by assumption, there is a point $x \in \partial X \cap B(x_0, \varepsilon)$ with $\|z - x\| = \delta(z, \partial X) < \varepsilon$. As in the proof of (4.5.9) we bound $|1 - \Phi_{x.z_0}(z)|$:

$$|1 - \Phi_{x, z_0}(z)| \leq M \|z - x\|.$$

In the proof of (4.5.9), M was independent of z_0 as long as z_0 stays in a fixed compact set K. This time, we claim that M can be chosen independent of z_0 as long as z_0 stays outside $B(x_0, 2\varepsilon)$. In fact, since $x \in \partial X \cap B(x_0, \varepsilon)$, by (4.5.4) there is a constant $a > 0$ such that $|\Psi(x, z_0)| \leq 1 - a$ for $z_0 \in X - B(x_0, 2\varepsilon)$. Since ∂X is compact, a can be chosen independent of $x \in \partial X$. Hence,

$$|\Phi(x, z_0, z)| \leq \left| \frac{1 - \overline{\Psi(x, z_0)}}{1 - \Psi(x, z_0)} \right| \leq \frac{1 + a}{1 - a},$$

proving our claim. The rest of the argument proceeds as in case (i) in the proof of (4.5.9). $\qquad\square$

The following is a weaker version of Forstneric-Rosay [1] who assumed strong pseudoconvexity only at x_1 and x_2.

(4.5.12) Corollary. *Let $X \subset \mathbf{C}^n$ be a strongly pseudoconvex bounded domain with C^2 boundary. Given two boundary points $x_1, x_2 \in \partial X$ with $x_1 \neq x_2$, there exist constants $\varepsilon > 0$ and $c \in \mathbf{R}$ depending only on X, x_1 and x_2 such that*

$$d_X(z_1, z_2) \geq c - \log \delta(z_1, \partial X) - \log \delta(z_2, \partial X)$$

for any $z_1 \in X \cap B(x_1, \varepsilon)$ and $z_2 \in X \cap B(x_2, \varepsilon)$.

Proof. Let $\varepsilon(x_j)$ and $c(x_j)$ be given by (4.5.11) for $j = 1, 2$, and choose $\varepsilon < \varepsilon(x_j)$, $j = 1, 2$, so small that $B(x_1, 2\varepsilon) \cap B(x_2, 2\varepsilon) = \emptyset$. Let σ be any curve from $z_1 \in B(x_1, \varepsilon)$ to $z_2 \in B(x_2, \varepsilon)$. Then part of σ is outside both $B(x_1, 2\varepsilon)$ and $B(x_2, 2\varepsilon)$. By (4.5.11), the length $L(\sigma)$ of σ measured by d_X must satisfy the following inequality:

$$L(\sigma) \geq c(x_1) + c(x_2) - \log \delta(z_1, \partial X) - \log \delta(z_2, \partial X).$$

Since d_X is an inner distance, the corollary follows. $\qquad\square$

Finally, the behavior of $d_X(z_1, z_2)$ as z_1 and z_2 approach the same boundary point is given by the following theorem of Forstneric-Rosay [1] (who actually assumed only that ∂X is of class $C^{1+\varepsilon}$).

(4.5.13) Theorem. *Let $X \subset \mathbf{C}^n$ be a bounded domain with C^2 boundary, and $x_0 \in \partial X$. Then there exist $\varepsilon > 0$ and $C \in \mathbf{R}$ depending only on X and x_0 such that*

$$d_X(z_1, z_2) \leq C - \sum_{j=1}^{2} \log \delta(z_j, \partial X) + \sum_{j=1}^{2} \log(\delta(z_j, \partial X) + \|z_1 - z_2\|)$$

for any $z_1, z_2 \in X \cap B(x_0, \varepsilon)$

Proof. For every $x \in \partial X$, let $\mathbf{n}_x$ denote the inward unit normal vector to ∂X at x. Choose $\varepsilon > 0$ so small that $\partial X \cap B(x_0, 4\varepsilon)$ is connected and for any point $z \in B(x_0, \varepsilon)$ the nearest point on ∂X is in $B(x_0, 2\varepsilon)$. We may further assume

(i) $\|\mathbf{n}_x - \mathbf{n}_{x_0}\| < 1/8$ for all $x \in \partial X \cap B(x_0, \varepsilon)$;

(ii)   for every $r \in [0, 4\varepsilon]$, $z \in \bar{X} \cap B(x_0, \varepsilon)$ and $x \in \partial X \cap B(x_0, 4\varepsilon)$, we have

$$z + r\mathbf{n}_x \in X \quad \text{and} \quad \delta(z + r\mathbf{n}_x, \partial X) > 3r/4.$$

Let $z_1, z_2 \in B(x_0, \varepsilon) \cap X$, and for $j = 1, 2$ let x_j be any point on ∂X nearest to z_j so that $\|z_j - x_j\| = \delta(z_j, \partial X)$. Then $x_j \in B(x_0, 2\varepsilon)$, and

$$\|z_j - x_j\| \le \|z_j - x_0\| < \varepsilon \quad \text{and} \quad z_j = x_j + \|z_j - x_j\|\mathbf{n}_{x_j}.$$

Set

$$z_j' = z_j + \|z_1 - z_2\|\mathbf{n}_{x_j} = x_j + (\|z_j - x_j\| + \|z_1 - z_2\|)\mathbf{n}_{x_j}.$$

Then $z_j' \in X$ by (ii) and

$$d_X(z_1, z_2) \le d_X(z_1', z_2') + \sum_{j=1}^{2} d_X(z_j, z_j').$$

In order to find an upper bound for $d_X(z_1', z_2')$, we note that $\|z_1 - z_2\| < 2\varepsilon$ implies $\delta(z_j', \partial X) > 3\|z_1 - z_2\|/4$ by (ii) and $\|z_1' - z_2'\| < 5\|z_1 - z_2\|/4$ by (i). Define an open set $\Omega \subset \mathbf{C}$ by

$$\Omega = \left\{ \zeta \in \mathbf{C};\ \min\{|\zeta|, |\zeta - 1|\} < \frac{3}{5} \right\},$$

and a map $\varphi \colon \Omega \to \mathbf{C}^n$ by

$$\varphi(\zeta) = z_1' + \zeta(z_2' - z_1').$$

Clearly, $\varphi(0) = z_1'$, and $\varphi(1) = z_2'$. Moreover, $\varphi(\Omega) \subset X$. This is because if $|\zeta| \le |\zeta - 1|$, then

$$\|\varphi(\zeta) - z_1'\| = |\zeta|\|z_2' - z_1'\| < \frac{5}{4}|\zeta|\|z_2 - z_1\| < \frac{3}{4}\|z_2 - z_1\|,$$

while if $|\zeta - 1| \le |\zeta|$, then

$$\|\varphi(\zeta) - z_2'\| = |\zeta - 1|\|z_2' - z_1'\| < \frac{5}{4}|\zeta - 1|\|z_2 - z_1\| < \frac{3}{4}\|z_2 - z_1\|.$$

Hence,

$$d_X(z_1', z_2') \le d_\Omega(0, 1).$$

Next, we shall bound $d_X(z_j, z_j')$ from above. Let $\varphi_j \in \mathrm{Hol}(\mathbf{C}, \mathbf{C}^n)$ be defined by

$$\varphi_j(\zeta) = x_j + \zeta\mathbf{n}_{x_j}.$$

Then $\varphi_j(0) = x_j$, $\varphi_j(\|z_j - x_j\|) = z_j$, and $\varphi_j(\|z_j - x_j\| + \|z_1 - z_2\|) = z_j'$. Set

$$\Omega_0 = \{\zeta = \xi + i\eta \in \mathbf{C}; \ |\zeta| < 4\varepsilon, \ \xi > K|\eta|^2\}.$$

If K is large enough, then $\varphi_j(\Omega_0) \subset X \cap B(x_0, 5\varepsilon)$. For convenience, fix a domain $\Omega_1 \subset \Omega_0$, symmetric with respect to the real axis, obtained by smoothing $\partial\Omega_0$ at its two corners. We have

$$d_X(z_j, z_j') \leq d_{\Omega_1}(\|z_j - x_j\|, \ \|z_j - x_j\| + \|z_1 - z_2\|).$$

It remains to show that if a and b are real numbers satisfying $0 < a < b < 3\varepsilon$, then

$$d_{\Omega_1}(a, b) \leq (\log b - \log a) + C',$$

where C' is a constant which depends only on X and x_0. Let $\tau : \Omega_1 \to D$ be a biholomorphic map such that $\tau(0) = 1$ and τ is real on the real axis. Since $\partial\Omega_1$ is of class C^2, τ extends to a diffeomorphism between $\bar{\Omega}_1$ and $\bar{D}$. Therefore, there are $K > 1$ and $\theta \in (-1, 1)$ such that

$$\max\{\theta, \ 1 - Kc\} \leq \tau(c) \leq 1 - c/K \qquad \text{for} \quad c \in (0, 3\varepsilon).$$

Then

$$
\begin{aligned}
d_{\Omega_1}(a, b) \ &= \ \rho(\tau(a), \tau(b)) = \rho(0, \tau(a)) - \rho(0, \tau(b)) \\
&\leq \ \log \frac{2}{a/K} - \log \frac{1+\theta}{Kb}.
\end{aligned}
$$

$\square$

As an application of these estimates, we prove that every biholomorphism between two strongly pseudoconvex bounded domains with C^2 boundary extends to a homeomorphism of their closures.

(4.5.14) Lemma. *Let $U \subset \mathbf{R}^N$ be a bounded domain with C^2 boundary. Let $f : \bar{U} \to \mathbf{r}$ be subharmonic on U and C^1 on $\bar{U}$. If f has a local maximum at $x_0 \in \partial U$ and if $\mathbf{n} = \mathbf{n}_{x_0}$ is the inner unit normal to ∂U at x_0, then*

$$\frac{\partial f}{\partial \mathbf{n}}(x_0) < 0.$$

Proof. Let $\varepsilon > 0$ be such that there exists a ball B of radius ε internally tangent to ∂U at x_0 so that $f(x_0) > f(x)$ for all $x \in B$. We may assume that the center of B is the origin 0. Let B_1 be a ball of radius $\varepsilon_1 < \varepsilon$ centered at x_0, and let $B' = B \cap B_1$. Then $\partial B'$ is the union of $S' = \partial B \cap \bar{B}_1$ and $S_1' = \partial B_1 \cap \bar{B}$.

Define $h : \mathbf{R}^N \to \mathbf{R}$ by

$$h(x) = e^{-\alpha\|x\|^2} - e^{-\alpha\varepsilon^2},$$

where $\alpha > 0$. Then $h > 0$ on $B' \subset B$ and

$$\triangle h = e^{-\alpha\|x\|^2}(4\alpha^2\|x\|^2 - 2\alpha N).$$

In particular, if α is large enough, then $\triangle h > 0$ on B'. Set

$$v(x) = f(x) + \delta h(x).$$

If δ is small enough, then $v(x) < f(x_0)$ on S_1'; moreover $v(x) = f(x) < f(x_0)$ for $x \in S' - \{x_0\}$. Since v is subharmonic on B', we have

$$\max_{x \in \overline{B}'} v(x) = f(x_0).$$

Therefore,

$$\frac{\partial v}{\partial \mathbf{n}}(x_0) = \frac{\partial f}{\partial \mathbf{n}}(x_0) + \delta \frac{\partial h}{\partial \mathbf{n}}(x_0) \leq 0.$$

But $\partial h / \partial \mathbf{n}(x_0) = 2\alpha\varepsilon e^{-\alpha\varepsilon^2} > 0$, and so $\partial f / \partial \mathbf{n}(x_0) < 0$. $\square$

(4.5.15) Lemma. *Let X, $X' \subset \mathbf{C}^n$ be bounded domains with C^2 boundary. Assume that X is strongly pseudoconvex. Let $f: X \to X'$ be a biholomorphic mapping. Then there exists a constant $C > 0$ such that*

$$\delta(f(z), \partial X') \leq C \cdot \delta(z, \partial X), \qquad z \in X.$$

Proof. We have

$$\delta(f(z), \partial X') \leq ce^{-d_{X'}(f(z_0), f(z))} = ce^{-d_X(z_0, z)} \leq ce^{-c_X(z_0, z)} \leq C\delta(z, \partial X),$$

where the first inequality comes from (4.5.8) while the last inequality is from (4.5.9). $\square$

(4.5.16) Theorem. *Let X, $X' \subset \mathbf{C}^n$ be strongly pseudoconvex bounded domains with C^2 boundary. Then every biholomorphic mapping $f: X \to X'$ extends to a homeomorphism of $\bar{X}$ onto $\bar{X}'$.*

Proof. Let $x_0 \in \partial X$. Assume that there are two sequences $\{z_j^1\}$, $\{z_j^2\} \subset X$ both converging to x_0 such that $f(z_j^1) \to y^1 \in \partial X'$ and $f(z_j^2) \to y^2 \in \partial X'$ with $y^1 \neq y^2$. We shall show that this leads to a contradiction. By (4.5.13) we have eventually

$$d_X(z_j^1, z_j^2) \leq C - \sum_{\mu=1}^{2} \log \delta(z_j^\mu, \partial X) + \sum_{\mu=1}^{2} \log(\delta(z_j^\mu, \partial X) + \|z_j^1 - z_j^2\|).$$

On the other hand, (4.5.12) yields

$$d_{X'}(f(z_j^1), f(z_j^2)) \geq C' - \sum_{\mu=1}^{2} \log \delta(f(z_j^\mu), \partial X').$$

But, $d_{X'}(f(z_j^1), f(z_j^2)) = d_X(z_j^1, z_j^2)$. Hence, using (4.5.15) we obtain

$$-\sum_{\mu=1}^{2} \log(\delta(z_j^\mu, \partial X) + \|z_j^1 - z_j^2\|) \leq C''.$$

Letting $j \to \infty$ we obtain a contradiction.

Hence, f extends to a continuous map $\bar{X} \to \bar{X}'$. Applying the same argument to f^{-1}, we see that f extends to a homeomorphism of $\bar{X}$ to $\bar{X}'$. $\square$

This topological extension theorem is due to Margulis [1] who used the Bergman metric as well as Henkin [2] and Vormoor [1] who used the boundary behavior of the Carathéodory distance. Using the Bergman kernel function and the Bergman metric, Fefferman [1] obtained a smooth extension theorem for biholomorphic maps between strongly pseudoconvex bounded domains with C^∞ boundary, and his proof was greatly simplified by Webster [1], Bell and Ligocka [1], and Forstnerič [2]. The proof by Lempert [4] which makes use of his theory of extremal discs is more in line with the content of this book. For other simplified proofs, see a survey by Forstnerič [1].

Continuous extension theorems for proper holomorphic mappings have been obtained by various people including Pinchuk [1], Alexander [1], Range [1], Diederich-Fornaess [1], Forstnerič-Rosay [1]. The earlier proofs relied on the boundary behavior of the Carathéodory distance. Diederich-Fornaess [1] and then Forstnerič-Rosay [1] used the Kobayashi distance to simplify the proof. The proof of (4.5.16) given here is based on Forstnerič-Rosay [1]. The proof shows that it suffices to prove (4.5.15) for proper holomorphic maps, for which (4.5.14) can be used (see Forstnerič-Rosay [1]). As we have shown, (4.5.14) is actually unnecessary for biholomorphic maps. More recently, smooth extension theorems for proper holomorphic mappings have been obtained by Bell-Catlin [1] and Diederich-Fornaess [3]. For a survey of the subject, see Diederich-Lieb [1], Bedford [1], Bell-Narasimhan [1], Forstnerič [1].

On the boundary behavior of the infinitesimal metric F_X for strongly pseudoconvex domains X, see Forstnerič-Rosay [1], Aladro [3], Pang [3], Fu [1, 2]. For the case of weakly pseudoconvex domains, see Cho [1, 2], Herbort [1], Krantz [4], Jarnicki-Pflug [10; Misc. E].

6 Extremal Discs and Complex Geodesics

Let X be a complex space. Given two points x and y of X, we say that a holomorphic map $f: D \to X$ is an **extremal disc** for the pair $\{x, y\}$ if there exist points $a, b \in D$ such that $f(a) = x$, $f(b) = y$ and

$$\rho(a, b) = d_X(x, y).$$

Given a point $x \in X$ and a tangent vector ξ at x, we say that a holomorphic map $f \in \mathrm{Hol}(D, X)$ is an **extremal disc** for $\xi \in T_x X$ if there exists a tangent vector $u \in T_0 D$ such that $f_*(u) = \xi$ and $\|u\| = F_X(\xi)$, where $\|u\|$ denotes the Poincaré length of u.

Replacing d_X or F_X in the above definitions by the Carathéodory pseudo-distance c_X or the infinitesimal Carathéodory metric E_X, we define the concept of **C-extremal disc**.

Following Vesentini [1, 2, 3] we say that a holomorphic map $f \in \mathrm{Hol}(D, X)$ is a **complex geodesic** (resp. **complex C-geodesic**) if it is extremal (resp. C-extremal) for every pair of points in its image $f(D)$. In other words, f is a complex geodesic (resp. complex C-geodesic) if and only if

$$\rho(a, b) = d_X(f(a), f(b)) \quad (\text{resp.} \quad \rho(a, b) = c_X(f(a), f(b)))$$

for all $a, b \in D$.

Now we prove basic results on complex geodesics by Vesentini. See also Abate [2].

(4.6.1) Theorem. *Let X be a complex space, and $f \in \mathrm{Hol}(D, X)$.*

(1)   If f is a C-extremal disc for one pair of distinct points $\{x, y\}$, then it is a complex C-geodesic and also a complex geodesic;

(2)   If f is a C-extremal disc for one nonzero vector $\xi \in T_x X$, then it is a complex C-geodesic and also a complex geodesic.

Proof. (1)   Take $a, b \in D$ such that $x = f(a)$ and $y = f(b)$. By definition, there is a map $h \in \mathrm{Hol}(X, D)$ such that

$$\rho(h(f(a)), h(f(b))) = c_X(f(a), f(b)) = \rho(a, b).$$

By Schwarz-Pick lemma, $h \circ f$ is an automorphism of D. Therefore, for any $c \in D$ we have

$$\rho(a, c) \geq c_X(f(a), f(c)) \geq \rho(h(f(a)), h(f(c))) = \rho(a, c),$$

which shows that f is C-extremal for the pair $\{f(a), f(c)\}$. Replacing a by an arbitrary point of D and repeating the same argument once more, we see that f is C-extremal for any two points of $f(D)$, i.e., f is a complex C-geodesic.

To see that f is a complex geodesic, take any pair $a', b' \in D$. Since

$$\rho(a', b') = c_X(f(a'), f(b')) \leq d_X(f(a'), f(b')) \leq \rho(a', b'),$$

f is extremal for the pair $\{f(a'), f(b')\}$., showing that f is a complex geodesic.

(2)   Take $u \in T_0 D$ such that $f_*(u) = \xi$ and $\|u\| = E_X(\xi)$. Then there is a map $h \in \mathrm{Hol}(X, D)$ such that $\|h_*(f_*(u))\| = E_X(f_*(u)) = \|u\|$. By Schwarz-Pick lemma, $h \circ f$ is an automorphism of D. The remainder of the proof is the same as in (1). $\square$

(4.6.2) Corollary. *Let X be a complex space. If $f \in \mathrm{Hol}(D, X)$ is extremal for one pair of points $\{x, y\}$ and if $c_X(x, y) = d_X(x, y)$, then it is a complex geodesic.*

(4.6.3) Proposition. *If $f \in \mathrm{Hol}(D, X)$ is a complex geodesic, it is a proper injective map of D into X.*
Proof. Let $K \subset X$ be compact. It is bounded with respect to d_X. If $f^{-1}(K)$ is not compact, it would be unbounded with respect to ρ, and this is a contradiction. Clearly, f is injective. $\square$

A complex geodesic is essentially determined by its image:

(4.6.4) Corollary. *If f and g are two complex geodesics in a complex space X and if $f(D) = g(D)$, then there is an automorphism α of D such that $g = f \circ \alpha$.*

Proof. By (4.6.3), $\alpha = f^{-1} \circ g$ is a homeomorphism of D onto itself and is holomorphic outside of a discrete subset of D since $df \neq 0$ outside of a discrete subset. By Riemann's extension theorem, α is holomorphic everywhere.   □

Let $p: \mathbf{C}^n \to \mathbf{R}^+$ be a norm (not necessarily the Euclidean norm), and B the unit ball for this norm. We say that a point $x_0 \in \partial B$ is a **complex extreme point** of $\bar{B}$ if there is no nonzero vector y such that $x_0 + ty \in \bar{B}$ for all $t \in \mathbf{C}$, $|t| < 1$. For example, if B is the usual Euclidean ball, then every boundary point is complex extreme. However, if B is a polydisc D^n, then the distinguished boundary $(\partial D)^n$ is exactly the set of complex extreme points.

(4.6.5) **Theorem**. *Let p be an arbitrary norm on $\mathbf{C}^n$, and B the open unit ball with respect to p. Let $z \in B$. Then $f_z(t) = tz/p(z)$, $t \in D$, is a complex geodesic, and it is the unique complex geodesic for the pair $\{0, z\}$ if and only if $z/p(z)$ is a complex extreme point of $\bar{B}$.*

Proof. From Example (3.1.24) we see that f_z is a complex geodesic.

Suppose that $x = z/p(z) \in \partial B$ is not a complex extreme point, and let y be a nonzero element of $\mathbf{C}^n$ such that $x + ty \in \bar{B}$ for all $t \in D$. For each $s \in D$, define $g_s \in \mathrm{Hol}(D, B)$ by

$$g_s(t) = t\left[x + s\frac{t - p(z)}{1 - p(z)t} y \right], \qquad t \in B.$$

Then $g_s(0) = 0$ and $g_s(p(z)) = z$. Since $\rho(0, p(z)) = c_B(0, z)$ by (3.1.24), each g_s is C-extremal for the pair $\{0, z\}$. By (4.6.1) g_s is a complex geodesic through 0 and z.

Conversely, let $f \in \mathrm{Hol}(D, B)$ be a complex geodesic through 0 and z, and assume that $x = z/p(z)$ is a complex extreme point of ∂B. Since $d_B(0, f(t)) = \rho(0, t)$ and since $d_B(0, f(t)) = \rho(0, p(f(t)))$ by (3.1.24), we have $p(f(t)) = |t|$. By reparametrization, we may assume that $f(0) = 0$ and $f(p(z)) = z$.

Define $h \in \mathrm{Hol}(D, \mathbf{C}^n)$ by $h(t) = f(t)/t$. Then $h(D) \subset \partial B$, and $h(p(z)) = z/p(z) = x$. We claim that h is a constant map, i.e., $h(t) = z/p(z)$ for all $t \in D$. If this claim is granted, then $f(t) = h(t)t = tz/p(z)$, which proves the uniqeness.

As the first step for proving our claim, we establish a few inequalities. Let $f \in \mathrm{Hol}(D, D)$. From the Schwarz-Pick lemma,

$$\left| \frac{f(t) - f(0)}{1 - \overline{f(0)}f(t)} \right| \leq |t|, \qquad t \in D.$$

On the other hand, the elementary inequalities

$$|1 - \bar{a}b| \leq 1 - |a|^2 + |\bar{a}b - |a|^2| \leq 2(1 - |a|) + |a||b - a|, \qquad a, b \in D$$

applied to $f(0)$ and $f(t)$ yields

$$|1 - \overline{f(0)}f(t)| \leq 2(1 - |f(0)|) + |f(0)||f(t) - f(0)|.$$

Hence,

$$|f(t) - f(0)| \leq 2|t|(1 - |f(0)|) + |tf(0)||f(t) - f(0)|.$$

This may be rewritten as

$$(4.6.6) \qquad 2|t||f(0)| + (1 - |tf(0)|)|f(t) - f(0)| \leq 2|t|.$$

(4.6.7) Lemma. *Let $B \subset \mathbf{C}^n$ be the unit ball with respect to a norm $p: \mathbf{C}^n \to \mathbf{R}^+$, and let $h \in \mathrm{Hol}(D, \mathbf{C}^n)$ be such that $h(D) \subset \bar{B}$. Then*

$$p(h(0) + s(h(t) - h(0))) \leq 1$$

for all $t \in D^*$ and $s \in \mathbf{C}$ such that $2|st| \leq 1 - |t|$.

Proof. For $n = 1$, $p(\cdot) = r|\cdot|$ for some $r > 0$ and Lemma follows from (4.6.6).

For $n > 1$, assume that Lemma does not hold for some $t \in D^*$ and $s \in \mathbf{C}$ such that $2|st| \leq 1 - |t|$. But there is a complex linear functional $\lambda: \mathbf{C}^n \to \mathbf{C}$ such that $|\lambda(z)| \leq p(z)$ for all $z \in \mathbf{C}^n$ and

$$\lambda(h(0) + s(h(t) - h(0))) = p(h(0) + s(h(t) - h(0))) > 1.$$

If we set $g = \lambda \circ h$, this means that $g(D) \subset \bar{D}$ and $|g(0) + s(g(t) - g(0))| > 1$, contradicting the case $n = 1$ of Lemma. $\quad\square$

We are now in a position to prove our claim:

(4.6.8) Lemma. *Let B be the unit ball with respect to a norm p in $\mathbf{C}^n$, and let $h \in \mathrm{Hol}(D, \mathbf{C}^n)$ be such that $h(D) \subset \bar{B}$. If $h(D)$ contains a complex extreme point of $\bar{B}$, then h is constant.*

Proof. If $h(D)$ contains a complex extreme point, which we can assume to be $h(0)$, (4.6.7) implies $h(t) - h(0) = 0$ for all $t \in D$. $\quad\square$

If B^n is the Euclidean ball defined by the Euclidean norm of $\mathbf{C}^n$, it is homogeneous and its boundary points are all complex extreme. Hence (4.6.5) determines all complex geodesics of B^n:

(4.6.9) Corollary. *Let B^n be the Euclidean unit ball in $\mathbf{C}^n$. For any pair of points $\{z_0, z_1\}$ in B^n the unique complex geodesic through z_0 and z_1 is the intersection of B^n with the affine complex line joining z_0 and z_1.*

Proof. If $z_0 = 0$, this follows from (4.6.5). The general case follows from the fact that every automorphism of B^n sends affine lines into affine lines. $\quad\square$

(4.6.10) Remark. In (3.1.3) we defined a function d'_X on $X \times X$ by setting

$$d'_X(x, x') = \inf \rho(a, a'),$$

where the infimum is taken over all holomorphic maps $f: D \to X$ and all pairs of points $a, a' \in D$ such that $f(a) = x$ and $f(a') = x'$. (If no such map f exists, then $d'_X(x, x') = \infty$ by definition). We then pointed out that d'_X may not satisfy the triangular inequality. If it satisfies the triangular inequality, then $d'_X = d_X$ since d_X is the largest pseudo-distance bounded by d'_X, (see Section 1 of Chapter 3).

Following Pang [1], we say that a point $x \in X$ is d_X-**simple** if x has a neighborhood U such that $d_X(x, y) = d_X'(x, y)$ for all $y \in U$. A complex space X is said to be d_X-**simple** if all points of X are d_X-simple.

As we shall see later in (4.8.6), every convex bounded domain X in $\mathbf{C}^n$ is d_X-simple. It is clear that a complex space X is d_X-simple if and only if its universal covering space $\tilde{X}$ is $d_{\tilde{X}}$-simple.

7 Extremal Problems and Extremal Discs

The purpose of this and next sections is to prove the theorem of Lempert (4.8.13) that on a bounded convex domain in $\mathbf{C}^n$ the Carathéodory distance and the Kobayashi distance coincide. We follow largely the functional analytic approach by Royden and Wong [1]. See also Abate [2] and Jarnicki-Pflug [10].

In general, let L be a complex Banach space. Let $P: L \to \mathbf{R}$ be a **Minkowski functional** on L, i.e., a non-negative real valued map having the properties (i), (ii) and (iii) below:

(i) $P(f + g) \leq P(f) + P(g)$, $f, g \in L$;

(ii) $P(rf) = r P(f)$, $f \in L$, $r \geq 0$;

(iii)   there exists $c > 0$ such that $c^{-1} \| f \| \leq P(f) \leq c \| f \|$ for all $f \in L$.

Since we do not assume $P(-f) = P(f)$, P is not a norm. But, $P(g - f)$ may be considered as a "non-symmetric distance" from f to g.

Given a linear subspace $M \subset L$ and a point f_0 not in the closure of M, we define

$$(4.7.1) \qquad m = \inf_{f \in M} P(f_0 + f).$$

Clearly, m depends only on $f_0 + M$, not on f_0 itself; m is the "distance" from the origin 0 to the affine subspace $f_0 + M$. A **linear extremal problem** is to find $f \in f_0 + M$ such that $m = P(f)$.

Let L^* be the dual Banach space. On L^* we define the dual Minkowski functional P^* by

$$(4.7.2) \qquad P^*(u) = \sup_{f \neq 0} \mathrm{Re}(u(f))/P(f).$$

To the linear extremal problem above, we associate the dual extremal problem as follows. Let M^0 be the annihilator of M, i.e.,

$$M^0 = \{ u \in L^*; \ u(f) = 0 \ \ \forall f \in M \},$$

and define

$$(4.7.3) \qquad m^* = \inf\{ P^*(u); \ u \in M^0 \ \ \text{and} \ \ \mathrm{Re}(u(f_0)) = 1 \}.$$

Again, m^* depends only on $f_0 + M$, not on f_0. Then the **dual extremal problem** is to find $u \in M^0$ such that

$$\operatorname{Re}(u(f_0)) = 1 \quad \text{and} \quad m^* = P^*(u).$$

We prove now the following **principle of duality**.

(4.7.4) Theorem *Let L be a complex Banach space, $P\colon L \to \mathbf{R}$ a Minkowski functional on L, M a linear subspace of L, and f_0 a point of L not in the closure of M. Then*

(1) $mm^* = 1$;

(2)   *there is always a solution u to the dual extremal problem*;

(3)   *if $f \in f_0 + M$ and $u \in M^0$ are such that* $\operatorname{Re}(u(f)) = P(f)P^*(u) = 1$, *then* $P(f) = m$ *and* $P^*(u) = m^*$, *i.e., f and u are respectively solutions of the linear extremal and dual extremal problems.*

Proof. On the linear span $\mathbf{R}f_0 + M$ of f_0 and M, define a real linear functional φ by setting

$$\varphi(af_0 + f) = a \qquad \text{for} \quad a \in \mathbf{R}, \quad f \in M.$$

Since

$$P(af_0 + f) = aP(f_0 + \frac{1}{a}f) \ge am \qquad \text{for} \quad a > 0$$

and since $P(af_0 + f) \ge 0$ for all $a \in \mathbf{R}$, we conclude that

$$\varphi(f) \le \frac{1}{m}P(f) \qquad \text{for} \quad f \in \mathbf{R}f_0 + M.$$

By the Hahn-Banach theorem φ can be extended to a real linear functional φ on L such that $\varphi(f) \le \frac{1}{m}P(f)$ for $f \in L$. Define a complex linear functional u by setting

$$u(f) = \varphi(f) - i\varphi(if).$$

Then

$$|u(f)| \le (\varphi(f)^2 + \varphi(if)^2)^{1/2} \le \frac{\sqrt{2}c}{m}\|f\|,$$

where c is the constant appearing in property (iii) of a Minkowski functional. In particular, $u \in L^*$. If $f \in M$, then $u(f) = \varphi(f) - i\varphi(if) = 0$ because M is a complex linear subspace and φ annihilates M by construction. Hence, $u \in M^0$. Furthermore, $\operatorname{Re}(u(f_0)) = \varphi(f_0) = 1$.

From $\operatorname{Re}(u(f)) = \varphi(f) \le \frac{1}{m}P(f)$ and from the definition of m^* we obtain

$$m^* \le P^*(u) \le \frac{1}{m};$$

in particular, $mm^* \le 1$.

On the other hand, for any $v \in M^0$ with $\operatorname{Re}(v(f_0)) = 1$ and any $f \in M$, from the definition of P^* we have

$$P(f_0 + f)P^*(v) \geq \mathrm{Re}(v(f_0 + f)) = \mathrm{Re}(v(f_0)) = 1;$$

in particular, $mm^* \geq 1$. Hence, $mm^* = 1$ and

$$m^* = P^*(u) = \frac{1}{m},$$

completing the proof of (1) and (2).

As for (3), if $f \in f_0 + M$ and $u \in M^0$ are such that $\mathrm{Re}(u(f)) = P(f)P^*(u) = 1$, then $\mathrm{Re}(u(f_0)) = \mathrm{Re}(u(f)) = 1$ and so $P^*(u) \geq m^*$ and $P(f) \geq m$. On the other hand,

$$mm^* \leq mP^*(u) \leq P(f)P^*(u) = 1 = mm^*.$$

Hence, $P^*(u) = m^*$ and $P(f) = m$. $\square$

Let D be the open unit disk in $\mathbf{C}$. Let $C(\bar{D})$ be the space of continuous complex functions on $\bar{D}$, and $H(D)$ denote the space of holomorphic functions on D. We shall need the following Banach spaces of holomorphic functions and subspaces.

$$(4.7.5) \qquad H_1(D) = \{f \in H(D);\ \|f\|_1 = \sup_{r \in (0,1)} \int_0^{2\pi} |f(re^{i\theta})|d\theta < \infty\};$$

$$(4.7.6) \qquad H_\infty(D) = \{f \in H(D);\ \|f\|_\infty = \sup_{t \in D} |f(t)| < \infty\};$$

$$(4.7.7) \qquad A(D) = H(D) \cap C(\bar{D}).$$

Thus, $A(D)$ is the space of holomorphic functions on D that extend continuously to the boundary ∂D.

Let $C(\partial D)$ be the space of continuous complex functions on ∂D with the supremum norm. While the dual space of $L_1(\partial D)$ is $L_\infty(\partial D)$, the dual space $L_\infty^*(\partial D)$ of $L_\infty(\partial D)$ is the space of absolutely continuous complex valued finitely additive set functions of bounded variation defined for all measurable subsets of ∂D. The dual space of $C^*(\partial D)$ of $C(\partial D)$ is the space of Radon measures on ∂D (i.e., complex valued completely additive set functions of bounded variation defined for all Borel subsets of ∂D).

There is a natural isometric immersion $L_1(\partial D) \to C^*(\partial D)$; each function $h \in L_1(\partial D)$ gives rise to a Radon measure $\frac{1}{2\pi}hd\theta$, that is, the linear functional associated to h is given by

$$(4.7.8) \qquad h(f) = \frac{1}{2\pi} \int_0^{2\pi} h(e^{i\theta})f(e^{i\theta})d\theta \qquad \text{for} \quad f \in C(\partial D).$$

The elements of $A(D)$ are completely determined by their values on ∂D; hence we can identify $A(D)$ with a closed subspace of $C(\partial D)$.

The following theorem of F. and M. Riesz says that the same holds for $H_1(D)$, see Duren [1].

(4.7.9) Theorem. (1)   *For any $h \in H_1(D)$ the limit*

$$h^*(e^{i\theta}) = \lim_{r \to 1} h(re^{i\theta})$$

exists for almost all θ, and $h \mapsto h^*$ is an isometric imbedding of $H_1(D)$ into $L_1(\partial D)$, and thus onto a closed subspace of $C^*(\partial D)$;
   (2)   *The annihilator of $A(D) \subset C(\partial D)$ is given by*

$$tH_1(D) = \{tf(t); \ f \in H_1(D)\},$$

where t is the coordinate function of $D = \{|t| < 1\}$;
   (3)   *If $h \in H_1(D)$, then*

$$h(0) = \frac{1}{2\pi} \int_0^{2\pi} h^*(e^{i\theta})d\theta.$$

In summary, we have

$$
\begin{array}{ccccccc}
A(D) & \subset & H_\infty(D) & \subset & H_1(D) & & \\
\cap & & \cap & & \cap & & \\
C(\partial D) & \subset & L_\infty(\partial D) & \subset & L_1(\partial D) & \subset & C^*(\partial D).
\end{array}
$$

We shall denote the product of n copies of any of the above spaces by a superscript n such as $C(\partial D)^n$, $L_1(\partial D)^n$, $H_1(D)^n$, $A(D)^n$, $C^*(\partial D)^n$, etc.; these are spaces of mappings from ∂D or D into $\mathbf{C}^n$.
   For $z = (z^1, \ldots, z^n)$, $w = (w^1, \ldots, w^n) \in \mathbf{C}^n$, write

$$(4.7.10) \qquad\qquad \langle w, z \rangle = \sum w^j z^j,$$

with no complex conjugate bar over z^j. Then (4.7.8) yields a natural map $L_1(\partial D)^n \to C^*(\partial D)^n$. Namely, for $h \in L_1(\partial D)^n$, we have

$$(4.7.11) \qquad h(f) = \frac{1}{2\pi} \int_0^{2\pi} \langle h(e^{i\theta}), f(e^{i\theta}) \rangle d\theta, \qquad f \in C(\partial D)^n.$$

Now, let $X \subset \mathbf{C}^n$ be a bounded convex domain containing 0. Then we define the Minkowski function $p: \mathbf{C}^n \to \mathbf{R}$ of X by

$$(4.7.12) \qquad\qquad p(z) = \inf\{r^{-1} \in \mathbf{R}; \ rz \in X, r > 0\}.$$

Clearly, $X = \{z \in \mathbf{C}^n; \ p(z) < 1\}$, and

(i) $p(z + w) \le p(z) + p(w)$ for $z, w \in \mathbf{C}^n$,

(ii) $p(rz) = rp(z)$ for $z \in \mathbf{C}^n$, $r \ge 0$.

Since X is not necessarily symmetric about 0, $p(-z) \ne p(z)$ in general.
   Using the notation $\langle w, z \rangle$ defined in (4.7.10), we define the dual Minkowski function $p^*: \mathbf{C}^n \to \mathbf{R}$ by

$$(4.7.13) \qquad p^*(w) = \sup_{z \neq 0} \frac{\mathrm{Re}\langle w, z \rangle}{p(z)} = \sup_{p(z)=1} \mathrm{Re}\langle w, z \rangle.$$

(4.7.14) Remark. A point z on the boundary ∂X where the supremum in (4.7.13) is attained can be found as follows. If we identify $\mathbf{C}^n$ with $\mathbf{R}^{2n}$ by writing

$$z^j = x^j + iy^j, \quad w^j = u^j + iv^j,$$

then $\mathrm{Re}\langle w, z \rangle = \sum(u^j x^j - v^j y^j)$ so that $\mathrm{Re} < w, z >$ is the usual inner product of $\bar{w}$ with z as real vectors in $\mathbf{R}^{2n}$. So we want to maximize the inner product with $\bar{w}$ on ∂X. Consider the supporting real hyperplanes of X (i.e.,tangent planes of ∂X if ∂X is smooth) that are perpendicular to the vector $\bar{w} \in \mathrm{R}^{2n}$. Since X is convex, there are two of them. We choose the one that is on the same side as $\bar{w}$ with respect to the origin 0. A point z where this hyperplane touches X is where the supremum is attained. In general, such a point z is not unique. But if X is strictly convex in the direction of $\bar{w}$, then z is unique (by definition). We note that X is strictly convex in almost all (in the sense of measure) directions $\bar{w}$. Hence, the map

$$w \in \mathbf{C}^n \to z \in \partial X \subset \mathbf{C}^n$$

is in $L_\infty(\partial D)^n$ in general but continuous if X is strictly convex.

Let L be a Banach space of bounded holomorphic mappings from D into $\mathbf{C}^n$ or (essentially) bounded mappings from ∂D into $\mathbf{C}^n$ such as $H_\infty(D)^n$, $A(D)^n$, $L_\infty(\partial D)^n$, or $C(\partial D)^n$. Then the Minkowski function p associated to X induces a Minkowski functional P on L by

$$(4.7.15) \qquad P(f) = \sup_t p(f(t)) \qquad \text{for} \quad f \in L,$$

where t varies over D or ∂D depending on the case. The definition is consistent with the inclusion $A(D)^n \subset C(\partial D)^n$ or $H_\infty(D)^n \subset L_\infty(\partial D)^n$ in the sense that for $f \in A(D)^n$ or $f \in H_\infty(D)^n$ the following equality holds:

$$(4.7.16) \qquad \sup_{t \in D} p(f(t)) = \sup_{t \in \partial D} p(f(t)).$$

In fact, let LH and RH denote the left and right hand sides of (4.7.16). Clearly, LH $\geq$ RH. Let $X_r = \{x \in X; p(x) < r\}$. Assume LH $>$ RH. Then there is $t_0 \in D$ such that $f(t_0) \notin \bar{X}_r$. Let x_0 be the point of ∂X_r closest to $f(t_0)$. Take a linear functional $\lambda \colon \mathbf{C}^n \to \mathbf{C}$ such that

$$\mathrm{Re}(\lambda(x)) \leq \mathrm{Re}(\lambda(x_0)) < \mathrm{Re}(\lambda(f(t_0))), \qquad x \in \bar{X}_r.$$

Define $g \in \mathrm{Hol}(\mathbf{C}^n, \mathbf{C})$ by $g(z) = e^{\lambda(z - x_0)}$. Then $|g \circ f| \leq 1$ on ∂D and $|g \circ f(t_0)| > 1$, thus contradicting the maximum principle.

(4.7.17) Lemma. *Let $X \in \mathbf{C}^n$ be a convex bounded domain with Minkowski function p.*

(1)   *Let P be the induced Minkowski functional on $L = L_\infty(\partial D)^n$ and P^* the dual Minkowski functional on $L_\infty^*(\partial D)^n$. Then for any $h \in L_1(\partial D)^n \subset L_\infty^*(\partial D)^n$, we have*

$$(*) \qquad P^*(h) = \frac{1}{2\pi} \int_0^{2\pi} p^*(h(e^{i\theta}))d\theta.$$

If X is strictly convex, then there is a unique $f \in L_\infty(\partial D)^n$ such that

$$(**) \qquad P^*(h) = \frac{\mathrm{Re}(h(f))}{P(f)} = \frac{1}{2\pi P(f)} \int_0^{2\pi} \mathrm{Re}\langle h(e^{i\theta}), f(e^{i\theta})\rangle d\theta.$$

(2)   *If X is strictly convex, we may use the induced Minkowski functional P on $L = C(\partial D)^n$ and the dual Minkowski functional P^* on $C^*(\partial D)^n$ to obtain the equality $(*)$ above for any $h \in L_1(\partial D)^n \subset C^*(\partial D)^n$, and moreover, there is a unique $f \in C(\partial D)^n$ satisfying $(**)$.*

Proof. We prove both (1) and (2) at the same time. We have

$$
\begin{aligned}
P^*(h) &= \sup_{f \in L} \frac{\mathrm{Re}(h(f))}{P(f)} \leq \frac{1}{2\pi} \int_0^{2\pi} \sup_f \frac{\mathrm{Re}\langle h(e^{i\theta}), f(e^{i\theta})\rangle}{P(f)} d\theta \\
&\leq \frac{1}{2\pi} \int_0^{2\pi} \sup_f \frac{\mathrm{Re}\langle h(e^{i\theta}), f(e^{i\theta})\rangle}{p(f(e^{i\theta}))} d\theta \\
&\leq \frac{1}{2\pi} \int_0^{2\pi} \sup_{z \in \mathbf{C}^n} \frac{\mathrm{Re}\langle h(e^{i\theta}), z\rangle}{p(z)} d\theta \\
&= \frac{1}{2\pi} \int_0^{2\pi} p^*(h(e^{i\theta}))d\theta.
\end{aligned}
$$

We shall now prove the opposite inequality. By Remark (4.7.14) there is a map $f \in L$ so that for (almost) all θ

$$p(f(e^{i\theta})) = 1 \quad \text{and} \quad \mathrm{Re}\langle h(e^{i\theta}), f(e^{i\theta})\rangle = p^*(h(e^{i\theta})).$$

Then

$$P^*(h) \geq \frac{1}{2\pi P(f)} \int_0^{2\pi} \mathrm{Re}\langle h(e^{i\theta}), f(e^{i\theta})\rangle d\theta = \frac{1}{2\pi} \int_0^{2\pi} p^*(h(e^{i\theta}))d\theta,$$

which establishes the desired equality.

Also by (4.7.14), if X is strictly convex, such an f can be found uniquely in $C(\partial D)^n$. $\qquad\square$

Let

$$\Delta = m_1 a_1 + \ldots + m_k a_k, \qquad a_i \in D, \quad m_i \in \mathbf{Z}$$

be a divisor on the unit disc D. We consider only positive divisors, i.e., divisors with $m_i > 0$. To the divisor Δ we associate the following bounded holomorphic function on D:

$$(4.7.18) \qquad \gamma_\Delta(t) = \prod_{i=1}^{k} \left(\frac{t - a_i}{1 - \bar{a}_i t} \right)^{m_i}.$$

Then γ_Δ maps D into itself and its zeros are exactly $a_1, \ldots, a_k$ with multiplicities $m_1, \ldots, m_k$.

In order to define the space of bounded holomorphic maps with prescribed derivatives of order up to $m_i - 1$ at a_i, $i = 1, \ldots, k$, we consider a set A of **data** consisting of $\sum m_i$ elements of $\mathbf{C}^n$:

$$A = \{\alpha_{i,\mu_i} \in \mathbf{C}^n; \ 1 \le i \le k, \ 0 \le \mu_i \le m_i - 1\}.$$

Let

$$(4.7.19) \qquad L(A, \Delta) = \{f \in H_\infty(D)^n; \ f^{(\mu_i)}(a_i) = \alpha_{i,\mu_i}\};$$

it is an affine subspace of $H_\infty(D)^n$ consisting of f with μ_i-th derivative at a_i equal to the prescribed value α_{i,μ_i}. Let M_Δ be the space of mappings $f \in A(D)^n$ that vanish at $a_1, \ldots, a_k \in D$ with multiplicities $m_1, \ldots, m_k$; it is a linear subspace of finite codimension in $A(D)^n$, and can be written as

$$(4.7.20) \qquad M_\Delta = \gamma_\Delta A(D)^n = \{\gamma_\Delta f; \ f \in A(D)^n\}.$$

Let t denote the coordinate function in D so that $D = \{|t| < 1\}$. Since the annihilator of $A(D)$ in $C(\partial D)$ is $t H_1(D)$ (see (4.7.9)), the annihilator $M_\Delta^0 \subset C^*(\partial D)^n$ of M_Δ is given by

$$(4.7.21) \qquad M_\Delta^0 = \frac{t}{\gamma_\Delta} H_1(D)^n;$$

here $1/\gamma_\Delta$ should be considered as an element of $C(\partial D) \subset C^*(\partial D)$.

To the Minkowski function $p: X \to \mathbf{R}$ of a bounded convex domain X we associate the Minkowski functional $P: H_\infty(D)^n \to \mathbf{R}$ following the definition (4.7.15):

$$P(f) = \sup_{t \in D} p(f(t)) \qquad f \in H_\infty(D)^n,$$

and then define

$$(4.7.22) \qquad m(A, \Delta) = \inf_{f \in L(A,\Delta)} P(f).$$

(4.7.23) **Theorem.** *Let X be a convex bounded domain in $\mathbf{C}^n$ with Minkowski function p, $\Delta = m_1 a_1 + \ldots + m_k a_k$ a divisor on D and A a set of data associated to Δ. Then for any map $f \in L(A, \Delta)$ the following are equivalent:*

(a) *f is extremal, i.e., $P(f) = m(A, \Delta)$;*

(b) *there exists $h \in M_\Delta^0$ such that for almost all $\theta \in \mathbf{R}$, we have*

$$p(f(e^{i\theta})) = m(A, \Delta)$$

and

$$\text{Re}\langle h(e^{i\theta}), f(e^{i\theta})\rangle = p(f(e^{i\theta}))p^*(h(e^{i\theta}));$$

(c)   *there exists $h \in M_\Delta^0$ such that*

$$\frac{1}{2\pi} \int_0^{2\pi} \text{Re}\langle h(e^{i\theta}), f(e^{i\theta})\rangle d\theta = P(f)P^*(h).$$

Geometrically, f and h are related as follows. For almost all θ, $\overline{h(e^{i\theta})}$ is perpendicular to a supporting real hyperplane to the convex domain $X_{m(A,\Delta)} = \{z \in \mathbf{C}^n; \ p(z) < m(A, \Delta)\}$ at the point $f(e^{i\theta})$. Consequently, if X is strictly convex, then an extremal map $f \in L(A, \Delta)$ is uniquely determined by h. On the other hand, if ∂X is smooth of class C^1, then h is unique.

Proof. Fix $f_0 \in L(A, \Delta) \cap A(D)^n$. Set

$$m^*(A, \Delta) = \inf\{P^*(h); \ h \in M_\Delta^0, \quad \text{Re}(h(f_0)) = 1\}.$$

We shall apply the principle of duality (4.7.4) to the following situation:

$$L = C(\partial D)^n, \quad M = M_\Delta, \quad m = m(A, \Delta), \quad m^* = m^*(A, \Delta).$$

By (1) and (2) of (4.7.4),

$$m(A, \Delta)m^*(A, \Delta) = 1,$$

and there is $h \in M_\Delta^0$ such that

$$\text{Re}(h(f_0)) = 1 \quad \text{and} \quad P^*(h) = m^*(A, \Delta).$$

Take $f \in L(A, \Delta)$, not yet assumed to be extremal. Since $f - f_0$ is bounded on D and vanishes at $a_1, \ldots, a_k$ with multiplicities $m_1, \ldots, m_k$ and since

$$h \in M_\Delta^0 = \frac{t}{\gamma_\Delta} H_1(D)^n,$$

it follows that $\langle h, f - f_0\rangle \in tH_1(D)$. Since $tH_1(D)$ is the annihilator of $A(D)$ (see (4.7.9)), $\langle h, f - f_0\rangle$ annihilates the constant function 1:

$$h(f - f_0) = \frac{1}{2\pi} \int_0^{2\pi} \langle h(e^{i\theta}), f(e^{i\theta}) - f_0(e^{i\theta})\rangle d\theta = 0.$$

In particular, $\text{Re}(h(f)) = \text{Re}(h(f_0)) = 1$. Making use of (4.7.11), (4.7.13) and (4.7.17), we obtain

$$\begin{aligned}
1 &= \frac{1}{2\pi} \int_0^{2\pi} \text{Re}\langle h(e^{i\theta}), f(e^{i\theta})\rangle d\theta \\
&\leq \frac{1}{2\pi} \int_0^{2\pi} p(f(e^{i\theta}))p^*(h(e^{i\theta})) d\theta \\
&\leq P(f)P^*(h) = P(f)m(A, \Delta)^{-1}.
\end{aligned}$$

Now we shall prove the implications:

(a) $\Rightarrow$ (b). If f is extremal, i.e., $P(f) = m(A, \Delta)$, then all the inequalities above are equalities. In particular, (a) implies (b).

(b) $\Rightarrow$ (c). This is evident from (4.7.17).

(c) $\Rightarrow$ (a). Replacing h by $(P(f)P^*(h))^{-1}h$ in (c), we have

$$\mathrm{Re}(h(f)) = P(f)P^*(h) = 1.$$

Now, (3) of (4.7.4) implies (a).

The geometrical assertion on f and h follows from (4.7.14). $\square$

(4.7.24) **Remark.** If X is strictly convex, then (4.7.23) and its proof are valid when $L(A, \Delta)$ is replaced by $A(D)^n \cap L(A, \Delta)$.

(4.7.25) **Corollary.** *Let X be as in (4.7.23). Let Δ and Δ' be two divisors on D with associated sets of data A and A', respectively. Assume $\deg \Delta' \geq \deg \Delta$. Let $f \in L(A, \Delta) \cap L(A', \Delta')$. If f is extremal for $L(A, \Delta)$, it is extremal for $L(A', \Delta')$, i.e., if $P(f) = m(A, \Delta)$, then $P(f) = m(A', \Delta')$.*

Proof. We claim that since $\deg \Delta' \geq \deg \Delta$, there is a meromorphic function φ on $\mathbf{C}$ with the same zeros and poles as $\gamma_\Delta / \gamma_{\Delta'}$ and which is positive on ∂D. Using $\mathrm{Aut}(D)$ the proof of this can be reduced to considering combinations of the following two cases: i) simple pole at the origin, and ii) simple pole at the origin and simple zero at $t = 1/2$. In case i), take

$$\varphi(t) = 3 + t + \frac{1}{t}.$$

In case ii), take

$$\varphi(t) = \frac{5}{2} - t - \frac{1}{t}.$$

Now, let $h \in M_\Delta^0$ be the map in (4.7.23), and set $h' = \varphi h$. Then

$$h' \in \frac{t}{\gamma_{\Delta'}} H_1(D)^n = M_{\Delta'}^0,$$

and for almost all $\theta \in \mathbf{R}$ we have

$$
\begin{aligned}
\mathrm{Re}\langle h'(e^{i\theta}), f(e^{i\theta})\rangle &= \varphi(e^{i\theta})\mathrm{Re}\langle h(e^{i\theta}), f(e^{i\theta})\rangle \\
&= \varphi(e^{i\theta})p(f(e^{i\theta}))p^*(h(e^{i\theta})) \\
&= p(f(e^{i\theta}))p^*(h'(e^{i\theta})).
\end{aligned}
$$

Since $p(f(e^{i\theta})) = m(A, \Delta) = P(f)$ for almost all θ, we have

$$\frac{1}{2\pi}\int_0^{2\pi} \mathrm{Re}\langle h'(e^{i\theta}), f(e^{i\theta})\rangle d\theta = P(f)P^*(h').$$

By (4.7.23) we conclude that f is extremal for $L(A', \Delta')$. $\square$

8 Intrinsic Distances on Convex Domains

In this section we shall complete the proof of the theorem of Lempert we stated at the beginning of the preceding section.

In (3.1.3) we defined a function d'_X on $X \times X$ by setting

$$d'_X(x, x') = \inf \rho(a, a'),$$

where the infimum is taken over all holomorphic maps $f \colon D \to X$ and all pairs of points $a, a' \in D$ such that $f(a) = x$ and $f(a') = x'$. If there is no such map f, then $d'_X(x, x') = \infty$ by definition. We then pointed out that d'_X may not satisfy the triangular inequality. If it satisfies the triangular inequality, then $d'_X = d_X$ since d_X is the largest pseudo-distance bounded by d'_X, (see Section 1 of Chapter 3).

We prove first the following theorem due to Lempert [1].

(4.8.1) **Theorem.** *If X is a convex domain in $\mathbf{C}^n$, then $d'_X = d_X$.*

Proof. It suffices to show that d'_X satisfies the triangular inequality. Given $x, y, z \in X$ and $\varepsilon > 0$, let $f, g \in \mathrm{Hol}(D, X)$ and $a, b, a', b' \in D$ be such that

$$f(a) = x, \quad f(b) = g(a') = y, \quad g(b') = z,$$

and

$$d'_X(x, y) > \rho(a, b) - \varepsilon, \quad d'_X(y, z) > \rho(a', b') - \varepsilon.$$

By composing f and g with suitable automorphisms of D, we may assume that $a, b, a', b' \in D$ are all real numbers such that $0 = a < b = a' < b' < 1$. Also by composing f and g with a homothety $\mu_r \colon z \in D \to rz \in D$ with $r < 1$ but sufficiently close to 1, we may assume that f and g are continuously extendable to $\bar{D}$.

We define $\lambda \colon \mathbf{C} - \{b, b^{-1}\} \to \mathbf{C}$ by setting

$$\lambda(z) = \frac{(z - b')(z - b'^{-1})}{(z - b)(z - b^{-1})}.$$

Then define $h \colon \bar{D} \to \mathbf{C}^n$ by

$$h(z) = \lambda(z)f(z) + (1 - \lambda(z))g(z), \qquad z \in \bar{D}.$$

Except for a simple pole at b, the function λ is holomorphic in $\bar{D}$, $\lambda(0) = 1$, $\lambda(b') = 0$, and takes real values between 0 and 1 on the boundary ∂D, and h is holomorphic on D (since $f(b) = g(b)$). Moreover, $h(0) = f(0) = x$, $h(b') = g(b') = z$ and $h(\partial D) \subset \bar{X}$. By the maximum principle (applied to the composition of h with a peak function for X at each boundary point of X, see the proof of (4.1.10)), we obtain $h(D) \subset X$.

In particular,

$$d'_X(x, z) \leq \rho(0, b') = \rho(0, b) + \rho(b, b') \leq d'_X(x, y) + d'_X(y, z) + 2\varepsilon.$$

Since ε is arbitrary, the assertion follows. $\square$

(4.8.2) Corollary. *Let $X \subset \mathbf{C}^n$ be a bounded convex domain. Then, given any pair of points $x, y \in X$, there is an extremal disc for the pair $\{x, y\}$, i.e., there exist a map $f \in \mathrm{Hol}(D, X)$ and a point $a \in D$ such that $f(0) = x$, $f(a) = y$ and*

$$d_X(x, y) = \rho(0, a).$$

Similarly, given a vector $\xi \in T_x X$, there is an extremal disc for ξ, i.e., there exists a map $f \in \mathrm{Hol}(D, X)$ and a vector $\tau \in T_0 D$ such that

$$f_*(\tau) = \xi, \qquad F_X(\xi) = \|\tau\|,$$

where $\|\tau\|$ denotes the Poincaré length of τ.

Proof. By (4.8.1), for each $n > 0$ there exist $f_n \in \mathrm{Hol}(D, X)$ and $a_n \in D$ such that $f_n(0) = x$, $f_n(a_n) = y$ and

$$d_X(f_n(0), f_n(a_n)) \leq \rho(0, a_n) < d_X(f_n(0), f_n(a_n)) + \frac{1}{n}.$$

We may further assume that all a_n are real and positive. Then $\{a_n\}$ converges to a point $a \in D$ such that $\rho(0, a) = d_X(x, y)$. Since X is (strongly) complete with respect to d_X by (4.1.10), $\{f_n\}$ has a convergent subsequence by (1.3.3). Then the limit map f has the desired property. $\qquad\qquad\square$

(4.8.3) Corollary. *Let $X \subset \mathbf{C}^n$ be a bounded convex domain. Then the closed balls of X with respect to d_X are compact convex subsets of X.*

Proof. Given $x \in X$ and $r > 0$, let $K = \{y \in X;\ d_X(x, y) \leq r\}$. By (4.1.10) K is compact. Let $y, z \in K$ with $d_X(x, z) \leq d_X(x, y)$. Given $\varepsilon > 0$ there exist $f, g \in \mathrm{Hol}(D, X)$ and $a, b \in D$ such that $f(0) = g(0) = x$, $f(a) = y$, $g(b) = z$ and

$$\rho(0, a) < d_X(x, y) + \varepsilon \quad \text{and} \quad \rho(0, b) < d_X(x, z) + \varepsilon.$$

We may assume that a and b are real and $0 < b \leq a$. Define $h_\lambda \in \mathrm{Hol}(D, X)$, $0 \leq \lambda \leq 1$, by

$$h_\lambda(t) = \lambda f(t) + (1 - \lambda) g(bt/a), \quad t \in D.$$

Then

$$d_X(x, \lambda y + (1 - \lambda)z) = d_X(x, h_\lambda(a)) \leq \rho(0, a) < d_X(x, y) + \varepsilon \leq r + \varepsilon.$$

Letting $\varepsilon \to 0$, we obtain the desired result. $\qquad\qquad\square$

In (4.8.2) we proved that given any pair of points x, y in a bounded convex domain X there is an extremal disc $f \in \mathrm{Hol}(D, X)$ for the pair. Lempert [1] proved also regularity and boundary smoothness of f. In Chang-Hu-Lee [1], this result of Lempert was extended to the situation where x and y are allowed to lie on the boundary of X.

We make use of the notation in the preceding section. In particular, if $p: \mathbf{C}^n \to \mathbf{R}$ is the Minkowski function for a convex domain X, i.e.,

$$X = \{z \in \mathbf{C}^n;\ p(z) < 1\},$$

then P denotes the Minkowski functional on $H_\infty(D)^n$. Given a divisor $\Delta = a + b$, $(a, b \in D)$, on D and the data $A = \{x, y\}$, $(x, y \in \mathbf{C}^n)$, we have (see (4.7.19))

$$L(A, \Delta) = \{f \in H_\infty(D)^n;\ f(a) = x \quad \text{and} \quad f(b) = y\}.$$

We recall also (see (4.7.22))

$$m(A, \Delta) = \inf_{f \in L(A,\Delta)} P(f).$$

The following theorem of Royden and Wong [1] relates the linear extremal problem of finding $f \in L(A, \Delta)$ such that $P(f) = m(A, \Delta)$ to that of finding extremal discs.

(4.8.4) **Theorem.** *Let $X \subset \mathbf{C}^n$ be a bounded convex domain with Minkowski function p. Let P be the induced Minkowski functional on $H_\infty(D)^n$. Then for a holomorphic map $f: D \to X$ with $f(a) = x$ and $f(b) = y$, the following are equivalent*:

(a)   *f is extremal for x, y in the sense that $d_X(x, y) = \rho(a, b)$;*

(b)   *For the divisor $\Delta = a + b$ and the data $A = \{x, y\}$, f satisfies*

$$m(A, \Delta) = P(f) = 1.$$

Proof. We may assume that $a = 0$ and $b > 0$.

(b) $\Rightarrow$ (a). Suppose that $m(A, \Delta) = P(f) = 1$ and that f is not extremal for x, y, i.e., $d_X(x, y) < \rho(0, b)$. Then there exists a holomorphic map $g: D \to X$ such that $g(0) = x$ and $g(c) = y$ with $0 < c < b$. The map $h(t) = g(ct/b)$ satisfies $h(0) = x$ and $h(b) = y$, i.e., $h \in L(A, \Delta)$. Since

$$\overline{h(D)} \subset g(D) \subset X,$$

we have $P(h) < 1 = m(A, \Delta)$, which is a contradiction.

(a) $\Rightarrow$ (b). Since $f(D) \subset X$, $f(a) = x$ and $f(b) = y$, we have

$$m(A, \Delta) \le P(f) \le 1.$$

If $m(A, \Delta) < 1$, there is $g \in L(A, \Delta)$ with $P(g) < 1$. Then $\overline{g(D)} \subset X$. For each $r > 1$, we define $g_r: D_{1/r} \to X$ by

$$g_r(t) = g(rt),$$

where $D_{1/r} = \{t \in \mathbf{C};\ |t| < 1/r\}$. Then $g_r(0) = x$ and $\varepsilon(r) = y - g_r(b) \to 0$ as $r \to 1$. Now define $h_r: D_{1/r} \to \mathbf{C}^n$ by

$$h_r(t) = g_r(t) + \frac{\varepsilon(r)}{b} t.$$

Then $h_r(0) = x$, $h_r(b) = y$ and we can choose r_0 so close to 1 that

$$\overline{h_{r_0}(D_{1/r_0})} \subset X.$$

Finally, we define $h \in \mathrm{Hol}(D, X)$ by

$$h(t) = h_{r_0}(t/r_0).$$

Then $h(0) = x$ and $h(r_0 b) = y$, contradicting the extremality of f. Hence, $m(A, \Delta) = P(f) = 1$. $\qquad\qquad\square$

Let t be the coordinate system for $\mathbf{C}$ so that D is given by $|t| < 1$. Let $\tau_0 \in T_0 D$ be the tangent vector given by $\tau_0 = (d/dt)_0$. For $f \in \mathrm{Hol}(D, X)$ we use both $f'(0)$ and $f_*(\tau_0)$ to denote the derivative of f at 0.

(4.8.5) Theorem. *Let X, p, and P be as in (4.8.4). Then for a holomorphic map $f: D \to X$ with $f(0) = x \in X$ and $f_*(\tau_0) = \xi \in T_x X$, where $\tau_0 = (d/dt)_0 \in T_0 D$, the following are equivalent:*

(a) *f is extremal for a nonzero vector $\xi \in T_x X$ in the sense that $F_X(\xi) = \|\tau_0\|$, where $\|\tau_0\|$ denotes the Poincaré length of the vector τ_0;*

(b) *For the divisor $\Delta = 2 \cdot 0$ with $0 \in D$ and the data $A = \{x, \xi\}$, f satisfies*

$$m(A, \Delta) = P(f) = 1.$$

Proof. The proof is similar to that of (4.8.4). We note that (see (4.7.19))

$$L(A, \Delta) = \{f \in H_\infty(D)^n; \ f(0) = x, \ f'(0) = \xi\}.$$

(b) $\Rightarrow$ (a). If f is not extremal for ξ, then there exists a map $g \in \mathrm{Hol}(D, X)$ with $g(0) = x$ and $g'(0) = r\xi$ for some $r > 1$. Then the map $h(t) = g(t/r)$ is in $L(A, \Delta)$, and

$$\overline{h(D)} \subset g(D) \subset X,$$

which implies $P(h) < 1 = m(A, \Delta)$. This is a contradiction.

(a) $\Rightarrow$ (b). If $P(f) < 1$ or if $m(A, \Delta) < P(f)$, there exists a map $g \in L(A, \Delta)$ with $P(g) < 1$. Then $\overline{g(D)} \subset X$. For each $r > 1$ we define $g_r: D_{1/r} \to X$ as in the proof of (4.8.4). Then $g_r(0) = x$ and $g_r'(0) = r\xi$. Define $h_r: D_{1/r} \to \mathbf{C}^n$ by

$$h_r(t) = g_r(t) + t(1 - r)\xi.$$

Then $h_r(0) = x$ and $h_r'(0) = \xi$, and we can choose r_0 so close to 1 that $\overline{h_{r_0}(D_{1/r_0})} \subset X$. Finally, we define $h \in \mathrm{Hol}(D, X)$ by

$$h(t) = h_{r_0}(t/r_0).$$

Then $h(0) = x$ and $h'(0) = \xi/r_0$. Hence, f is not extremal for ξ. $\qquad\square$

(4.8.6) Theorem. *Let X be a bounded convex domain in $\mathbf{C}^n$. If $f \in \mathrm{Hol}(D, X)$ is extremal for a pair of distinct points $x, y \in X$, then it is a complex geodesic, i.e., it is extremal for any other pair of distinct points x', y' in $f(D)$.*

Similarly, if f is extremal for a nonzero vector $\xi \in T_x X$, then it is a complex geodesic.

In particular, given any pair $x, y \in X$ (resp. $\xi \in T_x X$) there is a complex geodesic $f \in \mathrm{Hol}(D, X)$ passing through x, y (such that $f'(0) = \xi$).

Proof. We prove the first statement. Let

$$f(a) = x, \quad f(b) = y, \quad f(a') = x', \quad f(b') = y',$$

$$\Delta = a + b, \quad A = \{x, y\}, \quad \Delta' = a' + b', \quad A' = \{x', y'\}.$$

Then $\deg \Delta = \deg \Delta'$, and $f \in L(A, \Delta) \cap L(A'\Delta')$. By (4.8.4), $m(A, \Delta) = P(f) = 1$. By (4.7.25), $m(A', \Delta') = P(f) = 1$. By (4.8.4) f is extremal for the pair x', y'.

In order to prove the second statement, let

$$f(0) = x, \quad f'(0) = \xi, \quad f(a') = x', \quad f(b') = y',$$

$$\Delta = 2 \cdot 0, \quad A = \{x, \xi\}, \quad \Delta' = a' + b', \quad A' = \{x', y'\},$$

and use (4.8.5) in place of (4.8.4).

Now the last assertion follows from (4.8.2). $\qquad\square$

(4.8.7) Corollary. *Let $X \subset \mathbb{C}^n$ be a bounded convex domain. Then $f \in \mathrm{Hol}(D, X)$ is a complex geodesic if and only if it is extremal for every $\xi \in T_x X$, $x \in X$.*

Proof. If f is extremal for one ξ, then it is a complex geodesic by (4.8.6). The converse follows from (4.8.5) and (4.7.25); see the proof of (4.8.6). $\qquad\square$

Let $f \in \mathrm{Hol}(D, X)$ be a complex geodesic. For the divisor $\Delta = 2 \cdot 0$ and the data $A = \{f(0), f'(0)\}$, we have (by (4.8.5) and (4.8.7))

$$m(2 \cdot 0, A) = P(f) = 1.$$

For the divisor $\Delta = 2 \cdot 0$, the function γ_Δ is given by (see (4.7.18))

$$\gamma_{2 \cdot 0}(t) = t^2.$$

Let

$$h \in M_{2 \cdot 0}^0 = \frac{t}{\gamma_{2 \cdot 0}} H_1(D)^n = \frac{1}{t} H_1(D)^n$$

be the map given by (4.7.23), see (4.7.21). Since $m(2 \cdot 0, A) = 1$, (b) of (4.7.23) says that

$$\mathrm{Re}\langle h(e^{i\theta}), f(e^{i\theta})\rangle = p^*(h(e^{i\theta}))$$

for almost all θ. This, together with (4.7.13), implies

$$(4.8.8) \qquad \mathrm{Re}\langle h(e^{i\theta}), f(e^{i\theta}) - z\rangle > 0 \qquad \text{for} \quad z \in X$$

for almost all θ. Geometrically, this simply says that $\overline{h(e^{i\theta})}$ is perpendicular to a supporting real hyperplane of X at $f(e^{i\theta})$ for almost all θ, which we already know from (4.7.23).

We define a **dual map** $\tilde{f}$ of the complex geodesic $f \in H_1(D)^n$ by

$$(4.8.9) \qquad \tilde{f}(t) = \gamma_0(t)h(t) = th(t).$$

(4.8.10) **Lemma.** *Let $X \subset \mathbf{C}^n$ be a bounded convex domain, $f \in \mathrm{Hol}(D, X)$ a complex geodesic and $\tilde{f} = \gamma_0 h \in H_1(D)^n$ the dual map of f. Then $\tilde{f}$ never vanishes in D and*

$$\mathrm{Re}\langle \tilde{f}(0), f'(0)\rangle > 0.$$

Proof. Since $f(0) \in X$, (4.8.8) and (4.8.9) imply

$$\mathrm{Re}\langle \tilde{f}(t), \frac{1}{t}(f(t) - f(0))\rangle > 0$$

for almost all $t = e^{i\theta} \in \partial D$. Since the left hand side is the real part of a function in $H_1(D)$, (3) of (4.7.9) implies $\mathrm{Re}\langle \tilde{f}(0), f'(0)\rangle > 0$, and, in particular, $\tilde{f}(0) \neq 0$. Since $\tilde{f} \circ \alpha$ is a dual map of $f \circ \alpha$ for any $\alpha \in \mathrm{Aut}(D)$, $\tilde{f}$ never vanishes in D. $\square$

Since $\tilde{f}(t) \neq 0$ for $t \in D$ by (4.8.10), for each $t \in D$ we can define a complex hyperplane H_t in $\mathbf{C}^n$ by the equation

$$(4.8.11) \qquad \langle \tilde{f}(t), z - f(t)\rangle = 0,$$

where $z \in \mathbf{C}^n$.

We claim that each $z \in X$ lies exactly in one hyperplane H_t, i.e., the equation (4.8.11) has a unique solution $t = \tilde{p}(z)$ in D for each $z \in X$. Fixing z, let $g_z(t)$ denote the left hand side of (4.8.11):

$$g_z(t) = \langle \tilde{f}(t), z - f(t)\rangle = t\langle h(t), z - f(t)\rangle.$$

It is a holomorphic function on D. The number of its zeros is the winding number of g_z, i.e., the number of times $g_z(t)$ goes around the origin as t goes around (a circle slightly smaller than) the unit circle. This number is the sum of the winding numbers of two functions $\langle h(t), z - f(t)\rangle$ and t. The winding number of the latter is clearly 1 while that of the former is 0 because of (4.8.8). This proves our claim.

By the implicit function theorem, $\tilde{p}$ is holomorphic. Since the left hand side of (4.8.11) vanishes for $z = f(t)$, we have $\tilde{p}(f(t)) = t$. In summary, we have

(4.8.12) **Theorem.** *Let $X \subset \mathbf{C}^n$ be a bounded convex domain. Given a complex geodesic $f \in \mathrm{Hol}(D, X)$, let $\tilde{p}(z) \in D$ be the unique point such that $t = \tilde{p}(z)$ satisfies the equation (4.8.11). Then $\tilde{p}: X \to D$ is a holomorphic fibration with section f, i.e.,*

$$\tilde{p} \circ f = \mathrm{id}_D.$$

Clearly, $p = f \circ \tilde{p}: X \to f(D) \subset X$ is a holomorphic retract.

Finally, we have the following theorem of Lempert [1].

(4.8.13) **Theorem.** *If X is a bounded convex domain in $\mathbf{C}^n$, then*

$$d_X = c_X \quad and \quad F_X = E_X.$$

Proof. Let $x, y \in X$. By (4.8.6) there is a complex geodesic $f \in \mathrm{Hol}(D, X)$ with $x = f(a)$ and $y = f(b)$ for some $a, b \in D$. Then

$$d_X(x, y) \le \rho(a, b) = \rho(p \circ f(a), p \circ f(b)) \le c_X(f(a), f(b)).$$

Since the opposite inequality is always true, this proves $d_X = c_X$. The proof for the equality $F_X = E_X$ is similar. $\qquad\square$

This result of Lempert has been generalized to more general domains (called "strongly linearly convex domains") in Lempert [3]. For an operator theoretic approach to Lempert's theorem, see Agler [1], Salinas [1] and Meyer [1]. It has been generalized also to convex domains in a locally convex vector space of infinite dimension by Dinen-Timoney-Vigué [1].

9 Product Property for Carathéodory Distance

The purpose of this section is to establish the following result on the product property for the Carathéodory distance by Jarnicki and Pflug [4, 7, 8, 10].

(4.9.1) **Theorem.** *For any complex spaces X and Y, we have*

$$c_{X\times Y}((x, y), (x', y')) = \max\{c_X(x, x'), c_Y(y, y')\}, \qquad x, x' \in X, \ y, y' \in Y.$$

In (3.1.9) we established a similar formula for the Kobayashi pseudo-distance. The proof of (4.9.1) will make use of (3.1.9) as well as Lempert's theorem (4.8.13).

It is convenient to use the following Möbius pseudo-distance c'_X defined by

$$(4.9.2) \qquad c'_X(x_0, x) = \sup\{|f(x)|; \ f \in \mathrm{Hol}(X, D), \ f(x_0) = 0\},$$

where $x_0, x \in X$.

Since an automorphism of D sending $a \in D$ to $0 \in D$ is given by

$$z \mapsto \frac{z - a}{1 - \bar{a}z},$$

we may define c'_X by

$$(4.9.3) \qquad c'_X(x, x') = \sup_{f \in \mathrm{Hol}(X, D)} \left| \frac{f(x') - f(x)}{1 - \overline{f(x)}f(x')} \right|.$$

Since $\log(|z' - z|/|1 - \bar{z}z'|)$ is harmonic in z as well as in z', it follows that $c'_X(x, x')$ is plurisubharmonic in x as well as in x'.

The Carathéodory pseudo-distance c_X and the Möbius pseudo-distance c'_X are related by (see the explicit expression for ρ in Section 1 of Chapter 2)

$$(4.9.4) \qquad c_X = \log \frac{1 + c'_X}{1 - c'_X} = 2\tanh^{-1} c'_X.$$

Then the desired product formula is equivalent to

$$(4.9.5) \qquad c'_{X \times Y}((x, y), (x', y')) = \max\{c'_X(x, x'), \ c'_Y(y, y')\}.$$

Since we already know (see (3.1.11)) that the left hand side is greater than or equal to the right hand side, it suffices to prove the opposite inequality. In view of the definition (4.9.2), this amounts to showing the following inequality

$$(4.9.6) \qquad |f(x, y)| \leq \max\{c'_X(x_0, x), \ c'_Y(y_0, y)\}, \qquad x_0, x \in X, \ y_0, y' \in Y$$

for all $f \in \mathrm{Hol}(X \times Y, D)$ with $f(x_0, y_0) = 0$.

We fix $x_0 \in X$, $y_0 \in Y$ and a map $f \in \mathrm{Hol}(X \times Y, D)$ with $f(x_0, y_0) = 0$. Let $\{X_m\}$ (resp. $\{Y_m\}$) be a monotone increasing sequence of relatively compact domains in X (resp. Y) such that $X = \bigcup X_m$ (resp. $Y = \bigcup Y_m$). We consider only large m so that $x_0 \in X_m$ and $y_0 \in Y_m$. In view of (3.1.20) it now suffices to show the following inequality for all large m:

$$|f(x, y)| \leq \max\{c'_{X_m}(x_0, x), \ c'_{Y_m}(y_0, y)\} \qquad x \in X_m, \ y \in Y_m.$$

Now we need the following lemma.

(4.9.7) Lemma. *Any holomorphic function $f(x, y)$ on $X \times Y$ can be approximated by functions of the form*

$$f_s(x, y) = \sum_i g_{s,i}(x) h_{s,i}(y), \qquad s = 1, 2, \ldots,$$

where $g_{s,i}(x)$ and $h_{s,i}(y)$ are holomorphic functions on X and Y, respectively, and the sum is finite.

By approximation we mean a uniform convergence of f_s on compact sets.

Temporarily assuming the lemma, we shall complete the proof of (4.9.1). If no nonconstant bounded holomorphic functions exist on X or Y, i.e., if $c_X \equiv 0$ or $c_Y \equiv 0$, then (4.9.1) is evidently true. So we assume that both c_X and c_Y are nontrivial. Then in (4.9.7) we may assume that $g_{s,i}$ and $h_{s,i}$ are nonconstant (by adding small nonconstant holomorphic functions if necessary). Since $f(x_0, y_0) = 0$, in approximating $f(x, y)$ by $f_s(x, y)$ we may also assume that $f_s(x_0, y_0) = 0$ by replacing $f_s(x, y)$ with $f_s(x, y) - f_s(x_0, y_0)$. Fixing $m \geq 1$, set

$$M_s = \max\{1, \ \|f_s\|_{X_m \times Y_m}\}.$$

By replacing f_s with f_s/M_s we may further assume that f_s maps $X_m \times Y_m$ into D. Now it suffices to prove

$$|f_s(x, y)| \leq \max\{c'_{X_m}(x_0, x), \ c'_{Y_m}(y_0, y)\}, \qquad x \in X_m, \ y \in Y_m.$$

If the sum $\sum_i g_{s,i} h_{s,i}$ runs from $i = 1$ to $i = N$, set

$$g_s = (g_{s,1}, \ldots, g_{s,N}) : X \to \mathbf{C}^N,$$

$$h_s = (h_{s,1}, \ldots, h_{s,N}) : Y \to \mathbf{C}^N.$$

Define $\varphi \colon \mathbf{C}^N \times \mathbf{C}^N \to \mathbf{C}$ by

$$\varphi(\xi, \eta) = \sum_i \xi_i \eta_i, \qquad \xi = (\xi_1, \ldots, \xi_N), \ \eta = (\eta_1, \ldots, \eta_N).$$

Then put

$$U = \{\xi \in \mathbf{C}^N; \ |\xi_i| < \|g_{s,i}\|_{X_m}, \ |\varphi(\xi, h_s(y))| < 1, \ y \in Y_m\},$$

$$V = \{\eta \in \mathbf{C}^N; \ |\eta_i| < \|h_{s,i}\|_{Y_m}, \ |\varphi(g_s(x), \eta)| < 1, \ x \in X_m\}.$$

Then both U and V are bounded convex domains in $\mathbf{C}^N$. Since $g_{s,i}$ and $h_{s,i}$ are nonconstant and take the values $\|g_{s,i}\|_{X_m}$ and $\|h_{s,i}\|_{Y_m}$ only on the boundaries ∂X_m and ∂Y_m, we have $g_s(X_m) \subset U$ and $h_s(Y_m) \subset V$. Since $d_U = c_U$ and $d_V = c_V$ for convex domains U and V, the product property (3.1.9) for the Kobayashi pseudo-distance implies the product property for the Carathéodory pseudo-distances c_U, c_V, and $c_{U \times V}$. In other words, an inequality similar to (4.9.6) must hold for U and V. More explicitly, consider the holomorphic map

$$\frac{\varphi}{\|\varphi\|_{U \times V}} \colon U \times V \to D.$$

Then in this case, (4.9.6) reads as follows:

$$|\varphi(\xi, \eta)| \leq \|\varphi\|_{U \times V} \cdot \max\{c'_U(\xi_0, \xi), \ c'_V(\eta_0, \eta)\}$$

for all $(\xi_0, \eta_0), (\xi, \eta) \in U \times V$ with $\varphi(\xi_0, \eta_0) = 0$. Hence,

$$
\begin{aligned}
|f_s(x, y)| &= |\varphi(g(x), h(y))| \\
&\leq \|\varphi\|_{U \times V} \cdot \max\{c'_U(g(x_0), g(x)), \ c'_V(h(y_0), h(y))\} \\
&\leq \|\varphi\|_{U \times V} \cdot \max\{c'_{X_m}(x_0, x), \ c'_{Y_m}(y_0, y)\}
\end{aligned}
$$

where the last inequality comes from the distance-decreasing property for the Möbius pseudo-distance c'. This completes the proof of (4.9.1) except for the proof of Lemma (4.9.7).

Proof of (4.9.7). This is from Narasimhan [2]. Choose volume elements dv_X and dv_Y on X and Y, respectively. We fix a holomorphic function $f(x, y)$ on $X \times Y$. Then there is a positive function ζ on $X \times Y$ such that

$$\int_{X \times Y} |f|^2 \zeta \, dv_X dv_Y < \infty.$$

We claim that there are positive functions ξ and η on X and Y, respectively, such that $\xi(x)\eta(y) \leq \zeta(x, y)$. To construct such functions ξ and η, we consider increasing sequences of compact subsets $\{K_i\}$ and $\{L_i\}$ of X and Y, respectively, such that

$$X = \bigcup K_i, \qquad Y = \bigcup L_i.$$

For each i, choose δ_i, $0 < \delta_i < 1$, such that

$$\delta_i \leq \zeta(x, y) \qquad \text{for} \quad (x, y) \in K_i \times L_i.$$

There exist positive functions $\xi(x)$ and $\eta(y)$ such that

$$\xi(x) \leq \delta_i \qquad \text{for} \quad x \in K_i - K_{i-1},$$

$$\eta(y) \leq \delta_i \qquad \text{for} \quad y \in L_i - L_{i-1}.$$

Then for any

$$(x, y) \in K_i \times L_i - K_{i-1} \times L_{i-1} = (K_i \times (L_i - L_{i-1})) \cup ((K_i - K_{i-1}) \times L_i),$$

we have

$$\xi(x)\eta(y) \leq \delta_i \leq \zeta(x, y).$$

Having found ξ and η, we replace the volume elements dv_X and dv_Y by ξdv_X and ηdv_Y. We denote the new volume elements by dv_X and dv_Y. Then

$$\int_{X \times Y} |f|^2 dv_X dv_Y < \infty.$$

Let $\{a_j(x)\}$ be a complete orthonormal system for the square-integrable holomorphic functions on (X, dv_X); thus

$$\int_X a_j(x)\overline{a_k(x)}dv_X = \delta_{jk},$$

and every square-integrable holomorphic function on X is a linear combination of $\{a_j(x)\}$. Similarly, let $\{b_k(y)\}$ be a complete orthonormal system for the square-integrable holomorphic functions on (Y, dv_Y). Then $\{a_j(x)b_k(y)\}$ is a complete orthonormal system for the square-integrable holomorphic functions on $(X \times Y, dv_X dv_Y)$. Hence, $f(x, y)$ is a linear combination of $\{a_j(x)b_k(y)\}$. This completes the proof of the lemma. $\qquad\square$

(4.9.8) **Corollary**. *For any complex spaces X and Y, we have*

$$c^i_{X \times Y}((x, y), (x', y')) = \max\{c^i_X(x, x'), \ c^i_Y(y, y')\}, \qquad x, x' \in X, \ y, y' \in Y.$$

10　Bergman Metric

Let X be an n-dimensional complex manifold. Unless otherwise stated, in this section $\bar{X}$ denotes the conjugate complex manifold to X, not the closure of X. Thus, if J defines the complex structure of X, then $-J$ defines that of $\bar{X}$.

Let W be the space of square-integrable holomorphic n-forms ω on X. With the inner product

$$(\omega, \theta) = \int_X i^{n^2} \omega \wedge \bar{\theta}, \qquad \omega, \theta \in W$$

and the norm

$$\|\omega\| = \sqrt{(\omega, \omega)},$$

W is a separable complex Hilbert space. Let $\omega_0, \omega_1, \ldots$ be an orthonormal basis for W. (By an orthonormal basis, we always mean a complete orthonormal basis). Then the non-negative (n, n)-form

$$(4.10.1) \qquad B_X = \sum_{j=0}^{\infty} i^{n^2} \omega_j \wedge \bar{\omega}_j$$

is independent of the orthonormal basis chosen. We call B_X the **Bergman kernel form**. The notation B_X is a little ambiguous. Sometimes we consider it as a holomorphic $2n$-form

$$B_X(z, \bar{w}) = \sum_j i^{n^2} \omega_j(z) \wedge \bar{\omega}_j(\bar{w})$$

on $X \times \bar{X}$, with z denoting a point of X and $\bar{w}$ a point of the conjugate manifold $\bar{X}$. (We note that $\bar{\omega}_j(\bar{w})$ is holomorphic on $\bar{X}$ although it is conjugate holomorphic on X). By sending $z \in X$ to $(z, \bar{z}) \in X \times \bar{X}$ we can identify X with the "diagonal" of $X \times \bar{X}$. Then $B_X(z, \bar{z})$ will be considered as a real, non-negative (n, n)-form on X.

In terms of a local coordinate system $z^1, \ldots, z^n$ in X, we can write every $\omega \in W$ locally as

$$\omega = f dz^1 \wedge \ldots \wedge dz^n,$$

where f is a locally defined holomorphic function. In particular, by setting

$$\omega_j = f_j dz^1 \wedge \ldots \wedge dz^n, \quad j = 0, 1, \ldots,$$

we can express B_X locally

$$B_X(z, \bar{w}) = b_X(z, \bar{w}) i^{n^2} dz^1 \wedge \ldots \wedge dz^n \wedge d\bar{w}^1 \wedge \ldots \wedge d\bar{w}^n$$

with

$$(4.10.2) \qquad b_X(z, \bar{w}) = \sum_{j=0}^{\infty} f_j(z) \bar{f}_j(\bar{w}).$$

If X is a domain in $\mathbf{C}^n$, then using the natural coordinate system in $\mathbf{C}^n$, we can write ω, ω_j, B_X and b_X as above. Thus, we can identify W with the space of square-integrable holomorphic functions on the domain X. The function b_X is the **Bergman kernel function** of the domain X.

However, even for a domain X in $\mathbf{C}^n$, we prefer to use the kernel form B_X since it is invariantly defined on any complex manifold. In particular, B_X is invariant under any holomorphic automorphism of X.

(4.10.3) **Proposition.** *Fix a point z on a complex manifold X, and let W' be the subspace of W consisting of forms that vanish at z. Then*

$$B_X(z, \bar{z}) = \max_{\|\omega\| \leq 1} i^{n^2} \omega(z) \wedge \bar{\omega}(\bar{z}),$$

where the maximum is taken over all $\omega \in W$ with $\|\omega\| \leq 1$.

If $B_X(z, \bar{z}) \neq 0$, or equivalently, if $W' \neq W$, then the maximum is attained by a form $\omega_0 \in W$ perpendicular to W' such that $\|\omega_0\| = 1$. Moreover, such a form ω_0 is unique up to a constant factor c with $|c| = 1$.

Proof. If $W' = W$, there is nothing to prove. If $W' \neq W$, then W' is of codimension 1 in W, and we can choose an orthonormal basis $\omega_0, \omega_1, \ldots$ in such a way that ω_0 is perpendicular to W' while $\omega_1, \omega_2, \ldots$ are in W'. By expressing ω in terms of this basis, we easily obtain the proposition. $\square$

(4.10.4) Corollary. *If X' is a connected open subset of a complex manifold X, then*

$$B_X(z, \bar{z}) \leq B_{X'}(z, \bar{z}), \qquad z \in X'.$$

(4.10.5) Proposition. *If A is a closed complex subspace of codimension ≥ 1 in a complex manifold X, then*

$$B_{X-A} = B_X.$$

Proof. It suffices to show that every square-integrable holomorphic n-form ω on $X - A$ extends holomorphically to X. This is a purely local problem, and it can be easily reduced to the situation where $X = D^n$ and $X - A = D^* \times D^{n-1}$. Then we can write

$$\omega = f\, dz^1 \wedge \ldots \wedge dz^n,$$

with f holomorphic in $D^* \times D^{n-1}$. Then f can be written as a Laurent series in z^1:

$$f(z^1, z^2, \ldots, z^n) = \sum_{q=-\infty}^{\infty} a_q(z^2, \ldots, z^n)(z^1)^q,$$

where each $a_q(z^2, \ldots, z^n)$ is holomorphic in $z^2, \ldots, z^n$. Now the problem is to show that $a_q(z^2, \ldots, z^n) = 0$ for $q < 0$, and this is reduced to proving a simple lemma that a square-integrable holomorphic function $f(z) = \sum_{q=-\infty}^{\infty} a_q z^q$ on D^* is holomorphic at $z = 0$, i.e., $a_q = 0$ for $q < 0$. Since we prove a more general lemma in (7.5.8), we omit its proof here. $\square$

Using a local coordinate system $z^1, \ldots, z^n$, we write

$$B_X(z, \bar{z}) = b_X(z, \bar{z})i^{n^2} dz^1 \wedge \ldots \wedge dz^n \wedge d\bar{z}^1 \wedge \ldots \wedge d\bar{z}^n.$$

Assume that $B_X(z, \bar{z}) > 0$ for every z, that is, at every z there is a square-integrable holomorphic n-form ω such that $\omega(z) \neq 0$. Then we can define $\log b_X(z, \bar{z})$ and

$$(4.10.6) \qquad ds_X^2 = 2 \sum h_{j\bar{k}} dz^j\, d\bar{z}^k, \qquad h_{j\bar{k}} = \frac{\partial^2 \log b_X(z, \bar{z})}{\partial z^j \partial \bar{z}^k}.$$

Although $b_X(z, \bar{z})$ depends on the local coordinate system $z^1, \ldots, z^n$, ds_X^2 is independent of the coordinate system. Using the local expression $b_X = \sum |f_j|^2$, we can write ds_X^2 as follows:

$$(4.10.7) \qquad ds_X^2 = 2\frac{(\sum |f_j|^2)(\sum df_k d\bar{f}_k) - (\sum \bar{f}_j df_j)(\sum f_k d\bar{f}_k)}{(\sum |f_j|^2)^2}.$$

In general, ds_X^2 is a positive semi-definite Hermitian form, called the **Bergman pseudo-metric** of X.

As an immediate consequence of (4.10.5) we have

(4.10.8) Proposition. *Let X be a complex manifold with Bergman kernel form $B_X(z, \bar{z}) > 0$ everywhere. If A is a closed complex subspace of codimension ≥ 1 in X, then*

$$ds_X^2 = ds_{X-A}^2.$$

We fix a point $z \in X$ and also a tangent vector v at z, and choosing a suitable orthonormal basis $\{\omega_j\}$ we want to evaluate ds_X^2 on v. As before, let W' be the hyperplane of W consisting of forms ω vanishing at z. Let W'' be the subspace of W' consisting of forms $\omega = f dz^1 \wedge \ldots \wedge dz^n$ such that $df(v) = 0$. For $\omega \in W'$, the condition $df(v) = 0$ does not depend on the local coordinate system $z^1, \ldots, z^n$ and is well defined. (For an element $\omega \in W$, not belonging to W', this condition is not well defined). Let $ds_X(v)$ denote the length of v with respect to the Bergman pseudo-metric ds_X^2. There are two cases:

(a) $W'' = W'$. In this case, we choose ω_0 perpendicular to W', and $\omega_1, \omega_2, \ldots$ from W'. From (4.10.7) we see that $ds_X(v) = 0$.

(b) $W'' \neq W'$. In this case, W'' is of codimension 1 in W'. So we choose ω_0 perpendicular to W', $\omega_1 \in W'$ perpendicular to W'', and $\omega_2, \omega_3, \ldots$ in W''. Then, from (4.10.7) we obtain

$$(4.10.9) \qquad ds_X(v) = \frac{|df_1(v)|}{|f_0(z)|}.$$

The proof of the following proposition is similar to that of (4.10.3).

(4.10.10) Proposition. *Let X be a complex manifold. Fix a point $z \in X$ and a vector v at z, and define $W'' \subset W' \subset W$ as above. Let $\omega_0 = f_0 dz^1 \wedge \ldots \wedge dz^n$ be an element of W perpendicular to W' such that $B_X(z, \bar{z}) = i^{n^2} \omega_0(z) \wedge \bar{\omega}_0(\bar{z})$; (by (4.10.3) such an ω_0 is unique up to a constant factor c with $|c| = 1$).*

Then

$$ds_X(v) = \max \frac{|df(v)|}{|f_0(z)|},$$

where the maximum is taken over all $\omega = f dz^1 \wedge \ldots \wedge dz^n \in W'$ with $\|\omega\| \leq 1$. If $ds_X(v) > 0$, or equivalently, if $W'' \neq W'$, then the maximum is attained by an element $\omega_1 \in W'$ perpendicular to W'' such that $\|\omega_1\| = 1$, and, moreover, such a form ω_1 is unique up to a constant factor c with $|c| = 1$.

We have proved also the following

(4.10.11) Proposition. *For each $z \in X$ and $v \in T_z X$ of a complex manifold X, let $W'' \subset W' \subset W$ be as above. Then*

(1) $B_X(z, \bar{z}) > 0$ *if and only if $W' \neq W$;*

(2) *Under the assumption that $B_X(z, \bar{z}) > 0$, we have $ds_X(v) > 0$ if and only if $W'' \neq W'$.*

In order to explain the geometric meaning of ds_X^2, we consider the dual Hilbert space W^* of W and the projective space $P(W^*)$ of lines in W^* (or equivalently the projective space of hyperplanes in W). In general, $P(W^*)$ is an infinite dimensional Hilbertian manifold. Corresponding to an orthonormal basis $\omega_0, \omega_1, \ldots$ for W, we have a homogeneous coordinate system $(\zeta^0, \zeta^1, \ldots)$ for $P(W^*)$. Write

$$\|\zeta\|^2 = \sum_{j=0}^{\infty} |\zeta^j|^2.$$

Then the Fubini-Study metric for $P(W^*)$ is given by

$$(4.10.12) \qquad ds_{P(W^*)}^2 = 2 \sum_{j,k=0}^{\infty} g_{j\bar{k}} d\zeta^j d\bar{\zeta}^k, \qquad g_{j\bar{k}} = \frac{\partial^2 \log \|\zeta\|^2}{\partial \zeta^j \partial \bar{\zeta}^k}.$$

If X is a complex manifold with Bergman kernel form $B_X > 0$, then we can define a natural map

$$\iota: X \rightarrow P(W^*)$$

by assigning to each point $z \in X$ the hyperplane of W consisting of forms ω vanishing at z. In terms of the basis $\{\omega_j\}$ for W and the corresponding homogeneous coordinate system for $P(W^*)$ defined above, the map ι is given by

$$z \mapsto [\omega_0(z) : \omega_1(z) : \omega_2(z) : \ldots].$$

From the construction of ds_X^2, we immediately see

$$(4.10.13) \qquad \iota^* ds_{P(W^*)}^2 = ds_X^2.$$

If ds_X^2 is positive definite everywhere, or equivalently, if ι defines an immersion of X into $P(W^*)$, then we say that X admits the **Bergman metric** ds_X^2. In this case,

$$\iota: (X, ds_X^2) \rightarrow (P(W^*), ds_{P(W^*)}^2)$$

is an isometric immersion.

For X to admit the Bergman metric, W must be "very ample". If X is a bounded domain in $\mathbf{C}^n$, then W contains all polynomial functions, and X admits the Bergman metric.

(4.10.14) **Example**. For the unit disk D, a simple orthonormal basis $\{\omega_j\}$ is given by

$$\omega_j = \sqrt{\frac{j+1}{2\pi}} z^j dz, \qquad j = 0, 1, 2, \ldots,$$

and the kernel form B_D by

$$B_D(z, \bar{z}) = \frac{i dz \wedge d\bar{z}}{2\pi (1 - |z|^2)^2}.$$

Hence, the Bergman metric of D is given by

$$ds_D^2 = \frac{4dz\,d\bar{z}}{(1 - |z|^2)^2},$$

which agrees with the Poincaré metric of curvature -1, see (2.1.3).

This generalizes to the n-dimensional unit ball B^n as follows.
(4.10.15) **Example.** Let

$$B_n = \{z = (z^1, \ldots, z^n) \in \mathbf{C}^n; \ \|z\|^2 = |z^1|^2 + \ldots + |z^n|^2 < 1\}.$$

Then its Bergman kernel form is given by

$$B_{B^n} = \frac{n!}{(2\pi)^n} \frac{i^{n^2} dz^1 \wedge \ldots \wedge dz^n \wedge d\bar{z}^1 \wedge \ldots \wedge d\bar{z}^n}{(1 - \|z\|^2)^{n+1}}.$$

We note that $\pi^n/n!$ is the Euclidean volume of the unit ball B^n. The Bergman
metric $ds_{B^n}^2 = 2 \sum g_{j\bar{k}} dz^j d\bar{z}^k$ is given by

$$\begin{aligned}
g_{j\bar{k}} &= -\frac{\partial^2}{\partial z^j \partial \bar{z}^k} \log(1 - \|z\|^2)^{n+1} \\
&= \frac{n+1}{(1 - \|z\|^2)^2}[(1 - \|z\|^2)\delta_{jk} + z^k \bar{z}^j].
\end{aligned}$$

If X is a compact complex manifold, it admits no nonconstant holomorphic
functions. But it may admits sufficiently many holomorphic n-forms to produce
the Bergman metric. This is the case when X is a projective manifold with very
ample canonical bundle. Postponing the general discussion on manifolds with
ample canonical bundle to Chapter 7, we give only a simple example here.

(4.10.16) **Example.** Let X be a nonsingular hypersurface of degree d in $P_{n+1}\mathbf{C}$
defined by the polynomial equation $f(z) = f(z^0, \ldots, z^{n+1}) = 0$. If $d \geq n + 3$,
then X has a very ample canonical bundle and admits the Bergman metric. Let
H be the hyperplane line bundle of $P_{n+1}\mathbf{C}$, and H_X its restriction to X. Then the
canonical line bundle K_X of X is given by

$$K_X = H_X^{d-n-2},$$

where H_X^{d-n-2} denotes the $(d-n-2)$-th tensor power of H_X. Since H_X^d is the
normal bundle of X in $P_{n+1}\mathbf{C}$ and since $H^1(P_{n+1}; H^m) = 0$ for all m by Bott's
vanishing theorem, we have the following exact sequence:

$$0 \to H^0(P_{n+1}; H^{m-d}) \to H^0(P_{n+1}; H^m) \to H^0(X; H_X^m) \to 0.$$

Since $H^0(P_{n+1}; H^{m-d}) = 0$ for $m - d < 0$, we have a natural isomorphism

$$H^0(P_{n+1}; H^{d-n-2}) \cong H^0(X; H_X^{d-n-2}) \cong H^0(X; K_X).$$

The space $H^0(P_{n+1}; H^{d-n-2})$ is naturally isomorphic to the space $Q_{d-n-2}(n+2)$ of homogeneous polynomials of degree $d-n-2$ in $n+2$ variables $z^0, z^1 \ldots, z^{n+1}$. If $g(z) = g(z^0, \ldots, z^{n+1})$ is such a polynomial, then the corresponding holomorphic n-form on X is given by

$$\mathrm{Res}_X\left(\frac{g(z) \sum (-1)^j dz^0 \wedge \ldots \wedge \widehat{dz^j} \wedge \ldots \wedge dz^{n+1}}{f(z)} \right),$$

where Res_X denotes the Poincaré residue map.

In the simplest case of $d = n+3$, $K_X = H_X$ and $H^0(M; K_X)$ is dual to $\mathbf{C}^{n+2}$. In this case, the map $\iota \colon X \to P_n\mathbf{C}$ is the given imbedding we started with.

The following product formulas for the Bergman kernel form and the Bergman metric are well known and easy to prove.

(4.10.17) **Proposition.** *Let X and Y be complex manifolds. Then*

$$B_{X \times Y} = B_X \wedge B_Y, \qquad ds_{X \times Y}^2 = ds_X^2 + ds_Y^2.$$

This follows from the fact that if $\{\omega_j\}$ (resp. $\{\varphi_k\}$ is a complete orthonormal basis for the space W_X (resp. W_Y) of square integrable holomorphic forms of the highest degree on X (resp. Y), then $\{\omega_j \wedge \varphi_k\}$ is a complete orthonormal basis for the space $W_{X \times Y}$.

The following comparison between the Bergman metric ds_X^2 and the Carathéodory metric is due to Hahn [2], see also Hahn [1], Burbea [1], Matsuura [2].

(4.10.18) **Theorem.** *Let X be a complex manifold with Bergman kernel form $B_X(z, \bar{z}) > 0$ everywhere. Then the infinitesimal Carathéodory pseudo-metric E_X is bounded by the Bergman pseudo-metric ds_X^2:*

$$\frac{1}{4} E_X^2 \leq ds_X^2.$$

Proof. Fix a point $o \in X$ and a vector $v \in T_z X$. Choose $\omega_0 \in W$ and $\omega_1 \in W'$ as in (4.10.10) so that the Bergman-length $ds_X(v)$ of v is given by

$$ds_X(v) = \frac{|df_1(v)|}{|f_0(o)|}.$$

Take any map $\varphi \in \mathrm{Hol}(X, D)$ such that $\varphi(o) = 0$. Then $\varphi\omega_0 \in W'$. By (4.10.10), we have

$$|df_1(v)| \geq |d(\varphi f_0)(v)| = |d\varphi(v) \cdot f_0(o)|.$$

Hence,

$$ds_X(v) = \frac{|df_1(v)|}{|f_0(o)|} \geq |d\varphi(v)| = \frac{1}{2}\|d\varphi(v)\|,$$

where $\| \ \|$ is the length in the Poincaré metric. Since this holds for all $\varphi \in \mathrm{Hol}(X, D)$ with $\varphi(z) = 0$, we have

$$ds_X(v) \geq \frac{1}{2} E_X(v). \qquad \square$$

An inequality such as $ds_X \leq cF_X$ is false in general, (see Diederich-Fornaess [2]). For special domains, see Hahn-Pflug [1], Masaaki Suzuki [2, 3, 4].

(4.10.19) Corollary. *Let X be a complex manifold with $B_X(z, \bar{z}) > 0$ everywhere.*

(1)   If it is Carathéodory-hyperbolic, (i.e., c_X is a distance and induces the complex space topology of X), then it admits the Bergman metric ds_X^2.

(2)   If it is Carathéodory-hyperbolic and strongly complete with respect to c_X, then it is complete with respect to the Bergman metric ds_X^2.

This combined with (4.1.7), (4.1.9) and (4.1.10) yields

(4.10.20) Corollary. *If $X \subset \mathbf{C}^n$ is a bounded domain such that there is a weak peak function for X at each point of ∂X, then it is complete with respect to its Bergman metric ds_X^2.*

In particular, every generalized analytic polyhedron is complete with respect to its Bergman metric. So is every bounded convex domain in $\mathbf{C}^n$.

On the other hand, we have the following result due to Bremermann [1].

(4.10.21) Theorem. *If a bounded domain X in $\mathbf{C}^n$ is complete with respect to its Bergman metric, it must be a domain of holomorphy.*

Proof. It suffices to show that the Bergman metric ds_X^2 of X extends to a metric on the envelop of holomorphy $\mathcal{E}(X)$ of X. The functions f_j of (4.10.2) extend to holomorphic functions on $\mathcal{E}(X)$, which will be still denoted f_j. Since $b_X(z, \bar{w})$ is holomorphic in $(z, \bar{w}) \in X \times \bar{X}$, it extends to a holomorphic function, still denoted $b_X(z, \bar{w})$, on the envelop of holomorphy $\mathcal{E}(X \times \bar{X}) = \mathcal{E}(X) \times \mathcal{E}(\bar{X})$. Then (4.10.2) holds on $\mathcal{E}(X)$. Since the Hilbert space W of square-integrable holomorphic functions contains all polynomials, in particular, all linear functions as well as the constant functions, ds_X^2 given by (4.10.7) is defined and positive on $\mathcal{E}(X)$. $\qquad\square$

The converse is not true. Let X be a bounded domain of holomorphy. Let A be the closed complex subspace defined by a holomorphic function $f = 0$. Then $X - A$ is also a domain of holomorphy, but it cannot be complete with respect to its Bergman metric ds_{X-A}^2 since by (4.10.5) ds_{X-A}^2 is the restriction of ds_X^2. The simplest example is the punctured disk D^* in $\mathbf{C}$.

Given a bounded domain $X \subset \mathbf{C}^n$, its **outerhull** $\mathcal{A}(X)$ is defined to be the intersection of all domains of holomorphy containing the closure of X. Then $\mathcal{A}(X)$ is a domain of holomorphy and contains the envelop of holomorpy $\mathcal{E}(X)$; in general, it can be strictly greater than $\mathcal{E}(X)$. The first systematic study on extension of the Bergman's kernell function and metric was made by Bremermann [1]. The following example is due to him.

(4.10.22) Example. The **Hartogs triangle**

$$X = \{(z, w) \in \mathbf{C}^2;\ |w| < |z| < 1\}$$

is a domain of holomorphy with outerhull $\mathcal{A}(X) = D^2$, and the kernel function b_X becomes infinite everywhere at the boundary of X. In fact, the transformation

$$h\colon (z, w) \mapsto (s, t) = (z, w/z)$$

maps X biholomorphically onto $D^* \times D$. By (4.10.5), (4.10.14) and (4.10.17), we have

$$B_{D^* \times D} = B_{D \times D} = \frac{ds \wedge dt \wedge d\bar{s} \wedge d\bar{t}}{4\pi^2(1 - |s|^2)(1 - |t|^2)}.$$

By the invariance property of the kernel form, we have $h^* B_{D^* \times D} = B_X$. Hence,

$$b_X = \frac{1}{4\pi^2(1 - |z|^2)(|z|^2 - |w|^2)}.$$

Sommer and Mehring [1] proved that the kernel form B_X cannot be extended beyond $\mathcal{A}(X)$. (The maximum domain to which B_X can be extended may be strictly smaller than $\mathcal{A}(X)$). It seems reasonable to ask whether every bounded domain X with $\mathcal{A}(X) = X$ is complete with respect to its Bergman metric.

For a bounded domain X in $\mathbf{C}^n$, the Bergman kernel function b_X is globally defined. The boundary behavior of b_X has been extensively studied. However, a condition such as $\lim_{z \to \partial X} b_X(z, \bar{z}) = \infty$ is not intrinsic; it depends on how X is imbedded in $\mathbf{C}^n$, and it does not make sense if X is an abstract complex manifold. Therefore we consider the following slightly stronger, but intrinsic growth condition for the kernel form B_X of a complex manifold X with $B_X > 0$.

(C)   For every sequence S of points in X with no accumulation points and for every $\omega \in W$, there is a subsequence S' of S such that

$$\lim_{S'} \frac{i^{n^2} \omega(z) \wedge \overline{\omega(z)}}{B_X(z, \bar{z})} = 0.$$

The following condition, slightly weaker than (C), is also useful.

(C$'$)   There is a dense subset V of W such that for every sequence S of points in X and for every $\omega \in V$, there is a subsequence S' of S such that

$$\lim_{S'} \frac{i^{n^2} \omega(z) \wedge \overline{\omega(z)}}{B_X(z, \bar{z})} = 0.$$

The primary reason for introducing conditions (C) and (C$'$) lies in the following theorem (see Kobayashi [1], [3]).

(4.10.23). **Theorem.** *Let X be a complex manifold with Bergman metric ds_X^2. If it satisfies condition (C$'$), then it is complete with respect to ds_X^2.*

Proof. We make use of the natural isometric immersion $\iota\colon X \to P(W^*)$. Let S be a Cauchy sequence in X with no accumulation points in X. Then $\iota(S)$ is a Cauchy sequence in $P(W^*)$. Since $P(W^*)$ is complete, there is a limit point $x_0 \in P(W^*)$. Choose an orthonormal basis $\omega_0, \omega_1, \ldots$ for W in such a way that x_0 is represented by a point $\zeta_0 \in W^*$ with homogeneous coordinates $\zeta_0 = (1, 0, 0, \ldots)$. Let ω be an arbitrary element of V. Write $\omega = \sum_{j=0}^{\infty} a_j \omega_j$.

For each $z \in S$, represent $\iota(z) \in P(W^*)$ by a point in the unit sphere of W^*; such a representative is unique up to a constant factor of absolute value 1. Explicitly, we can represent it by

$$\frac{1}{\sqrt{B_X}} (\omega_0(z), \omega_1(z), \ldots),$$

where by $\sqrt{B_X}$ we mean $\sqrt{b_X} dz^1 \wedge \ldots \wedge dz^n$. Up to a multiplicative factor of absolute value 1, this has to converge to $(1, 0, 0, \ldots)$. Hence,

$$\lim_{z \in S} \frac{i^{n^2}}{B_X(z, \bar{z})} (\omega_0(z) \wedge \overline{\omega_0(z)}, \omega_1(z) \wedge \overline{\omega_1(z)}, \ldots) = (1, 0, 0, \ldots).$$

This implies

$$\lim_{z \in S} \frac{i^{n^2} \omega(z) \wedge \overline{\omega(z)}}{B_X(z, \bar{z})} = |a_0|^2.$$

By condition (C'), we have $a_0 = 0$. This means that every element ω of V is perpendicular to ω_0, contradicting the assumption that V is dense in W. $\qquad\square$

A refinement of the theorem above by Pflug [2] and Ligocka was used by Ohsawa [2] to prove the following

(4.10.24) Theorem. *Every bounded pseudoconvex domain in $\mathbf{C}^n$ with a C^1-smooth boundary is complete with respect to its Bergman metric.*

A simpler proof was given by Diederich-Fornaess-Herbort [1] as an application of the following localization theorem.

(4.10.25) Theorem. *Let X be a bounded domain of holomorphy in $\mathbf{C}^n$, and $x_0 \in \partial X$. Let $U \subset\subset V$ be small open neighborhoods of x_0. Then there are positive constants C and C' such that*

$$C \cdot ds^2_{X \cap V} \leq ds^2_X \leq C' \cdot ds^2_{X \cap V} \qquad \text{on} \quad X \cap U.$$

For pseudoconvex domains with C^2 boundary, Diederich-Ohsawa [1] has given an estimate for the Bergman metric implying the completeness of the metric.

Ohsawa [3] proved also that if X is a hyperconvex bounded domain, then $\lim_{z \to \partial X} b_X(z, \bar{z}) = \infty$. We conjecture that X actually satisfies condition (C').

(4.10.26) Remark. Let F be a continuous Minkowski function on $\mathbf{C}^n$ as in (4.1.14), and $X_F = \{v \in \mathbf{C}^n;\ F(v) < 1\}$. Then X_F is complete with respect to its Bergman metric, (Jarnicki-Pflug [3]). As we stated in (4.1.14), such a domain need not be complete hyperbolic, (Jarnicki-Pflug [9]).

In this section we discussed the question of completeness with respect to the Bergman metric. A related problem is to characterize those domians which admit complete Kähler metrics. This problem was first considered by Grauert [1], who proved that every Stein manifold admits a complete Kähler metric and that if a bounded domain with real analytic boundary admits a complete Kähler metric, it must be Stein. The first statement is now a consequence of Remmert's imbedding theorem that every Stein manifold is a closed complex subspace in $\mathbf{C}^N$. The second statement has been generalized by Ohsawa [1] to domains with C^1-boundary. On recent results on the Bergman kernel function metric, see a survey paper by Ohsawa [4].

A Pseudoconvexity

We summarize here various definitions of pseudoconvexity. We shall consider only domains in $\mathbf{C}^n$ although some of the definitions and results apply to domains in Stein spaces or even to more general complex spaces. Let X be a domain in $\mathbf{C}^n$. We give five definitions of pseudoconvexity for X.

(i)   A domain X is said to be δ-**pseudoconvex** if the distance $\delta_X(z)$ from $z \in X$ to the boundary ∂X has the property that $-\log \delta_X(z)$ is a plurisubharmonic function in X.

(ii)   A domain X is said to be **Lelong-pseudoconvex** if it admits a continuous plurisubharmonic function u such that for every real number a the set $\{z \in X; u(z) \le a\}$ is compact. This means that u becomes ∞ at every boundary point of X.

(iii) Let $\mathcal{P}$ denote the set of plurisubharmonic functions on X. Given a compact subset $K \subset X$, let

$$\hat{K} = \{z \in X;\ u(z) \le \sup_{K} u \quad \text{for} \quad u \in \mathcal{P}\}.$$

Since some $u \in \mathcal{P}$ may not be continuous, $\hat{K}$ is not necessarily closed. Let K^* be the closure of $\hat{K}$ in X. A domain X is said to be P-**pseudoconvex** if, for every compact set $K \subset X$, K^* is compact.

(iv) A domain G is said to be **strongly pseudoconvex** if there is a C^∞ strongly plurisubharmonic function v defined on a domain U containing $\bar{G}$ such that G is one of the connected components of $\{z \in U;\ v(z) < 0\}$. A domain X is said to be **Grauert-pseudoconvex** if it is a union of increasing sequence of strongly pseudoconvex domains G_i.

(v) Finally, a domain X is said to be **Oka-pseudoconvex** if it has the property that if $f: \bar{D} \times [0, 1] \to \mathbf{C}^n$ is a continuous map such that

(a)   for each $t \in [0, 1]$ the map $f_t: D \to \mathbf{C}^n$ defined by $f_t(z) = f(z, t)$ is a holomorphic immersion and

(b) $f_t(\bar{D}) \subset X$ for $0 \le t < 1$ and $f_1(\partial D) \subset X$,

then $f(\bar{D} \times [0, 1]) \subset X$, i.e., $f_1(D) \subset X$.

Before we show that these five definitions are equivalent, we prove

(4.A.1) Lemma. *Every domain X in $\mathbf{C}$ is δ-pseudoconvex.*

Proof. For each boundary point $\zeta \in \partial X$, $-\log|z - \zeta|$ is a (sub)harmonic function of $z \in X$. Since $\delta_X(z) = \inf_{\zeta \in \partial X} |z - \zeta|$, it follows that $-\log \delta_X(z) = \sup_{\zeta \in \partial X}(-\log|z - \zeta|)$ is subharmonic function. $\qquad\square$

(4.A.2) Theorem. *The five concepts of pseudoconvexity (δ-pseudoconvexity, Lelong-pseudoconvexity, P-pseudoconvexity, Grauert-pseudoconvexity, and Oka-pseudoconvexity) defined above are all equivalent, and every domain of holomorphy is P-pseudoconvex.*

Proof. (i) $\Rightarrow$ (ii). Let $u = -\log \delta_X$.

(ii) $\Rightarrow$ (iii). Let u be the plurisubharmonic function u in the definition of Lelong-pseudoconvexity. Then $\{z \in X; \ u(z) \le \sup_K u\}$ is compact, and K^* is a closed subset of this compact set.

holomorphy $\Rightarrow$ (iii). Let $\mathcal{F}$ be the family of holomorphic functions on X. Since X is holomorph-convex, for every compact $K \subset X$ its holomorph-convex hull $\{z \in X; \ |f(z)| \le \sup_K |f|, \ f \in \mathcal{F}\}$ is compact. Since $\{|f|; \ f \in \mathcal{F}\} \subset \mathcal{P}$, K^* is compact.

(iii) $\Rightarrow$ (iv). Given a compact set $K \subset X$, choose a domain U such that $\bar{U}$ is compact and $K^* \subset U \subset \bar{U} \subset X$. We claim that there is a domain R, $K^* \subset R \subset \bar{R} \subset U$, which is given as a connected component of a set of the form

$$\{z \in U; \ u_1(z) < 0, \ldots, u_k(z) < 0 \quad \text{for} \quad u_i \in \mathcal{P}\}.$$

(R is a plurisubharmonic analogue of an analytic polyhedron). To find such plurisubharmonic functions $u_1, \ldots, u_k$, for each boundary point a of U we choose $u_a \in \mathcal{P}$ such that $\sup_K u_a < u_a(a)$. By the following approximation theorem of Lelong we may assume that u_a is continuous, (Lelong [1]).

Approximation theorem of Lelong: *given a compact subset $K \subset X$ every plurisubharmonic function u is written as the limit of a monotone decreasing sequence of plurisubharmonic functions which are C^∞ and strongly plurisubharmonic in a neighborhood of K.*

We may further assume that $\sup_K u_a < 0 < u_a(a)$. Let $V(a)$ be a neighborhood of a such that $u_a > 0$ on $V(a)$. Since ∂U is compact, it is covered by a finite number of these neighborhoods, say $V(a_1), \ldots, V(a_k)$. Set $u_1 = u_{a_1}, \ldots, u_k = u_{a_k}$.

Having constructed the desired functions $u_1, \ldots, u_k$ and hence the domain R, we define a plurisubharmonic function u by setting

$$u(z) = \max(u_1(z), \ldots, u_k(z)) \quad \text{for} \quad z \in X.$$

Now, R is given as a connected component of the set defined by $u < 0$.

Let $\sup_K u = -3\varepsilon < 0$. By the approximation theorem of Lelong, we can find a function v which is C^∞ strognly plurisubharmonic in a neighborhood of K and satisfies $|u - v| < \varepsilon$. Let G be the component of the set $\{z \in X;\ v(z) + \varepsilon < 0\}$ that contains K. Then G is strongly pseudoconvex.

(iv) $\Rightarrow$ (v). If X is Grauert-pseudoconvex, then $X = \bigcup G_i$, where $G_1 \subset G_2 \subset \dots$ are strongly pseudoconvex. Let $f: \bar{D} \times [0, 1] \to \mathbf{C}^n$ and $f_t(z) = f(z, t)$ be as in the definition of Oka-pseudoconvexity. Since $f(\partial D \times [0, 1]) \cup f(D \times \{0\})$ is compact and is contained in X, it is contained in some G_i. Assume that $f_1(D) \not\subset G_i$. There is a unique $t_0 \in (0, 1]$ such that $f_t(D) \subset G_i$ for $t < t_0$ and $f_{t_0}(D) \not\subset G_i$. Then $f_{t_0}(D) \subset \bar{G}_i$. Let v be a continuous plurisubharmonic function defined in a neighborhood U of $\bar{G}_i$ such that G_i is a connected component of $\{z \in U;\ v(z) < 0\}$. Since $v \circ f_{t_0}$ is subharmonic and is negative on the boundary ∂D, it is negative in D by the maximum modulus principle. Hence the $f_{t_0}(D)$ is also in G_i. This is a contradiction.

(v) $\Rightarrow$ (i). Let a be a unit vector in $\mathbf{C}^n$, and through each $z \in X$ we consider a complex line $l(z, a)$ in the direction of a. Thus, $l(z, a) = \{z + \tau a \in \mathbf{C}^n;\ \tau \in \mathbf{C}\}$. Let $\delta_a(z)$ be the Euclidean distance from z to the boundary of the open set $X \cap l(z, a)$ in $l(z, a)$. Since $\delta_X(z) = \inf_a \delta_a(z)$, it follows that $-\log \delta_X(z) = \sup_a(-\log \delta_a(z))$ is plurisubharmonic if $-\log \delta_a(z)$ is plurisubharmonic for every a. We fix a and set $u_a(z) = -\log \delta_a(z)$. The problem is to show that u_a is a plurisubharmonic function in X.

We fix $z_0 \in X$ and a unit vector b in $\mathbf{C}^n$. It suffices to prove that $u_a(z_0 + \zeta b)$ is subharmonic in $\zeta \in \mathbf{C}$, $|\zeta| < \varepsilon$. If a and b are proportional, then $\delta_a(z)$ is the distance from z to the boundary of the open set $X \cap l(z, a)$ in $l(z, a)$. By (4.A.1), $u_a(z_0 + \zeta b)$ is subharmonic in ζ.

We therefore assume that a and b are not proportional. If $u_a(z_0 + \zeta b)$ is not subharmonic, there exist a closed disc $\bar{D} \subset \mathbf{C}$ and a harmonic function $h(z)$ on $\bar{D}$ such that

$$(*) \qquad\qquad u_a(z_0 + \zeta b) \le h(\zeta) \qquad \text{for} \quad \zeta \in \bar{D},$$

$$(**) \qquad\qquad u_a(z_0 + \zeta b) < h(\zeta) \qquad \text{for} \quad \zeta \in \partial D,$$

$$(***) \qquad\qquad u_a(z_0 + \zeta_0 b) = h(\zeta_0) \qquad \text{for some} \quad \zeta_0 \in D.$$

Let $l(z_0 + \zeta_0 b, a)$ be the complex line through $z_0 + \zeta_0 b$ in the direction of a, and $z_1 = z_0 + \zeta_0 b + \tau_0 a$ a boundary point of $X \cap l(z_0 + \zeta_0 b, a)$ in $l(z_0 + \zeta_0 b, a)$ nearest to $z_0 + \zeta_0 b$. Then $|\tau_0| = \delta_a(z_0 + \zeta_0 b)$. Let $\tilde{h}$ be the harmonic function conjugate to h determined by the normalizing condition $\tilde{h}(\zeta_0) = -\arg(\tau_0)$, i.e., $e^{-(h+i\tilde{h})(\zeta_0)} = \tau_0$.

We define $f: \bar{D} \times [0, 1] \to \mathbf{C}^n$ by

$$f(\zeta, t) = z_0 + \zeta b + e^{-(h+i\tilde{h})(\zeta)-(1-t)} a.$$

Since a and b are not proportional, we have $\partial f/\partial \zeta \ne 0$, i.e., each f_t defines an immersion of D into $\mathbf{C}^n$. Since

$$e^{-h(\zeta)-(1-t)} \le \delta_a(z_0 + \zeta b) \qquad \text{for} \quad (\zeta, t) \in \bar{D} \times [0, 1]$$

by ($*$) and since the strict inequality holds except at $(\zeta, 1) \in D \times [0, 1]$ by ($*$) and ($**$), we have $f(\zeta, t) \in X$ except at $(\zeta, 1) \in D \times [0, 1]$. Since $f(\zeta_0, 1) = z_1 \in \partial X$ by ($***$), X is not Oka-pseudoconvex. $\qquad\square$

If any one of these pseudoconvexity conditions is satisfied, X is said to be **pseudoconvex**.

In (4.A.2) we showed that every domain of holomorphy is pseudoconvex. The famous problem of Levi was to prove the converse. For reference, we state its solution (by Oka-Bremermann-Norguet-Grauert) as

(4.A.3) **Theorem**. *Every pseudoconvex domain X in $\mathbf{C}^n$ is a domain of holomorphy.*

Chapter 5. Holomorphic Maps into Hyperbolic Spaces

1 Normality, Tautness and Hyperbolicity

Let X and Y be complex spaces. Let $C(X, Y)$ denote the family of continuous maps from X into Y with compact-open topology. Let $\mathcal{D}(X, Y)$ be the subfamily of distance-decreasing maps from X into Y with respect to their intrinsic pseudo-distances d_X and d_Y. Then $\mathcal{D}(X, Y)$ is closed in $C(X, Y)$. The family $\mathrm{Hol}(X, Y)$ of holomorphic maps from X into Y is a closed subset of $\mathcal{D}(X, Y)$.

From (1.3.2) we obtain

(5.1.1) Theorem. *Let X be an arbitrary complex space and Y a hyperbolic complex space. Then a family $\mathcal{F} \subset \mathrm{Hol}(X, Y)$ is relatively compact in $\mathrm{Hol}(X, Y)$ if and only if for every $x \in X$, the set $\{f(x);\ f \in \mathcal{F}\}$ is relatively compact in Y.*

In particular, if Y is compact hyperbolic, then $\mathrm{Hol}(X, Y)$ is compact.

From (1.3.3) we obtain the following corollary.

(5.1.2) Corollary. *Let X be an arbitrary complex space and Y a complete hyperbolic complex space. Then a family $\mathcal{F} \subset \mathrm{Hol}(X, Y)$ is relatively compact in $\mathrm{Hol}(X, Y)$ if and only if the set $\{f(x_0);\ f \in \mathcal{F}\}$ is relatively compact in Y for some $x_0 \in X$.*

We recall (see (1.3.5)) that a family $\mathcal{F} \subset \mathrm{Hol}(X, Y)$ is **normal** if every sequence in $\mathcal{F}$ either has a convergent subsequence or is compactly divergent; see Section 3 of Chapter I for the definition of *compactly divergent*. If $Y^* = Y \cup \{\infty\}$ denotes the one-point compactification of Y and if ∞ denotes also the constant map which sends X to ∞, then $\mathcal{F} \subset \mathrm{Hol}(X, Y)$ is normal if and only if $\mathcal{F} \cup \{\infty\}$ is a relatively compact subset of $C(X, Y^*)$, see (1.3.5). A systematic study of normal families of holomorphic maps into complex manifolds were initiated by Grauert-Reckziegel [1], Wu [1] and Kaup [3, 4].

We say that a complex space Y is **taut** if $\mathrm{Hol}(D, Y)$ is normal, i.e., if for every sequence $\{f_n\}$ in $\mathrm{Hol}(D, Y)$, either

(a) there is a subsequence $\{f_{n_k}\}$ which converges in $\mathrm{Hol}(D, Y)$, or

(b) for each compact set $K \subset D$ and each compact set $L \subset Y$, there is an integer n_0 such that $f_n(K) \cap L = \emptyset$ for $n > n_0$.

We formulate this definition slightly differently. Namely, Y is taut if every sequence $\{f_n\}$ in $\mathrm{Hol}(D, Y)$ has a subsequence $\{f_{n_k}\}$ for which one of the following holds:

(a′) $\{f_{n_k}\}$ converges in $\mathrm{Hol}(D, Y)$;

(b′)   for each compact set $K \subset D$ and each compact set $L \subset Y$ there is an integer k_0 such that $f_{n_k}(K) \cap L = \emptyset$ for $k > k_0$.

This seemingly weaker second definition is actually equivalent to the first definition. In fact, assuming that Y is not taut in the sense of the first definition, let $\{f_n\}$ be a sequence for which neither (a) nor (b) holds. Since (b) does not hold, there exist a compact set $K \subset D$, a compact set $L \subset Y$ and a subsequence $\{f_{n_k}\}$ such that $f_{n_k}(K) \cap L \neq \emptyset$ for all k. Then for any subsequence of $\{f_{n_k}\}$ neither (a′) nor (b′) holds. Hence, Y is not taut even in the sense of the second definition.

Since $\mathrm{Hol}(D, Y)$ is closed in $\mathcal{C}(D, Y)$, Y being taut amounts to saying that $\mathrm{Hol}(D, Y) \cup \{\infty\} \subset \mathcal{C}(D, Y^*)$ is compact.

The concept of taut complex space was introduced by Wu [1] and Kaup [3]; Kaup used the term "hyperbolicity", which has a different meaning in this book.

Given a closed subset $\Delta \subset Y$, we say that a family $\mathcal{F} \subset \mathrm{Hol}(X, Y)$ is **normal modulo** Δ if for every sequence $\{f_n\}$ in $\mathcal{F}$ one of the following holds:

(a) $\{f_n\}$ has a convergent subsequence;

(b)   for each compact set $K \subset X$ and each compact set $L \subset Y - \Delta$, there exists an integer n_0 such that $f_n(K) \cap L = \emptyset$ for all $n > n_0$.

When Δ is empty, this definition reduces to that of normal family. We say that Y is **taut modulo** Δ if $\mathrm{Hol}(D, Y)$ is normal modulo Δ. Again, we may say that Y is taut modulo Δ if every sequence $\{f_n\}$ in $\mathrm{Hol}(D, Y)$ has a subsequence $\{f_{n_k}\}$ for which one of the following holds:

(a′) $\{f_{n_k}\}$ converges in $\mathrm{Hol}(D, Y)$;

(b′)   for each compact set $K \subset D$ and each compact set $L \subset Y - \Delta$, there exists an integer k_0 such that $f_{n_k}(K) \cap L = \emptyset$ for $k > k_0$.

We say that Y is **strongly taut modulo** Δ if for each compact set $K \subset D$ and each compact set $L \subset Y - \Delta$, there exist compact subsets $L_1, \ldots, L_m$ of L and taut open subsets $U_1, \ldots, U_m$ of Y such that

(a) $L = \bigcup_j L_j$ and $L_j \subset U_j$,

(b)   if $f: D \to Y$ is holomorphic and $f(0) \in L_j$, then $f(K) \subset U_j$.

When Δ is empty, we simply say that Y is **strongly taut**. From (1.3.6) we obtain

(5.1.3) **Theorem.** *For a complex space Y with a closed subset Δ (possibly empty), we have the following implications*:

complete hyperbolic $\Rightarrow$ taut $\Rightarrow$ hyperbolic, strongly taut modulo Δ $\Rightarrow$ taut modulo Δ $\Rightarrow$ hyperbolic modulo Δ.

Proof. (1)   complete hyperbolic $\Rightarrow$ taut, (Kiernan [2], Eisenman [2]). By (1.3.6), if Y is complete hyperbolic, then $\mathrm{Hol}(X, Y)$ is normal for any complex space X.

(2)   taut modulo Δ $\Rightarrow$ hyperbolic modulo Δ, (Kiernan [2]). We start with the proof of the following lemma.

(5.1.4) **Lemma.** *Let U, V, W, and U' be open subsets of a complex space Y such that $\bar{U} \cap \bar{U}' = \emptyset$ and $W \subset\subset V \subset\subset U$ and that U is hyperbolic. Assume that there exists a positive number $\delta < 1$ such that, for every $f \in \mathrm{Hol}(D, Y)$ with $f(0) \in V$, we have $f(D_\delta) \subset U$. Then $d_Y(W, U') > 0$.*

Proof. Let U, V, W, U', and δ be as above. Choose a constant $c > 0$ such that $d_D(0, b) \geq c \cdot d_{D_\delta}(0, b)$ for all $b \in D_{\delta/2}$. Let $p \in W$ and $q \in U'$, and

$$\alpha = \{p = p_0, p_1, \ldots, p_l = q; a_1, b_1, \ldots, a_l, b_l; f_1, \ldots, f_l\}$$

be a chain of holomorphic discs from p to q. Without loss of generality we may assume that $a_1 = \ldots = a_l = 0$ and $b_1, \ldots, b_l \in D_{\delta/2}$. Let k be the integer such that $p_1, \ldots, p_{k-1} \in V$ but $p_k \notin V$. By taking a refinement of α we may further assume that $p_k \in U$. Since $f_1(0) = p_0, \ldots, f_k(0) = p_{k-1}$ are all in V, by our assumption $f_1(D_\delta), \ldots, f_k(D_\delta)$ are all contained in U. Hence, the length $l(\alpha)$ of α may be estimated as follows:

$$\begin{aligned} l(\alpha) &\geq \sum_{i=1}^{k} d_D(0, b_i) \geq c \sum_{i=1}^{k} d_{D_\delta}(0, b_i) \geq c \sum_{i=1}^{k} d_U(p_{i-1}, p_i) \\ &\geq c \cdot d_U(p, p_k) \geq c \cdot d_U(W, U - V). \end{aligned}$$

In order to derive the third inequality, we used the fact that $f_i(D_\delta) \subset U$ for $i = 1, \ldots, k$. Hence, $d_Y(W, U') \geq c \cdot d_U(W, U - V)$. $\qquad\square$

We resume the proof of (2). Assume that Y is not hyperbolic modulo Δ. Then there exist distinct points $p \notin \Delta$ and q with $d_Y(p, q) = 0$. Take a complete hyperbolic neighborhood U of p such that $q \notin U$ and $\Delta \cap U = \emptyset$. Let V be a neighborhood of p such that $V \subset\subset U$. By Lemma above, for each integer $n > 0$ there exists a holomorphic map $f_n \in \mathrm{Hol}(D, Y)$ such that $f_n(0) \in V$ and $f_n(D_{1/n}) \not\subset U$. Then $\{f_n\}$ has no convergent subsequence (since $f_n(D_{1/n}) \not\subset U$) nor satisfies condition (b) for normality modulo Δ (since $f_n(0) \in V$).

(3)   strongly taut modulo Δ $\Rightarrow$ taut modulo Δ. Let $\{f_n\}$ be a sequence in $\mathrm{Hol}(D, Y)$. Assume that we are not in case (b) in the definition of taut modulo Δ, i.e., that there exist compact sets $K \subset D$ and $L \subset Y - \Delta$ such that $f_n(K) \cap L \neq \emptyset$ for infinitely many n. We must show that we are in case (a) in the definition of taut modulo Δ. Namely, we want to produce a convergent subsequence of $\{f_n\}$. By taking a subsequence, we may assume that $f_n(K) \cap L \neq \emptyset$ for all n. For each n, take an automorphism g_n of D such that $g_n(0) \in K$ and $f_n(g_n(0)) \in L$. Since $g_n(0)$ is in a fixed compact set, taking a subsequence we may assume that g_n converges to an automorphism g of D. Set $h_n = f_n \circ g_n$.

If we prove that a subsequence of $\{h_n\}$ converges to a map $h \in \mathrm{Hol}(D, Y)$, it will follow that the corresponding subsequence of $\{f_n\}$ converges to $f = h \circ g^{-1}$. Therefore it suffices to show that, for each fixed $r < 1$, $\{h_n\}$ has a subsequence which converges uniformly on $\bar{D}_r$. From the defintion of strong tautness modulo Δ, we have compact subsets $L_1, \ldots, L_m$ of L and taut open subsets $U_1, \ldots, U_m$ of Y such that (a) $L = \cup L_j$ and $L_j \subset U_j$, and (b) if $f : D \to Y$ is holomorphic and $f(0) \in L_j$, then $f(\bar{D}_r) \subset U_j$. Since $h_n(0) \in L = \bigcup L_j$ for all n, by taking a subsequence we may assume that all $h_n(0)$ lie in the same L_j. Then $h_n(\bar{D}_r) \subset U_j$. Since U_j is taut and all $h_n(0)$ are in the fixed compact set L_j, $\{h_n\}$ converges uniformly on $\bar{D}_r$. $\qquad\square$

Some authors call Y taut when $\mathrm{Hol}(X, Y)$ is normal for all X. This is justified by the following result, (see Kobayashi [4], Barth [1]).

(5.1.5) Theorem. *If a complex space Y is taut, then the family $\mathrm{Hol}(X, Y)$ is normal for every complex space X.*

Proof. By (5.1.3) Y is hyperbolic. Let $f_n \in \mathrm{Hol}(X, Y)$ be a sequence which is not compactly divergent. Then there exist compact subsets $K \subset X$ and $L \subset Y$ such that $f_n(K) \cap L \neq \emptyset$ for infinitely many n, i.e., there exists a sequence $x_n \in K$ such that $f_n(x_n) \in L$ for infinitely many n. By taking a subsequence of $\{x_n\}$, we may assume that $x_n \to p \in K$. Fix a relatively compact neighborhood V of L. Since $d_Y(f_n(x_n), f_n(p)) \leq d_X(x_n, p) \to 0$ and since $f_n(x_n) \in L$, there exists n_0 such that $f_n(p) \in V$ for all $n \geq n_0$. Thus $\{f_n(p)\}$ is relatively compact in Y. We shall show that $\{f_n(q)\}$ is relatively compact in Y for every $q \in X$. We consider a chain of holomorphic discs from p to q given by

$$p = p_0, p_1, \ldots, p_k = q \in X; a_1, b_1, \ldots, a_k, b_k \in D; h_1, \ldots, h_k \in \mathrm{Hol}(D, X).$$

Since $f_n(h_1(a_1)) = f_n(p) \in L$ for all n and $\mathrm{Hol}(D, Y)$ is normal, $\{f_n(h_1(b_1))\}$ is relatively compact in Y. Since $f_n(h_1(b_1)) = f_n(h_2(a_2))$ and $\mathrm{Hol}(D, Y)$ is normal, $\{f_n(h_2(b_2))\}$ relatively compact in Y. Continuing this construction we see that $\{f_n(q)\}$ is relatively compact in Y. Since this holds for all $q \in Y$, (1.3.2) implies that $\mathrm{Hol}(X, Y)$ is normal. $\qquad\square$

The following result (independently obtained by Abate [5]) clarifies the implication "taut $\Rightarrow$ hyperbolic". However, we do not have a similar characterization of complete hyperbolicity.

(5.1.6) Theorem. *Let X be a complex space, and $X^* = X \cup \{\infty\}$ be its one-point compactification. (If X is compact, we let $X^* = X$). Let $\mathcal{C}(D, X^*)$ denote the space of continuous maps from D into X^*. Then*

(1) X is hyperbolic if and only if $\mathrm{Hol}(D, X)$ is relatively compact in $\mathcal{C}(D, X^*)$, i.e., every sequence $f_n \in \mathrm{Hol}(D, X)$ has a subsequence which converges in $\mathcal{C}(D, X^*)$;

(2) X is taut if and only if $\mathrm{Hol}(D, X) \cup \{\infty\}$ is a compact subset of $\mathcal{C}(D, X^*)$.

Proof. (1)   Suppose that X is not hyperbolic, and take two distinct points x_0 and y_0 with $d_X(x_0, y_0) = 0$. Choose relatively compact neighborhoods $V \subset\subset U$ of x_0 such that $y_0 \notin \bar{U}$.

We claim that for any positive integer n there is a map $f_n \in \mathrm{Hol}(D, X)$ such that $f_n(0) \in \bar{V}$ but $f_n(D_{1/n}) \not\subset U$. If this claim is granted, we can complete the proof as follows. Let $t_n \in D_{1/n}$ be such that $f_n(t_n) \notin U$. If $\mathrm{Hol}(D, X)$ were relatively compact in $\mathcal{C}(D, X^*)$, then $\{f_n\}$ would have a subsequence $\{f_{n_i}\}$ converging to $f \in \mathcal{C}(D, X^*)$; but then $\{f_{n_i}(t_{n_i})\}$ would converges to $f(0) \in \bar{V}$. This is impossible.

We shall now prove our claim. Assume, to the contrary, that there is a positive intger n such that $f(0) \in \bar{V}$ implies $f(D_{1/n}) \subset U$ for any $f \in \mathrm{Hol}(D, X)$. Then from the definition of d_X it follows that $d_X(x_0, y_0) \geq \rho(0, 1/n) > 0$, contradicting our supposition.

The converse follows from the first part of (1.3.11).

(2)   As explained at the beginning of this section, this is immediate from the definition of tautness. $\square$

(5.1.7) **Examples**. There are taut complex spaces which are not complete hyperbolic. The following example is from Barth [6]. We construct a taut complex space Y by attaching a countable number of unit discs, say $D_1, D_2, D_3, \ldots$, together in the following manner. In the n-th unit disc D_n we choose a point a_n such that its Poincaré distance $d_{D_n}(0, a_n)$ from the origin is equal to $1/2^n$. We attach the second disc D_2 to the first disc D_1 by identifying $a_1 \in D_1$ with the origin 0 of D_2. Then we attach the third disc D_3 to D_2 by identifying $a_2 \in D_2$ with the origin 0 of D_3. In general, we identify $a_n \in D_n$ with the origin 0 of the next disc D_{n+1}. We denote the resulting complex space by Y. The discs $D_n, n = 1, 2, 3, \ldots$ are the irreducible components of Y. Since every holomorphic map $f : D \to Y$ sends D into one of the irreducible components D_n, the family $\mathrm{Hol}(D, Y)$ is a union of subfamilies $\mathrm{Hol}(D, D_n)$. Let $\{f_j\}$ be sequence in $\mathrm{Hol}(D, Y)$. Assume that it has no convergent subsequence. Let $\{f_{nj}\}$ be the subsequence consisting of those f_j which map D in D_n. Since $\mathrm{Hol}(D, D_n)$ is normal, the sequence $\{f_{nj}\}$ must be compactly divergent. Each compact subset $L \subset Y$ is contained $D_1 \cup \ldots \cup D_k$ for some k. Then, for each compact set $K \subset D$ and for each fixed $n \leq k$ we have $f_{nj}(K) \cap L = \emptyset$ for all but a finite number of f_{nj}. For $n > k$, $f_{nj}(K) \cap L = \emptyset$ since $D_n \cap L = \emptyset$. Hence, $\{f_j\}$ is compactly divergent, proving that $\mathrm{Hol}(D, Y)$ is normal. Let p_n be the point of Y corresponding to $a_n \in D_n$. Then the sequence $\{p_n\}$ is divergent in Y but is Cauchy since

$$d_Y(p_{n-1}, p_n) \leq d_{D_n}(0, a_n) = 1/2^n.$$

Later, Rosay [2] gave an example of taut bounded domain in $\mathbf{C}^3$ which is not complete hyperbolic.

Unlike (5.1.5), for a complex space Y which is taut modulo Δ, the family $\mathrm{Hol}(X, Y)$ may not be normal modulo Δ. Consider the following example by Barth [6]. Let $Y \subset \mathbf{C}^2$ be the 1-dimensional complex subspace having two irreducible components $\Delta = \{0\} \times \mathbf{C}$ and $D \times \{0\}$. Then Y is complete hyperbolic modulo Δ and taut modulo Δ, but $\mathrm{Hol}(Y, Y)$ is not normal modulo Δ. In fact, define $f_n \in \mathrm{Hol}(Y, Y)$ by $f_n(z, w) = (z, nw)$. Then every subsequence of $\{f_n\}$ diverges in $\mathrm{Hol}(Y, Y)$. Moreover, for any nonempty compact subset $K \subset Y - \Delta$, we have $f_n(K) \cap K = K \neq \emptyset$ for all n, showing that $\{f_n\}$ is not normal modulo Δ. It does not seem to be known if there is an irreducible example.

We do not know the relationship between "complete hyperbolic modulo Δ" and "(strongly) taut modulo Δ".

In spite of these examples, a taut complex space seems to be very close to being complete hyperbolic. Many of the results for complete hyperbolic complex spaces hold also for taut complex spaces, see Kaup [4] and Wu [1]. We have the following analogue of (3.11.2), (see Do Duc Thai [1]).

(5.1.8) **Theorem.** *Let* $\pi\colon X \to T$ *be a proper holomorphic map of complex spaces such that every* $t \in T$ *has an open neighborhood* U *such that* $\pi^{-1}(U)$ *is taut. If* T *is taut, then* X *is also taut.*

Proof. Let $\{U_i\}$ be a countable open cover of T such that each $\pi^{-1}(U_i)$ is taut. Let $\{f_n\} \subset \mathrm{Hol}(D, X)$, and assume that $\{f_n\}$ is not compactly divergent. Then $\{\pi \circ f_n\}$ is not compactly divergent. Since T is taut, by taking a subsequence we may assume that $\{\pi \circ f_n\}$ converges to $g \in \mathrm{Hol}(D, T)$. Put $V_i = g^{-1}(U_i) \subset D$. Take an open cover $\{W_i\}$ of D such that $W_i \subset\subset V_i$.

Consider the sequence $\{f_n|_{W_1}\}$, and take an integer n_1 such that $f_n(W_1) \subset \pi^{-1}(U_1)$ for $n \geq n_1$. Since $\pi \circ f_n|_{W_1} \to g|_{W_1}$, it follows that $\{f_n|_{W_1}\}$ is not compactly divergent. Hence, there exists a subsequence $\{f_n^{(1)}\}$ of $\{f_n\}$ converging in $\mathrm{Hol}(W_1, \pi^{-1}(U_1)) \subset \mathrm{Hol}(W_1, X)$.

Considering the sequence $\{f_n^{(1)}|_{W_2}\}$, we obtain a subsequence $\{f_n^{(2)}\}$ such that $\{f_n^{(2)}|_{W_2}\}$ converges in $\mathrm{Hol}(W_2, \pi^{-1}(U_2)) \subset \mathrm{Hol}(W_2, X)$. Continuing this process, we obtain a subsequence $\{f_n^{(k)}\} \subset \{f_n^{(k-1)}\}$ such that $\{f_n^{(k)}\}$ converges in $\mathrm{Hol}(W_k, \pi^{-1}(U_k)) \subset \mathrm{Hol}(W_k, X)$. Then the diagonal subsequence $\{f_n^{(n)}\}$ converges in $\mathrm{Hol}(D, Z)$. $\qquad\square$

The following corollary should be compared with (3.2.11) and (3.2.12).

(5.1.9) **Corollary.** *Let* $\pi\colon \tilde{X} \to X$ *be a proper finite map between complex spaces. If* X *is taut, so is* $\tilde{X}$.

In particular, if X *is taut, its normalization is also taut.*

Proof. Given $x \in X$, let U be a complete hyperbolic open neighborhood of x. Then $\pi^{-1}(U)$ is complete hyperbolic by (3.2.11) and hence taut by (5.1.3). Then $\tilde{X}$ is taut by (5.1.8). $\qquad\square$

A relatively compact complex subspace Y of a complex space Z is said to be **tautly imbedded** in Z if $\mathrm{Hol}(D, Y)$ is relatively compact in $\mathrm{Hol}(D, Z)$, i.e., every sequence $f_n \in \mathrm{Hol}(D, Y)$ has a subsequence which converges in $\mathrm{Hol}(D, Z)$. (In applications, Y is usually a relatively compact open domain in Z).

Let Δ be a closed subset of Z. A relatively compact complex subspace Y of Z is said to be **tautly imbedded modulo Δ in Z** if for each sequence $\{f_n\}$ in $\mathrm{Hol}(D, Y)$ one of the following holds:

(a) $\{f_n\}$ has a subsequence $\{f_{n_k}\}$ which converges in $\mathrm{Hol}(D, Z)$;

(b)   for each compact set $K \subset D$ and each compact set $L \subset Z - \Delta$ there exists an integer n_0 such that $f_n(K) \cap L = \emptyset$ for $n > n_0$.

Here again, we may say that Y is tautly imbedded modulo Δ in Z if each sequence $\{f_n\}$ in $\mathrm{Hol}(D, Y)$ has a subsequence $\{f_{n_k}\}$ for which one of the following holds:

(a$'$) $\{f_{n_k}\}$ converges in $\mathrm{Hol}(D, Z)$;

(b$'$)   for each compact set $K \subset D$ and each compact set $L \subset Z - \Delta$ there exists an integer k_0 such that $f_{n_k}(K) \cap L = \emptyset$ for $k > k_0$.

We say that Y is **strong-tautly imbedded modulo Δ in Z** if for each compact set $K \subset D$ and each compact set $L \subset Z - \Delta$, there exist compact subsets $L_1, \ldots, L_m$ of L and taut open subsets $U_1, \ldots, U_m$ of Z such that

(a) $L = \bigcup_j L_j$ and $L_j \subset U_j$;

(b)   if $f: D \to Y$ is holomorphic and $f(0) \in L_j$, then $f(K) \subset U_j$.

Reasoning in the same way as in (3) of the proof of (5.1.3), we see that if Y is strong-tautly imbedded modulo Δ in Z, it is taulty imbedded modulo Δ in Z, (see Kiernan-Kobayashi [2] for details).

The following theorem is essentially the same as the main result in Joseph-Kwack [2].

(5.1.10) **Theorem.** *Given a relatively compact complex subspace Y of Z and a closed subset Δ of Z, let $\pi: \bar{Y} \to \bar{Y}/\Delta$ be the projection which collapses $\Delta \cap \bar{Y}$ into a single point. Then Y is hyperbolically imbedded modulo Δ in Z if and only if the family $\pi \circ \mathrm{Hol}(D, Y) = \{\pi \circ f;\ f \in \mathrm{Hol}(D, Y)\}$ is relatively compact in $\mathcal{C}(D, \bar{Y}/\Delta)$.*

Proof. Assume that $\pi \circ \mathrm{Hol}(D, Y)$ is relatively compact in $\mathcal{C}(D, \bar{Y}/\Delta)$. If Y is not hyperbolically imbedded modulo Δ in Z, then there is a non-hyperbolic point $p \in \bar{Y} - (\Delta \cap Y)$. This means that there is an open neighborhood U of p in Z such that for any open neighborhood V of p with $\bar{V} \subset U$ we have

$$d_Y(\bar{V} \cap Y, Y - (U \cap Y)) = 0.$$

Take U small enough so that $U \cap \Delta = \emptyset$.

As in the proof of (1) of (5.1.6), there is a sequence of maps $f_n \in \mathrm{Hol}(D, Y)$ such that $f_n(0) \in \bar{V}$ but $f_n(D_{1/n}) \not\subset U$. Let $t_n \in D_{1/n}$ be such that $f_n(t_n) \notin U$. Since $\pi \circ \mathrm{Hol}(D, Y)$ is relatively compact in $\mathcal{C}(D, \bar{Y}/\Delta)$, $\{\pi \circ f_n\}$ has a subsequence $\{\pi \circ f_{n_i}\}$ converging to $g \in \mathcal{C}(D, \bar{Y}/\Delta)$. Then $\{\pi(f_{n_i}(t_{n_i}))\}$ converges to $g(0) \in \pi(\bar{V})$. This is a contradiction.

Conversely, assume that Y is hyperbolically imbedded modulo Δ in Z. Let K be a compact neighborhood of $\bar{Y}$. Given a length function F on K, let φ be a continuous nonnegative function on K satisfying conditions (a) and (b) of (3.3.13). Let δ_K be the pseudo-distance on K defined by the pseudo-length function φF. Then $\delta_K(p, q) > 0$ for $p \neq q$ unless both p and q are in Δ. Then δ_K induces a distance function $\delta_{K/\Delta}$ on the quotient space K/Δ and hence on $\bar{Y}/\Delta$. By (1.3.9), $\pi \circ \mathrm{Hol}(D, Y)$ is relatively compact in $\mathcal{C}(D, \bar{Y}/\Delta)$. $\square$

Let $\mathcal{F}_{Y,Z} \subset \mathrm{Hol}(D, Z)$ denote the family of holomorphic maps $f : D \to Z$ which map the whole disc D, except possibly one point, into Y.

The following result is due partly to Kiernan [6] and partly to Joseph-Kwack [1].

(5.1.11) **Theorem.** *For a relatively compact complex subspace Y of a complex space Z, the following conditions are mutually equivalent:*
   (a)   *Y is hyperbolically imbedded in Z;*
   (b)   *Y is tautly imbedded in Z;*
   (c)   *$\mathrm{Hol}(X, Y)$ is relatively compact in $\mathrm{Hol}(X, Z)$ for all complex spaces X;*
   (d)   *$\mathcal{F}_{Y,Z}$ is relatively compact in $\mathrm{Hol}(D, Z)$;*
   (e)   *Given a length function F on Z, there is a constant $c > 0$ such that*

$$f^*(c^2 F^2) \leq ds_D^2 \qquad \text{for all} \quad f \in \mathrm{Hol}(D, Y),$$

where ds_D^2 denotes the Poincaré metric of D;
   (f)   *Given a length function F on Z, there is a constant $c > 0$ such that*

$$f^*(c^2 F^2) \leq ds_D^2 \qquad \text{for all} \quad f \in \mathcal{F}_{Y,Z}.$$

   (g)   *for $p, q \in \bar{Y}$, $p \neq q$,*

$$d_{Y,Z}(p, q) > 0.$$

Proof. (a) $\Leftrightarrow$ (e). This was proved in (3.3.3).
   (e) $\Rightarrow$ (c). Let δ_Z be the distance function on Z defined by the length function cF. Now apply (1.3.9) to (Y, d_Y) and (Z, δ_Z).
   (c) $\Rightarrow$ (b). This is trivial.
   (b) $\Rightarrow$ (e). Assume that (e) is false. Then for each integer $n > 0$ there exists a holmorphic map $f_n : D \to Y$ such that $f_n^*(F^2) > n \cdot ds_D^2$ at some point $a_n \in D$. Since D is homogeneous, we may assume that $a_n = 0$. This means that $\lim F(f_{n*}(e)) = \infty$, where e denotes the tangent vector $(\partial/\partial z)_0$ of D at 0. Since Y is relatively compact, we may assume that $\lim f_n(0) = p \in \bar{Y}$.
   If (b) is true, taking a subsequence we may assume that $\{f_n\}$ converges to a holomorphic map $f : D \to Z$. Then $f_{n*}(e)$ would have to converge to $f_*(e)$. This is a contradiction.
   (f) $\Leftrightarrow$ (g). This was proved in (3.4.11).
   (a) $\Leftrightarrow$ (g). This was proved in (3.4.11) and (3.6.20).

(f) $\Rightarrow$ (d). Every element of $\mathcal{F}_{Y,Z}$ is distance-decreasing with respect to the Poincaré metric of D and the length function cF. The desired implication follows from (1.3.2).

(d) $\Rightarrow$ (b). This is obvious since $\mathrm{Hol}(D, Y) \subset \mathcal{F}_{Y,Z}$. $\qquad\square$

(5.1.12) **Corollary**. *Let Z be a complex space and A a Cartier divisor. If $Y = Z - A$ is tautly imbedded in Z, then Y is complete hyperbolic and hence taut.*

Proof. By (5.1.11) Y is hyperbolically imbedded in Z. It is complete hyperbolic by (3.3.6) and taut by (5.1.3). $\qquad\square$

(5.1.13) **Theorem**. *Let Δ be a closed subset of a complex space Z. If a relatively compact complex subspace Y of Z is tautly imbedded modulo Δ in Z, then it is hyperbolically imbedded modulo Δ in Z.*

Proof. Let p, q be two distinct points of $\bar{Y}$ such that $p \notin \Delta$. Let U and U' be neighborhoods of p and q in Z such that $\bar{U} \cap \bar{U}' = \emptyset$ and $\bar{U} \cap \Delta = \emptyset$. Taking U small, we may assume that U is hyperbolic. Let V be a smaller neighborhood of p such that $\bar{V}$ is compact and is contained in U. We claim that there is a positive number $\delta < 1$ such that if $f : D \to Y$ is holomorphic and $f(0) \in V$, then $f(D_\delta) \subset U$. Assume the contrary. Then for each positive integer n, there exists a map $f_n \in \mathrm{Hol}(D, Y)$ such that $f_n(0) \in V$ and $f_n(D_{1/n}) \not\subset U$. Then the sequence $\{f_n\}$ has no subsequence converging to, say g in $\mathrm{Hol}(D, Z)$. (For g would have the property that $g(0) \in \bar{V} \subset U$. But then $f_n(D_{1/n})$ would be contained in U for a large n.) This means that condition (a) for taut imbedding modulo Δ is violated. By taking $K = \{0\}$ and $L = \bar{V}$, we see that condition (b) is not satisfied either. Hence, our claim must be true. Take a neighborhood W of p such that $\bar{W} \subset V$, and apply (5.1.4) to $U \cap Y$, $V \cap Y$, $W \cap Y$ and $U' \cap Y$. Then $d_Y(U \cap Y, U' \cap Y) > 0$.
$\qquad\square$

We do not know if the converse holds when Δ is nonempty.

In parallel to the concept of local complete hyperbolicity introduced in Section 2 of Chapter 3, we define the notion of local tautness. Let Y be a complex subspace of a complex space Z with compact closure $\bar{Y}$. We say that Y is **locally taut** in Z if every point $p \in \bar{Y}$ has a neighborhood V such that $V \cap Y$ is taut. Note that the condition is trivially satisfied by all points p of Y.

From (5.1.3) we have

(5.1.14) **Proposition**. *If a complex subspace Y of a complex space Z is locally complete hyperbolic, then it is locally taut.*

From (3.2.18) we obtain

(5.1.15) **Proposition**. *Let Z be a complex space and A a Cartier divisor. Then $Y = Z - A$ is locally taut.*

(5.1.16) **Theorem**. *Let Δ be a closed subset of a complex space Z. If a relatively compact complex subspace Y is tautly imbedded modulo Δ in Z and is locally taut in Z, then it is taut modulo Δ.*

Proof. Let $\{f_n\}$ be a sequence in $\mathrm{Hol}(D, Y)$. Assume that condition (b) for tautness modulo Δ is not satisfied, i.e., there exist compact sets $K \subset D$ and $L \subset Y - \Delta$ such that $f_n(K) \cap L \neq \emptyset$ for infinitely many n. Then condition (b) for taut imbedding modulo Δ would not be satisfied. Hence, condition (a) for taut imbedding modulo Δ must be satisfied, i.e., (a subsequence of) $\{f_n\}$ converges to a map $f \in \mathrm{Hol}(D, Z)$. We have to show that $f(D) \subset Y$. Assume $f(D) \not\subset Y$. Then the open subset $f^{-1}(Y)$ of D is distinct from D. It is nonempty since $f_n(K) \cap L \neq \emptyset$ implies $f(K) \cap L \neq \emptyset$. Let a be a boundary point of $f^{-1}(Y)$ in D, and set $p = f(a) \in \partial Y$. Let V be a neighborhood of p in Z such that $V \cap Y$ is taut. Let W be a neighborhood of a in D such that $f(\bar{W}) \subset V$. By taking a subsequence we may assume that $f_n(\bar{W}) \subset V$ for all n. Let b be a point in $W \cap f^{-1}(Y)$. Since $f_n(W) \subset V \cap Y$, we consider $\{f_n\}$ as a sequence in $\mathrm{Hol}(W, V \cap Y)$. Since $V \cap Y$ is taut, it converges in $\mathrm{Hol}(W, V \cap Y)$ or it is compactly divergent. But

$$\lim f_n(a) = f(a) = p \in \partial Y \quad \text{and} \quad \lim f_n(b) = f(b) \in Y.$$

This is a contradiction. $\qquad\square$

(5.1.17) **Remark**. Let Z be a complex space and A a Cartier divisor. If $Y = Z - A$ is tautly imbedded in Z, then by (5.1.15) and (5.1.16) Y is taut. But we already know this from (5.1.12).

Since the proof of the following theorem is similar to that of (5.1.16), it is omitted. Details can be found in Kiernan-Kobayashi [2].

(5.1.18) **Theorem.** *If Y is strong-tautly imbedded modulo Δ in Z and is locally taut in Z, then it is strongly taut modulo Δ*

We summarize some of the results above in the following chart:

(5.1.19) **Theorem.**

$$
\begin{array}{ccccc}
\text{s-taut imbd mod } \Delta & \Rightarrow & \text{taut imbd mod } \Delta & \Rightarrow & \text{hyp imbd mod } \Delta \\
\downarrow & & \downarrow & & \Downarrow \\
\text{s-taut mod } \Delta & \Rightarrow & \text{taut mod } \Delta & \Rightarrow & \text{hyp mod } \Delta
\end{array}
$$

where the implications indicated by the two single down-arrows are valid under the assumption that Y is locally taut in Z, while the double arrows are valid without such an assumption.

The concept of a normal meromorphic function was introduced by Lehto and Virtanen [1] for the purpose of generalizing the classical Picard theorem. Recently, several authors, including Cima-Krantz [1], Funahashi [1], Hahn [5, 6, 7, 8, 9], Järvi [1], Zaidenberg [11], and Joseph-kwack [3], extended the higher dimensional Picard theorem to normal holomorphic mappings. For normal functions in several variables, see also Aladro [1] and Aladro-Krantz [1]. For a survey, see Kwack [6]. We adopt here the definition due to Hahn [3, 7].

Let X and Z be complex spaces. A holomorphic map $f: X \to Z$ is said to be **normal** if the image $f(X)$ is relatively compact and if the family $\mathcal{F}_f = \{f \circ \varphi; \ \varphi \in$

$\mathrm{Hol}(D, X)\}$ is relatively compact in $\mathrm{Hol}(D, Z)$. Clearly, if f is normal, then $\mathcal{F}_f$ is a normal family. But the converse is not true.

(5.1.20) Proposition. *Let Y be a relatively compact complex subspace of a complex space Z. Then Y is hyperbolically imbedded in Z if and only if the natural inclusion map $j: Y \to Z$ is normal.*

Proof. By definition, Y is tautly imbedded in Z if and only if j is a normal map. The proposition follows from (5.1.11). $\qquad\square$

(5.1.21) Proposition. *Let X be a complex space and Z a hyperbolic complex space. If a holomorphic map $f: X \to Z$ has a relatively compact image, then f is normal.*

Proof. For every $a \in D$, the set $\{f(\varphi(a)); \ \varphi \in \mathrm{Hol}(D, X)\}$ is contained in $f(X)$ and hence is relatively compact. By (1.3.2), $\mathcal{F}_f$ is relatively compact in $\mathrm{Hol}(D, Z)$. $\qquad\square$

A normal holomorphic map $f: X \to Z$ behaves like a mapping into a hyperbolic complex space even when Z is not hyperbolic. The following result of Joseph-Kwack [1] explains this behavior, cf. (5.1.11).

(5.1.22) Theorem. *Let $f: X \to Z$ be a holomorphic mapping between complex spaces with relatively compact image $f(X)$. Then the following are equivalent:*
   (a)   *f is normal;*
   (b)   *Given a length function F on Z, there is a constant $c > 0$ such that*

$$(f \circ \varphi)^*(c^2 F^2) \le ds_D^2 \qquad \text{for all} \quad \varphi \in \mathrm{Hol}(D, X).$$

Proof. Set

$$|df \circ d\varphi| = \sup_{|v|=1} F(df(d\varphi(v))),$$

where the supremum is taken over all tangent vectors $v \in TD$ of unit length with respect to ds_D^2.

If (b) does not hold, there is a sequence $\varphi_n \in \mathrm{Hol}(D, X)$ such that $|df \circ d\varphi_n| > n$. Then $\{f \circ \varphi_n\}$ cannot be relatively compact in $\mathrm{Hol}(D, Z)$, and f is not normal.

Assume (b). Apply (1.3.2) to the family $\mathcal{F}_f = \{f \circ \varphi; \ \varphi \in \mathrm{Hol}(D, X)\}$. Since $f \circ \varphi$ is a distance-decreasing map of (D, ds_D^2) into $(Z, c^2 F^2)$ and since, for each $a \in D$, the set $\{f(\varphi(a)); \ \varphi \in \mathrm{Hol}(D, X)\}$ is contained in the relatively compact set $f(X)$, $\mathcal{F}_f$ is relatively compact in $\mathrm{Hol}(D, Z)$. $\qquad\square$

(5.1.23) Corollary. *If Y is hyperbolically imbedded in Z, then every map $f \in \mathcal{F}_{Y,Z}$ is normal.*

Proof. By (5.1.11), we have $f^*(c^2 F^2) \le ds_D^2$. Since $\varphi^*(ds_D^2) \le ds_D^2$ for every $\varphi \in \mathrm{Hol}(D, D)$, we obtain

$$(f \circ \varphi)^*(c^2 F^2) \le ds_D^2 \qquad \varphi \in \mathrm{Hol}(D, D).$$

By (5.1.22), f is normal. $\qquad\square$

We rewrite the inequality in (b) of (5.1.22):

$$\varphi^*(f^*(c^2 F^2)) \leq ds_D^2 \qquad \text{for all} \quad \varphi \in \text{Hol}(D, X).$$

Then by (3.5.19), $f^*(cF) \leq F_X$. In other words,

(5.1.24) Corollary. *Let $f: X \to Z$ be a normal holomorphic mapping between complex spaces. Let δ_Z be the distance function on Z defined by the length function cF of (5.1.22). Then f is distance-decreasing with respect to d_X and δ_Z.*

The following theorem of Zaidenberg [11] generalizes that of Eastwood [1] who assumed that Z is (complete) hyperbolic.

(5.1.25) Theorem. *Let $f: X \to Z$ be a normal mapping between complex spaces. If there is an open cover $\{V_\alpha\}$ of Z such that each $f^{-1}(V_\alpha)$ is either empty or (complete) hyperbolic, then X is (complete) hyperbolic.*

Proof. We make use of the distance δ_Z defined in (5.1.24). Let $p \in X$, and $f(p) \in V_\alpha$. Choose $\varepsilon > 0$ such that the ε-neighborhood $V(f(p), \varepsilon)$ of $f(p)$ with respect to δ_Z is contained in V_α. Let $U(p, \varepsilon)$ be the ε-neighborhood of p with respect to d_X. Then by (5.1.24),

$$U(p, \varepsilon) \subset f^{-1}(V(f(p), \varepsilon)) \subset f^{-1}(V_\alpha).$$

Hence, by our assumption $U(p, \varepsilon)$ is hyperbolic. Now, by (3.2.6) X is hyperbolic.

Let $\{p_n\}$ be a Cauchy sequence in X with respect to d_X. Then $\{f(p_n)\}$ is a Cauchy sequence in Z with respect to δ_Z. Since $f(X)$ is relatively compact, it converges to a point, say $z_0 \in \overline{f(X)}$. Let $z_0 \in V_\alpha$.

Choose $\varepsilon > 0$ and ρ such that the $(3\rho + 2\varepsilon)$-neighborhood $V(z_0, 3\rho + 2\varepsilon)$ of z_0 with respect to δ_Z is contained in V_α. By dropping a finite number of p_n, we may assume that all $f(p_n)$ are in $V(z_0, \varepsilon)$ and that $d_X(p_m, p_n) < \rho$ for all m, n. Let $U(p_1, 3\rho + \varepsilon)$ be the $(3\rho + \varepsilon)$-neighborhood of p_1 with respect to d_X. Since $f(p_1) \in V(z_0, \varepsilon)$, we have

$$f(U(p_1, 3\rho + \varepsilon)) \subset V(f(p_1), 3\rho + \varepsilon) \subset V(z_0, 3\rho + 2\varepsilon) \subset V_\alpha.$$

Since $p_m, p_n \in U(p_1, \rho)$, by (3.1.19) we find a constant $C > 1$ such that

$$d_{U(p_1, 3\rho+\varepsilon)}(p_m, p_n) \leq C \cdot d_X(p_m, p_n) \qquad \text{for all} \quad m, n,$$

showing that $\{p_n\}$ is a Cauchy sequence with respect to $d_{U(p_1, 3\rho+\varepsilon)}$. Since $U(p_1, 3\rho + \varepsilon) \subset f^{-1}(V_\alpha)$, the sequence $\{p_n\}$ is Cauchy with respect to $d_{f^{-1}(V_\alpha)}$. Since, by assumption, $f^{-1}(V_\alpha)$ is complete hyperbolic, $\{p_n\}$ converges. $\qquad\square$

The following result generalizes (3.11.2).

(5.1.26) Corollary. *Let $f: X \to Z$ be a proper, normal holomorphic map between complex spaces. If, for every $z \in Z$, (each connected component of) $f^{-1}(z)$ is hyperbolic, then X is complete hyperbolic.*

Proof. By (3.11.1) every point $z \in Z$ has a neighborhood V_z such that $f^{-1}(V_z)$ is hyperbolic. Choose V_z to be complete hyperbolic. Then by (3.11.2) $f^{-1}(V_z)$ is also complete hyperbolic. By (5.1.25) X is complete hyperbolic. $\qquad\square$

(5.1.27) **Remark.** Let X be a compact complex space, and Y be a compact complex space with a C-hyperbolic regular covering space $\tilde{Y}$ (so that $Y = \tilde{Y}/\Gamma$, where Γ is a covering group and Y is hyperbolic). In this case, several rigidity results on $\mathrm{Hol}(X, Y)$ are known, see Borel-Narasimhan [1], Imayoshi [3, 4], Kalka-Shiffman-Wong [1], and Noguchi-Sunada [1].

2 Taut Domains

In the preceding section we considered taut complex spaces. In this sections we consider tautness for domains in $\mathbf{C}^n$. In Appendix A of Chapter 4, we gave several equivalent definitions of pseudoconvexity.

We begin with the following result by Wu [1].

(5.2.1) **Theorem.** *Every taut domain X in $\mathbf{C}^n$ is pseudoconvex.*

Proof. We prove that X is Oka-pseudoconvex (see Appendix A of Chapter 4). Let $f: \bar{D} \times [0, 1] \to \mathbf{C}^n$ be a continuous map such that

(a)   for each $t \in [0, 1]$, the map f_t defined by $f_t(z) = f(z, t)$ holomorphically immerses D into $\mathbf{C}^n$ and
(b) $f(\bar{D} \times [0, 1) \cup \partial D \times \{1\}) \subset X$.

Since $f(\partial D \times [0, 1])$ is a compact subset of X, it has a compact neighborhood K in X. Then we can find point $z_0 \in D$ sufficiently close to the boundary such that $f(z_0, t) \in K$ for $t \in [0, 1]$. Let $\mathcal{F}(z_0, K) = \{f \in \mathrm{Hol}(D, X); f(z_0) \in K\}$. Since X is taut, $\mathcal{F}(z_0, K)$ is normal. Clearly, no sequences in $\mathcal{F}(z_0, K)$ can be compactly divergent. Since each f_t is in $\mathcal{F}(z_0, K)$ for $t \in [0, 1)$, the limit mapping f_1 must be also in $\mathcal{F}(z_0, K)$. In particular, f_1 maps D into X. $\qquad\square$

Before we prove a partial converse to (5.2.1), we recall (see Section 4 of Chapter 4) that a domain $X \subset \mathbf{C}^n$ is **hyperconvex** if there exists a continuous bounded plurisubharmonic function $u < 0$ such that the set $X_c = \{z \in X; u(z) \leq c\}$ is compact for every $c < 0$. We express the last property of u by calling u an **exhaustion function** since X_c exhausts X as $c \to 0$. Thus, $u(z) \to 0$ as z approaches the boundary ∂X.

(5.2.2) **Proposition.** *Every hyperconvex bounded domain X in $\mathbf{C}^n$ is taut.*

Proof. Let u be a function on X with the property described above. Let Y be a bounded domain, e.g., a ball, such that $\bar{X} \subset Y \subset \mathbf{C}^n$. Given a sequence $\{f_i\} \subset \mathrm{Hol}(D, X)$, there is a subsequence, still denoted $\{f_i\}$, which converges to $f \in \mathrm{Hol}(D, Y)$. Since $u \circ f_i$ is subharmonic in D, its limit $u \circ f$ is also subharmonic in D. Since $f(D) \subset \bar{X}$, $u \circ f \leq 0$ on D. If there is a point $\zeta_0 \in D$ such that $f(\zeta_0) \in \partial X$ so that $u \circ f(\zeta_0) = 0$, then by the maximum principle $u \circ f(\zeta) = 0$ for all $\zeta \in D$, and hence $f(\zeta) \in \partial X$ for all $\zeta \in D$. This means that $\{f_i\}$ is compactly divergent. $\qquad\square$

From (5.2.1) and (5.2.2) we see that every hyperconvex domain X is pseudoconvex:

$$\text{hyperconvex} \Rightarrow \text{taut} \Rightarrow \text{pseudoconvex}$$

However, the implication "hyperconvex $\Rightarrow$ pseudoconvex" can be seen more directly. Let $u < 0$ be the continuous bounded plurisubharmonic exhaustion function on X which appears in the definition of hyperconvexity. Set $v = -1/u$. Then v is a continuous plurisubharmonic function which tends to infinity at the boundary. Hence, X is pseudoconvex. We do not have a complete converse. The following example is due to Diederich and Fornaess [1]:

(5.2.3) **Example**. The Hartogs triangle

$$X = \{(z, w) \in \mathbf{C}^2; \ |z| < |w| < 1\}$$

is complete hyperbolic and pseudoconvex but not hyperconvex. In fact, under the map $(z, w) \mapsto (z/w, w)$, X is biholomorphic to $D \times D^*$, and hence it is complete hyperbolic and pseudoconvex. Assume that there is a continuous bounded plurisubharmonic exhaustion function u on X with $\sup_X u = 0$. Set $u_0(w) = u(0, w)$. Then u_0 is subharmonic on $D^* = \{w; \ 0 < |w| < 1\}$, $u \leq 0$, and $\lim_{w \to 0} u_0(w) = 0$. Therefore, u_0 can be extended to a subharmonic function on D by $u_0(0) = 0$. Because of the maximum principle $u_0 \equiv 0$. This contradicts the exhaustion property of u. We note that the boundary ∂X is not smooth in this example.

If the boundary ∂X is smooth, then the converse holds.

(5.2.4) **Theorem**. *Let $X \subset \mathbf{C}^n$ be a pseudoconvex bounded domain with Lipschitz boundary. Then X admits a bounded C^∞ strongly plurisubharmonic exhaustion function $u < 0$. In particular, X is hyperconvex.*

This theorem was first proved by Diederich and Fornaess [1] when ∂X is of class C^2 and by Kerzman and Rosay [1] when ∂X is of class C^1. The result with the weakest assumption on the boundary is due to Demailly [1], who gives also very precise estimates on u.

Referring the reader to the original papers for the proof of this theorem, we shall prove the following theorem of Kerzman [1] (see also Kerzman-Rosay [1]) using only part of the proof of (5.2.4). It is a partial converse to (5.2.1).

(5.2.5) **Theorem**. *Let $X \subset \mathbf{C}^n$ be a pseudoconvex bounded domain with boundary ∂X of class C^1. Then X is taut.*

As a first step we prove that X is locally hyperconvex following Kerzman-Rosay [1].

(5.2.6) **Lemma**. *Let X be a bounded pseudoconvex domain in $\mathbf{C}^n$ with boundary ∂X of class C^1, and $x \in \partial X$. Set $B(x, r) = \{z \in \mathbf{C}^n; \ |z - x| < r\}$. Then there is a* *positive number r such that the domain $X \cap B(x, r)$ admits a bounded continuous plurisubharmonic exhaustion function $u < 0$.*

Proof. As in Appendix A of Chapter 4, we denote the Euclidean distance from z to the boundary ∂X by $\delta_X(z)$.

Let $\mathbf{n}_x$ be the inward unit normal to ∂X at x. Then there exist positive numbers a, $\varepsilon_0 < a$, and $C < 1$ such that

$$\delta_X(z + \varepsilon\mathbf{n}_x) \geq \delta_X(z) + C\varepsilon \qquad \text{for} \quad z \in \bar{X} \cap \bar{B}(x, a), \quad 0 < \varepsilon < \varepsilon_0.$$

(As Kerzman and Rosay point out, it is only this property of $\mathbf{n}_x$ that is used in the proof.)

Let $X - \varepsilon\mathbf{n}_x$ be the translate of X by $-\varepsilon\mathbf{n}_x$. We choose $r > 0$ small enough so that $r + \varepsilon_0 < a$. For $0 < \varepsilon < \varepsilon_0$, we set

$$U = X \cap B(x, r) \quad \text{and} \quad U_\varepsilon = (X - \varepsilon\mathbf{n}_x) \cap B(x, r + \varepsilon).$$

We claim

$$C\varepsilon \leq \delta_{U_\varepsilon}(z) \leq \varepsilon \qquad \text{for} \quad z \in \partial U.$$

The second inequality is obvious. To prove the first inequality, we observe that $\partial U = (\partial X \cap \bar{B}(x, r)) \cup (X \cap \partial B(x, r))$. If $z \in \partial X \cap \bar{B}(x, r)$, then

$$\delta_{U_\varepsilon}(z) = \delta_{X - \varepsilon\mathbf{n}_x}(z) = \delta_X(z + \varepsilon\mathbf{n}_x) \geq C\varepsilon.$$

Consider the case $z \in X \cap \partial B(x, r)$. If a point, say w, of ∂U_ε nearest to z is on $\partial B(x, r + \varepsilon)$, then $\delta_{U_\varepsilon}(z) = \varepsilon$. If it is on $\partial(X - \varepsilon\mathbf{n}_x)$ rather than on $\partial B(x, r + \varepsilon)$, then the line segment from z to w interesects the boundary ∂X, say at z'. Then

$$\delta_{U_\varepsilon}(z) = |w - z| > |w - z'| = \delta_{X - \varepsilon\mathbf{n}_x}(z') = \delta_X(z' + \varepsilon\mathbf{n}_x) \geq C\varepsilon.$$

Since $\varepsilon > C\varepsilon$, we have $\delta_{U_\varepsilon}(z) \geq C\varepsilon$ in either case. This proves our claim.

Set

$$A_\varepsilon = \sup_{z \in \partial U} \log \frac{1}{\delta_{U_\varepsilon}(z)}.$$

Then

$$\log \frac{1}{\varepsilon} \leq A_\varepsilon \leq \log \frac{1}{C\varepsilon}.$$

We set

$$u_\varepsilon(z) = \frac{1}{A_\varepsilon}(\log \frac{1}{\delta_{U_\varepsilon}(z)} - A_\varepsilon) < 0 \qquad \text{for} \quad z \in U$$

and

$$u(z) = \sup_{\varepsilon < \varepsilon_0}(u_\varepsilon(z) - \frac{1}{\log \frac{1}{\varepsilon}}) \qquad \text{for} \quad z \in U.$$

Being an intersection of two pseudoconvex domains, U_ε is pseudoconvex. Hence, $-\log \delta_{U_\varepsilon}(z)$ is plurisubharmonic. Hence, $u(z)$ is plurisubharmonic.

We want establish the following estimate:

$$(*) \qquad \frac{-B}{\log \frac{1}{\delta_U(z)}} \leq u(z) \leq \frac{-A}{\log \frac{1}{\delta_U(z)}} < 0$$

for $z \in U$ satisfying $\delta_U(z) < \varepsilon_0$, where A and B are positive constants. (We use here only the lower bound of $u(z)$.)

For this purpose we set $\varepsilon = \delta_U(z)$. Then $\delta_{U_\varepsilon}(z) \leq \delta_U(z) + \varepsilon = 2\varepsilon$ and

$$u_\varepsilon(z) \geq \frac{1}{\log \frac{1}{\varepsilon}} (\log \frac{1}{2\varepsilon} - \log \frac{1}{C\varepsilon}) = \frac{-\log \frac{2}{C}}{\log \frac{1}{\delta_U(z)}},$$

and hence

$$u(z) \geq \frac{-1 - \log \frac{2}{C}}{\log \frac{1}{\delta_U(z)}} = \frac{-B}{\log \frac{1}{\delta_U(z)}},$$

where $B = 1 + \log \frac{2}{C}$. This gives the desired lower bound for $u(z)$.

To obtain an upper bound for $u(z)$ we have to consider two cases. We fix z.

(i)   For $\delta_U(z)^2 \leq \varepsilon$ we have

$$u_\varepsilon(z) - \frac{1}{\log \frac{1}{\varepsilon}} < -\frac{1}{\log \frac{1}{\varepsilon}} \leq -\frac{1}{2 \log \frac{1}{\delta_U(z)}}.$$

(ii)   For $\varepsilon \leq \delta_U(z)^2$ we have

$$u_\varepsilon(z) \leq \frac{1}{\log \frac{1}{C\varepsilon}} (\log \frac{1}{\sqrt{\varepsilon}} - \log \frac{1}{\varepsilon}) = \frac{-\frac{1}{2}\log \frac{1}{\varepsilon}}{\log \frac{1}{C\varepsilon}} \leq -D \leq \frac{-E}{\log \frac{1}{\delta_U(z)}}.$$

Hence,

$$u_\varepsilon(z) - \frac{1}{\log \frac{1}{\varepsilon}} \leq \frac{-E}{\log \frac{1}{\delta_U(z)}}.$$

By setting $A = \min(\frac{1}{2}, E)$, we obtain the desired upper bound for $u(z)$.

From the lower bound given by $(*)$ we see that $\{z \in U; u(z) \leq -c < 0\}$ is a closed subset of $\{z \in U; -\log \delta_U(z) \leq B/c\}$, which is compact since U is pseudoconvex. This shows that $u(z)$ is an exhaustion function. $\qquad \square$

(5.2.7) **Corollary.** *Let X be a pseudoconvex bounded domain in $\mathbf{C}^n$ with boundary ∂X of class C^1. Then X is locally hyperconvex in the sense that every boundary point $x \in \partial X$ has a neighborhood $B(x, r)$ such that $X \cap B(x, r)$ is hyperconvex. It is also locally taut in the sense that $X \cap B(x, r)$ is taut.*

We note that the second assertion in (5.2.7) is a consequence of the first assertion and (5.2.2).

Now (5.2.5) will follow from the following lemma.

(5.2.8) **Lemma.** *If a bounded domain $X \subset \mathbf{C}^n$ is locally taut, it is taut.*

Proof. Let Y be a large ball containing $\bar{X}$. In the proof we shall use only the following properties of X and Y:

"X is a complex subspace of a complex space Y such that (i) $\bar{X}$ is compact and (ii) every sequence $\{f_k\}$ in $\mathrm{Hol}(D, X)$ has a subsequence convergent in $\mathrm{Hol}(D, Y)$."

Let $\{f_k\}$ be a sequence in $\mathrm{Hol}(D, X)$. Assume that it is not compactly divergent. Then there exist compact sets $K \subset D$ and $L \subset X$ such that $f_k(K) \cap L \neq \emptyset$ for infinitely many k. By taking a subsequence we may assume that this is the case for all k. We shall show that $\{f_k\}$ has a subsequence which converges in $\mathrm{Hol}(D, X)$.

Since Y is taut, taking a subsequence we may assume that $\{f_k\}$ converges to a map f in $\mathrm{Hol}(D, Y)$. Since $f_k(K) \cap L \neq \emptyset$ for all k, we have $f(K) \cap L \neq \emptyset$.

We must show that $f \in \mathrm{Hol}(D, X)$. Assume the contrary. Then the open subset $f^{-1}(X)$ of D is distinct from D. It is nonempty since $f(K) \cap L \neq \emptyset$. Let a be a boundary point of $f^{-1}(X)$ in D, and set $p = f(a)$. Then $p \notin X$. Let V be a neighborhood of p in Y such that $V \cap X$ is taut. Let W be a neighborhood of a in D such that $f(\bar{W}) \subset V$. By taking a subsequence we may assume that $f_k(\bar{W}) \subset V$ for all k. Since $V \cap X$ is taut, $\{f_k\} \subset \mathrm{Hol}(D, V \cap X)$ is either compactly divergent or has a convergent subsequence. Since $\lim f_k(a) = f(a) = p \notin X$, it cannot have a convergent subsequence. It cannot be compactly divergent either because, for any point $b \in W \cap f^{-1}(X)$, we have $\lim f_k(b) = f(b) \in V \cap X$. This is a contradiction. $\qquad\square$

The converse to (5.2.1) does not hold in general for domains with non-smooth boundary. The following example is due to Kerzman [1], (see Kerzman-Rosay [1] and also Barth [6]).

(5.2.9) Example. Let D be the unit disk in $\mathbf{C}$ and $u\colon D \to (0, 1)$ a subharmonic function that is discontinuous at 0. Then the domain

$$X = \{(z, w) \in D \times \mathbf{C};\ |w| < \exp(-u(z))\}$$

is pseudoconvex. Since u is not lower semicontinuous at 0, there exist a constant c and a sequence $\{a_n\}$ coverging to 0 in D such that $0 < u(a_n) < c < u(0)$ for all n. Defining $f_n(w) = (a_n, e^{-c}w)$, we obtain a sequence $\{f_n\}$ in $\mathrm{Hol}(D, X)$ with $\lim f_n(0) = (0, 0) \in X$ but

$$\lim f_n(\exp(c - u(0))) = (0, \exp(-u(0))) \notin X,$$

which shows that X is not taut.

Another criterion for tautness can be stated in terms of peak functions, (see Section 1 of Chapter 4 for the definition of peak function). Although the following is a direct consequence of (4.1.11), we shall give a direct proof following Abate [2].

(5.2.11) Theorem. *If $X \subset \mathbf{C}^n$ is a bounded domain such that there is a local weak peak function for X at each point of ∂X, then it is taut.*

Proof. In view of (5.2.8) we may assume that there is a weak peak function for X at each boundary point. Since X is bounded, every sequence $h_k \in \mathrm{Hol}(D, X)$ has a subsequence which converges to a map $h \in \mathrm{Hol}(D, \mathbf{C}^n)$. Clearly, $h(D) \subset \bar{X}$. It suffices to show that either $h(D) \subset X$ or $h(D) \subset \partial X$.

Assume that there is a point $z_0 \in D$ such that $h(z_0) \in \partial X$. Let f be a weak peak function for X at $h(z_0)$. Then $f \circ h$ is a holomorphic function on D which attains its maximum at z_0. Hence, $f(h(z)) \equiv f(h(z_0))$ on D, which implies that $h(D) \cap X = \emptyset$. Hence, $h(D) \subset \partial X$. $\qquad\square$

(5.2.12) Corollary. *Every bounded convex domain $X \subset \mathbf{C}^n$ is taut.*

Proof. For each $x \in \partial X$ there is a complex linear functional $\varphi \colon \mathbf{C}^n \to \mathbf{C}$ such that $\mathrm{Re}(\varphi(z)) < \mathrm{Re}(\varphi(x))$ for all $z \in X$. Then $f = e^\varphi$ is a weak peak function for X at x. $\qquad\square$

3 Spaces of Holomorphic Mappings

We recall first the basic result of Douady [1], (see also Kaup [5]). Let X and Y be complex spaces. Let $\mathcal{F}$ be a subfamily of $\mathrm{Hol}(X, Y)$. We say that a complex structure on $\mathcal{F}$ (if it exists) is **universal** if:

(i)   the evaluation map $\Phi \colon X \times \mathcal{F} \to Y$ is holomorphic, and

(ii)   if T is a complex space and $\varphi \colon X \times T \to Y$ is a holomorphic map such that $\varphi(\cdot, t) \in \mathcal{F}$, then the map $\tilde{\varphi} \colon T \to \mathcal{F}$ defined by $\tilde{\varphi}(t) = \varphi(\cdot, t)$ is holomorphic.

If X is compact, then for any complex space Y, $\mathrm{Hol}(X, Y)$ admits a universal complex structure, Douady [1]. In the case X is nonsingular, a direct proof is given in Kaup [5]. We shall describe its tangent space at $f \in \mathrm{Hol}(X, Y)$. Let $\mathcal{F}$ be the irreducible component of $\mathrm{Hol}(X, Y)$ containing f. Following Kodaira [1], we define $\sigma_f \colon T_f \mathcal{F} \to H^0(X, f^*TY)$ as follows. If f_t is a curve in $\mathcal{F}$ such that $f_0 = f$ and if $\zeta = (df_t/dt)_0$, then

$$(\sigma_f(\zeta))(x) = (df_t(x)/dt)_0 \in T_{f(x)}Y.$$

More formally, using the differential

$$\Phi_* \colon TX \times T\mathcal{F} \to TY$$

of the evaluation map Φ, we set

$$(\sigma_f(\zeta))(x) = \Phi_*(0_x, \zeta) \in T_{f(x)}Y, \qquad x \in X, \ \zeta \in T_f\mathcal{F},$$

where 0_x stands for the zero vector at x. Then σ_f is injective, and it is bijective if $H^1(X, f^*TY) = 0$, see (Namba [1]). So we can identify the tangent space $T_f\mathcal{F}$ with a subspace of $H^0(X, f^*TY)$.

In this section, we shall first review general results on $\mathrm{Hol}(X, Y)$. In the following lemma (due to Urata [2] and Horst [1]), $\mathcal{F}$ is a general complex space although what we have in mind is a subfamily of $\mathrm{Hol}(X, Y)$.

(5.3.1) Lemma. *Let X, Y, and $\mathcal{F}$ be connected complex spaces with X compact. Let $\Phi\colon X \times \mathcal{F} \to Y$ be a holomorphic map. If $\Phi(\cdot, f_0)\colon X \to Y$ is a constant map for one element $f_0 \in \mathcal{F}$, then $\Phi(\cdot, f)$ is a constant map for every $f \in \mathcal{F}$.*

Proof. Let $\mathcal{F}_0$ be the set of $f \in \mathcal{F}$ such that $\Phi(\cdot, f)$ is a constant map. Then $\mathcal{F}_0$ is nonempty and closed in $\mathcal{F}$. We want to show that $\mathcal{F}_0$ is also open in $\mathcal{F}$. Let $f_1 \in \mathcal{F}_0$ and put $y_1 = \Phi(X, f_1)$. Let V be a coordinate neighborhood of y_1 in Y. If $f \in \mathcal{F}$ is sufficiently close to f_1, then $\Phi(X, f) \subset V$ since X is compact. By the maximum principle, every holomorphic map from a compact connected complex space into V is constant. Hence, $\Phi(\cdot, f)$ is constant. $\square$

We recall the **Stein factorization** of a proper holomorphic map, (see Grauert-Remmert [3; p. 213]). Every proper holomorphic map $f\colon X \to Y$ admits a unique factorization

$$f\colon X \xrightarrow{\ p_f\ } X'_f \xrightarrow{\ f'\ } Y$$

through a complex space X'_f with the following properties:

(i) p_f is a proper surjective holomorphic map and $p_*(\mathcal{O}_X) = \mathcal{O}_{X'_f}$, in particular, all fibers of p_f are connected,

(ii) f' is a finite map.

Intuitively, X'_f is obtained by collapsing each connected component of every fiber $f^{-1}(y)$ to a single point. The following corollary due to Horst [1] is a simultaneous Stein factorization for $\Phi(\cdot, f)$.

(5.3.2) Corollary. *Let X and Y be connected complex spaces, and let $\mathcal{F} \subset \mathrm{Hol}(X, Y)$ be a subfamily with a connected universal complex structure such that each $f \in \mathcal{F}$ is a proper map from X into Y. Then $\mathcal{F}$ admits a simultaneous factorization*

$$f\colon X \xrightarrow{\ p\ } X' \xrightarrow{\ f'\ } Y, \qquad f \in \mathcal{F}$$

through a common complex space X' so that p is a proper surjective holomorphic map with connected fibers and f' is a finite map.

Assume that X is compact. Let $f_0 \in \mathrm{Hol}(X, Y)$, and $f'_0 \in \mathrm{Hol}(X', Y)$ the corresponding map so that $f_0 = f'_0 \circ p$. Let $\mathcal{F}$ (resp. $\mathcal{F}'$) be the connected component of $\mathrm{Hol}(X, Y)$ (resp. $\mathrm{Hol}(X', Y)$) containing f_0 (resp. f'_0). Then the map $f \mapsto f'$ sends $\mathcal{F}$ biholomorphically onto $\mathcal{F}'$.

Proof. Given the Stein factorizations of two maps $f, g \in \mathcal{F}$:

$$f\colon X \xrightarrow{\ p_f\ } X'_f \xrightarrow{\ f'\ } Y, \qquad g\colon X \xrightarrow{\ p_g\ } X'_g \xrightarrow{\ g'\ } Y,$$

there is a unique biholomorphic map $\alpha\colon X'_f \to X'_g$ such that $p_g = \alpha \circ p_f$.

In fact, let $x' \in X'_f$. Since $p_f^{-1}(x')$ is compact and connected, we can apply (5.3.1) to $\Phi\colon p_f^{-1}(x') \times \mathcal{F} \to Y$ and conclude that $g(p_f^{-1}(x'))$ is a singleton, say $y \in Y$. Since $p_g(p_f^{-1}(x'))$ is connected, it is one of the points in the finite set $g'^{-1}(y)$, say x''. Define $\alpha(x') = x''$.

Fix one member $f \in \mathcal{F}$ as a reference map, and set $X' = X_f$ and $p = p_f$. Since

$$g = (g' \circ \alpha) \circ (\alpha^{-1} \circ p_g) = (g' \circ \alpha) \circ p,$$

write g' for $g' \circ \alpha$.

The second statement is clear. $\qquad\square$

(5.3.3) Corollary. *Let X be a compact connected complex space, and $\pi: \tilde{X} \to X$ be a resolution of the singularities of X. Let Y be a connected complex space. Let $f_0 \in \mathrm{Hol}(X, Y)$, and set $\tilde{f}_0 = f_0 \circ \pi \in \mathrm{Hol}(\tilde{X}, Y)$. Let $\mathcal{F}$ (resp. $\tilde{\mathcal{F}}$) be the connected component of $\mathrm{Hol}(X, Y)$ (resp. $\mathrm{Hol}(\tilde{X}, Y)$) containing f_0 (resp. $\tilde{f}_0$). Then the map $f \mapsto f \circ \pi$ sends $\mathcal{F}$ biholomorphically onto $\tilde{\mathcal{F}}$.*

Proof. All we have to show is that the map $\mathcal{F} \to \tilde{\mathcal{F}}$ is surjective. Let $x \in X$. Since $\tilde{f}_0(\pi^{-1}(x)) = f_0(x)$, we apply (5.3.1) to the evaluation map $\Phi: \pi^{-1}(x) \times \tilde{\mathcal{F}} \to Y$. Let $\tilde{f} \in \tilde{\mathcal{F}}$. Then by (5.3.1), for each fixed x, $\tilde{f}(\pi^{-1}(x))$ is a singleton. Define $f(x) = \tilde{f}(\pi^{-1}(x))$. Then $f \in \mathcal{F}$, and $\tilde{f} = f \circ \pi$. $\qquad\square$

As another consequence of (5.3.1) we obtain the following result of Urata [2].

(5.3.4) Corollary. *Let X and Y be complex spaces. Assume that $\mathrm{Hol}(X, Y)$ has a compact subfamily $\mathcal{F}$ with a universal complex structure. Then, for each fixed $x_0 \in X$, the map $\mathcal{F} \to Y$ given by $f \mapsto f(x_0)$ is a finite map. In other words, for any pair of points $x_0 \in X$ and $y_0 \in Y$ the family $\mathcal{F}_0 = \{f \in \mathcal{F};\ f(x_0) = y_0\}$ is finite. In particular, $\dim \mathcal{F} \leq \dim Y$.*

Proof. $\mathcal{F}_0$ is a compact complex subspace of $\mathcal{F}$. Let $\mathcal{H}$ be a connected component of $\mathcal{F}_0$. It suffices to prove that $\mathcal{H}$ is a singleton set. We apply (5.3.1) to $\Phi: X \times \mathcal{H} \to Y$. Here, $\mathcal{H}$ (which is compact) plays the role of X in (5.3.1), and X that of $\mathcal{F}$. (Therefore we need not assume here compactness of X). Since the map $\Phi(x_0, \cdot): \mathcal{H} \to Y$ is a constant map to y_0, $\Phi(x, \cdot)$ is also a constant map, i.e., for any $f, g \in \mathcal{H}$, we have $f(x) = g(x)$ for all $x \in X$. This means that $\mathcal{H}$ contains only one element. $\qquad\square$

We define the **rank** of a holomorphic mapping $f: X \to Y$ between complex spaces by

$$\mathrm{rank}\, f = \max_{x \in X}\{\dim_x X - \dim_x f^{-1}(f(x))\},$$

and, for each integer $k = 0, 1, \ldots, \dim X$, we set

$$\mathrm{Hol}(X, Y, k) = \{f \in \mathrm{Hol}(X, Y);\ \mathrm{rank}\, f = k\}.$$

We denote the family of surjective $f \in \mathrm{Hol}(X, Y)$ by $\mathrm{Sur}(X, Y)$, and the family of finite maps $f \in \mathrm{Hol}(X, Y)$ by $\mathrm{Fin}(X, Y)$. Then If X is compact and Y is irreducible, then $\mathrm{Sur}(X, Y) = \mathrm{Hol}(X, Y, m)$, where $m = \dim Y$.

Following Horst [2] we prove the following. The statement on $\mathrm{Hol}(X, Y, k)$ was obtained by Noguchi [10] by a different method.

(5.3.5) Corollary. *Let X and Y be connected complex spaces with X compact. Then $\mathrm{Fin}(X, Y)$, $\mathrm{Sur}(X, Y)$ and $\mathrm{Hol}(X, Y, k)$, $0 \leq k \leq m = \dim X$, are all open* *and closed in* $\mathrm{Hol}(X, Y)$. *If X is moreover normal, then* $\mathrm{Aut}(X)$ *is open and closed in* $\mathrm{Hol}(X, X)$.

Proof. Let $\mathcal{F}$ be a connected component of $\mathrm{Hol}(X, Y)$. We may assume that X and Y are irreducible.

(a)   Suppose $\mathcal{F} \cap \mathrm{Fin}(X, Y) \neq \emptyset$. We want to show $\mathcal{F} \subset \mathrm{Fin}(X, Y)$. Suppose there is a map $f_0 \in \mathcal{F}$ which is not finite. Then there is a point $y_0 \in Y$ such that $f_0^{-1}(y_0)$ has a connected component X_0 with $\dim X_0 > 0$. Apply (5.3.1) to $\Phi: X_0 \times \mathcal{F} \to Y$. Then for every $f \in \mathcal{F}$, $f(X_0)$ is a singleton. This is a contradiction.

(b)   Since $\mathrm{Hol}(X, Y)$ is a disjoint union of $\mathrm{Hol}(X, Y, k)$, $k = 0, 1, \ldots, m$, it suffices to show that $\mathcal{F}$ is contained in $\mathrm{Hol}(X, Y, k)$ for some k. Let k be the largest integer such that $\mathcal{F} \cap \mathrm{Hol}(X, Y, k) \neq \emptyset$. We want to show $\mathcal{F} \subset \mathrm{Hol}(X, Y, k)$. Every $f \in \mathcal{F}$ is of rank at most k. Suppose there is a map $f_0 \in \mathcal{F}$ which is of rank less than k. Then $m - \dim f_0^{-1}(f_0(x)) < k$ for all $x \in X$. Let X_x be the connected component of $f_0^{-1}(f_0(x))$ containing x. The inequality above amounts to saying $\dim X_x > m - k$. Apply (5.3.1) to $\Phi: X_x \times \mathcal{F} \to Y$. Since $f_0(X_x)$ is a singleton $\{f_0(x)\}$, (5.3.1) says that $f(X_x)$ is a singleton for every $f \in \mathcal{F}$, i.e., $X_x \subset f^{-1}(f(x))$. Then $\mathrm{rank}\, f \leq \max_x\{m - \dim X_x\} < k$ for all $f \in \mathcal{F}$. This is a contradiction.

(c)   Let $\mathcal{F}$ be a connected component of $\mathrm{Hol}(X, X)$ containing an element f_0 of $\mathrm{Aut}(X)$. We want to show $\mathcal{F} \subset \mathrm{Aut}(X)$. Since $f_0 \in \mathrm{Fin}(X, X) \cap \mathrm{Sur}(X, X)$, (a) and (b) imply $\mathcal{F} \subset \mathrm{Fin}(X, X) \cap \mathrm{Sur}(X, X)$. Since f_0 is a mapping of degree 1, every $f \in \mathcal{F}$ is of degree 1.

By the open mapping theorem (cf. Grauert-Remmert [3; p. 107]), every $f \in \mathcal{F}$ is an open map. We claim that f is a homeomorphism. To prove this, we only need to show that f is injective. If there is a point $p \in X$ such that $f^{-1}(p) = \{q_1, q_2, \ldots, q_s\}$, $s > 1$, take mutually disjoint open neighborhoods $U_1, U_2, \ldots, U_s$ of $q_1, q_2, \ldots, q_s$. Then $f(U_1) \cap f(U_2) \cap \ldots \cap f(U_s)$ is an open neighborhood of p. Clearly, the cardinality of the inverse image of every point in this neighborhood is at least s, thus contradicting the fact that f has degree 1.

Since X is normal, f^{-1} is also holomorphic. $\square$

Given a holomorphic map p from a complex space X into another complex space X' and a holomorphic vector bundle E over X', let p^*E denote the induced vector bundle over X. Then there is a natural linear map

$$p^*: H^0(X', E) \to H^0(X, p^*E).$$

If p is surjective, the map p^* above is injective. If p is a proper surjective map with connected fibers, then the map p^* is an isomorphism. In fact, for each $x' \in X'$, $p^*E|_{p^{-1}(x')}$ is naturally isomorphic to a product bundle $p^{-1}(x') \times E_{x'}$. Since $p^{-1}(x')$ is compact, every holomorphic section of p^*E is constant on $p^{-1}(x')$ and determines a point in the fiber $E_{x'}$, thus defining a section s' of E over X'. Clearly, $f^*s' = s$.

We apply this to the following two situations. They may be considered as infinitesimal versions of (5.3.2) and (5.3.3).

(5.3.6) Proposition. *Let X and Y be connected complex spaces, and $f: X \to Y$ a proper holomorphic map.*

(1) *If $f: X \xrightarrow{p} X' \xrightarrow{f'} Y$ is the Stein factorization of f, then*

$$p^*: H^0(X', f'^*TY) \to H^0(X, f^*TY)$$

is an isomorphism;

(2) *If $p: \tilde{X} \to X$ is a resolution of the singularities of X, then*

$$p^*: H^0(X, f^*TY) \to H^0(\tilde{X}, p^*f^*TY)$$

is an isomorphism.

Define a pseudo-distance function δ on $\mathrm{Hol}(X, Y)$ by setting

$$(5.3.7) \qquad \delta(f, g) = \sup_{x \in X} d_Y(f(x), g(x)) \qquad \text{for} \quad f, g \in \mathrm{Hol}(X, Y).$$

In general, $\delta(f, g)$ can be infinite.

If X is compact, the supremum exists and the pseudo-distance δ is well-defined. Even if X is not compact, $\delta(f, g)$ is finite for f, g belonging to a subfamily $\mathcal{F} \subset \mathrm{Hol}(X, Y)$ with a complex structure. In fact, for each fixed $x \in X$ the distance-decreasing property of a holomorphic map $\mathcal{F} \to Y$ sending f to $f(x)$ implies

$$d_Y(f(x), g(x)) \le d_{\mathcal{F}}(f, g) \qquad \text{for} \quad f, g \in \mathcal{F}.$$

Hence,

(5.3.8) Proposition. *Let X and Y be complex spaces, and $\mathcal{F} \subset \mathrm{Hol}(X, Y)$ a connected subfamily with a universal complex structure. Then*

$$\delta(f, g) \le d_{\mathcal{F}}(f, g) \qquad \text{for} \quad f, g \in \mathcal{F}.$$

If Y is (complete) hyperbolic, then $\mathcal{F}$ is a (complete) hyperbolic complex space.

In particular, if X is compact and Y is (complete) hyperbolic, then every connected component $\mathcal{F}$ of $\mathrm{Hol}(X, Y)$ is a (complete) hyperbolic complex space.

(5.3.9) Theorem. *Let X and Y be complex spaces. If Y is complete hyperbolic (more generally, taut), then for all compact subsets $K \subset X$ and $L \subset Y$ the family*

$$\mathcal{F}_{K,L} = \{f \in \mathrm{Hol}(X, Y);\ f(K) \cap L \ne \emptyset\}$$

is compact.

In particular, if Y is compact hyperbolic, then $\mathrm{Hol}(X, Y)$ itself is compact.

Proof. Since $\mathrm{Hol}(X, Y)$ is a normal family by (5.1.5), $\mathcal{F}_{K,L}$ which is closed in $\mathrm{Hol}(X, Y)$ is also a normal family. The theorem follows from (1.3.5). $\square$

We have now the following theorem of Urata [2].

(5.3.10) Theorem. *Let X be a compact complex space and Y a complete hyperbolic complex space, (more generally , a taut complex space). Then for each fixed point*

$x_0 \in X$, *the map* $\mathrm{Hol}(X, Y) \to Y$ *given by* $f \mapsto f(x_0)$ *is a finite map. In other words, for any pair of points* $x_0 \in X$ *and* $y_0 \in Y$ *the family*

$$\mathcal{F}_0 = \{ f \in \mathrm{Hol}(X, Y); \ f(x_0) = y_0 \}$$

is finite. In particular, $\dim \mathrm{Hol}(X, Y) \leq \dim Y$.

Proof. Since $\mathcal{F}_0$ is a complex subspace of $\mathrm{Hol}(X, Y)$ and is compact by (5.3.9), the theorem follows directly from (5.3.4). $\square$

(5.3.11) Proposition. *Let* X *and* X' *be compact complex space with a surjective holomorphic map* $\alpha: X \to X'$. *Then for any complex space* Y, *the natural map* $\alpha^*: \mathrm{Hol}(X', Y) \to \mathrm{Hol}(X, Y)$ *sends* $\mathrm{Hol}(X', Y)$ *biholomorphically onto the complex subspace* $\alpha^*(\mathrm{Hol}(X', Y))$ *of* $\mathrm{Hol}(X, Y)$.

Proof. By condition (i) for the universality of a complex structre, the composed map

$$X \times \mathrm{Hol}(X', Y) \to X' \times \mathrm{Hol}(X', Y) \to Y$$

is holomorphic. Then by condition (ii), the map α^* is holomorphic. Since α is surjective, α^* is injective. See Kaup [5] for details. $\square$

The following theorem of H. Cartan [9] yields the situation where (5.3.11) is applicable.

(5.3.12) Theorem. *Let* X *be a complex space and* R *the equivalence relation on* X *defined by a family of holomorphic maps* $f_i: X \to Y_i$, *i.e.,* $x \sim_R x'$ *if and only if* $f_i(x) = f_i(x')$ *for all* i. *If* R *is proper (i.e., the inverse image of every compact subset of* X/R *by the natural projection* $X \to X/R$ *is compact), then* X/R *is a complex space and the projection* $X \to X/R$ *is holomorphic.*

Using (5.3.12) we prove the following

(5.3.13) Proposition. *Let* X *be a compact complex space and* Y *an arbitrary complex space. Let* $\mathcal{F}$ *be a compact complex subspace of a complex space* $\mathrm{Hol}(X, Y)$. *Consider the equivalence relation* R *on* X *defined by* $\mathcal{F}$. *Let* $X' = X/R$ *and* $\alpha: X \to X'$ *the natural projection. Then*

(1)   *There is a unique compact complex subspace* $\mathcal{F}'$ *of* $\mathrm{Hol}(X', Y)$ *such that the natural map* $\alpha^*: \mathrm{Hol}(X', Y) \to \mathrm{Hol}(X, Y)$ *sends* $\mathcal{F}'$ *biholomorphically onto* $\mathcal{F}$;

(2)   *For every* $f' \in \mathcal{F}'$ *and for every* $y \in Y$, $f'^{-1}(y)$ *is a finite subset of* X';

(3)   *If* Y *is hyperbolic, so is* X'.

Proof. (1)   We set $\mathcal{F}' = (\alpha^*)^{-1}(\mathcal{F})$; from the construction of $X' = X/R$ it is clear that for each $f \in \mathcal{F}$ there is a unique $f' \in \mathrm{Hol}(X', Y)$ such that $f = f' \circ \alpha$.

(2)   Let

$$\Phi: \mathcal{F}' \times X' \to Y$$

be the restriction of the canonical map $\mathrm{Hol}(X', Y) \times X' \to Y$. Define a holomorphic map $\varphi: X' \to \mathrm{Hol}(\mathcal{F}', Y)$ by setting

$$\varphi(x') = \Phi(\cdot, x') \in \mathrm{Hol}(\mathcal{F}', Y) \qquad \text{for} \quad x' \in X'.$$

From the construction of X' we see that the family $\mathcal{F}'$ separates the points of X' and hence that the map $\varphi\colon X' \to \mathrm{Hol}(\mathcal{F}', Y)$ is injective. Considering X' as a compact complex subspace of $\mathrm{Hol}(\mathcal{F}', Y)$ we use (5.3.1) to see that for each fixed pair $f' \in \mathcal{F}'$ and $y \in Y$ the set $\{x' \in X'; \ f'(x') = y\}$ is finite.

(3)   Since $\varphi\colon X' \to \mathrm{Hol}(\mathcal{F}', Y)$ is injective and since $\mathrm{Hol}(\mathcal{F}', Y)$ is hyperbolic by (5.3.6), X' is hyperbolic. We may also prove hyperbolicity of X' by applying (3.2.11) to a finite map $f'\colon X' \to Y$. $\square$

4 Automorphisms of Hyperbolic Complex Spaces

We apply the following theorem of Dantzig and Van der Waerden [1] to hyperbolic complex spaces. For its proof, see also Kobayashi-Nomizu [1; pp. 46–50].

(5.4.1) **Theorem**. *The group $I(X)$ of isometries of a connected, locally compact metric space X is locally compact with respect to the compact-open topology, and for any point $x \in X$ and any compact subset $K \subset X$, the subset $\{f \in I(X); \ f(x) \in K\}$ is compact. In particular, at any point $x \in X$ the isotropy subgroup $I_x(X)$ is compact. If X is moreover compact, then $I(X)$ is compact.*

Given a complex space X, we denote the group of automorphisms (i.e., bi-holomorphic mappings of X onto itself by $\mathrm{Aut}(X)$ and its identity component by $\mathrm{Aut}^0(X)$.

The following theorem (see Kaup [2], Kobayashi [7]) generalizes the classical theorem of H. Cartan [4], [6] for bounded domains.

(5.4.2) **Theorem**. *Let X be a hyperbolic complex space of dimension n. Then*

(1)   With respect to the compact-open topology, $\mathrm{Aut}(X)$ is a real Lie group of dimension $\leq n(n+2)$;

(2)   For any point $x \in X$ and any compact subset $K \subset X$, the subset $\{f \in \mathrm{Aut}(X); \ f(x) \in K\}$ is compact. In particular, the isotropy subgroup of $\mathrm{Aut}(X)$ at each point $x \in X$ is compact;

(3)   the Lie algebra $\mathrm{aut}(X)$ of $\mathrm{Aut}(X)$ consists of complete (i.e., globally integrable) holomorphic vector fields. If $v \in \mathrm{aut}(X)$, then $\sqrt{-1}\,v$ is not in $\mathrm{aut}(X)$, i.e., it is not globally integrable. In other words, no complex Lie group of positive dimension acts on X effectively as a holomorphic transformation group.

Proof. (1)   Let X' be the open subset of X consisting of all regular points of X. Since every automorphism of X restricts to an automorphism of X', the group $\mathrm{Aut}(X)$ may be regarded as a closed subgroup of $\mathrm{Aut}(X')$. Applying (5.4.1) to the metric space $(X', d_{X'})$, we see that $\mathrm{Aut}(X')$ is locally compact. By a theorem of Bochner and Montgomery [1], a locally compact group of differentiable transformations of a manifold is a Lie transformation group. Hence, $\mathrm{Aut}(X')$ is a Lie transformation group acting on X'. Being a closed subgroup of $\mathrm{Aut}(X')$, $\mathrm{Aut}(X)$ is also a Lie group.

Let $\mathrm{Aut}_x(X)$ be the isotropy subgroup at $x \in X'$. Since it is compact by (5.4.1), there is a Hermitian inner product on T_xX invariant by $\mathrm{Aut}_x(X)$. We see that the linear isotropy representation of $\mathrm{Aut}_x(X)$ on the tangent space $T_x X$ is faithful and unitary with respect to the inner product. Hence, $\dim \mathrm{Aut}_x(X) \le \dim U(n) = n^2$ and $\dim \mathrm{Aut}(X) \le \dim_{\mathbf{R}} X + \dim \mathrm{Aut}_x(X) \le 2n + n^2$.

(2)   This is immediate from (5.4.1).

(3)   Suppose that both v and $\sqrt{-1}v$ are complete holomorphic vector fields. Then they would generate a complex 1-parameter group $\mathbf{C}$ acting on X. By (3.1.23) X cannot be hyperbolic. $\qquad\square$

It is not difficult to verify that when $\mathrm{Aut}(X)$ attains the maximum real dimension $n(n+2)$ in (5.4.2), X is biholomorphic to the unit ball B_n in $\mathbf{C}^n$. For further results on $\dim \mathrm{Aut}(X)$, see Kaup [2].

In contrast to (3) of (5.4.2) we have the following theorem of Bochner-Montgomery [1], [2] (in the nonsingular case) and Gunning [1], Kerner [1], Kaup [1] (in the singular case):

(5.4.3) **Theorem.** *If X is a compact complex space, then* $\mathrm{Aut}(X)$ *is a complex Lie group and its Lie algebra* $\mathrm{aut}(X)$ *consists of all holomorphic vector fields.*

We note that on a compact complex space, every vector field is complete so that every holomorphic vector field is in the Lie algebra of $\mathrm{Aut}(X)$.

If X is a complex manifold which is neither compact nor hyperbolic, $\mathrm{Aut}(X)$ can be too large to be a Lie group as in the case of $X = \mathbf{C}^n$, $n \ge 2$.

(5.4.4) **Theorem.** *If X is a compact hyperbolic complex space, then* $\mathrm{Aut}(X)$ *is finite.*

Proof. By (5.4.3) and (3) of (5.4.2) $\mathrm{Aut}(X)$ is 0-dimensional. By (2) of (5.4.2) it is compact. $\qquad\square$

In generalizing (5.4.2), Urata [2] proved that for any compact complex space X and any compact hyperbolic complex space Y the set of surjective holomorphic maps from X to Y with connected fibers is finite. His proof has been simplified by Simha [1].

Following Urata we introduce a concept which would clarify some aspects of hyperbolcity. We say that a complex space X is **immobile** (see Condition C in Urata [4]) if every holomorphic map $f: X \times D \to X$ with the property that $f(x, 0) = x$ for all $x \in X$ satisfies the equality $f(x, t) = x$ for all $t \in D$ and $x \in X$. (By writing $f_t(x) = f(x, t)$ we may say that X is immobile if every holomorphic family of self-maps $f_t: X \to X$ parametrized by $t \in D$ such that f_0 is the identity automorphism of X is the identity automorphism for all $t \in D$.)

If X is immobile, then for every complex space Y and for every $f \in \mathrm{Hol}(X \times Y, X)$ such that $f(x, y_0) = x$ for some $y_0 \in Y$ and all $x \in X$ we have $f(x, y) = x$ for all $x \in X$ and $y \in Y$. This is obvious since any pair of points in Y can be joined by a chain of holomorphic discs.

If X is immobile, then for every Y and every map $f \in \mathrm{Hol}(X \times Y, X)$ such that, for some $y_0 \in Y$, $f(\cdot, y_0): X \to X$ is an automorphism of X we have

$$f(x, y) = f(x, y_0) \qquad \text{for} \quad x \in X, \ y \in Y.$$

This follows from the case above by composing f with the inverse of the automorphism $f(\cdot, y_0)$.

The following result of Royden [5] (see also Urata [4]) goes back essentially to H. Cartan [8]. It may be considered as a generalization of (3) of (5.4.2).

(5.4.5) Theorem. *Every hyperbolic complex space X is immobile.*

Proof. Let $f \in \mathrm{Hol}(X \times D, X)$ such that $f(x, 0) = x$ for all $x \in X$. Fix any regular point $x_0 \in X$, and define $h_n \in \mathrm{Hol}(D, X)$ by setting

$$h_1(t) = f(x_0, t), \qquad h_n(t) = f(h_{n-1}(t), t) \qquad \text{for} \quad t \in D.$$

Then $h_m(0) = x_0$ for all m. Since X is hyperbolic, the family $\{h_m\}$ is equicontinuous, and for every positive integer k there is a positive constant A_k such that

$$\left\| \frac{d^k h_m}{dt^k}(0) \right\| \leq A_k, \qquad m = 1, 2, \ldots,$$

where $\| \cdot \|$ is defined in terms of a local coordinate system around x_0.

Since $f(x, 0) = x$ for all $x \in X$, f has a power series expansion of the form

$$f(x, t) = x + a(x)t^k + O(t^{k+1})$$

around the point $(x_0, 0) \in X \times D$. Then h_m has an expansion of the form

$$h_m(t) = x_0 + ma(x_0)t^k + O(t^{k+1})$$

around $0 \in D$. Then

$$k! m \|a(x_0)\| = \left\| \frac{d^k h_m}{dt^k}(0) \right\| \leq A_k, \qquad m = 1, 2, \ldots,$$

which implies $a(x_0) = 0$. Hence, $f(x, t) = x$. $\qquad\qquad \square$

(5.4.6) Theorem. *A compact complex space X is immobile if and only if $\mathrm{Aut}(X)$ is discrete.*

Proof. The group $\mathrm{Aut}(X)$ is open in $\mathrm{Hol}(X, X)$ since it consists of holomorphic self-maps of degree 1. Suppose that $\mathrm{Aut}(X)$ is discrete. Then the identity automorphism of X is isolated in $\mathrm{Hol}(X, X)$. Given a holomorphic map $f: X \times D \to X$ with $f(x, 0) = x$, consider the family $f_t = f(\cdot, t) \in \mathrm{Hol}(X, X)$. Since f_0 is isolated in $\mathrm{Hol}(X, X)$, we have $f_t = f_0$. Hence, X is immobile.

Since $\mathrm{Aut}^0(X)$ is a complex space, the natural map $X \times \mathrm{Aut}^0(X) \to X$ would violate the condition of immobility unless $\mathrm{Aut}^0(X)$ is trivial. This proves the converse. $\qquad\qquad \square$

As we shall see later (see (7.1.17)), if X is measure hyperbolic, then $\mathrm{Aut}(X)$ is discrete. Therefore, by considering the immobility condition we can sometimes treat hyperbolic complex spaces and mesaure hyperbolic complex spaces at the same.

The following is due to Urata [4].

(5.4.7) Proposition. *Complex spaces X and Y are immobile if and only if $X \times Y$ are.*

Proof. Suppose that X and Y are immobile. Let $f: X \times Y \times D \to X \times Y$ be a holomorphic mapping such that $f(x, y, 0) = (x, y)$ for all $(x, y) \in X \times Y$. Write $f = (g, h)$, where g and h are holomorphic mappings from $X \times Y \times D$ into X and Y, respectively. Since $f(\cdot, \cdot, 0)$ is the identity transformation of $X \times Y$, it follows that, for any x and y, $g(\cdot, y, 0)$ (resp. $h(x, \cdot, 0)$) is the identity transformation of X (resp. Y). Since X and Y are immobile, this implies that $g(x, y, t) = x$ and $h(x, y, t) = y$ for all $t \in D$. Hence, $f(x, y, t) = (x, y)$.

Conversely, suppose that $X \times Y$ are immobile. Let $f \in \mathrm{Hol}(X \times D, X)$ such that $f(x, 0) = x$ for all $x \in X$. Define $g \in \mathrm{Hol}(X \times Y \times D, X \times Y)$ by

$$g(x, y, t) = (f(x, t), y) \qquad (x, y, t) \in X \times Y \times D.$$

Then $g(x, y, 0) = (x, y)$, and hence $g(x, y, t) = (x, y)$, which implies $f(x, t) = x$. This shows that X is immobile. Similarly, Y is immobile. $\qquad\qquad\square$

We call a complex subspace A of a complex space X a **direct factor** of X if there exists a complex space B together with a biholomorphic mapping $f: A \times B \to X$ such that, for some $b_0 \in B$, $f(a, b_0) = a$ for all $a \in A$.

(5.4.8) Lemma. *Let X, Y, V and W be complex spaces. Let*

$$h: X \times V \to Y \times W$$

be a biholomorphic mapping. Let π_X and π_V be the projections from $X \times V$ onto X and V, respectively. Similarly, let π_Y and π_W be the projections from $Y \times W$ onto Y and W, respectively. Fix a point $x_0 \in X$ and define mappings $f \in \mathrm{Hol}(V, Y)$ and $g \in \mathrm{Hol}(V, W)$ by

$$f(v) = \pi_Y h(x_0, v) \quad and \quad g(v) = \pi_W h(x_0, v)$$

so that

$$h(x_0, v) = (f(v), g(v)).$$

Set $Y' = f(V) \subset Y$ and $W' = g(V) \subset W$. Then

(1)   If X is immobile, then Y' and W' are complex subspaces of Y and W, respectively, and $h(x_0, \cdot) = (f, g)$ maps V biholomorphically onto $Y' \times W'$;

(2)   If X and Y (resp. X and W) are immobile, then W' (resp. Y') is a direct factor of W (resp. Y);

(3)   If X, Y, V and W are immobile and g maps V biholomorphically onto W, then

(a) $\pi_W h(x, v) = \pi_W h(x_0, v) = g(v)$, $x \in X$, $v \in V$,

(b) $\pi_Y h(x, v) = \pi_Y h(x, v_0)$, $x \in X$, $v_0, v \in V$,

and the mapping $x \mapsto \pi_Y h(x, v)$ maps X biholomorphically onto Y.

Proof. (1)   Define a map $\varphi: X \times (V \times V) \to X$ by

$$\varphi(x, v, v') = \pi_X h^{-1}(\pi_Y h(x, v), \pi_W h(x, v')).$$

Then $\varphi(x, v, v) = x$. Since X is immobile, $\varphi(x, v, v') = x$ for all $v, v' \in V$. For $x = x_0$, this means

$$\pi_X h^{-1}(f(v), g(v')) = x_0,$$

i.e., $h^{-1}(Y' \times W') \subset \{x_0\} \times V$. Thus, $Y' \times W' \subset h(\{x_0\} \times V)$. The opposite inclusion being trivial, we have

$$Y' \times W' = h(\{x_0\} \times V).$$

This shows also that $Y' \times W'$ is a complex subspace of $Y \times W$.

(2)   Define a map $h': Y \times W \to (X \times Y') \times W'$ by

$$h'(y, w) = (\pi_X h^{-1}(y, w), f\pi_V h^{-1}(y, w), g\pi_V h^{-1}(y, w)).$$

Fix $y \in Y' \subset Y$. Then

$$h'(y, w) = (f'(w), g'(w)),$$

where

$$f'(w) = (\pi_X h^{-1}(y, w), f\pi_V h^{-1}(y, w)) \quad \text{and} \quad g'(w) = g\pi_V h^{-1}(y, w).$$

We set $W'' = f'(W)$. On the other hand, $W' = g'(W)$ because every element $v \in V$ can be written as $\pi_V h^{-1}(y, w)$ for some $w \in W$.

Now apply (1) (replacing X by Y, V by W, Y by $X \times Y'$, W by W', and h by h'). Then we see that the mapping $w \mapsto h'(y, w) = (f'(w), g'(w))$ maps W biholomorphically onto $W'' \times W'$.

Take w in W'. By (1), there is an element $v \in V$ such that $y = f(v)$ and $w = g(v)$ so that $h^{-1}(y, w) = (x_0, v)$. Then

$$h'(y, w) = (\pi_X(x_0, v), f\pi_V(x_0, v), g\pi_V(x_0, v)) = (x_0, y, w).$$

This shows that the isomorphism $h': W \to W'' \times W'$ maps $W' \subset W$ to $\{(x_0, y)\} \times W'$. Thus, W' is a direct factor of W. The proof for Y' is similar.

(3a)   The map $g^{-1} \circ \pi_W \circ h: X \times V \to V$ sends (x_0, v) to v. Since V is immobile, it sends (x, v) to v for all $x \in X$. In other words, $\pi_W h(x, v) = g(v)$.

(3b)   Fix $v_0 \in V$. Since

$$h: (x, v_0) \mapsto (\pi_Y h(x, v_0), \pi_W h(x, v_0))$$

and since $\pi_W h(x, v_0) = \pi_W h(x_0, v_0)$ by (a), h maps $X \times \{v_0\}$ onto $Y \times \{\pi_W h(x_0, v_0)\}$. We denote the corresponding isomorphism from X onto Y by η, i.e.,

$$\eta(x) = \pi_Y h(x, v_0).$$

Consider the map $h': X \times V \to X$ defined by

$$h'(x, v) = \eta^{-1} \pi_Y h(x, v).$$

Then $h'(x, v_0) = \eta^{-1}\eta(x) = x$. Since X is immobile, $h'(x, v) = x$, i.e., $\eta^{-1}\pi_Y h(x, v) = x$, or $\pi_Y h(x, v) = \pi_Y h(x, v_0)$. $\qquad\square$

We say that a complex space X is **primary** if it has no direct factor of positive dimension different from X.

(5.4.9) Lemma. *As in (5.4.8), let $h: X \times V \to Y \times W$ be a biholomorphic mapping, and let $g(v) = \pi_W h(x_0, v)$. Assume that X and Y are immobile and that V and W are primary. If $g: V \to W$ is nonconstant, then it is biholomorphic.*

Proof. Let $W' = g(V)$ be as in (5.4.8). Since g is nonconstant, $\dim W' > 0$. By (5.4.8) W' is a direct factor of W. Since W is primary, $W' = W$. By (5.4.8) V is biholomorphic to $Y' \times W$. Since V is primary, $\dim Y' = 0$. This means that $f: V \to Y$ is constant and that $g: V \to W$ is biholomorphic. $\qquad\square$

We are now in a position to prove the following theorem of Urata [4].

(5.4.10) Theorem. *If a complex space X is immobile, then it is a direct product of primary complex spaces $X_1 \times \ldots \times X_m$, and this decomposition is unique up to an ordering.*

Proof. Let $X_1 \times \ldots \times X_m$ and $Y_1 \times \ldots \times Y_n$ be two primary decompositions of X. By (5.4.7), $X_1, \ldots, X_m, Y_1, \ldots, Y_n, X_1 \times \ldots \times X_i$ ($i = 2, \ldots, m$), $Y_1 \times \ldots \times Y_j$ ($j = 2, \ldots, n$) are all immobile. Let

$$f = (f_1, \ldots, f_n): X_1 \times \ldots \times X_m \to Y_1 \times \ldots \times Y_n$$

be any biholomorphic mapping. By reordering $Y_1, \ldots, Y_n$ we may assume that the mapping $f_n(x_1, \ldots, x_{m-1}, \cdot): X_m \to Y_n$ is nonconstant for some fixed $(x_1, \ldots, x_{m-1}) \in X_1 \times \ldots \times X_{m-1}$. This map is independent of $(x_1, \ldots, x_{m-1})$ by (3a) of (5.4.8) and is biholomorphic by (5.4.9). Moreover, by (3b) of (5.4.8), the map

$$(f_1(\cdot, x_m), \ldots, f_{n-1}(\cdot, x_m)): X_1 \times \ldots \times X_{m-1} \to Y_1 \times \ldots \times Y_{n-1}$$

is biholomorphic and is independent of x_m. Now the theorem follows by induction on m. $\qquad\square$

As a consequence we obtain the following generalization of the theorem of H. Cartan [8] (proved for bounded domains) and the theorem of Konrad Peters [1] (proved for hyperbolic complex spaces).

(5.4.11) Corollary. *Following (5.4.10), decompose an immobile complex space X in the following form:*

$$X = X_1 \times \ldots \times X_m, \quad X_i = (V_i)^{n_i} = V_i \times \ldots \times V_i,$$

where $V_1, \ldots, V_m$ are primary and mutually non-isomorphic. Then

$$\mathrm{Aut}(X) = \mathrm{Aut}(X_1) \times \ldots \times \mathrm{Aut}(X_m),$$

and, for each i, the following natural sequence is exact:

$$1 \to (\mathrm{Aut}(V_i))^{n_i} \to \mathrm{Aut}(X_i) \to S_{n_i} \to 1,$$

where S_{n_i} is the symmetry group of degree n_i.

(5.4.12) **Corollary**. *If complex spaces X and Y are immobile, then the natural injection $\mathrm{Aut}(X) \times \mathrm{Aut}(Y) \to \mathrm{Aut}(X \times Y)$ induces an isomorphism*

$$\mathrm{Aut}^0(X) \times \mathrm{Aut}^0(Y) \cong \mathrm{Aut}^0(X \times Y).$$

(5.4.13) **Remark**. As we explained in (4.1.14), every Siegel domain of the second kind is complete hyperbolic. Nakajima [1] has shown that every homogeneous hyperbolic complex manifold is biholomorphic to a Siegel domain of the second kind. For earlier partial results, see Kodama [4]. For the Kobayashi pseudodistance on homogeneous complex manifolds, see Winkelmann [1].

5 Self-Mappings of Hyperbolic Complex Spaces

The following theorem may be regarded as a generalization of the classical lemma of Schwarz to higher dimensions.

(5.5.1) **Theorem**. *Let X be a hyperbolic complex space and o a non-singular point of X. Let $f: X \to X$ be a holomorphic mapping such that $f(o) = o$, and $df_o: T_o X \to T_o X$ the differential of f at o. Then*
   (1)   *The eigenvalues of df_o have absolute value ≤ 1;*
   (2)   *If df_o is the identity transformation of $T_o X$, then f is the identity transformation of X;*
   (3)   *If $|\det(df_o)| = 1$, then f is a biholomorphic mapping.*

Proof. We take a small $r > 0$ so that the open ball $U(o, r) = \{x \in X; \; d_X(o, x) < r\}$ has compact closure $B = \overline{U(o, r)}$. Let $\mathcal{D}$ be the set of distance-decreasing maps f from B into itself with respect to $d_X|_B$. By (1.3.1) $\mathcal{D}$ is compact.

   (1)   Given $f \in \mathrm{Hol}(X, X)$ with $f(o) = o$, let λ be an eigenvalue of df_o. For each positive integer k, the iterated mapping f^k restricted to B belongs to $\mathcal{D}$ and its differential $(df_o)^k$ has an eigenvalue λ^k. Since $\mathcal{D}$ is compact, we must have $|\lambda| \leq 1$.

   (2)   For the sake of simplicity, we denote by $d^m f_o$ all partial derivatives of order m at o. We want to show that if df_o is the identity transformation of $T_o X$, then $d^m f_o = 0$ for all $m \geq 2$. Let m be the least integer ≥ 2 such that $d^m f_o \neq 0$. Then $d^m (f^k)_o = k \cdot d^m f_o$ for all positive integers k. As k goes to infinity, $d^m (f^k)_o$ also goes to infinity, thus contradicting the compactness of $\mathcal{D}$.

(3)   Assume $|\det(df_o)| = 1$. From (1), it follows that all eigenvalues of df_o have absolute value 1. Putting df_o in Jordan's canonical form, we claim that df_o is then in diagonal form. If not, it must have a diagonal block of the following form:

$$\begin{pmatrix} \lambda & 1 & 0 & \cdot & 0 \\ 0 & \lambda & 1 & \cdot & 0 \\ \cdot & \cdot & \cdot & \cdot & \cdot \\ \cdot & \cdot & \cdot & \cdot & \cdot \\ 0 & \cdot & 0 & 0 & \lambda \end{pmatrix}$$

with $|\lambda| = 1$. The corresponding diagonal block of $(df_o)^k$ is then of the form

$$\begin{pmatrix} \lambda^k & k\lambda^{k-1} & * & \cdot & * \\ 0 & \lambda^k & k\lambda^{k-1} & \cdot & * \\ \cdot & \cdot & \cdot & \cdot & \cdot \\ \cdot & \cdot & \cdot & \cdot & \cdot \\ 0 & \cdot & 0 & 0 & \lambda^k \end{pmatrix}.$$

This contradicts the compactness of $\mathcal{D}$ since the entries $k\lambda^{k-1}$ go to infinity as k goes to infinity,

We prove that a subsequence $\{f^{k_i}\}$ of $\{f^k\}$ which converges to the identity transformation of X. Since df_o is a diagonal matrix whose diagonal entries have absolute value 1, there is a subsequence $\{(df)_o^{k_i}\}$ of $\{(df_o)^k\}$ which converges to the identity matrix. Since $\mathcal{D}$ is compact, by taking a subsequence if necessary we may assume that $\{f^{k_i}\}$ converges to a map $h_{U(o,r)}$ $U(o,r)$ into itself. The differential of $h_{U(o,r)}$ at o, being equal to $\lim d(f^{k_i})_o$, is the identity transformation of T_oX. By (2), $h_{U(o,r)}$ must be the identity transformation of $U(o,r)$.

Let W be the largest open subset of X with the property that some subsequence of $\{f^{k_i}\}$ converges to the identity transformation of W. (To obtain such W, consider the union $W = \bigcup W_j$ of all open sets W_j of X such that on each W_j some subsequence of $\{f^{k_i}\}$ converges to the identity transformation. A countable number of W_j's already cover W. We consider the corresponding countable number of subsequences of $\{f^{k_i}\}$ and extract a desired subsequence by the standard argument using the diagonal subsequence.) Without loss of generality, we may assume that $\{f^{k_i}\}$ itself converges to the identity transformation on W. Since $U(o,r) \subset W$, W is nonempty. If $W \neq X$, take $x \in \partial W$ and choose s small so that $U(x,s) = \{y \in X;\ d_X(x,y) < s\}$ has compact closure. Since $\lim f^{k_i}(x) = x$ and f is distance-decreasing, there is a neighborhood U_x of x such that $f^{k_i}(U_x) \subset U(x,s)$ for $i \geq i_0$. let $\mathcal{F}$ be the set of all distance-decreasing maps from U_x into $\overline{U(x;s)}$. By (1.3.1) $\mathcal{F}$ is compact. We extract a subsequence from $\{f^{k_i}\}$ that is convergent on U_x. Since it converges to the identity transformation on $W \cap U_x$, it must converges to the identity transformation on U_x. Since this contradicts maximality of W, we must have $W = X$, thus proving our assertion. We may assume that $\{f^{k_i}\}$ itself converges to id_X.

Now we consider the sequence $\{(f)^{k_i-1}\}$ and show that it has a subsequence which converges to the inverse of f. By the same argument as above, by taking a subsequence if necessary we may assume that $\{(f)^{k_i-1}\}$ converges to a holomorphic map $g_{U(o,r)}$ of $U(o,r)$ into itself.

Let V be the largest open subset of X with the property that some subsequence of $\{(f)^{k_i-1}\}$ converges to a holomorphic transformation g_V of V. The existence of such V is proved in the same way as the existence of W above. From maximality of V we obtain $V = X$ again by the same argument as above. By taking a subsequence we may assume that $\{(f)^{k_i-1}\}$ converges to a holomorphic map g of X into itself. Then

$$f \circ g = f \circ (\lim(f)^{k_i-1}) = \lim f^{k_i} = \mathrm{id}_X.$$

Similarly, $g \circ f = \mathrm{id}_X$. Thus, g is the inverse of f. $\qquad\square$

For a bounded domain, (5.5.1) is due to H. Cartan [2, 4] and also to Carathéodory [3]. For a taut manifold, it is due to Wu [1]. It was proved independently by Kaup [2] under a weaker condition. It was proved in this form, i.e., under the hyperbolicity assumption in Kobayashi [7]. Eisenman [2] generalized Wu's result to tight manifolds (see (3.2.23) for the definition of "tight complex space"). Since tightness is equivalent to hyperbolicity (see (3.2.23)), Eisenman's result is equivalent to (5.5.1).

A domain X in $\mathbf{C}^n$ is said to be **circular** if $e^{i\theta}z \in X$ for all $\theta \in \mathbf{R}$ and $z \in X$. Then we have the following result of H. Cartan [2].

(5.5.2) **Corollary.** *If $X \subset \mathbf{C}^n$ is a bounded circular domain containing 0 and if f is a holomorphic automorphism of X such that $f(0) = 0$, then f is linear.*

Proof. For each $\theta \in \mathbf{R}$, define $h_\theta \in \mathrm{Aut}(X)$ by setting

$$h_\theta(z) = f^{-1}(e^{-i\theta} f(e^{i\theta}z)).$$

Then $h_\theta(0) = 0$ and $dh_\theta(0)$ is the identity transformation of $T_0 X$. By (5.5.1), h_θ is the identity transformation of X, that is,

$$f(e^{i\theta}z) = e^{i\theta} f(z) \qquad \text{for} \quad \theta \in \mathbf{R},\ z \in X.$$

Therefore all higher order terms in the power series expansion of f vanish. $\qquad\square$

A **holomorphic retraction** of a complex space X is a holomorphic map $\rho \in \mathrm{Hol}(X, X)$ such that $\rho^2 = \rho$. Then its image $\rho(X)$ is called a **holomorphic retract** of X. Clearly, $\rho(X)$ coincides with the set of fixed points of ρ. Hence, we have the first statement of the following proposition, (Abate [2]).

(5.5.3) **Proposition.** *If ρ is a holomorphic retraction of a complex space X, then $\rho(X)$ is a closed connected complex subspace of X. If $x \in \rho(X)$ is a nonsingular point of X, then $\rho(X)$ is nonsingular at x.*

Proof. In order to prove the last statement, let U be a coordinate neighborhood of x in X and set $V = \rho^{-1}(U \cap \rho(X)) \cap U$. Then V is an open neighborhood of x contained in a local chart U and $\rho(V) \subset V$. Replacing X by V we may assume that X is a bounded domain in $\mathbf{C}^n$.

Set $P = d\rho_x: \mathbf{C}^n \to \mathbf{C}^n$, and define $\varphi: X \to \mathbf{C}^n$ by

$$\varphi = \mathrm{id}_X + (2P - \mathrm{id}_X) \circ (\rho - P).$$

Since $d\varphi_x = \mathrm{id}$, φ defines a local chart in a neighborhood of x. Using $\rho^2 = \rho$ and $P^2 = P$, we have

$$\varphi \circ \rho = \rho + (2P - \mathrm{id}_X) \circ (\rho^2 - P \circ \rho) = P \circ \rho = P + P \circ (2P - \mathrm{id}_X) \circ (\rho - P) = P \circ \varphi.$$

This shows that in terms of the local chart given by φ, ρ is represented by a linear map P. Hence, $\rho(X)$ is nonsingular at x. $\qquad\qquad\square$

Thus, if $\dim X = 1$, then a holomorphic retraction of X is either a constant map or id_X.

The following theorem is due to Abate [2].

(5.5.4) Theorem. *Let X be a taut complex space, and $f \in \mathrm{Hol}(X, X)$. Assume that the sequence of iterates $\{f^k\}$ is not compactly divergent. Then*

(1)   there exists a unique holomorphic retraction ρ of X such that every limit map $h \in \mathrm{Hol}(X, X)$ of $\{f^n\}$ is of the form

$$h = \gamma \circ \rho,$$

where γ is an automorphism of $\rho(X)$;

(2)   moreover, a suitable subsequence of $\{f^k\}$ converges to ρ;

(3) $f|_{\rho(X)}$ is an automorphism of $\rho(X)$.

Proof. (1) and (2)   Let $\{f^{k_j}\}$ be a subsequence of $\{f^k\}$ converging to $h \in \mathrm{Hol}(X, X)$. Taking a subsequence if necessary, we may assume that both $p_j = k_{j+1} - k_j$ and $q_j = p_j - k_j = k_{j+1} - 2k_j$ tend to ∞ as $j \to \infty$, and that $\{f^{p_j}\}$ and $\{f^{q_j}\}$ are either convergent or compactly divergent. Since

$$\lim_{j \to \infty} f^{p_j}(f^{k_j}(x)) = \lim_{j \to \infty} f^{k_{j+1}}(x) = h(x), \qquad x \in X,$$

$\{f^{p_j}\}$ cannot be compactly divergent and hence converges to a map $\rho \in \mathrm{Hol}(X, X)$ such that

$$h \circ \rho = \rho \circ h = h.$$

Since

$$\lim_{j \to \infty} f^{q_j}(f^{k_j}(x)) = \lim_{j \to \infty} f^{p_j}(z) = \rho(x), \qquad x \in X,$$

$\{f^{q_j}\}$ cannot be compactly divergent and hence coverges to a map $g \in \mathrm{Hol}(X, X)$ such that

$$g \circ h = h \circ g = \rho.$$

Hence,

$$\rho^2 = g \circ h \circ \rho = g \circ h = \rho,$$

which shows that ρ is a holomorphic retraction of X. Since $h = \rho \circ h$ and $\rho = h \circ g$, we have $h(X) = \rho(X)$. Since $g \circ \rho = g \circ h \circ g = \rho \circ g$, we have $g(\rho(X)) \subset \rho(X)$. Since $g \circ h = h \circ g = \rho$, we have

$$g \circ h|_{\rho(X)} = h \circ g|_{\rho(X)} = \mathrm{id}_{\rho(X)}.$$

Setting $\gamma = h|_{\rho(X)}$, we obtain $h = \gamma \circ \rho$.

In order to see that $\rho(X)$ does not depend on the choice of convergent subsequence, let $\{f^{k'_j}\}$ be another subsequence of $\{f^k\}$ converging to a map $h' \in \mathrm{Hol}(X, X)$. Arguing as before, we may assume that both $s_j = k'_j - k_j$ and $t_j = k_{j+1} - k'_j$ tend to ∞ as $j \to \infty$, and that $\{f^{s_j}\}$ and $\{f^{t_j}\}$ converge, respectively, to maps $\alpha \in \mathrm{Hol}(X, X)$ and $\beta \in \mathrm{Hol}(X, X)$ such that

$$(*) \qquad \alpha \circ h = h \circ \alpha = h' \quad \text{and} \quad \beta \circ h' = h' \circ \beta = h.$$

Then $h(X) = h'(X)$, and hence $\rho(X) = \rho'(X)$. We write $M = h(X) = h'(X) = \rho(X) = \rho'(X)$.

Finally we shall show that ρ itself does not depend on the chosen sequence. Write

$$h = \gamma_1 \circ \rho_1, \quad h' = \gamma_2 \circ \rho_2, \quad \alpha = \gamma_3 \circ \rho_3, \quad \beta = \gamma_4 \circ \rho_4,$$

where ρ_1, ρ_2, ρ_3 and ρ_4 are holomorphic retractions of X onto M, and $\gamma_1, \gamma_2, \gamma_3$ and γ_4 are automorphisms of M. Then $h \circ h' = h' \circ h$ and $\alpha \circ \beta = \beta \circ \alpha$ together with $(*)$ give

$$\gamma_1 \circ \gamma_2 \circ \rho_2 = \gamma_2 \circ \gamma_1 \circ \rho_1,$$

$$\gamma_3 \circ \gamma_4 \circ \rho_4 = \gamma_4 \circ \gamma_3 \circ \rho_3,$$

$$\gamma_3 \circ \gamma_1 \circ \rho_1 = \gamma_1 \circ \gamma_3 \circ \rho_3 = \gamma_2 \circ \rho_2,$$

$$\gamma_4 \circ \gamma_2 \circ \rho_2 = \gamma_2 \circ \gamma_4 \circ \rho_4 = \gamma_1 \circ \rho_1.$$

From the first and the third equations, we obtain

$$\gamma_2 \circ \gamma_1 \circ \rho_1 = \gamma_1 \circ \gamma_1 \circ \rho_2 = \gamma_1 \circ \gamma_3 \circ \gamma_1 \circ \rho_1,$$

and hecne $\gamma_2 = \gamma_1 \circ \gamma_3$. Similarly, from the first and the fourth equations we obtain $\gamma_1 = \gamma_2 \circ \gamma_4$. Hence, $\gamma_3 = \gamma_4^{-1}$. This with the second equation yields $\rho_3 = \rho_4$. Then using the third and the fourth equations we obtain

$$\rho_2 = \gamma_3^{-1} \circ \gamma_1^{-1} \circ \rho_2 = \rho_3 = \rho_4 = \gamma_4^{-1} \circ \gamma_2^{-1} \circ \gamma_1 \circ \rho_1 = \rho_1.$$

(3)   Taking a subsequence of a sequence $\{f^{k_j}\}$ which converges to h say, we may assume that $\{f^{k_j+1}\}$ also converges (to $h' = f \circ h = h \circ f$). Then

$$f \circ \gamma \circ \rho = f \circ h = h' = \gamma' \circ \rho,$$

where γ' is an automorphism of $\rho(X)$. Hence, $f \circ \gamma = \gamma'$ on $\rho(X)$ and $f = \gamma' \circ \gamma^{-1}$ on $\rho(X)$. $\qquad \square$

(5.5.5) **Corollary.** *Let X be a taut complex space containing no compact complex subspaces of positive dimension. Let $f \in \mathrm{Hol}(X, X)$ be such that $f(X)$ has compact closure in X. Then f has a unique fixed point $x_0 \in X$, and the sequence of iterates of f converges to x_0.*

Proof. Since the sequence $\{f^k\}$ cannot be compactly divergent, (5.5.4) yields a holomorphic retraction ρ with the property described in (5.5.4). Since $\rho(X) \subset f(X)$, the retract $\rho(X)$ is a compact connected complex subspace of X, which is a point x_0 of X. By (5.5.4), the unique limit point of $\{f^k\}$ is the constant map x_0.
$\qquad\square$

The corollary above goes back to Wavre [1].

(5.5.6) Theorem. *Let X be a taut complex space, and $f \in \mathrm{Hol}(X, X)$. If the sequence $\{f^k\}$ converges, it converges to a holomorphic retraction ρ (so that $\rho(X)$ is pointwise fixed by f), and at every nonsingular point $z_0 \in \rho(X)$ the differential df_{z_0} has eigenvalues in the set $D \cup \{1\}$. Conversely, if f has a nonsingular fixed point $z_0 \in X$ such that df_{z_0} has eigenvalues in $D \cup \{1\}$, then $\{f^k\}$ converges.*

Proof. Assume that $\{f^k\}$ converges. Then by (5.5.4) it converges to a retraction ρ and

$$f(\rho(z)) = \lim_{k \to \infty} f(f^k(z)) = \lim_{k \to \infty} f^{k+1}(z) = \rho(z),$$

so that $\rho(X)$ is pointwise fixed by f. Let $z_0 \in \rho(X)$ be a nonsingular point and λ an eigenvalue of df_{z_0}. Then $\{\lambda^k\}$ tends to an eigenvalue of $d\rho_{z_0}$, i.e., 0 or 1. Hence, $\lambda \in D \cup \{1\}$.

Conversely, assuming that f has a nonsingular fixed point z_0 such that df_{z_0} has eigenvalues in $D \cup \{1\}$, put df_{z_0} in Jordan's canonical form. Then (see the proof of (5.5.1))

$$A = \begin{pmatrix} I_r & 0 \\ 0 & A \end{pmatrix},$$

where r is the multiplicity of 1 as eigenvalues of df_{z_0}, and A is a matrix such that $\lim A^k = 0$. Since f has a fixed point, $\{f^k\}$ is not compactly divergent and (5.5.4) can be applied. Let ρ be the retraction produced in (5.5.4). Let h be any limit point of $\{f^k\}$. Then h fixes z_0, sends $\rho(X)$ into itself and

$$dh_{z_0} = \begin{pmatrix} I_r & 0 \\ 0 & 0 \end{pmatrix} = d\rho_{z_0}.$$

Applying (2) of (5.5.1) to $h|_{\rho(X)}$ we see that h fixes every point of $\rho(X)$. Therefore γ in (5.5.4) is the identity automorphism of $\rho(X)$, and it follows that ρ is the unique limit point of $\{f^k\}$.
$\qquad\square$

The following theorem is due to W. Kaup [4].

(5.5.7) Theorem. *Let X be a compact hyperbolic complex space, and $f \in \mathrm{Hol}(X, X)$. Then there exists a positive integer m such that f^m is a holomorphic retraction. In particular, the sequence of iterates $\{f^k\}$ converges if and only if f itself is a holomorphic retraction.*

Proof. Let ρ be the limit retraction obtained in (5.5.4). Then $f|_{\rho(X)}$ is an automorphism of $\rho(X)$. In particular, $f^k(X) \supset \rho(X)$ for every k. Therefore, the descending chain of compact hyperbolic complex spaces $X \supset f(X) \supset f^2(X) \supset \ldots$

must stop, and there is an integer k_0 such that $f^{k+1}(X) = f^k(X)$ for all $k \geq k_0$. Since ρ is a limit map of $\{f^k\}$, we have $f^k(X) = \rho(X)$ for all $k \geq k_0$.

By (5.4.4) $\mathrm{Aut}(\rho(X))$ is finite. Therefore there exists an integer $m \geq k_0$ such that $(f|_{\rho(X)})^m = \mathrm{id}_{\rho(X)}$. Since $f^m(z) \in \rho(X)$ for all $z \in X$, we have

$$f^{2m}(z) = f^m(f^m(z)) = f^m(z), \qquad z \in X,$$

which shows that f^m is a retraction of X to $\rho(X)$. $\square$

The following is due to Vigué [8].

(5.5.8) **Theorem**. *Let X be a taut complex space, $f \in \mathrm{Hol}(X, X)$, and $\mathrm{Fix}(f)$ the set of fixed points of f. Then $\mathrm{Fix}(f)$ is a closed complex space of X and is singular only where X is singular. If $x \in \mathrm{Fix}(f)$ is a nonsingular point of X, then*

$$T_x(\mathrm{Fix}(f)) = \{\xi \in T_x X;\ f_*\xi = \xi\}.$$

Proof. Assume that $\mathrm{Fix}(f)$ is nonempty. Then $\{f^k\}$ cannot be compactly divergent. Let ρ be the holomorphic retraction constructed in (5.5.4). As we proved in (5.5.4), f is an isometry of $\rho(X)$ with respect to $d_{\rho(X)}$. Since $\mathrm{Fix}(f) \subset \rho(X)$ and since $\rho(X)$ is singular only at singular points of X (see (5.5.3)), the proof is reduced to the case where f is an isometry.

Let $x \in \mathrm{Fix}(f)$ be a nonsingular point. Since the group of isometries of X fixing x is compact (see (5.4.1)), there is a Hermitian metric (at least in a neighborhood of x) which is invariant by f. In other words, f is an isometry with respect to not only d_X but also suitable Hermitian metric. Hence, $\mathrm{Fix}(f)$ is a complex submanifold in a neighborhood of x.

In the last assertion of the theorem, the left hand side is clearly contained in the right hand side. We shall prove the reverse inclusion. If $f_*\xi = \xi$, then $\rho_*\xi = \xi$ so that

$$\{\xi \in T_x X;\ f_*\xi = \xi\} \subset T_x(\rho(X)).$$

Hence, the proof is again reduced to the case where f is an automrophism of X. Then the argument using an invariant Hermitian metric yields the desired result.

$\square$

On the fixed point set $\mathrm{Fix}(f)$, see also Abate [3], Abate-Vigué [1] and Vigué [3, 6, 8, 9, 11].

The classical theorem of Denjoy-Wolff states that a holomorphic self-map $f \in \mathrm{Hol}(D, D)$ of the unit disc D has no fixed point if and only if the sequence of its iterates $\{f^k\}$ converges to a boundary point. Theorem (5.5.4) points in the same direction although $\rho(X)$ is not the fixed point set of f, nor is the sequence $\{f^k\}$ being compactly divergent quite the same as convergence to a boundary point. The theorem of Denjoy-Wolff has been generalized by Abate [2] (to strongly convex domains in $\mathbf{C}^n$), Ma [2] (to strongly pseudoconvex contractible domains in $\mathbf{C}^2$) and Huang [1] (to strongly pseudoconvex contractible domains in $\mathbf{C}^n$). We state here Huang's result:

(5.5.9) **Theorem**. *Let X be a topologically contractible bounded strongly pseudo-convex domain in $\mathbf{C}^n$ with C^3 boundary. Then a holomorphic self-map f of X has no fixed point in X if and only if the sequence of its iterates $\{f^k\}$ converges to a boundary point uniformly on compacta.*

Chapter 6. Extension and Finiteness Theorems

1 The Classical Big Picard Theorem

The classical little Picard theorem states that every entire function f missing two values must be constant. In (3.10.2) we stated E. Borel's generalization to a system of entire functions. One of its geometric consequences is that given $n + p$ hyperplanes $H_1, \ldots, H_{n+p}$ in $P_n\mathbf{C}$ in general position, every $f \in \mathrm{Hol}(\mathbf{C}, P_n\mathbf{C} - \bigcup_{i=1}^{n+p} H_i)$ has its image in a linear subspace of dimension $\leq [n/p]$, see (3.10.7). For $p = n + 1$, this means that every $f \in \mathrm{Hol}(\mathbf{C}, P_n\mathbf{C} - \bigcup_{i=1}^{2n+1} H_i)$ must be constant, (see (3.10.8)). This has been further strengthened to the statement that $P_n\mathbf{C} - \bigcup_{i=1}^{2n+1} H_i$ is complete hyperbolic and is hyperbolically imbedded in $P_n\mathbf{C}$, (see (3.10.9)), which is a statement on $\mathrm{Hol}(D, P_n\mathbf{C} - \bigcup_{i=1}^{2n+1} H_i)$ rather than on $\mathrm{Hol}(\mathbf{C}, P_n\mathbf{C} - \bigcup_{i=1}^{2n+1} H_i)$ since the hyperbolicity and hyperbolic imbeddedness of $X = P_n\mathbf{C} - \bigcup_{i=1}^{2n+1} H_i$ is defined in terms of d_X which, in turn, is constructed by $\mathrm{Hol}(D, P_n\mathbf{C} - \bigcup_{i=1}^{2n+1} H_i)$. Thus, (3.10.9) is a statement *en termes finis* in the sense of Bloch [1].

The classical big Picard theorem is usually stated as follows:

If a function $f(z)$ holomorphic in the punctured disc $0 < |z| < R$ has an essential singularity at $z = 0$, then there is at most one value a $(\neq \infty)$ such that the equation $f(z) = a$ has only a finite number of solutions in the disc.

If $f(z) = a$ has only a finite number of solutions in the punctured disc above, then it has no solutions in a smaller punctured disc $0 < |z| < R'$, $R' \leq R$. Hence, the big Picard theorem may be rephrased as follows:

If a function $f(z)$ holomorphic in the punctured disc $0 < |z| < R$ misses two values a, b $(\neq \infty)$, then the origin $z = 0$ is either a removable singularity or a pole so that f extends through $z = 0$ as a meromorphic function.

We regard a holomorphic function which misses two values a, b as a holomorphic mapping into $P_1\mathbf{C} - \{\infty, a, b\}$, and we consider a meromorphic function as a holomorphic map into $P_1\mathbf{C}$. Then the big Picard theorem may be stated as an extension theorem for holomorphic maps. (Since any triple of points of $P_1\mathbf{C}$ can be mapped into $\{\infty, 0, 1\}$ by an automorphism of $P_1\mathbf{C}$, we may assume that the missing values are 0 and 1). Thus:

Every holomorphic map from the punctured disc $0 < |z| < R$ into $P_1\mathbf{C} - \{\infty, 0, 1\}$ extends to a holomorphic map from the disc $|z| < R$ into $P_1\mathbf{C}$.

Since $P_1\mathbf{C} - \{\infty, 0, 1\}$ is hyperbolic, we ask the following question.

Given a (compact) complex space Z and a domain, or more generally, a complex subspace $Y \subset Z$ which is hyperbolic and relatively compact in Z, does every holomorphic map from the punctured disc $0 < |z| < R$ into Y extends to a holomorphic map from the disc $|z| < R$ into Z ?

The answer to this general question is negative as shown by the following example (see Kiernan [5, 7]).

(6.1.1) **Example.** Let $Z = P_2\mathbf{C}$ with homogeneous coordinate system (u, v, w) and
$$Y = \{(1, v, w) \in P_2\mathbf{C};\ 0 < |v| < 1, |w| < |e^{1/v}|\}.$$

Then the mapping
$$(1, v, w) \mapsto (v, we^{-1/v})$$

defines a biholomorphic isomorphism between Y and $D^* \times D$, (where $D^* = \{0 < |z| < 1\}$ and $D = \{|z| < 1\}$). Hence, Y is complete hyperbolic. The boundary of Y contains the line in $P_2\mathbf{C}$ defined by $v = 0$. Let $f: D^* \to Y$ be the mapping defined by
$$f(z) = \left(1, z, \frac{1}{2}e^{1/z}\right), \qquad z \in D^*.$$

For every neighborhood N of the line $v = 0$, there exists a small punctured disc $D_\delta^* = \{0 < |z| < \delta\}$ such that $f(D_\delta^*) \subset N$. This is to be expected since every complete hyperbolic space is taut (see (5.1.3)). However, f cannot be extended to a holomorphic map from D into $P_2\mathbf{C}$.

We can easily show that Y is *not* hyperbolically imbedded in Z. Let $p_n = (1, 1/n, 0)$, $q_n = (1, 1/n, 1) \in Y$ so that $\lim p_n = (1, 0, 0)$ and $\lim q_n = (1, 0, 1)$. Making use of a holomorphic map $f_n: D \to Y$ which sends $z \in D$ to $(1, 1/n, e^n z)$ we see that $d_Y(p_n, q_n) \le d_D(0, e^{-n}) \to 0$ as $n \to \infty$.

In Example (3.3.12) we gave another example where $Z = P_2\mathbf{C}$ and Y is the complement of three projective lines and two transcendental curves together with a holomorphic map $h: D^* \to Y$ which does not extend to a map $h: D \to Z$. In that example, Y is biholomorphic to $(\mathbf{C} - \{0, 1\})^2$ (and hence complete hyperbolic) but is *not* hyperbolically imbedded in Z.

In Section 3 we shall show that the holomorphic extension theorem holds if Y is hyperbolically imbedded in Z. More precisely (see (6.3.9)), we have:
If a complex space Y is hyperbolically imbedded in a complex space Z and if X is a nonsingular complex manifold and A is a closed complex subspace consisting of hypersurfaces with at most normal crossing singularities, then every $h \in \mathrm{Hol}(X - A, Y)$ extends to a map $\bar{h} \in \mathrm{Hol}(X, Z)$.
The assumption on the nature of sinularities of $A \subset X$ is in general essential.

From (6.3.9) and (3.10.8) we immediately obtain the following generalization of the big Picard theorem.

(6.1.2) Theorem. *Let $A \subset X$ be as above, and let $H_1, \ldots, H_{2n+1}$ be $2n + 1$ hyperplanes in $P_n\mathbf{C}$ in general position. Then every holomorphic map h from $X - A$ into $P_n\mathbf{C} - \bigcup_{i=1}^{2n+1} H_i$ extends to a holomorphic map $\bar{h}$ from X into $P_n\mathbf{C}$.*

As we remarked in Section 2 of Chapter 2, $P_1\mathbf{C} - \{\infty, 0, 1\}$ may be identified with the quotient $SL(2; \mathbf{Z})_2\backslash H$ of the upper halfplane H. The original proof of the Picard theorems depended on this fact. From this view point, the big Picard theorem can be generalized in the following form, Kobayashi-Ochiai [2]. Since the proof requires much preparation, the reader is referred to the original paper. See also A. Borel [1].

(6.1.3) Theorem. *Let $\mathcal{D} = G/K$ be a symmetric bounded domain, and $\Gamma \subset G$ a discrete arithmetic subgroup acting freely on $\mathcal{D}$. Then $\Gamma\backslash\mathcal{D}$ is hyperbolically imbedded in its Satake compactification $(\Gamma\backslash\mathcal{D})^*$.*

This means that the Satake compactification $(\Gamma\backslash\mathcal{D})^*$ is in some sense a minimal compactification of $\Gamma\backslash\mathcal{D}$.

(6.1.4) Corollary. *Let $A \subset X$ and $\Gamma\backslash\mathcal{D}$ be as above. Then every holomorphic map h from $X - A$ into $\Gamma\backslash\mathcal{D}$ extends to a map $\bar{h}$ from X into $(\Gamma\backslash\mathcal{D})^*$.*

We stated above that the assumption on $A \subset X$ about the nature of their singularities is essential in the generalized big Picard Theorem. However, there is a notable exception. Although $(\Gamma\backslash\mathcal{D})^*$ aquires bad singularities when compactified, we have the following extension theorem (Kiernan-Kobayashi [1]).

(6.1.5) Theorem. *Let $\Gamma\backslash\mathcal{D} \subset (\Gamma\backslash\mathcal{D})^*$ be as above. Let Y be a complex space hyperbolically imbedded in a complex space Z. Then every holomorphic map h from $\Gamma\backslash\mathcal{D}$ into Y extends to a map $(\Gamma\backslash\mathcal{D})^*$ into Z.*

Again, for the proof, see the original paper.

For extendability of holomorphic maps into complete hyperbolic-like manifolds, see Dolbeault-Ławrynowicz [1].

2 Extension through Subsets of Large Codimension

In this section we discuss the problem of extending a holomorphic map $f: X - A \to Y$ to X when A has codimension greater than 1. In Section 3, we shall consider the case where the codimension of A is 1. The following proposition strengthens (3.2.19). See also (3.2.22).

(6.2.1) Proposition. *Let A be a closed subset of the unit polydisc $D^m = D \times D^{m-1}$ of the form $A = \{0\} \times A'$, where A' is nowhere dense in D^{m-1}. Then the distance $d_{D^m - A}$ is the restriction of the distance d_{D^m} to $D^m - A$.*

Proof. Let $p, q \in D^m - A$. The problem is to prove the inequality $d_{D^m-A}(p, q) \leq d_{D^m}(p, q)$, the oppositie inequality being obvious, (see (3.1.6)). It suffices to establish the desired inequality for pairs (p, q) belonging to a dense subset of $(D^m - A) \times (D^m - A)$.

Let S be the subset of $(D^m - A) \times (D^m - A)$ consisting of pairs (p, q) for which there exist points $a, b \in D$ and a holomorphic map $f : D \to D^m - A$ such that $d_{D^m}(p, q) = d_D(a, b)$, $f(a) = p$, and $f(b) = q$. If $(p, q) \in S$, then

$$d_{D^m-A}(p, q) = d_{D^m-A}(f(a), f(b)) \leq d_D(a, b) = d_{D^m}(p, q).$$

It suffices therefore to prove that S is a dense subset.

Let $p = (a^1, \ldots, a^m)$ and $q = (b^1, \ldots, b^m)$ be arbitrary points of $D^m - A$. In order to show that every neighborhood of (p, q) in $(D^m - A) \times (D^m - A)$ contains a point of S, we may assume without loss of generality that a^1, b^1 and 0 are mutually distinct since the set of such pairs is dense in $(D^m - A) \times (D^m - A)$. The distance $d_{D^m}(p, q)$ is equal to the maximum of $d_D(a^j, b^j)$, $j = 1, \ldots, m$, say $d_D(a^k, b^k)$, (see (3.1.9)). We set $a = a^k$ and $b = b^k$ so that $d_{D^m}(p, q) = d_D(a, b)$. Since $d_D(a^j, b^j) \leq d_D(a, b)$ for $j = 1, \ldots, m$, there exist holomorphic mappings $f_j : D \to D$ such that $f_j(a) = a^j$ and $f_j(b) = b^j$ for $j = 1, \ldots, m$. Since $a^1 \neq b^1$, we may impose the additional condition that f_1 be injective. Then $f_1^{-1}(0)$ is either empty or a single point $c \in D$. If $f_1^{-1}(0)$ is empty, then the mapping $f : D \to D^m$ defined by $f(z) = (f_1(z), \ldots, f_m(z))$ sends D into $D^m - A$ since $f_1(z)$ never vanishes. In this case, (p, q) belongs to S. Assume $c = f_1^{-1}(0)$. Then the mapping $f : D \to D^m$ defined above sends D into $D^m - A$ if and only if $(f_2(c), \ldots, f_m(c))$ is not in A'. Since $f(D) \subset D^m - A$ implies $(p, q) \in S$, we have only to consider the case $(f_2(c), \ldots, f_m(c)) \in A'$.

We assert that given a positive number ε there exists a positive number δ such that for any points $c^j \in D$, $(j = 2, \ldots, m)$, with $d_D(c^j, f_j(c)) < \delta$ there exist automorphisms h_j of D such that

$$h_j(f_j(c)) = c^j, \quad d_D(a^j, h_j(a^j)) < \varepsilon, \quad \text{and} \quad d_D(b^j, h_j(b^j)) < \varepsilon.$$

We shall first complete the proof of the proposition and then come back to the proof of this assertion.

Given $\varepsilon > 0$, let $\delta > 0$ be as above. Since A' is nowhere dense in D^{m-1}, there exists a point $(c^2, \ldots, c^m) \in D^{m-1} - A'$ such that $d_D(c^j, f_j(c)) < \delta$ for $j = 2, \ldots, m$. Let h_j be automorphisms of D as above. We consider the points $p' = (a, h_2(a^2), \ldots, h_m(a^m))$ and $q' = (b^1, h_2(b^2), \ldots, h_m(b^m))$ of D^m and the holomorphic map $f' : D \to D^m$ defined by

$$f'(z) = (f_1(z), h_2(f_2(z)), \ldots, h_m(f_m(z))).$$

Since a^1, b^1 and 0 are mutually distinct, both p' and q' are in $D^m - A$. Since h_j is an automorphism, we have $d_D(h_j(a^j), h_j(b^j)) = d_D(a^j, b^j)$ and hence

$$d_{D^m}(p', q') = \max\{d_D(a^j, b^j); \ j = 1, \ldots, m\} = d_{D^m}(p, q) = d_D(a, b).$$

Clearly, $f'(a) = p'$ and $f'(b) = q'$. This shows that (p', q') belongs to S. Since $d_D(a^j, h_j(a^j)) < \varepsilon$, we have $d_{D^m}(p, p') < \varepsilon$. Similarly, $d_{D^m}(q, q') < \varepsilon$. Thus the ε-neighborhood of (p, q) in $(D^m - A) \times (D^m - A)$ with respect to $d_{D^m} \times d_{D^m}$ contains a point (p', q') of S. This completes the proof of the proposition except for the proof of the assertion above.

To simplify the notations in the assertion above, we denote $a^j, b^j, f_j(c), c^j$ and h_j by a, b, c, c' and h, respectively. Then the assertion we have to prove reads as follows:

(6.2.2) Lemma. *Given $a, b, c \in D$ and $\varepsilon > 0$, there exists $\delta > 0$ such that for any $c' \in D$ with $d_D(c, c') < \delta$ there exists an automorphism h of D satisfying*

$$h(c) = c' \quad d_D(a, h(a)) < \varepsilon, \quad and \quad d_D(b, h(b)) < \varepsilon.$$

In order to prove the lemma, it is more convenient to replace D by the upper half-plane H in $\mathbf{C}$. Given $\varepsilon > 0$, choose $\delta_1 > 0$ and $\delta_2 > 0$ such that

$$d_H(a, r(a + t)) < \varepsilon \quad and \quad d_H(b, r(b + t)) < \varepsilon$$

for any real numbers t and r such that $|t| < \delta_1$ and $|r - 1| < \delta_2$. The set $\{r(c + t); \ |t| < \delta_1, \ |r - 1| < \delta_2\}$ contains the δ-neighborhood of c for some $\delta > 0$. Given c' in the δ-neighborhood of c, we can find an automorphism h of H of the form $h(z) = r(z + t)$ such that $h(c) = c'$, $d_H(a, h(a)) < \varepsilon$ and $d_H(b, h(b)) < \varepsilon$. $\qquad\square$

As an application of (6.2.1) we prove

(6.2.3) Theorem. *Let Y be a complete hyperbolic space. Let X be a complex manifold of dimension m, and let A be a subset which is nowhere dense in a complex subspace $B \subset X$ of dimension $\leq m - 1$. Then every holomorphic map $f: X - A \to Y$ extends to a holomorphic map $f: X \to Y$.*

Proof. Let $S(B)$ denote the singular locus of B. First we note that we may assume that B is nonsingular; i.e, we extend the map f to $X - S(B)$ then to $X - S(S(B))$, and so on.

By localizing the map f, we may assume that $X = D^m$ and $B = \{0\} \times D^{m-1}$ so that A is of the form $A = \{0\} \times A'$, where A' is nowhere dense in D^{m-1}. Since $f: D^{m-1} - A \to Y$ is distance-decreasing, f extends to a continuous map from the completion of the metric space $D^m - A$ into Y. By (6.2.1) D^m is the completion of $D^m - A$ with respect to $d_{D^m - A}$. By the Riemann extension theorem, the extended continuous map f is holomorphic. $\qquad\square$

The theorem above contains the following result of Kwack [1], which was proved by a different method. This follows also from (3.2.19).

(6.2.4) Corollary. *Let Y be a complete hyperbolic space. Let X be a complex manifold of dimension m, and let A be a complex subspace of dimension $\leq m - 2$. Then every holomorphic mapping $f: X - A \to Y$ extends to a holomorphic map $f: X \to Y$.*

In particular, every meromorphic map f of a complex manifold X into a complete hyperbolic space Y is holomorphic. However, it will be shown in (6.3.19) that f is holomorphic if Y is only hyperbolic (complete or not), (Kodama [2]).

A similar extension theorem has been proved by Andreotti and Stoll [1] when Y has a Stein covering space:

(6.2.5) **Theorem**. *Let Y be a complex space which has a Stein covering space. Let X be a complex manifold of dimension m, and A a subset of topological dimension $\leq 2m - 3$ contained in a complex subspace $B \subset X$ of dimension $\leq m - 1$. Then every holomorphic map $f: X - A \to Y$ extends to a holomorphic map $f: X \to Y$.*

3 Generalized Big Picard Theorems and Applications

We know (see (2.2.3)) that the complete metric of constant curvature -1 on the punctured unit disc $D^* = \{0 < |z| < 1\}$ is given by

$$(6.3.1) \qquad ds_{D^*}^2 = \frac{4dz d\bar{z}}{|z|^2 (\log |z|^2)^2}.$$

and that the Kobayashi distance d_{D^*} of D^* is obtained by integration of $ds_{D^*}^2$.

For each positive number $r < 1$, let $L(r)$ denote the arc-length of the circle $|z| = r$ with respect to $ds_{D^*}^2$. Then

$$(6.3.2) \qquad L(r) = \frac{2\pi}{\log(1/r)}.$$

In particular,

$$(6.3.3) \qquad \lim_{r \to 0} L(r) = 0,$$

The following result is due to Kwack [1].

(6.3.4) **Theorem**. *Let $f: D^* \to Y$ be a holomorphic map from the punctured disc D^* into a complex space Y. It extends to a holomorphic map $\bar{f}: D \to Y$ if the following two conditions are satisfied:*

(a)   *there exists a distance function δ_Y on Y such that $f: (D^*, d_{D^*}) \to (Y, \delta_Y)$ is distance-decreasing, (e.g., Y is hyperbolic);*

(b)   *there exists a sequence of points $z_k \in D^*$ converging to the origin 0 such that $f(z_k)$ converges to a point $y_0 \in Y$.*

(6.3.5) **Corollary**. *If Y is a compact hyperbolic complex space, then every holomorphic map $f: D^* \to Y$ extends to a holomorphic map of D into Y.*

Proof of (6.3.4). We set

$$r_k = |z_k|, \qquad \gamma_k(t) = f(r_k e^{2\pi i t}).$$

Thus, γ_k is the image of the circle $|z| = r_k$ by f. We may assume that $\{r_k\}$ is monotone decreasing.

Let U be a neighborhood of y_0 in Y. We identify U with a complex subspace of a polydisc neighborhood

$$V = \{|w^1| < \varepsilon, \ldots, |w^n| < \varepsilon\}$$

of 0 in $\mathbf{C}^n$ and the point y_0 with the origin 0. Let W be the neighborhood of 0 defined by

$$W = \{|w^1| < \varepsilon/2, \ldots, |w^n| < \varepsilon/2\}.$$

We have to show that, for a suitable positive number δ, the small punctured disc $D^*_\delta = \{z \in D^*; \; |z| < \delta\}$ is mapped into V by f.

Since the diameter of γ_k approaches zero (by (6.3.3)), all (but a finite number of) γ_k's are contained in W. Consider the set of integers k such that the image of the annulus $r_{k+1} < |z| < r_k$ by f is not entirely contained in W. If this set of integers is finite, then f maps a small punctured disc D^*_δ into $\bar{W}$. Assuming that this set of integers is infinite, we shall obtain a contradiction.

By taking a subsequence we may assume that, for every k, the image of the annulus $r_{k+1} < |z| < r_k$ by f is not entirely contained in W. For each k, let

$$R_k = \{z \in D^*; \; a_k < |z| < b_k\}$$

be the largest open annulus such that (i) $a_k < r_k < b_k$ and (ii) f maps R_k into W. We set

$$\sigma_k(t) = a_k e^{2\pi it}, \qquad 0 \leq t \leq 1,$$
$$\tau_k(t) = b_k e^{2\pi it}, \qquad 0 \leq t \leq 1.$$

Thus, σ_k is the inner boundary of the annulus R_k while τ_k is the outer boundary of R_k. From the definition of a_k and b_k, it is clear that both $f(\sigma_k)$ and $f(\tau_k)$ are contained in $\bar{W}$ but not in W. By (6.3.3) the diameters of $f(\sigma_k)$ and $f(\tau_k)$ approach zero as k goes to infinity. By taking a subsequence we may assume that the sequences $\{f(\sigma_k)\}$ and $\{f(\tau_k)\}$ converge to points p and q of ∂W, respectively. Since y_0 is in W and both p and q are on the boundary of W, the points p and q are distinct from y_0. By a linear change of coordinates, we may assume that

$$w^1(p) \neq w^1(y_0) = 0, \quad w^1(q) \neq w^1(y_0) = 0.$$

If $f(z) = (f^1(z), \ldots, f_n(z))$ is the local expression of f, then

$$\lim_{k \to \infty} f^1(\sigma_k) = w^1(p),$$
$$\lim_{k \to \infty} f^1(\tau_k) = w^1(q),$$
$$\lim_{k \to \infty} f^1(z_k) = w^1(y_0) = 0.$$

It follows that if k is sufficiently large, then

$$f^1(z_k) \notin f^1(\sigma_k) \cup f^1(\tau_k).$$

If k is sufficiently large, we can find a simply connected open neighborhood G_k of $w^1(p)$ in $\mathbf{C}$ such that $f^1(\sigma_k) \subset G_k$ and $f^1(z_k) \notin G_k$. We apply Cauchy's theorem to the holomorphic function $1/(w^1 - f^1(z_k))$ and the closed curve $f^1(\sigma_k)$ in G_k. Then

$$\int_{f^1(\sigma_k)} \frac{dw^1}{w^1 - f^1(z_k)} = 0.$$

This may be rewritten as follows:

$$\int_{\sigma_k} \frac{df^1}{f^1(z) - f^1(z_k)} = 0.$$

Similarly, if k is sufficiently large, then

$$\int_{\tau_k} \frac{df^1}{f^1(z) - f^1(z_k)} = 0.$$

On the other hand, the principle of the argument applied to the function $f^1(z) - f^1(z_k)$ defined in a neighborhood of the annulus R_k with the boundary $\tau_k - \sigma_k$ yields the following equality:

$$\int_{\tau_k} \frac{df^1}{f^1(z) - f^1(z_k)} - \int_{\sigma_k} \frac{df^1}{f^1(z) - f^1(z_k)} = 2\pi i(N - P),$$

where N and P denote the numbers of zeros and poles of $f^1(z) - f^1(z_k)$ in R_k. In the present situation, $P = 0$ and $N \geq 1$. We have finally arrived at a contradiction.
$\square$

The winding number argument used here is due to Grauert and Reckziegel [1]. For related extension theorems, see Carleson [1], where T^*Y is assumed to be ample.

Example (6.1.1) falls into the second case of the following corollary.

(6.3.6) Corollary. *Let Y be a relatively compact hyperbolic complex subspace of a complex space Z. Then every holomorphic map $f \colon D^* \to Y$ satisfies one of the following two conditions:*

(a)   *f extends to a holomorphic map $\bar{f} \colon D \to Y$;*

(b)   *For every neighborhood N of the boundary ∂Y of Y, there exists a small punctured disc $D_\delta^* = \{0 < |z| < \delta\}$ such that $f(D_\delta^*) \subset N$.*

The big Picard theorem can be generalized if Y is not only hyperbolic but hyperbolically imbedded in Z, (Kobayashi [7]).

(6.3.7) Theorem. *If a complex space Y is hyperbolically imbedded in a complex space Z, then every holomorphic map $f \colon D^* \to Y$ extends to a holomorphic map $\bar{f} \colon D \to Z$.*

We shall often denote $\bar{f}$ by f.

Proof. Let $\{r_k\}, 0 < r_k < 1$, be a monotone decreasing sequence with $\lim r_k = 0$. We consider $\{r_k\}$ as a sequence of points in D^* converging to the origin 0. Since $\bar{Y}$ is compact, we may assume (by taking a subsequence) that $\{f(r_k)\}$ converges to a point p_0 of $\bar{Y}$.

In view of (6.3.4) all we have to do is to construct a distance function δ_Z on Z such that $f: (D^*, d_{D^*}) \to (Z, \delta_Z)$ is distance-decreasing. Let $\pi: D \to D^*$ be the covering map. According to (3.3.3), given a length function F on Z there is a constant $c > 0$ such that

$$\pi^* f^* (c^2 F^2) \leq ds_D^2 \qquad \text{for all} \quad f \in \text{Hol}(D^*, Y).$$

Since π is a local isometry in the sense that $\pi^* ds_{D^*}^2 = ds_D^2$, we obtain $\pi^* f^* (c^2 F^2) \leq \pi^* ds_{D^*}^2$ and $f^* (c^2 F^2) \leq ds_{D^*}^2$. Let δ_Z be the distance function defined by cF. $\qquad\square$

For a direct proof which does not make use of (3.3.3), see the original argument in Kobayashi [7].

If we set $Z = P_1 \mathbf{C}$ and $Y = P_1 \mathbf{C} - \{3 \text{ points}\}$, then we obtain the classical big Picard theorem.

In Section 1 of Chapter 5, we introduced the concept of normal holomorphic map and proved that if f is a holomorphic map from D^* into a hyperbolically imbedded space $Y \subset Z$, then f is normal, (see (5.1.23)). The following generalization of (6.3.7) is due to Joseph-Kwack [1].

(6.3.8) Theorem. *Every normal holomoprhic map f of the punctured disc D^* into a complex space Z extends to a holomorphic map $\bar{f}: D \to Z$.*

Proof. Since $f(D^*)$ is relatively compact (from the definition of normal map), condition (b) in (6.3.4) is satisfied. By (5.1.24) condition (a) of (6.3.4) is also satisfied. Now the theorem follows from (6.3.4). $\qquad\square$

If A is a nonsingular hypersurface of a complex manifold X, then in terms of a local coordinate system $(x^1, \ldots, x^n)$ of X it is locally given by $x^1 = 0$. Thus, X is locally biholomorphic to a polydisc D^n, and $X - A$ is then identified with the product $D^* \times D^{n-1}$ of a punctured disc D^* and an $(n-1)$-disc D^{n-1}. More generally, if A consists of hypersurfaces with no worse than normal crossing singularities, then it is locally defined by $x^1 x^2 \ldots x^k = 0$. In this case, $X - A$ is locally given by $(D^*)^k \times D^{n-k}$.

Now, we state a generalization of (6.3.7) to higher dimensional domain spaces, (Kiernan [6]). The proof given here is significantly simpler than the original one.

(6.3.9) Theorem. *Let X be an m-dimensional complex manifold and A a closed complex subspace consisting of hypersurfaces with normal crossing singularities. If a complex space Y is hyperbolically imbedded in a complex space Z, then every map $h \in \text{Hol}(X - A, Y)$ extends to a map $\bar{h} \in \text{Hol}(X, Z)$.*

Proof. By localizing f, we may assume that $X = D^n$ and $X - A = (D^*)^k \times D^{n-k}$. Now the theorem follows from (6.3.7), (3.6.20) and (3.4.19). $\qquad\square$

If Y is compact, the condition on the nature of singularity of A is unnecessary.

(6.3.10) **Theorem.** *Let X be a complex manifold and A a closed complex subspace. If Y is a compact hyperbolic complex space, then every map $h \in \mathrm{Hol}(X - A, Y)$ extends to a map $\bar{h} \in \mathrm{Hol}(X, Y)$.*

Proof. We first note that Y is hyperbolically imbedded in Y itself. Let $S(A)$ be the singular locus of A. Since $A - S(A)$ is a nonsingular submanifold of $X - S(A)$, by (6.3.9) h extends to a holomorphic map $\bar{h} \colon X - S(A) \to Y$. Apply the same argument to the singular locus $S(S(A))$ of $S(A)$ and repeat the process, or apply (6.2.4) since $S(S(A))$ has codimension ≥ 2. $\qquad\square$

The following corollaries are due to Kiernan [6].

(6.3.11) **Corollary.** *Let A be a closed complex subspace of a complex manifold X such that $d_X \equiv 0$, and Y a compact hyperbolic complex space. Then every holomorphic map $h \colon X - A \to Y$ is necessarily constant.*

(6.3.12) **Corollary.** *Let X be a complex manifold such that $d_X \equiv 0$. If A is a closed complex subspace such that $X - A$ is hyperbolic, then $X - A$ cannot be a covering space of a compact complex manifold.*

Proof. Suppose that $X - A$ is a covering space of a compact complex manifold Y. Then Y is hyperbolic by (3.2.8). Now, by (6.3.11) the projection $X - A \to Y$ is constant. $\qquad\square$

In (5.1.11) we proved that $\mathrm{Hol}(D, Y)$ is relatively compact in $\mathrm{Hol}(D, Z)$ if (and only if) Y is hyperbolically imbedded in Z. Making use of (6.3.7) and (5.1.11) we obtain the following results due to Noguchi [10] (see also Lang [3]). The proofs given here are simpler than those of Noguchi.

(6.3.13) **Theorem.** *If a complex space Y is hyperbolically imbedded in Z, then $\mathrm{Hol}(D^*, Y)$ is relatively compact in $\mathrm{Hol}(D, Z)$.*

In (6.3.13) $\mathrm{Hol}(D^*, Y)$ is regarded as a subset of $\mathrm{Hol}(D, Z)$ under the injection $f \in \mathrm{Hol}(D^*, Y) \mapsto \bar{f} \in \mathrm{Hol}(D, Z)$. So (6.3.13) means that for any sequence $\{f_k\} \subset \mathrm{Hol}(D^*, Y)$ a suitable subsequence of $\{\bar{f}_k\} \subset \mathrm{Hol}(D, Z)$ converges.

Proof. For every $f \in \mathrm{Hol}(D^*, Y)$ the extended map $\bar{f} \in \mathrm{Hol}(D, Z)$ belongs to the family $\mathcal{F}_{Y,Z}$ in (5.1.11). Thus, $\mathrm{Hol}(D^*, Y) \subset \mathcal{F}_{Y,Z} \subset \mathrm{Hol}(D, Z)$. But, by (5.1.11), $\mathcal{F}_{Y,Z}$ is relatively compact in $\mathrm{Hol}(D, Z)$. $\qquad\square$

In particular,

(6.3.14) **Corollary.** *Let Y be hyperbolically imbedded in Z. If a sequence $\{f_k\} \subset \mathrm{Hol}(D^*, Y)$ converges to g in $\mathrm{Hol}(D^*, Z)$, then g extends to a holomorphic map $\bar{g} \in \mathrm{Hol}(D, Z)$ and the sequence $\{\bar{f}_k\} \subset \mathrm{Hol}(D, Z)$ converges to $\bar{g}$ in $\mathrm{Hol}(D, Z)$.*

We generalize (6.3.13) to higher dimensional domains.

(6.3.15) **Theorem.** *Let Y be hyperbolically imbedded in Z. Let X be a complex manifold and A a closed complex subspace consisting of hypersurfaces with* *only normal crossing singularities. Then* $\mathrm{Hol}(X - A, Y)$ *is relatively compact in* $\mathrm{Hol}(X, Z)$.

Proof. It suffices to consider the case $X = D^n$ and $X - A = (D^*)^k \times D^{n-k}$. By (6.3.9), every $f \in \mathrm{Hol}(X - A, Y)$ extends to $\bar{f} \in \mathrm{Hol}(X, Z)$. In view of (1.3.2) it suffices to construct a distance function δ_Z on Z such that, for every $f \in \mathrm{Hol}(X - A, Y)$, $\bar{f}$ is a distance-decreasing map of (X, d_X) into (Z, δ_Z), i.e.,

$$\delta_Z(\bar{f}(p), \bar{f}(q)) \leq d_X(p, q) \qquad p, q \in X.$$

It suffices to verify this inequality for $p, q \in X - A$. Given two points

$$p = (a_1, \ldots, a_n), q = (b_1, \ldots, b_n) \in X - A,$$

we consider a chain of points $p = p_0, p_1, \ldots, p_{n-1}, p_n = q$ given by

$$p_i = (b_1, \ldots, b_i, a_{i+1}, \ldots, a_n) \in X - A.$$

Let f_i be the restriction of $\bar{f}$ to the disc

$$D_i = \{(b_1, \ldots, b_{i-1}, z, a_{i+1}, \ldots, a_n); \ z \in D\};$$

by identifying D_i with D, we consider f_i as a map from D in Z such that $f_i(D^*) \subset Y$. Let δ'_Z be the distance function on Z defined by the length function cF in (f) of (5.1.11) so that f_i is a distance-decreasing map of (D, d_D) into (Z, δ'_Z). Now we have

$$\delta'_Z(f(p), f(q)) \ \leq \ \sum_{i=1}^{n} \delta'_Z(f(p_{i-1}), f(p_i)) = \sum_{i=1}^{n} \delta'_Z(f_i(p_{i-1}), f_i(p_i))$$

$$\leq \ \sum_{i=1}^{n} d_D(p_{i-1}, p_i) \leq n \cdot d_{D^n}(p, q).$$

Therefore, it suffices to set $\delta_Z = n\delta'_Z$. $\qquad\qquad\square$

(6.3.16) **Remark.** In the special case where Y is the complement of a Cartier divisor B in Z, if a sequence $\{f_k\} \subset \mathrm{Hol}(X - A, Z - B)$ converges to $g \in \mathrm{Hol}(X, Z)$, then either $g(X - A) \subset Z - B$ or $g(X) \subset B$. This is an immediate consequence of the generalized Hurwicz theorem (3.6.11).

(6.3.17) **Corollary.** *Let Y be hyperbolically imbedded in Z, and let $A \subset X$ be as in (6.3.13). If a sequence $\{f_k\} \subset \mathrm{Hol}(X - A, Y)$ converges to g in $\mathrm{Hol}(X - A, Z)$, then g extends to a holomorphic map $\bar{g} \in \mathrm{Hol}(X, Z)$ and the extended sequence $\{\bar{f}_k\} \subset \mathrm{Hol}(X, Z)$ converges to $\bar{g}$ in $\mathrm{Hol}(X, Z)$.*

(6.3.18) **Remark.** The corollary above implies that $\mathrm{Hol}(X - A, Y)$ is homeomorphic to its image in $\mathrm{Hol}(X, Z)$ under the natural injection $f \mapsto \bar{f}$. We may therefore consider $\mathrm{Hol}(X - A, Y)$ as a subfamily of $\mathrm{Hol}(X, Z)$.

Let f be a meromorphic map from a nonsingular complex manifold X into a complex space Y. Since the singularity set S of f is a closed complex subspace of codimension ≥ 2 in X, the holomorphic map $f: X - S \to Y$ extends to a holomorphic map $X \to Y$ if Y is complete hyperbolic, (see (6.2.4)). In other words, a meromorphic map $f: X \to Y$ is actually holomorphic if X is nonsingular and Y is complete hyperbolic. However, Kodama [2] removed the completeness assumption by utilizing the original proof of (6.3.9) in Kiernan [6].

(6.3.19) **Theorem**. *Every meromorphic map from a nonsingular complex manifold X into a hyperbolic complex space Y is holomorphic.*

Proof. As in the proof of (6.2.3) we may assume that $X = D^m$ and that the singularity set S of f is contained in the subset $\{0\} \times D^{m-1}$. For each $t \in D^{m-1}$, we define a holomorphic maps $f_t: D^* \to Y$ by $f_t(z) = f(z, t)$, $z \in D^*$. Fix t. In order to extend f_t to the origin of D, let $\{z_k\}$ be a sequence of points in D^* converging to 0. Let $\Gamma_f \subset X \times Y$ be the graph of f. Since the projection $\pi: \Gamma_f \to X$ is proper, a suitable subsequence of $\{((z_k, t), f(z_k, t))\} \subset \Gamma_f$ converges to a point, say $((0, t), y_t) \in \Gamma_f$. In other words, a subsequence of $\{f_t(z_k)\}$ converges to $y_t \in Y$. By (6.3.4), f_t extends to a holomorphic map of D into Y. We denote the extended map by the same symbol f_t (so that $f_t(0) = y_t$).

Now, we extend the map $f|_{X-S}$ to $g: X \to Y$ by setting $g(0, t) = f_t(0) = y_t$. It suffices to prove that g is continuous at $(0, t) \in D \times D^{m-1}$. Since D^{m-1} is homogeneous, we may assume that $t = 0$. Put $y_0 = g(0, 0) \in Y$, and let ε be any positive number. Let V (resp. W) be the ε-neighborhood of 0 in D (resp. 0 in D^{m-1}) with respect to d_D (resp. $d_{D^{m-1}}$). We shall show that g maps $V \times W$ into the 3ε-neighborhood of y_0 in Y with respect to d_Y. Let $(z, t) \in V \times W$. Since the restriction of $g|_{D \times \{0\}} = f_0$ is holomorphic, we have

(i) $$d_Y(g(0, 0), g(z, 0)) \leq d_D(0, z) < \varepsilon.$$

First, we consider the case $z \neq 0$. Since $g|_{D^* \times D^{m-1}} = f$ is holomorphic and distance-decreasing, we have

(ii) $$d_Y(g(z, 0), g(z, t)) \leq d_{D^* \times D^{m-1}}((z, 0), (z, t)) = d_{D^{m-1}}(0, t) < \varepsilon,$$

where the equality is a consequence of (3.1.9). Combining (i) and (ii) we have

(iii) $$d_Y(g(0, 0), g(z, t)) < 2\varepsilon,$$

provided that $z \neq 0$.

Now, we consider a point $(0, t) \in V \times W$. Choose any $z \in V$ different from 0. Since $g|_{D \times \{t\}} = f_t$ is holomorphic, we have

(iv) $$d_Y(g(z, t), g(0, t)) \leq d_D(z, 0) < \varepsilon.$$

From (iii) and (iv) we obtain

$$d_Y(g(0, 0), g(0, t)) < 3\varepsilon.$$

This proves that g is continuous at (0,0). $\qquad\square$

Following Weil ([1; p. 27]) we say that a complex space Y is **strongly minimal** if every meromorphic map from a complex space X into Y is holomorphic at every simple (i.e., nonsingular) point of Y. Hence,

(6.3.20) **Corollary.** *Every hyperbolic complex space Y is strongly minimal.*

Although the following result is not realted to the big Picard theorem, it shows that there is another important class of strongly minimal complex spaces.

(6.3.21) **Theorem.** *If Y is a complex space which has a Stein space $\tilde{Y}$ as a covering space, then Y is strongly minimal.*

Proof. We may assume that $\tilde{Y}$ is a closed complex subspace of $\mathbf{C}^N$. Let f be a meromorphic map from a complex manifold X into Y, and let $S \subset X$ be the singularity set of f. We may assume that X is a ball. Since $\mathrm{codim}\,S \geq 2$, $X - S$ is simply connected and the holomorphic map $f : X - S \to Y$ lifts to a holomorphic map $\tilde{f} : X - S \to \tilde{Y} \subset \mathbf{C}^N$. Since $\tilde{f}$ is given by N holomorphic functions, it extends to a holomorphic map $\tilde{f} : X \to \mathbf{C}^N$ by Hartogs' theorem. Since $\tilde{Y}$ is closed, $\tilde{f}$ maps X into $\tilde{Y}$. Now compose $\tilde{f}$ with the projection $\pi : \tilde{Y} \to Y$. $\square$

For a complex space X, let $\mathrm{Aut}(X)$ and (resp. $\mathrm{Bim}(X)$) denote the group of biholomorphic (resp. bimeromorphic) automorphisms of X. In Section 4 of Chapter 5 we discussed properties of $\mathrm{Aut}(X)$. We state another consequence of (6.3.19), (Kodama [2]).

(6.3.22) **Corollary.** *If X is a nonsingular hyperbolic complex manifold, then $\mathrm{Aut}(X) = \mathrm{Bim}(X)$.*

(6.3.23) **Remark.** The equality above does not hold, in general, if X is not hyperbolic even if X is compact and nonsingular. For example if X is the blow-up of $P_n\mathbf{C}$ at a finite number of points, then $\mathrm{Aut}(X)$ is a proper subset of $\mathrm{Bim}(X) \cong \mathrm{Aut}(P_n\mathbf{C})$.

It is essential, in (6.3.19), that X is nonsingular as the following example shows. Let Y be a compact hyperbolic manifold imbedded in $P_N\mathbf{C}$ in such a way that the affine cone $C(Y)$ is a normal complex space. (By definition, $C(Y) \subset \mathbf{C}^{N+1}$ is the union of all complex lines through the origin of $\mathbf{C}^{N+1}$ representing the points of Y.) Then $C(Y)$ is singular only at 0. The natural map $\pi : C(Y) - \{0\} \to Y$ does not extend holomorphically through 0. Resolve the singularity of $C(Y)$, let $C(Y)^*$ be the resulting nonsingular manifold. By (6.3.10), π extends to a holomorphic map from $C(Y)^*$ into Y. Hence, π extends meromorphically to $C(Y)$.

Making use of this example, Kodama [2] constructed an example of complete hyperbolic complex space X such that $\mathrm{Bim}(X) \neq \mathrm{Aut}(X)$. Let U be an open complete hyperbolic neighborhood of 0 in $C(Y)$. Let $\Gamma_\pi \subset C(Y) \times Y$ be the graph of the meromorphic map $\pi : C(Y) \to Y$, and $\pi : \Gamma_\pi \to C(Y)$ the natural projection. We set $V = p^{-1}(U)$; being a closed complex subspace of $U \times Y$ it is complete hyperbolic. We set $X = U \times V$ and define a meromorphic map $f \in \mathrm{Bim}(X)$ by $f(u, v) = (p(v), p^{-1}(u))$, $(u, v) \in U \times V$. Then $f \notin \mathrm{Aut}(X)$.

(6.3.24) Theorem. *Let Y be a complex space hyperbolically imbedded in another complex space Z. Let A be a closed complex subspace of a complex space X. Then every meromorphic map $f \in \mathrm{Mer}(X - A, Y)$ extends to a meromorphic map $\bar{f} \in \mathrm{Mer}(X, Z)$.*

Proof. Let $p: (\tilde{X}, \tilde{A}) \rightarrow (X, A)$ be a resolution of the singularities, where $\tilde{A}$ consists of hypersurfaces with only normal crossing singularities. By (6.3.19) $f \circ p$ is holomorphic. By (6.3.9) $f \circ p$ extends to a holomorphic map $\tilde{X} \rightarrow Z$, which induces a meromorphic map of X into Z. $\qquad\square$

4 Moduli of Maps into Hyperbolically Imbedded Spaces

We start this section by recalling some results from Chapter 5. In Section 3 of Chapter 5 we explained that if X is a compact complex space, then for any complex space Y the family $\mathrm{Hol}(X, Y)$ admits the structure of a complex space with all expected natural properties, called a universal complex structure. From (5.1.1), (5.3.4) and (5.3.8) we have

(6.4.1) Theorem. *Let X and Y be compact complex spaces, and $\mathcal{F}$ a connected component of $\mathrm{Hol}(X, Y)$. Assume that Y is hyperbolic. Then*
   (1) $\mathrm{Hol}(X, Y)$ *is compact.*
   (2) $\mathcal{F}$ *is a compact hyperbolic complex space.*
   (3)   *The evaluation map $\Phi: X \times \mathcal{F} \rightarrow X \times Y$ which maps (x, f) to $(x, f(x))$ is a finite map. In particular, $\dim \mathcal{F} \leq \dim Y$.*

This can be generalized as follows.

(6.4.2) Theorem. *Let X and Z be compact complex spaces. Let B be a Cartier divisor of Z, and set $Y = Z - B$. Let $\mathcal{F}$ be a connected component of $\mathrm{Hol}(X, Y)$, and $\partial \mathcal{F} = \bar{\mathcal{F}} - \mathcal{F}$ its boundary. Assume that Y is hyperbolically imbedded in Z. Then*
   (1)   *The closure $\bar{\mathcal{F}}$ of $\mathcal{F}$ in $\mathrm{Hol}(X, Z)$ is a compact complex subspace of $\mathrm{Hol}(X, Z)$, and the boundary $\partial \mathcal{F}$ is contained in $\mathrm{Hol}(X, B)$.*
   (2)   *$\mathcal{F}$ is complete hyperbolic and hyperbolically imbedded in $\bar{\mathcal{F}}$.*
   (3)   *The boundary $\partial \mathcal{F}$ is a Cartier divisor of $\bar{\mathcal{F}}$.*
   (4)   *The evaluation map $\Phi: X \times \bar{\mathcal{F}} \rightarrow X \times Z$ is a finite map. In particular, $\dim \mathcal{F} \leq \dim Y$.*

Proof. (1)   By (5.1.11), $\mathrm{Hol}(X, Y)$ is relatively compact in $\mathrm{Hol}(X, Z)$. Hence, $\bar{\mathcal{F}}$ is compact.

Let $\mathcal{G}$ be the connected component of $\mathrm{Hol}(X, Z)$ such that $\mathcal{F} \subset \mathcal{G}$. We set

$$\mathrm{Hol}(X, Z)_B = \mathrm{Hol}(X, Z) - \mathrm{Hol}(X, B), \qquad \mathcal{G}_B = \mathcal{G} \cap \mathrm{Hol}(X, Z)_B.$$

We prove that $\mathcal{F}$ coincides with one of the connected components of $\mathcal{G}_B$. Since X is compact, $\mathrm{Hol}(X, Y)$ is open in $\mathrm{Hol}(X, Z)$. Hence, $\mathcal{F}$ is open in $\mathrm{Hol}(X, Z)$, which implies that $\mathcal{F}$ is open in $\mathcal{G}_B$. In order to show that $\mathcal{F}$ is closed in $\mathcal{G}_B$, we consider the boundary $\partial\mathcal{F}$. Let $f \in \partial\mathcal{F}$. By Hurwitz' theorem (3.6.11), we have either $f(X) \subset Y$ or $f(X) \subset B$. Since f is not in $\mathcal{F}$, we must have $f(X) \subset B$. Thus, $\partial\mathcal{F} \subset \mathrm{Hol}(X, B)$. This shows that $\mathcal{F}$ is closed in $\mathrm{Hol}(X, Z)_B$, proving that $\mathcal{F}$ is closed in $\mathcal{G}_B$. Hence, $\mathcal{F}$ coincides with one of the connected components of $\mathcal{G}_B$, and its closure is Zariski closed in $\mathcal{G}$.

(2)   The relative pseudo-distance $d_{Y,Z}$ on Z defined by (3.4.1) is a distance since Y is hyperbolically imbedded in Z, see (3.4.11) and (3.6.20). We define

$$(6.4.3) \qquad \delta_{\mathcal{F}}(f, g) = \sup_{x \in X} d_Y(f(x), g(x)), \qquad f, g \in \mathcal{F},$$

$$(6.4.4) \qquad \delta_{\mathcal{F}, \bar{\mathcal{F}}}(f, g) = \sup_{x \in X} d_{Y,Z}(f(x), g(x)) \qquad f, g \in \bar{\mathcal{F}}.$$

Since d_Y is a distance on Y and $d_{Y,Z}$ is a distance on Z, it follows that $\delta_{\mathcal{F}}$ is a distance on $\mathcal{F}$ and $\delta_{\mathcal{F}, \bar{\mathcal{F}}}$ is a distance on $\bar{\mathcal{F}}$. At each $x \in X$, the evaluation map $\mathcal{F} \to Y$ which sends f to $f(x)$ is holomorphic and distance-decreasing. Hence

$$d_Y(f(x), g(x)) \le d_{\mathcal{F}}(f, g), \qquad f, g \in \mathcal{F},$$

from which

$$(6.4.5) \qquad \delta_{\mathcal{F}}(f, g) \le d_{\mathcal{F}}(f, g), \qquad f, g \in \mathcal{F}.$$

The evaluation at each $x \in X$ maps $\bar{\mathcal{F}}$ into Z, and $\mathcal{F}$ into Y. By (3.4.4) we have

$$(6.4.6) \qquad \delta_{\mathcal{F}, \bar{\mathcal{F}}}(f, g) \le d_{\mathcal{F}, \bar{\mathcal{F}}}(f, g), \qquad f, g \in \bar{\mathcal{F}}.$$

By (3.3.6), Y is complete hyperbolic. Hence, $\delta_{\mathcal{F}}$ is a complete distance from its construction (6.4.3). Hence, (6.4.5) implies that $d_{\mathcal{F}}$ is a complete distance.

Since $d_{Y,Z}$ is a distance on Z by (3.4.20), $\delta_{\mathcal{F}, \bar{\mathcal{F}}}$ is a distance on $\bar{\mathcal{F}}$. By (6.4.6) $d_{\mathcal{F}, \bar{\mathcal{F}}}$ is a distance on $\bar{\mathcal{F}}$. By (3.4.11) $\mathcal{F}$ is hyperbolically imbedded in $\bar{\mathcal{F}}$.

(3)   Fix $f_0 \in \partial\mathcal{F}$. Fix a point $x_0 \in X$, and let $z_0 = f_0(x_0)$. By (1), $z_0 \in B$. Since B is a Cartier divisor, it is locally defined by a single holomorphic function, say φ, in a neighborhood of z_0. Set $\psi(f) = \varphi(\Phi(x_0, f))$ for f belonging to a neighborhood of f_0 in $\bar{\mathcal{F}}$. From (1) we see that $\partial\mathcal{F}$ is locally defined by the holomorphic function ψ.

(4)   Apply (5.3.4) to the compact family $\bar{\mathcal{F}}$. $\square$

Let Z be a compact complex space with a Cartier divisor B such that $Y = Z - B$ is hyperbolically imbedded in Z. Let X be a compact nonsingular complex manifold and A a divisor with only normal crossing singularities. Consider $\mathrm{Hol}(X - A, Y)$ as a subset of $\mathrm{Hol}(X, Z)$ by the natural imbedding

$$\iota : \mathrm{Hol}(X - A, Y) \subset \mathrm{Hol}(X, Z),$$

sending each $f : X - A \to Y$ to its extension $\bar{f} : X \to Z$, (see (6.3.9)). Let $\mathcal{F}$ be a connected component of $\mathrm{Hol}(X - A, Y)$ and $\bar{\mathcal{F}}$ the closure of $\iota(\mathcal{F})$ in $\mathrm{Hol}(X, Z)$.

We write A as a union of irreducible hypersurfaces A_i with only normal crossing singularities:

$$A = \bigcup_{i=1}^{m} A_i .$$

Fixing a map $f \in \mathcal{F}$, we partition the index set $\{1, \ldots, m\}$ into a disjoint union $I \cup J$ as follows:

$$\begin{aligned} \bar{f}(A_i) &\subset B &&\text{for} \quad i \in I, \\ \bar{f}(A_j) &\not\subset B &&\text{for} \quad j \in J. \end{aligned}$$

Set

$$A_I = \bigcup_{i \in I} A_i \quad \text{and} \quad A_J = \bigcup_{j \in J} A_j .$$

We note that although this partition depends on $\mathcal{F}$, it does not depend on the choice of f from $\mathcal{F}$ since $\mathcal{F}$ is connected.

(6.4.7) Lemma. *Let $\mathcal{F}$ and A_I be as above. Then*

$$\bar{f}^{-1}(Y) = X - A_I \quad \text{and} \quad \bar{f}^{-1}(B) = A_I \quad \text{for} \quad f \in \mathcal{F},$$

$$f(X) \subset B \quad \text{for} \quad f \in \partial \mathcal{F} = \bar{\mathcal{F}} - \mathcal{F}.$$

Proof. Let $f \in \mathcal{F}$. Clearly, we have

$$\bar{f}^{-1}(B) = A_I \cup (A_J \cap \bar{f}^{-1}(B)).$$

Since $\bar{f}(A_j) \not\subset B$ for $j \in J$, it follows that $A_J \cap \bar{f}^{-1}(B)$ has codimension ≥ 2 in X. Since Y is complete hyperbolic by (3.3.6), the extension $\bar{f}$ of the map

$$f : X - A = (X - A_I) - (A_{I'} \cap \bar{f}^{-1}(B)) \to Y$$

sends $X - A_I$ into Y by (6.2.4). Thus, $\bar{f}(X - A_I) \subset Y$. Hence, $\bar{f}^{-1}(B) \subset A_I$, and the opposite inclusion follows from $\bar{f}(A_I) \subset B$. Finally, from $\bar{f}(X - A_I) \subset Y$ and $\bar{f}^{-1}(B) = A_I$, we obtain $\bar{f}^{-1}(Y) = X - A_I$.

Let $f \in \partial \mathcal{F}$. Since f is the limit of a sequence of maps belonging to $\mathcal{F}$ and since every element of $\mathcal{F}$ maps $X - A$ into Y, by Hurwitz' theorem (3.6.11) f maps $X - A$ into either Y or B. If it maps $X - A$ into Y, then it must be already in $\mathcal{F}$. Hence, $f(X - A) \subset B$. By continuity, $f(X) \subset B$. $\qquad \square$

(6.4.8) Theorem. *Let Z be an n-dimensional compact complex space with a Cartier divisor B such that $Y = Z - B$ is hyperbolically imbedded in Z. Let X be a compact nonsingular complex manifold and A a divisor with only normal crossing singularities. Let $\mathrm{Hol}(X - A, Y, n)$ be the family of holomorphic maps of $X - A$ to Y of rank n.*

Then $\mathrm{Hol}(X - A, Y, n)$ is compact, and every connected component $\mathcal{F}$ of $\mathrm{Hol}(X - A, Y, n)$ is a compact hyperbolic complex space.

Making use of this proposition, in (6.6.9) we shall show that $\mathcal{F}$ reduces to a single point.

Proof. First, we prove that $\text{Hol}(X - A, Y, n)$ is compact. The family $\text{Hol}(X, Z, n)$ is (open and) closed in $\text{Hol}(X, Z)$ (see (5.3.5)), the closure of $\iota(\text{Hol}(X - A, Y, n))$ in $\text{Hol}(X, Z)$ is contained in $\text{Hol}(X, Z, n)$. Since $\iota(\text{Hol}(X - A, Y))$ is relatively compact in $\text{Hol}(X, Z)$ by (6.3.15), it suffices to show that $\iota(\text{Hol}(X - A, Y, n))$ is closed in $\text{Hol}(X, Z, n)$. Let f be an element of the closure of $\iota(\text{Hol}(X - A, Y))$. By Hurwitz' theorem (3.6.11), we have either $f(X - A) \subset Y$ or $f(X - A) \subset B$. In the first case, f is in $\iota(\text{Hol}(X - A, Y, n))$. In the second case, $f(X) \subset B$ by continuity and the rank of f is at most $n - 1 \geq \dim B$, contradicting $f \in \text{Hol}(X, Z, n)$.

Define $\delta_{\mathcal{F}}$ as in (6.4.3), but taking the supremum over $X - A$ instead of over X. Then we see that $\mathcal{F}$ is hyperbolic. $\qquad\square$

When B is in the singular locus of Z, $\dim B$ can be strictly less than $n - 1$. For example, if Y is a quotient of a symmetric domain G/K by an arithmetic discrete subgroup of G and Z is its Satake compactification, the codimension of the boundary B is usually greater than 1. It is clear from the proof that in such a case, the theorem above holds for maps of rank $k > \dim B$. We state this fact as follows.

(6.4.9) Theorem. *Let Z be an n-dimensional compact complex space with a Cartier divisor B such that $Y = Z - B$ is hyperbolically imbedded in Z. Let X be a compact nonsingular complex manifold and A a divisor with only normal crossing singularities. If k is greater than the dimension of every irreducible component of B, then $\text{Hol}(X - A, Y, k)$ is compact, and every connected component $\mathcal{F}$ of $\text{Hol}(X - A, Y, k)$ is a compact hyperbolic complex space.*

Proof. Much of the argument in the proof of (6.4.8) is still valid in this case. It suffices to show that $\iota(\text{Hol}(X - A, Y, k))$ is closed in $\text{Hol}(X, Z, k)$. Let f be an element of the closure of $\iota(\text{Hol}(X - A, Y, k))$ with $f = \lim \iota(f_\nu)$, $f_\nu \in \text{Hol}(X - A, Y, k)$. Since $k = \text{rank}(f) > \dim B$, it follows that there is a point $x_0 \in X$ such that $f(x_0) \notin B$. Choosing a nearby point if necessary, we may assume that $x_0 \in X - A$. Take a compact neighborhood K of $f(x_0)$ in Y. Then $f_\nu(x_0) \in K$ for all large ν. Since Y is complete hyperbolic, by (5.1.2) $\{f_\nu\}$ is relatively compact in $\text{Hol}(X - A, Y)$. $\qquad\square$

Although (6.4.8) suffices for the proof of (6.6.9), the following result of Noguchi [10] is needed in the proof of (6.5.17), which in turn will be used in proving (6.9.5).

(6.4.10) Theorem. *Let Z be a compact complex space with a Cartier divisor B such that $Y = Z - B$ is hyperbolically imbedded in Z. Let X be a compact nonsingular complex manifold and A a divisor with only normal crossing singularities. Let $\mathcal{F}$ be a connected component of $\text{Hol}(X - A, Z - B)$, $\bar{\mathcal{F}}$ its closure in $\text{Hol}(X, Z)$, and $\partial \mathcal{F} = \bar{\mathcal{F}} - \mathcal{F}$ its boundary. Then*

(1) $\bar{\mathcal{F}}$ is a compact complex subspace of $\text{Hol}(X, Z)$, and $\mathcal{F}$ is a Zariski open subset of $\bar{\mathcal{F}}$ with $\partial \mathcal{F} \subset \text{Hol}(X, B)$.

(2) $\mathcal{F}$ is complete hyperbolic and hyperbolically imbedded in $\bar{\mathcal{F}}$.

(3)   The boundary $\partial \mathcal{F}$ is a Cartier divisor of $\bar{\mathcal{F}}$.

(4)   *The evaluation map $\Phi: X \times \bar{\mathcal{F}} \to X \times Z$ is a finite map. In particular,*
$\dim \bar{\mathcal{F}} \leq \dim Z$.

Part (4) of the theorem will be improved in (6.6.11).

Proof. (1)   Let $\mathcal{G}$ be the connected component of $\mathrm{Hol}(X, Z)$ such that $\iota(\mathcal{F}) \subset \mathcal{G}$.
Set

$$\mathcal{G}_B = \mathcal{G} \cap \mathrm{Hol}(X, Z)_B.$$

Let A_I be as above so that $\bar{f}^{-1}(B) = A_I$ for all $\in \mathcal{F}$. This is a set theoretic
equality. If $\bar{f}$ maps A_i into B with multiplicity m_i, we write

$$\bar{f}^* B = \sum_{i \in I} m_i A_i.$$

This effective divisor $\sum_{i \in I} m_i A_i$ with support in $\bar{f}^{-1}(B) = A_I$ does not depend
on the choice of f from $\mathcal{F}$ because $\mathcal{F}$ is connected.

Let $\mathcal{H}$ be the family of maps $g \in \mathrm{Hol}(X, Z)$ which map each A_i into B with
multiplicity at least m_i for $i \in I$, namely,

$$\mathcal{H} = \{g \in \mathrm{Hol}(X, Z); \ g^* B = \sum_{i \in I} m_i A_i + \ldots\},$$

where the dots indicate an effective divisor (or zero). Then H is Zariski closed in
$\mathrm{Hol}(X, Z)$. Set

$$\mathcal{H}_B = \mathcal{H} \cap \mathrm{Hol}(X, Z)_B.$$

Then $\mathcal{G}_B \cap \mathcal{H}_B$ is Zariski open in $\mathcal{G} \cap \mathcal{H}$. Since $\bar{f} \in \mathcal{H}_B$, we have $\iota(\mathcal{F}) \subset \mathcal{G}_B \cap \mathcal{H}_B$.

We claim that $\iota(\mathcal{F})$ is one of the connected components of $\mathcal{G}_B \cap \mathcal{H}_B$. Let $f \in \mathcal{F}$
and $g \in \mathcal{G} \cap \mathcal{H}_B$. Then

$$g^* B - \bar{f}^* B = g^* B - \sum_{i \in I} m_i A_i$$

is either zero or an effective divisor. It has to be zero if $\bar{f}$ and g are in the same
connected component of $\mathcal{G}_B \cap \mathcal{H}_B$. Then $g^{-1}(B) \subset A_I$ and hence $g(X - A) \subset Y$,
showing that g is an extension of a map beloning to $\mathrm{Hol}(X - A, Y)$. This implies
$g \in \iota(\mathcal{F})$, proving our claim. Hence, its closure $\overline{\iota(\mathcal{F})}$ in $\mathrm{Hol}(X, Z)$ is a Zariski
closed in $\mathcal{G} \cap \mathcal{H}$. By (6.3.15) it is compact.

If $f \in \partial \mathcal{F}$, then by Hurwitz's theorem (3.6.11), we have either $f(X - A) \subset Y$
or $f(X - A) \subset B$. Since f is not in $\mathcal{F}$, we must have $f(X - A) \subset B$. By
continuity, $f(X) \subset B$.

(2)   The proof is the same as in (2) of (6.4.2) except for the following minor
change. We define $\delta_{\mathcal{F}}$ and $\delta_{\mathcal{F}, \bar{\mathcal{F}}}$ as in (6.4.3) and (6.4.4), but taking the supremum
over $X - A$ instead of X. Then as in (6.4.5) and (6.4.6) we have

$$\delta_{\mathcal{F}} \leq d_{\mathcal{F}}(f, g) \quad \text{and} \quad \delta_{\mathcal{F}, \bar{\mathcal{F}}} \leq d_{\mathcal{F}, \bar{\mathcal{F}}}.$$

(3) and (4)   The proof is the same as in (3) and (4) of (6.4.2). $\square$

5 Hyperbolic and Hyperbolically Imbedded Fiber Spaces

A complex fiber space (Y, π, X) consists of complex spaces X, Y and a surjective holomorphic map $\pi : Y \to X$. Throughout this section we assume that both X and Y are irreducible. For a point $x \in X$ and an open set $U \subset X$, we put

$$Y_x = \pi^{-1}(x), \qquad \text{and} \qquad Y_U = \pi^{-1}(U),$$

$$\Gamma(U, Y) = \{ f \in \text{Hol}(U, Y); \ \pi(f(x)) = x \in U \}.$$

A complex fiber space (Y, π, X) is said to be (complete) hyperbolic at $x \in X$ if x has a neighborhood U such that Y_U is (complete) hyperbolic. It is called **(complete) hyperbolic** if it is (complete) hyperbolic at every $x \in X$. The concept of taut fiber space is closely related to that of relative hyperbolicity in Ohgai [1]. Similarly, a fiber space (Y, π, X) is said to be taut at $x \in X$ if x has a neighborhood U such that Y_U is taut. It is called **taut** if it is taut at every $x \in X$.

The concept of hyperbolic or taut imbedding can be also generalized to fiber spaces. Let (Z, π, X) be a complex fiber space Let Y be a complex subspace (often a domain) of Z such that $X = \pi(Y)$. Then (Y, π, X) is also complex fiber space over X, called a **subfiber space** of (Z, π, X). The subfiber space (Y, π, X) is **hyperbolically** (resp. **tautly) imbedded in** (Z, π, X) if every $x \in X$ has a neighborhood U such that Y_U is hyperbolically (resp. tautly) imbedded in Z_U.

If a complex space Y is (complete) hyperbolic (resp. taut), then the product fiber space $(Y \times X, \pi, X)$ is (complete) hyperbolic (resp. taut) in the sense defined above. Similarly, if Y is hyperbolically (resp. tautly) imbedded in a complex space Z, then the product fiber space $(Y \times X, \pi, X)$ is hyperbolically (resp. tautly) imbedded in the product fiber space $(Z \times X, \pi, X)$.

Let A be a closed subset of a compact complex space X. (We are mostly interested in the case where A is a closed complex subspace of X). We say that compact complex fiber space (Y, π, X) is **hyperbolic over** (X, A) if every point $x \in X$ has a neighborhood U such that $Y_{U-(A \cap U)}$ is hyperbolically imbedded in Y_U. We note that in this case $(Y_{X-A}, \pi, X - A)$ is a hyperbolic fiber space but the the fiber Y_x over $x \in A$ may not be hyperbolic. If A is empty, this amounts to saying that (Y, π, X) is hyperbolic.

We say that a subfiber space (Y, π, X) of a complex fiber space (Z, π, X) is **hyperbolically imbedded in** (Z, π, X) **over** (X, A) if every point $x \in X$ has a neighborhood U such that $Y_{U-(A \cap U)}$ is hyperbolically imbedded in Z_U. If A is empty, this simply means that (Y, π, X) is hyperbolically imbedded in (Z, π, X).

Clearly, if (Y, π, X) is (complete) hyperbolic (resp. taut), then every fiber Y_x is (complete) hyperbolic (resp. taut). If (Y, π, X) is hyperbolically (resp. tautly) imbedded in (Z, π, X), then every fiber Y_x is hyperbolically (resp. tautly) imbedded in Z_x.

In (3.11.1) we proved that if (Y, π, X) is a complex fiber space with compact fibers, then hyperbolicity of Y_x implies hyperbolicity of (Y, π, X) at x. For a similar statement on hyperbolically imbedded fiber spaces, see (3.9.4).

From (5.1.11) we have

(6.5.1) Theorem. *For a subfiber space (Y, π, X) of a complex fiber space (Z, π, X) the following conditions (a) and (b) are equivalent:*

    (a)   *(Y, π, X) is tautly imbedded in (Z, π, X);*
    (b)   *(Y, π, X) is hyperbolically imbedded in (Z, π, X).*

In order to reduce problems to the case where X is nonsingular, we shall later use the following proposition when $p: X' \to X$ is a resolution of singularities

(6.5.2) Proposition. *Let (Y, π, X) and (Z, π, X) be complex fiber spaces, and $p: X' \to X$ be a holomorphic map. Let A be a (possibly empty) closed subset of X.*

    (1)   *If (Y, π, X) is hyperbolic over (X, A), then its pull-back $(p^{-1}Y, \pi', X')$ is hyperbolic over $(X', p^{-1}(A))$.*

    (2)   *If (Y, π, X) is hypebolically imbedded in (Z, π, X) over (X, A), then its pull-back $(p^{-1}Y, \pi', X')$ is hyperbolically imbedded in the pull-back $(p^{-1}Z, \pi', X')$ over $(X', p^{-1}(A))$.*

Proof. (1)   For the sake of notational simplicity, we assume that A is empty. There is essentially no difference in the proof whether A is empty or not. Set $Y' = p^{-1}Y$. Given a point $x' \in X'$, let U be a neighborhood of $p(x')$ in X such that Y_U is hyperbolic. Let U' be a hyperbolic neighborhood of x' in X' such that $U' \subset p^{-1}(U)$. We shall show that $Y'_{U'}$ is hyperbolic. Let $y, y' \in Y'_{U'}$ be distinct points. If $\pi'(y) \neq \pi'(y')$, then $d_{Y'_{U'}}(y, y') \geq d_{U'}(\pi'(y), \pi'(y')) > 0$. If $\pi'(y) = \pi'(y')$, then $d_{Y'_{U'}}(y, y') \geq d_{Y_U}(p(y), p(y')) > 0$. This shows that $Y'_{U'}$ is hyperbolic.

The proof for (2) is similar.   □

The following proposition will be used when $q: \tilde{Y} \to Y$ is the normalization of Y.

(6.5.3) Proposition. *Let A be a (possibly empty) closed subset of X.*

    (1)   *Let (Y, π, X) and $(\tilde{Y}, \tilde{\pi}, X)$ be complex fiber spaces with a surjective finite map $q: \tilde{Y} \to Y$ such that $\tilde{\pi} = \pi \circ q$. If (Y, π, X) is a hyperbolic over (X, A), so is $(\tilde{Y}, \tilde{\pi}, X)$.*

    (2)   *Let $(Y, \pi, X) \subset (Z, \pi, X)$ and $(\tilde{Y}, \tilde{\pi}, X) \subset (\tilde{Z}, \tilde{\pi}, X)$ be subfiber spaces. Let $q: \tilde{Z} \to Z$ be a proper finite map such that $\tilde{\pi} = \pi \circ q$. If (Y, π, X) is hyperbolically imbedded in (Z, π, X) over (X, A), then $(\tilde{Y}, \tilde{\pi}, X)$ is hyperbolically imbedded in $(\tilde{Z}, \tilde{\pi}, X)$ over (X, A).*

Proof. (1) follows from (3.2.11) while (2) follows from (3.3.7).   □

(6.5.4) Theorem. *Let (Z, π, X) be a fiber space with compact fibers. Let B be a Cartier divisor in Z. Let $Y = Z - B$. If (Y, π, X) is hyperbolically imbedded in (Z, π, X), then it is complete hyperbolic.*

Proof. Although we cannot directly apply (3.3.15) to the present situation since Z here may not be compact, we can use the proof of (3.3.15). Namely, we observe the fiber space version of (3.3.14) combined with (3.2.18) implies (6.5.4). $\qquad\square$

(6.5.5) Theorem. (1) *If (Y, π, X) is hyperbolic with compact fibers, then $\Gamma(X, Y)$ is compact.*

(2) *If (Y, π, X) is hyperbolically imbedded in (Z, π, X), then $\Gamma(X, Y)$ is relatively compact in $\Gamma(X, Z)$.*

We emphasize that X need not be compact here.

Proof. It suffices to prove (2). Let $\{U_j\}$ be a countable open cover of X such that each U_j is complete hyperbolic and each Y_{U_j} is hyperbolically imbedded in Z_{U_j}. By (5.1.11) $\Gamma(U_j, Y)$ is relatively compact in $\Gamma(U_j, Z)$. Given an infinite sequence $\{f_k\} \subset \Gamma(X, Y)$, we extract a subsequence which converges on U_1 to a section of Z over U_1. Then extract a subsequence from this subsequence which converges on U_2 to a section of Z over U_2, and so on. Then the diagonal subsequence converges to a section of Z over X. $\qquad\square$

The following example due to Zaidenberg [7] shows that the theorem would fail if even one fiber is not hyperbolic.

(6.5.6) Example. Let $X = P_1\mathbf{C}$, W a compact Riemann surface of genus ≥ 2, and $X \times W$ a product bundle over X with fiber W. Let Y be the space obtained by blowing up a point $(x_0, w_0) \in X \times W$, and $\sigma: Y \to X \times W$ be the blow-up. Then we have a compact complex fiber space (Y, π, X) with a singular fibre at x_0. The singular fiber is connected but reducible, with irreducible components $W_0 \cong W$ and $E = \sigma^{-1}((x_0, w_0)) \cong P_1\mathbf{C}$; E may be considered as the projective space consisting of all tangential directions of $X \times W$ at (x_0, w_0), and W_0 intersects E at the point corresponding to the fiber direction at (x_0, w_0). The constant section $f_0: X - \{x_0\} \to Y$ given by $f_0(x) = w_0$ extends to a section $f_0 \in \Gamma(Y)$, and $f_0(x_0)$ is the point of E corresponding to the horizontal direction of $X \times W$ at (x_0, w_0), which is different from the point $E \cap W_0$. It follows that $\Gamma(Y)$ consists of an isolated point $\{f_0\}$ and a noncompact set $\mathrm{Hol}(X, W - \{w_0\}) \cong W - \{w_0\}$. We note that although there is a natural one-to-one correspondence $\Gamma(X, Y) \to \Gamma(X, X \times W)$ (actually a holomorphic map), these two spaces differ as complex space.

(6.5.7) Theorem. *Let X be a nonsingular complex manifold, and A a divisor with only normal crossing singularities.*

(1) *If (Y, π, X) has compact fibers and is hyperbolic over (X, A), then $\Gamma(X - A, Y) \cong \Gamma(X, Y)$ in a natural way and is compact.*

(2) *If (Y, π, X) is hyperbolically imbedded in (Z, π, X) over (X, A), then $\Gamma(X - A, Y)$ is relatively compact in $\Gamma(X, Z)$.*

As in (6.5.5) we are not assuming that X is compact here.

Proof. It suffices to prove (2). Let $f \in \Gamma(X - A, Y)$. By (6.3.9), f extends to a map $\bar{f} \in \mathrm{Hol}(X, Z)$. Clearly, $\bar{f} \in \Gamma(X, Z)$.

Let $\{U_j\}$ be a countable open cover of X such that each U_j is complete hyperbolic and $Y_{U_j-(A\cap U_j)}$ is hyperbolically imbedded in Z_{U_j}. By (6.3.15), $\Gamma(U_j - (A \cap U_j), Y)$ is relatively compact in $\Gamma(U_j, Z)$. Given a sequence $\{f_k\}$ in $\Gamma(X - A, Y)$, as in the proof of (6.5.5) we can extract a converging subsequence from the sequence $\{\bar{f}_k\} \subset \Gamma(X, Z)$ of extended sections. $\qquad\square$

By Douady [1] (see Section 3 of Chapter 5), if X is compact, then $\mathrm{Hol}(X, Y)$ has a universal complex structure. In this case, $\Gamma(X, Y)$ is a closed complex subspace of $\mathrm{Hol}(X, Y)$. If X is not compact, neither $\mathrm{Hol}(X, Y)$ nor $\Gamma(X, Y)$ may admit complex structures. However, $\Gamma(X, Y)$ may contain a connected subset S which carries a **universal** complex structure, i.e., a complex structure such that

(i)   the evaluation map $\Phi: X \times S \to Y$ is holomorphic and

(ii)   if T is a complex space and $\varphi: X \times T \to Y$ is a holomorphic map such that $\varphi(\cdot, t) \in S$, then the map $\tilde{\varphi}: T \to S$ defined by $\tilde{\varphi}(t) = \varphi(\cdot, t)$ is holomorphic.

Instead of assuming that X is compact, we consider the situation where $\Gamma(X, Y)$ has a subfamily S with a universal complex structure. Let $S \subset \Gamma(X, Y)$ be a connected subfamily with a universal complex structure. We define a pseudo-distance δ_S on S by setting

$$(6.5.8) \qquad \delta_S(f, g) = \sup_{x \in X} d_{Y_x}(f(x), g(x)), \qquad f, g \in S.$$

For each fixed $x \in X$ the distance-decreasing property of a holomorphic map $S \to Y_x$ sending f to $f(x)$ implies

$$d_{Y_x}(f(x), g(x)) \le d_S(f, g), \qquad f, g \in S.$$

Hence,

$$(6.5.9) \qquad\qquad\qquad \delta_S(f, g) \le d_S(f, g).$$

(6.5.10) Proposition. *Let (Y, π, X) be a complex fiber space, and $S \subset \Gamma(X, Y)$ a connected subfamily with a universal complex structure.*

(1)   *If (Y, π, X) is hyperbolic at a point $x_0 \in X$, then S is hyperbolic;*

(2)   *If (Y, π, X) is complete hyperbolic, then S is complete hyperbolic.*

Proof. (1)   Let $f, g \in S$ be such that $d_S(f, g) = 0$. By (6.5.9), $d_{Y_x}(f(x), g(x)) = 0$ for all $x \in X$. Since Y_x is hyperbolic for all x in a neighborhood U of x_0, we have $f(x) = g(x)$ for $x \in U$. Hence, $f = g$.

(2)   Let $\{f_n\}$ be a Cauchy sequence in S with respect to d_S. Then

$$d_{Y_x}(f_m(x), f_n(x)) \le d_S(f_m, f_n) \to 0 \qquad \text{as} \quad m, n \to \infty.$$

Hence, $f_n \to g \in S$, where $g(x) = \lim f_n(x)$. $\qquad\square$

(6.5.11) Proposition. *Let (Y, π, X) be a compact hyperbolic complex fiber space. Then, for any $y_0 \in Y$, there are only finitely many sections passing through y_0.*

In particular, $\dim \Gamma(X, Y) \le \dim Y_x$ *for all $x \in X$.*

Proof. Since $\Gamma(X, Y)$ is compact by (6.5.5), this follows from (5.3.4). $\qquad\square$

If X is compact, then by Douady's theorem $\Gamma(X, Y)$ has a universal complex structure. If (Y, π, X) is a compact hyperbolic complex fiber space, then $\Gamma(X, Y)$ is compact by (6.5.5). Now, making use of (6.5.10) and (6.5.11) we obtain

(6.5.12) Theorem. *Let (Y, π, X) be a compact hyperbolic complex fiber space. Then*

(1)   *$\Gamma(X, Y)$ is a compact complex space.*

(2)   *Each connected components S of $\Gamma(X, Y)$ is a compact hyperbolic complex space.*

(3)   *The evaluation map $\Phi\colon X \times \Gamma(X, Y) \to Y$ is a finite map. In particular, $\dim \Gamma(X, Y) \le \dim Y_x$ for any $x \in X$.*

The following theorem generalizes (6.5.12).

(6.5.13) Theorem. *Let X be a nonsingular compact complex manifold, and A a divisor with only normal crossing singularities. Let (Y, π, X) be a compact complex fiber space hyperbolic over (X, A). Then*

(1)   *$\Gamma(X - A, Y) \cong \Gamma(X, Y)$ is a compact complex space.*

(2)   *Each connected component S of $\Gamma(X, Y)$ is a compact hyperbolic complex space.*

(3)   *The evaluation map $\Phi\colon X \times \Gamma(X, Y) \to Y$ is a finite map. In particular, $\dim \Gamma(X, Y) \le \dim Y_x$ for any $x \in X$.*

Proof. (1)   By (6.5.7), $\Gamma(X - A, Y)$ is naturally isomorphic to $\Gamma(X, Y)$ and is compact. Since X is compact, $\Gamma(X, Y)$ has a universal complex structure.

(2)   By (6.5.10), S is hyperbolic.

(3)   This follows from (5.3.4). $\square$

(6.5.14) Theorem. *Let (Z, π, X) be a compact complex fiber space, and B a Cartier divisor of Z transversal to the fibers in the sense that, at each $x \in X$, $B \cap Z_x$ is a Cartier divisor of the fiber Z_x. Set $Y = Z - B$. Assume that (Y, π, X) is hyperbolically imbedded in (Z, π, X). Let S be a connected component of $\Gamma(X, Y)$. Then*

(1)   *The closure $\bar{S}$ of S in $\Gamma(X, Z)$ is a compact complex space.*

(2)   *The boundary $\partial S = \bar{S} - S$ is a Cartier divisor of $\bar{S}$.*

(3)   *S is complete hyperbolic and hyperbolically imbedded in $\bar{S}$.*

(4)   *The evaluation map $\Phi\colon X \times \bar{S} \to Z$ is finite. In particular, $\dim \bar{S} \le \dim Z_x$ for any $x \in X$.*

Proof. (1)   By (6.5.7), $\bar{S}$ is compact. Let $\mathcal{T}$ be the connected component of the complex space $\Gamma(X, Z)$ such that $S \subset \mathcal{T}$. We set

$$\Gamma(X, Z)_B = \Gamma(X, Z) - \Gamma(X, B), \qquad \mathcal{T}_B = \mathcal{T} \cap \Gamma(X, Z)_B.$$

We prove that S coincides with one of the connected components of $\mathcal{T}_B$. Since X is compact, $\Gamma(X, Y)$ is open in $\Gamma(X, Z)$. Hence, S is open in $\Gamma(X, Z)$, which implies that S is an open subset of $\mathcal{T}_B$. In order to show that S is closed in $\mathcal{T}_B$, we consider the boundary ∂S of S. Let $f \in \partial S$. By Hurwitz' theorem (3.6.11), either

$f(X) \subset Y$ or $f(X) \subset B$. Since f is not in $\mathcal{S}$, $f(X) \subset B$. Thus, $\partial \mathcal{S} \subset \mathrm{Hol}(X, B)$. This shows that $\mathcal{S}$ is closed in $\Gamma(X, Z)_B$, proving that $\mathcal{S}$ is closed in $\mathcal{T}_B$. Hence, $\mathcal{S}$ coincides with one of the connected components of $\mathcal{T}_B$, and its closure is Zariski closed in $\mathcal{T}$.

(2)   The proof is almost the same as that of (6.4.7).

(3)   Since Y_x is hypebolically imbedded in Z_x, using the relative intrinsic distance d_{Y_x, Z_x} on Z_x, we define a distance function $\delta_{\mathcal{S}, \bar{\mathcal{S}}}$ on $\bar{\mathcal{S}}$ by

$$(6.5.15) \qquad \delta_{\mathcal{S}, \bar{\mathcal{S}}}(f, g) = \sup_{x \in X} d_{Y_x, Z_x}(f(x), g(x)), \qquad f, g \in \bar{\mathcal{S}}.$$

The argument giving (6.4.5) and (6.4.6) gives also the following:

$$(6.5.16) \qquad \delta_{\mathcal{S}, \bar{\mathcal{S}}}(f, g) \leq d_{\mathcal{S}, \bar{\mathcal{S}}}(f, g), \qquad f, g \in \bar{\mathcal{S}},$$

showing that $d_{\mathcal{S}, \bar{\mathcal{S}}}$ is a distance. By (3.4.11), $\mathcal{S}$ is hyperbolically imbedded in $\bar{\mathcal{S}}$. Now, (2) and (3.3.6) imply that $\mathcal{S}$ is complete hyperbolic. $\square$

Finally, we consider the most general case.

(6.5.17) **Theorem.** *Let X be a compact nonsingular complex manifold and A a divisor with only normal crossing singularities. Let (Z, π, X) be a compact complex fiber space, and B a Cartier divisor of Z transversal to the fibers. Set $Y = Z - B$. Let $\mathcal{S}$ be a connected component of $\Gamma(X - A, Y)$. Assume that (Y, π, X) is hyperbolically imbedded in (Z, π, X) over (X, A). By the natural imbedding*

$$\iota \colon \Gamma(X - A, Y) \subset \Gamma(X, Z),$$

sending each $f \in \Gamma(X - A, Y)$ to its extension $\bar{f} \in \Gamma(X, Z)$, we consider $\mathcal{S}$ as a subset of $\Gamma(X, Z)$. Let $\bar{\mathcal{S}}$ be its closure in $\Gamma(X, Z)$. Then

(1)   *$\bar{\mathcal{S}}$ is a compact complex subspace of $\Gamma(X, Z)$, and $\mathcal{S}$ is a Zariski open subset of $\bar{\mathcal{S}}$.*

(2)   *$\mathcal{S}$ is complete hyperbolic and hyerpbolic imbedded in $\bar{\mathcal{S}}$.*

(3)   *The boundary $\partial \mathcal{S} = \bar{\mathcal{S}} - \mathcal{S}$ is a Cartier divisor of $\bar{\mathcal{S}}$.*

(4)   *The evaluation map $\Phi \colon X \times \bar{\mathcal{S}} \to Z$ is finite. In particular, $\dim \bar{\mathcal{S}} \leq \dim Z_x$ for any $x \in X$.*

Proof. (1)   The proof of (1) is almost identical to that of (1) of (6.4.10). Now we shall sketch the proof, leaving the details to the proof of (6.4.10). Let $A = \bigcup_{i=1}^{m} A_i$ be the decomposition into irreducible components. Using a section $f \in \mathcal{S} \subset \Gamma(X - A, Y)$, we partition the index set $\{1, \ldots, m\}$ into a disjoint union $I \cup J$ in such a way that

$$\begin{aligned} \bar{f}(A_i) &\subset B &&\text{for} \quad i \in I, \\ \bar{f}(A_j) &\not\subset B &&\text{for} \quad j \in J, \end{aligned}$$

and set

$$A_I = \bigcup_{i \in I} A_i \quad \text{and} \quad A_J = \bigcup_{j \in I'} A_j.$$

This partition depends only on $\mathcal{S}$, not on f.

With the same proof as in (6.4.7) we have

$$\bar{f}^{-1}(Y) = X - A_I \quad \text{and} \quad \bar{f}^{-1}(B) = A_I \quad \text{for} \quad f \in \mathcal{S},$$

$$f(X) \subset B \qquad \text{for} \quad f \in \partial\mathcal{S} = \bar{\mathcal{S}} - \mathcal{S}.$$

We set

$$\Gamma(X, Z)_B = \Gamma(X, Z) - \Gamma(X, B).$$

Since $\Gamma(X, B)$ is Zariski closed in $\Gamma(X, Z)$, $\Gamma(X, Z)_B$ is Zariski open.

Exactly as in the proof of (6.4.10), we show that $\mathcal{S}$ is a closed complex subspace of $\Gamma(X, Z)_B$. Then its closure $\bar{\mathcal{S}}$ in $\Gamma(X, Z)$ is a closed complex subspace of $\Gamma(X, Z)$ because $\Gamma(X, Z)_B$ is Zariski open in $\Gamma(X, Z)$.

The proof for (2), (3) and (4) is almost the same as that of (6.4.10). $\square$

Let (Y, π, X) with $Y = Z - B$ be as in (6.5.16). Given a connected component $\mathcal{S}$ of $\Gamma(X - A, Y)$, we define a distance function $\delta_{\mathcal{S}}$ on $\mathcal{S}$ in the same way as (6.4.3) (the only difference from $\delta_{\mathcal{S}}$ in (6.5.8) being that $X - A$ is used as the base space here):

$$(6.5.18) \qquad \delta_{\mathcal{S}}(f, g) = \sup_{x \in X - A} d_{Y_x}(f(x), g(x)). \qquad f, g \in \mathcal{S}.$$

Since Y_x is hypebolically imbedded in Z_x, using the relative intrinsic distance d_{Y_x, Z_x} on Z_x, we define a distance function $\delta_{\mathcal{S}, \bar{\mathcal{S}}}$ on $\bar{\mathcal{S}}$ by

$$(6.5.19) \qquad \delta_{\mathcal{S}, \bar{\mathcal{S}}}(f, g) = \sup_{x \in X - A} d_{Y_x, Z_x}(f(x), g(x)), \qquad f, g \in \bar{\mathcal{S}}.$$

The argument giving (6.4.5) and (6.4.6) gives also the following:

$$(6.5.20) \qquad \delta_{\mathcal{S}}(f, g) \le d_{\mathcal{S}}(f, g), \qquad f, g \in \mathcal{S},$$

$$(6.5.21) \qquad \delta_{\mathcal{S}, \bar{\mathcal{S}}}(f, g) \le d_{\mathcal{S}, \bar{\mathcal{S}}}(f, g), \qquad f, g \in \bar{\mathcal{S}}.$$

When the base space X has singularities, it is more natural to consider not only holomorphic but meromorphic sections. A **meromorphic section** f of a complex fiber space (Y, π, X) is a correspondence satisfying the following conditions:

(i) For each $x \in X$, $f(x)$ is a nonempty compact complex subspace of Y_x;

(ii) The graph $f(X) = \bigcup_{x \in X} f(x)$ is a connected complex subspace of Y with $\dim f(X) = \dim X$;

(iii) If we set

$$S = \{x \in X; \ \dim f(x) > 0\} \quad \text{and} \quad E = \pi^{-1}(S),$$

then S is a closed complex subspace of codimension ≥ 2 in X, and E is a closed complex subspace of codimension ≥ 1 in Y.

It is clear that f is single valued and holomorphic on $X - S$. We call S the **singular locus** of f. This generalizes the concept of meromorphic map explained in Section 4 of Chapter 2. (We changed the definition here a little for the sake of simplicity).

Let $\Gamma^*(Y) = \Gamma^*(X, Y)$ denote the set of meromorphic sections of (Y, π, X). If X is compact and n-dimensional, $\Gamma^*(X, Y)$ is a closed complex subspace of the Douady space of compact n-dimensional complex subspaces of Y.

(6.5.23) **Proposition.** *Let (Y, π, X) be a hyperbolic fiber space. Let $p: X' \to X$ be a resolution of singularities of X, and $(p^{-1}Y, \pi', X')$ the pull-back of (Y, π, X) by p. Then the set $\Gamma(X', p^{-1}Y)$ of holomorphic sections and the set $\Gamma^*(X, Y)$ of meromorphic sections are in a natural one-to-one correspondence.*
Proof. Let $f \in \Gamma^*(X, Y)$. Then $f \circ p$ defines a meromorphic section of $(p^{-1}Y, \pi', X')$. Since $(p^{-1}Y, \pi', X')$ is a hyperbolic fiber space by (6.5.4), the meromorphic section $f \circ p$ is holomorphic by (6.3.19).

Conversely, if $f' \in \Gamma(X, p^{-1}Y)$, then by setting $f(x) = p(f'(p^{-1}(x))) \subset Y_x$ for each $x \in X$ we obtain the corresponding meromorphic section $f \in \Gamma^*(X, Y)$.
$\square$

(6.5.24) **Proposition.** *Let (Y, π, X) be a complex fiber space with X normal. Let $q: \tilde{Y} \to Y$ be the normalization of Y. Let N be the locus of nonnormal points of Y. Consider the complex fiber space $(\tilde{Y}, \tilde{\pi}, X)$ with $\tilde{\pi} = \pi \circ q$. Every section $\tilde{f} \in \Gamma(\tilde{Y})$ induces a section $q \circ \tilde{f} \in \Gamma(Y)$. Conversely, if $f \in \Gamma(Y)$ is a section such that $f(X) \not\subset N$, then there is a section $\tilde{f} \in \Gamma(\tilde{Y})$ such that $f = q \circ \tilde{f}$.*

Proof. The first assertion is trivial. Let $X_0 = \{x \in X; \ f(x) \notin N\}$, which is Zariski open in X. Then $q^{-1} \circ f$ is a holomorphic map of X_0 into $\tilde{Y}$ and is weakly holomorphic on X. Since X is normal, it extends to a holomorphic map $\tilde{f}$ of X into $\tilde{Y}$.
$\square$

6 Surjective Maps to Hyperbolic Spaces

After this section, the reader may proceed to Section 9 skipping Sections 7 and 8.

A meromorphic map f of a complex space X into another complex space Y is said to be **dominant** if its graph $G_f \subset X \times Y$ maps onto Y under the natural projection $X \times Y \to Y$. If f is a holomorphic map, this means that f is surjective. If X and Y are compact and irreducible, f is dominant if and only if there is a regular point $x \in X$ such that f is holomorphic at x, $f(x)$ is a regular point of Y, and the differential $f_*: T_x X \to T_{f(x)} Y$ is surjective.

The classical theorem of de Franchis [1] (see also Samuel [1] and Martens [1], and for a generalization to finite Riemann surfaces, Imayoshi [1]) states

(6.6.1) **Theorem.** *Let X and Y be compact Riemann surfaces. If the genus $g(Y)$ of Y is greater than or equal to 2, then the number of surjective holomorphic maps from X to Y is finite.*

The theorem of Severi allows Y to vary. Namely, it says that for a fixed X the number of pairs (Y, f) consisting of compact Riemann surfaces Y with $g(Y) \geq 2$

and surjective holomorphic maps $f: X \to Y$ is finite, (see Samuel [1], Imayoshi [2], Howard-Sommese [2], and Kani [1]). This will be discussed in Section 6 of Chapter 7.

The first generalization of (6.6.1) to higher dimensional varieties was in the case where Y is a compact complex space of general type, (Kobayashi-Ochiai [3]), and it will be proved in Section 6 of Chapter 7.

In this section we shall first prove the following generalization (6.6.2) conjectured by Lang [1]. Theorem (6.6.2) was proved first under additional assumptions by several authors: by Urata [3] (when Y is Carathéodory-hyperbolic or carries a Hermitian metric of negative holomorphic sectional curvature), by Kalka-Shiffman-Wong [1] (also when Y admits a Hermitian metric of negative holomorphic sectional curvature), by Noguchi [7] (when Y is a hyperbolic Kähler manifold with semi-positive canonical bundle), by Horst [4] (when Y is hyperbolic Kähler), and finally by Noguchi [13] with no extra conditions. After we prove (6.6.2) we shall present also differential geometric results of Urata and Kalka-Shiffman-Wong mentioned above. Mappings of lower ranks will be discussed in Sections 7 and 8.

Let X and Y be compact complex spaces. The set of dominant meromorphic maps from X to Y will be denoted $\mathrm{Dom}(X, Y)$ while $\mathrm{Sur}(X, Y) \subset \mathrm{Dom}(X, Y)$ stands for the set of surjective holomorphic maps from X to Y.

(6.6.2) **Theorem.** *Let X and Y be compact, irreducible complex space. If Y is hyperbolic, then* $\mathrm{Dom}(X, Y)$ *is finite.*

We shall first prove that $\mathrm{Sur}(X, Y)$ is finite.

(6.6.3) **Lemma.** *Let X and Y be compact, irreducible complex spaces. If Y is hyperbolic, then* $\mathrm{Sur}(X, Y)$ *is a compact complex space.*

Proof. Since X is compact, $\mathrm{Hol}(X, Y)$ has a natural complex structure. Since Y is compact and hyperbolic, $\mathrm{Hol}(X, Y)$ is compact by (5.3.9). By (5.3.5) $\mathrm{Sur}(X, Y)$ is (open and) closed in $\mathrm{Hol}(X, Y)$. $\qquad\square$

Let $\mathcal{S}$ be a connected component of $\mathrm{Sur}(X, Y)$. It suffices to show that $\mathcal{S}$ is a singleton. We apply the simultaneous Stein factorization (5.3.2) to the family $\mathcal{S}$:

$$ X \xrightarrow{\ p\ } X' \xrightarrow{\ f'\ } Y, \qquad f \in \mathcal{S}. $$

Since the map $f \mapsto f'$ injects $\mathcal{S}$ into $\mathrm{Sur}(X', Y)$ and since f' is a finite map, it suffices to show that $\mathrm{Sur}(X', Y)$ contains only a finite number of finite maps. Thus, the proof of (6.6.2) is reduced to showing the following.

If X and Y are irreducible and compact and if Y is hyperbolic, then there are only a finite number of finite surjective holomorphic maps of X onto Y.

If there is a finite surjective holomorphic map from X to Y, then $\dim X = \dim Y$ and by (3.2.11) Y is also hyperbolic. We may also replace X and Y by their normalization. We shall therefore assume that *X and Y are compact irreducible hyperbolic normal complex spaces with* $\dim X = \dim Y$.

Let $\mathcal{F}$ be an irreducible closed complex subspace of $\mathrm{Hol}(X, Y)$ consisting of surjective finite maps. Assuming that $\dim \mathcal{F} > 0$, we shall derive a contradiction. Let

$$n = \dim X = \dim Y, \quad r = \dim \mathcal{F}.$$

We use the natural projections π_X, $\pi_{\mathcal{F}}$, and the evaluation map Φ:

$$\pi_X : X \times \mathcal{F} \to X, \qquad \pi_{\mathcal{F}} : X \times \mathcal{F} \to \mathcal{F}, \qquad \Phi : X \times \mathcal{F} \to Y.$$

We apply the Stein factorization theorem to Φ. Thus, there exist a finite map $p : Y' \to Y$ and a surjective map $\Phi' : X \times \mathcal{F} \to Y'$ with connected fibres such that $\Phi = p \circ \Phi'$. Since Y' is hyperbolic by (3.2.11), we may replace Y and Φ by Y' and Φ', respectiely. Hence, *we may assume that $\Phi : X \times \mathcal{F} \to Y$ has connected fibres.*

Let TX be the Zariski tangent space of X. Let X_{sing} denote the singular locus of X, and $X_{\mathrm{reg}} = X - X_{\mathrm{sing}}$ the set of regular points.

For each $f \in \mathcal{F}$ we write its differential $f_* : TX \to TY$ and its Jacobian $Jf : \bigwedge^n TX \to \bigwedge^n TY$. We define the degeneracy locus $\mathcal{D}$ by setting

$$\mathcal{D}_0 = \{(x, g) \in X \times \mathcal{F}; \; x \in X_{\mathrm{reg}}, \; g(x) \in Y_{\mathrm{reg}}, \; (Jg)(x) = 0\},$$

$$\mathcal{D} = \bar{\mathcal{D}}_0 \subset X \times \mathcal{F}.$$

(6.6.4) Lemma.   (a) $\pi_{\mathcal{F}}(\mathcal{D}) = \mathcal{F}$, *and*   (b) $\pi_X(\mathcal{D}) = X$.

Proof. (a)   We fix $f \in \mathcal{F}$. At the beginning of Section 3 of Chapter 5, we defined an injective map $\sigma_f : T_f \mathcal{F} \to H^0(X, f^*TY)$. For each tangent vector $\zeta \in T_f \mathcal{F}$, we set $\tilde{\zeta} = \sigma_f(\zeta) \in H^0(X, f^*TY)$.

Assume $\pi_{\mathcal{F}}(\mathcal{D}) \neq \mathcal{F}$. By choosing f in $\mathcal{F} - \pi_{\mathcal{F}}(\mathcal{D})$, we have $Jf(x) \neq 0$ at $x \in X_{\mathrm{reg}} \cap f^{-1}(Y_{\mathrm{reg}})$, and we can define

$$v(x) = f_*^{-1}(\tilde{\zeta}(x)) \in T_x X, \qquad x \in X_{\mathrm{reg}} \cap f^{-1}(Y_{\mathrm{reg}}).$$

Thus we have a nonzero holomorphic vector field v on $X_{\mathrm{reg}} \cap f^{-1}(Y_{\mathrm{reg}})$. It is undefined on $X_{\mathrm{sing}} \cup f^{-1}(Y_{\mathrm{sing}})$, which is of codimension at least 2 since both X and Y are normal and f is a finite map. Since X is normal, it extends to a holomorphic vector field on X, which generates a 1-parameter group of automorphisms of X. This contradicts the hyperbolicity of X, (see (5.4.4)).

(b)   Fix $f \in \mathcal{F}$ and a nonzero vector $\zeta \in T_f \mathcal{F}$. The section $\tilde{\zeta}$ of f^*TY defines a multi-valued section $\hat{\zeta}$ of TY, i.e.,

$$\hat{\zeta}(y) = \{\tilde{\zeta}(x) \in T_y Y; \; x \in f^{-1}(y)\}.$$

Let $\hat{Y} = \hat{\zeta}(Y) \subset TY$. Then the natural projection $\hat{Y} \to Y$ is a finite surjective map. Let $\tilde{\Phi} : \Phi^*TY \to TY$ be the natural map (which identifies the fibre $T_{g(x)}Y$ of Φ^*TY at $(x, g) \in X \times \mathcal{F}$ with the fibre $T_{g(x)}Y$ of TY at $g(x) \in Y$). Then $\tilde{\Phi}^{-1}(\hat{Y}) \to X \times \mathcal{F}$ is a finite surjective map. We may view $\tilde{\Phi}^{-1}(\hat{Y})$ as a multi-valued section of Φ^*TY over $X \times \mathcal{F}$.

On the other hand, viewing f^*TY as the restriction of the pull-back bundle Φ^*TY to $X \times \{f\}$ we consider $\tilde{\zeta}$ as a section of Φ^*TY over $X \times \{f\}$, and we shall extend $\tilde{\zeta}$ to a holomorphic section $\tilde{\zeta}$ of Φ^*TY over $(X - A) \times \mathcal{F}$, where

$$A = \{x \in X_{\mathrm{reg}};\ f(x) \in Y_{\mathrm{reg}},\ (Jf)(x) = 0\} \cup X_{\mathrm{sing}} \cup f^{-1}(Y_{\mathrm{sing}}).$$

We define $v(x) = f_*^{-1}(\tilde{\zeta}(x)) \in T_x X$ for $x \in X - A$ and

$$(6.6.5) \qquad \tilde{\zeta}(x, g) = g_*(v(x)) \in T_{g(x)}Y, \qquad (x, g) \in (X - A) \times \mathcal{F}.$$

Thus we have a single valued section $\tilde{\zeta}$ of Φ^*TY defined only on $(X - A) \times \mathcal{F}$ as well as a multi-valued section $\tilde{\Phi}^{-1}(\hat{Y})$ of Φ^*TY defined on all of $X \times \mathcal{F}$.

Now, suppose that $\pi_X(\mathcal{D}) \neq X$, and set

$$B = \pi_X(\mathcal{D}) \cup X_{\mathrm{sing}} \cup f^{-1}(Y_{\mathrm{sing}}).$$

Clearly, $A \subset B$. Then every $g \in \mathcal{F}$ has a nonvanishing Jacobian Jg at $x \in X - B$ and maps a neighborhood U of x biholomorphically onto a neighborhood V of $g(x)$. Hence, the differential $g_*: T_x X \to T_{g(x)}Y$ is an isomorphism. Since $T_{g(x)}Y$ is the fibre of the pull-back bundle Φ^*TY at $(x, g) \in X \times \mathcal{F}$, we see that $(\Phi^*TY)|_{\{x\} \times \mathcal{F}}$ is isomorphic to the product bundle over $\{x\} \times \mathcal{F}$ with fibre $T_x X$, i.e.,

$$(6.6.6) \qquad \Phi^*TY|_{\{x\} \times \mathcal{F}} \cong (\{x\} \times \mathcal{F}) \times T_x X, \qquad x \in X - B.$$

Fix $x \in X - B$. Since the bundle Φ^*TY restricted to $\{x\} \times \mathcal{F}$ is a product bundle $(\{x\} \times \mathcal{F}) \times T_x X$ by (6.6.6) and since $\{x\} \times \mathcal{F}$ is compact, the restriction of the mutli-valued section $\tilde{\Phi}^{-1}(\hat{Y})$ to $\{x\} \times \mathcal{F}$ consists of constant sections $(\{x\} \times \mathcal{F}) \times \sigma_i(x)$, $i = 1, \ldots, m$, where $\sigma_i(x) \in T_x X$ is independent of $g \in \mathcal{F}$.

On the other hand, since $\tilde{\zeta}(x) \subset \hat{\zeta}(f(x))$, $\tilde{\zeta}(X)$ is contained in the mutli-valued section $\tilde{\Phi}^{-1}(\hat{Y})$. So by renumbering $\sigma_1, \ldots, \sigma_m$ we may assume that

$$\tilde{\zeta}(x) = (x, f, \sigma_1(x)) \in \{x\} \times \mathcal{F} \times T_x X, \qquad x \in X - B.$$

Since the extension $\tilde{\zeta}$ of $\tilde{\zeta}$ to $(X - B) \times \mathcal{F}$ defined by (6.6.5) is nothing but the extension of $\tilde{\zeta}$ by the trivialization (6.6.6), we have

$$\tilde{\zeta}(x, g) = (x, g, \sigma_1(x)) \in \{x\} \times \mathcal{F} \times T_x X, \qquad x \in X - B.$$

Since $\tilde{\zeta}((X - B) \times \mathcal{F})$ is contained in $\tilde{\Phi}^{-1}(\hat{Y})$, $\tilde{\zeta}$ extends to $X \times \mathcal{F}$.

We denote this extended section of Φ^*TY over $X \times \mathcal{F}$ by the same symbol $\tilde{\zeta}$. For each $y \in Y$, we consider a mapping $\Phi^{-1}(y) \to T_y Y$ which sends $(x, g) \in \Phi^{-1}(y)$ to $g_*(\tilde{\zeta}(x, g)) \in T_y Y$. This map is constant since $\Phi^{-1}(y)$ is connected and compact. Thus we obtain a non-trivial holomorphic vector field on Y, which contradicts the hyperbolicity of Y. This completes the proof of Lemma (6.6.4).

We continue the proof of (6.6.2). Our next step is to reduce the proof to the case where X and Y are Moishezon and $\dim \mathcal{F} = 1$. Let

$$\mathcal{D}' = \{(x, g) \in \mathcal{D};\ g(x) \in Y_{\text{reg}}\}.$$

Then $\mathcal{D}'$ is an open dense subset of $\mathcal{D}$. To see this, let $\Psi: X \times \mathcal{F} \to Y \times \mathcal{F}$ be defined by $\Psi(x, g) = (g(x), g)$. Define

$$C = \Psi(\mathcal{D}), \qquad C' = C - (Y_{\text{sing}} \times \mathcal{F}).$$

Since $\dim C = r + n - 1 > r + n - 2 \geq \dim(Y_{\text{sing}} \times \mathcal{F})$, C' is open and dense in C. Hence, $\mathcal{D}' = \Psi^{-1}(C') \cap \mathcal{D}$ is also open and dense in $\mathcal{D}$.

Since $\{x \in X;\ \{x\} \times \mathcal{F} \subset \mathcal{D}\}$ is a closed complex subspace of X of dimension $\leq n - 1$, there is a point $x_0 \in X_{\text{reg}}$ such that $\{x_0\} \times \mathcal{F} \not\subset \mathcal{D}$. By (b) of (6.6.4), there is a map $g_0 \in \mathcal{F}$ such that $(x_0, g_0) \in \mathcal{D}$. Since $\mathcal{D}'$ is open and dense, we may choose (x_0, g_0) in $\mathcal{D}'$ so that $g_0(x_0) \notin Y_{\text{sing}}$.

Put

$$\mathcal{F}_0 = \{g \in \mathcal{F};\ (Jg)_{x_0} = 0\}.$$

Then $\mathcal{F}_0$ contains g_0 and is of pure dimension $r - 1$. Let $\mathcal{F}_1$ be an irreducible component of $\mathcal{F}_0$. Now we repeat the above argument with $\mathcal{F} = \mathcal{F}_1$. By induction, we may assume that $\dim \mathcal{F} = 1$.

Fix a point $x \in X$ and consider the curve $\Phi(\{x\} \times \mathcal{F})$ in Y. Let Y_0 be a maximal dimensional irreducible Moishezon complex subspace of Y that contains this curve. Fix $f \in \mathcal{F}$, and let X_0 be an irreducible component of $f^{-1}(Y_0)$. Since $f: X_0 \to Y_0$ is finite and surjective, X_0 is also Moishezon, and so is $X_0 \times \mathcal{F}$. Therefore $\Phi(X_0 \times \mathcal{F})$ is also Moishezon and $\Phi(X_0 \times \mathcal{F}) \supset Y_0$. By the maximality of $\dim Y_0$, we have $\Phi(X_0 \times \mathcal{F}) = Y_0$. Thus we may assume that X and Y are Moishezon and $\mathcal{F}$ is a curve.

Let $\alpha: \tilde{Y} \to Y$ be a proper modification with center $I \supset Y_{\text{sing}}$, codim $I \geq 2$, such that $\tilde{Y}$ is a nonsingular projective algebraic manifold, (see Moishezon [1]). Since $\dim \mathcal{F} = 1$, we have $\dim p_X(\Phi^{-1}(I)) \leq n - 1$. Hence, there exist a point $x \in X_{\text{reg}} - p_X(\Phi^{-1}(I))$ and a map $f \in \mathcal{F}$ such that $f(x) \in Y_{\text{reg}}$ and $(Jf)_x \neq 0$. We fix such a pair (x, f).

With this fixed x we define a map $\varphi: \mathcal{F} \to Y$ by setting $\varphi(g) = g(x)$. Fix a basis $\{e_1, \ldots, e_n\}$ of $T_x X$. Then the Jacobian map $(Jg)_x: \bigwedge^n T_x X \to \bigwedge^n T_{g(x)} Y$ yields a non-trivial holomorphic section σ of the pull-back line bundle $\varphi^*(\bigwedge^n TY)$ over the curve $\mathcal{F}$:

$$\sigma(g) = (Jg)_x(e_1 \wedge \ldots \wedge e_n), \qquad g \in \mathcal{F}.$$

By (b) of (6.6.4), the section σ has zeros. Hence, the degree of the line bundle $\varphi^*(\bigwedge^n TY)$ is positive. Let $K_Y = (\bigwedge^n TY)^{-1}$ be the canonical bundle of Y. Then

$$(K_Y, \varphi(\mathcal{F})) = (\varphi^* K_Y, \mathcal{F}) = \deg \varphi^* K_Y = -\deg \varphi^*(\bigwedge^n TY) < 0.$$

Let $K_{\tilde{Y}}$ be the canonical bundle of $\tilde{Y}$. Since we have chosen x in such a way that $\Phi(\{x\} \times \mathcal{F}) \cap I = \emptyset$, we have $\varphi(\mathcal{F}) \cap I = \emptyset$, and hence

$$(K_{\tilde{Y}}, \varphi(\mathcal{F})) < 0.$$

Now we make use of the following theorem of Miyaoka-Mori [1]:

(6.6.7) Theorem. *Let X be a non-singular projective algebraic manifold, C a closed curve on X, and x a general point of C. If $(K_X, C) < 0$, then there exists a rational curve L through x.*

By this theorem, there is a rational curve L in $\tilde{Y}$ passing through a general point of $\varphi(\mathcal{F})$. Then $\alpha(L)$ is a rational curve on Y. This contradicts the hyperbolicity of Y, showing that $\mathrm{Sur}(X, Y)$ is finite.

Let $\mu \colon \tilde{X} \to X$ be a desingularization of X. If $f \colon X \to Y$ is meromorphic, then by (6.3.19) $f \circ \mu \colon \tilde{X} \to Y$ is holomorphic. This gives a one-to-one correspondence between $\mathrm{Dom}(X, Y)$ and $\mathrm{Sur}(\tilde{X}, Y)$, proving that $\mathrm{Dom}(X, Y)$ is finite. $\square$

(6.6.8) Corollary. *Let X and Y be compact, irreducible complex spaces. If Y is hyperbolic, then for any compact complex subspaces $A \subset X$ and $B \subset Y$, the family $\{f \in \mathrm{Hol}(X, Y);\ f(A) = B\}$ is finite.*

Proof. Let $\mathcal{F}$ be the family in question. We may assume that A and B are irreducible. Since Y is hyperbolic, so is B. By (6.6.2) $\mathrm{Sur}(A, B)$ is finite. So it suffices to show that for each $f_0 \in \mathcal{F}$ the set $\{f \in \mathcal{F};\ f = f_0 \ \text{on}\ A\}$ is finite. But this is obvious from (5.3.10). $\square$

We prove now a generalization of (6.6.2) by Makoto Suzuki [2]:

(6.6.9) Theorem. *Let X be a compact complex space with a Cartier divisor A, and Z be a compact complex space with a Cartier divisor B such that $Y = Z - B$ is hyperbolically imbedded in Z. Then $\mathrm{Dom}(X - A, Y)$ is a finite set.*

Proof. By resolving the singularities of X, we may assume that X is nonsingular and A is a divisor with only normal crossing singularities.

By (3.3.6) Y is complete hyperbolic. Let f be a meromorphic map from $X - A$ to Y. Since the singularity set of f has codimension 2 in X, by (6.2.4) f is holomorphic. By (6.3.9) f extends to a holomorphic map $\bar{f} \in \mathrm{Hol}(X, Z)$.

If $n = \dim Z$, then

$$\mathrm{Dom}(X - A, Y) = \mathrm{Hol}(X - A, Y, n),$$

where $\mathrm{Hol}(X - A, Y, n)$ denotes the set of maps $f \in \mathrm{Hol}(X - A, Y)$ with $\mathrm{rank}\, f = n$. By (6.4.8) $\mathrm{Hol}(X - A, Y, n)$ is compact, and each irreducible component $\mathcal{F}$ is a compact hyperbolic complex space. The problem is reduced to showing $\dim \mathcal{F} = 0$.

We use the evaluation map

$$\Phi \colon X \times \mathcal{F} \to Z, \quad \Phi(x, f) = \bar{f}(x).$$

Take a point $x_0 \in X - A$. Then $\Phi(x_0, \mathcal{F}) = \{f(x_0);\ f \in \mathcal{F}\}$ is a compact complex subspace of Y. Among all compact irreducible complex subspaces of Y containing $\Phi(x_0, \mathcal{F})$, let Y' be one with the maximum dimension. Since Y is hyperbolic, so is Y'.

Fix $f_0 \in \mathcal{F}$, and let $\bar{f}_0 \in \mathrm{Hol}(X, Z)$ be its extension. Since Y' is compact, $\bar{f}_0^{-1}(Y')$ is a compact complex subspace of X. Since $\bar{f}^{-1}(Y) = X - A_I$ for all

$f \in \mathcal{F}$ by (6.4.7) (see (6.4.7) for the definition of A_I), it follows that $\bar{f}_0^{-1}(Y') \subset X - A_I$. Then at least one of the irreducible components of $\bar{f}_0^{-1}(Y')$ is mapped surjectively onto Y' by f_0. Call it X'. Since $X' \subset X - A_I$, again by (6.4.7) $\bar{f}(X') \subset Y$ for all $f \in \mathcal{F}$. Hence, $\Phi(X', \mathcal{F})$ is a compact irreducible complex subspace of Y. Since $f_0(X') = Y'$, $\Phi(X', \mathcal{F})$ contains Y'. By the maximality of Y', we have

$$\Phi(X', \mathcal{F}) = Y'.$$

We consider two elements $f, f' \in \mathcal{F}$ to be equivalent if they induce the same map on X', i.e., if $\Phi(x, f) = \Phi(x, f')$ for all $x \in X'$. Diving $\mathcal{F}$ by this equivalence relation R, set $\mathcal{F}' = \mathcal{F}/R$ with the projection $\pi : \mathcal{F} \to \mathcal{F}'$. By (5.3.10), π is a finite map. In particular, $\dim \mathcal{F}' = \dim \mathcal{F} > 0$.

Now, $\mathcal{F}'$ is an irreducible complex subspace of $\mathrm{Hol}(X', Y')$. For every $f \in \mathcal{F}$, the map $\pi(f) : X' \to Y'$ is of maximal rank $n' = \dim Y'$. Thus, $\mathcal{F}' \subset \mathrm{Hol}(X', Y', n')$. By (6.6.2), $\mathrm{Hol}(X', Y', n')$ is a finite set, contradicting $\dim \mathcal{F}' > 0$ $\qquad\square$

(6.6.10) Corollary. *Let X be a compact complex space with a Cartier divisor A such that $X - A$ is hyperbolically imbedded in X. Then the group $\mathrm{Bim}(X - A)$ of bimeromorphic automorphisms of $X - A$ is finite.*

The following corollary, also due to Suzuki, improves part of (6.4.10).

(6.6.11) Corollary. *Let $A \subset X$ and $B \subset Z$ be as in (6.6.9). Then every irreducible component of $\mathrm{Hol}(X - A, Z - B)$, except the one consisting of all constant maps, has dimension at most $\dim Z - 1$.*

Proof. Set $Y = Z - B$. Let $\mathcal{F}$ be an irreducible component of $\mathrm{Hol}(X - A, Y)$, and $\bar{\mathcal{F}}$ its closure in $\mathrm{Hol}(X, Z)$. By (6.4.10), $\bar{\mathcal{F}}$ is a compact complex subspace of $\mathrm{Hol}(X, Z)$, and for each $x \in X$, the map

$$\varphi_x : \bar{\mathcal{F}} \to Z, \quad \varphi_x(f) = \bar{f}(x)$$

is a finite map. In particular, $\dim \bar{\mathcal{F}} \leq \dim Z$. Assume $\dim \bar{\mathcal{F}} = \dim Z$. Then $\mathrm{rank}\varphi_x = \dim Z = n$. If $x \in X - A$, then $\varphi_x \in \mathrm{Hol}(\bar{\mathcal{F}}, Y, n)$. Since $\mathrm{Hol}(\bar{\mathcal{F}}, Y, n)$ is a finite set of by (6.6.9), φ_x is independent of x, showing that every $f \in \bar{\mathcal{F}}$ is a constant map. $\qquad\square$

Although the following result, which is due, except for some technical improvement, to Kalka-Shiffman-Wong [1] and Urata [3], is not as strong as (6.6.2), the method of the proof is sufficiently different that it would be of some interest.

We recall from Section 3 of Chapter 2 that for an upper semicontinuous pseudo-length function F the holomorphic sectional curvature K_F can be defined, (see (2.1.9)) .

(6.6.12) Theorem. *Let X and Y be compact complex spaces. Assume that Y admits an upper semicontinuous length function F bounded below by a continuous length function and with holomorphic sectional curvature $K_F \leq -1$. Then the set $\mathrm{Dom}(X, Y)$ is finite.*

Proof. Since Y is hyperbolic by (3.7.1), the argument at the very end of the proof of (6.6.2) shows that it suffices to prove finiteness of $\mathrm{Sur}(X, Y)$. If $v\colon \tilde{Y} \to Y$ is the normalization of Y, then $\tilde{F} := v^* F$ satisfies the same curvature condition as F. Now, as in the beginning part of the proof of (6.6.2), the proof can be reduced to the situation where both X and Y are compact irreducible hyperbolic normal complex spaces with $\dim X = \dim Y$. Let $\mathcal{F}$ be an irreducible closed complex subspace of $\mathrm{Hol}(X, Y)$ consisting of surjective finite maps. Assuming that $\dim \mathcal{F} > 0$, we shall derive a contradiction.

Fix a nonzero tangent vector ζ of $\mathcal{F}$ at a regular point f of $\mathcal{F}$, and set $\tilde{\zeta} = \sigma_f(\zeta) \in H^0(X, f^* TY)$. Using the length function F we measure the length of $\tilde{\zeta}(x)$. Since F is upper semicontinuous, $F(\tilde{\zeta}(x))$ is an upper semicontinuous function on X. Let $x_0 \in X$ be a maximum point of this function. Set $y_0 = f(x_0)$.

Let U be a neighborhood of x_0 such that $U \cap f^{-1}(y_0)$ contains only one point, namely x_0. Set $V = f(U) \subset Y$. Then $f\colon U \to V$ is a ramified finite (say r-sheeted) covering. Using the section $\tilde{\zeta}$ and the map f we define a holomorphic vector field v on V as follows. If $y \in V$ is a point outside the ramification locus, we set

$$v(y) = \sum_{i=1}^{r} \tilde{\zeta}(x_i), \qquad \text{where} \quad \{x_i\} = U \cap f^{-1}(y).$$

By assumption, there is a continuous length function F' such that $F' \leq F$. Since $F'(v(y))$ is bounded by a constant $r \cdot F(\tilde{\zeta}(x_0))$, the vector field v extends through the ramification locus. The extended holomorphic vector field v on V satisfies

$$v(y_0) = r \cdot \tilde{\zeta}(x_0).$$

The existence of a nonzero holomorphic vector field v on V such that $F(v)$ attaining a maximum at y_0 contradicts (2.3.13). $\qquad\square$

It is not known if every hyperbolic complex space Y admits a Finsler metric F satisfying the condition of (6.6.12).

The curvature condition on F was used to show that, for any holomorphic vector field η defined on an open set V of Y, $F(\eta)$ cannot attain a maximum in the interior of V. It is not known if for a hyperbolic complex space Y the intrinsic metric F_Y has such a property.

We shall now replace the assumption on holomorphic sectional curvature of F by a negativity condition on the *mean curvature*.

In general, let E be a holomorphic vector bundle of rank r over a complex manifold Y, and h a Hermitian structure in E. In terms of a local coordinate system $z^1, \ldots, z^n$ of Y and a local holomorphic frame field $s_1, \ldots, s_r$ of E, we can express h and its curvature R by their components $h_{i\bar{j}}$ and $R^i_{j\alpha\bar{\beta}}$. If $\xi = \sum \xi^i s_i$ is a local holomorphic section of E, then we have (see, for example, Kobayashi [21; p. 50])

$$(6.6.13) \qquad \frac{\partial^2 h(\xi, \bar{\xi})}{\partial z^\alpha \partial \bar{z}^\beta} = \sum h_{i\bar{j}} \nabla_\alpha \xi^i \nabla_{\bar{\beta}} \bar{\xi}^j - \sum h_{i\bar{k}} R^i_{j\alpha\bar{\beta}} \xi^j \bar{\xi}^k.$$

Let $g = 2 \sum g_{\alpha\bar\beta} dz^\alpha d\bar z^\beta$ be a Hermitian metric on Y. As usual, let $(g^{\alpha\bar\beta})$ be the inverse matrix of $(g_{\alpha\bar\beta})$. Taking the trace of (6.6.13) with respect to g, we have

$$(6.6.14) \qquad \sum g^{\alpha\bar\beta} \frac{\partial^2 h(\xi,\bar\xi)}{\partial z^\alpha \partial \bar z^\beta} = \sum g^{\alpha\bar\beta} \sum h_{i\bar j} \nabla_\alpha \xi^i \nabla_{\bar\beta} \bar\xi^j - \sum K_{j\bar k} \xi^j \bar\xi^k,$$

where

$$K_{j\bar k} = \sum g^{\alpha\bar\beta} h_{i\bar k} R^i_{j\alpha\bar\beta}.$$

The Hermitian form K on E with components $K_{j\bar k}$ is what we call the **mean curvature** of (E, h) with respect to g.

In the special case where E is the tangent bundle TY of Y, a Hermitian structure in TY is nothing but a Hermitian metric on Y. It is then possible to use h as a metric g. In particular, if $h = g$ is a Kähler metric, then the mean curvature K is the usual Ricci curvature. Although we shall later specialize to this situation, at the moment we do not assume $g = h$ even when $E = TY$.

(6.6.15) Lemma. *Let X be a compact complex space and Y a complex manifold. Let h and g be Hermitian metrics on Y, and K the mean curvature of (TY, h) with respect to g. Assume that K is negative semi-definite. Let $f \in \mathrm{Hol}(X, Y, n)$, where $n = \dim Y$, and let $\mathcal{F}$ an irreducible component of $\mathrm{Hol}(X, Y, n)$ containing f. Let $\zeta \in T_f\mathcal{F}$, and $\tilde\zeta \in H^0(X, f^*TY)$ the induced section. We denote the induced Hermitian structure on f^*TY by the same letter h. Then*

(1)   The function $h(\tilde\zeta, \tilde\zeta)$ is constant on X;

(2)   For every $x \in X$, the map $\zeta \mapsto \tilde\zeta(x)$ sends $T_f\mathcal{F}$ injectively into $T_{f(x)}Y$. In particular, $\dim \mathcal{F} \leq \dim Y$;

(3)   The section $\tilde\zeta$ of f^*TY is parallel with respect to h over the nonsingular locus X_{reg} of X.

In order not get involved with connections in a singular variety, we consider in (3) only the regular points of X.

Proof. (1)   Using the Stein factorization theorem and (5.3.6) we may assume that f is a finite map. In particular, $\dim X = \dim Y$. Consider the set X' of points of X where $h(\tilde\zeta, \tilde\zeta)$ attains its maximum value, say a^2. Since X' is obviously closed, all we have to show is that X' is open. Let $x_0 \in X'$. Set $y_0 = f(x_0)$, and let U, V and v be as in the proof of (6.6.12).

We apply (6.6.14) to the vector field v. Since K is negative semi-definite, it follows that the right hand side is nonnegative. Hence, $h(v, v)$ is subharmonic on V. We write $\|v\| = h(v, v)^{1/2}$ and $\|\tilde\zeta\| = h(\tilde\zeta, \tilde\zeta)^{1/2}$. Since $\|v\|$ attains a maximum value ra at an interior point y_0, it must be constant. If $y \in Y$ is a point outside the ramification locus of $f : U \to V$ so that $v(y) = \sum_{i=1}^r \tilde\zeta(x_i)$ as in the proof of (6.6.12), then

$$\sum_{i=1}^r \|\tilde\zeta(x_i)\| \geq \|v(y)\| \equiv \|v(y_0)\| = ra.$$

On the other hand, $\|\tilde\zeta(x_i)\| \leq a$. Hence, $\|\tilde\zeta\| \equiv a$ on U.

(2)   By (5.3.3) we may assume that X is nonsingular. Let $\zeta \in T_f \mathcal{F}$ be a nonzero tangent vector. Since $\sigma_f : T_f \mathcal{F} \to H^0(X, f^*TY)$ is injective (see Section 3 of Chapter 5), $\tilde{\zeta} = \sigma_f(\zeta)$ is nonzero. By (1), $\tilde{\zeta}(x) \neq 0$ at every $x \in X$.

(3)   Again, by the Stein factorization theorem and (5.3.6), we may assume that f is a finite map. (In fact, if

$$f : X \xrightarrow{\ p\ } X' \xrightarrow{\ f'\ } Y$$

is a Stein factorization of f, then, for each $x' \in X'$, the bundle $f^*TY|_{p^{-1}(x')}$ is naturally isomorphic to a product bundle $p^{-1}(x') \times T_{f'(x')}Y$. The section $\tilde{\zeta}$ of f^*TY is constant along $p^{-1}(x')$ and hence parallel along $p^{-1}(x')$. It suffices therefore to show that the section $\tilde{\zeta}'$ of f'^*TY induced by ζ is parallel.)

Now, assuming that f is a finite map, Let X_0 denote the set of regular points of X where f is of maximal rank $n = \dim Y$. Then f^*g is a Hermitian metric on X_0. We pull back the local coordinate system $z^1, \ldots, z^n$ of Y and the local frame field $s_1, \ldots, s_n$ of TY to X_0 and f^*TY by f. Then the mean curvature K of (TY, h) pulls back to the mean curvature f^*K of (f^*TY, h) with respect to f^*g. Apply (6.6.13) to the constant function $\tilde{h}(\tilde{\zeta}, \tilde{\zeta})$. Since the left hand side vanishes and since f^*K is negative semi-definite, it follows that $\tilde{\zeta}$ is parallel on X_0. Since X_0 is dense, $\tilde{\zeta}$ is parallel on X. $\square$

(6.6.16) **Theorem.** *Let X be a compact complex space, and Y a compact complex manifold. Let h and g be Hermitian metrics on Y. Assume that the mean curvature K of (TY, h) with respect to g is negative semi-definite. Then the family $\mathrm{Sur}(X, Y)$ is discrete under either one of the following additional conditions:*
(a)   *K is negative definite at some point of Y;*
(b)   *at least one of the Chern numbers of Y is nonzero;*

Proof. By taking a resolution of the singularities $p : \tilde{X} \to X$ and pulling back everything by p, we may assume that X is nonsingular. Using the notation in the proof of (6.6.15), assume that $\mathcal{F}$ has a positive dimension. Let $\zeta \in T_f \mathcal{F}$ be a nonzero vector and the $\tilde{\zeta}$ be the section of f^*TY defined by ζ. By (6.6.15), $\tilde{\zeta}$ is a parallel section and $h(\tilde{\zeta}, \tilde{\zeta})$ is a constant function on X.

(a)   Since K is negative definite on a nonempty open subset of Y, there is a point x_0 of X_0 where f^*K is negative definite. Applying (6.6.13) to the constant function $h(\tilde{\zeta}, \tilde{\zeta})$ at x_0 yields a contradiction unless $\tilde{\zeta} = 0$.

(b)   For simplicity of the notation, consider the case $c_n(TY)[X] \neq 0$. Then we have

$$c_n(f^*TY)[X] = (f^*c_n(TY))[X] = \deg_f \cdot c_n(TY)[Y] \neq 0,$$

where $\deg_f$ is the degree of f, i.e., the number of points in $f^{-1}(y)$ for a generic point $y \in Y$. Then f^*TY cannot admit the nonzero holomorphic section $\tilde{\zeta}$. $\square$

Before we derive a corollary, we need to explain a differential geometric characterization of semistable vector bundles. Let Y be a compact complex manifold with an ample line bundle H, and E a holomorphic vector bundle over Y. We choose a Kähler metric g in such a way that its Kähler form ω represents $c_1(H)$. Let h be a Hermitian structure in E, and K the mean curvature of (E, h) with respect to g. We define a real constant c by equating the integral of the trace of $K - ch$ to zero:

$$\int_Y (\sum h^{i\bar{j}} K_{i\bar{j}} - cr)\omega^n = 0.$$

This constant is determined by $c_1(H)$ and $c_1(E)$ as we can see from the following formula (see Kobayashi [21; pp. 103–104]):

$$\int_Y c_1(E) \wedge c_1(H)^{n-1} = \frac{cr}{2n\pi} \int_Y c_1(H)^n.$$

(The left hand side is called the **degree** of E with respect to H.) In particular, c does not depend on h. Define the norm $\|K - ch\|$ by

$$\|K - ch\|^2 = \max_Y |K - ch|^2 = \max_Y \sum h^{i\bar{k}} h^{l\bar{j}} (K_{i\bar{j}} - ch_{i\bar{j}})(\bar{K}_{k\bar{l}} - ch_{k\bar{l}}).$$

Then (see Kobayashi [21; p. 234])

The vector bundle E is H-semistable if and only if, given $\varepsilon > 0$, there is a Hermitian structure h such that $\|K - ch\| < \varepsilon$.

This implies that if $c < 0$ and E is H-semistable, there is a Hermitian structure h such that K is negative definite.

Specializing to the situation where $E = TY$, we obtain

(6.6.17) **Corollary.** *Let X be a compact complex space. Let Y be a compact complex manifold with an ample line bunlde H. If the tangent bundle TY is H-semistable and $\deg(TY) = c_1(Y) \wedge c_1(H)^{n-1} < 0$, then $\mathrm{Sur}(X, Y)$ is discrete.*

In deriving (6.6.17) it is essential not to assume $g = h$ in (6.6.16).

We generalize a result of Kalka-Shiffman-Wong [1], where it is assumed that a particular Chern number $c_n(Y)[Y]$ or $c_1(Y)^n[Y]$ is nonzero.

(6.6.18) **Theorem.** *Let X be a compact complex space, and Y a compact Kähler manifold with negative semi-definite Ricci tensor. If one of the Chern numbers of Y is nonzero, then $\mathrm{Sur}(X, Y)$ is discrete.*

Proof. We let $g = h$ be the Kähler metric on Y with negative semi-definite Ricci tensor. As we noted earlier, the mean curvature K in the Kähler case is nothing but the Ricci tensor. By (3) of (6.6.15), $\tilde{\zeta}$ is a parallel section of f^*TY. Let $X_0 \subset X$ be as in the proof of (3) of (6.6.15), and let $Y_0 = f(X_0) \subset Y$. Given $y_0 \in Y_0$, choose a point $x_0 \in X_0$ such that $y_0 = f(x_0)$. Let U be a neighborhood of x_0 and V a neighborhood of y_0 such that f is a biholomorphic map from U to V. Then, corresponding to $\tilde{\zeta}$ we have a parallel section ζ_V of TY over V; it is a holomorphic parallel vector field on V. Taking V sufficiently small, we can decompose V into a direct product $V' \times V''$ as a Kähler manifold, where V' is flat and V'' is a product of (holonomy) irreducible Kähler manifolds, (see Kobayashi-Nomizu [1; I, p. 187, and II, p. 171]). Since we have a parallel vector field on V, V' is nontrivial. We choose a coordinate system $z^1, z^2, \ldots, z^n$ in V in such a way that $z^1, \ldots, z^k$ is the coordinate system for V' while $z^{k+1}, \ldots, z^n$ is the coordinate system for V''. Since V' is flat, the curvature form of V does not involve $dz^1, \ldots, dz^k$. Any form of bidegree greater than $(n-k, n-k)$ involving only $dz^{k+1}, \ldots, dz^n$ must vanish identically on V. In particular, every characteristic form of degree (n, n) is identically zero on V. Since this holds for a neighborhood of every $y_0 \in Y_0$ and Y_0 is dense in Y, every characteristic (n, n)-form of Y is identically zero on Y.

$\square$

(6.6.19) Corollary. *Let X be a compact complex space, and Y a compact Kähler manifold with $c_1(Y) = 0$. If one of the Chern numbers of Y is nonzero, then* Sur(X, Y) *is discrete.*

In (6.6.19) we cannot conclude that Sur(X, Y) is finite. It was shown in Matsumura-Monsky [1] that there exist examples of nonsingular quartic surfaces in $P_3 C$ with infinite number of automorphisms.

We conclude this section by proving the following theorem of Kaup [4].

(6.6.20) Theorem. *If X is a compact hyperbolic complex space, then every surjective holomorphic self-map $f : X \to X$ is an automorphism of X.*

Proof. Let Hol$_m(X, X)$ be the set of holomorphic self-maps of degree m. Since Hol(X, X) is compact, Hol$_m(X, X)$ is empty for large m. Let f be a surjective holomorphic map of X onto itself. If its degree were $m > 1$, then f^k would have degree m^k. Hence, the f must be of degree 1 and is an automorphism of X.

$\square$

(6.6.21) Remark. The condition on Y in each of the finiteness theorems (6.6.2), (7.6.1) and (6.6.17) for Sur(X, Y) has its differential geometric counterpart on the curvature of Y. Thus,

$$
\begin{array}{lcl}
\text{hyperbolic} & \Leftarrow & \text{holomorphic sectional curvature} < 0 \\
\text{general type} & \Leftarrow & \text{Ricci curvature} < 0 \\
\deg(TY) < 0 & \Leftarrow & \int \text{scalar curvature} < 0
\end{array}
$$

7 Holomorphic Maps into Spaces of Nonpositive Curvature

Let X and Y be irreducible complex spaces with X compact, $\dim X = m$ and $\dim Y = n$. Fix an integer k, $0 \leq k \leq n$, and let Hol(X, Y, k) be the space of holomorphic maps $f : X \to Y$ of rank k, i.e., maps f such that $\dim f(X) = k$. By (5.3.5) Hol(X, Y, k) is open and closed in Hol(X, Y) and hence carries a universal complex structure. In the preceding section, we studied Hol(X, Y, n). Following Sunada [1], Urata [3], Kalka-Shiffman-Wong [1], and Noguchi-Sunada [1], we study Hol(X, Y, k) for an arbitrary k.

As we explained at the beginning of Section 3 of Chapter 5, if $\mathcal{F}$ is an irreducible component of Hol(X, Y), at each $f \in \mathcal{F}$ there is a natural injective map

$$\sigma_f : T_f \mathcal{F} \to H^0(X, f^* TY).$$

Therefore we look for conditions for vanishing of $H^0(X, f^* TY)$. For this reason, we consider various negativity for holomorphic vector bundles in general.

Let E be a holomorphic vector bundle of rank r over a complex manifold Y, and let $E^{\times}$ denote E minus its zero section. We say that E is p-**negative** if there is a nonnegative continuous function h on E such that

(i) $h^{-1}(0)$ is the zero section of E;
(ii)   everywhere on $E^{\times}$, h is of class C^2 and its complex Hessian $\partial\bar{\partial}h$ has at least $r + p$ positive eigenvalues. (In local coordinates, the Hessian $\partial\bar{\partial}h$ is an $(r + n) \times (r + n)$ Hermitian matrix).

If E is p-negative, it is clearly q-negative for all $q < p$.

(6.7.1) **Theorem.** *Let X be a compact irreducible complex space, and Y a complex manifold of dimension n. Let $0 \le p \le n$. If the tangent bundle TY is p-negative, then $\mathrm{Hol}(X, Y, k)$ is discrete for $k > n - p$.*

Proof. Let $\mathcal{F}$ be a connected component of $\mathrm{Hol}(X, Y, k)$. Assuming that $\dim \mathcal{F} > 0$, we shall obtain a contradiction. Let $\Phi : X \times \mathcal{F} \to Y$ be the evaluation map. We fix $\zeta \in T_f \mathcal{F}$, $f \in \mathcal{F}$, such that $\Phi_*(x_0, \zeta) \neq 0$ for some x_0. The differential Φ_* of Φ defines map $\Phi_*(\cdot, \zeta) : X \to TY$ sending $x \in X$ to $\Phi_*(x, \zeta) \in T_{f(x)}Y$. Its image $S = \Phi_*(X, \zeta)$ is a compact irreducible complex subspace of TY. It is not contained in the zero section $h^{-1}(0)$ of TY. Since S projects onto the k-dimensional complex subspace $f(X) \subset Y$, it follows that $\dim S \ge k$.

Let $\eta_0 \in S$ be such that $h(\eta_0) = \max_{\eta \in S} h(\eta)$. Since $\partial\bar{\partial}h$ has at least $n + p$ positive eigenvalues at η_0, there is an $(n + p)$-dimensional local complex submanifold $N \subset TY$ through η_0 such that $\partial\bar{\partial}(h|_N)$ is positive-definite. Then $\dim(S \cap N) > 0$ if $\dim S + \dim N > \dim TY$, i.e., if $k > n - p$. Since $h|_{S \cap N}$ is strongly plurisubharmonic and attains a maximum at ζ_0, we have a contradiction. $\square$

Let E be a holomorphic vector bundle of rank r over an n-dimensional complex manifold Y. Let $L(E)$ be the tautological line bundle over $P(E)$ with projection p; it is the blow-up of E along the zero section of E. Let $L^{\times}(E)$ be $L(E)$ minus its zero section. Then $E^{\times}$ is naturally identified with $L^{\times}(E)$, see (6.A.1).

We consider the situation where the function h in the definition of p-negativity is a Hermitian structure in $L(E)$. In general, let L be a holomorphic line bundle over a complex manifold P with projection p. Let $z = (z^1, \ldots, z^m)$ be a local coordinate system for P. Let s be a non-vanishing local holomorphic section of $L^{\times}$ over $U \subset P$. A point of L over $z \in U$ can be uniquely written as $\zeta s(z)$ with $\zeta \in \mathbf{C}$. We use $z^1, \ldots, z^m, \zeta$ as a local coordinate system for $L|_U$.

Let h be a Hermitian structure in L. It is positive on $L^{\times}$. Set $g = s^* h$. Then g is a positive function on U, and $h(z, \zeta) = g(z)|\zeta|^2$. The connection form of h is given by

$$\omega = d' \log h = d' \log g + d \log \zeta.$$

Its curvature is given by

$$\Omega = d''d'\log h = d''d'\log g = \frac{1}{g}d''d'g - \frac{1}{g^2}d''g \wedge d'g.$$

Given a point $z_0 \in U$, we can find a non-vanishing local section s such that $dg(z_0) = 0$. In fact, from any non-vanishing local section t of $L^\times$ we can obtain such a section s by a transformation of the form

$$t(z) \mapsto s(z) = f(z)t(z), \quad \text{where} \quad f(z) = 1 + \sum_{j=1}^{m} a_j z^j.$$

With respect to such a section s, at z_0 the curvature looks as follows:

$$\Omega = -\frac{1}{g}d''d'g = \frac{1}{g}\sum g_{i\bar{j}}dz^i \wedge d\bar{z}^j, \quad \text{where} \quad g_{i\bar{j}} = \partial^2 g/\partial z^i \partial \bar{z}^j.$$

On the other hand, the complex Hessian of h at (z_0, ζ) is given by

$$\partial\bar{\partial}h = \begin{pmatrix} g_{i\bar{j}}(z_0)|\zeta|^2 & 0 \\ 0 & g \end{pmatrix}.$$

Hence, the number of positive eigenvalues of the complex Hessian $\partial\bar{\partial}h$ is equal to the number of negative eigenvalues of the curvature $\sqrt{-1}\Omega$ plus 1.

Applying this to $L = L(E)$ and $P = P(E)$ yields the following

(6.7.2) Lemma. *If the line bundle $L(E)$ admits a Hermitian structure h such that its curvature $\sqrt{-1}\Omega$ has at least $r-1+p$ negative eigenvalues, then E is p-negative.*

Hence,

(6.7.3) Corollary. *Let X be a compact irreducible complex space, and Y a complex manifold of dimension n. Let $0 \leq p \leq n$. If the line bundle $L(TY)$ over $P(TY)$ admits a Hermitian structure h whose curvature $\sqrt{-1}\Omega$ has at least $n-1+p$ negative eigenvalues, then $\mathrm{Hol}(X, Y, k)$ is discrete for $k > n - p$.*

Under the identification $L^\times(E) \cong E^\times$, let G be the positive function on $E^\times$ corresponding to h. (By setting $G = 0$ on the zero section, we extend it to all of E). Then its square root $F = \sqrt{G}$ is the complex Finsler structure on E corresponding to the Hermitian structure h of $L(E)$, see Appendix A of this chapter. If F is strongly pseudo-convex (see (6.A.22) for the corresponding condition on h), then the curvature of F is defined, see (6.A.26) and (6.A.31). We say that the curvature of F is p-**negative** if $\Psi(z, \zeta, \zeta)$ of (6.A.31) has at least p negative eigenvalues, i.e., for every nonzero ζ, the $(n \times n)$-matrix $(\sum R_{i\bar{j}\alpha\bar{\beta}}\zeta^i\bar{\zeta}^j)_{\alpha\bar{\beta}}$ has at least p negative eigenvalues. Then (see (6.A.42))

(6.7.4) Lemma. *The curvature of a strongly pseudo-convex complex Finsler structure F in E is p-negative if and only if the complex Hessian of $G = F^2$ on $E^\times$ has at least $r + p$ positive eigenvalues.*

(6.7.5) Corollary. *Let X be a compact irreducible complex space, and Y a complex manifold of dimension n. Let $0 \leq p \leq n$. If the tangent bundle TY admits a strongly*

pseudo-convex complex Finsler structure F whose curvature is p-negative, then $\mathrm{Hol}(X, Y, k)$ *is discrete for* $k > n - p$.

This is a technical extension to the Finsler case of the corresponding result in the Hermitian case by Urata [3] (for $p = n$) and by Kalka-Shiffman-Wong [1] (for all p). See also Goloff-To [1].

The generalization to the Finsler case makes it possible to draw an algebraic geometric conclusion. In general, the dual bundle E^* of a holomorphic vector bundle E is ample if and only if the dual line bundle $L(E)^*$ of $L(E)$ is ample; this is one of the definitions for ampleness for a vector bundle. Hence, if E^* is ample, then $L(E)$ admits a Hermitian structure h with negative curvature. Thus, if E^* is ample E is n-negative.

In particular, if Y is compact and its cotangent bundle T^*Y is ample, then $\mathrm{Hol}(X, Y, k)$ is discrete for all $k > 0$. On the other hand, if T^*Y is ample, then Y is hyperbolic (see (3.6.21)), and $\mathrm{Hol}(X, Y)$ is compact (see (5.1.1)). Hence,

(6.7.6) Corollary. *Let X be a compact irreducible complex space, and Y a compact complex manifold with ample cotangent bundle. Then the set of nonconstant holomorphic mappings from X into Y is finite.*

In order to obtain finer results, we specialize now to the Hermitian situation. Let E be a holomorphic vector bundle or rank r over a complex manifold M, and h a Hermitian structure in E. At each point $x \in M$, the curvature R defines a map

$$R: T_x M \times \overline{T}_x M \to \mathrm{End}(E_x).$$

Thus, $R(u, \bar{v})s \in E_x$ for $u, v \in T_x M$, $s \in E_x$.

For each linear subspace $V_x \subset T_x M$, we set

$$(6.7.7) \qquad N(V_x, E) = \{s \in E_x;\ R(u, \bar{v})s = 0 \quad \text{for all} \quad u, v \in V_x\}.$$

By letting $\lambda = 1$ and then $\lambda = i$ in the following identity

$$R(u + \lambda v, \bar{u} + \bar{\lambda}\bar{v})s - R(u, \bar{u})s - |\lambda|^2 R(v, \bar{v})s = \lambda R(v, \bar{u})s + \bar{\lambda} R(u, \bar{v})s,$$

we see that

$$(6.7.8) \qquad N(V_x, E) = \{s \in E_x;\ R(u, \bar{u})s = 0 \quad \text{for all} \quad u \in V_x\}.$$

Then, for each integer k, $0 < k \le \dim M$, we define k-**nullity** of (E, h) **at** x to be

$$(6.7.9) \qquad \nu_x(k, E) = \max_{\dim V_x = k} \dim N(V_x, E),$$

where the maximum is taken over all k-dimensional subspaces V_x of $T_x M$. We set

$$\nu_x(0, E) = \mathrm{rank}(E).$$

Finally, we define the k-**nullity** $\nu(k, E)$ of (E, h) by

(6.7.10) $$\nu(k, E) = \max_{x \in M} \nu_x(k, E).$$

If E is the tangent bundle TM of a Hermitian manifold (M, h), we write

$$N(V_x) = N(V_x, TM), \qquad \nu_M(k) = \nu(k, TM),$$

and call $\nu_M(k)$ simply the k-nullity of (M, h). Clearly, we have

$$0 \le \nu_M(n) \le \nu_M(n-1) \le \ldots \le \nu_M(1) \le \nu_M(0) = n.$$

We say that the curvature R is **negative semi-definite** if, for every tangent vector $v \in T_x M$, $R(v, \bar{v})$ is a negative semi-definite endormorphism of E, i.e., if $h(R(v, \bar{v})\xi, \bar{\xi}) \le 0$ for all $v \in T_x M$ and $\xi \in E_x$. In the special case where E is the tangent bundle of a Hermitian manifold M, for two unit tangent vectors $v, w \in T_x M$, we call $h(R(v, \bar{v})w, \bar{w})$ the **bisectional curvature** in the direction of v and w. So in this case, to say that R is negative semi-definite amounts to saying that the bisectional curvature is nonpositive. In the special case where $v = w$, $h(R(v, \bar{v})v, \bar{v})$ is the holomorphic sectional curvature in the direction of v.

(6.7.11) **Lemma.** *Let (E, h) be a Hermitian vector bundle over a compact complex manifold X. Assume that the curvature R of h is negative semi-definite. Then every holomorphic section ξ of E satisfies the following equations:*

(a) $\nabla \xi = 0$;

(b) $R(u, \bar{v})\xi = 0$,   *for* $u, v \in T_x X$, $x \in X$.

We note that any C^∞ section ξ of E satisfying (a) is holomorphic.
Proof. From (6.6.13) we see that $h(\xi, \bar{\xi})$ is plurisubharmonic on X. Since X is compact, it must be constant. As in Section 6, in terms of a local coordinate system $z^1, \ldots, z^n$ of X and a local holomorphic frame field $s_1, \ldots, s_r$, we express h and R by their components $h_{i\bar{j}}$ and $R^i_{j\alpha\bar{\beta}}$. Then

$$0 = \sum h_{i\bar{j}} \nabla_\alpha \xi^i \overline{\nabla_\beta \xi^j} - \sum h_{i\bar{k}} R^i_{j\alpha\bar{\beta}} \xi^j \bar{\xi}^k.$$

Since R is negative semi-definite, each of the two sums above must vanish. $\square$

(6.7.12) **Corollary.** *Let (E, h) be as in (6.7.11). Let $k \le \dim X$. Then for any $x \in X$ and any k-dimensional subspace V_x of $T_x X$, we have*

$$\dim H^0(X, E) \le \dim N(V_x, E).$$

Proof. Since ξ is parallel by (a) of (6.7.11), the map $\xi \mapsto \xi(x)$ is an injection $H^0(X, E) \to E_x$. By (b) of (6.7.11), $R(u, v)\xi(x) = 0$ for all $u, v \in V_x$. Hence, $\xi(x) \in N(V_x, E)$, and $\dim H^0(X, E) \le \dim N(V_x, E)$. $\square$

We have now the main result of Sunada [1].

(6.7.13) **Theorem.** *Let X be a compact complex space, and (Y, h) a Hermitian manifold with nonpositive bisectional curvature. Let $\nu_Y(k)$ be the k-nullity of (Y, h). Then*

$$\dim H^0(X, f^*TY) \leq \nu_Y(k) \qquad \text{for} \quad f \in \text{Hol}(X, Y, k).$$

Consequently,

$$\dim \text{Hol}(X, Y, k) \leq \nu_Y(k).$$

Proof. Replacing X by its nonsingular model and applying (2) of (5.3.6), we may assume that X is nonsingular. Let (f^*TY, f^*h) be the induced Hermitian vector bundle over X. Its curvature f^*R is negative semi-definite.

Consider $\dim N(V_x, f^*TY)$ as a function on the Grassmann bundle of k-planes in TX. Let $x_0 \in X$ and $V_{x_0} \subset T_{x_0}X$, $\dim V_{x_0} = k$, be where $\dim N(V_x, f^*TY)$ attains its minimum. Then in a neighborhood of V_{x_0}, $\dim N(V_x, f^*TY)$ remains constant since the negative eigenvalues of $(f^*R)(u, \bar{v})$ remains negative. We may therefore choose x_0 and V_{x_0} in such a way that $f_*: T_{x_0} \to T_{f(x_0)}Y$ is of rank k and $\dim f_*(V_{x_0}) = k$. For $u, v \in V_{x_0}$, $(f^*R)(u, \bar{v}) \in \text{End}((f^*TY)_{x_0})$ coincides with $R(f_*u, f_*\bar{v}) \in \text{End}(T_{f(x_0)}Y)$ under the natural identification $(f^*TY)_{x_0} \cong T_{f(x_0)}Y$. By (6.7.12),

$$\dim H^0(X, f^*TY) \leq \dim N(V_{x_0}, f^*TY)$$
$$= \dim N(f_*(V_{x_0}), TY) \leq \nu_{f(x_0)}(k, TY) \leq \nu_Y(k). \qquad \square$$

(6.7.14) Lemma. *Let (Y, h) be an n-dimensional Hermitian manifold with nonpositive bisectional curvature and negative holomorphic sectional curvature. Then*

(1) $N(V_y) \cap V_y = 0$ *for any linear subspace $V_y \subset T_yY$;*

(2) $\nu_Y(k) \leq n - k$.

Proof. (1)   If $V_y \cap N(V_y)$ contains a nonzero vector, say v, then $R(v, v)v = 0$ and the holomorphic sectional curvature $h(R(v, \bar{v})v, \bar{v})$ would vanish, contradicting our assumption.

(2)   By the very definition of $\nu_y(k, TY)$ and by (1),

$$\nu_y(k, TY) = \max_{\dim V_y = k} \dim N(V_y) \leq n - k,$$

where the maximum is taken over all k-dimensional linear subspace of T_yY. $\square$

If (Y, h) satisfies the assumption in the lemma above, then $\nu_Y(n) = 0$ and $\nu_Y(1) \leq n - 1$. We consider the largest l such that $\nu_Y(l) > 0$. Thus

$$(6.7.15) \qquad 0 = \nu_Y(n) = \ldots = \nu_Y(l + 1) < \nu_Y(l) \leq \ldots \leq \nu_Y(1) < n.$$

(6.7.16) Theorem. *Let (Y, h) be an n-dimensional Hermitian manifold with nonpositive bisectional curvature. Let l be the largest integer such that $\nu_Y(l) > 0$. Then for any compact complex space X, we have*

(1) $\dim \text{Hol}(X, Y, k) = 0$ *for all* $k > l$;

If, in addition, Y is compact hyperbolic, then $\text{Hol}(X, Y, k)$ is finite for $k > l$.

(2)   *Assume, in addition, that (Y, h) has negative holomorphic sectional curvature. Then*

$$\dim \mathrm{Hol}(X, Y, k) \le n - k \quad \text{for all} \quad k \ge 0;$$

Proof. (1)   The first part is immediate from (6.7.13). If Y is compact hyperbolic, then $\mathrm{Hol}(X, Y)$ is compact by (5.1.1), and $\mathrm{Hol}(X, Y, k)$ is finite.

(2)   This follows from (6.7.13) and (6.7.14).   □

We examine $N(V_x)$ (see (6.7.7)) when E is the tangent bundle of a Kähler manifold M with metric h. Since

$$h(R(u, \bar{v})s, \bar{t}) = h(R(s, \bar{t})u, \bar{v}), \qquad u, v, s, t \in T_x M,$$

we have, in particular,

$$h(R(u, \bar{u})s, \bar{s}) = h(R(s, \bar{s})u, \bar{u}), \qquad u, s \in T_x M.$$

From (6.7.8) we have

$$(6.7.17) \qquad\qquad V_x \subset N(N(V_x)), \qquad V_x \subset T_x M.$$

Replacing V_x by $N(V_x)$ in the above, we obtain $N(V_x) \subset N(N(N(V_x)))$. On the other hand, applying N to both sides of (6.7.17) yields the reversed inclusion. Hence,

$$(6.7.18) \qquad\qquad N(V_x) = N(N(N(V_x))), \qquad V_x \subset T_x M.$$

We can refine (6.7.15) in the Kähler case.

(6.7.19) **Lemma.** *Let M be an n-dimensional Kähler manifold, and let l be the largest integer such that $\nu_M(l) > 0$. Then*

$$\nu_M(l) \le \nu_M(l - 1) \le \ldots \le \nu_M(\nu_M(l)) = \ldots = \nu_M(1) = l.$$

Proof. Fix an integer i, $1 \le i \le n$, and let $V_x \subset T_x M$ be an i-dimensional subspace such that $\nu_M(i) = \dim N(V_x)$. By (6.7.17), $\dim N(N(V_x)) \ge i$. Hence,

$$(*) \qquad\qquad i \le \dim N(N(V_x)) \le \nu_M(\nu_M(i)).$$

Letting $i = 1$ in $(*)$, we have $0 < \nu_M(\nu_M(1))$. Hence,

$$\nu_M(1) \le l.$$

Letting $i = l$ in $(*)$, we have

$$l \le \nu_M(\nu_M(l)).$$

Hence,

$$l \le \nu_M(\nu_M(l)) \le \ldots \le \nu_M(1) \le l. \qquad \text{Q.E.D.}$$

From (6.7.13) and (6.7.19) we obtain (Sunada [1])

(6.7.20) Theorem. *Let (Y, h) be an n-dimensional Kähler manifold with nonpositive bisectional curvature. Let l be the largest integer such that $v_Y(l) > 0$. Let X be a compact complex space, and $k > 0$. Then*

$$\dim \operatorname{Hol}(X, Y, k) \begin{cases} \le l & \text{if } k \le l; \\ = 0 & \text{if } k > l. \end{cases}$$

If, in addition, (Y, h) has negative holomorphic sectional curvature, then

$$\dim \operatorname{Hol}(X, Y, k) \begin{cases} \le \min\{l, \; n - k\} & \text{if } k \le l; \\ = 0 & \text{if } k > l. \end{cases}$$

Proof. The last statement follows from the first part and (6.7.16). $\qquad\square$

(6.6.21) Remark. It is possible to draw the same conclusion from a slightly different assumption. Namely,

Let (Y, h) be an n-dimensional compact Hermitian manifold with nonpositive bisectional curvature and negative holomorphic sectional curvature, and X a compact complex space. Then for any $k > 0$ we have

$$\dim \operatorname{Hol}(X, Y, k) \begin{cases} \le \min\{l, \; n - k\} & \text{if } k \le l; \\ = 0 & \text{if } k > l. \end{cases}$$

In order to prove this, suppose that there is an irreducible component $\mathcal{F}$ of $\operatorname{Hol}(X, Y, k)$ with $\dim \mathcal{F} > l$. By (3.7.1) Y is hyperbolic, and by (5.1.1) $\operatorname{Hol}(X, Y)$ is compact. Consider the evaluation map

$$\Phi \colon X \times \mathcal{F} \to Y.$$

Each point $x \in X$ defines a holomorphic map $\Phi(x, \cdot) \in \operatorname{Hol}(\mathcal{F}, Y)$. By the universal property of the complex structure of $\mathcal{F}$ (see Section 3 of Chapter 5), the map $\alpha \colon X \to \operatorname{Hol}(\mathcal{F}, Y)$ defined by $\alpha(x) = \Phi(x, \cdot)$ is holomorphic. Then $\dim \alpha(X) > 0$; for otherwise, every $f \in \mathcal{F}$ would be a constant map. Since the map $\mathcal{F} \to Y$ given by $f \mapsto \Phi(x, f)$ is a finite map by (5.3.4), it has rank equal to $\dim \mathcal{F}$. Hence, $\alpha(x) \in \operatorname{Hol}(\mathcal{F}, Y, r)$, where $r = \dim \mathcal{F}$. This implies $\dim \operatorname{Hol}(\mathcal{F}, Y) \ge \dim \alpha(X) > 0$. Since $r > l$, this contradicts (6.7.16).

In order to relate the integer l to the negativity of $\bigwedge^k TM$, we again consider a general Hermitian vector bundle (E, h) over a complex manifold M. We denote the induced Hermitian structure in $\bigwedge^k E$ also by h and its curvature also by R:

$$R \colon T_x M \times \overline{T}_x M \to \operatorname{End}\left(\bigwedge^k E_x\right).$$

If $s_1, \ldots, s_k \in E_x$, then

$$R(u, \bar{v})(s_1 \wedge \ldots \wedge s_k) = \sum_{i=1}^{k} s_1 \wedge \ldots \wedge R(u, \bar{v})s_i \wedge \ldots \wedge s_k.$$

For a fixed nonzero vector $v \in T_x M$, we choose an orthonormal basis $s_1, \ldots, s_r$ of E_x such that

$$R(v, \bar{v})s_i = \lambda_i s_i,$$

where $\lambda_1, \ldots, \lambda_r$ are the eigenvalues of $R(v, \bar{v})$ operating on E_x with eigenvectors $s_1, \ldots, s_r$. Then $R(v, \bar{v})$ acting on $\bigwedge^k E_x$ has eigenvalues

$$\lambda_{i_1} \ldots \lambda_{i_k}, \quad 1 \leq i_1 < \ldots < i_k \leq r$$

with eigenvectors

$$s_{i_1} \wedge \ldots \wedge s_{i_k}, \quad 1 \leq i_1 < \ldots < i_k \leq r.$$

We assume that the curvature of (E, h) is negative semi-definite, i.e., all $\lambda_i \leq 0$. Let $V_x \subset T_x M$ be the 1-dimensional subspace spanned by v. The number of those eigenvalues λ_i that are equal to 0, being $v(V_x)$ by (6.7.8), is bounded by $v(1, E)$. Let $k > v(1, E)$. Then, given $\lambda_{i_1}, \ldots, \lambda_{i_k}$, at least one of them is negative, and hence

$$\lambda_{i_1} + \ldots + \lambda_{i_k} < 0.$$

Let $k \leq v(1, E)$. Choose $x \in M$ and a nonzero $v \in T_x M$ such that $v(1, E) = v(V_x)$. Then $v(1, E)$ is the multiplicity of the eigenvalue 0 for $R(v, \bar{v})$. Hence

$$\lambda_{i_1} + \ldots + \lambda_{i_k} = 0$$

for some $i_1 < \ldots < i_k$. We have shown

(6.7.22) Proposition. *Let (E, h) be a Hermitian vector bundle with negative semi-definite curvature. Then the curvature of $(\bigwedge^k E, h)$ is negative definite if and only if $k > v(1, E)$.*

Combined with (6.7.20), this implies (see Noguchi-Sunada [1])

(6.7.23) Corollary. *Let M be an n-dimensional Kähler manifold with non-positive bisectional curvature, and let l be the smallest integer such that $v_M(l) > 0$. Then the curvature of $\bigwedge^k TM$ is negative definite if and only if $k > l$.*

From (6.7.20) and (6.7.23) we obtain

(6.7.24) Corollary. *Let X be a compact complex space. Let (Y, h) be an n-dimensional Kähler manifold with nonpositive bisectional curvature. If the curvature of $(\bigwedge^k TY, h)$ is negative definite, then $\mathrm{Hol}(X, Y, k)$ is discrete.*

We have often used (5.3.4) in the above discussion. When the bisectional curvature of Y is nonpositive, (5.3.4) can be strengthened as follows.

(6.7.25) Theorem. *Let X be a compact complex manifold, and Let (Y, h) a Hermitian manifold with nonpositive bisectional curvature. Assume that $\mathrm{Hol}(X, Y)$ has a compact subfamily $\mathcal{F}$ with a universal irreducible complex structure. Then, for each fixed $x_0 \in X$, the map $\mathcal{F} \to Y$ given by $f \mapsto f(x_0)$ is not only a finite map but also an immersion.*

Proof. In (5.3.4), we proved that the map in question is finite. As we explained at the beginning of Section 3 of Chapter 5, the natural injection $\sigma_f : T_f \mathcal{F} \to$

$H^0(X, f^*TY)$ can be expressed in terms of the differential Φ_* of the evaluation map $\Phi: X \times \mathcal{F} \to Y$ as follows:

$$(\sigma_f(\zeta))(x) = \Phi_*(0_x, \zeta) \in T_{f(x)}Y, \qquad x \in X, \ \zeta \in T_f\mathcal{F}.$$

If $\zeta \neq 0$, then the section $\sigma_f(\zeta)$ is not identically equal to zero since σ_f is injective. Since the curvature of the Hermitian vector bundle (f^*TY, f^*h) is negative semi-definite, the section $\sigma_f(\zeta)$ is parallel by (6.7.11) and hence vanishes nowhere on X. In particular, $\Phi(0_{x_0}, \zeta) \neq 0$, showing that the differential of the map $f \mapsto f(x_0)$ is nondegenerate. $\qquad\square$

In order to generalize some of the precedings results to the noncompact space X, we begin with the following technical generalization of (6.7.11).

(6.7.26) **Lemma.** *Let X be a compact normal complex space, and A a proper closed complex subspace such that $X - A$ is nonsingular. Let (E, h) be a Hermitian vector bundle over $X - A$. Assume that the curvature R of h is negative semi-definite. If ξ is a holomorphic section of E over $X - A$ such that $h(\xi, \bar{\xi})$ is bounded, then it satisfies the following equations:*

(a) $\nabla\xi = 0$;

(b) $R(u, \bar{v})\xi = 0$, $\quad$ for $\quad u, v \in T_x(X - A)$, $x \in X - A$.

Proof. By (6.6.13), $h(\xi, \bar{\xi})$ is plurisubharmonic on $X - A$. Since it is bounded on $X - A$, it extends to a plurisubharmonic function on X. Now the rest of the argument is the same as in the proof of (6.7.11). $\qquad\square$

With the help of the lemma above, we obtain the following generalization of (6.7.20) with no essential change in the proof.

(6.7.27) **Theorem.** *Let Z be an n-dimensional compact complex space with a Cartier divisor B such that $Y = Z - B$ is hyperbolically imbedded in Z. Assume that Y carries a Kähler metric h with bisectional curvature nonpositive and holomorphic sectional curvature bounded by a negative constant. Let l be the largest integer such that $v_Y(l) > 0$. Let X be a compact complex manifold, and A a divisor with only normal crossing singularities. Then*

$$\dim \mathrm{Hol}(X - A, Y, k) \begin{cases} \leq \min\{l, \ n - k\} & \text{if } k \leq l; \\ = 0 & \text{if } k > l. \end{cases}$$

Proof. Let $\mathcal{F}$ be a connected component of $\mathrm{Hol}(X - A, Y, k)$, and $\bar{\mathcal{F}}$ its closure in $\mathrm{Hol}(X, Z, k)$. By (6.4.10), $\bar{\mathcal{F}}$ is a compact complex subspace of $\mathrm{Hol}(X, Z, k)$, and $\mathcal{F}$ is a Zariski open subset of $\bar{\mathcal{F}}$. Let $\Phi: X \times \bar{\mathcal{F}} \to Z$ be the evaluation map.

Given $f \in \mathcal{F}$, we extend it to $\bar{f}$ by (6.3.9). As we explained in Section 3 of Chapter 5, each tangent vector $\zeta \in T_{\bar{f}}\bar{\mathcal{F}}$ induces a holomorphic section $\sigma_{\bar{f}}(\zeta) \in H^0(X, \bar{f}^*TZ)$ by

$$\sigma_{\bar{f}}(\zeta)(x) = \Phi_*(x, \zeta) \in T_{\bar{f}(x)}Z, \qquad x \in X.$$

Since $\bar{f}(X - A) \subset Y$, restricting $\sigma_{\bar{f}}(\zeta)$ to $X - A$, we obtain a holomorphic section $\sigma_f(\zeta) \in H^0(X - A, f^*TY)$.

Each $x \in X - A$ defines a holomorphic map from $\mathcal{F}$ to Y sending f to $f(x)$ and ζ to $\sigma_f(\zeta)(x)$. Let F_Y and $F_{\mathcal{F}}$ be the intrinsic length functions of Y and $\mathcal{F}$. By the length-decreasing property of a holomorphic map, we have

$$F_Y(\sigma_f(\zeta)(x)) \leq F_{\mathcal{F}}(\zeta).$$

Since the right hand side is a number independent of x, this shows that the intrinsic length $F_Y(\sigma_f(\zeta))$ of the holomorphic section $\sigma_f(\zeta)$ is bounded on $X - A$. Since the holomorphic sectional curvature of (Y, h) is bounded above by a negative constant, by suitably normalizing h we may assume that $h \leq F_Y^2$ by (2.3.5) and (3.5.19). Hence, $h(\sigma_f(\zeta), \overline{\sigma_f(\zeta)})$ is bounded on $X - A$. Now, we apply Lemma (6.7.26) to the holomorphic section $\xi = \sigma_f(\zeta)$.

The compactness of X was needed in (6.7.12) and (6.7.13) only to the extent that the compactness of X was used in (6.7.11). Since we are interested only in those sections of f^*TY which come from $T_f\mathcal{F}$, i.e., sections of the form $\sigma_f(\zeta)$, $\zeta \in T_f\mathcal{F}$, we use (6.7.26) instead of (6.7.11) to obtain $\dim T_f\mathcal{F} \leq \nu_Y(k)$ with no changes in the proof of (6.7.13). Hence, $\dim \mathcal{F} \leq \nu_Y(k)$. Since (6.7.14) and (6.7.19) have nothing to do with X, we are free to use them here and obtain the stated result. $\qquad\square$

The theorem above can be applied to an arithmetic quotient $Y = \Gamma\backslash\mathcal{D}$ of a symmetric domain and its Satake compactification Z. For further related results, see Noguchi [10].

8 Holomorphic Maps into Quotients of Symmetric Domains

Most of the results in the preceding section can be applied and sharpened when Y is the quotient $\Gamma\backslash(G/K)$ of a symmetric bounded domain G/K by a torsion-free discrete subgroup of G. We follow Sunada [1] and Noguchi-Sunada [1].

Let $M = G/K$ be a bounded symmetric domain, where G is a connected noncompact semisimple Lie group with Lie algebra $\mathbf{g}$, and K is a maximal compact subgroup with Lie algebra $\mathbf{k}$. Let $Y = \Gamma\backslash M$, where Γ is a torsion-free discrete subgroup of G. In order to apply results in preceding section to $\mathrm{Hol}(X, Y)$, we need to calculate the k-nullity $\nu_Y(k)$. Clearly, $\nu_Y(k) = \nu_M(k)$.

Let β be the Killing form of $\mathbf{g}$. Let

$$\mathbf{g} = \mathbf{k} + \mathbf{p}$$

be the Cartan decomposition of $\mathbf{g}$ so that $\mathbf{k}$ and $\mathbf{p}$ are perpendicular with respect to β. Then $\mathbf{p}$ is identified with the tangent space of M at the origin. The Killing form β is negative definite on $\mathbf{k}$ and positive definite on $\mathbf{p}$.

Let $\mathbf{a} \subset \mathbf{p}$ be a linear subspace of maximal dimension such that $[\mathbf{a}, \mathbf{a}] = 0$. Its dimension $r = \dim \mathbf{a}$ is called the **rank** of the symmetric space G/K.

Let $J: \mathbf{p} \to \mathbf{p}$ be the endomorphism defined by the complex structure J of M. There is a unique element z in the center of $\mathbf{k}$ such that

$$(6.8.1) \qquad\qquad J x = [z, x], \qquad x \in \mathbf{p}.$$

The extension of J to the complexification $\mathbf{p}_{\mathbf{C}}$ of $\mathbf{p}$ gives an eigenspace decomposition

$$\mathbf{p}_{\mathbf{C}} = \mathbf{p}_{+} + \mathbf{p}_{-}, \qquad \mathbf{p}_{-} = \bar{\mathbf{p}}_{+},$$

and $\mathbf{p}_{+}$ is identified with the holomorphic tangent space of M at the origin. From (6.8.1) we obtain

$$(6.8.2) \qquad\qquad [\mathbf{p}_{+}, \mathbf{p}_{+}] = [\mathbf{p}_{-}, \mathbf{p}_{-}] = 0.$$

It is sometimes convenient to extend J to $\mathbf{g}$ (and $\mathbf{g}_{\mathbf{C}}$) by setting $J = 0$ on $\mathbf{k}$ (and $\mathbf{k}_{\mathbf{C}}$). Then the integrability condition for J is given by (see Kobayashi-Nomizu [1; vol. II, p. 217])

$$(6.8.3) \qquad [Jx, Jy] - [x, y] - J[x, Jy] - J[Jx, y] = 0, \qquad x, y \in \mathbf{g}_{\mathbf{C}}.$$

We extend the Killing form β to $\mathbf{g}_{\mathbf{C}}$ as complex bilinear form. Then

$$\beta(\mathbf{p}_{+}, \mathbf{p}_{+}) = \beta(\mathbf{p}_{-}, \mathbf{p}_{-}) = 0,$$
$$(6.8.4) \qquad \beta(u, \bar{u}) < 0 \quad \text{for nonzero} \quad u \in \mathbf{k}_{\mathbf{C}},$$
$$\beta(u, \bar{u}) > 0 \quad \text{for nonzero} \quad u \in \mathbf{p}_{\mathbf{C}}.$$

The curvature $R: \mathbf{p}_{+} \times \mathbf{p}_{-} \to \operatorname{End}(\mathbf{p}_{+})$ of M is given by (see, for example, Kobayashi-Nomizu [1; vol. II; p. 193])

$$(6.8.5) \qquad\qquad R(u, \bar{v}) w = -[[u, \bar{v}], w], \qquad u, v, w \in \mathbf{p}_{+}.$$

Then

$$\begin{aligned}
\beta(R(u, \bar{v}) w, \bar{t}) &= -\beta([[u, \bar{v}], w], \bar{t}) = \beta([[\bar{v}, w], u], \bar{t}) \\
&= \beta([\bar{v}, w], [u, \bar{t}]) = \beta([u, \bar{t}], \overline{[v, \bar{w}]}), \qquad u, v, w, t \in \mathbf{p}_{+}.
\end{aligned}$$

In particular,

$$(6.8.6) \qquad\qquad \beta(R(u, \bar{u}) w, \bar{w}) = \beta([u, \bar{w}], \overline{[u, \bar{w}]}) \leq 0,$$

which shows that, for a symmetric bounded domain, the bisectional curvature is nonpositive and the holomorphic sectional curvature is negative.

To each linear subspace $\mathbf{v}$ of $\mathbf{p}_+$ we associate a subspace $\mathbf{n}(\mathbf{v})$ of $\mathbf{p}_+$ as follows (see (6.7.7) and (6.8.5)):

$$(6.8.7) \qquad \mathbf{n}(\mathbf{v}) = \{w \in \mathbf{p}_+;\ [[u, \bar{v}], w] = 0 \quad \text{for} \quad u, v \in \mathbf{v}\}.$$

We prove the following two characterizations of $\mathbf{n}(\mathbf{v})$.

$$(6.8.8), \qquad \mathbf{n}(\mathbf{v}) = \{w \in \mathbf{p}_+;\ [w, \bar{u}] = 0 \quad \text{for} \quad u \in \mathbf{v}\}$$

$$(6.8.9) \qquad \mathbf{n}(\mathbf{v}) = \{w \in \mathbf{p}_+;\ [[u, \bar{v}], w] = 0 \quad \text{for} \quad u \in \mathbf{p}_+,\ v \in \mathbf{v}\}.$$

In fact, if $w \in \mathbf{n}(\mathbf{v})$, then (6.8.5) together with (6.8.6) implies $[w, \bar{u}] = 0$ for all $u \in \mathbf{v}$. Conversely, if $[w, \bar{v}] = 0$ for all $v \in \mathbf{v}$, then

$$[[u, \bar{v}], w] = [u, [\bar{v}, w]] = 0, \qquad u \in \mathbf{p}_+,\ v \in \mathbf{v},$$

showing (6.8.8) and (6.8.9) at the same time.

(6.8.10) **Lemma.** *For any $\mathbf{v} \subset \mathbf{p}_+$, we have*

    (1) $\mathbf{v} \perp \mathbf{n}(\mathbf{v})$,   i.e., $\beta(\mathbf{v}, \bar{\mathbf{n}}(\mathbf{v})) = 0$;

    (2) $\mathbf{v} \subset \mathbf{n}(\mathbf{n}(\mathbf{v}))$;

    (3) $\mathbf{n}(\mathbf{v}) = \mathbf{n}(\mathbf{n}(\mathbf{n}(\mathbf{v})))$.

    (4) $\mathbf{n}(\mathbf{v}) \perp \mathbf{n}(\mathbf{n}(\mathbf{v}))$;

    (5) $[[\mathbf{n}(\mathbf{v}), \overline{\mathbf{n}(\mathbf{v})}], \mathbf{n}(\mathbf{v})] \subset \mathbf{n}(\mathbf{v})$.

Proof. (1)   Let $u \in \mathbf{v}$, $w \in \mathbf{n}(\mathbf{v})$, and let z be the element in the center of $\mathbf{k}$ giving J. Then

$$\beta(iu, \bar{w}) = \beta([z, u], \bar{w}) = \beta(z, [u, \bar{w}]) = 0$$

by (6.8.8).

    (2)   This is immediate from (6.8.8).

    (3)   Replacing $\mathbf{v}$ by $\mathbf{n}(\mathbf{v})$ in (2), we have $\mathbf{n}(\mathbf{v}) \subset \mathbf{n}(\mathbf{n}(\mathbf{n}(\mathbf{v})))$. On the other hand, applying $\mathbf{n}$ to both sides of (2) we obtain the reverse inclusion.

    (4)   Replace $\mathbf{v}$ by $\mathbf{n}(\mathbf{v})$ in (1), and apply (3).

    (5)   Use the Jacobi identity togehter with (6.8.2) and (6.8.8). $\square$

Under the natural identification of $\mathbf{p}$ with $\mathbf{p}_+$ ($x \in \mathbf{p} \to x - iJx \in \mathbf{p}_+$), the J-invariant subspaces of $\mathbf{p}$ are in one-to-one correspondence with the complex subspaces of $\mathbf{p}_+$. Given a complex subspace of $\mathbf{p}_+$, we denote the corresponding J-invariant subspace of $\mathbf{p}$ by adding a subscript $\mathbf{R}$, e.g., $\mathbf{v}_{\mathbf{R}}$ and $\mathbf{n}(\mathbf{v})_{\mathbf{R}}$.

If $u = x - iJx$ and $w = y - iJy$ with $x, y \in \mathbf{p}$, then

$$(6.8.11) \qquad [w, u] = [w, \bar{u}] = 0 \iff [y, x] = [y, Jx] = 0.$$

Hence,

$$(6.8.12) \qquad \mathbf{n}(\mathbf{v})_{\mathbf{R}} = \{y \in \mathbf{p};\ [y, x] = [y, Jx] = 0 \quad \text{for} \quad x \in \mathbf{v}_{\mathbf{R}}\}.$$

Now, (5) of (6.8.10) may be restated as follows.

(6.8.13) **Lemma.** *For any complex subspace $\mathbf{v}$ of $\mathbf{p}_+$, $\mathbf{n}(\mathbf{v})_\mathbf{R}$ is a J-invariant Lie triple system, i.e.,*

$$[[\mathbf{n}(\mathbf{v})_\mathbf{R}, \mathbf{n}(\mathbf{v})_\mathbf{R}], \mathbf{n}(\mathbf{v})_\mathbf{R}] \subset \mathbf{n}(\mathbf{v})_\mathbf{R}.$$

In general, if $\mathbf{p}' \subset \mathbf{p}$ is a J-invariant Lie triple system, then

$$\mathbf{g}' = \mathbf{k}' + \mathbf{p}', \quad \text{where} \quad \mathbf{k}' = [\mathbf{p}', \mathbf{p}']$$

is the Cartan decomposition of a Hermitian symmetric subspace $G'/K' \subset G/K$.

Summarizing the construction so far, we state

(6.8.14) **Theorem.** *Given $\mathbf{v} \subset \mathbf{p}_+$, set*

$$\begin{aligned}
\mathbf{p}' &= \mathbf{n}(\mathbf{n}(\mathbf{v}))_\mathbf{R}, & \mathbf{k}' &= [\mathbf{p}', \mathbf{p}'], & \mathbf{g}' &= \mathbf{k}' + \mathbf{p}', \\
\mathbf{p}'' &= \mathbf{n}(\mathbf{v})_\mathbf{R}, & \mathbf{k}'' &= [\mathbf{p}'', \mathbf{p}''], & \mathbf{g}'' &= \mathbf{k}'' + \mathbf{p}''.
\end{aligned}$$

Then $\mathbf{g}' \perp \mathbf{g}''$ with respect to the Killing form β, and $[\mathbf{g}', \mathbf{g}''] = 0$. Thus, we have mutually perpendicular totally geodesic Hermitian symmetric subspaces G'/K' and G''/K'' of $M = G/K$:

$$(G'/K') \times (G''/K'') \subset G/K.$$

Let $r = \operatorname{rank}(G, K)$, $r' = \operatorname{rank}(G'/K')$, and $r'' = \operatorname{rank}(G''/K'')$. Then (6.8.14) implies

$$r' + r'' \leq r.$$

However, in the special case $\dim \mathbf{v} = 1$, we have

(6.8.15) **Corollary.** *If $\dim \mathbf{v} = 1$, then*

$$\operatorname{rank}(G'/K') = 1, \qquad \operatorname{rank}(G''/K'') = r - 1.$$

Proof. We have a basis of the form $x - iJx$ for $\mathbf{v}$, where $x \in \mathbf{p}$. Let $\mathbf{a} \subset \mathbf{p}$ be a real vector subspace of maximal dimension containing x such that $[\mathbf{a}, \mathbf{a}] = 0$. Then $r = \dim \mathbf{a}$ is the rank of $M = G/K$. Let $\mathbf{a}' \subset \mathbf{a}$ be the real 1-dimensional space spanned by x. Let $\mathbf{a}''$ be the orthogonal complement of $\mathbf{a}'$ in $\mathbf{a}$ with respect to β. Thus,

$$\mathbf{a} = \mathbf{a}' + \mathbf{a}'', \qquad \mathbf{a}_\mathbf{C} = \mathbf{a}_\mathbf{C}' + \mathbf{a}_\mathbf{C}''.$$

Decomposing $\mathbf{a}_\mathbf{C}$ according to the eigenvalues of J, we have

$$\mathbf{a}_\mathbf{C} = \mathbf{a}_+ + \mathbf{a}_-, \qquad \mathbf{a}_\mathbf{C}' = \mathbf{a}_+' + \mathbf{a}_-', \qquad \mathbf{a}_\mathbf{C}'' = \mathbf{a}_+'' + \mathbf{a}_-''.$$

From our construction, we have $\mathbf{v} = \mathbf{a}'^+$. We claim

$$(6.8.16) \qquad\qquad\qquad [\mathbf{a}_+'', \mathbf{a}_-'] = 0.$$

Let $u = x - iJx \in \mathbf{a}_+'$ and $w = y - iJy \in \mathbf{a}_+''$ with $y \in \mathbf{a}''$. Now, our assertion follows from (6.8.11).

From (6.8.16) we obtain $\mathbf{a}''_+ \subset \mathbf{n}(\mathbf{v})$, which implies that $\mathbf{a}'' \subset \mathbf{n}(\mathbf{v})_{\mathbf{R}} = \mathbf{p}''$.

Since $\mathbf{a}'' \subset \mathbf{p}''$, G''/K'' has rank $\geq r - 1$ and hence $r - 1$. Then G'/K' must have rank at most 1 and hence 1. $\qquad\square$

Even under the constraint that $\dim \mathbf{v} = 1$, the dimension of $\mathbf{n}(\mathbf{v})$ varies with $\mathbf{v}$. By (6.7.19),

$$l = v_M(1) = \max_{\dim \mathbf{v}=1} \dim \mathbf{n}(\mathbf{v}),$$

where the maximum is taken over all 1-dimensional subspace $\mathbf{v}$ of $\mathbf{p}_+$.

(6.8.17) **Theorem.** *The integer l above coincides with the maximum dimension of proper boundary components of $M = G/K$.*

Proof. Let l^* be the maximum dimension of proper boundary components of M. In (6.8.14) let $\mathbf{v}$ be the 1-dimensional subspace of $\mathbf{p}_+$ such that $l = \dim \mathbf{n}(\mathbf{v})$. Then in the notation of (6.8.14), G''/K'' is a totally geodesic Hermitian symmetric subspace of dimension l. If a is any boundary point of G'/K', then $\{a\} \times (G''/K'')$ is a boundary component of M. Hence, $l \leq l^*$.

On the other hand, every proper boundary component of M can be obtained as the Cayley transform of a totally geodesic Hermitian symmetric subspace whose tangent space (considered as a subspace of $\mathbf{p}_+$) is of the form $\mathbf{n}(\mathbf{v})$ for some $\mathbf{v} \subset \mathbf{p}_+$, (see Wolf [1; p. 287] or Wolf-Korányi [1]). In particular, if $\mathbf{n}(\mathbf{v})$ is the tangent space to a proper boundary component of maximum dimension, then $l^* = \dim \mathbf{n}(\mathbf{v})$. Since $\mathbf{n}(\mathbf{n}(\mathbf{v})) \neq 0$ by (6.8.10) and since l is the largest integer such that $v_M(l) > 0$ (see (6.7.19)), we have $\dim \mathbf{n}(\mathbf{v}) \leq l$, thus proving $l^* \leq l$.

$\qquad\square$

In order to determine the integer $l = v(1)$, we have to find a 1-dimensional $\mathbf{v}$ that maximizes the $\dim \mathbf{n}(\mathbf{v})$. For this purpose we quickly review root systems for symmetric bounded domains. Let $\mathbf{h}$ be a maximal abelian subalgebra of $\mathbf{k}$. In the Hermitian case, it is a Cartan subalgebra of $\mathbf{g}$. Let $\mathbf{h}_{\mathbf{C}}$ be its complexification. Let Δ be the set of nonzero roots of $\mathbf{g}_{\mathbf{C}}$ with respect to $\mathbf{h}_{\mathbf{C}}$. For $\alpha \in \Delta$, let $\mathbf{g}^\alpha$ denote the root subspace for α. Then either $\mathbf{g}^\alpha \subset \mathbf{k}_{\mathbf{C}}$ (then α is called a compact root), or $\mathbf{g}^\alpha \subset \mathbf{p}_{\mathbf{C}}$ (then α is called a noncompact root). Thus,

$$\mathbf{k}_{\mathbf{C}} = \mathbf{h} + \sum_\alpha \mathbf{g}^\alpha, \qquad \mathbf{p}_{\mathbf{C}} = \sum_\beta \mathbf{g}^\beta,$$

where α runs over all compact roots while β runs over all noncompact roots. Let $\mathbf{c}$ be the center of $\mathbf{h}$. Then α is a compact root if and only if it vanishes on $\mathbf{c}$.

Since each root α is real valued on $i\mathbf{h}$, we can introduce a lexicographic ordering in Δ by choosing a basis $X_1, \ldots, X_r$ for $i\mathbf{h}$ in such a way that $X_1, \ldots, X_k$ form a basis for $i\mathbf{c}$. Thus, $\alpha > \beta$ if $\alpha(X_1) = \beta(X_1), \ldots, \alpha(X_{i-1}) = \beta(X_{i-1})$ and

$\alpha(X_i) = \beta(X_i)$. Let Δ^+ be the set of positive roots, and Q_+ the set of noncompact positive roots. Then

$$\mathbf{p}_+ = \sum_{\beta \in Q_+} \mathbf{g}^\beta, \qquad \mathbf{p}_- = \sum_{-\beta \in Q_+} \mathbf{g}^\beta.$$

Let $\alpha, \beta \in Q_+$. Since

$$[\mathbf{g}^\beta, \mathbf{g}^{-\alpha}] = \begin{cases} \mathbf{g}^{\beta - \alpha} & \text{if } \beta - \alpha \in \Delta, \\ 0 & \text{if } \beta - \alpha \notin \Delta, \end{cases}$$

it follows from (6.8.8) that the maximum of $\dim \mathbf{n}(\mathbf{v})$ is achieved by some $\alpha \in Q_+$, and

$$(6.8.18) \qquad \nu_M(1) = \max_{\alpha \in Q_+} \#\{\beta \in Q_+; \ \beta - \alpha \notin \Delta\}.$$

(6.8.19) Example. Let $G/K = SU(p, q)/S(U(p) \times U(q))$. Then $\mathbf{g_C} = \mathbf{sl}(p+q; \mathbf{C})$ and

$$\mathbf{k_C} = \left\{ \begin{pmatrix} * & 0 \\ 0 & * \end{pmatrix} \right\}, \qquad \mathbf{p_C} = \left\{ \begin{pmatrix} 0 & * \\ * & 0 \end{pmatrix} \right\}.$$

Let $e_{i,j}$ denote the $(p+q) \times (p+q)$-matrix with 1 at the (i, j)-th place (i.e., the i-th row and j-th column) and 0 elsewhere. Set

$$e' = \frac{1}{p}(e_{1,1} + \ldots + e_{p,p}), \qquad e'' = \frac{1}{q}(e_{p+1,p+1} + \ldots + e_{p+q,p+q}).$$

Then $e' - e''$ spans the center of $\mathbf{k_C}$, and the following elements form a basis for a Cartan subalgebra $\mathbf{h_C}$:

$$e' - e'', \ e_{i,i} - e_{i+1,i+1}, \ i = 1, \ldots, p - 1, p + 1, \ldots, p + q - 1.$$

Then

$$\mathbf{p}_+ = \left\{ \begin{pmatrix} 0 & * \\ 0 & 0 \end{pmatrix} \right\}, \qquad \mathbf{p}_- = \left\{ \begin{pmatrix} 0 & 0 \\ * & 0 \end{pmatrix} \right\}.$$

Thus $\mathbf{p}_+$ is identified with the space of $p \times q$-matrices. The maximum of $\dim \mathbf{n}(\mathbf{v})$ is achieved by the space $\mathbf{v}$ spanned by $e_{1,p+1}$, and $\mathbf{n}(\mathbf{v})$ is then spanned by $e_{i,p+j}$, $2 \leq i \leq p$, $2 \leq j \leq q$ and is identified with the space of $(p-1) \times (q-1)$-matrices. Hence,

$$l = \nu_M(1) = (p - 1)(q - 1).$$

Furthermore (see (6.8.14)),

$$G''/K'' = SU(p - 1, q - 1)/S(U(p - 1) \times U(q - 1)).$$

The following is a complete list of l for the irreducible symmetric bounded domains, (see Wolf-Korányi [1]):

	Domain	Dimension	l
I	$SU(p, q)/S(U(p) \times U(q))$	pq	$(p-1)(q-1)$
II	$Sp(m, \mathbf{R})/U(m)$	$m(m+1)/2$	$(m-1)m/2$
III	$SO^*(2m)/U(m)$	$m(m-1)/2$	$(m-2)(m-3)/2$
IV	$SO_0(m, 2)/SO(m) \times SO(2)$	m	1
V	$E_6/SO(10) \cdot SO(2)$	16	1
VI	$E_7/E_6 \cdot SO(2)$	27	8

Most of the results in Section 7 can be applied to holomorphic mappings into arithmetic quotients of symmetric domains with the integer l given in the table above. Part of (6.7.27) can be strengthened in such cases.

(6.8.20) Theorem. *Let $Y = \Gamma \backslash (G/K)$ be the quotient of a Hermitian symmetric domain G/K by a torsion-free arithmetic subgroup Γ of G. Let l be the maximum dimension of proper boundary components of G/K. Let X be a compact complex manifold, and A a divisor with only normal crossing singularities.*

(1) *If $k > l$, then $\mathrm{Hol}(X - A, Y, k)$ is finite.*

(2) *If $k \leq l$, then $\dim \mathrm{Hol}(X - A, Y, k) \leq \min\{l, n - k\}$. Moreover, if $\mathcal{F}$ is a connected component of $\mathrm{Hol}(X - A, Y, k)$, $\bar{\mathcal{F}}$ its closure in $\mathrm{Hol}(X, Z)$ and $\partial \mathcal{F} = \bar{\mathcal{F}} - \mathcal{F}$ its boundary, then*
   (a) *$\bar{\mathcal{F}}$ is a compact complex space,*
   (b) *$\mathcal{F}$ is a complete hyperbolic and hyperbolically imbedded Zariski open subset of $\bar{\mathcal{F}}$,*
   (c) *$\partial \mathcal{F}$ is a Cartier divisor of $\bar{\mathcal{F}}$ and $\partial \mathcal{F} \subset \mathrm{Hol}(X, B)$.*

Proof. (1) Let Z be the Satake compactification of Y, (see (6.1.3)). By (6.7.27) and (6.8.17), $\dim \mathrm{Hol}(X - A, Y, k) = 0$. By (6.4.9), $\mathrm{Hol}(X - A, Y, k)$ is compact.
   (2) This follows from (6.4.10) and (6.7.27). $\square$

For further results on maps into quotients of symmetric domains, see Noguchi [10] and Miyano-Noguchi [1].

9 Finiteness Theorems for Sections of Hyperbolic Fibre Spaces

The function field analogue of Mordell's conjecture states if $\pi: Y \to X$ is a holomorphic map of a nonsingular algebraic surface Y onto a compact Riemann surface X such that the fiber $\pi^{-1}(x)$ over a generic point $x \in X$ is a compact Riemann surface of geneus ≥ 2, then there are only finitely many meromorphic sections of the projection π unless the fibering is bimeromorphically trivial. This was proved, independently, by Grauert [2] and Manin [1]. Other proofs and generalizations have been obtained by Samuel [2], Parshin [1, 2] and Raynaud [1]. See Lang [6] on comments on these various proofs.

The theorem of Manin and Grauert has been generalized to families of *noncompact* curves of general type by Zaidenberg [9].

In this section we consider the higher dimensional Mordell conjecture over function fields as formulated by Lang [1]:

If an algebraic family of compact hyperbolic complex spaces admits infinitely many sections, then the family contains split subfamilies, and all but a finite number of sections are trivial, i.e., constant sections of these split subfamilies.

This conjecture of Lang was verified under additional assumptions on fibers by several authors: by Riebesehl [1] when the fibers admit a complex Finsler metric with negative holomorphic sectional curvature, by Martin-Deschamps [1] and Noguchi [5] when the fibers have ample cotangent bundle, and then by Noguchi [7] when the fibers are hyperbolic and admit only finitely many surjective holomorphic maps from any compact complex space. Finally, Noguchi [13] proved de Franchis' theorem (6.6.2) for compact hyperbolic complex spaces, thus removing the last condition. For a survey on these results on the conjecture, see Miyano-Noguchi [1]. Recently, Makoto Suzuki [3] extended results of Noguchi and Zaidenberg to the case of hyperbolically imbedded fiber space. The Mordell conjecture for an algebraic family of projective varieties of general type has been studied by Maehara [3, 4].

In this section we shall prove Noguchi's result and Suzuki's generalization.

We use the notation explained in Section 4. In particular, given a holomorphic fiber space $\pi: Y \to X$, we denote its fiber over $x \in X$ by $Y_x = \pi^{-1}(x)$ and its restriction to $U \subset X$ by $Y_U = \pi^{-1}U$. The space of holomorphic sections over U is written $\Gamma(U, Y)$ and the space of global holomorphic sections $\Gamma(Y) = \Gamma(X, Y)$.

By Douady [1] (see Section 3 of Chapter 5), if X is compact, then $\mathrm{Hol}(X, Y)$ has a universal complex structure and the evaluation map

$$\Phi: X \times \mathrm{Hol}(X, Y) \to Y, \qquad \Phi(x, f) = f(x),$$

is holomorphic, and $\Gamma(Y)$ is a closed complex subspace of $\mathrm{Hol}(X, Y)$. If X is not compact, $\Gamma(Y)$ need not be a complex space. But, in certain cases, $\Gamma(Y)$ may contain a connected subset $\mathcal{S}$ which carries a *universal* complex structure.

We are now in a position to prove the following theorem of Noguchi [7, 13]. The proof uses (6.6.2) in an essential way.

(6.9.1) Theorem. *Let X be a nonsingular compact complex manifold, and $A \subset X$ a divisor with only normal crossing singularities. Let (Y, π, X) be a compact complex fiber space hyperbolic over (X, A). If Y is irreducible and normal and if there is a point $x_0 \in X - A$ such that*

$$(*) \qquad\qquad Y_{x_0} = \{f(x_0);\ f \in \Gamma(X - A, Y)\},$$

then Y is biholomorphic to a product bundle $X \times Y_{x_0}$.

Proof. By (6.5.13), $\Gamma(X - A, Y) \cong \Gamma(X, Y)$ is a compact complex space. We set $\Gamma(Y) = \Gamma(X, Y)$. Since $\Phi(X \times \Gamma(Y))$ contains the fiber Y_{x_0} by $(*)$ and is the union of sections and since $X \times \Gamma(Y)$ is compact, we have $\Phi(X \times \Gamma(Y)) = Y$. Hence, there is an irreducible component $\mathcal{S}$ of $\Gamma(Y)$ such that $\Phi(X \times \mathcal{S}) = Y$.

By (6.5.13) the evaluation map $\Phi: X \times \mathcal{S} \to Y$ is a finite map, and hence it is a finite ramified covering with, say, k-sheets. Since $\Phi: \{x\} \times \mathcal{S} \to Y_x$ is a finite map, we have $\dim \mathcal{S} = \dim Y_x$. Let

$$\mathcal{S}_y = \{f \in \mathcal{S};\ f(\pi(y)) = y\}.$$

For a generic y, $\mathcal{S}_y$ has k elements.

Given a holomorphic vector field v on an open set U of X, we lift it to a "horizontal" holomorphic vector field $\tilde{v}$ on $Y_U = \pi^{-1}(U)$ in the following manner. First, we restrict the differential

$$\Phi_*: TX \times TS \to TY$$

of Φ to $TX \times \mathcal{S}$. Then $\Phi_*(v, \mathcal{S})$ is, roughly speaking, a multivalued holomorphic vector field on Y_U; at generic points (i.e., at unramified points) it gives k holomorphic vector fields. We define $\tilde{v}$ to be the average of these k vector fields. More precisely, at each point $y \in Y_U$ outside the ramification locus of the covering map Φ, we set

$$\tilde{v}(y) = \frac{1}{k} \sum_{f \in \mathcal{S}_y} \Phi_*(v(\pi(y)), f).$$

Since $\mathcal{S}$ is compact, $\Phi_*(v, \mathcal{S})$ is bounded. Since Y is normal, $\tilde{v}$ extends through the ramification locus.

With respect to a local coordinate system in U, take the coordinate vector fields in U. Then lift these vector fields to horizontal vector fields on Y_U as above. We can use these horizontal holomorphic vector fields to translate a fiber over $x_0 \in U$ to nearby fibers, and we see that Y is a holomorphic fiber bundle over X with all fibers isomorphic to the fiber Y_{x_0}. By (5.4.4) its structure group $\mathrm{Aut}(Y_{x_0})$ is finite.

Let $p: Y_U \cong U \times Y_{x_0} \to Y_{x_0}$ be the projection to the standard fiber. Each $x \in U$ defines a surjective holomorphic map

$$\lambda_x: \mathcal{S} \to Y_{x_0}, \quad \text{where} \quad \lambda_x(f) = p(f(x)).$$

Since $\mathcal{S}$ is compact and Y_{x_0} is compact hyperbolic, $\mathrm{Sur}(\mathcal{S}, Y_{x_0})$ is finite by (6.6.2). Since U is connected, λ_x has to be independent of x. In other words, the sections $f \in \mathcal{S}$, restricted to U, are all constant with respect to the local trivialization $Y_U \cong U \times Y_{x_0}$. Hence, if $y \in Y_U$ with $x = \pi(y)$ and if $f, f' \in \mathcal{S}_y$ so that $f(x) = f'(x) = y$, then $f = f'$ on U, and consequently, $f = f'$. This means that $k = 1$ and $\Phi: X \times \mathcal{S} \to Y$ is an isomorphism. $\square$

(6.9.2) **Corollary.** *Let X be a compact complex space, and $A \subset X$ a closed complex subspace such that $X - A$ is nonsingular. Let (Y, π, X) be a compact complex fiber space hyperbolic over (X, A). Assume that Y is irreducible and $Y_{X-A} = \pi^{-1}(X - A)$ is normal and that there is a point $x_0 \in X - A$ such that*

$$(*) \qquad Y_{x_0} = \bigcup_{f \in \Gamma(X-A, Y)} f(x_0)$$

Then there is a holomorphic map of $X \times Y_{x_0}$ onto Y which is bimeromorphic and induces an isomorphism from $(X - A) \times Y_{x_0}$ onto Y_{X-A}.

Proof. Let $p: (X', A') \to (X, A)$ be a resolution of singularities of X such that A' is a divisor with only normal crossing singularities and $X' - A' \cong X - A$ under p. By replacing (Y, π, X) by its pull-back $(p^{-1}Y, \pi', X')$ and making use of (6.5.2), we may assume that X is nonsingular and A has only normal crossing singularities.

Let $q: \tilde{Y} \to Y$ be the normalization of Y. Since Y_{X-A} is normal, $\tilde{Y}_{X-A} \cong Y_{X-A}$ under q. By (6.9.1), $\tilde{Y} \cong X \times Y_{x_0}$. $\quad\square$

(6.9.3) Corollary. *Let X be a compact complex space, and $A \subset X$ a closed complex subspace. Let (Y, π, X) be a compact complex fiber space hyperbolic over (X, A). Let $\Gamma^*(X - A, Y)$ denote the set of meromorphic sections of Y over $X - A$. If there is a point $x_0 \in X - A$ such that*

$$(*) \qquad\qquad Y_{x_0} = \bigcup_{f \in \Gamma^*(X-A,Y)} f(x_0),$$

then Y is bimeromorphic to a product bundle $X \times Y_{x_0}$.

Proof. Let $p: (X', A') \to (X, A)$ be a resolution of singularities of X such that A' is a divisor with only normal crossing singularities. By replacing (Y, π, X) by its pull-back $(p^{-1}Y, \pi', X')$ and making use of (6.5.2) and (6.3.19), we may assume that X is nonsingular and A has only normal crossing singularities. (Now all meromorphic sections are holomorphic).

As shown in the proof of (6.9.1), there is an irreducible component $\mathcal{S}$ of $\Gamma(Y)$ such that $\Phi(X \times \mathcal{S}) = Y$. In particular, condition $(*)$ is satisfied by any point x_0 of X.

Let $q: \tilde{Y} \to Y$ be the normalization of Y. The complex fiber space $(\tilde{Y}, \tilde{\pi}, X)$ with $\tilde{\pi} = \pi \circ q$ is hyperbolic over (X, A) by (6.5.3). Let $N \subset Y$ denote the locus of nonnormal points of Y. Since $(*)$ is satisfied by any point x_0 of $X - A$, we take x_0 such that Y_{x_0} is not contained in N.

Every section $\tilde{f} \in \Gamma(\tilde{Y})$ induces a section $q \circ \tilde{f} \in \Gamma(Y)$. Conversely, if a section $f \in \Gamma(Y)$ is such that $f(x_0) \notin N$, then there is a section $\tilde{f} \in \Gamma(\tilde{Y})$ such that $f = q \circ \tilde{f}$, see (6.5.24). Hence, $\{\tilde{f}(x_0); \ \tilde{f} \in \Gamma(X - A, \tilde{Y})\}$ covers $\tilde{Y}_{x_0}$ since the only points which may not be covered are points belonging to $q^{-1}(N) \cap \tilde{Y}_{x_0}$, which has a lower dimension than $\tilde{Y}_{x_0}$. By (6.9.1), $\tilde{Y}$ is biholomorphic to a product bundle $X \times \tilde{Y}_{x_0}$. $\quad\square$

We now consider the case where condition $(*)$ in (6.9.1) may not be satisfied.

(6.9.4) Corollary. *Let X be a compact complex space, and $A \subset X$ a closed complex subspace such that $X - A$ is nonsingular. Let (Y, π, X) be a compact complex fiber space hyperbolic over (X, A). Let $\mathcal{S}$ be an irreducible closed complex subspace of $\Gamma(X, Y)$. Then $\mathcal{S}$ is a compact hyperbolic complex space. Put*

$$Y(\mathcal{S}) = \Phi(X \times \mathcal{S}) = \{f(x); \ x \in X, \ f \in \mathcal{S}\} \subset Y.$$

Then the normalization $\tilde{Y}(\mathcal{S})$ of $Y(\mathcal{S})$ is biholomorphic to a product bundle.

Proof. By (6.5.13), $\mathcal{S}$ is a compact hyperbolic complex space. Hence, $Y(\mathcal{S})$ is a compact complex subspace of Y. The fiber space $(Y(\mathcal{S}), \pi, X)$ is hyperbolic along A. Let $\tilde{Y}(\mathcal{S})$ be the normalization of $Y(\mathcal{S})$. By (6.5.3), the fiber space $(\tilde{Y}(\mathcal{S}), \tilde{\pi}, X)$ is also hyperbolic along A. It satisfies condition $(*)$ in (6.9.1); see the argument at the end of the proof of (6.9.3). Now, apply (6.9.1) to $(\tilde{Y}(\mathcal{S}), \tilde{\pi}, X)$. $\square$

Now we shall prove the result of Makoto Suzuki [3] which generalizes (6.9.1) to hyperbolically imbedded fiber spaces. We work in the same set-up as in (6.5.17).

(6.9.5) Theorem. *Let X be a compact nonsingular complex manifold and A a divisor with only normal crossing singularities. Let (Z, π, X) be a compact complex fiber space, and B a Cartier divisor of Z transversal to the fibers in the sense that, at each $x \in X$, $B \cap Z_x$ is a Cartier divisor of the fiber Z_x. Setting $Y = Z - B$, assume that (Y, π, X) is hyperbolically imbedded in (Z, π, X). If Z is irreducible and normal and if there is a point $x_0 \in X - A$ such that*

$$(*) \qquad Y_{x_0} = \{f(x_0); \ f \in \Gamma(X - A, Y)\},$$

then there is an irreducible component $\mathcal{S}$ of $\Gamma(X - A, Y)$ such that

$$Y \cong X \times \mathcal{S}, \qquad Z \cong X \times \bar{\mathcal{S}},$$

where $\bar{\mathcal{S}}$ is the closure of $\mathcal{S}$ in $\Gamma(X, Z)$, and the biholomorphic isomorphisms are given by the evaluation map Φ.

Proof. We consider $\Gamma(X - A, Y)$ as a subset of $\Gamma(X, Z)$ by the natural imbedding

$$\iota: \Gamma(X - A, Y) \subset \Gamma(X, Z),$$

sending each $f \in \Gamma(X - A, Y)$ to its extension $\bar{f} \in \Gamma(X, Z)$.

Since $\Phi(X - A, \Gamma(X - A, Y))$ contains Y_{x_0} by assumption $(*)$ and is the union of sections $f(X - A)$ with $f \in \Gamma(X - A, Y)$, it follows that $\Phi(X - A, \Gamma(X - A, Y))$ covers a neighborhood of Y_{x_0} in Y. Then there is an irreducible component $\mathcal{S}$ of $\Gamma(X - A, Y)$ such that $\Phi(X - A, \mathcal{S})$ covers a neighborhood of Y_{x_0}.

By (6.5.17), $\bar{\mathcal{S}}$ is a compact complex subspace of $\Gamma(X, Z)$. Since $\Phi(X, \bar{\mathcal{S}})$ is a compact complex subspace of Z and covers a neighborhood of Y_{x_0}, it covers the entire Z, i.e., $\Phi(X, \bar{\mathcal{S}}) = Z$.

By (5.3.4) the evaluation map $\Phi: X \times \bar{\mathcal{S}} \to Z$ is a finite map. Hence, it is a ramified finite, say k-fold, covering map. In particular, $\dim \mathcal{S} = \dim Z_x$ for $x \in X$.

Let $A = \bigcup_{i=1}^{m} A_i$ be the decomposition into irreducible components. Using a section belonging to $\mathcal{S}$, as in the proof of (6.5.17) we partition the index set $\{1, \ldots, n\}$ and define A_I. We prove that A_I is empty, i.e., the partition is trivial. Since $\bar{f}(A_I) \subset B$ for $f \in \mathcal{S}$, we have $\Phi(A_I, \mathcal{S}) \subset B$. By continuity, $\Phi(A_I, \bar{\mathcal{S}}) \subset B$. Hence, $\Phi(x, \bar{\mathcal{S}}) \subset B \cap Z_x$ for any $x \in A_I$. Since $B \cap Z_x$ is a divisor in Z_x, we have $\dim(B \cap Z_x) < \dim Z_x$. This, together with $\dim \mathcal{S} = \dim Z_x$ implies $\dim(B \cap Z_x) < \dim \mathcal{S}$, contradicting the fact that $\Phi: \{x\} \times \bar{\mathcal{S}} \to B \cap Z_x$ is a finite map. This proves that A_I is empty. Hence,

$$\Phi(X, \mathcal{S}) \subset Y.$$

Given a holomorphic vector field v on an open set U of X, we lift it to a holomorphic vector field $\tilde{v}$ on Z_U by averaging the multivalued vector field $\Phi_*(v, \bar{\mathcal{S}})$ exactly as in the proof of (6.9.1). More precisely, let

$$\bar{\mathcal{S}}_z = \{f \in \bar{\mathcal{S}};\ f(\pi(y)) = z\}.$$

If $z \in Z$ is not in the ramification locus of the covering map Φ, $\bar{\mathcal{S}}_z$ has k elements, and $\tilde{v}(z)$ is given by

$$\tilde{v}(z) = \frac{1}{k} \sum_{f \in \bar{\mathcal{S}}_z} \Phi_*(v(\pi(z)), f).$$

Since $\bar{\mathcal{S}}$ is compact, $\Phi_*(v, \bar{\mathcal{S}})$ is bounded. Since Z is normal, $\tilde{v}$ extends through the ramification locus. We note that since $f(X) \subset B$ for $f \in \partial\mathcal{S}$, we have $\bar{\mathcal{S}}_y \subset \mathcal{S}$ for $y \in Y$.

As in the proof of (6.9.1) we obtain a holomorphic local isomorphism

$$Z_U \cong U \times Z_{x_0}$$

with $x_0 \in U$. This isomorphism induces an isomorphism

$$Y_U \cong U \times Y_{x_0}.$$

Hence, Z is a holomorphic fiber bundle with standard fiber Z_{x_0}, and Y is a holomorphic subbundle with standard fiber Y_{x_0}. These two bundles share a common set of transition functions and have the same structure group $\mathrm{Aut}(Y_{x_0})$, which is known to be finite by (6.6.10).

Let $p\colon Z_U \cong U \times Z_{x_0} \to Z_{x_0}$ be the projection to the standard fiber. Then p induces the projection $Y_U \cong U \times Y_{x_0} \to Y_{x_0}$. At each $x \in U$, define a surjective holomorphic map

$$\lambda_x\colon \bar{\mathcal{S}} \to Z_{x_0}, \quad \lambda_x(f) = p(f(x)).$$

Then λ_x induces a surjective holomorphic map $\lambda_x\colon \mathcal{S} \to Y_{x_0}$. Since $\partial\mathcal{S}$ is a Cartier divisor of $\bar{\mathcal{S}}$ by (6.5.17) and since $B \cap Z_{x_0}$ is a Cartier divisor in Z_{x_0} with the hyperbolically imbedded complement Y_{x_0}, $\mathrm{Sur}(\mathcal{S}, Y_{x_0})$ is finite by (6.6.9). Since U is connected, λ_x has to be independent of x. In other words, the sections $f \in \mathcal{S}$, restricted to U, are all constant with respect to the local trivialization $Y_U \cong U \times Y_{x_0}$. Hence, if $y \in Y_U$ with $x = \pi(y)$ and if $f, f' \in \mathcal{S}_y$ so that $f(x) = f'(x) = y$, then $f = f'$ on U, and hence $f = f'$. This means that $k = 1$, and both $\Phi\colon X \times \bar{\mathcal{S}} \to Z$ and $\Phi\colon X \times \mathcal{S} \to Y$ are isomorphisms. $\qquad\square$

A Complex Finsler Vector Bundles

Let E be a holomorphic vector bundle of rank r over a complex manifold M of dimension n with projection π. We identify M with the zero section of E. Let $E^\times$ be E minus its zero section. Then $\mathbf{C}^*$ acts on $E^\times$ by scalar multiplication. The projective bundle $P(E)$ is defined by $P(E) = E^\times/\mathbf{C}^*$ with projection $p\colon P(E) \to M$. The pull-back $\tilde{E} = p^{-1}E$ is a vector bundle of rank r over $P(E)$. Let $L(E)$ be the tautological line subbundle of $\tilde{E}$.

We summarize the construction in the following diagram:

$$
\text{(6.A.1)} \qquad
\begin{array}{ccccc}
L(E) & \subset & \tilde{E} & \xrightarrow{\ \tilde{p}\ } & E \\
 & & \downarrow \tilde{\pi} & & \downarrow \pi \\
 & & P(E) & \xrightarrow{\ p\ } & M.
\end{array}
$$

Let $L^\times(E)$ be $L(E)$ minus its zero section. There is a natural map $L(E) \to E$, which maps $L^\times(E)$ biholomorphiclly to $E^\times$ and collapses the zero section $P(E)$ of $L(E)$ to the zero section M of E by p; thus, $L(E)$ is a blow-up of E along the zero section M of E.

We explain local coordinate systems associated to the bundles in (6.A.1). Let $z = (z^1, \ldots, z^n)$ be a local coordinate system in M, and $\zeta = (\zeta^1, \ldots, \zeta^r)$ the local fibre coordinate system defined by a local holomorphic frame field $s = (s_1, \ldots, s_r)$ of E. Then $(z, \zeta) = (z^1, \ldots, z^n, \zeta^1, \ldots, \zeta^r)$ is a local coordinate system for E. This may be considered also as a local coordinate system for $P(E)$ as long as $\zeta^1, \ldots, \zeta^r$ is considered as a homogenenous coordinate system for fibres. Setting

$$
Z^i = \zeta^i \circ \tilde{p},
$$

we take $(z, \zeta, Z) = (z^1, \ldots, z^n, \zeta^1 : \ldots : \zeta^r, Z^1, \ldots, Z^r)$ as a local coordinate system for $\tilde{E} = p^{-1}(E)$ with the understanding that $(\zeta^1 : \ldots : \zeta^r)$ is a homogenenous coordinate system. Then the line subbundle $L(E) \subset \tilde{E}$ is defined by

$$
(Z^1 : \ldots : Z^r) = (\zeta^1 : \ldots : \zeta^r).
$$

A **complex Finsler structure** F in E is a real function on E satisfying the following conditions (a), (b) and (c).

(a) F is smooth outside of the zero section of E;

(b) $F(z, \zeta) \geq 0$ and $= 0$ if and only if $\zeta = 0$;

(c) $F(z, \lambda\zeta) = |\lambda| F(z, \zeta)$ for all $\lambda \in \mathbf{C}$.

There is a natural correspondence between the Hermitian structures h on $L(E)$ and the Finsler structures F on E; namely $F^2(\zeta) = h(\zeta)$ for any nonzero element $\zeta \in E^\times = L^\times(E)$.

Since we use $F^2(z, \zeta)$ more often than $F(z, \zeta)$, we set

$$
G(z, \zeta) = F^2(z, \zeta).
$$

Then

$$(6.A.2) \qquad G(z, \lambda\zeta) = \lambda\bar{\lambda}G(z, \zeta).$$

We shall do our local calculation on $E^\times$ rather than on $P(E)$. We write

$$G_i = \partial G/\partial \zeta^i, \quad G_{\bar{j}} = \partial G/\partial \bar{\zeta}^j, \quad G_{i\bar{j}} = \partial^2 G/\partial \zeta^i \partial \bar{\zeta}^j,$$

$$G_{i\alpha} = \partial G_i/\partial z^\alpha, \quad G_{i\bar{j}\bar{\beta}} = \partial G_{i\bar{j}}/\partial \bar{z}^\beta, \quad \text{etc.,}$$

denoting differentiation in ζ^i, $\bar{\zeta}^j$, z^α, $\bar{z}^\beta$ by subscripts i, $\bar{j}$, α, $\bar{\beta}$, respectively.
Differentiating (6.A.2) with respect to λ and $\bar{\lambda}$, we obtain

$$(6.A.3) \qquad \sum G_i(z, \lambda\zeta)\zeta^i = \bar{\lambda}G(z, \zeta), \qquad \sum G_{\bar{j}}(z, \lambda\zeta)\bar{\zeta}^j = \lambda G(z, \zeta).$$

Differentiating the first equation of (6.A.3) by $\bar{\zeta}^j$ and the second equation by ζ^i yields

$$(6.A.4) \qquad \sum G_{i\bar{j}}(z, \lambda\zeta)\zeta^i = G_{\bar{j}}(z, \zeta), \qquad \sum G_{i\bar{j}}(z, \lambda\zeta)\bar{\zeta}^j = G_i(z, \zeta).$$

Differentiating the first equation of (6.A.3) by ζ^k and the second equation by $\bar{\zeta}^l$ yields

$$\sum G_{ik}(z, \lambda\zeta)\zeta^i\lambda + G_k(z, \lambda\zeta) = \bar{\lambda}G_k(z, \zeta),$$

$$\sum G_{\bar{j}\bar{l}}(z, \lambda\zeta)\bar{\zeta}^j\bar{\lambda} + G_{\bar{l}}(z, \lambda\zeta) = \lambda G_{\bar{l}}(z, \zeta),$$

Setting $\lambda = 1$ and then plowing back the resulting equations into the above, we obtain

$$(6.A.5) \qquad \sum G_{ik}(z, \zeta)\zeta^i = 0, \qquad \sum G_{\bar{j}\bar{l}}(z, \zeta)\bar{\zeta}^j = 0,$$

$$(6.A.6) \qquad G_i(z, \lambda\zeta) = \bar{\lambda}G_i(z, \zeta), \qquad G_{\bar{j}}(z, \lambda\zeta) = \lambda G_{\bar{j}}(z, \zeta).$$

Differentiating the first equation of (6.A.4) by ζ^k and the second equation by $\bar{\zeta}^l$ yields

$$\sum G_{i\bar{j}k}(z, \lambda\zeta)\lambda\zeta^i + G_{k\bar{j}}(z, \lambda\zeta) = G_{k\bar{j}}(z, \zeta),$$

$$\sum G_{i\bar{j}\bar{l}}(z, \lambda\zeta)\lambda\bar{\zeta}^j + G_{i\bar{l}}(z, \lambda\zeta) = G_{i\bar{l}}(z, \zeta).$$

Setting $\lambda = 1$, we obtain

$$(6.A.7) \qquad \sum G_{i\bar{j}k}(z, \zeta)\zeta^i = 0, \qquad \sum G_{i\bar{j}\bar{l}}(z, \zeta)\bar{\zeta}^j = 0,$$

$$(6.A.8) \qquad G_{k\bar{j}}(z, \lambda\zeta) = G_{k\bar{j}}(z, \zeta), \qquad G_{i\bar{l}}(z, \lambda\zeta) = G_{i\bar{l}}(z, \zeta).$$

On the other hand, differentiating the second equation of (6.A.4) by ζ^k and the first equation by $\bar{\zeta}^l$ yields

$$(6.A.9) \quad G_{ik}(z, \zeta) = \sum G_{i\bar{j}k}(z, \lambda\zeta)\lambda\bar{\zeta}^j, \qquad G_{\bar{j}\bar{l}}(z, \zeta) = \sum G_{i\bar{j}\bar{l}}(z, \lambda\zeta)\bar{\lambda}\zeta^i.$$

Differentiating the first equation of (6.A.6) by ζ^k and the second equation by $\bar\zeta^l$ gives

(6.A.10) $\qquad G_{ik}(z, \lambda\zeta)\lambda = \bar\lambda G_{ik}(z, \zeta), \qquad G_{\bar j\bar l}(z, \lambda\zeta)\bar\lambda = \lambda G_{\bar j\bar l}(z, \zeta).$

From (6.A.3) and (6.A.4) we obtain

(6.A.11) $\qquad\qquad \sum G_{i\bar j}(z, \lambda\zeta)\zeta^i\bar\zeta^j = G(z, \zeta).$

We set

(6.A.12) $\qquad\qquad \tilde G(z, \zeta, Z) = \sum G_{i\bar j}(z, \zeta)Z^i\bar Z^j.$

Then

(6.A.13) $\qquad\qquad G(z, \zeta) = \tilde G(z, \zeta, \zeta),$

which says that, restricted to $L(E)^\times = E^\times$, $\tilde G$ coincides with G.

Given a complex Finsler structure F in E, we consider now the corresponding Hermitian structure h in $L(E)$, and relate its curvature to that of F. We recall that $h = F^2$ on $E^\times = L^\times(E)$. The connection form φ and the curvature form Φ of h are given by

(6.A.14) $\qquad\qquad \varphi = d' \log h, \qquad \Phi = d''d' \log h.$

We write

(6.A.15)
$$\begin{aligned} \Phi \;=\; & \sum K_{\alpha\bar\beta}dz^\alpha \wedge d\bar z^\beta + \sum K_{\alpha\bar j}dz^\alpha \wedge d\bar\zeta^j \\ & + \sum K_{i\bar\beta}d\zeta^i \wedge d\bar z^\beta + \sum K_{i\bar j}d\zeta^i \wedge d\bar\zeta^j. \end{aligned}$$

Then

(6.A.16) $\qquad K_{\alpha\bar\beta} = -\dfrac{1}{G}\sum G_{i\bar j\alpha\bar\beta}\zeta^i\bar\zeta^j + \dfrac{1}{G^2}\sum G_{i\bar j\alpha}G_{k\bar l\bar\beta}\zeta^i\bar\zeta^j\zeta^k\bar\zeta^l,$

(6.A.17) $\qquad K_{\alpha\bar j} = \bar K_{j\bar\alpha} = -\dfrac{1}{G}\sum G_{i\bar j\alpha}\zeta^i + \dfrac{1}{G^2}\sum G_{\bar j}G_{k\bar l\alpha}\zeta^k\bar\zeta^l,$

(6.A.18) $\qquad K_{i\bar j} = -\dfrac{1}{G}G_{i\bar j} + \dfrac{1}{G^2}\sum G_{i\bar l}G_{k\bar j}\zeta^k\bar\zeta^l.$

Utilizing (6.A.7) and (6.A.4) simplifies these equations to

(6.A.19) $\qquad\qquad K_{\alpha\bar\beta} = -\dfrac{1}{G}G_{\alpha\bar\beta} + \dfrac{1}{G^2}G_\alpha G_{\bar\beta},$

(6.A.20) $\qquad\qquad K_{\alpha\bar j} = \bar K_{j\bar\alpha} = -\dfrac{1}{G}G_{\alpha\bar j} + \dfrac{1}{G^2}G_\alpha G_{\bar j},$

$$(6.A.21) \qquad K_{i\bar{j}} = -\frac{1}{G}G_{i\bar{j}} + \frac{1}{G^2}G_i G_{\bar{j}}.$$

From (6.A.11) and (6.A.18) we have

$$\sum K_{i\bar{j}} Z^i \bar{Z}^j = -\frac{1}{G}\left(\left(\sum G_{i\bar{j}} Z^i \bar{Z}^j\right)\left(\sum G_{k\bar{l}}\zeta^k \bar{\zeta}^l\right) - \left(\sum G_{i\bar{l}} Z^i \bar{\zeta}^l\right)\left(\sum G_{k\bar{j}}\zeta^k \bar{Z}^j\right)\right).$$

This shows that $\sum G_{i\bar{j}} Z^i \bar{Z}^j > 0$ if and only if $\sum K_{i\bar{j}} Z^i \bar{Z}^j > 0$ except when $Z^i = c\zeta^i$, $i = 1, \ldots, r$. (The curvature form Φ degenerates in the fiber direction of the line bundle $L(E)$ and is actually a form on $P(E)$; the exceptional direction $(Z^1, \ldots, Z^r)$ above is precisely the fiber direction of $L(E)$.)

We say that a Finsler structure F is **strongly pseudo-convex** if $(G_{i\bar{j}})$ is positive definite. What we proved, together with the comments in the parenthesis, may be summarized as follows.

(6.A.22) **Proposition.** *Let E be a holomorphic vector bundle over M. Let F be a complex Finsler structure in E and h the corresponding Hermitian structure in $L(E)$. Then F is strongly pseudo-convex if and only if the curvature $\sqrt{-1}\Phi$ of h, restricted to each fiber of $P(E)$ is negative-definite.*

Assume that F is strongly pseudo-convex. Then by (6.A.12) $\tilde{G}(z, \zeta, Z)$ is a Hermitian metric in the vector bundle $\tilde{E}$ over $P(E)$, and we can apply results from Hermitian vector bundles to $(\tilde{E}, \tilde{G})$.

Let $(G^{i\bar{j}})$ be the inverse matrix of $(G_{i\bar{j}})$. We consider the connection form $\omega = (\omega^i_j)$ defined by

$$(6.A.23) \qquad \omega^i_j = \sum G^{i\bar{k}}\partial G_{j\bar{k}} = \sum \Gamma^i_{j\alpha}dz^\alpha + \sum C^i_{jk}d\zeta^k,$$

where

$$C^i_{jk} = C^i_{kj} = \sum G^{i\bar{h}}G_{j\bar{h}k}, \qquad \Gamma^i_{j\alpha} = \sum G^{i\bar{h}}G_{j\bar{h}\alpha}.$$

From (6.A.5) we have

$$(6.A.24) \qquad \sum C^i_{jk}\zeta^j = \sum C^i_{jk}\zeta^k = 0.$$

The curvature form $\Omega = (\Omega^i_j)$ of the connection $\omega = (\omega^i_j)$ is given by

$$(6.A.25) \qquad \Omega^i_j = \bar{\partial}\omega^i_j.$$

We can write

$$(6.A.26) \qquad \begin{aligned} \Omega^i_j = {}& \sum R^i_{j\alpha\bar{\beta}}dz^\alpha \wedge d\bar{z}^\beta + \sum P^i_{j\alpha\bar{l}}dz^\alpha \wedge d\bar{\zeta}^l \\ & + \sum P^i_{jk\bar{\beta}}d\zeta^k \wedge d\bar{z}^\beta + \sum Q^i_{jk\bar{l}}d\zeta^k \wedge d\bar{\zeta}^l, \end{aligned}$$

where

$$(6.A.27) \qquad \begin{aligned} R^i_{j\alpha\bar{\beta}} &= -\partial\Gamma^i_{j\alpha}/\partial\bar{z}^\beta, & P^i_{j\alpha\bar{l}} &= -\partial\Gamma^i_{j\alpha}/\partial\bar{\zeta}^l, \\ P^i_{jk\bar{\beta}} &= -\partial C^i_{jk}/\partial\bar{z}^\beta, & Q^i_{jk\bar{l}} &= -\partial C^i_{jk}/\partial\bar{\zeta}^l. \end{aligned}$$

Setting $R_{i\bar{j}\alpha\bar{\beta}} = \sum G_{k\bar{j}} R^{k}_{i\alpha\bar{\beta}}$, etc., we obtain

$$
\begin{aligned}
R_{i\bar{j}\alpha\bar{\beta}} &= -G_{i\bar{j}\alpha\bar{\beta}} + \sum G^{k\bar{l}} G_{k\bar{j}\bar{\beta}} G_{i\bar{l}\alpha}, \\
P_{i\bar{j}\alpha\bar{l}} &= -G_{i\bar{j}\alpha\bar{l}} + \sum G^{k\bar{h}} G_{k\bar{j}\bar{l}} G_{i\bar{h}\alpha}, \\
P_{i\bar{j}k\bar{\beta}} &= -G_{i\bar{j}k\bar{\beta}} + \sum G^{h\bar{l}} G_{h\bar{j}\bar{\beta}} G_{i\bar{l}k}, \\
Q_{i\bar{j}k\bar{l}} &= -G_{i\bar{j}k\bar{l}} + \sum G^{h\bar{m}} G_{h\bar{j}\bar{l}} G_{i\bar{m}k}.
\end{aligned}
$$

(6.A.28)

Utilizing (6.A.5), (6.A.12) and (6.A.8) yields

$$
\begin{aligned}
\sum R_{i\bar{j}\alpha\bar{\beta}} \zeta^{i}\bar{\zeta}^{j} &= -G_{\alpha\bar{\beta}} + \sum G^{k\bar{l}} G_{k\bar{\beta}} G_{\bar{l}\alpha}, \\
\sum P_{i\bar{j}\alpha\bar{l}}\bar{\zeta}^{j} &= \sum P_{i\bar{j}\alpha\bar{l}}\bar{\zeta}^{l} = 0, \\
\sum P_{i\bar{j}k\bar{\beta}} \zeta^{i} &= \sum P_{i\bar{j}k\bar{\beta}} \zeta^{k} = 0, \\
\sum Q_{i\bar{j}k\bar{l}} \zeta^{i} &= \sum Q_{i\bar{j}k\bar{l}} \zeta^{k} = 0.
\end{aligned}
$$

(6.A.29)

To each vector (z, ζ, Z) in $\tilde{E}$, we associate the following Hermitian form on $P(E)$:

$$
\begin{aligned}
\Psi(z, \zeta, Z) &= \sum R_{i\bar{j}\alpha\beta} Z^{i}\bar{Z}^{j} dz^{\alpha} d\bar{z}^{\beta} + \sum P_{i\bar{j}\alpha\bar{l}} Z^{i}\bar{Z}^{j} dz^{\alpha} d\bar{\zeta}^{l} \\
&\quad + \sum P_{i\bar{j}k\bar{\beta}} Z^{i}\bar{Z}^{j} d\zeta^{k} d\bar{z}^{\bar{\beta}} + \sum Q_{i\bar{j}k\bar{l}} Z^{i}\bar{Z}^{j} d\zeta^{k} d\bar{\zeta}^{l}.
\end{aligned}
$$

(6.A.30)

Restricting it to $L(E)$ and using (6.A.29), we have

$$
\begin{aligned}
\Psi(z, \zeta, \zeta) &= \sum R_{i\bar{j}\alpha\bar{\beta}} \zeta^{i}\bar{\zeta}^{j} dz^{\alpha} d\bar{z}^{\beta} \\
&= \sum (-G_{\alpha\bar{\beta}} + \sum G^{k\bar{l}} G_{k\bar{\beta}} G_{\bar{l}\alpha}) dz^{\alpha} d\bar{z}^{\beta}.
\end{aligned}
$$

(6.A.31)

(6.A.32) **Remark**. If G is a Hermitian structure in E, then $F = \sqrt{G}$ is a strongly pseudo-convex Finsler structure in E. In this case C^{i}_{jk} in (6.A.23) and $P^{i}_{j\alpha l}$, $P^{i}_{jk\bar{\beta}}$ and $Q^{i}_{jk\bar{l}}$ in (6.A.26) vanish. Even in the general Finsler case, $R^{i}_{j\alpha\bar{\beta}}$ are the only important components of the curvature.

In order to facilitate calculation, it is useful to have analogue of normal coordinate systems. Given a point (z_0, ζ_0) in E, we can find a local frame field $s_1, \ldots, s_r$ such that

(6.A.33) $\qquad G_{i\bar{j}}(z_0, \zeta_0) = \delta_{ij}, \qquad G_{i\bar{j}\alpha}(z_0, \zeta_0) = G_{i\bar{j}\bar{\beta}}(z_0, \zeta_0) = 0.$

Such a local frame field $s_1, \ldots, s_r$ is called a **normal frame field** at (z_0, ζ_0). For simplicity, take a local coordinate system $z^1, \ldots, z^n$ such that z_0 is its origin. Then from a given local frame field $t_1, \ldots, t_r$ we can obtain such a frame field $s_1, \ldots, s_r$ by a transformation of the the type

$$t_i(z) \mapsto s_i(z) = \sum f_i^j(z) t_j(z),$$

where $f_i^j(z) = a_i^j + \sum b_{i\alpha}^j z^\alpha$. A normal frame field is not unique.

With respect to a normal frame field, we have

$$(6.A.34) \qquad G(z_0, \zeta_0) = \sum |\zeta_0^i|^2, \qquad \text{(by (6.A.11))}$$

$$(6.A.35) \qquad G_i(z_0, \zeta_0) = \bar\zeta_0^i, \qquad G_{\bar j}(z_0, \zeta_0) = \zeta_0^j. \qquad \text{(by (6.A.4))}$$

Differentiating (6.A.4) by z^α and $\bar z^\beta$, we obtain

$$(6.A.36) \qquad G_{i\alpha}(z_0, \zeta_0) = G_{i\bar\beta}(z_0, \zeta_0) = G_{\bar j\alpha}(z_0, \zeta_0) = G_{\bar j\bar\beta}(z_0, \zeta_0) = 0.$$

Differentiating (6.A.11) by z^α and $\bar z^\beta$, we have

$$(6.A.37) \qquad G_\alpha(z_0, \zeta_0) = G_{\bar\beta}(z_0, \zeta_0) = 0.$$

With respect to a normal frame field, (6.A.16), (6.A.17) and (6.A.18) reduce to the following:

$$(6.A.38) \qquad K_{\alpha\bar\beta}(z_0, \zeta_0) = -\frac{1}{G(z_0, \zeta_0)} G_{\alpha\bar\beta}(z_0, \zeta_0),$$

$$(6.A.39) \qquad K_{\alpha\bar j}(z_0, \zeta_0) = \bar K_{j\bar\alpha}(z_0, \zeta_0) = 0,$$

$$(6.A.40) \qquad K_{i\bar j}(z_0, \zeta_0) = -\frac{1}{G(z_0, \zeta_0)} \delta_{ij} + \frac{1}{G(z_0, \zeta_0)^2} \zeta_0^j \bar\zeta_0^i.$$

From (6.A.40) we see that the number of negative eigenvalues for the $n \times n$-matrix $(K_{\alpha\bar\beta}(z_0, \zeta_0))$ is equal to the number of positive eigenvalues of the matrix $(G_{\alpha\bar\beta}(z_0, \zeta_0))$, which we denote by p.

Since $(G_{i\bar j})$ is positive-definite, the number of positive eigenvalues of the $(n+r) \times (n+r)$-matrix

$$\begin{pmatrix} G_{\alpha\bar\beta} & G_{\alpha\bar j} \\ G_{i\bar\beta} & G_{i\bar j} \end{pmatrix}$$

is equal to $r + p$.

On the other hand, as we saw in (6.A.22), the $r \times r$-matrix $(K_{i\bar j}(z_0, \zeta_0))$ has $r - 1$ negative eigenvalues with the remaining eigenvalue equal to

Hence, the number of negative eigenvalues of $(n+r) \times (n+r)$-matrix

$$\begin{pmatrix} K_{\alpha\bar\beta} & K_{\alpha\bar j} \\ K_{i\bar\beta} & K_{i\bar j} \end{pmatrix}$$

is equal to $r - 1 + p$.

From (6.A.26) we have

$$(6.A.41) \qquad \sum R_{i\bar j\alpha\bar\beta}(z_0, \zeta_0) \zeta_0^i \bar\zeta_0^j = -G_{\alpha\bar\beta}(z_0, \zeta_0) = G(z_0, \zeta_0) K_{\alpha\bar\beta}(z_0, \zeta_0).$$

From (6.A.41) we see that the number of negative eigenvalues of the $n \times n$-matrix $(K_{\alpha\bar\beta}(z_0, \zeta_0))$, can be stated in terms of $R_{i\bar j\alpha\bar\beta}$. In summary, we have

(6.A.42) Proposition. *Let F be a strongly pseudo-convex Finsler structure on E, and $G = F^2$. Then the following are equivalent:*

(a) *the complex Hessian of G on $E^\times$ has $r + p$ positive eigenvalues:*
(b) *the curvature form $\sqrt{-1}\Phi$ of the line bundle $L(E)$ has $r - 1 + p$ negative eigenvalues*
(c) *the number of negative eigenvalues of $\Psi(z, \zeta, \zeta)$ is p.*

In particular, the complex Hessian of G is positive-definite if and only if $\sqrt{-1}\Phi$ is neagative-definite.

Chapter 7. Manifolds of General Type

1 Intrinsic Volume Forms

In general, given a topological space X with a pseudo-distance d and a non-negative real number k, the k-dimensional **Hausdorff measure** m_k is defined as follows. For a subset $E \subset X$, we set

$$m_k(E) = \sup_{\varepsilon > 0} \inf\{\sum_{i=1}^{\infty} (\delta(E_i))^k; \ E = \bigcup_{i=1}^{\infty} E_i, \ \delta(E_i) < \varepsilon\},$$

where $\delta(E_i)$ denotes the diameter of E_i. If X is a complex space, then the pseudo-distances c_X and d_X induce Hausdorff measures on X. Since every holomorphic map is distance-decreasing with respect to these intrinsic pseudo-distances, it is also measure-decreasing with respect to the Hausdorff measures they define. There are other intrinsic measures on complex spaces. For a systematic study of intrinsic measures on complex manifolds, see Eisenman [1]. In Section 2 we shall discuss the intrinsic mesaures which may be considered as direct generalizations of c_X and d_X. In this section we discuss their infinitesimal forms.

Let B^n be the unit ball in $\mathbf{C}^n$ with the invariant volume form $\mu = \mu_{B^n}$ defined by (see (2.4.7))

$$(7.1.1) \qquad \mu = \frac{2^n}{(1 - \|\mathbf{z}\|^2)^{n+1}} \prod_{j=1}^{n} i\, dz^j \wedge d\bar{z}^j$$

with respect to the natural coordinate system $\mathbf{z} = (z^1, \ldots, z^n)$ of $\mathbf{C}^n$. At the origin 0 it reduces to

$$\mu_0 = 2^n \prod i(dz^j \wedge d\bar{z}^j)_0.$$

Let X be a complex space of dimension n. We define an intrinsic pseudo-volume form Φ_X analogous to the Carathéodory pseudo-metric E_X by setting

$$(7.1.2) \qquad (\Phi_X)_x = \sup_{f}(f^*\mu)_x \qquad \text{for} \quad x \in X,$$

where the supremum is taken over all holomorphic maps $f \in \mathrm{Hol}(X, B^n)$. (Because of homogeneity of B^n, we may take the supremum over only those f which send x to the origin 0 of B^n.) It would be prudent to consider Φ_X as a form defined only at the nonsingular points of X since the tangent space at a singular point may have dimension higher than n.

To obtain an intrinsic pseudo-volume form Ψ_X^h analogous to the intrinsic pseudo-metric F_X, we set

$$(7.1.3) \qquad (\Psi_X^h)_x = \inf_f (f^{-1})^*(\mu_0) \qquad \text{for} \quad x \in X,$$

where the infimum is taken over all holomorphic maps $f : B^n \to X$ which send the origin 0 to x and are non-degenerate at 0. (Again, Ψ_X^h is defined only at the regular points of X. The inverse f^{-1} is defined at least in a neighborhood of x provided that x is a nonsingular point of X and f is non-degenerate at 0.) The superscript h stands for *holomorphic*.

Clearly,

(a) $f^* \Psi_X^h \leq \mu$ for all $f \in \text{Hol}(B^n, X)$.

This together with the following characterizes $(\Psi_X^h)_x$ when $(\Psi_X^h)_x \neq 0$:

(b)   Given a positive number $r < 1$, there is a map $f \in \text{Hol}(B^n, X)$ such that $f(0) = x$ and $r\mu < f^* \Psi_X^h$ at $0 \in B^n$.

In (7.1.3), by taking the infimum over all meromorphic maps $f : B^n \to X$ which are holomorphic and non-degenerate at 0, we obtain another intrinsic pseudo-volume form, which will be denoted Ψ_X^m, (see Yau [2]). (The superscript m stands for *meromorphic*). Since every meromorphic map $f : X \to B^n$ is holomorphic, by allowing f to be meromorphic in the definition of Φ_X we would obtain nothing new.

We make two remarks. Since B^n is homogeneous, the role of the origin 0 may be played by any other point. Since the map $B^n \to B_a^n$ given by $z \mapsto az$ pulls back μ_a to μ, in defining Ψ_X^h and Ψ_X^m we may replace B^n and μ by B_a^n and μ_a of (2.4.7). These remarks will be used in the proof of (7.1.5).

The following theorems summarize basic properties of three pseudo-volume forms Φ_X, Ψ_X^h and Ψ_X^m; the proofs of these assertions are similar to those of the corresponding statements for E_X and F_X.

(7.1.4) **Theorem.** *Let X and Y be complex spaces of dimension n.*

(1)   *For the unit ball B^n,*

$$\mu = \Phi_{B^n} = \Psi_{B^n}^h = \Psi_{B^n}^m;$$

(2)   *If $f : X \to Y$ is holomorphic, then*

$$f^* \Phi_Y \leq \Phi_X \quad \text{and} \quad f^* \Psi_Y^h \leq \Psi_X^h;$$

(3)   *If $f : X \to Y$ is meromorphic, then*

$$f^* \Phi_Y \leq \Phi_X \quad \text{and} \quad f^* \Psi_Y^m \leq \Psi_X^m;$$

(4)   *If ω_X is any pseudo-volume form of X such that $f^*\omega_X \leq \mu$ for all holomorphic (resp. meromorphic) maps $f: B^n \to X$, then*

$$\omega_X \leq \Psi_X^h \quad (resp. \ \omega_X \leq \Psi_X^m);$$

(5)   *If ω_X is any pseudo-volume form of X such that $f^*\mu \leq \omega_X$ for all holomorphic maps $f: X \to B^n$, then*

$$\Phi_X \leq \omega_X;$$

(6)   *The three intrinsic pseudo-volume forms are related as follows:*

$$\Phi_X \leq \Psi_X^m \leq \Psi_X^h;$$

(7)   *If $\pi: \tilde{X} \to X$ is a covering projection, then*

$$\Psi_{\tilde{X}}^h = \pi^*\Psi_X \quad and \quad \Psi_{\tilde{X}}^m = \pi^*\Psi_X^m;$$

(8)   *Both Ψ_X^h and Ψ_X^m are upper semi-continuous, and Φ_X is continuous. If X is complete hyperbolic, then Ψ_X^h and Ψ_X^m are also continuous.*

Proof. The definitions of these intrinsic pseudo-volume forms imply immediately (2), (3), (4) and (5).

By Schwarz' lemma (2.4.16), we have $f^*\mu \leq \mu$ for all holomorphic maps $f: B^n \to B^n$, and from (4) we obtain $\mu \leq \Psi_{B^n}^h$. The reverse inequality can be obtained by setting f to be the identity transformation of B^n in the definition (7.1.3). The other equalities $\mu = \Phi_{B^n}$ and $\mu = \Psi_{B^n}^m$ in (1) can be verified in the same way.

The first inequality in (6) follows from Schwarz' lemma applied to the composed map $B^n \xrightarrow{f} X \xrightarrow{g} B^n$. The second inequality is trivial.

In (7) the inequality in one direction follows from (2) and (3). The inequality in the reverse direction follows from the fact that any map $B^n \to X$ lifts to a map $B^n \to \tilde{X}$.

The proof for continuity of Φ_X is similar to that of (4.2.6), i.e., continuity of the infinitesimal Carathéodory pseudo-metric. The proof for upper semicontinuity of Ψ_X^h and Ψ_X^m is similar to but simpler than that of (3.5.27). It will appear as part of the proof of (7.1.5) in which we show that Ψ_X^h and Ψ_X^m are upper semicontinuous under deformations of the complex structure of X. The proof for continuity of Ψ_X^h and Ψ_X^m for a complete hyperbolic complex space X is essentially the same as that of (3.5.38). Incidentally, as we shall see in (7.1.6), Ψ_X^m coincides with Ψ_X^h when X is hyperbolic. $\qquad\qquad\square$

Let X be a complex manifold of dimension $n + r$, R a complex manifold of dimension r, and $\pi: X \to R$ a surjective holomorphic map of maximal rank r everywhere. Set

$$X_r = \pi^{-1}(r) \qquad for \quad r \in R.$$

Then each X_r is a complex submanifold of dimension n in X. We consider $\{X_r\}$ as a family of complex manifolds parametrized by $r \in R$.

Let $T^v X$ denote the subbundle of the tangent bundle TX consisting of vertical vectors, i.e., vectors which are annihilated by π; it is a vector bundle of rank n over X. Let $T^{v*}X$ be its dual bundle, and $\bigwedge^{n,n} T^{v*}X$ the bundle of (n,n)-forms along the fibers.

Let $\Psi^h_{X_r}$ be the intrinsic pseudo-volume form of X_r; it is a real nonnegative (n,n)-form on X_r. As it varies with $r \in R$, it defines a (possibly discontinuous) section of $\bigwedge^{n,n} T^{v*}X$. We shall show that not only is $\Psi^h_{X_r}$ upper semicontinuous on X_r for each fixed r, it is upper semicontinuous in $r \in R$. More precisely,

(7.1.5) Theorem. *Let $X = \{X_r\}_{r \in R}$ be a family of complex manifolds parametrized by R. Then as sections of $\bigwedge^{n,n} T^{v*}X$, the intrinsic pseudo-volume forms $\{\Psi^h_{X_r}\}_{r \in R}$ and $\{\Psi^m_{X_r}\}_{r \in R}$ are upper semi-continuous on X.*

Proof. Since the same proof works for both Ψ^h_X and Ψ^m_X, we shall drop the superscripts h and m in the proof.

We fix a point $x_0 \in X$ and put $o = \pi(x_0) \in R$ so that $x_0 \in X_o$. Let $f: B^n \to X_o$ be a holomorphic map which sends the origin $0 \in B^n$ to x_0 and is non-degenerate at 0. As in the proof of (3.11.5), it suffices to show that, for $a < 1$, arbitrarily close to 1, there exist a neighborhood U of o in R and a holomorphic map $F: B^n_a \times U \to X$ such that

$$\pi(F(z,r)) = r \quad \text{and} \quad F(z,o) = f(z) \qquad \text{for} \quad z \in B^n_a, \ r \in U.$$

In fact, if $x \in X$ is near x_0 so that $r = \pi(x)$ is near o, then there is a unique point $z \in B^n_a$ near 0 such that $F(z,r) = x$. Put $f_r(\cdot) = F(\cdot, r)$. Then $(f_r^{-1})^*(\mu_a)_z$ is close to $(f^{-1})^*(\mu)_0$. Now the upper semicontinuity of $\{\Psi_{X_r}\}_{r \in R}$ follows from the definition of Ψ_X, (see the two remarks following the definitions of Ψ^h_X and Ψ^m_X).

The rest of the proof is essentially the same as that of (3.11.5), the map F constructed here playing the role of the map φ in (3.11.6). $\square$

Although $\Psi^m_X \leq \Psi^h_X$ in general, the equality holds in some cases. For example, if X is strongly minimal, by definition every meromorphic map from B^n into X is holomorphic, and the equality holds. In particular, from (6.3.20) and (6.3.21) we obtain

(7.1.6) Proposition. *The equality $\Psi^m_X = \Psi^h_X$ holds in the following cases;*

(a)   *X has a Stein space $\tilde{X}$ as a covering space;*
(b)   *X is hyperbolic.*

When $\Psi^h_X = \Psi^m_X$, we drop the superscripts and set

$$\Psi_X = \Psi^h_X = \Psi^m_X.$$

Since Ψ^h is upper semi-continuous, for any open set $B \subset X$ we can define its Ψ^h_X-measure

$$\Psi^h_X[B] = \int_B \Psi^h.$$

Similarly, the Ψ^m_X-measure $\Psi^m_X[B]$ and the Φ_X-measure $\Phi_X[B]$ can be defined.

We say that X is Φ-**measure hyperbolic** (resp. Ψ^h-**measure hyperbolic**, Ψ^m-**measure hyperbolic**) if $\Phi_X[B] > 0$ (resp. $\Psi^h[B] > 0$, $\Psi^m[B] > 0$) for every nonempty open subset $B \subset X$. If $\Phi_X > 0$ on a dense open subset of X, then X is Φ-measure hyperbolic. Similarly for Ψ^h_X and Ψ^m_X.

We say that X is **strongly Φ-measure hyperbolic** if there is a continuous positive volume form v on X such that $\Phi_X \geq v$. Similarly, for Ψ^h and Ψ^m.

(7.1.7) Proposition. (1) *Let X and Y be complex spaces of equal dimension n, and $f: X \to Y$ a meromorphic map which is holomorphic and non-degenerate at some regular point $x \in X$. Then each of the following properties for Y is inherited by X:*

(a) *Φ-measure hyperbolic;*

(b) *Ψ^m-measure hyperbolic;*

(c) *Ψ^h-measure hyperbolic;*

(d) *Φ is positive outside a proper analytic subset;*

(e) *Ψ^m is positive outside a proper analytic subset;*

(f) *Ψ^h is positive outside a proper analytic subset.*

(2) *If f is an unbranched covering projection, then each of the following properties for Y is inherited by X:*

(a) *$\Phi > 0$ everywhere;*

(b) *$\Psi^m > 0$ everywhere;*

(c) *$\Psi^h > 0$ everywhere.*

If X has property (b) *or* (c)*, so does Y.*

Proof. (1)   Let X' be the set of regular points of X where f is holomorphic and non-degenerate. Then its complement $S = X - X'$ is a proper closed analytic subset of X. Then $f^*\Phi_Y$ is no more degenerate than Φ_Y except possibly along S. All our assertions concerning Φ follows from $f^*\Phi_Y \leq \Phi_X$. Similarly, for Ψ^m and Ψ^h.

(2)   This is trivial. $\qquad\qquad\qquad\qquad\qquad\qquad\qquad\qquad\qquad\qquad$ □

We shall now prove the following product formula of Graham-Wu [1].

(7.1.8) Theorem. *For two complex spaces X and Y we have*

$$\Psi^h_{X \times Y} = \Psi^h_X \wedge \Psi^h_Y \quad \text{and} \quad \Psi^m_{X \times Y} = \Psi^m_X \wedge \Psi^m_Y.$$

Proof. Let $m = \dim X$ and $n = \dim Y$. Let $x \in X$ and $y \in Y$ be regular points. Let $p: X \times Y \to X$ and $q: X \times Y \to Y$ be the natural projections.

We first prove $\Psi^h_{X \times Y} \leq \Psi^h_X \wedge \Psi^h_Y$ at $(x, y) \in X \times Y$. Let $f: B^m \to X$ and $g: B^n \to Y$ be holomorphic maps such that $f(0) = x$ and $g(0) = y$ and non-degenerate at the origin. Composing (f, g) with the natural injection $B^{m+n} \subset B^m \times B^n$, we obtain a holomorphic map $h = (f, g)|_{B^{m+n}}: B^{m+n} \to X \times Y$, which sends the origin to (x, y) and is non-degenerate at the origin. Then

$$(h^{-1})^*(\mu_{B^{m+n}})_{(0,0)} = (f^{-1})^*(\mu_{B^m})_0 \wedge (g^{-1})^*(\mu_{B^n})_0.$$

From the definition (7.1.3) we obtain the asserted inequality.

Now we prove $\Psi_X^h \wedge \Psi_Y^h \le \Psi_{X \times Y}^h$ at (x, y). Given a holomorphic map $h: B^{m+n} \to X \times Y$ which sends the origin to (x, y) and is non-degenerate at the origin, we construct holomorphic maps $f: B^m \to X$ and $g: B^n \to Y$ as follows. Under the natural identification $T_{(x,y)}(X \times Y) \simeq T_x X \times T_y Y$ we regard $T_x X$ and $T_y Y$ as subsapces of $T_{(x,y)}(X \times Y)$. Then $h_*^{-1}(T_x X)$ and $h_*^{-1}(T_y Y)$ are transversal subspaces of $T_0 B^{m+n} \simeq \mathbf{C}^{m+n}$. Intersecting B^{m+n} with $h_*^{-1}(T_x X)$ (resp. with $h_*^{-1}(T_y Y)$) in $\mathbf{C}^{m+n}$ we obtain a unit ball B^m (resp. B^n). We define maps $f: B^m \to X$ and $g: B^n \to Y$ by setting

$$f = p \circ h|_{B^m} \quad \text{and} \quad g = q \circ h|_{B^n}.$$

Then

$$(h^{-1})^*(\mu_{B^{m+n}})_{(0,0)} = (f^{-1})^*(\mu_{B^m})_0 \wedge (g^{-1})^*(\mu_{B^n})_0.$$

The asserted inequality follows from the definition (7.1.3).

The proof for $\Psi_{X \times Y}^m$ is similar. $\square$

As an application of (7.1.8) we determine Ψ_{D^n} explicitly. For $n = 1$, we have

$$\Psi_D = \mu = \frac{2i \, dz \wedge d\bar{z}}{(1 - |z|^2)^2}.$$

Hence, by (7.1.8)

$$(7.1.9) \qquad \Psi_{D^n} = 2^n \prod_{j=1}^{n} \frac{i \, dz^j \wedge d\bar{z}^j}{(1 - |z^j|^2)^2},$$

which shows that Ψ_{D^n} agrees with the invariant volume element ν introduced in (2.4.11).

Another application is to the punctured polydisc $(D^*)^n = D^* \times \ldots \times D^*$. Applying (7) of (7.1.4) to the covering projection $D \to D^*$ given by

$$z \mapsto w = e^{-(1+z)/(1-z)},$$

we see that

$$(7.1.10) \qquad \Psi_{D^*} = \frac{2i \, dw \wedge d\bar{w}}{(|w| \log |w|^2)^2}.$$

Hence, by (7.1.8) we have

$$(7.1.11) \qquad \Psi_{(D^*)^n} = \prod_{j=1}^{n} \frac{2i \, dw^j \wedge d\bar{w}^j}{(|w^j| \log |w^j|^2)^2}.$$

We consider the simplest example.

(7.1.12) **Example.** For a ball $B_a^n = \{\mathbf{z} = (z^1, \ldots, z^n); \|\mathbf{z}\| < a\}$ of radius a in $\mathbf{C}^n$, we have

$$\mu_a = \Phi_{B_a^n} = \Psi_{B_a^n},$$

where μ_a is the invariant volume element defined in (2.4.7).

We know this for $a = 1$, (see (7.1.4)). If $f_a : B^n \to B_a^n$ is the biholomorphic mapping sending $\mathbf{z} \in B^n$ to $a\mathbf{z} \in B_a^n$, then $f_a^* \mu_a = \mu$, $f^* \Phi_{B_a^n} = \Phi_{B^n}$, etc..

The following *theorem of Landau-Shottky type* is essentially contained in the definition of Ψ_X. By Ψ_X we mean either Ψ_X^h or Ψ_X^m. Correspondingly, by a map f we mean either a holomorphic or meromorphic map.

(7.1.13) **Theorem.** *Let X be an n-dimensional complex space, and B_a^n the ball of radius a in $\mathbf{C}^n$. If $f: B_a^n \to X$ is a map such that*

$$c^{2n} \prod_j i \, dz^j \wedge d\bar{z}^j \leq f^* \Psi_X \quad \text{at} \quad 0 \in B_a^n,$$

then $a \leq \sqrt{2}/c$.

Conversely, if $a < \sqrt{2}/c$, then for $x \in X$ such that $(\Psi_X)_x \neq 0$ there is a map $f: B_a^n \to X$ satisfying $f(0) = x$ and the inequality above.

The inequality in the theorem implies that $\Psi_X > 0$ at $f(0)$. If $f(0)$ is a regular point, then we can take a local coordinate system $w^1, \ldots, w^n$ around $f(0)$ such that $\Psi_X = \prod \sqrt{-1} \, dw^j \wedge d\bar{w}^j$ at $f(0)$. Then the assumption says that the absoulte value of the Jacobian of f is bounded below by c^n.

Proof. By (7.1.4) and (7.1.12) we have $f^* \Psi_X \leq \mu_a$. Hence.

$$c^{2n} \prod i \, dz^j \wedge d\bar{z}^j \leq f^* \Psi_X \leq \frac{2^n}{a^{2n}} \prod i \, dz^j \wedge d\bar{z}^j \quad \text{at} \quad 0 \in B_a^n.$$

This implies $a \leq \sqrt{2}/c$.

Assume $a < \sqrt{2}/c$, and set $r = (a^2 c^2 / 2)^n < 1$. By properties (a) and (b) characterizing Ψ_X (see (a) and (b) following (7.1.3)), there is a map $h: B^n \to X$ such that $h(0) = x$ and

$$r 2^n \prod i \, dz^j \wedge d\bar{z}^j < (h^* \Psi_X)_0 \leq 2^n \prod i \, dz^j \wedge d\bar{z}^j.$$

Let $f: B_a^n \to X$ be the map defined by $f(z) = h(z/a)$. Then f has the desired property. $\qquad\square$

We shall now examine intrinsic pseudo-volume forms for simple domains.

(7.1.14) **Proposition.** *If $X \subset \mathbf{C}^n$ is a bounded domain contained in a ball B_r^n of radius r, then*

$$\mu_r = \Phi_{B_r^n} \leq \Phi_X \leq \Psi_X^m = \Psi_X^h.$$

In particular, X is strongly Φ-measure hyperbolic.

Proof. This is immediate from (7.1.4) and (7.1.6). $\qquad\square$

(7.1.15) **Proposition.** (1) $\quad \Phi_{\mathbf{C}^n} = \Psi_{\mathbf{C}^n}^m = \Psi_{\mathbf{C}^n}^h = 0$;

(2) *For any complex space X,* $\quad \Phi_{\mathbf{C} \times X} = \Psi_{\mathbf{C} \times X} = 0$.

Proof. (1) Let a tend to infinity in μ_a (2.4.7) and use (7.1.12).

(2)   By (1), $\Psi_\mathbf{C}^h = 0$. By (7.1.8), $\Psi_{\mathbf{C}\times X}^h = 0$. By (6) of (7.1.4), $\Psi_{\mathbf{C}\times X}^m = 0$ and $\Phi_{\mathbf{C}\times X}^m = 0$. $\square$

The following theorem resulted from a discussion with Bun Wong.

(7.1.16) Theorem. *If the group* $\mathbf{C}$ *acts holomorphically on a complex space* X, *then* Φ_X, Ψ_X^m *and* Ψ_X^h *vanish at any regular point* x *that is not fixed by the group* $\mathbf{C}$.

Proof. Imbed a ball B^{n-1}, $(n = \dim X)$, into X in such a way that the imbedded B^{n-1} is transversal to the orbit of $\mathbf{C}$ through x. Translating the imbedded B^{n-1} along the orbit by the group action, we obtain a holomorphic map $f : \mathbf{C} \times B^{n-1} \to X$ which is non-degenerate at x. By (2) of (7.1.4) and (2) of (7.1.15) we have $f^*\Psi_X^h \le \Psi_{\mathbf{C}\times B^{n-1}}^h = 0$ at x. Since $f^*\Phi_X$ and $f^*\Psi_X^m$ are dominated by $f^*\Psi_X^h$, this completes the proof. $\square$

(7.1.17) Corollary. *Let* X *be a compact* Ψ^h*-measure hyperbolic complex space. Then its group* $\mathrm{Aut}(X)$ *of biholomorphic automorphisms is discrete, and* X *is immobile.*

Proof. For a compact complex space X, $\mathrm{Aut}(X)$ is a complex Lie group. By (7.1.16), $\dim \mathrm{Aut}(X) = 0$. By (5.4.6), X is immobile. $\square$

The following example is from Graham-Wu [1].

(7.1.18) Example. Let $X = \{(z, w) \in \mathbf{C}^2;\ |zw| < 1\}$. Since X is a domain of holomorphy, $\Psi_X^m = \Psi_X^h$ by (7.1.6), and

(1) $\Psi_X = 0$ on $X - \{(0, 0)\}$,   (2) $\Psi_X > 0$ at $(0,0)$,   (3) $\Phi_X = 0$ on X.

To see (1), we use the $\mathbf{C}$-action on X defined by

$$t : (z, w) \mapsto (e^t z, e^{-t} w), \quad t \in \mathbf{C},$$

and apply (7.1.15). Since Φ_X is continuous, (3) follows from (1). To see (2), let $f : B^2 \to X$ be a holomorphic map sending the origin to the origin. After a rotation of B^2, the map $f = (f_1, f_2)$ is of the form

$$
\begin{aligned}
f_1(s, t) &= as + bt + O(2), \\
f_2(s, t) &= ct + O(2).
\end{aligned}
$$

On the disc $D_{1/\sqrt{2}}$ of radius $1/\sqrt{2}$, we define

$$g(t) = f_1(t, t) f_2(t, t) = c(a + b)t^2 + O(3).$$

Since $|f_1 f_2| < 1$, we have $|g(t)| < 1$. Using Cauchy's integral formula for $g(t)$, we estimate the second derivative of $g(t)$ at $t = 0$ and we obtain $|c(a + b)| \le 2$. Similarly, using the function

$$h(t) = f_1(t, -t) f_2(t, -t) = -c(a - b)t^2 + O(3),$$

we obtain $|c(a - b)| \le 2$. Hence,

$$|ac| \leq \frac{1}{2}|(a+b)c)| + \frac{1}{2}|(a-b)c| \leq 2.$$

From (7.1.3) we obtain

$$(\Psi_X)_0 \geq \inf_f \frac{2^2}{|ac|^2} dz \wedge dw \wedge d\bar{z} \wedge d\bar{w} \geq dz \wedge dw \wedge d\bar{z} \wedge d\bar{w}.$$

The following example is related to (7.1.18).

(7.1.19) **Example.** Blow up the origin (0,0) of $D \times \mathbf{C}$, and let Y be the resulting space. Then $\Phi_Y = \Psi_Y = 0$.

Since $\Psi_{D \times \mathbf{C}} = 0$, Ψ_Y also vanishes except possibly at the exceptional curve obtained by blowing up the origin. To see that it vanishes at the exceptional curve, let $\mathbf{C}$ act on $D \times \mathbf{C}$ by

$$t : (z, w) \mapsto (z, e^t w) \qquad \text{for} \quad t \in \mathbf{C}, \ (z, w) \in D \times \mathbf{C}.$$

The induced action on Y has only one fixed point, namely the point on the exceptional curve that corresponds to the complex direction $w = 0$. By (7.1.16), Ψ_Y vanishes on Y except at this one point. Consider next the action of $\mathbf{C}$ given by

$$t : (z, w) \mapsto (z, tz + w).$$

Then the induced action on Y has only one fixed point, namely the point on the exceptional curve that corresponds to the complex direction $z = 0$. By (7.1.6), Ψ_Y vanishes on Y except at this one point. Hence, Ψ_Y vanishes everywhere.

Let X be the domain given in (7.1.18), and let $p : X \to D \times \mathbf{C}$ be the map defined by $p(z, w) = (zw, w)$; the map p collapses the complex line $\mathbf{C} \times \{0\}$ to the origin (0,0). It is not difficult to see that X can be realized as a domain in Y in such a way that the line $\mathbf{C} \times \{0\}$ becomes part of the exceptional curve of Y. Since $\Psi_X > 0$ at 0, it is not completely trivial that $\Psi_Y \equiv 0$.

In (7.1.5) we proved that Ψ_X^h is upper semi-continuous under deformations. We shall sharpen an estimate in an example by Graham-Wu [1].

(7.1.20) **Example.** For each $s \in D$ let

$$X_s = \{(z, w) \in \mathbf{C}^2; \ |z| < 1, \ |zw| < 1\} \cup \{(z, w) \in \mathbf{C}^2; \ |z| < |s|\},$$

with the understanding that $X_0 = \{(z, w) \in \mathbf{C}^2; \ |z| < 1, \ |zw| < 1\}$. Then $X = \{X_s\}_{s \in D}$ is a family of domains parametrized by $s \in D$.

Since the holomorphic map $f : X_0 \to D \times D$ defined by $f(z, w) = (z, zw)$ is non-degenerate outside the locus $z = 0$, by (7.1.7) Ψ_{X_0} is positive outside the locus $z = 0$. In fact, by (2) of (7.1.4) and (7.1.9) we have

$$\Psi_{X_0} \geq \frac{2^2 |z|^2 dz \wedge dw \wedge d\bar{z} \wedge d\bar{w}}{(1 - |z|^2)^2 (1 - |zw|^2)^2}.$$

From (7.1.7) we know that $\Psi_{X_0} > 0$ at the origin $(0,0)$. We shall show that $\Psi_{X_0} > 0$ on the entire locus $z = 0$. Let $f: B^2 \to X_0$ be a holomorphic map sending the origin to $(0, w_0) \in X_0$. After a rotation of B^n, the map $f = (f_1, f_2)$ is of the form

$$
\begin{aligned}
f_1(s, t) &= as + bt + \alpha s^2 + \beta st + \gamma t^2 + O(3), \\
f_2(s, t) &= w_0 + ct + O(2).
\end{aligned}
$$

As in (7.1.18) we want to find an upper bound for $|ac|$ independent of f. On the disc $D_{1/\sqrt{2}}$ of radius $1/\sqrt{2}$ we have

$$
f_1(t, t) = (a + b)t + (\alpha + \beta + \gamma)t^2 + O(3).
$$

Since $|f_1| < 1$, from Cauchy's integral formula we obtain

$$
|\alpha + \beta + \gamma| \le 2.
$$

On $D_{1/\sqrt{2}}$ we set

$$
g(t) = f_1(t, t)f_2(t, t) = (a + b)w_0 t + ((a + b)c + (\alpha + \beta + \gamma)w_0)t^2 + O(3).
$$

Since $|f_1 f_2| < 1$, Cauchy's integral formula yields

$$
|(a + b)c + (\alpha + \beta + \gamma)w_0| \le 2.
$$

Hence,

$$
|(a + b)c| = |(a + b)c + (\alpha + \beta + \gamma)w_0 - (\alpha + \beta + \gamma)w_0| \le 2 + 2|w_0|.
$$

By a similar calculation using $h(t) = f_1(t, -t)f_2(t, -t)$ we obtain

$$
|(a - b)c| \le 2 + 2|w_0|.
$$

Hence,

$$
|ac| \le \frac{1}{2}|(a + b)c| + \frac{1}{2}|(a - b)c| \le 2 + 2|w_0|.
$$

From (7.1.3) we obtain

$$
(\Psi_{X_0})_{(0, w_0)} = \inf_f \frac{2^2}{|ac|^2} dz \wedge dw \wedge d\bar{z} \wedge d\bar{w} \ge \frac{1}{(1 + |w_0|)^2} dz \wedge dw \wedge d\bar{z} \wedge d\bar{w}.
$$

On the other hand, for $s \ne 0$, by (7.1.15) Ψ_{X_s} vanishes on $D_s \times \mathbf{C} = \{(z, w); |z| < |s|\}$.

In contrast to (7.1.20), in the following trivial example we have $\Psi_{X_0} \equiv 0$ while $\Psi_{X_s} > 0$ for all $s \ne 0$.

(7.1.21) **Example.** For each $s \in D$, let

$$
X_s = \{z \in \mathbf{C}; \ |zs| < 1\}.
$$

(7.1.22) **Example**. Let T be a complex torus of dimension n. Then $\Phi_T = \Psi_T = 0$. This follows directly from (7) of (7.1.4) and (7.1.15).

Let ι be the involution of T sending $t \in T$ to $-t$. Then the quotient space T/ι has 2^{2n} singular points corresponding to the fixed points of ι, and $\pi: T \to T/\iota$ is a double covering ramified at these singular points. By (2) of (7.1.4), $\Psi_{T/\iota}^h = 0$ at all regular points of T/ι.

Let X be the complex manifold obtained by blowing up the singular points of T/ι. (For $n = 2$, the resulting surface X is called the **Kummer surface** associated to T.) Since Ψ_X^m is meromorphically invariant, we see that Ψ_X^m vanishes except possibly at the exceptional divisors. It is not clear what happens at the exceptional divisors.

(7.1.23) **Remarks**. On the boundary behavior of the intrinsic volume Ψ_X for bounded domains X in $\mathbf{C}^n$, see Ma [1], Cheung-Wong [1], and T. G. Chen [1].

Given an n-dimensional complex manifold X, for any k, $1 \le k \le n$ we can introduce k-dimensional intrinsic area elements $\Phi_{X,k}$ and $\Psi_{X,k}$ as generalizations of Φ_X and Ψ_X. They are nonnegative functions defined on the set of decomposable elements of $\bigwedge^k TX$. For details, see Eisenman [1], Kobayashi [7], Graham-Wu [1], and Graham [2, 3]. See also Venturini [1], Kaliman [2], and Kaliman-Zaidenberg [2].

Bland-Graham [1] and Graham-Wu [1] used Φ_X and Ψ_X to characterize the ball in $\mathbf{C}^n$ and to give criteria for the indicatrices of the infinitesimal metrics E_X and F_X to be ellipsoid; see also Graham [2].

Chinak [1, 2] defined a variant of Ψ_X using only injective holomorphic maps $B_n \to X$. This is analogous to the pseudo-metric introduced by Hahn [3] using only injective holomorphic discs $D \to X$; see Remark (3.1.30). Chinak [3] defined also another variant of Ψ_X using only holomorphic maps $B_n \to X$ with "multiplicity" bounded by a given integer k.

Intrinsic volume forms have been used by Zaidenberg [10] and Kaliman [3] to construct exotic complex structures on $\mathbf{C}^n$.

2 Intrinsic Measures

Let B^n be the unit ball in $\mathbf{C}^n$ with center at 0, and $\mu = \mu_{B^n}$ the invariant volume form defined by (7.1.1). The Borel measure on B^n defined by μ will be denoted also by μ. By Schwarz' lemma (2.4.16), every holomorphic map $f: B^n \to B^n$ is measure-decreasing with respect to μ, i.e.,

$$(7.2.1) \qquad \mu[f(E)] \le \mu[E] \qquad \text{for every Borel set} \quad E \subset B^n.$$

Let X be an n-dimensional complex space. We define three intrinsic measures $\bar{\Phi}_X$, $\bar{\Psi}_X^h$ and $\bar{\Psi}_X^m$ corresponding to pseudo-volume forms Φ_X, Ψ_X^h and Ψ_X^m.

Given a Borel set $A \subset X$, choose holomorphic maps $f_i: B^n \to X$ and Borel sets $E_i \subset B^n$ for $i = 1, 2, \ldots$ such that $A \subset \bigcup f_i(E_i)$. Then the measure $\bar{\Psi}_X^h$ is defined by

(7.2.2)
$$\bar{\Psi}_X^h[A] = \inf \sum_i \mu[E_i],$$

where the infimum is taken over all possible choices of f_i and E_i.

Allowing f_i to be meromorphic in the definition above, we obtain a measure $\bar{\Psi}_X^m$ on X. The singularity set of f_i can be ignored since it is of μ-measure 0 in B^n.

Given a Borel set $A \subset X$, we set

(7.2.3)
$$\Phi_X^*[A] = \sup_f \mu[f(A)],$$

where the supremum is taken over all holomorphic maps $f: X \to B^n$. We note that Φ_X^* is not a measure since it is, in general, not additive. We write A as a union of countably many Borel sets A_i, and we define a measure $\bar{\Phi}_X$ by setting

(7.2.4)
$$\bar{\Phi}_X[A] = \sup \sum_i \Phi_X^*[A \cap B_i],$$

where the supremum is taken over all families of countably many disjoint Borel subsets B_i, $i = 1, 2, \ldots$ of X. More directly, we can write

(7.2.5)
$$\bar{\Phi}_X[A] = \sup \sum_i \mu[f_i(A \cap B_i)],$$

where the supremum is taken over all $f_i \in \mathrm{Hol}(X, B^n)$ and all disjoint Borel sets $B_i \subset X$, $i = 1, 2, \ldots$. The proof of the following theorem is similar to that of (7.1.4).

(7.2.6) **Theorem.** *Let X and Y be complex spaces of dimension n.*

    (1)   *For the unit ball B^n,*

$$\mu = \bar{\Phi}_{B^n} = \bar{\Psi}_{B^n}^h = \bar{\Psi}_{B^n}^m;$$

    (2)   *If $f: X \to Y$ is holomorphic, then for any Borel set $A \subset X$ we have*

$$\bar{\Phi}_Y[f(A)] \le \bar{\Phi}_X[A] \quad \text{and} \quad \bar{\Psi}_Y^h[f(A)] \le \bar{\Psi}_X^h[A];$$

    (3)   *If $f: X \to Y$ is meromorphic, then for any Borel set $A \subset X$ we have*

$$\bar{\Phi}_Y[f(A)] \le \bar{\Phi}_X[A] \quad \text{and} \quad \bar{\Psi}_Y^m[f(A)] \le \bar{\Psi}_X^m[A];$$

    (4)   *If ν_X is any measure on X such that $f^*\nu_X \le \mu$ for all holomorphic (resp. meromorphic) maps $f: B^n \to X$, then*

$$\nu_X \le \bar{\Psi}_X^h \quad (\text{resp. } \nu_X \le \bar{\Psi}_X^m);$$

    (5)   *If ν_X is any measure on X such that $f^*\mu \le \nu_X$ for all holomorphic maps $f: X \to B^n$, then*

$$\bar{\Phi}_X \le \nu_X;$$

    (6)   *The three intrinsic measures are related as follows:*

$$\bar{\Phi}_X \le \bar{\Psi}_X^m \le \bar{\Psi}_X^h;$$

(7)   *If $\pi: \tilde{X} \to X$ is a covering projection, then*

$$\bar{\Psi}_{\tilde{X}}^h = \pi^* \bar{\Psi}_X \quad \text{and} \quad \bar{\Psi}_{\tilde{X}}^m = \pi^* \bar{\Psi}_X^m.$$

For the same reason as in the proof of (7.1.6), we have

(7.2.7) **Proposition**. *The equality $\bar{\Psi}_X^m = \bar{\Psi}_X^h$ holds in the following cases*:

(a)   *X has a Stein space $\tilde{X}$ as a covering space*;
(b)   *X is hyperbolic.*

When the equality above holds, we simply write $\bar{\Psi}_X$ for $\bar{\Psi}_X^m = \bar{\Psi}_X^h$.

Since Φ_X is continuous and since Ψ_X^h and Ψ_X^m are upper semicontinuous (see (8) of (7.1.4)), for any open set $A \subset X$ we can define the integrals:

$$\Phi_X[A] = \int_A \Phi_X, \quad \Psi_X^h[A] = \int_A \Psi_X^h \quad \text{and} \quad \Psi_X^m[A] = \int_A \Psi_X^m.$$

Since Ψ_X^h and Ψ_X^m are upper semicontinuous on X, Ψ_X^h and Ψ_X^m define Borel measures on X. Thus $\Psi_X^h[A]$ and $\Psi_X^m[A]$ are defined for any Borel set $A \subset X$. Similarly, $\Phi_X[A]$ is also defined for any Borel set $A \subset X$.

(7.2.8) **Theorem**. *For any Borel set $A \subset X$, we have*

$$\bar{\Phi}_X[A] = \Phi_X[A], \quad \bar{\Psi}_X^h[A] = \Psi_X^h[A] \quad and \quad \bar{\Psi}_X^m[A] = \Psi_X^m[A].$$

Proof. First, we prove the second equality. The proof for the third equality is essentially the same. For simplicity, we write Ψ_X and $\bar{\Psi}_X$ for Ψ_X^h and $\bar{\Psi}_X^h$, respectively.

Since every $f \in \mathrm{Hol}(B^n, X)$ is a measure-decreasing map of (B^n, μ) into (X, Ψ_X), by (4) of (7.2.6) we have $\Psi_X \le \bar{\Psi}_X$.

It suffices to verify the reverse inequality $\bar{\Psi}_X[A] \le \Psi_X[A]$ for a relatively compact open set A. Since the pseudo-volume form Ψ_X is upper semicontinuous, it is the limit of a monotone decreasing sequence of continuous pseudo-volume forms v_m, and by the Lebesgue convergence theorem we have

$$\Psi_X[A] = \lim v_m[A], \quad \text{where} \quad v_m[A] = \int_A v_m.$$

Fix an arbitrary $\varepsilon > 0$, and choose a large m such that

$$v_m[A] < \Psi[A] + \varepsilon.$$

We denote this v_m simply by v. We fix also a (positive) continuous volume form w. For each point x_0 of A we take a holomorphic map $f: B^n \to X$ such that $f(0) = x_0$ and

$$(f^{-1})^*(\mu_0) < v_{x_0} + \varepsilon w_{x_0}.$$

Then in a small neighborhood $U_0 \subset B^n$ of 0 and in the neighborhood $f(U_0)$ of x_0 in X, we have

$$(f^{-1})^*(\mu_b) < v_x + \varepsilon w_x \qquad \text{for} \quad b \in U_0, \ x = f(b) \in f(U_0).$$

Now we subdivide A into small domains A_i, $i = 1, 2, \ldots, N$, so that (i) $A \supset \bigcup A_i$ (disjoint), (ii) $A \subset \bigcup \overline{A}_i$, and (iii) for each i there is a holomorphic map $f_i \colon B^n \to X$ which maps a neighborhood E_i of $0 \in B^n$ biholomorphically onto A_i and satisfies

$$(f_i^{-1})^*(\mu_b) < v_x + \varepsilon w_x \qquad \text{for} \quad b \in E_i, \ x = f(b) \in A_i.$$

Integrating the above inequality and using $v[A] < \Psi_X[A] + \varepsilon$, we obtain

$$\sum \mu[E_i] < v[A] + \varepsilon w[A] < \Psi_X[A] + \varepsilon + \varepsilon w[A].$$

Since ε is arbitrary, we obtain the desired inequality $\bar{\Psi}_X[A] \leq \sum \mu[E_i] \leq \Psi_X[A]$.

Now we consider Φ_X. The inequality $\bar{\Phi}_X[A] \leq \Phi_X[A]$ follows from (5) of (7.2.6). In order to prove the reverse inequality, we fix a (positive) continuous volume form w as above. We may also assume that A is relatively compact. Given $\varepsilon > 0$, we subdivide A into small domains A_i, $i = 1, 2, \ldots, N$ and find holomorphic maps $f_i \colon X \to B^n$ such that

$$\Phi_X \leq f_i^* \mu + \varepsilon w \qquad \text{on} \quad A_i.$$

(Since Φ_X is continuous, the proof is simpler in this case than the case of Ψ_X considered above). Integrating the inequality above, we obtain

$$\Phi_X[A] = \sum \Phi_X[A_i] \leq \sum \mu[f_i(A_i)] + \varepsilon \sum w[A_i] \leq \Phi_X[A] + \varepsilon w[A].$$

$$\square$$

From now on, we can denote the measures $\bar{\Phi}_X$, $\bar{\Psi}_X^h$ and $\bar{\Psi}_X^m$ also by Φ_X, Ψ_X^h and Ψ_X^m, respectively.

Let X be a complex space with intrinsic pseudo-distance d_X. As we stated at the beginning of Section 1, we can associate Hausdorff measures to d_X. Let $n = \dim X$. Then we define the $2n$-dimensional **Hausdorff measure** $\mu_X^{(2n)}$ as follows. For a subset $E \subset X$, we set

$$(7.2.9) \qquad \mu_X^{(2n)}(E) = \sup_{\varepsilon > 0} \inf \Big\{ c_n \sum_{i=1}^{\infty} (\delta(E_i))^{2n}; \ E = \bigcup_{i=1}^{\infty} E_i, \ \delta(E_i) < \varepsilon \Big\},$$

where $\delta(E_i)$ is the diameter of E_i measured by d_X and c_n is a universal constant to normalize $\mu_{B^n}^{(2n)}$ for the unit ball $B^n \in \mathbf{C}^n$ as follows. Since d_{B^n} is invariant by the automorphism group $\mathrm{Aut}(B^n)$, so is $\mu_{B^n}^{(2n)}$. Since $\mathrm{Aut}(B^n)$ is transitive on B^n, the measured defined by the volume element (7.1.1) coincides with $\mu_{B^n}^{(2n)}$ with a suitable constant c_n. Thus, c_n is determined by the condition that $\mu = \mu_{B^n}^{(2n)}$.

Since every holomorphic map $f: X \to Y$ is distance-decreasing with respect to d_X and d_Y, it is also measure-decreasing with respect to $\mu_X^{(2n)}$ and $\mu_Y^{(2n)}$. From (4) of (7.2.6) we obtain

$$(7.2.10) \qquad\qquad \mu_X^{(2n)} \leq \Psi_X^h.$$

(7.2.11) **Proposition**. *If an n-dimensional complex space X is hyperbolic modulo a closed subset Δ, then $\mu_X^{(2n)}(E) > 0$ for every nonempty open subset $E \subset X$ with $E \cap \Delta = \emptyset$.*

Proof. Let F be a smooth length function on X. Then there is a nonnegative continuous function φ on X which is positive outside Δ such that $\varphi F \leq F_X$, (see (3.5.41)). Since the $2n$-dimensional Hausdorff measure defined by φF is positive on $X - \Delta$, $\mu_X^{(n)}$ is positive on $X - \Delta$. $\qquad\qquad\square$

From (7.2.10) and (7.2.11) we obtain

(7.2.12) **Corollary**. *If a complex space X is hyperbolic modulo a closed subset Δ which has no interior point, then X is Ψ^h-measure hyperbolic.*

For a complex space X, we consider its **intrinsic total volume** $\Psi_X^h[X] = \int_X \Psi_X^h$ as well as $\Psi_X^m[X] = \int_X \Psi_X^m$.

Let X and Y be compact complex spaces of dimension n, and $f: X \to Y$ a holomorphic map. Then the degree of f is a non-negative integer; f covers Y generically as many times as $\deg(f)$. Hence, if $\Psi_Y^h[Y] > 0$, then

$$\deg(f) = (f^* \Psi_Y^h)[X] / \Psi_Y^h[Y],$$

and the inequality $f^* \Psi_Y^h \leq \Psi_X^h$ implies

$$(7.2.13) \qquad\qquad \deg(f) \leq \Psi_X^h[X] / \Psi_Y^h[Y].$$

If X is a compact Riemann surface of genus $g \geq 1$, then the Gauss-Bonnet formula gives

$$\Psi_X[X] = 2\pi(2g - 2).$$

The inequality (7.2.13) generalizes the well known fact that there is no nonconstant holomorphic map from a compact Riemann surface X into another compact Riemann surface Y of larger genus.

In general, from (7.2.13) we obtain

(7.2.14) **Theorem**. *Let X and Y be compact complex spaces of dimension n, and $f: X \to Y$ a holomorphic map.*

(1) *If $0 < \Psi_X^h[X] < k\Psi_Y^h[Y]$, then $\deg(f) < k$. In particular, if $0 < \Psi_X^h[X] < \Psi_Y^h[Y]$, then f is degenerate everywhere on X.*

(2) *Assume that there exists a continuous (positive) volume form v on X such that $0 < v \leq \Psi_X^h$ on X. If $\Psi_X^h[X] = \Psi_Y^h[Y]$, then f is either degenerate everywhere or nondegenerate everywhere in the nonsigular part of X.*

Proof. (1)   The first assertion follows directly from (7.2.13). If $\deg(f)$ is 0, then f must be degenerate everywhere.

(2)   If $\Psi_X^h[X] = \Psi_Y^h[Y]$, then either $\deg(f) = 0$ (in which case f is degenerate everywhere) or $\deg(f) = 1$. In the latter case, we have

$$1 = \deg(f) = (f^*\Psi_Y^h)[X]/\Psi_Y^h[Y] \leq \Psi_X^h[X]/\Psi_Y^h[Y] = 1.$$

Hence, $(f^*\Psi_Y^h)[X] = \Psi_X^h[X]$. On the other hand, $f^*\Psi_Y^h \leq \Psi_X^h$ on X. Then f must be measure-preserving in the sense that $(f^*\Psi_Y^h)[A] = \Psi_X^h[A]$ or $\Psi_Y^h[f(A)] = \Psi_X^h[A]$ for all Borel sets $A \subset X$. Suppose that there is a regular point $x_0 \in X$ where f is degenerate. Since $f^*\Psi_Y^h$ vanishes at x_0 and since Ψ_Y^h is upper semicontinuous, there is a neighborhood U of x_0 such that $f^*\Psi_Y^h < v$ on U. Then $(f^*\Psi_Y^h)[U] < v[U] \leq \Psi_X^h[U]$, which is a contradiction.   □

If $X = Y$, we have the following result (Kobayashi [7], Yau [2]), which generalizes (6.6.20).

(7.2.15) Corollary. *Let X be a compact normal complex space with intrinsic total volume $\Psi_X^h[X] > 0$. Then every holomorphic mapping f of X into itself is either degenerate everywhere or biholomorphic.*

Proof. Clearly, $\deg(f) = 0$ or 1. If $\deg(f) = 0$, then f is everywhere degenerate. If $\deg(f) = 1$, f is biholomorphic by the following lemma. (The proof given here is by Yingchen Li).

(7.2.16) Lemma. *If a holomorphic map f of a compact complex space X into itself is of degree 1, then it is a homeomorphism.*

Proof. Using the Stein factorization theorem we can factor f uniquelly as the composition of g and h:

$$f: X \xrightarrow{g} Y \xrightarrow{h} X,$$

where g is a proper surjective holomorphic mapping with connected fibres, and h is a finite holomorphic mapping.

Clearly h is also surjective of degree 1. As in part (c) in the proof of (5.3.5), a finite holomorphic mapping h of degree 1 is a homeomorphism.

So from now on we assume that f has connected fibres. Let $A \subset X$ be the set of $x \in X$ such that $\dim(f^{-1}(f(x))) > 0$. Then A and its image $B = f(A)$ are closed complex subspaces of X. We shall show that A is empty. Assume that A is nonempty. Then $\dim B < \dim A$. Put $p = \dim A$ and $q = \dim B$. We form the commutative diagram of homology groups induced by f:

$$
\begin{array}{ccccccccc}
H_{2p+1}(A) & \to & H_{2p+1}(X) & \to & H_{2p+1}(X, A) & \to & H_{2p}(A) & \to & H_{2p}(X) \\
\downarrow & & \downarrow & & \downarrow & & \downarrow & & \downarrow \\
H_{2p+1}(B) & \to & H_{2p+1}(X) & \to & H_{2p+1}(X, B) & \to & H_{2p}(B) & \to & H_{2p}(X).
\end{array}
$$

Since any compact complex space can be triangulated, we may assume all spaces in the above diagram are finite polyhedra. And thus all homology groups appearing above are finite dimensional.

Since $H_{2p+1}(A) = H_{2p+1}(B) = H_{2p}(B) = H_{2p-1}(B) = 0$, we have $H_{2p+1}(X) \cong H_{2p+1}(X, B)$. On the other hand, we know from singular homology theory that $H_i(X, A) \cong H_i(X, B)$ for all i. So $H_{2p+1}(X)$ and $H_{2p+1}(X, A)$ have the same dimension. From the diagram we deduce that the map $H_{2p+1}(X) \to H_{2p+1}(X, A)$ in the diagram is an isomorphism.

Thus the right part of the diagram becomes

$$
\begin{array}{ccc}
0 \to & H_{2p}(A) & \to H_{2p}(X) \\
& \downarrow & \downarrow \\
0 \to & 0 & \to H_{2p}(X).
\end{array}
$$

For any complex space Z we use $H_{2l}^{\mathrm{anal}}(Z)$ to denote the subgroup of $H_{2l}(Z)$ generated by l-dimensional analytic cycles of Z. The last diagram now implies

$$
\begin{array}{ccc}
0 \to & H_{2p}^{\mathrm{anal}}(A) & \to H_{2p}^{\mathrm{anal}}(X) \\
& \downarrow & \downarrow \\
0 \to & 0 & \to H_{2p}^{\mathrm{anal}}(X).
\end{array}
$$

Now, since $f: X \to X$ is surjective and generically one-to-one, each p-dimensional analytic cycle of X is the image of a p-dimensional analytic cycle of X by f, i.e., the map $H_{2p}^{\mathrm{anal}}(X) \to H_{2p}^{\mathrm{anal}}(X)$ in the above diagram is surjective and, hence, is an isomorphism. Hence, $H_{2p}^{\mathrm{anal}}(A) = 0$. On the other hand, $H_{2p}^{\mathrm{anal}}(A)$ is nontrivial since it contains the fundamental cycle $[A]$. This is a contradiction, showing that A must be empty. This completes the proof of Lemma.

If X is normal, f^{-1} is also holomorphic.   □

While $\mathbf{C}$ can be compactified as a Zariski open subset of $P_1\mathbf{C}$, the disc D cannot be. We give a necessary condition for a complex space X to be compactified.

(7.2.17) **Theorem.** *Let Y be a compact complex space, A a closed complex subspace of lower dimension, and $X = Y - A$. Then the intrinsic total volume $\Psi_X[X]$ of X is finite.*

Proof. Let $\pi: (\hat{Y}, \hat{A}) \to (Y, A)$ be a resolution of singularities such that A has only normal crossing singularities. Since π decreases the intrinsic measure, it suffices to prove the theorem for $(\hat{Y}, \hat{A})$. We assume therefore that Y is nonsingular and that A has only normal crossing singularities.

We cover Y by a finite number of open sets U_i, each of which is biholomorphic to the unit polydisc D^n. We may assume that $X \cap U_i$ is biholomorphic to $(D^*)^k \times D^{n-k}$. Let r be a number slightly smaller than 1, and let D_r^n be the polydisc of radius r. Let $U_i(r)$ be the subset of U_i corresponding to D_r^n. If r is sufficiently close to 1, then $\{U_i(r)\}$ already covers Y, and $X \cap U_i(r)$ corresponds to $(D_r^*)^k \times D_r^{n-k}$.

Using (7.1.8), (7.1.9) and (7.1.11) we can calculate the integral

$$
\Psi_{(D^*)^k \times D^{n-k}}[(D_r^*)^k \times D_r^{n-k}] < \infty.
$$

Hence, $\Psi_X[U_i(r)] \le \Psi_{U_i}[U_i(r)] < \infty.$ □

3 Pseudo-ampleness and L-dimension

Results on line bundles L proved in this section will be applied to canonical line bundles in Section 4.

Let X be a compact complex space of dimension n, and $\mathbf{C}(X)$ its field of meromorphic functions. The transcendence degree of $\mathbf{C}(X)$, denoted $a(X)$, is called the **algebraic dimension** of X. Then $a(X) \leq n$. If $a(X) = n$, then X is called a **Moishezon space**. A Moishezon space does not differ very much from a projective variety. If X is a Moishezon space, there is a modification $\pi: X^* \to X$ with a smooth projective variety X^*.

Let L be a line bundle over a compact complex space X such that the space of holomorphic sections $\Gamma(L)$ is nonzero. Let $\psi_0, \ldots, \psi_N$ be a basis for $\Gamma(L)$. Then we can define a map $\Phi_L: X \to P_N\mathbf{C}$ by

$$(7.3.1) \qquad z \in X \to (\psi_0(z) : \ldots : \psi_N(z)) \in P_N\mathbf{C}.$$

If at each $z \in X$ at least one ψ_j does not vanish at z, then this map is holomorphic. Otherwise, that is, if the sections of L has common zeros (called **base points**), then Φ_L is not well defined at the base points and is merely a meromorphic map.

If the map Φ_L gives a holomorphic imbedding of X into $P_N\mathbf{C}$, then L is said to be **very ample**. If L^m is very ample for some integer $m > 0$, then L is said to be **ample**. A compact complex space X is a projective variety if and only if it has an ample line bundle L.

A little more generally, a line bundle L is said to be **very pseudo-ample** if Φ_L gives a meromorphic imbedding of X into $P_N\mathbf{C}$. In this case, Φ_L gives a holomorphic imbedding of a nonempty Zariski open subset of X in o $P_N\mathbf{C}$. If L^m is very pseudo-ample for some positive integer m, then L is said to be **pseudo-ample**.

The so-called L-**dimension** $\kappa(L, X)$ of a compact normal complex space X is defined by

$$(7.3.2) \qquad \kappa(L, X) = \max_{m>0} \dim \Phi_{L^m}(X),$$

provided $\Gamma(L^m) \neq 0$ for some $m > 0$. If $\Gamma(L^m) = 0$ for all $m > 0$, then by convention we set

$$\kappa(L, X) = -\infty.$$

If X is not normal, we take its normalization $\pi: X^* \to X$ and set

$$\kappa(L, X) = \kappa(\pi^*L, X^*).$$

More generally, if $\pi: X' \to X$ is a modification with X' normal, then $\kappa(L, X) = \kappa(\pi^*L, X')$, see Ueno [1, p. 51].

We have in general

$$(7.3.3) \qquad \kappa(L, X) \leq a(X) \leq \dim X.$$

The L-dimension $\kappa(L, X)$ is equal to the complex dimension n of X if and only if L is pseudo-ample. In such a case, X is clearly Moishezon.

Let

$$\mathbf{N}(L, X) = \{m > 0;\ \Gamma(L^m) \neq 0\}.$$

Then $\mathbf{N}(L, X)$ is a semigroup under addition. Let d denote the greatest common divisor of $\mathbf{N}(L, X)$. One of the fundamental theorems on L-dimension states

(7.3.4) Theorem. *Let X be a compact complex space, and L a line bundle over X. Then there exist positive numbers a, b and a positive integer m_0 such that*

$$am^{\kappa(L,X)} \leq \dim \Gamma(L^{md}) \leq bm^{\kappa(L,X)} \qquad \text{for} \quad m \geq m_0.$$

Iitaka [1] used the inequalities in (7.3.4) to define the L-dimension. For a systematic account on L-dimension and, in particular, for the proof of (7.3.4), see Ueno [1] or Iitaka [1].

Here we shall consider only the special case where $\kappa(L, X) = n$, so that X is Moishezon and can be resolved by a smooth algebraic variety $\pi: X^* \to X$. We shall therefore assume that X is projective.

As a first step, we prove the following two lemmas, which are contained in (7.3.4) as special cases.

(7.3.5) Lemma. *For any line bundle L over a smooth projective variety X of dimension n, we have*

$$\limsup_{m \to \infty} \frac{1}{m^n} \dim \Gamma(L^m) < \infty.$$

Proof. Choose an ample line bundle H such that LH is also ample. If m is large enough so that H^m is very ample, then we have an exact sequence

$$0 \to H^0(X, L^m) \to H^0(X, L^m H^m) \to H^0(S, (L^m H^m)_S) \to,$$

where S is a nonsingular positive divisor of X obtained as the zero set of a general holomorphic section of H^m, and $(L^m H^m)_S$ denotes the restriction of $L^m H^m$ to S, (see, for example, Hirzebruch [1; p. 130]). From this exact sequence we obtain

$$\dim \Gamma(L^m) \leq \dim \Gamma(L^m H^m) \qquad \text{for} \quad m \geq m_0.$$

Since $L^m H^m$ is ample, Kodaira's vanishing theorem implies

$$\dim \Gamma(L^m H^m) = \chi(X, L^m H^m) \qquad \text{for} \quad m \geq m_0.$$

On the other hand (Hirzebruch [1; p. 150]), we have

$$\chi(X, L^m H^m) = a_0 + a_1 m + \ldots + a_n m^n,$$

where $a_0, a_1, \ldots, a_n$ are rational numbers determined by characteristic classes of X and LH, (in particular, $n! a_n = (c_1(LH))^n[X]$). Hence,

$$\limsup_{m\to\infty} \frac{1}{m^n} \dim \Gamma(L^m) \le a_n.$$

$\square$

(7.3.6) Lemma. *Let X be an n-dimensional projective algebraic manifold with a very ample line bundle H and a line bundle L such that*

$$\limsup_{m\to\infty} \frac{1}{m^n} \dim \Gamma(L^m) > 0.$$

Then there exists a positive integer m such that $\Gamma(L^m H^{-1}) \ne 0$.

Proof. Let S be a non-singular positive divisor of X obtained as the zero set of a general holomorphic section of H. As in the proof of (7.3.5), we have an exact sequence

$$0 \to H^0(X, L^m H^{-1}) \to H^0(X, L^m) \to H^0(S, L_S^m).$$

Since $\dim H^0(X, L^m)$ is of order m^n by assumption and since $\dim H^0(S, L_S^m)$ is of order at most m^{n-1} by (7.3.5), it follows that $H^0(X, L^m H^{-1}) \ne 0$ if m is sufficiently large. $\square$

Conversely,

(7.3.7) Lemma. *Let H be a very ample line bundle and L a line bundle over X such that $\Gamma(L^m H^{-1}) \ne 0$ for some positive integer m. Then Φ_{L^m} gives a meromorphic imbedding of X into a projective space and, hence, L is pseudo-ample.*

Proof. We prove more than what is stated above. Let α be a non-trivial holomorphic section of $L^m H^{-1}$. Let Γ_X be the image of the injective map

$$\varphi \in \Gamma(H) \to \alpha\varphi \in \Gamma(L^m).$$

This means that although L^m is not very ample, we can still obtain a holomorphic projective imbedding of X by using only the subspace Γ_X, i.e., the sections of L^m that are divisible by α; the imbedding thus obtained is none other than the imbedding obtained by using the sections of H. If we use all the sections of L^m (divisible by α or not), then we obtain only a meromorphic imbedding of X into a projective space. $\square$

In the course of the proof of (7.3.7) we established

(7.3.8) Lemma. *Let $\Gamma_X \subset \Gamma(L^m)$ be as above, and Γ_X^* its dual space. Let $\varphi_0, \ldots, \varphi_N$ be a basis for $\Gamma(H)$, and $\alpha\varphi_0, \ldots, \alpha\varphi_N$ the corresponding basis for Γ_X. Then*

$$x \mapsto [\varphi_0(x) : \ldots : \varphi_N(x)] = [\alpha\varphi_0(x) : \ldots : \alpha\varphi_N(x)]$$

defines a holomorphic imbedding of X into $P(\Gamma_X^*)$.

This shows that if $\kappa(L, X) = n$, then L is pseudo-ample. Conversely, assume that L is a pseudo-ample line bundle over a smooth projective variety X. Let $\Phi_{L^k}: X \to P_N\mathbf{C}$ be the meromorphic imbedding defined by the sections of L^k. Let

z_0 be a point where Φ_{L^k} is holomorphic. We choose a basis $\psi_0, \psi_1, \ldots, \psi_n, \ldots$ of $H^0(X, L^k)$ in such a way that

$$\psi_0(z_0) \neq 0, \quad \psi_1(z_0) = \ldots = \psi_n(z_0) = \ldots = 0,$$

and $\psi_1/\psi_0, \ldots, \psi_n/\psi_0$ form a local coordinate system around z_0. Then, for any positive integer m, the set

$$\{\psi_{i_1}\psi_{i_2}\ldots\psi_{i_m}; \; 0 \leq i_1 \leq i_2 \leq \ldots \leq i_m \leq n\}$$

is a linearly independent subset of $H^0(X, L^{km})$. Hence,

$$\dim H^0(X, L^{km}) \geq \binom{n+m}{m} \geq am^n,$$

where a is a positive number.

Summarizing the results above we state

(7.3.9) Proposition. *Let L be a line bundle over an n-dimensional smooth projective variety X. Then the following are equivalent*:

(a) $\qquad\qquad\qquad\qquad$ *L is pseudo-ample*;

(b) $\qquad\qquad\qquad\qquad$ $\kappa(L, X) = n$;

(c) $\qquad\qquad\qquad\qquad$ $\displaystyle\limsup_{m \to \infty} \frac{1}{m^n} \dim \Gamma(L^m) > 0.$

Given a pseudo-metric h on a line bundle, we can associate a closed $(1,1)$-form $\mathrm{Ric}(h)$ called the Ricci form, (see (2.4.1)). In general, $\mathrm{Ric}(h)$ is defined only where h is positive. However, it can be defined even at zeroes of h if h degenerates holomorphically as we explained in Section 4 of Chapter 2. This situation arises in a natural way when we consider pseudo-ample line bundles.

Let L be an ample line bundle over X with L^m very ample, and let $\psi_0, \ldots, \psi_N$ be a basis for $H^0(X, L^m)$. Then $\sum |\psi_j|^2$ is a metric on L^{-m}, and $h = (\sum |\psi_j|^2)^{1/m}$ is a metric on L^{-1}. A direct calculation shows that the Ricci form $\mathrm{Ric}(h)$ is negative definite; in fact, $-\mathrm{Ric}(h)$ is the pull-back of (Kähler form of) the Fubini-Study metric of $P_N\mathbf{C}$ by the immersion $\Phi_{L^m}: X \to P_N\mathbf{C}$. Conversely, if L^{-1} admits a metric h with negative Ricci form $\mathrm{Ric}(h) < 0$ (or equivalently, if L admits a metric with positive Ricci form), then L is ample; this is the Kodaira imbedding theorem.

In summary,

(7.3.10) Theorem. *A line bundle L over a compact complex manifold X is ample if and only if L^{-1} admits a metric h with negative Ricci form $\mathrm{Ric}(h)$.*

More generally, let L be a pseudo-ample line bundle over a smooth projective variety X. Let H and m be as in (7.3.7) and α a non-trivial holomorphic section of $L^m H^{-1}$. Let $\Gamma_X \subset H^0(X, L^m)$ denote the image of the injection $H^0(X, H) \to H^0(X, L^m)$ given by (7.3.8). Let $\varphi_0, \varphi_1, \ldots, \varphi_N$ be a basis for $H^0(X, H)$, and

$\psi_0 = \alpha\varphi_0, \psi_1 = \alpha\varphi_1, \ldots, \psi_N = \alpha\varphi_N$ be the corresponding basis for Γ_X. As we noted in the proof of (7.3.7), the mapping

$$(7.3.11) \qquad x \mapsto (\psi_0(x) : \ldots : \psi_N(x)) = (\varphi_0(x) : \ldots : \varphi_N(x))$$

imbeds X holomorphically into $P_N\mathbf{C}$. Using the basis $\psi_0, \psi_1, \ldots, \psi_N$ we obtain a pseudo-metric $\sum |\psi_j|^2$ on the line bundle L^{-m}. Hence, we have a pseudo-metric

$$(7.3.12) \qquad h = \left(\sum_{j=0}^{N} |\psi_j|^2\right)^{1/m} = |\alpha|^{2/m}\left(\sum_{j=0}^{N} |\varphi_j|^2\right)^{1/m}$$

on the line bundle L^{-1}. Then α accounts for the degeneracy of h and, as we explained in Section 4 of Chapter 2, the Ricci form $\mathrm{Ric}(h)$ is well defined everywhere on X. A direct calculation shows that $\mathrm{Ric}(h)$ is negative definite everywhere; again, $-\mathrm{Ric}(h)$ is the pull-back of (the Kähler form of) the Fubini-Study metric of $P_N\mathbf{C}$ by the imbedding (7.3.11).

The following theorem is a generalization of (7.3.10). We have already established the "only if " part of the theorem. Its converse is due to Burt Totaro.

(7.3.13) **Theorem.** *Let L be a line bundle over a smooth projective variety X. Then L is pseudo-ample if and only if L^{-1} admits a pseudo-metric h with at most holomorphic degeneracy (i.e., locally, h is of the form $h = a^{2q}g$, where a is holomorphic, g is strictly positive and q is a rational number) such that $\mathrm{Ric}(h) < 0$.*

Proof. Since q is a rational number, $q = p/m$, we replace L by L^m and put the natural pseudo-metric h^m on L^m. Then h^m is locally of the form $a^{2p}g^m$. Since it suffices to show that L^m is pseudo-ample, we may and shall assume that q is a positive integer.

It suffices to prove that there is an ample line bundle F and a positive integer k such that $\Gamma(L^k F^{-1}) \neq 0$. Indeed, let α be a nonzero holomorphic section of $L^k F^{-1}$. Choose $l > 0$ so that F^l is very ample. Then α^l is a section of $L^{kl}F^{-l}$. By (7.3.7) L is pseudo-ample.

We shall construct an ample line bundle F in terms of its transition functions. Let $\{U_j\}$ be an open cover of X such that L has a non-vanishing holomorphic section s_j on each U_j. Then our assumption on h means:

$$h(s_j) = |a_j|^{2q}g_j,$$

where a_j is holomorphic and g_j is positive on U_j. If $s_j = \lambda_{jk}s_k$ on $U_j \cap U_k$, then $h(s_j) = |\lambda_{jk}|^2 h(s_k)$ so that

$$|a_j|^{2q}g_j = |\lambda_{jk}|^2 |a_k|^{2q}g_k.$$

We set

$$\varphi_{jk} = \lambda_{jk}(a_k a_j^{-1})^q$$

so that

$$g_j = |\varphi_{jk}|^2 g_k,$$

and define a line bundle F by transition functions $\{\varphi_{jk}\}$. (Notice that, although the holomorphic functions a_j and a_k may have zeroes, $a_k a_j^{-1}$ is a holomorphic function on $U_j \cap U_k$ without zeroes.) Then $\{g_j\}$ define a metric g on F^{-1}. Since $h(s_j) = |a_j|^{2q} g_j$, g has the same Ricci form as h. In particular, $\mathrm{Ric}(g)$ is negative. By (7.3.10) F is ample.

Since the transition functions for LF^{-1} are given by $\{\lambda_{jk} \varphi_{jk}^{-1}\}$ and since

$$\lambda_{jk} \varphi_{jk}^{-1} = (a_j a_k^{-1})^q,$$

$\{a_j^q\}$ represents a holomorphic section of LF^{-1}. This shows that $\Gamma(LF^{-1}) \neq 0$.
$\square$

4 Measure Hyperbolicity and Manifolds of General Type

As an application of Schwarz lemma for volume elements, we obtain a differential geometric criterion for measure hyperbolicity. This criterion will be applied to compact complex manifolds of general type.

The following geometric criterion follows immediately from (2.4.15) and (4) of (7.1.4).

(7.4.1) **Theorem.** *Let X be a complex space of dimension n. If X admits a pseudo-volume form v with negatively bounded Ricci form $\mathrm{Ric}(v) < 0$, normalized so that $K_v \leq -1$, then*

$$v \leq \Psi_X^m \leq \Psi_X^h.$$

For the definition of K_v, see (2.4.6).

In order to apply (7.4.1) to compact complex manifolds of general type, we specialize results of the preceding section on line bundles L to the canonical line bundle.

Given a compact complex manifold X of dimension n, let K_X be its canonical bundle. Let $m > 0$ be an integer, and put $N+1 = \dim \Gamma(K_X^m)$. The map $\Phi_{K_X^m} : X \to P_N C$ defined by (7.3.1) is called the m-th **pluri-canonical map** of X.

By specializing the concept of L-dimension to the case $L = K$, we obtain the **Kodaira dimension** $\kappa(X)$:

$$(7.4.2) \qquad\qquad \kappa(X) = \max_{m > 0} \dim \Phi_{K_X^m}(X).$$

For a smooth projective variety X, the following proposition is a special case of (7.3.9). In the general case, it follows from (7.3.4).

(7.4.3) **Proposition.** *For a compact complex manifold X of dimension n, the following are equivalent:*

(a) $\qquad\qquad\qquad$ *K is pseudo-ample;*

(b) $\qquad\qquad\qquad$ *$\kappa(X) = n$;*

$$\text{(c)} \qquad \limsup_{m \to \infty} \frac{1}{m^n} \dim \Gamma(K_X^m) > 0.$$

A compact complex manifold X is said to be **of general type** if one of the three equivalent conditions in (7.4.3) is satisfied. A compact complex space is said to be **of general type** if it has a nonsingular model which is of general type. As we explained in the preceding section, every compact complex space of general type is Moishezon and can be resolved by a smooth projective variety $\pi: X^* \to X$.

Let X be a compact complex manifold. A (pseudo-) volume form v on X is nothing but a (pseudo-) metric on K_X^{-1}, see Section 4 of Chapter 2. As we have seen in (7.3.10), K_X is ample if and only if X admits a volume form v with negative Ricci form $\mathrm{Ric}(v)$. As a special case of (7.3.13) we have

(7.4.4) Theorem. *A smooth projective variety X is of general type if and only if it admits a pseudo-volume form v with at most holomorphic degeneracy (i.e., v is locally of the form $v = |a|^{2q} w$, where a is a holomorphic function, w is a volume form and q is a rational number) such that* $\mathrm{Ric}(v) < 0$.

We shall exhibit the pseudo-volume form v a little more explicitly. Let H be a very ample line bundle over X and m a positive integer such that $\Gamma(K_X^m H^{-1}) \neq 0$, see (7.3.6) and (7.3.7). Let α be a nonzero element of $\Gamma(K_X^m H^{-1})$, and $\varphi_0, \varphi_1, \ldots, \varphi_N$ be a basis for $\Gamma(H)$. Define

$$\omega_j = \alpha \varphi_j \in \Gamma(K_X^m), \qquad j = 0, 1, \ldots, N.$$

As we explained in the preceding section, the map

$$x \mapsto (\omega_0 : \ldots : \omega_N) = (\varphi_0 : \ldots : \varphi_N)$$

imbeds X into $P_N(\mathbf{C})$. Write ω_j in terms of a local coordinate system $z^1, \ldots, z^n$ in the form

$$\omega_j = a(z) h_j(z) (dz^1 \wedge \ldots \wedge dz^n)^m,$$

where $a(z)$ and $h_j(z)$ are locally defined holomorphic functions representing α and φ_j. Define a non-negative (n, n)-form $|\omega_j \wedge \bar\omega_j|^{1/m}$ by

$$|\omega_j \wedge \bar\omega_j|^{1/m} = |a(z) h_j(z)|^{2/m} i^m dz^1 \wedge d\bar{z}^1 \wedge \ldots \wedge dz^n \wedge d\bar{z}^n.$$

Now, we define v by

$$\text{(7.4.5)} \qquad v = \sum_j |\omega_j \wedge \bar\omega_j|^{1/m} = |a|^{2/m} w,$$

where

$$\text{(7.4.6)} \qquad w = \left(\sum_j |h_j(z)|^{2/m} \right) i^m dz^1 \wedge d\bar{z}^1 \wedge \ldots \wedge dz^n \wedge d\bar{z}^n.$$

Then v vanishes precisely where α vanishes.

Pseudo-volume forms arise in a natural way. Let X and $\tilde{X}$ be complex manifolds of the same dimension n, and $\pi\colon \tilde{X} \to X$ a holomorphic map whose differential $\pi_*\colon T\tilde{X} \to TX$ is nondegenerate outside an analytic subset Δ of $\tilde{X}$. If v is a volume form on X, then π^*v is a pseudo-volume form on $\tilde{X}$ which is positive outside the degeneracy set Δ. Since $\mathrm{Ric}(\pi^*v) = \pi^*\mathrm{Ric}(v)$ and $K_{\pi^*v} = \pi^*K_v$, if v is negatively bounded so is π^*v. For example, if we blow up a compact complex manifold X with ample canonical bundle K_X, then the resulting manifold $\tilde{X}$ has a pseudo-volume form with negatively bounded Ricci form. A manifold $\tilde{X}$ which is obtained by blowing up a manifold X with ample canonical bundle K_X is an example of a projective variety of general type.

By (7.4.1) and (7.4.4), for a smooth projective variety X of general type X, we have $\Psi_X^m \geq v$ when v is suitably normalized, and in particular, $\Psi_X^m > 0$ outside an analytic subset Δ. We shall show that actually Ψ^m is positive everywhere, (Kodaira [1]; the proof given here is more direct and elemetary). Fix a point $x \in X$, and let f be a meromorphic map from B^n into X which is holomorphic and non-degenerate at the origin 0 and maps 0 to x. We want to find a positive lower bound for $(f^{-1})^*\mu_0$ that is independent of f. As above, let H be a very ample line bundle over X and m a positive integer such that $K_X^m H^{-1}$ has a nonzero section α. Let $\varphi_0, \varphi_1, \ldots, \varphi_N$ be a basis for $\Gamma(H)$. Define

$$\omega_j = \alpha\varphi_j \in \Gamma(K_X^m), \qquad j = 0, 1, \ldots, N.$$

Since all bundles over B^n, in particular, f^*H and $f^*(K_X^m H^{-1})$ are trivial, the sections $f^*\alpha, f^*\varphi_0, \ldots, f^*\varphi_N$ are identified with holomorphic functions $\tilde{a}, \tilde{h}_0, \ldots, \tilde{h}_N$ on B^n. Write $f^*\omega_j$ in terms of the natural coordinate system $z^1, \ldots, z^n$ of B^n in the form

$$f^*\omega_j = \tilde{a}(z)\tilde{h}_j(z)(dz^1 \wedge \ldots \wedge dz^n)^m.$$

Since f is non-degenerate at 0, we may consider $z^1, \ldots, z^n$ as a local coordinate system around x in X. Then the expression above may be considered as a local expression for ω_j. Consider a positive (n, n)-form:

$$\tilde{w} = \Big(\sum_j |\tilde{h}_j(z)|^{2/m}\Big) i^m dz^1 \wedge d\bar{z}^1 \wedge \ldots \wedge dz^n \wedge d\bar{z}^n.$$

This volume form $\tilde{w}$ on B^n has a negatively bounded Ricci form $\mathrm{Ric}(\tilde{w}) < 0$ and is related to the local volume form w of X in (7.4.6) by $\tilde{w} = f^*w$. After suitable normalization, we have (see (2.4.14))

$$\tilde{w} \leq \mu.$$

Now,

$$(f^{-1})^*\mu_0 \geq (f^{-1})^*\tilde{w}_0 = (f^{-1})^*f^*w_x = w_x.$$

This proves our assertion.

Summarizing results in this section, we state

(7.4.7) **Theorem**. *For a smooth projective variety X, consider the following conditions*:

(a)   *Its canonical line bundle K_X is ample*;

(b)   *It is of general type*;

(c) $\Psi_X^m > 0$ *everywhere, and X is strongly Ψ^m-measure hyperbolic*;

(d) $\Psi_X^h > 0$ *everywhere, and X is strongly Ψ^h-measure hyperbolic*;

(a')   *It admits a volume form v with* $\mathrm{Ric}(v) < 0$;

(b')   *It admits a pseudo-volume form v which is positive outside an analytic subset Δ and has negatively bounded* $\mathrm{Ric}(v)$ *(outside Δ)*;

(c') $\Psi_X^m > 0$ *outside an analytic subset Δ, and X is Ψ^m-measure hyperbolic*;

(d') $\Psi_X^h > 0$ *outside an analytic subset Δ, and X is Ψ^h-measure hyperbolic*.

Then we have the following implications bewteen these conditions:

$$
\begin{array}{ccccccc}
(a) & \Rightarrow & (b) & \Rightarrow & (c) & \Rightarrow & (d) \\
\Updownarrow & & \Downarrow & & \Downarrow & & \Downarrow \\
(a') & \Rightarrow & (b') & \Rightarrow & (c') & \Rightarrow & (d')
\end{array}
$$

The implications $(a) \Rightarrow (b)$, $(c) \Rightarrow (c')$ and $(d) \Rightarrow (d')$ are sharp. The equivalence $(a) \Leftrightarrow (a')$ is a special case of (7.3.10), (i.e., $L = K_X$). We do not know if any of the other horizontal implications is an equivalence. (7.4.4) does not quite say $(b') \Rightarrow (b)$, but it comes very close to it. The three conditions (b), (b') and (c') are bimeromorphically invariant. (We see the bimeromorphic invariance of (b') from the argument used in the proof of (2.4.15), i.e., extension of plurisubharmonic functions). But it is not clear if (d') is also bimeromorphically invariant.

(7.4.8) **Example**. Let X be a non-singular hypersurface of degree d in $P_{n+1}\mathbf{C}$. Let H be the hyperplane line bundle (restricted to X). The canonical line bundle K_X of X is isomorphic to H^{n+2-d}. So, if $d > n + 2$, then K_X is very ample. See (4.10.16) for details.

More generally, let X be a complete intersection of k non-singular hypersurfaces of degree $d_1, \ldots, d_k$ in $P_{n+k}\mathbf{C}$, then K_X is isomorphic to $H^{n+k+1-\sum d_i}$. So, if $\sum d_i > n + k + 1$, then K_X is very ample.

(7.4.9) **Remark**. It is known that every measure hyperbolic algebraic surface is of general type, see Green-Griffiths [1], Mori-Mukai [1].

In (3.8.28) we proved a reduction theorem for closed complex subspaces of complex tori. We supplement (3.8.28) by the following theorem.

(7.4.10) **Theorem**. *Let X be an n-dimensional closed complex subspace of a complex torus T, and let T' be the identity component of the group of translations of T leaving X invariant. Then the following are equivalent*:

(a) $T' = 0$;

(b) X is Ψ^h-measure hyperbolic;

(c) Ψ_X^h is non-trivial;

(d) X is of general type;

(e)   The Ricci tensor of X is negative definite on a dense open subset.

If X is nonsingular, then the conditions above are equivalent to each of the following:

(f) $c_1(X)^n \neq 0$;

(g) $c_n(X) \neq 0$.

Proof. (a) $\Leftrightarrow$ (e). This is in (3.8.22).
   (a) $\Leftrightarrow$ (d). This is in (3.8.28).
   (d) $\Rightarrow$ (b). This follows from (7.4.7).
   (b) $\Rightarrow$ (c). Trivial.
   (c) $\Rightarrow$ (a). Assume $T' \neq 0$. Since T' is a group of translations, it has no fixed points. Hence, $\Psi_X^h = 0$ everywhere by (7.1.16).

Now we assume that X is nonsingular.
   (e) $\Rightarrow$ (f). Up to a constant factor, $c_1(X)$ is represented by the Ricci form $\sum R_{i\bar{j}} dz^i \wedge d\bar{z}^j$. Then up to a constant factor, $c_1(X)^n$ is given by $\det(R_{i\bar{j}}) dz^1 \wedge d\bar{z}^1 \wedge \ldots \wedge dz^n \wedge d\bar{z}^n$. Now (f) follows immediately from (e).
   (f) $\Rightarrow$ (e). In general, for a complex submanifold X of a flat manifold T, the Ricci tensor is negative semi-definite. If $\det(R_{i\bar{j}}) = 0$ on a nonempty open set, then $\det(R_{i\bar{j}}) \equiv 0$ by real analyticity of the metric. That would imply $c_1(X)^n = 0$, (see the proof above). Hence, $\det(R_{i\bar{j}}) \neq 0$ on a dense open subset of X.
   (g) $\Rightarrow$ (a). If T' is non-trivial, then X is a torus bundle over X/T'. Hence, $c_n(X) = 0$.
   (a) $\Rightarrow$ (g). We shall not prove this implication, referring the reader to Smyth [1]. $\qquad\square$

(7.4.11) **Remark**. As we remarked in (3.8.29), X is projective algebraic if and only if T' is. In particular, if T' is trivial, then X is projective algebraic.

The implication (a) $\Rightarrow$ (d) says, in particular, that if X is a closed complex submanifold of a simple abelian variety, then X is of general type. In Hartshorne [2] it was shown that the canonical bundle K_X of such an X is ample. (In fact, he proved that the normal bundle of X is ample).

For more algebraic geometric aspects of the fibration $X \to X/T'$, see Ueno [1; p. 120].

(7.4.12) **Theorem**. *Let X be a compact complex manifold with ample canonical line bundle. Then*

$$\frac{(-4\pi)^n}{n!(n+1)^n} c_1(X)^n \leq \int_X \Psi_X^m.$$

Proof. This follows from (2.4.20) and (7.4.1). $\square$

We do not know if we have a similar topological lower bound for the intrinsic total volume when X is a manifold of general type.

(7.4.13) Remark. It is likely that every compact complex manifold of general type is hyperbolic modulo a proper closed complex subspace. Green conjectures that every surface of general type is hyperbolic modulo the union of all its rational and elliptic curves. This conjecture of Green has been verified by Grant [2] under some additional conditions. Namely, she has shown that if a surface of general type X admits a nonconstant holomorphic map $f: X \to T$ into a complex torus T of dimension at least two such that the image $f(X)$ contains no elliptic curves, then X is hyperbolic modulo the union of all its rational and elliptic curves.

Conversely, every compact Kähler hyperbolic manifold is conjectured to have ample canonical bundle; this conjecture is discussed in Campana [1].

5 Extension of Maps into Manifolds of General Type

Let Y be a compact complex space of general type, or equivalently, assume that the canonical line bundle of a nonsingular model is pseudo-ample, (see 7.3.9). In the equidimensional case, the following theorem proved by Griffiths [1] for smooth projective varieties with very ample canonical bundle, was generalized to the case of ample canonical bundle by Kobayashi-Ochiai [1] and to the case of pseudo-ample canonical bundle by Kodaira. We present it in the form generalized by Kobayashi-Ochiai [3].

(7.5.1) Theorem. *Let X be a complex space and A a complex subspace of X. Let Y be a compact complex space of general type with* $\dim Y \leq \dim X$. *Then every meromorphic map $f: X - A \to Y$ of maximal rank extends to a meromorphic map of X into Y.*

Before we begin the proof, we need to explain the assumption. We say that f is of maximal rank if it is holomorphic at some regular point x of $X - A$ and if its differential $f_*: T_x(X - A) \to T_{f(x)}Y$ is of maximal rank (i.e., surjective in the present case).

Proof. Since every compact complex space Y of general type has a non-singular projective model as explained in the preceding section, we may assume that Y is a smooth projective variety of general type.

Clearly, we can replace X also by a non-singular model. We may also assume that $f: X - A \to Y$ is holomorphic since the points of indeterminacy may be included in A. We shall show that we may assume that A is a non-singular hypersurface of X. Let B be the union of all irreducible components of A of codimension ≥ 2 and the singular locus of A. Then $A - B$ is a non-singular hypersurface of $X - B$. Suppose that $f: X - A \to Y$ extends to a meromorphic map

$f: X - B \to Y$. Fix an imbedding $Y \subset P_N\mathbf{C}$, and let $w^0, \ldots, w^N$ be a homogeneous coordinate system for $P_N\mathbf{C}$. The meromorphic functions $f^*(w^j/w^k)$ on $X - B$ extend to meromorphic functions on X since B has codimension at least 2 in X. Hence, f extends to a meromorphic map $f: X \to Y$.

By localizing f, we may further assume that X is a unit polydisc D^p and A is a subpolydisc $\{0\} \times D^{p-1}$ so that $X - A = D^* \times D^{p-1}$.

Now we use the particular imbedding $Y \subset P_N\mathbf{C}$ constructed in the preceding section. We recall its construction. Given a very ample line bundle H over Y, take a positive integer m such that $K_Y^m H^{-1}$ has a nonzero section, say α, (see (7.3.6)). Set

$$(7.5.2) \qquad \Gamma_Y = \{\alpha\varphi;\ \varphi \in \Gamma(H)\} \subset \Gamma(K_Y^m), \qquad N + 1 = \dim \Gamma_Y,$$

i.e., Γ_Y consists of holomorphic sections of K_Y^m divisible by α and is naturally isomorphic to $\Gamma(H)$. Let Γ_Y^* denote the dual space of Γ_Y. Using the sections in Γ_Y we obtain a holomorphic imbedding of Y into the projective space $\mathbf{P}(\Gamma_Y^*)$; this is essentially the natural imbedding of Y into $\mathbf{P}(\Gamma(H)^*)$, see (7.3.8).

We consider first the equidimensional case, i.e., $p = n$. In order to prove that f extends meromorphically to X, it suffices to show that, for every $s \in \Gamma_Y \subset \Gamma(K_Y^m)$, $f^*s \in \Gamma(K_{X-A}^m)$ extends to a meromorphic section of K_X^m over X.

We define a pseudo-volume form $v = |a|^{2/m} w$ on Y as in (7.4.6). Then its Ricci form $\mathrm{Ric}(v)$ is negative. We may assume that v is normalized so that $K_v \leq -1$.

For simplicity we set

$$X = D^n, \quad X^* = D^* \times D^{n-1}.$$

We need to construct an invariant volume element on X^*. We know that an invariant volume element for $D = \{|z| < 1\}$ is given by

$$\mu_D = \frac{i dz \wedge d\bar{z}}{(1 - |z|^2)^2}.$$

The upper half-plane is biholomorphic to the unit disc D. Using the natural covering projection from the upper half-plane to the punctured unit disc $D^* = \{0 < |z| < 1\}$ we constructed a complete metric $ds_{D^*}^2$ of curvature -1, see (2.2.3). Its area element is given by

$$\mu_{D^*} = \frac{i dz \wedge d\bar{z}}{|z|^2 (\log |z|^2)^2}.$$

Hence, an invariant volume form for $X^* = D^* \times D^{n-1}$ is given by

$$\mu_{X^*} = i^n \frac{dz^1 \wedge d\bar{z}^1}{|z^1|^2 (\log |z^1|^2)^2} \wedge \prod_{j=2}^{n} \frac{dz^j \wedge d\bar{z}^j}{(1 - |z^j|^2)^2}.$$

Although the origin 0 of D is at infinity with respect to the metric $ds_{D^*}^2$, the area around 0 with respect to μ_{D^*} is bounded. More explicitly, let $D_a^* = \{0 < |z| < a\}$ with $a < 1$. Then (see (2.2.4))

$$(7.5.3) \qquad \int_{D_a^*} \mu_{D^*} < \infty.$$

Let $X_a^* = D_a^* \times D_a^{n-1}$. Then (7.5.3) implies

$$(7.5.4) \qquad \int_{X_a^*} \mu_{X^*} < \infty.$$

For the pseudo-volume form v of Y and the volume form μ_{X^*} of X^* we have the following inequality:

$$(7.5.5) \qquad f^*(v) \le c\mu_{X^*} \qquad \text{for} \quad f \in \mathrm{Hol}(X^*, Y),$$

where c is a positive constant independent of f. (By a suitable normalization of μ_{X^*}, we may assume that $c = 1$.) This is nothing but the Schwarz lemma for volume elements. In fact, if H denotes the upper half-plane and if $\pi : H \times D^{n-1} \to X^*$ is the natural covering projection, then by identifying H with D we may consider π as a map from D^n onto X^*. Then $\pi^*\mu_{X^*} = \mu_{D^n}$, and the inequality (7.5.5) is equivalent to

$$(f \circ \pi)^* v \le c\mu_{D^n}.$$

But this is essentially (2.4.15). In (2.4.15) the unit ball B^n was used as a "model domain", but as we remarked after (2.4.16) a similar inequality holds when D^n is used as a model domain.

Fix $s \in \Gamma_Y$. We shall show that f^*s extends to a meromorphic section of K_X^m. We define a non-negative function g on Y by setting

$$(7.5.6) \qquad g = |s|^2/v^m.$$

Since both $|s|^2$ and v^m are sections of the same bundle $K_Y^m \otimes \bar{K}_Y^m$ and have a common factor $\alpha\bar{\alpha}$ and since this factor is the only contributing factor to the zeros of v^m, the function g is smoothly defined on Y. More explicitly, in terms of a local coordinate system $w^1, \ldots, w^n$ of Y, s and v can be expressed as follows (see (7.4.5)):

$$s = a(w)\sigma(w)(dw^1 \wedge \ldots \wedge dw^n)^m,$$

$$v = |a(w)|^{2/m}\left(\sum |h_j(w)|^{2/m}\right) i^m dw^1 \wedge d\bar{w}^1 \wedge \ldots \wedge dw^n \wedge d\bar{w}^n.$$

Then

$$g = \frac{|\sigma(w)|^2}{\sum |h_j(w)|^2}.$$

Let M be the maximum value of $g^{1/m}$ on Y. Using (7.5.5) and (7.5.4) we obtain

$$(7.5.7) \qquad \int_{X_a^*} f^*|s|^{2/m} \le \int_{X_a^*} f^*(g^{1/m}v) \le M \int_{X_a^*} f^*v \le cM \int_{X_a^*} \mu_{X^*} < \infty.$$

Using (7.5.7) we shall prove that f^*s extends meromorphically to $X = D^n$. We set

$$f^*s = \varphi \cdot (dz^1 \wedge \ldots \wedge dz^n)^m,$$

where φ is a holomorphic function on $X^* = D^* \times D^{n-1}$ and can be written in a Laurent series in z^1 as follows:

$$\varphi(z^1, \ldots, z^n) = \sum_{q=-\infty}^{\infty} A_q(z^2, \ldots, z^n)(z^1)^q,$$

where each A_q is holomorphic in $z^2, \ldots, z^n$. It suffices to show that $A_q = 0$ for $q \leq -m$. This can be reduced to proving the following lemma for functions of one variable.

(7.5.8) **Lemma.** If $f(z) = \sum_{q=-\infty}^{\infty} a_q z^q$ is holomorphic on the punctured disk $D^* = \{0 < |z| < 1\}$ and satisfies

$$\int_{D^*} |f(z)|^{2/m} dx dy < \infty$$

for a positive integer m, then $a_q = 0$ for $q \leq -m$.

Proof of (7.5.8). The proof can be easily reduced to the case where f is of the form

$$f(z) = \sum_{q \leq -m} a_q z^q.$$

Then we have to show that $f(z) \equiv 0$. Assuming the contrary, let $k \geq m$ be the least integer such that $a_{-k} \neq 0$. If we write

$$f(z) = z^{-k} \sum_{q \leq -k} a_q z^{q+k}$$

and put

$$g(w) = \sum_{q=0}^{\infty} a_{-q-k} w^q,$$

then

$$f(z) = z^{-k} g(\frac{1}{z})$$

so that $g(w)$ is an entire function with $g(0) \neq 0$. If we put

$$w = \frac{1}{z} \qquad \text{and} \qquad w = u + iv,$$

then

$$(*) \qquad \int_{D^*} |f(z)|^{2/m} dx dy = \int_{|w|>1} |g(w)|^{2/m} |w|^{(2k-4m)/m} du dv.$$

In terms of the polar coordinate system $w = re^{i\theta}$, this integral is written as

$$(**) \qquad \int_1^{\infty} \int_0^{2\pi} |g(re^{i\theta})|^{2/m} r^{(2k-3m)/m} d\theta dr.$$

Since $|g(w)|^{2/m} = \exp(\frac{2}{m} \log |g(w)|)$ is subharmonic, we have

$$\int_0^{2\pi} g(re^{i\theta})^{2/m} d\theta \geq |g(0)|^{2/m} > 0.$$

Hence, the integral $(\ast\ast)$ is greater than or equal to

$$|g(0)|^{2/m} \int_1^{\infty} r^{(2k-3m)/m} dr.$$

But this integral is infinite since $(2k - 3m)/m \geq -1$. This contradiction completes the proof of (7.5.8) and that of (7.5.1) in the equidimensional case.

We shall now consider the case where $p > n$ and show that every holomorphic map $f: D^* \times D^{p-1} \to Y$ extends to a meromorphic map $f: D^p \to Y$.

Since the second Cousin problem is solvable for $D^* \times D^{p-1}$, we can lift the holomorphic map $f: D^* \times D^{p-1} \to Y \subset P_N \mathbf{C}$ to a holomorphic map $\tilde{f}: D^* \times D^{p-1} \to \mathbf{C}^{N+1}$, (see the argument for (8.2.1)). Then $\tilde{f}$ is given by a system of $N + 1$ functions $\varphi^0(z^1, \ldots, z^p), \ldots, \varphi^N(z^1, \ldots, z^p)$ holomorphic in $D^* \times D^{p-1}$.

Let

$$\varphi^j(z^1, \ldots, z^p) = \sum_{q=-\infty}^{\infty} A_q^j(z^2, \ldots, z^p)(z^1)^q, \quad j = 0, 1, \ldots, N,$$

be the Laurent expansions with respect to the variable z^1 with holomorphic coefficients $A_q^j(z^2, \ldots, z^p)$. We claim that $A_q^j(z^2, \ldots, z^p) = 0$ for $q \leq -m$.

We already proved this claim in the equidimensional case $p = n$, and we are now going to verify the claim in the case $p > n$. Since $f: D^* \times D^{p-1} \to Y$ is of rank n, there exists an n-dimensional plane P in $\mathbf{C}^p$ (not necessarily through the origin) such that the restriction of f to the intersection $P \cap (D^* \times D^{p-1})$ is of rank n. By moving P slightly if necessary, we may assume that P intersects the hyperplane $z^1 = 0$ transversally. By a linear change of the coordinate system in $\mathbf{C}^p$ we may further assume that P is defined by

$$z^{n+1} = a^{n+1}, \ldots, z^p = a^p,$$

where $a^{n+1}, \ldots, a^p$ are constants. Setting $\mathbf{a} = (a^{n+1}, \ldots, a^p)$, we write

$$f_{\mathbf{a}}(z^1, \ldots, z^n) = f(z^1, \ldots, z^n, a^{n+1}, \ldots, a^p),$$

$$\varphi_{\mathbf{a}}^j(z^1, \ldots, z^n) = \varphi^j(z^1, \ldots, z^n, a^{n+1}, \ldots, a^p).$$

Then $(\varphi_{\mathbf{a}}^0, \ldots, \varphi_{\mathbf{a}}^N)$ gives a lift of $f_{\mathbf{a}}$. The Laurent expansions of $\varphi_{\mathbf{a}}^j$ are given by

$$\varphi_{\mathbf{a}}^j(z^1, \ldots, z^n) = \sum_{q=-\infty}^{\infty} A_q^j(z^2, \ldots, z^n, a^{n+1}, \ldots, a^p)(z^1)^q.$$

Since our claim is valid for $p = n$ and hence for $f_{\mathbf{a}}$, we obtain

$$A_q^j(z^2, \ldots, z^n, a^{n+1}, \ldots, a^p) = 0 \quad \text{for} \quad q \leq -m.$$

Since $f_{\mathbf{a}}$ remains to be of rank n when $\mathbf{a}$ moves slightly, we have

$$A_q^j(z^1, \ldots, z^n, z^{n+1}, \ldots, z^p) = 0 \quad \text{for} \quad q \leq -m$$

for $(z^{n+1}, \ldots, z^p)$ in a neighborhood of $(a^{n+1}, \ldots, a^p)$. By analyticity this holds for all $(z^{n+1}, \ldots, z^p)$, which completes the proof of our claim and that of (7.5.1). $\square$

If the canonical line bundle K_Y of Y is very ample, we can let $m = 1$ in the proof above and obtain

(7.5.9) Corollary. *Let X be a complex manifold and A a complex subspace of X. Let Y be a smooth variety with a very ample canonical line bundle such that $\dim Y \leq \dim X$. Then every holomorphic map $f: X - A \to Y$ of maximal rank extends to a holomorphic map $f: X \to Y$.*

Proof. Let A_1 be the singular locus of A so that $A - A_1$ is a non-singular subvariety of $X - A_1$. We may assume that $A - A_1$ is a hypersurface (by locally enlarging it if its codimension is greater than 1). From the proof of (7.5.1) we see that f extends to a holomorphic map $f: X - A_1 \to Y$. Let A_2 be the singular locus of A_1. By repeating the same argument we can extend f to a holomorphic map $f: X - A_2 \to Y$. $\square$

We know that every complete hyperbolic space is taut, see (5.1.3). The following theorem may be regarded as an analogue of (5.1.3) for smooth projective varieties of general type.

(7.5.10) Theorem. *Let Y be an n-dimensional smooth projective variety of general type with a pseudo-volume form v as in (7.4.5). Let B^n be the unit ball in $\mathbf{C}^n$ with an invariant volume form μ as in (7.1.1). For a positive number c, let*

$$\mathcal{F}_c = \{ f \in \mathrm{Mer}(B^n, Y); \ c\mu \leq f^* v \ \text{at} \ 0 \}.$$

Then every sequence $\{f_\lambda\} \subset \mathcal{F}_c$ has a subsequence which converges to an element of $\mathcal{F}_c$.

For any $\omega \in \Gamma(K_Y^m)$ and $f \in \mathrm{Mer}(B^n, Y)$, $f^*\omega$ which is holomorphic outside the indeterminacy set of f extends holomorphically to B^n by Hartogs' theorem. Hence, f^*v is well-defined on all of B^n.

Proof. Let $\Gamma_Y \subset \Gamma(K_Y^m)$ be as in (7.5.2). Let $\omega_0, \ldots, \omega_N$ be a basis for Γ_Y, and let $v = \sum |\omega_j \wedge \bar\omega_j|^{1/m}$ be as in (7.4.5). For each $f \in \mathrm{Mer}(B^n, Y)$, we write

$$f^*\omega_j = f_j (dz^1 \wedge \ldots \wedge dz^n)^m,$$

where $(z^1, \ldots, z^n)$ denotes the coordinate system for $\mathbf{C}^n$. When v is appropriately normalized, we have (see (2.4.15))

$$\sum |f_j|^{2/m} i^n dz^1 \wedge d\bar z^1 \wedge \ldots \wedge dz^n \wedge d\bar z^n = f^* v \leq \mu_{B^n}.$$

Hence (see (2.4.7))

$$(7.5.11) \qquad \sum |f_j|^{2/m} \leq \frac{2^n}{(1 - \|z\|^2)^{n+1}}.$$

Setting $f_\lambda^* \omega_j = f_{\lambda j}(dz^1 \wedge \ldots \wedge dz^n)^m$ and applying (7.5.11) to f_λ, we obtain

$$\sum_j |f_{\lambda j}|^{2/m} \leq \frac{2^n}{(1 - \|z\|^2)^{n+1}}$$

for all $\lambda = 1, 2, \ldots$. By taking a subsequence we may assume that for each $j = 0, \ldots, N$ the sequence $\{f_{\lambda j}\}$ converges to a holomorphic function f_j uniformly on compact sets of B^n. Then $(f_0, \ldots, f_N)$ defines a meromorphic map $f: B^n \to Y$ provided that $(f_0, \ldots, f_N)$ does not vanish identically on B^n. This condition is guaranteed if all f_λ belong to $\mathcal{F}_c$ for some fixed c. $\qquad \square$

6 Dominant Maps to Manifolds of General Type

Let X and Y be compact complex spaces. We recall that a meromorphic map $f: X \to Y$ is said to be **dominant** if its graph $G_f \subset X \times Y$ maps onto Y under the projection $X \times Y \to Y$. Since we are assuming that X and Y are compact, this is equivalent to saying that there is a nonsingular point $x \in X$ where f is regular and its differential $df: T_x X \to T_{f(x)} Y$ is surjective. We denote the family of dominant meromorphic maps $f: X \to Y$ by $\mathrm{Dom}(X, Y)$.

In Section 6 of Chapter 6 we generalized the theorem of de Franchis on curves of genus ≥ 2 to compact hyperbolic spaces. The following generalization to varieties of general type is due to Kobayashi-Ochiai [3].

(7.6.1) **Theorem.** *Let X and Y be compact complex spaces. If Y is of general type, then the set $\mathrm{Dom}(X, Y)$ of dominant meromorphic maps of X onto Y is finite.*

Proof. As we explained in Section 3, every compact complex space Y of general type has a non-singular projective model. Hence, the proof of the theorem is reduced to the case where Y is projective algebraic.

By the following argument the proof is further reduced to the case where X is also projective algebraic. Let $\mathbf{C}(X)$ and $\mathbf{C}(Y)$ denote the fields of meromorphic functions on X and Y, respectively. Let X' be a projective algebraic manifold with $\mathbf{C}(X) = \mathbf{C}(X')$. Then

$$
\begin{aligned}
\mathrm{Dom}(X, Y) \quad &\subset \quad \{\varphi: \mathbf{C}(Y) \to \mathbf{C}(X); \text{ injective morphism}\} \\
&= \quad \{\varphi: \mathbf{C}(Y) \to \mathbf{C}(X'); \text{ injective morphism}\} \\
&= \quad \mathrm{Dom}(X', Y).
\end{aligned}
$$

This shows that if $\mathrm{Dom}(X', Y)$ is finite, so is $\mathrm{Dom}(X, Y)$.

We assume that both X and Y are smooth projective varieties. Let $n = \dim Y$, and $K_Y = \bigwedge^n T^*Y$ be the canonical line bundle of Y. Given a very ample line bundle H over Y, take a positive integer m such that $K_Y^m H^{-1}$ has a nonzero section, say α, see (7.3.6). Set

$$(7.6.2) \qquad \Gamma_Y = \{\alpha\varphi;\ \varphi \in \Gamma(H)\} \subset \Gamma(K_Y^m), \qquad N + 1 = \dim \Gamma_Y.$$

Let Γ_Y^* denote the dual space of Γ_Y. Let $P^*(\Gamma_Y)$ denote the projective space of hyperplanes in Γ_Y while $P(\Gamma_Y^*)$ denotes the projective space of lines in Γ_Y^*. Evidently, $P^*(\Gamma_Y) \cong P(\Gamma_Y^*)$ in a natural manner. As we observed in (7.3.8), Γ_Y defines a holomorphic imbedding of Y into a projective space of dimension N:

$$(7.6.3) \qquad j: Y \to P^*(\Gamma_Y) = P(\Gamma_Y^*);$$

for each $y \in Y$ we set $j(y)$ to be the hyperplane of Γ_Y (or the corresponding line in Γ_Y^*) consisting of those $\alpha\varphi$ with $\varphi(y) = 0$. This is the same as the imbedding defined by $\Gamma(H)$.

We may use $\Gamma(K_Y^m)$ itself in place of Γ_Y and the meromorphic imbedding $j: Y \to P^*(\Gamma(K_Y^m))$ defined by $\Gamma(K_Y^m)$, but we prefer the holomorphic imbedding defined above.

Assume that $\dim X = \dim Y$. (The case where $\dim X > \dim Y$ will be considered later). Let K_X be the canonical bundle of X, and put

$$(7.6.4) \qquad \Gamma_X = \Gamma(K_X^m).$$

Every $f \in \mathrm{Dom}(X, Y)$ defines a homomorphism

$$(7.6.5) \qquad f^*: \mathcal{O}(K_Y^m) \to \mathcal{O}(K_X^m).$$

In fact, for $\eta \in \Gamma(K_Y^m)$, $f^*\eta$ is holomorphic outside the singularity set of f (which is of codimension ≥ 2). By Hartogs' theorem it is actually holomorphic everywhere. Thus, f defines a linear map

$$f^*: \Gamma_Y \to \Gamma_X.$$

Since f is dominant, f^* is injective.

Since f^* is injective, Γ_X contains sufficiently many sections so that the natural meromorphic map $i: X \to P^*(\Gamma_X)$ is non-trivial.

Let $\tilde{f}: P^*(\Gamma_X) \to P^*(\Gamma_Y)$ be the meromorphic map induced by $f^*: \Gamma_Y \to \Gamma_X$. Then we have the following commutative diagram:

$$(7.6.6) \qquad \begin{array}{ccc} P^*(\Gamma_X) & \xrightarrow{\tilde{f}} & P^*(\Gamma_Y) \\ \uparrow i & & \uparrow j \\ X & \xrightarrow{f} & Y. \end{array}$$

Let

$$E = \mathrm{Hom}(\Gamma_Y, \Gamma_X) = \mathrm{Hom}(\Gamma_X^*, \Gamma_Y^*),$$

and $P(E)$ the projective space defined by E. Each element of $P(E)$ induces, in a natural way, a meromorphic map $P^*(\Gamma_X) \to P^*(\Gamma_Y)$.

In particular, the injective map $f^* \in E = \mathrm{Hom}(\Gamma_Y, \Gamma_X)$ induced by a map $f \in \mathrm{Dom}(X, Y)$ induces a meromorphic map $\tilde{f}$ of $P^*(\Gamma_X)$ into $P^*(\Gamma_Y)$ as above. We set

$$M = \{\tilde{f} \in P(E); \ f \in \mathrm{Dom}(X, Y)\} \subset P(E).$$

Since $j: Y \to P^*(\Gamma_Y)$ is an imbedding, the commutative diagram (7.6.6) implies that M is in a natural one-to-one correspondence with $\mathrm{Dom}(X, Y)$.

Let Z be the set of elements of $[t] \in P(E)$, $t \in E$, such that the induced meromorphic mappings $\tilde{t}: P^*(\Gamma_X) \to P^*(\Gamma_Y)$ send $i(X)$ into $j(Y)$. Then Z is an algebraic subvariety of $P(E)$. Let $\xi = (\ldots, \xi_\alpha, \ldots)$ and $\eta = (\ldots, \eta_\lambda, \ldots)$ be homogeneous coordinate systems for $P^*(\Gamma_X)$ and $P^*(\Gamma_Y)$, respectively. Let $\zeta = (\ldots, \zeta_\lambda^\alpha, \ldots)$ be the naturally induced homogeneous coordinate system for $P(E)$. Let J be the ideal of homogeneous polynomials $Q(\eta)$ defining the variety $j(Y) \subset P^*(\Gamma_Y)$. Let $\xi(x) = (\ldots, \xi_\alpha(x), \ldots)$ be the homogeneous coordinates of $x \in i(X)$, and let $\zeta \cdot \xi(x)$ denote the product of a matrix $\zeta = (\zeta_\lambda^\alpha)$ with the vector $\xi(x)$, i.e., $\zeta \cdot \xi(x) = (\ldots, \sum \zeta_\lambda^\alpha \xi_\alpha(x), \ldots)$. If we put $Q_x(\zeta) = Q(\zeta \cdot \xi(x))$, then Z is defined by the homogeneous polynomial equations $\{Q_x(\zeta) = 0; \ Q \in J, \ x \in i(X)\}$ in ζ.

Clearly, M is the subset of Z consisting of those elements which define dominant maps of $i(X)$ onto $j(Y)$. This shows that M is an open subset of Z. We shall show that M is also closed in Z.

In both Γ_X and Γ_Y we define a quasi-norm $\| \cdot \|$:

$$\|s\|^2 = \int_X |s \wedge \bar{s}|^{1/m} \qquad \text{for} \quad s \in \Gamma_X,$$

$$\|t\|^2 = \int_Y |t \wedge \bar{t}|^{1/m} \qquad \text{for} \quad t \in \Gamma_Y.$$

where $|s \wedge \bar{s}|^{1/m}$ and $|t \wedge \bar{t}|^{1/m}$ should be interpreted as in (7.4.5). (We note that for $m > 1$, these quasi-norms do not satisfy the convexity condition, i.e., the triangular inequality).

For any meromorphic map f from X to Y and for any $t \in \Gamma_Y$, we have

$$\|f^*t\|^2 = \int_X f^*|t \wedge \bar{t}|^{1/m} = (\deg f) \int_Y |t \wedge \bar{t}|^{1/m} = (\deg f)\|t\|^2.$$

Hence, f is dominant if and only if $\|f^*t\| \geq \|t\|$. (For f to be dominant, it suffices to have this inequality for one nonzero t).

This implies that a meromorphic map is dominant if it is a limit of dominant meromorphic maps. Thus, M is a closed subvariety of a projective variety Z.

On the other hand, the imbedding $\mathrm{Dom}(X, Y) \cong M \subset P(E)$ lifts to a holomorphic map $\mathrm{Dom}(X, Y) \to E$ that sends f to f^*. Since the only compact subvarieties of the vector space E are finite sets, this completes the proof of the theorem when $\dim X = \dim Y$.

Now we shall reduce the general case to the equidimensional case. Suppose $m = \dim X > \dim Y = n$. Assuming that $\mathrm{Dom}(X, Y)$ is infinite, we choose a countable infinite subset $\{f_1, f_2, \ldots\}$ of $\mathrm{Dom}(X, Y)$ and fix it once and for all.

For each point x of X, let $G_x(n)$ be the Grassmannian of n-planes in the m-dimensional tangent space $T_x X$, and $G(n) = \bigcup_{x \in X} G_x(n)$ the Grassmann bundle over X with projection $\pi: G(n) \to X$.

We claim that at some point $x \in X$ there is an n-plane $\xi \in G_x(n)$ such that the differential $df_1|_\xi, df_2|_\xi, \ldots$ are mutually distinct isomorphisms. (Implicit in the statement is that $f_1, f_2, \ldots$ are all regular at x).

In order to prove this claim, let S_j be the singularity set of the meromorphic map f_j, and let N_j be the set of $\xi \in G_x(n)$ such that f_j is regular at x but $df_j: \xi \to T_{f_j(x)} Y$ is not an isomorphism. For each pair (i, j), let P_{ij} be the set of $\xi \in G_x(n)$ such that both f_i and f_j are regular at x and $df_i|_\xi = df_j|_\xi$.

Clearly, $\pi^{-1}(S_j)$ is a proper subvariety of $G(n)$, and N_j is a proper subvariety of $G(n) - \pi^{-1}(S_j)$. The set P_{ij} is a proper subvariety of $G(n) - \pi^{-1}(S_i \cup S_j)$. Hence

$$G := G(n) - ((\bigcup_j \pi^{-1}(S_j)) \cup (\bigcup_j N_j) \cup (\bigcup_{i,j} P_{ij}))$$

is dense in $G(n)$. (In fact, G is the complement of a union of countably many subvarieties of lower dimension). Any $\xi \in G$ has the desired property. This proves our claim.

Finally, let V be a subvariety of X passing through x with $T_x V = \xi$. (Such a V exists since X is projective algebraic). Then $\dim V = \dim Y$, and $f_1|_V, f_2|_V, \ldots$ are mutually distinct elements of $\mathrm{Dom}(V, Y)$. So it suffices to prove that $\mathrm{Dom}(V, Y)$ is finite. $\qquad\square$

In Kobayashi-Ochiai [3] an unnecessary use is made of the equidimensional Schwarz lemma. In the above, that part of the proof was eliminated. For a little more algebraic proof, see Martin-Deschamps and Lewin-Menegaux [1] and Kurke [1]. For generalizations of (7.6.1) to complex spaces of log-general type, see Sakai [4] and Tsushima [1]. Effective proofs of (7.6.1) will be discussed in the next section.

As an immediate consequence of (7.6.1) we obtain the following result of Matsumura [1].

(7.6.7) **Corollary**. *The group of bimeromorphic automorphisms of a compact complex space of general type is finite.*

The corollary above raises the question whether the group of bimeromorphic automorphisms of a Ψ^m-measure hyperbolic compact complex space is also finite. We know that it is discrete, (see (7.1.17)).

The proof of (7.6.1) yields also the following result of Klaus Peters [1].

(7.6.8) **Theorem**. *If X is a compact complex manifold with ample canonical line bundle K, then every dominant meromorphic map f of X onto itself is a biholomorphic automorphism.*

Proof. In the commutative diagram (7.6.6), let $X = Y$. Since f^* is an injection from $\Gamma_X = \Gamma(K_X^m)$ into itself, it is an isomorphism. Hence, $\tilde{f}$ is a projective automorphism of $P^*(\Gamma_X)$. Since $i = j$ in (7.6.6) is a holomorphic imbedding, (7.6.6) implies that f is a biholomorphic automorphism. $\qquad\square$

Using the theorem above we prove that a compact complex manifold with ample canonical line bundle is the minimal model in its class of bimeromorphically equivalent complex spaces, (Kobayashi [7]).

(7.6.9) **Corollary**. *Let Y be a compact complex manifold with ample canonical line bundle. Let X be a compact complex space with a bimeromorphic map $f: X \to Y$. Then f is holomorphic.*

Proof. Let $g = f^{-1}$, i.e., let $g: Y \to X$ be a meromorphic map such that

$$x \in g(f(x)) \quad \text{for} \quad x \in X \quad \text{and} \quad y \in f(g(y)) \quad \text{and} \quad y \in Y.$$

It suffices to show that for every $x \in X$ the set $f(x) \in Y$ is a singleton. Let $y \in f(x)$. Then $x \in g(y)$. By (7.6.8), $f \circ g: Y \to Y$ is biholomorphic and $f(g(y))$ is a singleton $\{y\}$. Hence, $f(x) \subset f(g(y)) = \{y\}$. $\qquad\square$

In general, if X and Y are complex manifolds of equal dimension with local coordinate systems $x^1, \ldots, x^n$ and $y^1, \ldots, y^n$, then for any holomorphic n-form $\psi(y)dy^1 \wedge \ldots \wedge dy^n$ of Y we have

$$f^*(\psi(y)dy^1 \wedge \ldots \wedge dy^n) = J_f \cdot \psi(f(x))dx^1 \wedge \ldots \wedge dx^n,$$

where $J_f = \det(\partial y^i / \partial x^j)$. This means that the canonical line bundles K_X and K_Y are related by

$$(7.6.10) \qquad\qquad K_X = D_f \otimes f^{-1}K_Y,$$

where D_f is the line bundle defined by the divisor (J_f).

The following related result is also due to Peters [1].

(7.6.11) **Theorem**. *Let X be a compact complex manifold such that $\Gamma(K_X^m) \neq 0$ for some positive integer m. Then every surjective holomorphic map f of X onto itself is a unramified covering projection.*

Proof. Assuming that the theorem is not true, let (J_f) be the divisor defined as the zeros of the Jacobian $J_f: \det(TX) \to \det(TX)$ of f. Let r be an arbitrary positive integer. Since the Jacobian matrix of the r-th iterate f^r of f is the r-th power of the Jacobian matrix of f, the divisor (J_{f^r}) of the Jacobian of f^r contains at least r irreducible components (with multiplicities counted).

Since f^r is surjective, it induces an automorphism of $\Gamma(K_X^m)$. Hence, every nonzero section $s \in \Gamma(K_X^m)$ can be written as $s = (f^r)^*(t)$ for some $t \in \Gamma(K_X^m)$, and the divisor (s) of s contains the divisor (J_{f^r}) of the Jacobian of f^r. In particular, the divisor (s) contains at least r irreducible components. Since r is arbitrary, this is a contradiction. $\qquad\square$

The following result of Bandman [1] strengthens (7.6.9).

(7.6.12) **Corollary**. *Let Y be a compact complex manifold with ample canonical line bundle. Let X be a compact complex space of an equal dimension. Then every dominant meromorphic map $f: X \to Y$ is holomorphic.*

Proof. There exist a compact complex manifold $\tilde{X}$ and a holomorphic map $\pi: \tilde{X} \to X$ such that (1) π is bimeromorphic and (2) the composition $\tilde{f} = f \circ \pi$ is holomorphic. Let

$$\tilde{X} \xrightarrow{p} X' \xrightarrow{f'} Y$$

be the Stein factorization of the map $\tilde{f}: \tilde{X} \to Y$, (see Section 3 of Chapter 5); X' is obtained by collapsing each connected component of $\tilde{f}^{-1}(y)$ to a point. Thus, p has connected fibers while f' is a finite map. Since $\dim Y = \dim X = \dim \tilde{X}$, $\tilde{f}^{-1}(y)$ is a finite set for a generic y, and the projection $p: \tilde{X} \to X'$ is bimeromorphic and has connected fibers. Hence, $p \circ \pi^{-1}: X \to X'$ is bimeromorphic.

$$
\begin{array}{ccc}
\tilde{X} & \xrightarrow{\pi} & X \\
\downarrow p & & \downarrow f \\
X' & \xrightarrow{f'} & Y
\end{array}
$$

Since the induced map $f': X' \to Y$ is holomorphic and finite, the induced bundle $f'^{-1}K_Y$ is ample, see Hartshorne [1]. Since $K_{X'} = f'^{-1}K_Y \otimes D_{f'}$ where $D_{f'}$ is the line bundle defined by the zeros of the Jacobian $J_{f'}$, it follows that $K_{X'}$ is also ample.

Applying (7.6.9) to $p \circ \pi^{-1}: X \to X'$, we see that $p \circ \pi^{-1}$ is holomorphic. Hence, $f = f' \circ p \circ \pi^{-1}$ is holomorphic. $\qquad\square$

The theorem of de Franchis (6.6.1) asserts finiteness of dominant rational maps $f: X \to Y$ for fixed X and Y. The theorem of Severi allows Y to vary. Namely (see, for example, Samuel [1], Howard-Sommese [2], Kani [1] and, for a generalization to finite Riemann surfaces, Imayoshi [2]),

(7.6.13) **Theorem**. *Let X be a compact Riemann surface. Then there are only finitely many pairs (Y, f) consisting of compact Riemann surfaces Y of genus ≥ 2 and surjective holomorphic maps $f: X \to Y$.*

We are interested here in higher dimensional generalizations of (7.6.13). Given a compact complex space X, let $\mathrm{Dom}(X)$ denote the set of pairs (Y, f) consisting of compact complex spaces Y of general type and dominant meromorphic maps $f: X \to Y$ modulo bimeromorphic equivalence. We state a basic conjecture.

Conjecture. *For any compact complex space X, $\mathrm{Dom}(X)$ is finite.*

Since we are dealing with bimeromorphic equivalence classes of pairs (Y, f), we may assume that X and Y are nonsingular. Furthermore, as we explained in the proof of (7.6.1), we may assume that both X and Y are projective algebraic. Obviously, we may fix the dimension of Y. Moreover, by the argument used in the proof of (7.6.1), we can reduce the problem to the case $\dim X = \dim Y$. Now, given a projective algebraic manifold X of dimension n, let $\mathcal{Y}_X$ denote the set of

n-dimensional projective algebraic manifolds Y of general type for which there exist dominant meromorphic maps $f: X \to Y$. Because of (7.6.1) the problem is reduced to showing that the set $\mathcal{Y}_X$ is finite modulo bimeromorphic equivalence. For each positive integer m, we consider the subset $\mathcal{Y}_X^m$ of $\mathcal{Y}_X$ consisting of those manifolds Y which are meromorphically imbedded by the m-th canonical map $\Phi_{K_Y^m}$, (see (7.3.1) for the definition of $\Phi_{K_Y^m}$). Let $\mathcal{Y}_X^+$ denote the subset of $\mathcal{Y}_X$ consisting of $Y \in \mathcal{Y}_X$ with semipositive canonical bundle K_Y (i.e., $K_Y \cdot C \geq 0$ for all curves C of Y).

We state a theorem of Maehara [2].

(7.6.14) **Theorem.** *For fixed X and $m > 0$, both $\mathcal{Y}_X^m$ and $\mathcal{Y}_X^+$ are finite modulo bimeromorphic equivalence.*

It is known that for a nonsingular curve Y of genus ≥ 2, $\Phi_{K_Y^3}$ gives a meromorphic imbedding and that for a nonsingular surface of general type Y, $\Phi_{K_Y^5}$ gives a meromorphic imbedding. Therefore, the theorem above solves the conjecture in dimensions 1 and 2. For similar results in dimension 2, see also Martin-Deschamps and Lewin-Menegaux [2].

7 Effective Finiteness Theorems on Dominant Maps

As in the preceding section, $\mathrm{Dom}(X, Y)$ denotes the set of dominant meromorphic maps $f: X \to Y$ between compact complex spaces. We consider two maps $f \in \mathrm{Dom}(X, Y)$ and $f' \in \mathrm{Dom}(X, Y')$ to be equivalent if there exists a bimeromorphic map $\varphi: Y \to Y'$ such that $f' = \varphi \circ f$. We set

$$\mathrm{Dom}(X) = \bigcup_Y \mathrm{Dom}(X, Y) \quad \text{modulo equivalence,}$$

where the union is taken over compact complex spaces Y of general type. In (7.6.1) we proved that if Y is of general type, then $\mathrm{Dom}(X, Y)$ is finite and stated a theorem of Maehara (7.6.14) that $\mathrm{Dom}(X)$ is finite for X with $\dim X \leq 2$. See Tsai [1] and Bandman-Dethloff [1] for more general results. When $\dim X = 1$, Howard-Sommese [2] and Kani [1] found an explicit bound for the cardinality $|\mathrm{Dom}(X)|$ of $\mathrm{Dom}(X)$ in terms of the genus g of X. Namely, let $N(g)$ denote the maximum of $|\mathrm{Dom}(X)|$ on all compact Riemann surfaces X of genus g. Then Kani gave the following lower and upper bounds.

(7.7.1) $$c^{(\log g)^2} \leq N(g) \leq (g-1)2^{2g^2-2}(2^{2g^2-1} - 1) + 1,$$

where c is a constant.

It is not known if the number $|\mathrm{Dom}(X, Y)|$ can be bounded by a number which depends only on X when $\dim X > 1$ and Y is of general type. In this section we shall prove a theorem of Bandman [1] which gives a bound on the number $|\mathrm{Dom}(X, Y)|$ when both X and Y are nonsingular manifolds with ample canonical bundle. The proof of (7.7.16) contains already the main idea for the proofs of more general results, (see Bandman [2], Bandman-Markushevich [1]),

We shall first consider the equidimensional case. We may assume that X and Y are nonsingular since we are dealing with bimeromorphic equivalence classes of maps. Let $X \subset P_N$ and $Y \subset P_{N'}$ be n-dimensional projective manifolds of degree d and d', repsectively. Let H and H' be the hyperplane bundles of P_N and $P_{N'}$, respectively. For an n-dimensional subvariety $Z \subset P_N \times P_{N'}$, we define its degree $\deg Z$ by

$$(7.7.2) \qquad \deg Z = \int_Z (p^* c_1(H)^n + p'^* c_1(H')^n),$$

where $p \colon P_N \times P_{N'} \to P_N$ and $p' \colon P_N \times P_{N'} \to P_{N'}$ are the projections. In particular, for a meromorphic map $f \in \mathrm{Mer}(X, Y)$, we can speak of the degree of its graph $G_f \subset X \times Y \subset P_N \times P_{N'}$. This is not the same as the degree of Z as a subvariety of $P_N \times P_{N'} \subset P_{NN'+N+N'}$.

Given a positive integer k, we set

$$\mathrm{Dom}^k(X, Y) = \{f \in \mathrm{Dom}(X, Y); \ \deg G_f \leq k\}.$$

Given positive integers (N, n, d), let $U = U^{N,n,d}$ be the Chow variety of n-dimensional subvarieties of degree d in P_N. Similarly, let $V = V^{N',n,d'}$ be the Chow variety of n-dimensional subvarieties of degree d' in $P_{N'}$. The subvariety in P_N (resp. $P_{N'}$) corresponding to $u \in U$ (resp. $v \in V$) will be denoted X_u (resp. Y_v). Define subvarieties $\mathcal{X} \subset P_N \times U$ and $\mathcal{Y} \subset P_{N'} \times V$ by

$$\mathcal{X} = \bigcup_{u \in U} X_u, \qquad \mathcal{Y} = \bigcup_{v \in V} Y_v.$$

Let

$$\pi \colon \mathcal{X} \times \mathcal{Y} \to U \times V$$

be the natural projection so that $X_u \times Y_v = \pi^{-1}(u, v)$.

Let $A^{n,r}$ be the variety of n-dimensional subvarieties $Z \subset P_N \times P_{N'}$ of degree r:

$$A^{n,r} = \{Z \subset P_N \times P_{N'}; \ \dim Z = n, \ \deg Z = r\}.$$

We define

$$A^{n,r}(X_u \times Y_v) = \{Z \in A^{n,r}; \ Z \subset X_u \times Y_v\},$$

$$A^{n,r}(\mathcal{X} \times \mathcal{Y}) = \bigcup_{(u,v) \in U \times V} A^{n,r}(X_u \times Y_v) \subset A^{n,r} \times (U \times V),$$

with natural projection $\pi \colon A^{n,r}(\mathcal{X} \times \mathcal{Y}) \to U \times V$.

Let $A_i^{n,r}(\mathcal{X} \times \mathcal{Y})$, $i = 1, \ldots, q$, be the irreducible components of $A^{n,r}(\mathcal{X} \times \mathcal{Y})$ such that the restrictions $\pi_i = \pi|_{A_i^{n,r}(\mathcal{X} \times \mathcal{Y})}$ are all generically finite maps. This means that, for each i, $\pi_i^{-1}(u, v)$ is finite for almost all $(u, v) \in U \times V$. Let

$$v_i^{n,r} = \deg \pi_i;$$

i.e., $v_i^{n,r}$ is the cardinality of $\pi_i^{-1}(u, v)$ for a generic $(u, v) \in U \times V$.

Let k be a positive integer, and let $f \in \mathrm{Dom}^k(X_u, Y_v)$. By (7.7.2) the graph G_f is an element of $A^{n,r}(X_u \times Y_v)$ for some $r \leq k$. An element $Z \in A^{n,r}(X_u \times Y_v)$ near G_f must be also the graph of a map belonging to $\mathrm{Dom}^k(X_u, Y_v)$.

Now assume that Y_v is of general type. Then $\mathrm{Dom}^k(X_u, Y_v)$ is finite by (7.6.6), so that G_f is an isolated element of $A^{n,r}(X_u \times Y_v)$. Hence, G_f belongs to $A_i^{n,r}(\mathcal{X} \times \mathcal{Y})$ for some i.

Since the number of isolated points of $A^{n,r}(X_u \times Y_v)$ does not exceed $\sum_{i=1}^{q} v_i^{n,r}$, the cardinality $|\mathrm{Dom}^k(X_u, Y_v)|$ is bounded as follows:

$$(7.7.3) \qquad |\mathrm{Dom}^k(X_u, Y_v)| \leq \sum_{r=1}^{k} \sum_{i=1}^{q} v_i^{n,r}.$$

The right hand side of the inequality above is determined by (U, V, n, k), and does not depend on X_u or Y_v. Since U (resp. V) is determined by the numbers N, n, d (resp. N', n, d'), we set

$$(7.7.4) \qquad \mu(N, n, d, N', d', k) = \sum_{r=1}^{k} \sum_{i=1}^{q} v_i^{n,r}.$$

In summary, we have

(7.7.5) Proposition. *Let $X \subset P_N$ and $Y \subset P_{N'}$ be n-dimensional closed submanifolds of degree d and d', respectively. If Y is of general type, then*

$$|\mathrm{Dom}^k(X, Y)| \leq \mu(N, n, d, N', d', k).$$

Now, assume that the canonical bundle K_Y of Y is ample. By (7.6.12), every $f \in \mathrm{Dom}(X, Y)$ is holomorphic. A partial solution of Fujita's conjecture by Demailly [2] says that if K_Y is ample, then K_Y^l is very ample for $l \geq 12n^2$. We use the pluri-canonical imbedding $\Phi_l : Y \to P_{N'}$ with $l = 12n^2$ and $N' + 1 = \dim H^0(Y, K_Y^l)$. Assuming that $\mathrm{Dom}(X, Y)$ is nonempty, let $f \in \mathrm{Dom}(X, Y)$. Then

$$f^* : H^0(Y, K_Y^l) \to H^0(X, K_X^l)$$

is injective. Hence

$$(7.7.6) \qquad N' < \dim H^0(X, K_X^l).$$

Since the imbedding $Y \subset P_{N'}$ is by Φ_l, the degree d' of Y is given by

$$d' = \int_Y c_1(H')^n = \int_Y c_1(K_Y^l)^n = l^n \int_Y (c_1(K_Y))^n,$$

By (7.4.12),

$$d' \leq C(n) \int_Y \Psi_Y^m,$$

where $C(n) = l^n n(n+1)^n/(4\pi)^n$ and Ψ_Y^m is the intrinsic volume element of Y defined in Section 1.

Since $(f^*\Psi_Y^m) \leq \Psi_X^m$ for $f \in \mathrm{Dom}(X, Y)$, we have

$$\int_Y \Psi_Y^m \leq \int_X f^*\Psi_Y^m \leq \int_X \Psi_X^m.$$

Hence,

$$(7.7.7) \qquad d' \leq C(n) \int_X \Psi_X^n.$$

Since N, n, d are fixed when $X \subset P_N$ is given, it follows from (7.7.6) and (7.7.7) that $\mu(N, n, d, N', d', k)$ and hence $|\mathrm{Dom}^k(X, Y)|$ are bounded by a number which depends only on X and k.

Finally, we shall show that if K_X is also ample, then $\mathrm{Dom}(X, Y) = \mathrm{Dom}^k(X, Y)$ for a sufficiently large k. Let $f \in \mathrm{Dom}(X, Y)$. Since f is holomorphic, $\deg G_f$ defined by (7.7.2) can be rewritten as follows:

$$\deg G_f = \int_X (c_1(H)^n + f^*c_1(H')^n).$$

By Demailly's result mentioned above, K_X^l is very ample with $l = 12n^2$. Using the pluri-canonical imbeddings of X and Y given by $H = K_X^l$ and $H' = K_Y^l$, we estimate $\deg G_f$. Since the projection $p: G_f \to X$ has degree 1 and the projection $p': G_f \to Y$ has degree equal to $\deg f$, we have

$$\begin{aligned} \deg G_f &= \int_X (c_1(K_X^l)^n + f^*c_1(K_Y^l)^n) \\ &= l^n \left(\int_X c_1(K_X)^n + \deg f \int_Y c_1(K_Y)^n \right). \end{aligned}$$

But, using (2.4.22) we have

$$\deg f \int_Y c_1(K_Y)^n \leq \int_X c_1(K_X)^n.$$

Hence,

$$\deg G_f \leq 2l^n \int_X c_1(K_X)^n = 2 \int_X c_1(K_X^l)^n = 2d.$$

So it suffices to take

$$(7.7.8) \qquad k = 2d.$$

When the canonical bundle of X is ample, the bound for d' given in (7.7.7) can be made more explicit. Since

$$d = c_1(K_X^l)^n = l^n c_1(K_X)^n, \qquad d' = c_1(K_Y^l)^n = l^n c_1(K_Y)^n$$

and

$$(\deg f) \cdot (c_1(K_Y))^n \leq (c_1(K_X))^n$$

by (2.4.22) and since $\deg f \geq 1$ for a dominant map f, we have

(7.7.9) $$d' \leq d.$$

Summarizing, (assuming that $\mathrm{Dom}(X, Y) \neq \emptyset$)

$$N' \leq N = \dim H^0(X, K_X^l) - 1, \quad l = 12n^2,$$

$$d' \leq d = l^n c_1(K_X)^n, \qquad k = 2d.$$

We have shown that if X and Y are n-dimensional projective manifolds with ample canonical bundle, then $|\mathrm{Dom}(X, Y)|$ is bounded by a number which depends only on the dimension n, the Chern number $c_1(K_X)^n$ and the pluri-genus $\dim H^0(X, K_X^{12n^2})$ of X.

In order to dispose of the pluri-genus in the statement above, we need to make use of the Hilbert polynomial. The **Hilbert polynomial** $\chi(X, F, t)$ of a line bundle F over X is defined by

$$\chi(X, F, t) = \sum_{i=1}^{n} (-1)^i \dim H^i(X, F^t).$$

By the Riemann-Roch-Hirzebruch formula for a line bundle F (see Hirzebruch [1, p. 150]), $\chi(X, F, t)$ can be expressed in terms of Chern classes $c_i(X)$ and $c_1(F)$. In fact, it can be expressed in terms of $c_1(X)$, $c_1(F)$ and polynomials $A_s(p_1, \ldots, p_s)$ of Pontrjagin classes $p_j = p_j(X)$:

(7.7.10) $$\chi(X, F, t) = \sum \frac{1}{2^{4s} r!} (t c_1(F) + \frac{1}{2} c_1(X))^r A_s(p_1, \ldots, p_s)[X],$$

where the summation is taken over all r, s with $r + 2s = n$. (For definition of A_s, see Hirzebruch [1]. All we need here is the fact that $A_0 = 1$).

Writing (7.7.10) as a polynomial in t, we obtain (omitting $[X]$, i.e., the integral $\int_X$)

(7.7.11) $$\chi(X, F, t) = \frac{1}{n!} c_1(F)^n t^n + \frac{1}{2(n-1)!} c_1(F)^{n-1} c_1(X) t^{n-1} + \ldots.$$

In the special case where $F = K_X$, we write

$$\chi(X, t) = \chi(X, K_X, t),$$

and call it the **Hilbert polynomial** of X. As a special case of (7.7.11) we have

(7.7.12) $$\chi(X, m) = a_n m^n + a_{n-1} m^{n-1} + \ldots,$$

where

$$a_n = \frac{(-1)^n}{n!} c_1(X)^n, \quad a_{n-1} = \frac{(-1)^{n-1}}{2(n-1)!} c_1(X)^n.$$

Letting $F = K_X$ in (7.7.10), we see that the Hilbert polynomial $\chi(X, t)$ is determined by the first Chern class $c_1(X)$ and the Pontrjagin classes $p_j(X)$. We see also that $\chi(X, t)$ is determined by the Chern numbers of X.

The following result by Kollár-Matsusaka [1] shows that the first two leading coefficients a_n and a_{n-1} of $\chi(X, t)$ play decisive roles in determining $\chi(X, t)$.

(7.7.13) **Lemma.** *For every n there is a polynomial $Q(x, y)$ in two variables such that if X is an n-dimensional projective manifold with a semi-ample line bundle F and if $\chi(X, F, t) = \sum_{i=0}^{n} a_i t^i$, then $|a_i| \leq Q(a_n, a_{n-1})$ for all i.*

We shall not go into the definition of semi-ampleness (which is a little stronger than pseudo-ampleness) since we need only the case where F is ample.

If K_X is ample, $H^i(X, K_X^t) = 0$ for $i > 0,\ t > 1$ by Kodaira's vanishing theorem. Hence,

$$(7.7.14) \qquad \dim H^0(X, K_X^t) = \chi(X, t) \quad \text{for} \quad t > 1.$$

(7.7.15) **Lemma.** *If the canonical bundle K_X is ample, then*

(1) $\chi(X, t) \in \mathbf{Z}$ *for all positive integers t;*

(2) $a_n = (-c_1(X))^n/n!, \quad$ *and* $\quad a_{n-1} = -(-c_1(X))^n/2(n-1)!;$

(3) $|\chi(X, t)| \leq (n+1)^{n+1} Q(a_n, a_{n-1}) \quad$ *for $t = 1, \ldots, n+1$.*

Proof. While (1) is trivial, (2) is in (7.7.12). Finally, (3) is immediate from (7.7.13). $\qquad\square$

Assume that K_X is ample. By (7.7.13) and (7.7.14), given n and $c_1(X)^n$, there are only a finite number of polynomials $\chi_1(t), \ldots, \chi_q(t)$ that can be the Hilbert polynomial of X, and a complete list of these polynomials is determined by $c_1(X)^n$. By (7.7.14), $\chi_1(l), \ldots, \chi_q(l)$ are the only possible values for the plurigenus $\dim H^0(X, K_X^l)$.

We have now established the following theorem of Bandman in the case $\dim X = \dim Y$. It remains to be proven when $\dim X > \dim Y$.

(7.7.16) **Theorem.** *Let X and Y be compact complex manifolds with ample canonical bundle. Then the number $|\mathrm{Dom}(X, Y)|$ of dominant maps from X to Y can be bounded by a number which depends only on $n = \dim X$ and $c_1(X)^n$.*

Proof. We shall now consider the case $\dim Y < \dim X$. Let $m = \dim Y$. We claim that if V is the intersection of $X \subset P_N$ with a generic linear subspace L of codimension $n - m$ in P_N, then V is nonsingular, $\dim V = \dim Y$, and $|\mathrm{Dom}(X, Y)| \leq |\mathrm{Dom}(V, Y)|$. We have only to verify the last property.

For each point x of X, let $G_x(m)$ be the Grassmannian of m-planes in the tangent space $T_x X$, and $G_m = \bigcup_{x \in X} G_x(m)$ the Grassmann bundle over X with projection $\pi : G(m) \to X$.

Let $\mathrm{Dom}(X, Y) = \{f_j\}$. Let S_j be the singularity set of the meromorphic map f_j, and let N_j be the set of m-planes $\xi \in G_x(m)$ such that f_j is regular at x but

$df_j: \xi \to T_{f_j(x)}Y$ is not an isomorphism. For each pair (i, j) with $i \neq j$, let P_{ij} be the set of $\xi \in G_x(m)$ such that both f_i and f_j are regular at x and $df_i|_\xi = df_j|_\xi$. Clearly, $\pi^{-1}(S_j)$ is a proper subvariety of $G(m)$, and N_j is a proper subvariety of $G(m) - \pi^{-1}(S_j)$. The set P_{ij} is a proper subvariety of $G(m) - \pi^{-1}(S_i \cup S_j)$. Hence

$$G := G(m) - ((\bigcup_j \pi^{-1}(S_j)) \cup (\bigcup_j N_j) \cup (\bigcup_{i,j} P_{ij}))$$

is Zariski open in $G(m)$. Take any $\xi \in G$, $\xi \subset T_x X$. Then $f_1, f_2, \ldots$ are all regular at x, and $df_1|_\xi, df_2|_\xi, \ldots$ are mutually distinct isomorphisms. Take a linear subspace L of P_N such that ξ is the tangent plane of $X \cap L$, proving our claim

Let H be the hyperplane bundle of P_N so that $H|_X = K_X^l$. By the adjunction formula we have

$$K_V = K_X|_V \otimes H^{n-m}|_V = K_X^{l(n-m)+1}|_V,$$

and consequently,

$$c_1(K_V) = (l(n - m) + 1)c_1(K_X)|_V.$$

In particular, K_V is ample. By Demailly's result, K_V^l is very ample. Furthermore,

$$
\begin{aligned}
c_1(K_V)^m[V] &= (l(n - m) + 1)^m c_1(K_X)^m c_1(H)^{n-m}[X] \\
&= l^{n-m}(l(n - m) + 1)^m c_1(K_X)^n[X] \\
&\leq n^n l^n c_1(K_X)^n[X].
\end{aligned}
$$

Since $|\mathrm{Dom}(V, Y)|$ is bounded by a number which depends only on m and $(c_1(V))^m$, the inequality above shows that it is bounded by a number which depends only on n and $(c_1(X))^n$. Since $|\mathrm{Dom}(X, Y)| \leq |\mathrm{Dom}(V, Y)|$, this completes the proof. $\square$

Howard and Sommese [1] estimated the order of the automorphism group of a compact complex manifold X with ample canonical bundle in terms of the Chern numbers of X. We shall use their method to reprove Bandman's result.

Given a subvariety X of P_N, we consider the set of hyperplanes of P_N which are tangent to X (i.e., which contain the tangent space of X) at regular points. This is a subset of the dual projective space P_N^*. The Zariski closure of this set is the **dual variety** $X^* \subset P_N^*$, (see J. Harris [1] for basic properties of dual varieties). Let $m = \dim X^*$. The **incidence variety** E of X is, by definition, the closure of

$$\{(x, \xi) \in X \times X^*; \; \xi \text{ tangent to } X \text{ at } x \in X_{\mathrm{reg}}\}$$

in $X \times X^*$. Let $p: E \to X$ and $\pi: E \to X^*$ be the projections. For a regular point $x \in X$, the set $p^{-1}(x)$ of hyperplanes ξ of P_N containing $T_x X$ forms an $(N - n - 1)$-dimensional linear subspace of P_N^*. Thus

(7.7.17) **Proposition.** *Let X be an n-dimensional projective variety in P_N, and $X^* \subset P_N^*$ be the dual variety of X. Let E be the incidence variety defined above. Then $p^{-1}(X_{\mathrm{reg}})$ is P_{N-n-1}-bundle over X_{reg}, and, in particular, $\dim E = N - 1$.*

(7.7.18) **Proposition.** (1) X *is the dual variety of* X^*, *i.e.,* $X^{**} = X$;
  (2)   *If* X *is irreducible, so is* X^*.

Proof. (1)   Fix a point $(x_0, \xi_0) \in E$. Let x_t, $|t| < \varepsilon$, be an arbitrary curve in X passing through x_0. Consider each x_t as an element of $\mathbf{C}^{N+1}$ and ξ_0 as an element of the dual space $(\mathbf{C}^{N+1})^*$. Then the condition that the hyperplane ξ_0 passes through the point x_0 is expressed by the equality

$$\langle \xi_0, x_0 \rangle = 0.$$

The condition that the hyperplane ξ_0 is tangent to X at x_0 is expressed by the equality

$$\langle \xi_0, (dx_t/dt)_{t=0} \rangle = 0.$$

Now, let (x_t, ξ_t) be an arbitrary curve in E through (x_0, ξ_0). Then

$$\langle \xi_t, x_t \rangle = 0.$$

Differentiating this equality with respect to t at $t = 0$, we obtain

$$\langle (d\xi_t/dt)_{t=0}, x_0 \rangle + \langle \xi_0, (dx_t/dt)_{t=0} \rangle = 0.$$

Since the second term vanishes as shown above, the first term also vanishes:

$$\langle (d\xi_t)_{t=0}, x_0 \rangle = 0.$$

Since $(d\xi_t/dt)_{t=0}$ represents an arbitrary tangent vector of X^* at ξ_0, this says that the hyperplane of P_N^* represented by x_0 is tangent to X^* at ξ_0, thus proving (1).
  (2)   This follows from (1). $\qquad\square$

Hence, E is the incidence variety of X^* as well. Interchanging the roles of X and X^* in (7.7.17), we obtain

(7.7.19) **Proposition.** *Let* X, X^* *and* E *be as in* (7.7.17). *Let* $m = \dim X^*$. *Then* $\pi^{-1}(X_{\mathrm{reg}}^*)$ *is a* P_{N-m-1}*-bundle over* X_{reg}^*. *The fibre* $\pi^{-1}(\xi)$ *over any regular point* $\xi \in X^*$ *consists of all hyperplanes of* P_N^* *(equivalently, all points of* P_N*) tangent to* X^* *at* ξ, *and it is imbedded into* X *by* p.

The dual variety X^* is usually a hypersurface of P_N^*. Namely,

(7.7.20) **Proposition.** *If* $X \subset P_N$ *is nonsingular and* $\Gamma(K_X^k) \neq 0$ *for some* $k > 0$, *then* $\dim X^* = N - 1$.

Proof. By (7.7.19) there is an $(n - (N - m - 1))$-dimensional polydisc $D^{n-(N-m-1)}$ imbedded in X_{reg}^* such that p maps $\pi^{-1}(D^{n-(N-m-1)})$ onto an open subset of X. If ω is a nonzero section of K_X^k, then $\pi^*\omega$ gives a nonzero pluricanonical section of $\pi^{-1}(D^{n-(N-m-1)}) \cong D^{n-(N-m-1)} \times P_{N-m-1}$, but this is possible only when $N - m - 1 = 0$. $\qquad\square$

Let X be and Y be nonsingular projective manifolds of equal dimension n. Assume that X has ample canonical bundle K_X and that Y is of general type. Fix a very ample line bundle H' on Y. Let $l > 0$ be an integer such that K_X^l is very ample and $\Gamma(K_Y^l H'^{-1}) \neq 0$, see (7.3.6). Let α be a nonzero section of $K_Y^l H'^{-1}$. Let $\Gamma_Y = \{\alpha\varphi;\ \varphi \in \Gamma(H')\}$ so that we have (see (7.6.3)) a holomorphic imbedding

$$j: Y \to P(\Gamma_Y^*).$$

Set $\Gamma_X = \Gamma(K_X^m)$ and let $i: X \to P(\Gamma_X^*)$ be the pluricanonical imbedding of X.

Let $f: X \to Y$ be a dominant meromorphic map. Then it induces (see (7.6.5)) an injective homomorphism

$$f^*: \Gamma_Y \to \Gamma_X.$$

and a surjective homomorphism

$$f_*: \Gamma_X^* \to \Gamma_Y^*.$$

Let $X^* \subset P(\Gamma_X)$ and $Y^* \subset P(\Gamma_Y)$ be the dual varieties of X and Y, respectively. Then we have the following diagrams:

$$(7.7.21) \qquad
\begin{array}{ccc}
P(\Gamma_X^*) & \xrightarrow{f_*} & P(\Gamma_Y^*) \\
\uparrow i & & \uparrow j \\
X & \xrightarrow{\ f\ } & Y
\end{array}
\quad \text{and} \quad
\begin{array}{ccc}
P(\Gamma_Y) & \xrightarrow{f^*} & P(\Gamma_X) \\
\uparrow j^* & & \uparrow i^* \\
Y^* & \xrightarrow{\ f^*\ } & X^*.
\end{array}$$

Since $f^*: \Gamma_Y \to \Gamma_X$ is injective, $f^*: P(\Gamma_Y) \to P(\Gamma_X)$ is a linear holomorphic injection. However, $f_*: \Gamma_X^* \to \Gamma_Y^*$ is surjective, and the induced linear map $f_*: P(\Gamma_X^*) \to P(\Gamma_Y^*)$ is meromorphic. Dualizing the second diagram in (7.7.21) we obtain the first diagram. So, instead of counting the number of maps f, we shall count the number of the corresponding linear maps f^*.

Set $N + 1 = \dim \Gamma_X$ and $N' + 1 = \dim \Gamma_Y$. Then $N' \leq N$. Let ξ and η denote homogeneous coordinates for points of $P(\Gamma_X)$ and $P(\Gamma_Y)$, respectively. Let A denote the homogeneous coordinate for points of $P(\mathrm{End}(\Gamma_Y, \Gamma_X))$. We consider ξ, η and A as an $(N+1)$-vector, an $(N'+1)$-vector and an $(N'+1) \times (N+1)$-matrix, and write $\xi = A\eta$. Let

$$M = \{A \in P(\mathrm{End}(\Gamma_Y, \Gamma_X));\ \mathrm{rank}(A) = N' + 1,\ A(Y^*) \subset X^*\}.$$

Then $f \in \mathrm{Dom}(X, Y) \to f^* \in M$ gives a one-to-one correspondence between $\mathrm{Dom}(X, Y)$ and M. We shall estimate the cardinality of M.

By (7.7.18) and (7.7.20), X^* is an irreducible hypersurface in $P(\Gamma_X)$ and hence is given as the zeros of an irreducible homogeneous polynomial $F(\xi)$. Then the condition $A(Y^*) \subset X^*$ is written as $F(A\eta) = 0$ for $\eta \in Y^*$. For each $\eta \in Y^*$, let $S(\eta)$ be the hypersurface in $P(\mathrm{End}(\Gamma_Y, \Gamma_X))$ defined by $F(A\eta) = 0$, i.e.,

$$S(\eta) = \{A;\ F(A\eta) = 0\}.$$

Then

$$M = \bigcap_{\eta \in Y^*} S(\eta).$$

We already know that M is a finite set, see (7.6.1). Let

$$r = \dim P(\mathrm{End}(\Gamma_Y, \Gamma_X)) = (N+1)(N'+1) - 1.$$

We shall find r points $\eta_1, \ldots, \eta_r$ of Y^* such that

$$(7.7.22) \qquad M = \bigcap_{i=1}^{r} S(\eta_i).$$

Let η_1 be an arbitrary point of $Y*$. Let $S_1^{r-1}, \ldots, S_p^{r-1}$ be the irreducible components of $S(\eta_1)$; they all have dimension $r-1$. For each j, $j = 1, \ldots, p$, let Y_j^* be the set of points $\eta \in Y^*$ such that $S(\eta) \cap S_j^{r-1}$ has dimension $r-2$; it is Zariski open and is nonempty since $\dim M < r - 1$. Let η_2 be any point of $\bigcap Y_j^*$. Let $S_1^{r-2}, \ldots, S_q^{r-2}$ be the irreducible components of $S(\eta_1) \cap S(\eta_2)$; they all have dimension $r-2$. Repeating the same argument, we find η_3 such that the irreducible components of $S(\eta_1) \cap S(\eta_2) \cap S(\eta_3)$ has dimension $r-3$, and so on.

From (7.7.22) and from Bezout's theorem we obtain

$$|M| \le (\deg X^*)^r.$$

Summarizing, we have

(7.7.23) Theorem. *Let X be an n-dimensional compact complex manifold with ample canonical bundle K_X, and Y an n-dimensional projective algebraic manifold of general type. Let m be a positive integer such that K_X^m is very ample and $\Gamma(K_Y^m H'^{-1}) \ne 0$ for some very ample line bundle H' over Y. Set $N+1 = \dim K_X^m$ and $N'+1 = \dim \Gamma(H')$. Then*

$$|\mathrm{Dom}(X, Y)| \le (\deg X^*)^{NN'+N+N'},$$

where the degree of of the dual hypersurface $X^* \subset P(\Gamma(K_X^m))$ is given by

$$\deg X^* = \sum_{j=0}^{n} (-1)^n m^j \int_X c_1(X)^j c_{n-j}(X).$$

The formula for $\deg X^*$ is a special case (i.e., $H = K_X^m$) of the following formula, (see Deligne-Katz [1; Exposé XVII]).

(7.7.24) Proposition. *If $X \subset P_{N+1}$ is an n-dimensional nonsingular manifold and if the dual variety X^* has dimension N, then*

$$\deg X^* = \sum_{j=0}^{n} (-1)^{n-j} (j+1) c_1(H)^j c_{n-j}(X),$$

where H denotes the hyperplane line bundle.

We proved in (7.6.7) that the group of bimeromorphic automorphisms of a compact complex space of general type is finite. A natural problem is to estimate the order of this group.

Let $\mathrm{Aut}(X)$ and $|\mathrm{Aut}(X)|$ denote the group of biholomorphic automorphisms of a complex space X and its order. The classical theorem of Hurwitz says that if X is a compact Riemann surface of genus $g \geq 2$, then

$$(7.7.25) \qquad |\mathrm{Aut}(X)| \leq 84(g - 1),$$

(see, for example, Kobayashi [8]), and this is also the best possible estimate.

Now, let X be an algebraic surface of general type. Andreotti [1] proved that $|\mathrm{Aut}(X)|$ has an upper bound which depends only on the Chern numbers of the surface. Huckleberry-Sauer [1] proved that it is bounded by a polynomial of degree 8 in $c_2(X)$. Recently, Xiao [1] obtained a sharp estimate

$$(7.7.26) \qquad |\mathrm{Aut}(X)| \leq (42c_1(X))^2$$

for a minimal surface of general type.

For higher dimensional manifolds of general type, we have no general estimates better than what we have for $|\mathrm{Dom}(X, Y)|$, namely (7.7.16) and (7.7.23). The latter, as already mentioned, is based on the proof of Howard-Sommese [1]. The natural goal would be an estimate of the form $|\mathrm{Aut}(X)| \leq a|c_1(X)^n|$ with a universal constant a.

Chapter 8. Value Distributions

1 Grassmann Algebra

We fix n, and for each k, $0 \le k \le n$, consider $\bigwedge^{k+1} \mathbf{C}^{n+1}$. Set

$$n(k) = \binom{n+1}{k+1} - 1$$

so that $\bigwedge^{k+1} \mathbf{C}^{n+1} \simeq \mathbf{C}^{n(k)+1}$. Let $G(n,k)$ be the Grassmannian of k-planes in $P_n\mathbf{C}$, i.e., the Grassmannian of $(k+1)$-dimensional subspaces in $\mathbf{C}^{n+1}$. Then $\dim G(n,k) = (n-k)(k+1)$. To a $(k+1)$-dimensional subspace spanned by $\mathbf{a}_0, \ldots, \mathbf{a}_k \in \mathbf{C}^{n+1}$, we assign a decomposable $(k+1)$-vector $A = \mathbf{a}_0 \wedge \ldots \wedge \mathbf{a}_k \in \bigwedge^{k+1} \mathbf{C}^{n+1}$, which is determined, up to a constant factor, by the subspace. Conversely, each decomposable $(k+1)$-vector A determines a k-plane in $P_n\mathbf{C}$, i.e., a $(k+1)$-dimensional vector subspace of $\mathbf{C}^{n+1}$, both of which will be denoted by the same symbol $[A]$. This correspondence defines the **Plücker imbedding**

$$(8.1.1) \qquad G(n,k) \subset P_{n(k)}\mathbf{C}.$$

Let $\mathbf{e}_0, \ldots, \mathbf{e}_n$ be the natural basis of $\mathbf{C}^{n+1}$. This defines an inner product in $\mathbf{C}^{n+1}$. For each $(k+1)$-dimensional subspace $[A]$ of $\mathbf{C}^{n+1}$, let $[A]^\perp$ denote the orthogonal complement of $[A]$ in $\mathbf{C}^{n+1}$ and also the corresponding $(n-k-1)$-plane in $P_n\mathbf{C}$, which is called the **polar space** of $[A]$. Clearly, $[A]$ is the polar space of $[A]^\perp$. The polar space of a point in $P_n\mathbf{C}$ is a hyperplane.

If $\mathbf{a}_j = \sum_i \alpha_{ji} e_i \in \mathbf{C}^{n+1}$ for $j = 0, 1, \ldots k$, then

$$(8.1.2) \qquad A = \mathbf{a}_0 \wedge \ldots \wedge \mathbf{a}_k = \sum_{i_0 < \ldots < i_k} \alpha_{i_0 \ldots i_k} \mathbf{e}_{i_0} \wedge \ldots \wedge \mathbf{e}_{i_k},$$

where the coefficients

$$(8.1.3) \qquad \alpha_{i_0 \ldots i_k} = \det \begin{pmatrix} \alpha_{0 i_0} & \cdots & \alpha_{0 i_k} \\ \cdots & \cdots & \cdots \\ \alpha_{k i_0} & \cdots & \alpha_{k i_k} \end{pmatrix}$$

are determined by the k-plane $[A]$ up to a common nonzero factor. They are the homogeneous **Plücker coordinates** of $[A]$.

The natural inner product in $\mathbf{C}^{n+1}$ induces an inner product in $\bigwedge^{k+1} \mathbf{C}^{n+1}$ so that $\{\mathbf{e}_{i_0} \wedge \ldots \wedge \mathbf{e}_{i_k};\ 0 \le i_0 < \ldots < i_k \le n\}$ form an orthonromal basis. The inner product between two decomposable $(k+1)$-vectors $A = \mathbf{a}_0 \wedge \ldots \wedge \mathbf{a}_k$ and $B = \mathbf{b}_0 \wedge \ldots \wedge \mathbf{b}_k$ is given by

$$(8.1.4) \qquad \langle A, B \rangle = \det \begin{pmatrix} \langle \mathbf{a}_0, \mathbf{b}_0 \rangle & \cdots & \langle \mathbf{a}_0, \mathbf{b}_k \rangle \\ \cdots & \cdots & \cdots \\ \langle \mathbf{a}_k, \mathbf{b}_0 \rangle & \cdots & \langle \mathbf{a}_k, \mathbf{b}_k \rangle \end{pmatrix}.$$

For $A \in \bigwedge^{k+1} \mathbf{C}^{n+1}$ we set

$$|A| = \sqrt{\langle A, A \rangle},$$

which represents the volume of the parallelepiped spanned by $\mathbf{a}_0, \ldots, \mathbf{a}_k$ if $A = \mathbf{a}_0 \wedge \ldots \wedge \mathbf{a}_k$. The Schwarz inequality holds for all $A, B \in \bigwedge^{k+1} \mathbf{C}^{n+1}$:

$$|\langle A, B \rangle| \le |A| \cdot |B|, \qquad A, B \in \bigwedge^{k+1} \mathbf{C}^{n+1}.$$

(8.1.5) Lemma. *For a decomposable $A \in \bigwedge^{k+1} \mathbf{C}^{n+1}$ and for any $B \in \bigwedge^{l+1} \mathbf{C}^{n+1}$, we have*

$$|A \wedge B| \le |A| \cdot |B|,$$

and the equality holds if and only if $B \in \bigwedge^{l+1}([A]^\perp)$.

Proof. Choose an orthonormal basis $\mathbf{e}_0, \ldots \mathbf{e}_n$ in such a way that $A = \alpha \mathbf{e}_0 \wedge \ldots \wedge \mathbf{e}_k$.
$\qquad\square$

We note that the inequality in Lemma may not hold for arbitrary A, B. Take, for example, $A = B = \mathbf{e}_0 \wedge \mathbf{e}_1 + \mathbf{e}_2 \wedge \mathbf{e}_3 + \mathbf{e}_4 \wedge \mathbf{e}_5$.

(8.1.6) Lemma. *Let $A, B \in \bigwedge^k \mathbf{C}^{n+1}$ be decomposable and $\mathbf{a}, \mathbf{a}', \mathbf{b}, \mathbf{b}' \in \mathbf{C}^{n+1}$. Then*

$$\langle A, B \rangle \langle A \wedge \mathbf{a} \wedge \mathbf{a}', B \wedge \mathbf{b} \wedge \mathbf{b}' \rangle$$
$$= \langle A \wedge \mathbf{a}, B \wedge \mathbf{b} \rangle \langle A \wedge \mathbf{a}', B \wedge \mathbf{b}' \rangle - \langle A \wedge \mathbf{a}', B \wedge \mathbf{b} \rangle \langle A \wedge \mathbf{a}, B \wedge \mathbf{b}' \rangle.$$

In particular,

$$|A|^2 |A \wedge \mathbf{a} \wedge \mathbf{a}'|^2 = |A \wedge \mathbf{a}|^2 |A \wedge \mathbf{a}'|^2 - |\langle A \wedge \mathbf{a}', A \wedge \mathbf{a} \rangle|^2.$$

Proof. (Wu [4]). Let

$$A = \mathbf{a}_1 \wedge \ldots \wedge \mathbf{a}_k \quad \text{and} \quad B = \mathbf{b}_1 \wedge \ldots \wedge \mathbf{b}_k,$$

and set

$$\mathbf{a}_{k+1} = \mathbf{a}, \quad \mathbf{a}_{k+2} = \mathbf{a}', \quad \mathbf{b}_{k+1} = \mathbf{b}, \quad \mathbf{b}_{k+2} = \mathbf{b}'.$$

Considering (8.1.4) we define a $(k+2) \times (k+2)$ matrix

$$M = (\langle \mathbf{a}_i, \mathbf{b}_j \rangle)_{i,j=1,\ldots,k+2}.$$

Let M' denote the $k \times k$ submatrix in the upper left corner of M, i.e.,

$$M' = (\langle \mathbf{a}_i, \mathbf{b}_j \rangle)_{i,j=1,\ldots,k}.$$

Let $M_{\alpha\beta}$ be the $(k+1) \times (k+1)$ submatrix of M obtained by deleting the α-th row and β-th column from M. Then the lemma to be proved can be formulated as follows:

$$(8.1.7) \qquad \det M \cdot \det M' = \det\begin{pmatrix} \det M_{k+1,k+1} & \det M_{k+1,k+2} \\ \det M_{k+2,k+1} & \det M_{k+2,k+2} \end{pmatrix}.$$

This is a special case of Sylvester's theorem on compound determinants. $\qquad\square$

Now, we explain the concept of interior product or contraction. Assuming $l \le k$, for a $(k+1)$-vector $A \in \bigwedge^{k+1} \mathbf{C}^{n+1}$ and for an $(l+1)$-vector $B \in \bigwedge^{l+1} \mathbf{C}^{n+1}$ we define a $(k-l)$-vector $A \vee B$, called the **contraction** of A by B, by

$$(8.1.8) \qquad \langle A \vee B, C \rangle = \langle A, B \wedge C \rangle \qquad \text{for} \quad C \in \bigwedge^{k-l} \mathbf{C}^{n+1}.$$

If $k = l$, then

$$A \vee B = \langle A, B \rangle.$$

From the definition we obtain the following associativity:

$$(8.1.9) \qquad (A \vee B_1) \vee B_2 = A \vee (B_1 \wedge B_2).$$

The definition of $A \vee B$ given by (8.1.8) is indirect. If we express A and B in terms of an orthonormal basis $\mathbf{e}_0, \ldots, \mathbf{e}_n$ for $\mathbf{C}^{n+1}$, then we can calculate $A \vee B$ explicitly. Let

$$\mathbf{e}_I = \mathbf{e}_{i_0} \wedge \ldots \wedge \mathbf{e}_{i_k} \qquad \mathbf{e}_J = \mathbf{e}_{j_0} \wedge \ldots \wedge \mathbf{e}_{j_l},$$

with $I = (i_0, \ldots, i_k)$ and $J = (j_0, \ldots, j_l)$. Then

$$(8.1.10) \qquad \mathbf{e}_I \vee \mathbf{e}_J = \begin{cases} 0 & \text{if } J \not\subset I; \\ \sigma \mathbf{e}_{I-J} & \text{if } J \subset I, \end{cases}$$

where σ is the sign of the permutation $I \to (J, I - J)$

We may also calculate $A \vee B$ without the use of an othonormal basis as follows. We consider the case A is decomposable, i.e., $A = \mathbf{a}_0 \wedge \mathbf{a}_1 \wedge \ldots \wedge \mathbf{a}_k$ and B is a vector $\mathbf{b}$. Then it is easy to verify the following formula:

$$(8.1.11) \qquad A \vee \mathbf{b} = \sum_j (-1)^j \langle \mathbf{a}_j, \mathbf{b} \rangle \mathbf{a}_0 \wedge \ldots \wedge \hat{\mathbf{a}}_j \wedge \ldots \wedge \mathbf{a}_k,$$

where $\hat{\mathbf{a}}_j$ indicates the missing element. Using the associativity (8.1.9), we can extend the formula to the case where B is a decomposable $(l+1)$-vector. In fact, we have

$$(8.1.12) \qquad A \vee B = \sum_{j_0 < \ldots < j_k} \delta_{j_0 \ldots j_k} \langle \mathbf{a}_{j_0} \wedge \ldots \wedge \mathbf{a}_{j_l}, B \rangle \mathbf{a}_{j_{l+1}} \wedge \ldots \wedge \mathbf{a}_{j_k}.$$

Let $A \in \bigwedge^{k+1} \mathbf{C}^{n+1}$ be decomposable, and $\mathbf{b} \in \mathbf{C}^{n+1}$. Then A represents a $(k+1)$-dimensional subspace $[A]$ of $\mathbf{C}^{n+1}$. Let $\mathbf{b} = \mathbf{b}^A + \mathbf{b}^{A^\perp}$, where $\mathbf{b}^A \in [A]$ and $\mathbf{b}^{A^\perp}$ is perpendicular to $[A]$. Choosing an orthonormal basis $\mathbf{e}_0, \ldots \mathbf{e}_n$ in such a way that $\mathbf{b}^A$ is a multiple of $\mathbf{e}_0$ and that $\mathbf{e}_0, \ldots \mathbf{e}_k$ form a basis for $[A]$, we see the following:

$$A \vee \mathbf{b} = A \vee \mathbf{b}^A.$$

At the same time we see that if A is decomposable, so is $A \vee \mathbf{b}$. If $B = \mathbf{b}_0 \wedge \ldots \wedge \mathbf{b}_l$ and $B^A = \mathbf{b}_0^A \wedge \ldots \wedge \mathbf{b}_l^A$, then by induction on l we obtain the following:

$$(8.1.13) \qquad\qquad A \vee B = A \vee B^A.$$

We see also that if both A and B are decomposable, so is $A \vee B$.

(8.1.14) Lemma. *Let $A \in \bigwedge^{k+1} \mathbf{C}^{n+1}$ and $B \in \bigwedge^{l+1} \mathbf{C}^{n+1}$ be both decomposable with $l \leq k$. Then*

$$A \vee B = 0 \quad \text{if and only if} \quad [B] \cap [A]^{\perp} \neq 0.$$

In particular, in case $k = l$,

$$\langle A, B \rangle = 0 \quad \text{if and only if} \quad [B] \cap [A]^{\perp} \neq 0.$$

Proof. $A \vee B = 0$ if and only if $A \vee B^A = 0$. Since $[B^A] \subset [A]$, $A \vee B^A = 0$ if and only if $B^A = 0$. If $B = \mathbf{b}_0 \wedge \ldots \wedge \mathbf{b}_l$, then $B^A = 0$ if and only if $\sum \beta_i \mathbf{b}_i^A = 0$ for some nontrivial $\beta_0, \ldots, \beta_l$. This latter is equivalent to $\sum \beta_i \mathbf{b}_i \in [A]^{\perp}$. $\qquad\square$

(8.1.15) Lemma. *For any $A \in \bigwedge^{k+1} \mathbf{C}^{n+1}$ and any decomposable $B \in \bigwedge^{l+1} \mathbf{C}^{n+1}$ with $l \leq k$, we have*

$$|A \vee B| \leq |A||B|.$$

Proof. Using (8.1.8) and (8.1.5) we have

$$|A \vee B|^2 = \langle A \vee B, A \vee B \rangle = \langle A, B \wedge (A \vee B) \rangle \leq |A||B \wedge (A \vee B)| \leq |A||B||A \vee B|.$$

$$\square$$

Finally,

(8.1.16) Lemma. *Let*

$$A, B \in \overset{k+1}{\bigwedge} \mathbf{C}^{n+1}$$

be decomposable. Then for $\mathbf{a}, \mathbf{a}', \mathbf{b}, \mathbf{b}', \mathbf{u}, \mathbf{v} \in \mathbf{C}^{n+1}$ we have

$$\langle A \vee \mathbf{u}, B \vee \mathbf{v} \rangle \langle (A \wedge \mathbf{a} \wedge \mathbf{a}') \vee \mathbf{u}, (B \wedge \mathbf{b} \wedge \mathbf{b}') \vee \mathbf{v} \rangle$$
$$= \langle (A \wedge \mathbf{a}) \vee \mathbf{u}, (B \wedge \mathbf{b}) \vee \mathbf{v} \rangle \langle (A \wedge \mathbf{a}') \vee \mathbf{u}, (B \wedge \mathbf{b}') \vee \mathbf{v} \rangle$$
$$- \langle (A \wedge \mathbf{a}) \vee \mathbf{u}, (B \wedge \mathbf{b}') \vee \mathbf{v} \rangle \langle (A \wedge \mathbf{a}') \vee \mathbf{u}, (B \wedge \mathbf{b}) \vee \mathbf{v} \rangle.$$

In particular,

$$|A \vee \mathbf{u}|^2 |(A \wedge \mathbf{a} \wedge \mathbf{a}') \vee \mathbf{u}|^2$$
$$= |(A \wedge \mathbf{a}) \vee \mathbf{u}|^2 |(A \wedge \mathbf{a}') \vee \mathbf{u}|^2 - |\langle (A \wedge \mathbf{a}) \vee \mathbf{u}, (A \wedge \mathbf{a}') \vee \mathbf{u} \rangle|^2.$$

Proof. This follows from (8.1.6) and (8.1.11). $\qquad\square$

2 Associated Curves

After preliminaries on Grassmann algebra, we are in a position to define associated curves of a holomorphic curve

$$f: D_R \to P_n \mathbf{C},$$

where D_R denotes the disc of radius R. By definition, $P_n \mathbf{C}$ is the space of lines through the origin of $\mathbf{C}^{n+1}$ so that

$$P_n \mathbf{C} = (\mathbf{C}^{n+1} - \{0\})/\mathbf{C}^*.$$

If $\tilde{f}: D_R \to \mathbf{C}^{n+1}$ is a holomorphic map which is not identically zero, then it induces a holomorphic map $f: D_R \to P_n \mathbf{C}$. In fact, if $\tilde{f} = (\tilde{f}_0, \ldots, \tilde{f}_n)$ and $\tilde{f}_0, \ldots, \tilde{f}_n$ vanish to order $\geq k$ simultaneously at a point $x_0 \in D_R$, and if z is a local coordinate around x_0, then $(\tilde{f}_0/z^k, \ldots, \tilde{f}_n/z^k)$ has a component which does not vanish at x_0 and determines a point of $P_n \mathbf{C}$.

Actually, any holomorphic map $f: D_R \to P_n \mathbf{C}$ can be lifted globally to a holomorphic map

$$(8.2.1) \qquad \tilde{f}: D_R \to \mathbf{C}^{n+1} - \{0\}.$$

To see this, cover D_R by open sets U_α with local lifts $\tilde{f}_\alpha: U_\alpha \to \mathbf{C}^{n+1} - \{0\}$, and define nowhere vanishing holomorphic functions $\varphi_{\alpha\beta}$ on $U_\alpha \cap U_\beta$ by $\tilde{f}_\alpha = \varphi_{\alpha\beta} \tilde{f}_\beta$. Then $\{\varphi_{\alpha\beta}\}$ defines a 1-cocycle with coefficients in the sheaf $\mathcal{O}^*_{D_R}$. Since $H^1(D_R, \mathcal{O}^*_{D_R}) = 0$, our assertion follows. We call such a lift $\tilde{f}: D_R \to \mathbf{C}^{n+1} - \{0\}$ a **reduced representation** of f.

A holomorphic curve $f: D_R \to P_n \mathbf{C}$ is said to be **degenerate** if $f(D_R)$ lies in a lower dimensional linear (i.e., projective) subspace of $P_n \mathbf{C}$, and **nondegenerate** otherwise. Clearly f is degenerate if and only if a lift $\tilde{f}: D_R \to \mathbf{C}^{n+1}$ is contained in a lower dimensional linear subspace of $\mathbf{C}^{n+1}$. We are primarily interested in nondegenerate holomorphic curves.

To simplify our notation, we shall denote a lift of f by the same symbol f whenever possible. Thus, sometimes we simply write

$$(8.2.2) \qquad f = \tilde{f} = (f_0, \ldots, f_n).$$

Given a holomorphic curve (8.2.2), we consider

$$(8.2.3) \qquad \tilde{F}_k(z) = \tilde{f}(z) \wedge \tilde{f}'(z) \wedge \ldots \wedge \tilde{f}^{(k)}(z) \in \bigwedge^{k+1} \mathbf{C}^{n+1},$$

where $\tilde{f}^{(j)} = (f_0^{(j)}, \ldots, f_n^{(j)})$ is the j-th derivative of $\tilde{f}$. Using the natural basis $\mathbf{e}_0, \mathbf{e}_1, \ldots, \mathbf{e}_n$ for $\mathbf{C}^{n+1}$, we can write

$$(8.2.4) \qquad \tilde{F}_k(z) = \sum_{0 \leq j_0 < \ldots < j_k \leq n} \det \begin{pmatrix} f_{j_0} & \cdots & f_{j_k} \\ \cdots & \cdots & \cdots \\ f_{j_0}^{(k)} & \cdots & f_{j_k}^{(k)} \end{pmatrix} \mathbf{e}_{j_0} \wedge \ldots \wedge \mathbf{e}_{j_k}.$$

Let $n(k) = \binom{n+1}{k+1} - 1$. Then $\tilde{F}_k(z): D_R \to \bigwedge^{k+1} \mathbf{C}^{n+1} = \mathbf{C}^{n(k)+1}$ induces a holomorphic curve, called the k-th **associated curve** of f:

$$(8.2.5) \qquad\qquad F_k: D_R \to P_{n(k)}\mathbf{C};$$

this latter mapping is independent of the choice of a lift of f. We note that $\tilde{F}_k$ need not be a reduced representation of F_k, i.e., $\tilde{F}_k$ may vanish at some points of D_R. We sometimes denote $\tilde{F}_k$ simply by F_k.

In particular, the n-th associated curve is given by

$$(8.2.6) \qquad F_n(z) = \tilde{F}_n(z) = W(f_0, \ldots, f_n)\mathbf{e}_0 \wedge \ldots \wedge \mathbf{e}_n,$$

where $W(f_0, \ldots, f_n)$ is the Wronskian of $f_0, \ldots, f_n$. Since $n(n) = 0$ and the range of F_n is a point, F_n is of little interest.

Since we assumed that f is nondegenerate, all associated curves are well defined.

(8.2.7) **Remark**. More generally, $f(D_R)$ lies in a k-dimensional linear subspace of $P_n\mathbf{C}$ but in no $(k-1)$-dimensional linear subspaces if and only if $\tilde{F}_k(z) \not\equiv 0$ but $\tilde{F}_{k+1}(z) \equiv 0$.

Since $\tilde{F}_k(z)$ in (8.2.3) is a decomposable $k+1$ vector, the associated curve F_k lies in the Grassmannian $G(n, k)$ of k-planes in $P_n\mathbf{C}$:

$$(8.2.8) \qquad\qquad F_k: D_R \to G(n, k) \subset P_{n(k)}\mathbf{C}.$$

We recall the notation

$$d^c = i(d'' - d') \quad \text{so that} \quad dd^c = 2id'd''.$$

Let Z denote the homogeneous coordinate for $P_{n(k)}\mathbf{C}$. Then the Fubini-Study metric of $P_{n(k)}\mathbf{C}$, or rather the associated Kähler form, is given by

$$(8.2.9) \qquad\qquad \Phi_k = \frac{1}{2\pi} dd^c \log |Z|.$$

We know that Φ_k is the Chern form of the hyperplane line bundle over $P_{n(k)}\mathbf{C}$ (with respect to the natural Hermitian inner product).

By pulling back Φ_k to D_R by F_k we set

$$(8.2.10) \qquad\qquad \omega_k = F_k^* \Phi_k = \frac{1}{2\pi} dd^c \log |\tilde{F}_k(z)|.$$

Then ω_k is positive definite except where F_k degenerates. We set

$$(8.2.11) \qquad\qquad \omega_k = \frac{i}{2\pi} \lambda_k(z) dz \wedge d\bar{z},$$

so that $\lambda_k(z) > 0$ except where F_k degenerates.

In order to establish the inter-relationship between these ω_k, we prove

(8.2.12) **Lemma**. *For $k = 1, \ldots, n-1$, we have*

$$\omega_k = \frac{i}{2\pi}\lambda_k dz \wedge d\bar{z}, \quad \text{where} \quad \lambda_k = \frac{|\tilde{F}_{k-1}|^2|\tilde{F}_{k+1}|^2}{|\tilde{F}_k|^4}.$$

Proof. We substitute

$$d\tilde{F}_k = \tilde{f} \wedge \tilde{f}' \wedge \ldots \wedge \tilde{f}^{(k-1)} \wedge \tilde{f}^{(k+1)} dz$$

into

$$\omega_k = \frac{i}{2\pi}\frac{\langle \tilde{F}_k, \tilde{F}_k \rangle \langle d\tilde{F}_k, d\tilde{F}_k \rangle - \langle d\tilde{F}_k, \tilde{F}_k \rangle \langle \tilde{F}_k, d\tilde{F}_k \rangle}{|\tilde{F}_k|^4}.$$

Setting

$$A = B = \tilde{F}_{k-1}, \quad \mathbf{a} = \mathbf{b} = \tilde{f}^{(k)}, \quad \mathbf{a}' = \mathbf{b}' = \tilde{f}^{(k+1)},$$

we apply (8.1.6) to obtain the desired formula. $\qquad\qquad\square$

As in (2.4.4), to each semi-positive (1,1)-form $\omega = i\lambda dz \wedge d\bar{z}$ on D_R, we associate the **Ricci form**:

$$(8.2.13) \qquad\qquad \mathrm{Ric}(\omega) = -dd^c \log \lambda.$$

It is defined where $\lambda > 0$. The geometric meaning of the Ricci form is explained by the fact that if K denotes the Gaussian curvature of the metric $2\lambda dz d\bar{z}$, then

$$(8.2.14) \qquad\qquad \mathrm{Ric}(\omega) = 2K\omega.$$

From (8.2.12) and (8.2.13) we obtain

$$\mathrm{Ric}(\omega_k) = -dd^c \log |\tilde{F}_{k-1}|^2 - dd^c \log |\tilde{F}_{k+1}|^2 + dd^c \log |\tilde{F}_k|^4.$$

Using (8.2.10) we obtain

$$(8.2.15) \qquad\qquad \frac{1}{4\pi}\mathrm{Ric}(\omega_k) = -\omega_{k-1} - \omega_{k+1} + 2\omega_k.$$

We shall now prove the complex analogue of the classical Frenet formula for a space curve. Given a holomorphic curve $f: D_R \to P_n\mathbf{C}$, let $\tilde{f}: D_R \to \mathbf{C}^{n+1} - \{0\}$ be its lift. The Frenet formula is expressed in terms of a moving frame, called the **Frenet frame**, instead of the fixed basis we used earlier.

For each $z \in D_R$, let $e_0(z)$ be a unit vector in $\mathbf{C}^{n+1}$ such that

$$(8.2.16) \qquad\qquad \tilde{f}(z) = \tau_0(z)\mathbf{e}_0(z),$$

where $\tau_0(z)$ is a smooth function on D_R. We may assume that $\mathbf{e}_0(z)$ is also smooth; take for example, $\tau_0 = |\tilde{f}|$ and $\mathbf{e}_0 = \tilde{f}/|\tilde{f}|$. We take a unit vector $\mathbf{e}_1(z)$ perpendicular to $\mathbf{e}_0(z)$ such that

$$d\mathbf{e}_0 = \theta_0^0 \mathbf{e}_0 + \theta_0^1 \mathbf{e}_1,$$

where θ_0^0 and θ_0^1 are 1-forms. Continuing this construction we obtain a unitary frame $\mathbf{e}_0, \mathbf{e}_1, \ldots, \mathbf{e}_n$ along the curve such that $d\mathbf{e}_j$ is a linear combination of $\mathbf{e}_0, \ldots, \mathbf{e}_{j+1}$. Then using

$$0 = d\langle \mathbf{e}_i, \mathbf{e}_j \rangle = \langle d\mathbf{e}_i, \mathbf{e}_j \rangle + \langle \mathbf{e}_i, d\mathbf{e}_j \rangle,$$

we see that the matrix (θ_j^i) is skew-Hermitian, i.e.,

$$\theta_i^j = -\bar{\theta}_j^i$$

and that

$$(8.2.17) \qquad d\mathbf{e}_j = \theta_j^{j-1}\mathbf{e}_{j-1} + \theta_j^j \mathbf{e}_j + \theta_j^{j+1}\mathbf{e}_{j+1},$$

with the understanding that $\theta_0^{-1} = \theta_n^{n+1} = 0$. Thus, the matrix (θ_j^i) is tri-diagonal, that is,

$$(8.2.18) \qquad \theta_j^i = 0 \quad \text{for} \quad |i - j| > 1.$$

From $dd\mathbf{e}_j = 0$ we obtain

$$(8.2.19) \qquad d\theta_j^i = -\sum \theta_k^i \wedge \theta_j^k.$$

Geometrically speaking, we pulled back the tangent bundle $T(\mathbf{C}^{n+1})$ by $\tilde{f}$ and constructed a unitary frame field $\mathbf{e}_0, \ldots, \mathbf{e}_n$ for the induced vector bundle over D_R. The matrix valued 1-form (θ_j^i) defines the connection induced in this bundle from the trivial flat connection of $T(\mathbf{C}^{n+1})$, and (8.2.19) expresses the trivial fact that the induced connection is also flat.

Starting with (8.2.16) and using (8.2.17) we see inductively that $\tilde{f}^{(j)}$ is a linear combination of $\mathbf{e}_0, \ldots, \mathbf{e}_j$. Since $\tilde{f}, \tilde{f}', \ldots, \tilde{f}^{(j)}$ are all holomorphic, $d\tilde{f}^{(j-1)}$ is a linear combination of $\mathbf{e}_0, \ldots, \mathbf{e}_j$ with $(1, 0)$-forms as coefficients. From this fact, we see inductively that θ_{j-1}^j is a $(1, 0)$-form, say

$$\theta_{j-1}^j = \tau_j dz, \quad j = 1, \ldots, n,$$

and that

$$(8.2.20) \qquad \tilde{f}^{(j)} \equiv \tau_0 \tau_1 \ldots \tau_j \mathbf{e}_j \quad (\text{mod } \mathbf{e}_0, \ldots, \mathbf{e}_{j-1}).$$

In the construction of the unitary frame above, each $\mathbf{e}_j$ is unique up to a multiplicative factor of e^{it}. So we can require each τ_j to be real and non-negative. This makes $\mathbf{e}_j$ unique whenever τ_j is nonzero. Thus

$$(8.2.21) \qquad \tau_0 = |\tilde{f}|, \qquad \theta_{j-1}^j = \tau_j dz \quad \text{with} \quad \tau_j \geq 0.$$

Since (θ_j^i) is skew-Hermitian,

$$(8.2.22) \qquad \theta_j^{j-1} = -\tau_j d\bar{z}.$$

The diagonal elements θ_i^i are, of course, purely imaginary, i.e., $\bar{\theta}_i^i = -\theta_i^i$.

Since

$$\tilde{F}_k = \tilde{f} \wedge \tilde{f}' \wedge \ldots \wedge \tilde{f}^{(k)} = \tau_0^{k+1}\tau_1^k \ldots \tau_k \mathbf{e}_0 \wedge \mathbf{e}_1 \wedge \ldots \wedge \mathbf{e}_k$$

with $\tau_0^{k+1}\tau_1^k \ldots \tau_k \geq 0$, we have $|\tilde{F}_k| = \tau_0^{k+1}\tau_1^k \ldots \tau_k$. Differentiating

$$(8.2.23) \qquad \frac{\tilde{F}_k}{|\tilde{F}_k|} = \mathbf{e}_0 \wedge \mathbf{e}_1 \wedge \ldots \wedge \mathbf{e}_k$$

yields

$$
\begin{aligned}
\frac{d\tilde{F}_k}{|\tilde{F}_k|} - \frac{\tilde{F}_k d|\tilde{F}_k|}{|\tilde{F}_k|^2} &= \sum_{j=0}^{k} \mathbf{e}_0 \wedge \ldots \wedge d\mathbf{e}_j \wedge \ldots \wedge \mathbf{e}_k \\
&= (\sum_{j=0}^{k} \theta_j^j)\mathbf{e}_0 \wedge \ldots \wedge \mathbf{e}_k + \theta_k^{k+1}\mathbf{e}_0 \wedge \ldots \wedge \mathbf{e}_{k-1} \wedge \mathbf{e}_{k+1}.
\end{aligned}
$$

We take the inner product of this with (8.2.23) to get

$$
\begin{aligned}
\sum_{j=0}^{k} \theta_j^j &= \frac{\langle d\tilde{F}_k, \tilde{F}_k \rangle}{|\tilde{F}_k|^2} - \frac{d|\tilde{F}_k|}{|\tilde{F}_k|} \\
&= \frac{\langle d\tilde{F}_k, \tilde{F}_k \rangle - \langle \tilde{F}_k, d\tilde{F}_k \rangle}{2|\tilde{F}_k|^2} \\
&= \frac{d' - d''}{2} \log |\tilde{F}_k|^2 = id^c \log |\tilde{F}_k|.
\end{aligned}
$$

Differentiating once more and using ω_k defined by (8.2.10) we have

$$\omega_k = \frac{1}{2\pi i} \sum_{j=0}^{k} d\theta_j^j.$$

By (8.2.18) and (8.2.19) we have

$$
\begin{aligned}
\sum_{j=0}^{k} d\theta_j^j &= -\sum_{j=1}^{k} \theta_{j-1}^j \wedge \theta_j^{j-1} - \sum_{j=0}^{k} \theta_{j+1}^j \wedge \theta_j^{j+1} = -\theta_{k+1}^k \wedge \theta_k^{k+1} \\
&= -\theta_k^{k+1} \wedge \bar{\theta}_k^{k+1}.
\end{aligned}
$$

Hence, by (8.2.21)

$$(8.2.24) \qquad \omega_k = \frac{i}{2\pi}\theta_k^{k+1} \wedge \bar{\theta}_k^{k+1} = \frac{i}{2\pi}\tau_{k+1}^2 dz \wedge d\bar{z}.$$

Comparing this with (8.2.11) and (8.2.12) we have

$$(8.2.25) \qquad \lambda_k = \tau_{k+1}^2 = \frac{|\tilde{F}_{k-1}|^2 |\tilde{F}_{k+1}|^2}{|\tilde{F}_k|^4}.$$

We recall that, in general, if

$$\omega = \frac{i}{2\pi}\theta \wedge \bar{\theta} = \frac{i}{2\pi}\tau^2 ds \wedge d\bar{z} \quad \text{with} \quad \theta = \tau dz$$

is a Kähler form on D_R, then the $(1,0)$-form $\theta/\sqrt{2\pi}$ is an orthonormal coframe and the connection form φ is uniquely determined by the skew-Hermitian condition $\varphi + \bar{\varphi} = 0$ and by the structure equation

$$d\theta = -\varphi \wedge \theta.$$

This connection form is given by

$$\varphi = i d^c \log \tau,$$

and its curvature form by $d\varphi = i d d^c \log \tau$.

In particular, consider ω_k. It is a Kähler form on D_R except where τ_{k+1} vanishes. The $(1,0)$-form $\theta_k^{k+1} = \tau_{k+1} dz$ is an orthonormal coframe (except for the factor of $\sqrt{2\pi}$). From (8.2.18) and (8.2.19) we obtain

$$(8.2.26). \qquad d\theta_k^{k+1} = -(\theta_{k+1}^{k+1} - \theta_k^k) \wedge \theta_k^{k+1}.$$

Since $\theta_j^j + \bar{\theta}_j^j = 0$, the equation above must be the structure equation for the Kähler structure given by ω_k. In other words, $\theta_{k+1}^{k+1} - \theta_k^k$ is the connection form for the natural connection. From the uniqueness of such a connection form, we have

$$(8.2.27) \qquad \theta_{k+1}^{k+1} - \theta_k^k = i d^c \log \tau_{k+1}.$$

Its curvature form is given by $i d d^c \log \tau_{k+1}$.

3 Contact Functions

The main purpose of this section is to establish the inequality in (8.3.12) which will be used in Section 5. The results of this section will not be used in Section 4.

Let $f: D_R \to P_n \mathbf{C}$ be a non-degenerate holomorphic curve with a reduced representation $\tilde{f}: D_R \to \mathbf{C}^{n+1}$. Let $F_k: D_R \to P_{n(k)}\mathbf{C}$ be the k-th associated curve represented by $\tilde{F}_k: D_R \to \bigwedge^{k+1} \mathbf{C}^{n+1}$, see (8.2.3) and (8.2.5). We make use of the Frenet frame $\mathbf{e}_0, \dots \mathbf{e}_n$ introduced in the preceding section; at each point $z \in D_R$, $\mathbf{e}_0(z), \dots, \mathbf{e}_n(z)$ form an orthonormal basis for $\mathbf{C}^{n+1}$.

Now, fix a unit vector $\mathbf{a} = (a^0, \dots, a^n) \in \mathbf{C}^{n+1}$, and let

$$H_{\mathbf{a}} = \{\mathbf{z} = (z^0, \dots, z^n);\ \langle \mathbf{z}, \mathbf{a} \rangle = \sum z^j \bar{a}^j = 0\}$$

be the hyperplane perpendicular to $\mathbf{a}$. At each $z \in D_R$, we express $\mathbf{a}$ in terms of the Frenet frame. Thus

$$\mathbf{a} = \sum_{i=0}^{n} \alpha_i \mathbf{e}_i, \qquad \sum |\alpha_i|^2 = 1.$$

For each $k = 0, \dots, n$, we define the **k-th contact function** $\varphi_k(z, \mathbf{a})$ with respect to the hyperplane $H_{\mathbf{a}}$ as follows:

$$(8.3.1) \qquad \varphi_k(\mathbf{a}, z) = \frac{|\tilde{F}_k(z) \vee \mathbf{a}|^2}{|\tilde{F}_k(z)|^2} = |(\mathbf{e}_0 \wedge \ldots \wedge \mathbf{e}_k) \vee \mathbf{a}|^2 = \sum_{j=0}^{k} |\alpha_j|^2,$$

where $\vee$ is the contraction symbol defined in (8.1.8). Then $\varphi_k(\mathbf{a}, z)$ is a function on D_R satisfying $0 \le \varphi_k(\mathbf{a}, z) \le 1$.

We note that $\varphi_k(\mathbf{a}, z)$ depends on $H_\mathbf{a}$, not on $\mathbf{a}$ itself and should be written $\varphi_k(H_\mathbf{a}, z)$. But for the sake of simplicity of notation, we write $\varphi_k(\mathbf{a}, z)$.

We note also that $\varphi_k(\mathbf{a}, z)$ is independent of the representation $\tilde{f}$ and that the zeros of $\tilde{F}_k$ is canceled by the zeros of $\tilde{F}_k \vee \mathbf{a}$. It is a simple matter to see that $\varphi_k(\mathbf{a}, z)$ vanishes at z_0 if and only if the k-plane in $P_n\mathbf{C}$ given by $F_k(z_0)$ is contained in the hyperplane $H_\mathbf{a}$. In particular, if $\varphi_k(\mathbf{a}, z_0) = 0$, then $\varphi_j(\mathbf{a}, z_0) = 0$ for $j < k$. We say that the curve f has a **contact of order** $k+1$ with the hyperplane $H_\mathbf{a}$ at z_0 if $\varphi_k(\mathbf{a}, z_0) = 0$.

Now, in order to calculate $d'\varphi_k(\mathbf{a}, z)$, we consider $d'\mathbf{e}_j$. In (8.2.17), $\theta_j^{j+1} = \tau_j dz$ is a $(1, 0)$-form (see (8.2.21)) while θ_j^j is of the form $\sigma_j dz - \bar{\sigma}_j d\bar{z}$ since it is purely imaginary. Hence,

$$\begin{aligned} d'\varphi_k(\mathbf{a}, z) &= \theta_k^{k+1} \langle (\mathbf{e}_0 \wedge \ldots \wedge \mathbf{e}_{k-1} \wedge \mathbf{e}_{k+1}) \vee \mathbf{a}, (\mathbf{e}_0 \wedge \ldots \wedge \mathbf{e}_k) \vee \mathbf{a} \rangle \\ &= \theta_k^{k+1} \langle \bar{\alpha}_{k+1} \mathbf{e}_0 \wedge \ldots \wedge \mathbf{e}_{k-1}, \bar{\alpha}_k \mathbf{e}_0 \wedge \ldots \wedge \mathbf{e}_{k-1} \rangle \\ &= \bar{\alpha}_{k+1} \alpha_k \theta_k^{k+1}. \end{aligned}$$

For simplicity, we put

$$\varphi_j = \varphi_j(\mathbf{a}, z).$$

Since $d''\varphi = \overline{d'\varphi}$, we have

$$d\varphi_k \wedge d^c\varphi_k = 2i d'\varphi_k \wedge d''\varphi_k = 2i |\alpha_k|^2 |\alpha_{k+1}|^2 \theta_k^{k+1} \wedge \bar{\theta}_k^{k+1}.$$

By (8.3.1),

$$|\alpha_k|^2 = \varphi_k - \varphi_{k-1}, \qquad |\alpha_{k+1}|^2 = \varphi_{k+1} - \varphi_k.$$

Hence, using (8.2.24) we have

$$(8.3.2) \qquad \frac{1}{4\pi} d\varphi_k \wedge d^c\varphi_k = (\varphi_{k+1} - \varphi_k)(\varphi_k - \varphi_{k-1})\omega_k.$$

Now, in order to calculate $dd^c \log \varphi_k$, we start with

$$\begin{aligned} dd^c \log |\tilde{F}_k \vee \mathbf{a}|^2 &= 2i d'd'' \log \langle \tilde{F}_k \vee \mathbf{a}, \tilde{F}_k \vee \mathbf{a} \rangle \\ &= 2i \frac{|d\tilde{F}_k \vee \mathbf{a}|^2 |\tilde{F}_k \vee \mathbf{a}|^2 - \langle d\tilde{F}_k \vee \mathbf{a}, \tilde{F}_k \vee \mathbf{a} \rangle \langle \tilde{F}_k \vee \mathbf{a}, d\tilde{F}_k \vee \mathbf{a} \rangle}{|\tilde{F}_k \vee \mathbf{a}|^4}. \end{aligned}$$

In the numerator, we make the following substitution:

$$\begin{aligned} \tilde{F}_k \vee \mathbf{a} &= (\tilde{f} \wedge \tilde{f}' \wedge \ldots \wedge \tilde{f}^{(k-1)} \wedge \tilde{f}^{(k)}) \vee \mathbf{a}, \\ d\tilde{F}_k \vee \mathbf{a} &= ((\tilde{f} \wedge \tilde{f}' \wedge \ldots \wedge \tilde{f}^{(k-1)} \wedge \tilde{f}^{(k+1)}) \vee \mathbf{a})dz, \end{aligned}$$

and then apply (8.1.16) to obtain

$$(8.3.3) \qquad dd^c \log |\tilde{F}_k \vee \mathbf{a}|^2 = \frac{2|\tilde{F}_{k-1} \vee \mathbf{a}|^2 |\tilde{F}_{k+1} \vee \mathbf{a}|^2}{|\tilde{F}_k \vee \mathbf{a}|^4} idz \wedge d\bar{z}.$$

Combined with (8.2.12) and (8.3.1), this yields

$$(8.3.4) \qquad \frac{1}{4\pi} dd^c \log |\tilde{F}_k \vee \mathbf{a}|^2 = \frac{\varphi_{k-1}\varphi_{k+1}}{\varphi_k^2}\omega_k.$$

From (8.2.10), (8.3.1) and (8.3.4) we obtain the formula:

$$(8.3.5) \qquad \frac{1}{4\pi} dd^c \log \varphi_k = \frac{\varphi_{k-1}\varphi_{k+1} - \varphi_k^2}{\varphi_k^2}\omega_k.$$

In particular,

$$(8.3.6) \qquad \frac{1}{4\pi} dd^c \log \varphi_0 = -\omega_0.$$

Since $\varphi_n \equiv 1$,

$$(8.3.7) \qquad \frac{1}{4\pi} dd^c \log \varphi_n = 0.$$

Let c be a large positive constant, yet to be determined. We have

$$dd^c \log \frac{1}{(\log(c/\varphi_k))^2} = 2\frac{dd^c \log \varphi_k}{\log(c/\varphi_k)} + 2\frac{d\varphi_k \wedge d^c \varphi_k}{\varphi_k^2 (\log(c/\varphi_k))^2}.$$

Using (8.3.2) and (8.3.5) yields

$$\frac{1}{4\pi} dd^c \log \frac{1}{(\log(c/\varphi_k))^2} = 2\left[\frac{\varphi_{k+1}}{\varphi_k (\log(c/\varphi_k))^2} - \left(\frac{1}{\log(c/\varphi_k)} + \frac{1}{(\log(c/\varphi_k))^2} \right) \right]\omega_k$$

$$+2\left[\frac{\varphi_{k-1}\varphi_{k+1}}{\varphi_k^2} \left(\frac{1}{\log(c/\varphi_k)} - \frac{1}{(\log(c/\varphi_k))^2} \right) + \frac{\varphi_{k-1}}{\varphi_k (\log(c/\varphi_k))^2} \right]\omega_k.$$

Since $0 \le \varphi_k \le 1$, for every $\varepsilon > 0$ there exists a constant $c(\varepsilon) > 0$ such that

$$\left(\frac{1}{\log(c/\varphi_k)} + \frac{1}{(\log(c/\varphi_k))^2} \right) < \varepsilon,$$

and

$$\left(\frac{1}{\log(c/\varphi_k)} - \frac{1}{(\log(c/\varphi_k))^2} \right) > 0.$$

for $c \ge c(\varepsilon)$ so that

$$(8.3.8) \qquad \frac{1}{4\pi} dd^c \log \frac{1}{(\log(c/\varphi_k))^2} \ge \frac{2\varphi_{k+1}}{\varphi_k (\log(c/\varphi_k))^2}\omega_k - \varepsilon\omega_k.$$

Let $\mathbf{a}_0, \ldots, \mathbf{a}_q \in \mathbf{C}^{n+1}$ be $q+1$ vectors in general position; this means that any $n+1$ of $\mathbf{a}_0, \ldots, \mathbf{a}_q$ are linearly independent.

Let $f: D_R \to P_n\mathbf{C}$ be a holomorphic map such that its image $f(D_R)$ does not lie on a proper linear subspace of $P_n\mathbf{C}$. We write

$$\varphi_k(\mathbf{a}_j) = \varphi_k(\mathbf{a}_j, z), \qquad z \in D_R.$$

We shall need the following "sums into products" formula (cf. Cowen-Griffiths [1]):

(8.3.9) Lemma. *There is a constant $C_k > 0$ such that*

$$\sum_{j=0}^{q} \frac{\varphi_{k+1}(\mathbf{a}_j)}{\varphi_k(\mathbf{a}_j)(\log(c/\varphi_k(\mathbf{a}_j)))^2} \geq \frac{C_k}{2} \prod_{j=0}^{q} \left(\frac{\varphi_{k+1}(\mathbf{a}_j)}{\varphi_k(\mathbf{a}_j)(\log(c/\varphi_k(\mathbf{a}_j)))^2} \right)^{1/(n-k)}.$$

Proof. Since $\mathbf{a}_0, \ldots, \mathbf{a}_q$ are in general position, for any unit $(k+1)$-vector E in $\mathbf{C}^{n+1}$ we have

$$E \vee \mathbf{a}_j \neq 0$$

for all but at most $n - k$ of the $\mathbf{a}_j$. We cover the Grassmann manifold $G(n, k)$ of $(k+1)$-planes in $\mathbf{C}^{n+1}$ by a finite number of open sets $\{U_\alpha\}$ so that for each α

$$|E \vee \mathbf{a}_j| \geq \delta > 0 \qquad \text{for} \quad E \in U_\alpha$$

for all but at most $n - k$ of the $\mathbf{a}_j$. We can take the same constant δ for all U_α.

From the definition (8.3.1) of $\varphi_k(\mathbf{a}_j) = \varphi_k(\mathbf{a}_j, z)$, we see that

$$\varphi_k(\mathbf{a}_j) \geq \delta \qquad \text{on} \quad \tilde{F}_k^{-1}(U_\alpha) \subset D_R$$

for all but at most $n - k$ of the $\mathbf{a}_j$. Set

$$\psi_k(\mathbf{a}_j) = \frac{\varphi_{k+1}(\mathbf{a}_j)}{\varphi_k(\mathbf{a}_j)(\log(c/\varphi_k(\mathbf{a}_j)))^2}.$$

Since $0 \leq \varphi_k \leq 1$, we have

$$\psi_k(\mathbf{a}_j) \leq \frac{1}{\varphi_k(\mathbf{a}_j)(\log c)^2} \leq M \qquad \text{on} \quad \tilde{F}_k^{-1}(U_\alpha)$$

for all but at most $n - k$ of the $\mathbf{a}_j$, where $M = \max\{1/\delta(\log c)^2, 1\}$.

Renumbering $\mathbf{a}_0, \ldots, \mathbf{a}_q$, we may assume that

$$\psi_k(\mathbf{a}_j) > M \qquad \text{at most for} \quad j = 0, \ldots, p \leq n - k.$$

Then on $\tilde{F}_k^{-1}(U_\alpha)$ we have

$$\sum_{j=0}^{q} \psi_k(\mathbf{a}_j) \geq \sum_{j=0}^{p} \psi_k(\mathbf{a}_j) \geq (p+1) \prod_{j=0}^{p} \psi_k(\mathbf{a}_j)^{1/p+1}$$

$$\geq (p+1) \prod_{j=0}^{p} \psi_k(\mathbf{a}_j)^{1/(n-k)}$$

$$\geq (p+1) \prod_{j=0}^{q} \psi_k(\mathbf{a}_j)^{1/(n-k)} M^{-(q-p)/(n-k)}$$

$$= \frac{C_k}{2} \prod_{j=0}^{q} \psi_k(\mathbf{a}_j)^{1/(n-k)},$$

where $C_k = 2(p+1)M^{-(q-p)/(n-k)}$. $\qquad\qquad\square$

Now we define non-negative 2-forms $\hat{\omega}_k$, $k = 0, 1, \ldots, n-1$, on D_R by

$$(8.3.10) \qquad \hat{\omega}_k = C_k \prod_{j=1}^{N} \left(\frac{\varphi_{k+1}(\mathbf{a}_j)}{\varphi_k(\mathbf{a}_j)(\log(c/\varphi_k(\mathbf{a}_j)))^2} \right)^{1/(n-k)} \omega_k,$$

where the constants C_k are given in (8.3.9). Then

$$-(n-k)\mathrm{Ric}(\hat{\omega}_k) = \sum_{j=1}^{N}(dd^c \log \varphi_{k+1}(\mathbf{a}_j) - dd^c \log \varphi_k(\mathbf{a}_j))$$

$$+ \sum_{j=1}^{N} dd^c \log \frac{1}{(\log(c/\varphi_k(\mathbf{a}_j)))^2} - (n-k)\mathrm{Ric}(\omega_k).$$

Now we sum the equation above from $k = 0$ to $n - 1$. By (8.3.6) and (8.3.7),

$$\sum_{k=0}^{n-1}(dd^c \log \varphi_{k+1}(\mathbf{a}_j) - dd^c \log \varphi_k(\mathbf{a}_j)) = 4\pi\omega_0.$$

By (8.2.15)

$$\sum_{k=0}^{n-1}(n-k)\mathrm{Ric}(\omega_k) = 4\pi(n+1)\omega_0.$$

Hence,

$$(8.3.11) \quad -\frac{1}{4\pi}\sum_{k=0}^{n-1}(n-k)\mathrm{Ric}(\hat{\omega}_k) = \sum_{j=0}^{q} dd^c \log \frac{1}{(\log(c/\varphi_k(\mathbf{a}_j)))^2} + (q-n)\omega_0.$$

Now, substituing (8.3.8) into (8.3.11), we see that for any given $\varepsilon > 0$, there is $c(\varepsilon) > 0$ such that

$$-\frac{1}{4\pi}\sum_{k=0}^{n-1}(n-k)\mathrm{Ric}(\hat{\omega}_k)$$

$$\geq \sum_{k=0}^{n-1}\sum_{j=0}^{q} \frac{2\varphi_{k+1}(\mathbf{a}_j)}{\varphi_k(\mathbf{a}_j)(\log(c/\varphi_k(\mathbf{a}_j)))^2}\omega_k + (q-n)\omega_0 - \frac{\varepsilon(q+1)}{4\pi}\sum_{k=0}^{n-1}\omega_k$$

for $c \geq c(\varepsilon)$. Using (8.3.9) and (8.3.10) and replacing $\varepsilon(q+1)/4\pi$ by ε in the inequality above, we have

(8.3.12) Lemma. *Let* $f: D_R \to P_n\mathbf{C}$ *be a holomorphic map such that its image* $f(D_R)$ *does not lie on a proper linear subspace of* $P_n\mathbf{C}$. *Let* $\mathbf{a}_0, \ldots, \mathbf{a}_q \in \mathbf{C}^{n+1}$ *be* $q + 1$ *vectors in general position. Then, given* $\varepsilon > 0$, *there is a constant* $c(\varepsilon)$ *such that*

$$-\frac{1}{4\pi} \sum_{k=0}^{n-1} (n-k)\mathrm{Ric}(\hat{\omega}_k) \geq \sum_{k=0}^{n-1} \hat{\omega}_k + (q-n)\omega_0 - \varepsilon \sum_{k=0}^{n-1} \omega_k,$$

where the forms $\hat{\omega}_k$ *are defined by* (8.3.10) *with* $c \geq c(\varepsilon)$.

4 First Main Theorem

The following local formula, due to Lelong [1], is known under the name of the **Poincaré-Lelong formula**. If φ is a holomorphic function on an n-dimensional manifold X and if Δ_φ is its divisor, then

$$(8.4.1) \qquad\qquad \frac{1}{2\pi} dd^c \log |\varphi| = \Delta_\varphi;$$

this formula should be taken in the sense of current, that is,

$$\frac{1}{2\pi} \int_X dd^c \log |\varphi| \wedge \eta = \int_{\Delta_\varphi} \eta$$

for all C^∞ form η of degree $(n-1, n-1)$ with compact support. This yields the following line bundle version of the Poincaré-Lelong formula.

(8.4.2) Theorem. *Let* L *be a Hermitian line bundle over* X *with Hermitian inner product* h. *Let* σ *be a holomorphic section of* L *and* Δ_σ *its divisor. Then*

$$\frac{1}{2\pi} dd^c \log |\sigma|_h = -c_h + \Delta_\sigma,$$

where $|\sigma|_h$ *is the length of* σ *measured by* h, *and* $c_h = -\frac{1}{4\pi} dd^c \log h$ *is the Chern form of* (L, h).

Proof. Cover X by coordinate neighborhoods U_α. Using local expressions h_α and σ_α for h and σ, we have $|\sigma|_h^2 = h_\alpha |\sigma_\alpha|^2$ on U_α. Apply $dd^c \log$ to both sides. $\qquad\square$

Let X be a compact complex manifold with a Hermitian holomorphic line bundle (L, h). In Appendix B of Chapter 3, X was $P_n\mathbf{C}$ and L was the hyperplane line bundle H so that the divisor Δ_σ defined by a section σ was a hyperplane in $P_n\mathbf{C}$ and c_h was (up to a positive constant factor) the Kähler form of the Fubini-Study metric of $P_n\mathbf{C}$.

We fix a holomorphic map

$$f: D_R = \{z \in \mathbf{C};\ |z| < R\} \to X.$$

By pulling back the formula in (8.4.2) by f, we have

$$(8.4.3) \qquad \frac{1}{2\pi} dd^c \log |\sigma \circ f|_h = -f^*(c_h) + f^* \Delta_\sigma.$$

By integrating this formula twice, first over the disk D_ρ and then with respect to $d\rho/\rho$ from 0 to r, we shall obtain the so-called first main theorem.

For $r < R$, we set

$$(8.4.4) \qquad t(r) = \int_{D_r} f^* c_h.$$

If c_h is positive definite, it can be used as a Kähler metric on X. Then $t(r)$ represents the area of $f(D_r)$ in X. As in (3.B.2), the **order function** or **characteristic function** of f is defined to be

$$(8.4.5) \qquad T(r) = \int_0^r t(\rho) \frac{d\rho}{\rho} = \int_0^r \frac{d\rho}{\rho} \int_{D_\rho} f^* c_h.$$

Since we will be considering here only one fixed map f, we write $T(r)$ instead of $T(r, f)$. To see that the integral in (8.4.5) is well-defined, we note that since $f^* c_h$ is of the form

$$f^* c_h = i\lambda dz \wedge d\bar{z} = 2\lambda r dr \wedge d\theta$$

with respect to the coordinate $z = re^{i\theta}$ in D_R, we have

$$t(\rho) = \int_{D_\rho} f^* c_h = \rho^2 \psi,$$

where ψ a smooth function. Hence, $T(r)$ is well defined.

Differentiating (8.4.5) yields

$$\frac{dT}{d\log r} = r\frac{dT}{dr} = \int_{D_r} f^* c_h = 2 \int_0^r \left(\int_0^{2\pi} \lambda(\rho e^{i\theta}) d\theta \right) \rho d\rho.$$

Differentiating once more, we obtain

$$(8.4.6) \qquad \frac{d^2 T}{(d\log r)^2} = 2r^2 \int_0^{2\pi} \lambda(re^{i\theta}) d\theta.$$

For each $\alpha \in D_R$, let

$$\nu(\alpha, \sigma) = \text{the order of zero of } f^*\sigma \text{ at } \alpha;$$

it is the multiplicity with which f maps α into the divisor Δ_σ. Set

$$n(r, \sigma) = \sum_{\alpha \in \bar{D}_r} \nu(\alpha, \sigma).$$

Thus, $n(r, \sigma)$ is the number (with multiplicity counted) of points of $\bar{D}_r$ that are mapped into the divisor Δ_σ. As in (3.B.5), we define the **counting function** by

$$(8.4.7) \qquad N(r, \sigma) = \int_0^r (n(\rho, \sigma) - \nu(0, \sigma))\frac{d\rho}{\rho} + \nu(0, \sigma)\log r.$$

If $\nu(0, \sigma) = 0$, i.e., if $\sigma(f(0)) \neq 0$, then (8.4.7) reduces to

$$N(r, \sigma) = \int_0^r n(\rho, \sigma)\frac{d\rho}{\rho}.$$

If $\nu(0, \sigma) \neq 0$ and if $r_0 > 0$ is sufficiently small so that $n(r_0, \sigma) = \nu(0, \sigma)$, (i.e., $f^*\sigma$ does not vanish in D_{r_0} except at the origin), then

$$N(r, \sigma) = \int_{r_0}^r (n(\rho, \sigma) - \nu(0, \sigma))\frac{d\rho}{\rho} + \nu(0, \sigma)\log r.$$

Hence, in all cases we have

$$(8.4.8) \qquad N(r, \sigma) = \int_{r_0}^r n(\rho, \sigma)\frac{d\rho}{\rho} + \nu(0, \sigma)\log r_0.$$

As in (3.B.6), the integral in the definition of the counting function $N(r, \sigma)$ can be written as a finite sum. Namely, if $\alpha_1, \ldots, \alpha_k$ are the zeros of σ in $\bar{D}_r - \{0\}$ with multiplicities $\nu(\alpha_1, \sigma), \ldots, \nu(\alpha_k, \sigma)$, then

$$(8.4.9) \qquad N(r, \sigma) = \sum_{0 < |\alpha_i| \leq r} \nu(\alpha_i, \sigma)\log\frac{r}{|\alpha_i|} + \nu(0, \sigma)\log r.$$

We set

$$(8.4.10) \qquad u_\sigma = \log\frac{1}{|\sigma|_h}.$$

Then u_σ is a function smooth outside the divisor Δ_σ and bounded below by $-\log\max_x |\sigma(x)|_h$. Moreover,

$$(8.4.11) \qquad \frac{1}{2\pi}dd^c u_\sigma = \frac{1}{2\pi}dd^c \log h^{-1/2} = c_h.$$

Finally, we define the **proximity function** of σ by

$$(8.4.12) \qquad m(r, \sigma) = \frac{1}{2\pi}\int_0^{2\pi}(f^*u_\sigma)(re^{i\theta})d\theta,$$

provided $f(\partial D_r) \cap \Delta_\sigma = \emptyset$.

Having defined all these functions, we integrate (8.4.3) over D_r. Then we obtain the **non-integrated first main theorem**:

$$(8.4.13) \qquad \frac{1}{2\pi}\int_{\partial D_r} d^c \log|\sigma \circ f|_h = -t(r) + n(r, \sigma).$$

In general, if v is a function of $z = x + iy = re^{i\theta}$, then

$$(8.4.14) \qquad d^c v = \frac{\partial v}{\partial x} dy - \frac{\partial v}{\partial y} dx = \frac{\partial v}{\partial r} r d\theta - \frac{\partial v}{\partial \theta} \frac{dr}{r}.$$

Apply (8.4.14) to $v = \log |\sigma \circ f|_h$ to calculate the left hand side of (8.4.13). Since $dr = 0$ on ∂D_r, we have

$$(8.4.15) \qquad \int_{\partial D_r} d^c \log |\sigma \circ f|_h = \int_0^{2\pi} \frac{\partial \log |\sigma(f(re^{i\theta}))|_h}{\partial r} r d\theta.$$

We integrate (8.4.13) with respect to dr/r from r_0 to r. Apply (8.4.15) to the left hand side and (8.4.8) to the right hand side. Then

$$(8.4.16) \qquad N(r, \sigma) + m(r, \sigma) = \int_{r_0}^{r} t(\rho) \frac{d\rho}{\rho} + v(0, \sigma) \log r_0 + m(r_0, \sigma).$$

In order to let $r_0 \to 0$, we have to combine the last two terms on the right hand side. Let U_α be a coordinate neighborhood of $f(0)$. As in the proof of (8.4.2) we have

$$f^* |\sigma|_h = f^* \sqrt{h_\alpha} \cdot |f^* \sigma_\alpha|$$

in a neighborhood of $0 \in D_r$. By the very definition of $v(0, \sigma)$, we have

$$f^* \sigma_\alpha = z^{v(0,\sigma)} \varphi_\sigma,$$

where φ is nonzero around 0. Setting $\psi_\sigma = \sqrt{f^* h_\alpha} \cdot |\varphi_\sigma|$, we write

$$f^* |\sigma|_h = |z|^{v(0,\sigma)} \psi_\sigma,$$

where ψ_σ is a smooth function positive in a neighborhood of $0 \in D_r$. Hence,

$$\begin{aligned} m(r_0, \sigma) &= -\frac{1}{2\pi} \int_0^{2\pi} \log |\sigma(f(r_0 e^{i\theta}))|_h d\theta \\ &= -n(0, \sigma) \log r_0 - \frac{1}{2\pi} \int_0^{2\pi} \log \psi_\sigma(r_0 e^{i\theta}) d\theta. \end{aligned}$$

Substituting this into (8.4.16) and letting $r_0 \to 0$, we obtain the **first main theorem** of Nevanlinna:

$$(8.4.17) \qquad N(r, \sigma) + m(r, \sigma) = T(r) - \log \psi_\sigma(0).$$

If we replace the section σ by a section $c\sigma$, $c \in \mathbf{C}^*$ or the Hermitian inner product h by ah, $a > 0$, then neither $N(r, \sigma)$ nor $T(r)$ is affected since both σ and $c\sigma$ define the same divisor and since $c_{ah} = c_h$. On the other hand,

$$m(r, c\sigma) = m(r, \sigma) - \log |c| \quad \text{and} \quad \log \psi_{c\sigma} = \log \psi_\sigma + \log |c|.$$

Therefore, it suffices to consider the formula (8.4.17) only for sections σ normalized by the condition

(8.4.18) $$\max_{x \in X} |\sigma(x)|_h = 1.$$

Under this restriction, we have

(8.4.19) $$m(r, \sigma) \geq 0.$$

Since $\psi_\sigma(0)$ is a constant independent of r, we set

$$C_\sigma = -\log \psi_\sigma(0).$$

Then from the first main theorem we obtain **Nevanlinna's inequality**:

(8.4.20) $$N(r, \sigma) \leq T(r) + C_\sigma, \qquad r < R.$$

As an example, we consider the projective space $P_n \mathbf{C}$ with the hyperplane line bundle $H = \mathcal{O}(1)$. By definition, $P_n \mathbf{C}$ is the space of lines through the origin of $\mathbf{C}^{n+1}$ so that $P_n \mathbf{C} = (\mathbf{C}^{n+1} - \{0\})/\mathbf{C}^*$. By attaching to each point of $P_n \mathbf{C}$ the line it represents, we obtain the dual bundle $H^{-1} = \mathcal{O}(-1)$, which is called the **tautological line bundle**. Since H^{-1} minus its zero section is naturally identified with $\mathbf{C}^{n+1} - \{0\}$, the natural inner product in $\mathbf{C}^{n+1}$ induces a Hermitian inner product in H^{-1} and hence in H. We denote this inner product in H by h. If $\mathbf{z} = (z^0, \ldots, z^n)$ is the natural homogeneous coordinate for $P_n \mathbf{C}$, then the Chern form for (H, h) is given by

(8.4.21) $$c_h = \frac{1}{2\pi} dd^c \log \|\mathbf{z}\|.$$

In the remainder of this section, we denote the point of $P_n \mathbf{C}$ with homogeneous coordinates $\mathbf{z} = (z^0, \ldots, z^n)$ by the same letter $\mathbf{z}$; from the context it should be clear whether $\mathbf{z}$ denotes a point of $\mathbf{C}^{n+1}$ or it represents a point of $P_n \mathbf{C}$.

The space $\Gamma(H)$ of holomorphic sections of H is naturally isomorphic to the dual space of $\mathbf{C}^{n+1}$, which is identified with $\mathbf{C}^{n+1}$ by the natural inner product. To each point $\mathbf{a} = (a^0, \ldots, a^n) \in \mathbf{C}^{n+1}$ we associate a section $\sigma_{\mathbf{a}}$ of H as follows. Since $\sigma_{\mathbf{a}}(\mathbf{z})$ should be an element of the fibre $H_{\mathbf{z}}$, i.e., a linear functional on $H_{\mathbf{z}}^{-1}$, considering $\mathbf{z}$ as a point of $H_{\mathbf{z}}^{-1} \subset \mathbf{C}^{n+1}$ we define $\sigma_{\mathbf{a}}(\mathbf{z})$ by

(8.4.22) $$(\sigma_{\mathbf{a}}(\mathbf{z}), \mathbf{z}) = \sum \bar{a}^i z^i,$$

where $(\cdot, \cdot)$ on the left hand side is the dual pairing $H_{\mathbf{z}} \times H_{\mathbf{z}}^{-1} \to \mathbf{C}$. Hence, the zeros of the section $\sigma_{\mathbf{a}}$ is the hyperplane defined by

$$\langle \mathbf{z}, \mathbf{a} \rangle = \sum \bar{a}^i z^i = 0.$$

Let

(8.4.23) $$f = (f_0, \ldots, f_n)$$

be a reduced representation of a nondegenerate holomorphic curve $f: D_R \to P_n \mathbf{C}$. For each $\mathbf{a} \in \mathbf{C}^{n+1}$ and $\alpha \in D_R$, let

$$(8.4.24) \qquad v(\alpha, \mathbf{a}) = \text{the order of zero of } \sum \bar{a}^i f_i \text{ at } \alpha,$$

and

$$(8.4.25) \qquad n(r, \mathbf{a}) = \sum_{\alpha \in \bar{D}_r} v(\alpha, \mathbf{a}).$$

Thus $n(r, \mathbf{a})$ is the number (multiplicity counted) of points of $\bar{D}_r$ that are mapped by f into the hyperplane $\langle \mathbf{z}, \mathbf{a} \rangle = 0$. We define the **counting function** $N(r, \mathbf{a})$ in the same way as (8.4.7).

Before constructing the proximity function, we note that since $\sigma_{\mathbf{a}}(\mathbf{z})$ lies in the 1-dimensional fibre $H_{\mathbf{z}}$, its length $|\sigma_{\mathbf{a}}(\mathbf{z})|_h$ is given by $|(\sigma_{\mathbf{a}}(\mathbf{z}), \mathbf{z})| = |\sigma_{\mathbf{a}}(\mathbf{z})|_h \cdot \|\mathbf{z}\|$. Hence, by (8.4.22)

$$|\sigma_{\mathbf{a}}(\mathbf{z})|_h \cdot \|\mathbf{z}\| = |\langle \mathbf{z}, \mathbf{a} \rangle|.$$

Now, for each nonzero $\mathbf{a} \in \mathbf{C}^{n+1}$, we define a function $u_{\mathbf{a}}$ on $P_n \mathbf{C}$ by setting (see (8.4.10))

$$(8.4.26) \qquad u_{\mathbf{a}} = \log \frac{1}{|\sigma_{\mathbf{a}}|_h} = \log \frac{\|\mathbf{z}\|}{|\langle \mathbf{z}, \mathbf{a} \rangle|}.$$

Then $u_{\mathbf{a}}$ is a function smooth outside the hyperplane $\langle \mathbf{z}, \mathbf{a} \rangle = 0$ and bounded below by $-\log |\mathbf{a}|$. Since $\langle \mathbf{z}, \mathbf{a} \rangle$ is holomorphic, we have

$$(8.4.27) \qquad c_h = \frac{1}{2\pi} dd^c u_{\mathbf{a}} = \frac{1}{2\pi} dd^c \log \|\mathbf{z}\|.$$

Then the proximity function defined by (8.4.12) is given in this case by

$$(8.4.28) \qquad m(r, \mathbf{a}) = \frac{1}{2\pi} \int_0^{2\pi} \log \frac{\|f(re^{i\theta})\|}{|\langle f(re^{i\theta}), \mathbf{a} \rangle|} d\theta,$$

provided $\langle f(re^{i\theta}), \mathbf{a} \rangle \neq 0$ for all θ. This agrees with definition (3.B.3). Then (8.4.17) reads as follows:

$$(8.4.29) \qquad N(r, \mathbf{a}) + m(r, \mathbf{a}) = T(r) - \log \psi_{\mathbf{a}}(0),$$

where $\psi_{\mathbf{a}}(0)$ is an unimportant constant. Set $C_a = -\log \psi_a(0)$. As a special case of (8.4.20) we have

$$(8.4.30) \qquad N(r, \mathbf{a}) \leq T(r) + C_{\mathbf{a}}, \qquad r < R.$$

Now we simply apply what we did above to associated curves. Let F_k be the k-th associated curve of a holomorphic curve $f: D_R \to P_n \mathbf{C}$.

Let ω_k be the pull-back of the Kähler form Φ_k of the Fubini-Study metric by F_k, see (8.2.10). Following (8.4.4) and (8.4.5) we define

$$(8.4.31) \qquad t_k(r) = \int_{D_r} \omega_k,$$

and the **k-th characteristic function**

$$(8.4.32) \qquad T_k(r) = \int_0^r t_k(\rho) \frac{d\rho}{\rho} = \int_0^r \frac{d\rho}{\rho} \int_{D_\rho} \omega_k.$$

If we write

$$\omega_k = i\lambda_k dz \wedge d\bar{z},$$

then (8.4.6) yields

$$(8.4.33) \qquad \frac{d^2 T_k}{(d \log r)^2} = 2r^2 \int_0^{2\pi} \lambda_k(re^{i\theta}) d\theta.$$

Given a decomposable $(k+1)$-vector A, we define $n_k(r, A)$ to be the number of points of $\bar{D}_r$ that are mapped by F_k into the hyperplane $\langle Z, A \rangle = 0$ in $P_{n(k)}\mathbf{C}$. Let $\tilde{F}_k$ be a reduced representation of F_k, see Section 2. Then

$$(8.4.34) \qquad n_k(r, A) = \text{the number of zeros of } \langle \tilde{F}_k, A \rangle \text{ in } \bar{D}_r,$$

where the zeros are counted with multiplicity. The **k-th counting function** $N_k(r, A)$ is defined in the same way as (8.4.8):

$$(8.4.35) \qquad N_k(r, A) = \int_0^r (n_k(\rho, A) - n_k(0, A)) \frac{d\rho}{\rho} + n_k(0, A) \log r.$$

Following (8.4.28) we define the **k-th proximity function** of A by

$$(8.4.36) \qquad m_k(r, A) = \frac{1}{2\pi} \int_0^{2\pi} \log \frac{|\tilde{F}_k(re^{i\theta})|}{|\langle \tilde{F}_k(re^{i\theta}), A \rangle|} d\theta.$$

Then from (8.4.29) we obtain the **first main theorem for the k-th associated curve** F_k:

$$(8.4.37) \qquad N_k(r, A) + m_k(r, A) = T_k(r) - \log \psi_A(0)$$

for any decomposable $(k+1)$-vector A. From (8.4.20) we have also

$$(8.4.38) \qquad N_k(r, A) \leq T_k(r) + C_A, \qquad r < R.$$

where $C_A = -\log \psi_A(0)$.

5 Second Main Theorem

Let X be a complex manifold with a Hermitian metric ds^2 and the associated Kähler form Φ. Fix a holomorphic mapping from a disc of radius R into X:

$$f: D_R \to X.$$

Let $\alpha \in D_R$. If $w^1, \ldots, w^n$ is a local coordinate about the point $f(\alpha) \in X$, then f is given locally by power series

$$(8.5.1) \qquad w^i = a^i_{s+1}(z-\alpha)^{s+1} + a^i_{s+2}(z-\alpha)^{s+2} + \dots, \qquad i = 1, \dots, n$$

with $(a^1_{s+1}, \dots, a^n_{s+1}) \neq (0, \dots, 0)$. Clearly, $s > 0$ if and only if df vanishes at α. Such a point α is called a **stationary point** of f with **stationary index** s. A more useful way of determining the index s is to pull-back the Kähler form Φ by f. We set $\omega = f^*\Phi$. Then

$$(8.5.2) \qquad \omega = f^*\Phi = i\lambda dz \wedge d\bar{z} \quad \text{with} \quad \lambda = |z-\alpha|^{2s}\mu, \quad \mu(\alpha) \neq 0.$$

For $r < R$, there are only finitely many stationary points of f in D_r, say $\alpha_1, \dots, \alpha_m$. Let the stationary indices at these points be $s_1, \dots, s_m$. We assume that there are no stationary points on the boundary ∂D_r. We call the sum

$$(8.5.3) \qquad s(r) = \sum_{j=1}^{m} s_j$$

the **stationary index** of f in D_r.

As in (8.2.13) we consider the Ricci form $\mathrm{Ric}(\omega) = -dd^c \log \lambda$ of ω; it is defined everywhere except at the stationary points. Since, by (8.5.2),

$$dd^c \log \lambda = dd^c \log \mu,$$

$\mathrm{Ric}(\omega)$ can be extended to all of D_R.

If we consider $dd^c \log \lambda$ as a current, then the Poincaré-Lelong formula (8.4.1) gives

$$(8.5.4) \qquad \frac{1}{4\pi} dd^c \log \lambda = -\frac{1}{4\pi} \mathrm{Ric}(\omega) + \Delta_s \quad \text{in} \quad D_r,$$

where $\Delta_s = \sum_{j=1}^{m} s_j \alpha_j$ is the divisor of stationary points of f in D_r.

We shall obtain the second main theorem by integrating (8.5.4) twice, first on the disc D_ρ and then with respect to $d\rho/\rho$ from 0 to r.

Integrating (8.5.4) on D_r yields

$$(8.5.5) \qquad \frac{1}{4\pi} \int_{\partial D_r} d^c \log \lambda = -\frac{1}{4\pi} \int_{D_r} \mathrm{Ric}(\omega) + s(r).$$

Applying (8.4.14) to $v = \log \lambda$, we may rewrite (8.5.5) as follows:

$$(8.5.6) \qquad \frac{1}{4\pi} \int_0^{2\pi} \frac{\partial \log \lambda(re^{i\theta})}{\partial r} r d\theta = -\frac{1}{4\pi} \int_{D_r} \mathrm{Ric}(\omega) + s(r),$$

which may be called the **non-integrated form of the second main theorem**.

Imitating the definition of $N(r, a)$ in (8.4.7) we define $S(r)$ as follows:

$$(8.5.7) \qquad S(r) = \int_0^r (s(\rho) - s(0)) \frac{d\rho}{\rho} + s(0) \log r.$$

We note that if the stationary index $s(0)$ of f at 0 vanishes, then

$$S(r) = \int_0^r s(\rho)\frac{d\rho}{\rho}.$$

If $s(0) \neq 0$ and if $r_0 > 0$ is sufficiently small so that there are no stationary points of f in D_{r_0} except the origin, then

$$S(r) = \int_{r_0}^r (s(\rho) - s(0))\frac{d\rho}{\rho} + s(0)\log r.$$

Hence, in all cases we have

$$(8.5.8) \qquad S(r) = \int_{r_0}^r s(\rho)\frac{d\rho}{\rho} + s(0)\log r_0.$$

Integrating (8.5.6) with respect to dr/r from r_0 to r yields

$$\frac{1}{4\pi}\int_0^{2\pi}(\log\lambda(re^{i\theta}) - \log\lambda(r_0 e^{i\theta}))d\theta$$

$$= -\frac{1}{4\pi}\int_{r_0}^r \frac{d\rho}{\rho}\int_{D_\rho}\mathrm{Ric}(\omega) + S(r) - s(0)\log r_0.$$

Before we let $r_0 \to 0$, we have to eliminate the term $s(0)\log r_0$ in the above equation. For this purpose, we write, as in (8.5.2), in a neighborhood of the origin:

$$\omega = i\lambda dz \wedge d\bar{z}, \quad \text{with} \quad \lambda(z) = |z|^{2s(0)}\mu(z) = r^{2s(0)}\mu(z),$$

where $s(0)$ is the stationary index of f at 0. Then

$$\frac{1}{4\pi}\int_0^{2\pi}\log\lambda(r_0 e^{i\theta})d\theta = s(0)\log r_0 + \frac{1}{4\pi}\int_0^{2\pi}\log\mu(r_0 e^{i\theta})d\theta.$$

Hence,

$$\frac{1}{4\pi}\int_0^{2\pi}\log\lambda(re^{i\theta})d\theta - \frac{1}{4\pi}\int_0^{2\pi}\log\mu(r_0 e^{i\theta})d\theta$$

$$= -\frac{1}{4\pi}\int_{r_0}^r \frac{d\rho}{\rho}\int_{D_\rho}\mathrm{Ric}(\omega) + S(r).$$

Now, letting $r_0 \to 0$ we obtain

(8.5.9) Integrated Second Main Theorem.

$$\frac{1}{4\pi}\int_0^{2\pi}\log\lambda(re^{i\theta})d\theta = -\frac{1}{4\pi}\int_0^r \frac{d\rho}{\rho}\int_{D_\rho}\mathrm{Ric}(\omega) + S(r) + \frac{1}{2}\log\mu(0).$$

We apply the following lemma on the **concavity of the logarithm** to the left hand side of (8.5.9).

(8.5.10) Lemma. *Let E be a measure space with a positive measure ν with $\nu(E) < \infty$. If f is a nonnegative ν-integrable function on E, then*

$$\frac{1}{v(E)} \int_E (\log f) dv \le \log\left(\frac{1}{v(E)} \int_E f dv\right).$$

Proof. We may assume that $v(E) = 1$. Set $c = \int_E f dv$. Then

$$\begin{aligned}
\int_E (\log f) dv &= \int_E (\log c) dv + \int_E \log\left(1 + \frac{f-c}{c}\right) dv \\
&\le \log c + \int_E \frac{f-c}{c} dv = \log c,
\end{aligned}$$

where the inequality is a consequence of the general inequality $\log(1+t) \le t$, $(t \ge -1)$, applied to $t = (f-c)/c$. $\qquad\square$

By this lemma, we have

$$(8.5.11) \qquad \frac{1}{2\pi} \int_0^{2\pi} \log \lambda(re^{i\theta}) d\theta \le \log\left(\frac{1}{2\pi} \int_0^{2\pi} \lambda(re^{i\theta}) d\theta\right).$$

So if we denote twice the right hand side of (8.5.9) by $\eta(r)$, i.e., if we set

$$(8.5.12) \qquad \eta(r) = -\frac{1}{2\pi} \int_0^r \frac{d\rho}{\rho} \int_{D_\rho} \mathrm{Ric}(\omega) + 2S(r) + \log \mu(0),$$

then

$$\eta(r) \le \log\left(\frac{1}{2\pi} \int_0^{2\pi} \lambda(re^{i\theta}) d\theta\right).$$

Hence,

$$(8.5.13) \qquad e^{\eta(r)} \le \frac{1}{2\pi} \int_0^{2\pi} \lambda(re^{i\theta}) d\theta.$$

Integrating (8.5.13) yields:

$$\begin{aligned}
\int_0^r e^{\eta(\rho)} \rho d\rho &\le \frac{1}{4\pi} \int_0^r 2\rho d\rho \int_0^{2\pi} \lambda(\rho e^{i\theta}) d\theta \\
&= \frac{1}{4\pi} \int_{D_r} i\lambda dz \wedge d\bar{z} = \frac{1}{4\pi} \int_{D_r} \omega.
\end{aligned}$$

Assume that, as in Section 4, the Kähler form Φ is the Chern form c_h of an ample Hermitain line bundle (L, h) over X. Then (see (8.4.4))

$$\int_{D_r} \omega = \int_{D_r} f^*\Phi = t(r),$$

and the inequality above can be written as follows:

$$(8.5.14) \qquad \int_0^r e^{\eta(\rho)} \rho d\rho \le \frac{1}{4\pi} t(r).$$

Integrating (8.5.14) with respect to r/dr gives

$$(8.5.15) \qquad \int_0^r \frac{ds}{s} \int_0^s e^{\eta(\rho)} \rho \, d\rho \le \frac{1}{4\pi} T(r).$$

We shall now apply what we have done to a holomorphic map $f: D_R \to P_n\mathbf{C}$ and its associated curves F_k. Replace f and ω in (8.5.9) by the k-th associated curve F_k and ω_k. Then substituting (8.2.15) into $\mathrm{Ric}(\omega_k)$ yields the following:

(8.5.16) **Integrated Second Main Theorem for Associated Curves**. *We denote functions T, S, λ, and μ for the k-th associated curve F_k by T_k, S_k, λ_k, and μ_k. Then*

$$\frac{1}{4\pi} \int_0^{2\pi} \log \lambda_k(re^{i\theta}) d\theta = T_{k-1}(r) - 2T_k(r) + T_{k+1}(r) + S_k(r) + \frac{1}{2} \log \mu_k(0).$$

Similarly, (8.5.15), applied to the k-th associated curve, yields the following:

$$(8.5.17) \qquad \int_0^r \frac{ds}{s} \int_0^s e^{\eta_k(\rho)} \rho \, d\rho \le \frac{1}{4\pi} T_k(r),$$

where

$$\eta_k(\rho) = 2[T_{k-1}(\rho) + T_{k+1}(\rho) - 2T_k(\rho) + S_k(\rho)] + \log \mu_k(0).$$

Let $\mathbf{a}_0, \ldots, \mathbf{a}_q$ be $q+1$ vectors in general position in $\mathbf{C}^{n+1}$. We consider $\hat{\omega}_k = i\hat{\lambda}_k dz \wedge d\bar{z}$ defined by (8.3.10), where

$$(8.5.18) \qquad \hat{\lambda}_k = C_k \lambda_k \prod_{j=0}^q \left(\frac{\varphi_{k+1}(\mathbf{a}_j)}{\varphi_k(\mathbf{a}_j)(\log(c/\varphi_k(\mathbf{a}_j)))^2} \right)^{1/n-k}.$$

Apply the Poincaré-Lelong formula (8.5.4) to $\hat{\lambda}_k$ to obtain the following equation of currents:

(8.5.19) **Lemma.** *We have the following equation of divisors in the disc D_r:*

$$\frac{1}{4\pi} dd^c \log \hat{\lambda}_k = -\frac{1}{4\pi} \mathrm{Ric}(\hat{\omega}_k) + s_k + \frac{1}{n-k} \sum_{j=0}^q \{[\varphi_{k+1}(\mathbf{a}_j)] - [\varphi_k(\mathbf{a}_j)]\},$$

where s_k is the divisor of stationary points of the associated curve F_k in D_r, and $[\varphi_{k+1}(\mathbf{a}_j)]$ and $[\varphi_k(\mathbf{a}_j)]$ denote the divisors defined by the contact functions $\varphi_{k+1}(\mathbf{a}_j)$ and $\varphi_k(\mathbf{a}_j)$.

Proof. From the definition of the Ricci form $\mathrm{Ric}(\hat{\omega}_k)$, as current we have

$$dd^c \log \hat{\lambda}_k = -\mathrm{Ric}(\hat{\omega}_k) + \{\text{singular part}\}.$$

We have to determine the singular part in the equation above. According to (8.5.18), this singular part is equal to the singular part of the current

$$dd^c \log \lambda_k + \frac{1}{n-k} \left[dd^c \log \varphi_{k+1}(\mathbf{a}_j) - dd^c \log \varphi_k(\mathbf{a}_j) - dd^c \log \left(\log \frac{c}{\varphi_k(\mathbf{a}_j)} \right)^2 \right].$$

By applying (8.5.4) to λ_k we see that $4\pi s_k$ is the singular part of the current $dd^c \log \lambda_k$.

From the definition (8.3.1) of the contact function $\varphi_k(\mathbf{a}_j)$ it follows that $\varphi_k(\mathbf{a}_j)$ is of the form

$$\varphi_k(\mathbf{a}_j) = |z - \alpha|^{2m}\mu \qquad \text{at} \quad \alpha \in D_r,$$

where $m = m(\alpha)$ is the order of zero of $\hat{F}_k(z) \vee \mathbf{a}_j$ at α, and μ is smooth and $\mu(\alpha) \neq 0$. In a neighborhood of α we have the following equation of currents (see (8.4.1)):

$$dd^c \log \varphi_k(\mathbf{a}_j) = dd^c \log \mu + 4\pi[m\alpha],$$

so that the C^∞ form $dd^c \log \mu$ and the divisor $4\pi \sum_\alpha m\alpha$ represents the smooth part and the singular part of the current $dd^c \log \varphi_k(\mathbf{a}_j)$. By definition,

$$[\varphi_k(\mathbf{a}_j)] = \sum_\alpha m\alpha.$$

It suffices to show that the term $\log^2(c/\varphi_k(\mathbf{a}_j))$ does not contribute to the singular part of the current $dd^c \log \hat{\lambda}_k$, i.e., that $dd^c \log \log^2(c/\varphi_k(\mathbf{a}_j))$ in the sense of currents is the same as $dd^c \log \log^2(c/\varphi_k(\mathbf{a}_j))$ in the sense of differential forms. This amounts to showing that, for every point $\alpha \in D_r$ and for every test function ψ around α, we have

$$\lim_{\varepsilon \to 0} \int_{|z-\alpha| \le \varepsilon} \psi \cdot dd^c \log\left(\log\left(\frac{c}{\varphi_k(\mathbf{a}_j)}\right)^2\right)^2 = 0.$$

Set $g = c/\varphi_k(\mathbf{a}_j)$ for simplicity, and write

$$g = \left(\frac{c}{\varphi_k(\mathbf{a}_j)}\right)^2 = |z - \alpha|^{-2m}h, \qquad m \ge 0,$$

where h is smooth and $h(\alpha) \neq 0$. Then

$$\frac{d'g}{g} = d'\log g = \frac{d'h}{h} - \frac{mdz}{z - \alpha},$$

$$\frac{d''g}{g} = d''\log g = \frac{d''h}{h} - \frac{md\bar{z}}{\bar{z} - \bar{\alpha}},$$

$$d'\left(\frac{d''g}{g}\right) = \frac{hd'd''h - d'h \wedge d''h}{h^2}.$$

We substitute these into the following formula for $dd^c \log(\log g)^2$ as differential forms:

$$dd^c \log(\log g)^2 = 4id'\left(\frac{1}{\log g}\frac{d''g}{g}\right)$$

$$= \frac{4i}{\log g}d'\left(\frac{d''g}{g}\right) - \frac{4i}{(\log g)^2}\left(\frac{d'g}{g} \wedge \frac{d''g}{g}\right).$$

Noting that $\log g$ goes to infinity at α and ignoring the terms that are bounded in a neighborhood of α, we obtain

$$dd^c \log(\log g)^2 = \frac{4i}{(\log g)^2}\left(\frac{d'h}{h}\wedge\frac{md\bar{z}}{\bar{z}-\bar{\alpha}} + \frac{mdz}{z-\alpha}\wedge\frac{d''h}{h} - \frac{m^2 dz\wedge d\bar{z}}{|z-\alpha|^2}\right) + \ldots,$$

where the dots indicate terms that are bounded in a neighborhood of α. When those terms are integrated against ψ over the disc $|z-\alpha|\le\varepsilon$, they approach zero as $\varepsilon\to 0$. Therefore, it suffices to examine the three terms inside the parenthesis on the right. In terms of the polar coordinate $\rho e^{i\theta} = z-\alpha$ around α, each of the three terms is a product of $d\log\rho\wedge d\theta$ with a smooth function around α. On the other hand, the common denominator $(\log g)^2$ is asymptotically equal to $4m^2(\log\rho)^2$. Therefore, the problem is to show

$$\lim_{\varepsilon\to 0}\int_0^{2\pi} d\theta\int_0^{\varepsilon}\eta\frac{d\log\rho}{(\log\rho)^2} = 0$$

for every smooth function η. But this follows from

$$\lim_{\varepsilon\to 0}\int_0^{\varepsilon}\frac{d\log\rho}{(\log\rho)^2} = \lim_{\varepsilon\to 0}\frac{1}{\log 1/\varepsilon} = 0.$$

$\square$

Integrating (8.5.19) over D_r and using the argument used in establishing (8.5.6) we obtain

(8.5.20) Lemma. *As functions of r we have the following equality:*

$$\frac{1}{4\pi}\int_0^{2\pi}\frac{\partial\log\hat{\lambda}_k}{\partial r}r\,d\theta$$

$$= -\frac{1}{4\pi}\int_{D_r}\mathrm{Ric}(\hat{\omega}_k) + s_k(r) + \frac{1}{n-k}\sum_{j=0}^{q}(n_{k+1}(r,\mathbf{a}_j) - n_k(r,\mathbf{a}_j)),$$

where $s_k(r)$ is the degree of the divisor s_k of stationary points for F_k in D_r, and $n_k(r,\mathbf{a}_j)$ is the degree of the divisor $[\varphi_k(\mathbf{a}_j)]$ in $\bar{D}_r$.

Integrating (8.5.20) with respect to dr/r and using the argument used in proving (8.5.9) we obtain

(8.5.21) Lemma.

$$\frac{1}{4\pi}\int_0^{2\pi}\log\hat{\lambda}_k(re^{i\theta})d\theta = -\frac{1}{4\pi}\int_0^r\frac{d\rho}{\rho}\int_{D_\rho}\mathrm{Ric}(\hat{\omega}_k) + S_k(r)$$

$$+\frac{1}{n-k}\sum_{j=0}^{q}(N_{k+1}(r,\mathbf{a}_j) - N_k(r,\mathbf{a}_j)) + \frac{1}{2}\hat{\mu}_k(0),$$

where

$$S_k(r) = \int_0^r (s_k(\rho) - s_k(0))\frac{d\rho}{\rho} + s_k(0)\log r,$$

$$N_k(r, a_j) = \int_0^r (n_k(\rho, a_j) - n_k(0, a_j))\frac{d\rho}{\rho} + n_k(0, a_j)\log r,$$

and $\hat{\mu}_k$ is defined by $\hat{\lambda}_k = |z|^{2s_k(0)}\hat{\mu}_k$.

We note that since $\varphi_n(\mathbf{a}_j) \equiv 1$, we have

(8.5.22) $$\qquad n_n(r, \mathbf{a}_j) = 0, \qquad N_n(r, \mathbf{a}_j) = 0.$$

On the other hand, the degree of the divisor $[\varphi_0(\mathbf{a}_j)]$ in $\bar{D}_r$ is $n(r, \mathbf{a}_j)$, in other words,

(8.5.23) $$\qquad n_0(r, \mathbf{a}_j) = n(r, \mathbf{a}_j), \qquad N_0(r, \mathbf{a}_j) = N(r, \mathbf{a}_j).$$

Multiplying (8.5.21) by $n - k$ and summing over k, we obtain

$$\sum_{j=0}^q N(r, \mathbf{a}_j) = -\sum_{k=0}^{n-1} \frac{n-k}{4\pi}\left(\int_0^{2\pi} \log \hat{\lambda}_k(re^{i\theta})d\theta + \int_0^r \frac{d\rho}{\rho}\int_{D_\rho} \mathrm{Ric}(\hat{\omega}_k)\right)$$
$$+ \sum_{k=0}^{n-1}(n-k)\left(+S_k(r) + \frac{1}{2}\hat{\mu}_k(0)\right).$$

Integrating the inequality in (8.3.12) we obtain

$$-\sum_{k=0}^{n-1}\frac{n-k}{4\pi}\int_0^r \frac{d\rho}{\rho}\int_{D_\rho}\mathrm{Ric}(\hat{\omega}_k) \geq \sum_{k=0}^{n-1}\hat{T}_k(r) + (q-n)T(r) - \varepsilon\sum_{k=0}^{n-1}T_k(r),$$

where

$$\hat{t}_k(r) = \int_{D_r}\hat{\omega}_k, \qquad \hat{T}_k(r) = \int_0^r \hat{t}_k(\rho)\frac{d\rho}{\rho}.$$

Substituing this inequality into the equality above, we have the following inequality:

(8.5.24) **Lemma.** *Let $\mathbf{a}_0, \dots, \mathbf{a}_q$ be $q + 1$ vectors in general position in $\mathbf{C}^{n+1}$. Then*

$$\sum_{j=0}^q N(r, \mathbf{a}_j) \geq (q-n)T(r) - \varepsilon\sum_{k=0}^{n-1}T_k(r) + \sum_{k=0}^{n-1}(n-k)S_k(r)$$
$$+ \sum_{k=0}^{n-1}(\hat{T}_k(r) - \frac{n-k}{4\pi}\int_0^{2\pi}\log\hat{\lambda}_k(re^{i\theta})d\theta + \frac{n-k}{2}\hat{\mu}_k(0)).$$

In the next section we shall consider the asymptotic behavior of this inequality as $r \to \infty$.

6 Entire Curves

So far, we have been dealing with holomorphic mappings of a disk D_R into complex manifolds, in particular, into $P_n\mathbf{C}$. In this section we consider holomorphic mappings from $\mathbf{C}$ into $P_n\mathbf{C}$. We begin with the following lemma due to Nevanlinna.

(8.6.1) **Lemma.** *Fix $r_0 > 0$. Let $A(r)$ be a positive increasing function in the interval $[r_0, \infty)$ with piecewise continuous derivative $A'(r)$, $\alpha(s)$ a positive increasing function in the interval $(0, \infty)$, and $B(r)$ a positive continuous function in $[r_0, \infty)$. Then*

$$A'(r) \leq \alpha(A(r))B(r)$$

outside an open set $E \subset [r_0, \infty)$ such that

$$\int_E B(r)\,dr \leq \int_{A(r_0)}^{\infty} \frac{d\rho}{\alpha(\rho)}.$$

Proof. Let

$$E = \{r \in [r_0, \infty);\ A'(r) > \alpha(A(r))B(r)\}.$$

Then

$$\int_E B(r)\,dr \leq \int_E \frac{A'(r)\,dr}{\alpha(A(r))} \leq \int_{A(r_0)}^{\infty} \frac{d\rho}{\alpha(\rho)}.$$

$\square$

Applying (8.6.1) to $\alpha(s) = s^{1+\varepsilon}$ and $B(r) = 1/r$ yields

(8.6.2) **Corollary.** *Let $A(r)$ be as in (8.6.1), and $\varepsilon > 0$. Then*

$$r \cdot A'(r) \leq A(r)^{1+\varepsilon}$$

outside an open set $E \subset [r_0, \infty)$ such that

$$\int_E \frac{dr}{r} < \infty.$$

We call E in (8.6.1) and (8.6.2) an **exceptional set**. Following H. Weyl, we use the notation

$$P(r) \leq Q(r) \qquad \|$$

to mean that the inequality holds outside an exceptional set E such that

$$\int_E \frac{dr}{r} < \infty.$$

Thus the conclusion of the corollary above may be expressed by

$$r \cdot A'(r) \leq A(r)^{1+\varepsilon} \qquad \|.$$

We note that the symbol $\|$ defined here is a little different from that of Appendix B of Chapter 3, where an exceptional set has a finite Lebesgue measure.

We repeat here the derivation of (8.4.33), i.e., the proof of (8.4.6). Differentiating (8.4.32) yields

$$(8.6.3) \qquad \frac{dT_k}{d\log r} = r\frac{dT_k}{dr} = 2\int_0^r \left(\int_0^{2\pi} \lambda_k(\rho e^{i\theta})d\theta\right)\rho \, d\rho.$$

Differentiating once more, we obtain

$$(8.6.4) \qquad \frac{d^2 T_k}{(d\log r)^2} = 2r^2\int_0^{2\pi} \lambda_k(re^{i\theta})d\theta.$$

Using the convexity of the logarithm (8.5.10), we obtain

$$(8.6.5) \qquad \int_0^{2\pi} \log \lambda_k(re^{i\theta})d\theta \leq \log\left(\int_0^{2\pi} \lambda_k(re^{i\theta}d\theta)\right) = \log\left(\frac{1}{2r^2}\frac{d^2 T_k}{(d\log r)^2}\right).$$

Applying (8.6.2) to $A(r) = T_k(r)$ yields

$$\frac{dT_k}{d\log r} = r\frac{dT_k}{dr} \leq (T_k(r))^{1+\varepsilon} \qquad \|.$$

Applying (8.6.2) to the function $dT_k/d\log r$, which is increasing by (8.6.3), we have

$$\frac{d^2 T_k}{(d\log r)^2} = r\frac{d}{dr}\left(\frac{dT_k}{d\log r}\right) \leq \left(\frac{dT_k}{d\log r}\right)^{1+\varepsilon} \qquad \|.$$

Substituting the preceding inequality into this yields

$$\frac{d^2 T_k}{(d\log r)^2} \leq (T_k(r))^{1+\varepsilon} \qquad \|,$$

where $2\varepsilon + \varepsilon^2$ was replaced by ε. This together with (8.6.5) gives

(8.6.6) **Theorem.** *Given an entire curve* $f: \mathbf{C} \to P_n\mathbf{C}$, *we have for* $k = 0, 1, \ldots, n-1$

$$\int_0^{2\pi} \log \lambda_k(re^{i\theta})d\theta \leq c\log T_k(r) \qquad \|,$$

where c is a constant independent of r.

From the second main theorem (8.5.16) and (8.6.6) we obtain

(8.6.7) **Corollary.** *Given a non-degenerate entire curve* $f: \mathbf{C} \to P_n\mathbf{C}$, *we have for* $k = 0, 1, \ldots, n-1$

$$T_{k-1}(r) - 2T_k(r) + T_{k+1}(r) \leq c\log T_k(r) \qquad \|.$$

We set

$$\overline{T}(r) = \max_{0\leq k\leq n-1} T_k(r).$$

Then by (8.6.7) we have for $k = 0, 1, \ldots, n-1$

$$(8.6.8)_k \qquad T_{k-1}(r) - 2T_k(r) + T_{k+1}(r) \leq c\log\overline{T}(r) \qquad \|.$$

From (8.6.8) we conclude by induction that for $k = 1, \ldots, n - 1$

$(8.6.9)_k \qquad\qquad kT_k(r) - (k+1)T_{k-1}(r) \leq c \log \overline{T}(r) \qquad \parallel,$

$(8.6.10)_k \qquad\qquad (n-k)T_{k-1}(r) - (n+1-k)T_k(r) \leq c \log \overline{T}(r) \qquad \parallel.$

(The constants c in (8.6.8), (8.6.9) and (8.6.10) are of course different).

To start the induction we note that $(8.6.9)_1$ is nothing but $(8.6.8)_0$. Since

$$(k+1) \times (8.6.8)_k + (8.6.9)_k = (8.6.9)_{k+1},$$

the induction is complete for (8.6.9). Similarly, $(8.6.10)_{n-1}$ is nothing but $(8.6.8)_{n-1}$. Since

$$(n+1-k) \times (8.6.8)_{k-1} + (8.6.10)_k = (8.6.10)_{k-1},$$

the induction for (8.6.10) on decreasing k is now complete.

From (8.6.9) and (8.6.10) we derive the following inequalities for $k, l = 0, 1, \ldots, n - 1$, (see Wu [1]).

$(8.6.11)_{k,l} \qquad (k+1)T_l(r) - (l+1)T_k(r) \leq c \log \overline{T}(r) \qquad \parallel \quad \text{for} \quad l \geq k,$

$(8.6.12)_{k,l} \qquad (n-k)T_l(r) - (n-l)T_k(r) \leq c \log \overline{T}(r) \qquad \parallel \quad \text{for} \quad l \leq k.$

The proof is by induction on l. The left hand side of $(8.6.11)_{k,k}$ is zero, and the inequality is trivial for $k = l$. $(8.6.11)_{k,k+1}$ is nothing but $(8.6.9)_{k+1}$. Since

$$(l+1) \times (8.6.11)_{k,l+1} = (k+1) \times (8.6.9)_{l+1} + (l+2) \times (8.6.11)_{k,l},$$

$(8.6.11)_{k,l}$ implies $(8.6.11)_{k,l+1}$, thus completing the induction. The proof for (8.6.12) is similar.

What (8.6.11) and (8.6.12) mean is that all associated curves F_k of f have the same order of growth as f. In fact, by (8.6.11) and (8.6.12), there are positive constants c_1 and c_2 such that

$$\overline{T}(r) \leq c_1 T_k(r) + c_2 \log \overline{T}(r) \qquad \parallel \quad \text{for} \quad k = 0, 1, \ldots, n - 1.$$

Since $t_k(r)$ is positive and monotone increasing, $T_k(r) \to \infty$ as $r \to \infty$. Hence,

$$\frac{\log \overline{T}(r)}{\overline{T}(r)} \to 0 \quad \text{as} \quad r \to \infty.$$

So we can bound the term $c_2 \log \overline{T}(r)$ by $\overline{T}(r)/2$ for large r and obtain

$(8.6.13) \qquad\qquad \overline{T}(r) \leq 2c_2 T_k(r) \qquad \parallel \quad \text{for} \quad k = 0, 1, \ldots n - 1.$

Now, in place of $\omega_k = i\lambda_k dz \wedge d\bar{z}$, we use the non-negative $(1, 1)$-form $\hat{\omega}_k = i\hat{\lambda}_k dz \wedge d\bar{z}$ given in (8.3.10); for the expression of $\hat{\lambda}_k$, see (8.5.18). Then the argument used in proving (8.6.6) yields also

(8.6.14) **Theorem.** *Given an entire curve* $f: \mathbf{C} \to P_n\mathbf{C}$, *we have for* $k = 0, 1, \ldots, n-1$

$$\int_0^{2\pi} \log \hat{\lambda}_k(re^{i\theta})\,d\theta \leq c \log \hat{T}_k(r) \qquad \|,$$

where c is a constant independent of r.

Let $\mathbf{a}_0, \ldots, \mathbf{a}_q$ be $q+1$ vectors in general position in $\mathbf{C}^{n+1}$. Substituting (8.6.14) into (8.5.24) we obtain

$$\sum_{j=0}^{q} N(r, \mathbf{a}_j) \geq (q-n)T(r) - \varepsilon \sum_{k=0}^{n-1} T_k(r) + \sum_{k=0}^{n-1}(n-k)S_k(r)$$

$$+ \sum_{k=0}^{n-1}\{\hat{T}_k(r) - c \log \hat{T}_k(r) + \frac{1}{2}\hat{\mu}_k(0)\} \qquad \|.$$

We can replace the second term $\varepsilon \sum T_k(r)$ on the right hand side by $\varepsilon c T_0(r)$ because

$$\sum T_k(r) \leq n\overline{T}(r) \leq cT_0(r)$$

by (8.6.13). Now replace εc by ε. Since $\hat{t}_k(r)$ is nonnegative and monotone increasing, $\sum(\hat{T}_k(r) - c \log \hat{T}_k(r))$ goes to infinity, in particular, bounds any positive constant eventually as $r \to \infty$. Hence,

(8.6.15) **Theorem.** *Let holomorphic map* $f: \mathbf{C} \to P_n\mathbf{C}$ *be non-degenerate in the sense that its image is not contained in a linear subspace. Let* $\mathbf{a}_0, \ldots, \mathbf{a}_q$ *be* $q + 1$ *vectors in general position in* $\mathbf{C}^{n+1}$. *Then, given any positive* ε *and any constant C, we have*

$$\sum_{j=0}^{q} N(r, \mathbf{a}_j) \geq (q-n-\varepsilon)T(r) + \sum_{k=0}^{n-1}(n-k)S_k(r) + C \qquad \|.$$

7 Defect Relation

Consider the projective space $P_n\mathbf{C}$ with Fubini-Study metric ds^2 and its Kähler form Φ. We normalize the metric so that Φ represents the Chern class of the hyperplane line bundle, i.e.,

$$(8.7.1) \qquad \Phi = \frac{1}{2\pi}dd^c \log \|z\|.$$

As in Section 4 we shall write $\mathbf{a}$ for the point of $P_n\mathbf{C}$ represented by a nonzero vector $\mathbf{a}$ of $\mathbf{C}^{n+1}$.

Let V be a relatively compact domain in a Riemann surface S, and let $f: S \to P_n\mathbf{C}$ a holomorphic map. For $\mathbf{a} = (a^0, \ldots, a^n) \in P_n\mathbf{C}$, let $n(V, \mathbf{a})$ be the number of times the holomorphic curve $f(V)$ intersects the hyperplane $\langle \mathbf{z}, \mathbf{a} \rangle = 0$. Then **Crofton's formula** states

$$(8.7.2) \qquad \int_{\mathbf{a} \in P_n \mathbf{C}} n(V, \mathbf{a}) \Phi^n = \int_V f^* \Phi.$$

The left hand side represents the average number of times the holomorphic curve f meets the hyperplanes while the right hand side is the area of $f(V)$ in $P_n \mathbf{C}$. If V is a closed Riemann surface, i.e., f is an algebraic curve, then $n(V, \mathbf{a})$ is the degree of f and is independent of a. In this case, Crofton's formula is a well known fact in algebraic geometry.

We sketch the proof of Crofton's formula. Let P_n^* be the dual of $P_n = P_n \mathbf{C}$; it is the space of hyperplanes in P_n. Let $I \subset P_n \times P_n^*$ be the incidence set, i.e.,

$$I = \{(x, \xi) \in P_n \times P_n^*; \ x \in \xi\}$$

with projections

$$p: I \to P_n, \quad \pi: I \to P_n^*.$$

The unitary group $U(n + 1)$ acts on P_n, P_n^* and $P_n \times P_n^*$ in a natural way, and the action leaves the incidence set I invariant. Regarding $\mathbf{a} = (a^0, \ldots, a^n)$ as the homogeneous coordinate of the hyperplane $\langle \mathbf{z}, \mathbf{a} \rangle = 0$, we identify P_n^* with P_n. Then the incidence set I may be given by

$$I = \{(\mathbf{z}, \mathbf{a}) \in P_n \times P_n; \ \langle \mathbf{z}, \mathbf{a} \rangle = 0\}.$$

We use the same Φ to denote the Kähler form of the Fubini-Study metric for P_n^* as well as for P_n. The essence of Crofton's formula is in the following formula:

$$(8.7.3) \qquad p_* \pi^* \Phi^n = \Phi,$$

where p_* denotes the integration along fibers of the fibration $p: I \to P_n$. To prove (8.7.3) we observe that $p_* \pi^* \Phi^n$ is a $U(n + 1)$-invariant $(1, 1)$-form on P_n, and hence must be a constant multiple of Φ. Since the integral of Φ^{n-1} on each fibre $p^{-1}(x)$ is 1, this constant is 1, thus establishing (8.7.3).

In order to prove an integral formula such as (8.7.2), we can ignore singular points of f. By subdividing V we can also assume that f is an imbedding. Now, assuming that $f: V \to P_n$ is an imbedding, define a subset M of I by setting

$$M = p^{-1}(f(V)).$$

Then $\pi: M \to P_n^*$ is a finite-to-one mapping; if $\mathbf{a}$ represents a hyperplane of P_n, then $\pi^{-1}(\mathbf{a})$ consists of $n(V, \mathbf{a})$ points. Hence,

$$\int_{f(V)} \Phi = \int_{f(V)} p_* \pi^* \Phi^n = \int_M \pi^* \Phi^n = \int_{\mathbf{a} \in P_n} n(V, a) \Phi^n,$$

which proves (8.7.2).

When S is a domain containing the disc D_R and V is a smaller disc D_r, Crofton's formula (8.7.2) can be written as

$$(8.7.4) \qquad \int_{\mathbf{a} \in P_n \mathbf{C}} n(r, \mathbf{a}) \Phi^n = t(r).$$

We integrate (8.7.4) with respect to dr/r from 0 to r. Recalling the definition (8.4.8) of $N(r, \mathbf{a})$ and the fact that $v(0, \mathbf{a}) = 0$ for all $[\mathbf{a}] \in P_n$ except for the obvious one point $f(0)$, we obtain

$$(8.7.5) \qquad \int_{[\mathbf{a}] \in P_n \mathbf{C}} N(r, \mathbf{a}) \Phi^n = T(r).$$

Now, assuming that f is defined on all of $\mathbf{C}$, we define the **defect** of $\mathbf{a}$ to be

$$(8.7.6) \qquad \delta(\mathbf{a}) = \liminf_{r \to \infty} \left(1 - \frac{N(r, \mathbf{a})}{T(r)} \right).$$

Since both $N(r, \mathbf{a})$ and $T(r)$ are positive, we have $\delta(\mathbf{a}) \leq 1$. On the other hand, in (8.4.30) we proved

$$N(r, \mathbf{a}) < T(r) + C_{\mathbf{a}}.$$

Since $t(r)$ is monotone increasing and hence $\lim_{r \to \infty} T(r) = \infty$, we have

$$(8.7.7) \qquad \frac{N(r, \mathbf{a})}{T(r)} < 1 + \frac{C_{\mathbf{a}}}{T(r)} \to 1 \quad \text{as} \quad r \to \infty.$$

Hence,

$$(8.7.8) \qquad 0 \leq \delta(\mathbf{a}) \leq 1.$$

We note that if a hyperplane $\langle \mathbf{z}, \mathbf{a} \rangle = 0$ does not meet the curve $f(\mathbf{C})$, then $N(r, \mathbf{a}) = 0$ and $\delta(\mathbf{a}) = 1$. On the other hand, $\delta(\mathbf{a}) = 0$ means that the hyperplane $\langle \mathbf{z}, \mathbf{a} \rangle = 0$ meets the curve $f(\mathbf{C})$ as often as any other hyperplane does. We claim

$$(8.7.9) \qquad \delta(\mathbf{a}) = 0 \quad \text{for almost all} \quad \mathbf{a} \in P_n \mathbf{C}.$$

In fact, from (8.7.5) we have

$$(8.7.10) \qquad \int_{\mathbf{a} \in P_n \mathbf{C}} \left(1 - \frac{N(r, \mathbf{a})}{T(r)} \right) \Phi^n = 0.$$

But, the integrand, as a family of functions on $P_n \mathbf{C}$ parametrized by r, is uniformly bounded because

$$1 > 1 - \frac{N(r, \mathbf{a})}{T(r)} > -\frac{C_{\mathbf{a}}}{T(r)} \to 0 \quad \text{as} \quad r \to \infty.$$

Hence, by Fatou's lemma,

$$\int_{\mathbf{a} \in P_n \mathbf{C}} \delta(\mathbf{a}) \Phi^n = \int_{\mathbf{a} \in P_n \mathbf{C}} \liminf_{r \to \infty} \left(1 - \frac{N(r, \mathbf{a})}{T(r)} \right) \Phi^n$$

$$\leq \liminf_{r \to \infty} \int_{\mathbf{a} \in P_n \mathbf{C}} \left(1 - \frac{N(r, \mathbf{a})}{T(r)} \right) \Phi^n = 0,$$

which proves (8.7.9).

The statement (8.7.9) means that almost all hyperplanes of $P_n\mathbf{C}$ meet the curve $f(\mathbf{C})$ equally often. The following **defect relation** says much more. (However, it is not as strong as the truncated defect relation proved in (3.B.42)).

(8.7.11) **Theorem.** *Let* $\mathbf{a}_0, \ldots, \mathbf{a}_q \in \mathbf{C}^{n+1}$ *be* $q + 1$ *vectors in general position. Then for any holomorphic map* $f : \mathbf{C} \to P_n\mathbf{C}$ *that is non-degenerate in the sense that its image is not contained in any linear subsapce, we have*

$$\sum_{j=0}^{q} \delta(\mathbf{a}_j) \leq n + 1$$

Proof. Dropping the non-negative term $\sum(n - k)S_k(r)$ from (8.6.15) we have

$$\sum_{j=0}^{q} N(r, \mathbf{a}_j) \geq (q - n - \varepsilon)T(r) + C \qquad \|.$$

Then there is an increasing sequence of positive numbers $r_\nu \to \infty$ such that

$$\sum_{j=0}^{q} N(r_\nu, \mathbf{a}_j) \geq (q - n - \varepsilon)T(r_\nu) + C,$$

which implies

$$\limsup_{r \to \infty} \sum_{j=0}^{q} \frac{N(r, \mathbf{a}_j)}{T(r)} \geq q - n - \varepsilon.$$

Hence,

$$\sum_{j=0}^{q} \delta(\mathbf{a}_j) = \sum_{j=0}^{q} \left(1 - \limsup_{r \to \infty} \frac{N(r, \mathbf{a}_j)}{T(r)}\right)$$

$$\leq q + 1 - \limsup_{r \to \infty} \sum_{j=0}^{q} \frac{N(r, \mathbf{a}_j)}{T(r)} \leq n + 1 + \varepsilon.$$

Since ε is arbitrary, we have the desired defect relation. $\qquad\square$

We have now a generalization of the little Picard Theorem:

(8.7.12) **Corollary.** *If a holomorphic map* $f : \mathbf{C} \to P_n\mathbf{C}$ *misses* $n + 2$ *hyperplanes in general position, then its image is contained in a hyperplane.*

This implies Borel's theorem (3.8.2), see (3.B.45).

Now, we consider the k-th associated curve F_k. Let $A \in \bigwedge^{k+1} \mathbf{C}^{n+1}$ be a decomposable $(k + 1)$-vector of $\mathbf{C}^{n+1}$, or the k-plane in $P_n\mathbf{C}$ it represents. On account of (8.4.38), the k-**th defect** $\delta_k(A)$ of A can be defined by

(8.7.13)
$$\delta_k(A) = \liminf_{r \to \infty} \left(1 - \frac{N_k(r, A)}{T_k(r)}\right).$$

Then the k-**th defect relation** is given by

(8.7.14) **Theorem.** *Let $f \colon \mathbf{C} \to P_n\mathbf{C}$ be a non-degenerate holomorphic map, and $A_0, A_1, \ldots, A_q$ be k-planes in general position in $P_n\mathbf{C}$. Then*

$$\sum_{j=0}^{q} \delta_k(A_j) \le \binom{n+1}{k+1}, \qquad K = 0, 1, \ldots, n-1.$$

For the proof, we refer the reader to Wu [4, p. 205]. As we mentioned in Appendix B of Chapter 3, the truncated version of (8.7.14) can be found in Fujimoto [11].

References

Abate, M.

[1] Boundary behavior of invariant distances and complex geodesics. Rend. Acc. Naz. Lincei **80** (1986) 100–106

[2] *Iteration Theory of Holomorphic Maps on Taut Manifolds*. Mediterranean Press, Cosenza, 1989

[3] Common fixed points of commuting holomorphic maps. Math. Ann. **283** (1989) 645–655

[4] The Lindelöf principle and the angular derivative in strongly convex domains. J. d'Analyse Math. **54** (1990) 189–228

[5] Iteration theory, compactly divergent sequences and commuting holomorphic maps. Annali Scuola Norm. Sup. Pisa **18** (1991) 167–191; correction p. 631

[6] A characterization of hyperbolic manifolds. Proc. Amer. Math. Soc. **117** (1993) 789–793

Abate, M., and L. Geatti

[1] Cohomogeneity two hyperbolic acyclcic Stein manifolds. Intern'l J. Math. **3** (1992) 591–608

Abate, M., and G. Patrizio

[1] Uniqueness of complex geodesics and characterization of circular domains. Manuscripta Math. **74** (1992) 277–297

[2] Complex Finsler metrics, *Geometry and Analysis on Complex Manifolds*, (Festschrift for S. Kobayashi), ed. Mabuchi et al., pp. 1–38. World Scientific, 1994

[3] *Finsler Metrics – A Global Approach*. Lecture Notes in Math., vol. 1591. Springer, Berlin Heidelberg 1994

Abate, M., and J-P. Vigué

[1] Common fixed points in hyperbolic Riemann surfaces and convex domains. Proc. Amer. Math. Soc. **112** (1991) 503–512

Adachi, Y.

[1] On hyperbolicity of projective plane with lacunary curves. J. Math. Soc. Japan **46** (1994) 185–193

[2] A generalization of the big Picard theorem. Kodai. Math. J. **18** (1995) 408–424

[3] On the relation between tautly imbedded spaces modulo S and hyperbolically imbedded spaces modulo S. Preprint (1996)

Adachi, Y., and M. Suzuki

[1] On the family of holomorphic mappings into projective space with lacunary hypersurfaces. J. Math. Kyoto Univ. **30** (1990) 451–458

[2] Degeneracy points of the Kobayashi pseudodistance on complex manifolds. Proc. Symp. Pure Math., Amer. Math. Soc. **52** (1991) Part 2, 41–51

Agler, J.

[1] Operator theory and the Carathéodory metric. Invent. Math. **101** (1990) 483–500

Ahern, P. R., and R. Schneider

[1] Isometries of H^∞. Duke Math. J. **42** (1975) 321–326

Ahlfors, L. V.

[1] An extension of Schwarz's lemma. Trans. Amer. Math. Soc. **43** (1938) 359–364
[2] The theory of meromorphic curves. Acta Soc. Sci. Fenn. **3** (1941) 171–183
[3] *Complex Analysis*. McGraw-Hill, 1953
[4] *Conformal Invariants*. McGraw-Hill, 1973

Aihara, Y., and S. Mori

[1] Algebraic degeneracy theorem for holomorphic mappings into smooth projective algebraic varieties. Nagoya Math. J. **84** (1981) 209–218

Aihara, Y., and J. Noguchi

[1] Value distribution of meromorphic mappings into compactified locally symmetric spaces. Kodai Math. J. **14** (1991) 320–334

Aladro, G.

[1] Applications of the Kobayashi metric to normal functions of several complex variables. Utilitas Math. **31** (1987) 13–24
[2] The compatibility of the Kobayashi approach region and the admissible approach region. Illinois J. Math. **33** (1989) 42–63
[3] Localization of the Kobayashi distance. J. Math. Anal. Appl. **181** (1994) 200–204

Aladro, G., and S. G. Krantz

[1] A criterion for normality in $\mathbf{C}^n$. J. Math. Anal. Appl. **161** (1991) 1–8

Alexander, H.

[1] Proper holomorphic mappings in $\mathbf{C}^n$. Indiana Univ. Math. J. **26** (1977) 137–146

Ancona, V., and J-P. Speder

[1] Espaces de Banach-Stein. Ann. Scuola Norm. Sup. Pisa **25** (1971) 683–690

Andreotti, A.

[1] Sopra le superficie che possegno transformazioni birationali in se. Univ. Roma Ist. Naz. Alta. Mat. Red. Mat. e Appl. **9** (1950) 255–279

Andreotti, A., and W. Stoll

[1] Extension of holomorphic maps. Ann. of Math. **72** (1960) 312–349

Arakelov, S. Ju.

[1] Families of algebraic curves with fixed degeneracies. Izv. Akad. Nauk SSSR Mat. **35** (1971) 1277–1302

Aubin, T.

[1] Equations du type Monge-Ampère sur les variétés kähleriennes compactes. C. R. Acad. Sci. Paris **283** (1976) 119–121
[2] *Nonlinear Analysis on Manifolds, Monge-Ampère Equations*. Grundlehren Math. Wiss. 252, Springer-Verlag, 1982

Azukawa, K.

[1] Hyperbolicity of circular domains. Tohoku Math. J. **35** (1983) 403–413
[2] Two intrinsic pseudo-metrics with pseudoconvex indicatrices and starlike circular domains. J. Math. Soc. Japan **38** (1986) 627–647

[3]   The invariant pseudometric related to negative pluri-subharmonic functions. Kodai Math. J. **10** (1987) 83–92

[4]   A note on Carathéodory and Kobayashi pseudodistances. Kodai Math. J. **14** (1991) 1–12

[5]   The ratio of invariant metrics on the annulus and theta functions. *Topics in Complex Analysis* (Warsaw, 1992), 53–60, Banach Center Publ. 31, 1995

[6]   The pluri-complex Green function and a covering map. Michigan Math. J. **42** (1995) 593–602

Azukawa, K., and M. Suzuki

[1]   Some examples of algebraic degeneracy and hyperbolic manifolds. Rocky Mountain J. Math. **10** (1980) 655–659

Ax, J.

[1]   Some topics in differential algebraic geometry II: On the zeros of theta functions. Amer. J. Math. **94** (1972) 1205–1213

Babets, V. A.

[1]   A theorem on hyperbolicity. Funct. Analysis Appl. **14** (1980), 294–295

[2]   Geometry of logarithmically tangent bundles and hyperbolic manifolds. Funct. Analysis Appl. **16** (1982), 127–128

[3]   Picard type theorems for holomorphic mappings. Siberian Math. J. **25** (1984), 195–200

[4]   Pseudoforms with negative curvature on CP_n. Math. USSR Izvestiya **29** (1987), 677–688

Ballico, E.

[1]   Algebraic hyperbolicity of generic high degree hypersurfaces. Arch. Math. **63** (1994), 282–283

Bandman, T. M.

[1]   Surjective holomorphic mappings of projective manifolds. Siberian Math. J. **22** (1981), 204–210

[2]   Topological invariants of variety and the number of its holomorphic mappings. Intern'l Symp. *Holomorphic Mappings, Diophantine Geometry and Related Topics*. RIMS, Kyoto (1992), 188–202

[3]   Toplogical estimates of the number of rational mappings for low-dimensional varieties of general type. Proc. Hirzebruch 65 Conf. Alg. Geom., 41–48, Israel Math. Conf. Proc. 9, Bar-Ilan Univ. Ramat Gan, 1996

Bandman, T. M., and G. Dethloff

[1]   Estimates of the number of rational mappings from a fixed variety to varieties of general type. Ann. Inst. Fourier (Grenoble) **47** (1997), 801–824

Bandman, T. M., and D. Markushevich

[1]   On the number of rational maps between varieties of general type. J. Math. Sci. Univ. Tokyo **1** (1994), 423–433

Barth, T.

[1]   Taut and tight complex manifolds. Proc. Amer. Math. Soc. **24** (1970), 429–431

[2]   Normality domains for families of holomorphic maps. Math. Ann. **190** (1970), 293–297

[3]   The Kobayashi distance induces the standard topology. Proc. Amer. Math. Soc. **35** (1972), 439–441

[4]   Families of holomorphic maps into Riemann surfaces. Trans. Amer. Math. Soc. **207** (1975), 175–187

[5]   Separate analyticity, separate normality and radial normality for mappings. Proc. Symp. Pure Math., Amer. Math. Soc. **30** (1976), 221–224

[6] Some counterexamples concerning intrinsic distances. Proc. Amer. Math. Soc. **66** (1977), 49–53

[7] Convex domains and Kobayashi hyperbolicity. Proc. Amer. Math. Soc. **79** (1980), 556–558

[8] The Kobayashi indicatrix at the center of a circular domains. Proc. Amer. Math. Soc. **88** (1983), 527–530

[9] Topologies defined by some invariant pseudodistances. *Topics in Complex Analysis* (Warsaw 1992), 69–76, Banach Center Publ. 31, 1995

Bedford, E.

[1] Proper holomorphic mappings. Bull. Amer. Math. Soc. **10** (1984), 157–175

Bell, S., and D. Catlin

[1] Boundary regularity of proper holomorphic mappings. Duke Math. J. **49** (1982), 385–396

Bell, S., and E. Ligocka

[1] A simplification and extension of Fefferman's theorem on biholomorphic mappings. Invent. Math. **57** (1980), 283–289

Bell, S., and R. Narasimhan

[1] Proper holomorphic mappings of complex spaces. *Encycl. Math. Sci. 69, Several Complex Variables VI*, 1990, 1–38

Bergman, S.

[1] *The Kernel Functions and Conformal Mappings*. Math. Surveys V, Amer. Math. Soc. 1950

Biancofiore, A., and W. Stoll

[1] Another proof of the lemma of the logarithmic derivative in several complex variables. Ann. Math. Studies No. 100 (1981), 29–45, Princeton Univ. Press

Bishop, E.

[1] Conditions for the analyticity of certain sets. Michigan Math. J. **11** (1964), 289–304

Blanchard, A.

[1] Sur les variétés analytiques complexes. Ann. Sci. École Nor, Sup. **73** (1956), 157–202

Bland, J., T. Duchamp and M. Kalka

[1] On the automorphism group of strictly convex domains in $\mathbf{C}^n$. *Complex Differential Geometry and Nonlinear Differential Equations* (Brunswick, Maine 1984), 19–30, Contemp. Math. **49** Amer. Math. Soc. 1986

Bland, J., and I. Graham

[1] On the Hausdorff measures associated to the Carathéodory and Kobayashi metrics. Annali Scuola Norm. Sup. Pisa **12** (1985), 503–514

Blank, B. E., D. S. Fan, D. Klein, S. G. Krantz, D. W. Ma and M-Y. Pang

[1] The Kobayashi metric of a complex ellipsoids in $\mathbf{C}^2$. Experimental Math. **1** (1992), 47–55

Bloch, A.

[1] Sur les systèmes de fonctions holomorphes à variétés linéaires lacunaires. Ann. École Norm. Sup. **43** (1926), 309–362

[2] Sur les systèmes de fonctions uniformes satisfaisant à l'équation d'une variété algébrique dont l'irrégularité dépasse la dimension. J. de Math. Pure et Appl. **5** (1926), 19–66

[3] La conception actuelle de la théorie des fonctions entières et méromorphes. Enseign. Math. **25** (1926), 83–103

[4] *Les Fonctions Holomorphes et Méromorphes dans le Cercle-Unité*. Mémorial Sci. Math. Acad. Sci. Paris, Fasc. XX, 1926

Bochner, S., and D. Montgomery

[1] Groups of differentiable and real or complex analytic transformations. Ann. of Math. **46** (1945), 685–694

[2] Groups on analytic manifolds. Ann. of Math. **48** (1947), 659–669

Bombieri, E.

[1] Canonical models of surfaces of general type. Publ. IHES **42** (1973), 171–219

Borel, A.

[1] Some metric properties of arithmetic quotients of symmetric spaces and an extension theorem. J. Diff. Geom. **6** (1972), 543–560

Borel, A., and R. Narasimhan

[1] Uniqueness conditions for certain holomorphic mappings. Invent. Math. **2** (1967), 247–255

Borel, E.

[1] Sir les zéros de fonctions entières. Acta Math. **20** (1897), 357–396

Bott, R., and S. S. Chern

[1] Hermitian vector bundles and the equidistribution of the zeros of their holomorphic sections. Acta Math. **114** (1965), 71–112

Bremermann, H.

[1] Holomorphic continuation of the kernel function and the Bergman metric in several complex variables. *Lectures on Functions of a Complex Variable*, Univ. Michigan Press 1955, pp. 349–383

Brody, R.

[1] Compact manifolds and hyperbolicity. Trans. Amer. Math. Soc. **235** (1978), 213–219

Brody, R., and M. L. Green

[1] A family of smooth hyperbolic hypersurfaces in P_3. Duke Math. J. **44** (1977), 873–874

Burbea, J.

[1] Inequalities between intrinsic metrics. Proc. Amer. Math. Soc. **67** (1977), 50–54

[2] The Carathéodory metric and its majorant metrics. Canadian J. Math. **29** (1977), 771–780

[3] On the hessian of the Carathéodory metric. Rocky Mtn. Math. J. **8** (1978), 555–559

Burns, D. M., and S. G. Krantz

[1] Rigidity of holomorphic mappings and a new Schwarz lemma at the boundary. J. Amer. Math. Soc. **7** (1994), 661–676

Busemann, H., and W. Mayer

[1] On the foundation of calculus of variations. Trans. Amer. Math. Soc. **49** (1941), 173–198

Campana, F.

[1] Twistor spaces and nonhyperbolicity of certain symplectic Kähler manifolds. *Complex Analysis* (Wuppertal, 1991), 64–69, Aspects of Math. E 17, 1991

Campbell, L. A., A. Howard and T. Ochiai
[1] Moving holomorphic disks off analytic subsets. Proc. Amer. Math. Soc. **60** (1976), 106–108

Campbell, L. A., and R. H. Ogawa,
[1] On preserving the Kobayashi pseudodistance. Nagoya Math. J. **57** (1975), 37–47

Carathéodory, C.
[1] Über das Schwarzsche Lemma bei analytischen Funktionen von zwei komplexen Veränderlichen. Math. Ann. 97 (1926), 76–98
[2] Über die Geometrie der analytischen Abbildungen, die durch analytische Funktionen von zwei Veränderlichen vermittelt werden. Abh. Math. Sem. Univ. Hamburg **6** (1928), 96–145
[3] Über die Abbildungen, die durch Systeme von analytische Funktionen von mehreren Veränderlichen erzeugt werden. Math. Z. **34** (1932), 758–792
[4] *Theory of Functions of a Complex Variable.* In two volumes, Chelsea

Carlson, J. A.
[1] Some degeneracy theorems for entire functions with values in an algebraic variety. Trans. Amer. Math. Soc. **168** (1972), 273–301

Carlson, J. A., and P. A. Griffiths
[1] A defect relation for equidimensional holomorphic mappings between algebraic varieties. Ann. Math. **95** (1972), 556–584

Cartan, H.
[1] Sur les systèmes de fonctions holomorphes à variétés linéaires lacunaires. Ann. Sci. Ecole Norm. Sup. **45** (1928), 255–346
[2] Les fonctions de deux variables complexes et le problème de la représentation analytique. J. Math. Pure Appl. **10** (1931), 1–114
[3] Sur les fonctions de deux variables complexs:les transformations d'un domaine borné D en un domaine intérieur à D. Bull. Soc. Math. France **58** (1930), 199–219
[4] Sur les fonctions de plusieurs variables complexes. L'itération des transformations intérieures d'un domain borné. Math. Z. **35** (1932), 760–773
[5] Sur les zéros des combinaisons linéaires de p fonctions holomorphes données. Mathematica (Cluj) **7** (1933), 5–31
[6] Sur l'itération des transformations conformes ou pseudo-conformes. Compositio Math. **1** (1934), 223–227
[7] Sur les groupes de transformations analytiques. Collection à la mémoire de Jaques Herbrand, Hermann, Paris (1936)
[8] Sur les fonctions de n variables complexes: les transformations du produit topologique de deux domaines bornés. Bull. Soc. Math. France **64** (1936), 37–48
[9] Quotient d'un espace analytique par un groupe d'automorphismes. *Algebraic Geometry and Topology*, Princeton U. Press (1957), 90–102
[10] Quotients of complex analytic spaces. Intern'l Colloquium on Function Theory, Tata Inst. (1960), 1–15

Catlin, D. W.
[1] Boundary behavior of holomorphic functions on pseudoconvex domains. J. Diff. Geometry **15** (1980), 605–625
[2] Estimates of invariant metrics on pseudoconvex domains of dimension two. Math. Z. **200** (1989), 429–466

Chang, C-H., M. C. Hu and H-P. Lee
[1] Extremal analytic discs with prescribed boundary data. Trans. Amer. Math. Soc. **310** (1988), 355–369

Chen, S-S.

[1] Bounded holomorphic functions in Siegel domains. Proc. Amer. Math. Soc. **40** (1973), 539–542

Chen, T. G.

[1] Boundary behavior of the Kobayashi-Eisenman volume form. Chinese J. Math. **23** (1995), 235–244

Chen, W. W.

[1] A proof of the generalized Picard's little theorem using matrices. Linear Alg. and its Appl. **214** (1995), 187–192

Chen, Z. H., S-Y., Cheng and Q-K. Lu

[1] On the Schwarz lemma for complete Kähler manifolds. Sci. Sinica **22** (1979), 1238–1247

Chen, Z. H., and H. C. Yang

[1] Estimation of the upper bound on the Levi form of the distance function on Hermitian manifolds and some of its applications. Acta Math. Sinica **27** (1984), 631–643

Chern, S. S.

[1] Complex analytic mappings of Riemann surfaces I. Amer. J. Math. **82** (1960), 323–337
[2] The integrated form of the first main theorem for complex analytic mappings in several complex variables. Ann. of Math. **71** (1960), 536–551
[3] On holomorphic mappings of Hermitian manifolds of the same dimension. Proc. Symp. Pure Math. Amer. Math. Soc. Vol. 11, *Entire Functions and Related Parts Analysis*, (1968), 157–170
[4] Holomorphic curves in the plane. *Differential Geometry in honor of K. Yano*, Kinokuniya, Tokyo, 1972, pp. 73–94

Chern, S. S., H. I. Levine and L. Nirenberg

[1] Intrinsic norms on a complex manifold. *"Global Analysis"* in honor of K. Kodaira, Princeton U. Press, 1969, 14–148

Cheung, C. K.

[1] Hermitian metrics of negative holomorphic sectional curvature on some hyperbolic manifolds. Math. Z. **201** (1989), 105–119

Cheung, C. K., and K. T. Kim

[1] Analysis of the Wu metric I: The case of convex Thullen domains. Trans. Amer. Math. Soc. **348** (1996), 1429–1457
[2] Analysis of the Wu metric II: The case of nonconvex Thullen domains. Proc. Amer. Math. Soc. **125** (1997), 1131–1142

Cheung, W. S., and B. Wong

[1] An integral inequality of an intrinsic measure on bounded domains in $\mathbf{C}^n$. Rocky Mt. J. Math. **22** (1992), 825–836

Chi, D. P., I. H. Lee, S. G. Lee and S. M. Kim

[1] On a comparison theorem for pseudometrics of Riemann surfaces. J. Korean Math. Soc. **20** (1983), 31–35

Chinak, M.

[1] Weakly hyperbolic manifolds. Siberian Math. J. **29** (1988), 161–163
[2] On meromorphic mappings which contract a pseudovolume form. Math. Zametki **46** (1989), 116–117

436   References

[3] The weak hyperbolicity and many-dimensional analogues of Picard's theorem. Siberian Math. J. **32** (1991), 127–132

Cho, S.

[1] A lower bound on the Kobayashi metric near a point of finite type. J. Geometric Analysis **2** (1992), 317–326
[2] Estimates of invariant metrics on some pseudoconvex domains in $\mathbf{C}^n$. J. Korean Math. Soc. **32** (1995), 661–678

Cima, J. A., and S. G. Krantz

[1] The Lindelöf principle and normal functions of several complex variables. Duke Math. J. **50** (1983), 303–328

Cowen, M. J.

[1] Families of negatively curved Hermitian manifolds. Proc. Amer. Math. Soc. **39** (1973), 362–366
[2] The Kobayashi metric on $P_n - (2^{n+1})$ hyperplanes. *Value Distribution Theory*, Part A, Dekker, 1974, pp. 205–223
[3] The method of negative curvature: the Kobayashi metric on P_2 minus 4 lines. Trans. Amer. Math. Soc. **319** (1990), 729–745

Cowen, M. J., and P. A. Griffiths

[1] Holomorphic curves and metrics of negative curvature. J. Analyse Math. **29** (1976), 93–153

D'Angelo, J.

[1] *Several Complex Variables*. CRC Press, 1992

Dahlberg, B.

[1] Holomorphic mappings into smooth projective varieties. Ark. Mat. **12** (1974), 59–71

Davidov, J. T.

[1] Note on the extendability of holomorphic mappings into complex manifolds. Math. and Math. Education, 208–215, Bulgar. Acad. Sci. Sofia, 1986

Dektyarev, I. M.

[1] Test for equivalence of hyperbolic manifolds. Funct. Anal. Appl. **15** (1981), 292–293
[2] Multidimensional value distribution theory. Encyclopaedia Math. Sci. vol. 9, *Several Complex Variables* III, 31–61, Springer, 1989

Deligne, P.

[1] Théorie de Hodge. II, IHES Publ. Math. **40** (1971), 5–58

Deligne, P., and N. Katz

[1] *Groupes de Monodoromie en Géométrie Algébrique*. Lecture Notes in Math. **340** (1973)

Demailly, J-P.

[1] Mesures de Monge-Ampère et mesures pluriharmoniques. Math. Z. **194** (1987), 519–564
[2] A numerical criterion for very ample line bundle. J. Diff. Geom. **37** (1993), 323–374
[3] Algebraic criteria for Kobayashi hyperbolic varieties and jet differentials. Proc. Symp. Pure Math. Amer. Math. Soc. (Summer Inst. 1995), to appear
[4] Variétés hyperboliques et équations différentielles algébriques. Gazette Math. **73** (1997), 3–23

Demailly, J-P., and J. El Goul

[1] Connexions méromorphes projectives et variétés hyperboliques. C.R. Acad. Sci. Paris **324** (1997), 1385–1390

Demailly, J-P., L. Lempert and B. Shiffman

[1] Algebraic approximations of holomorphic maps from Stein domains to projective manifolds. Duke Math. J. **76** (1994), 333–363

Dethloff, G., G. Schumacher, and P. M. Wong

[1] Hyperbolicity of the complements of plane algebraic curves. Mathematica Göttingensis, Heft 31 (1992), 1–38; Amer. J. Math. **117** (1995), 573–599
[2] On the hyperbolicty of the complement of curves in algebraic surfaces: the three component case. Duke Math. J. **78** (1995), 193–212

Dethloff, G., and M. Zaidenberg

[1] Examples of plane curves of low degrees with hyperbolic and C-hyperbolic complements. *Geometric Complex Analysis* (Hayama 1995), 163–181, World Sci. Publ.

Diederich, K.

[1] Das Randverhalten der Bergmanschen Kernfunktion und Metrik in streng pseudokonvexen Gebieten. Math. Ann. **187** (1970), 9–36

Diederich, K., and J. E. Fornaess

[1] Proper holomorphic maps onto pseudoconvex domains with real analytic boundary. Ann. of Math. **110** (1979), 575–592
[2] Comparison of the Bergmann and the Kobayashi metric. Math. Ann. **254** (1980), 257–262
[3] Boundary regularity of proper holomorphic mappings. Invent. Math. **67** (1982), 363–384

Diederich, K., J. E. Fornaess, and G. Herbort

[1] Boundary behavior of the Bergman metric. Proc. Symp. Pure Math. Amer. Math. Soc. **41** (1984), 59–75

Diederich, K., and I. Lieb

[1] *Konvexität in der Komplexen Analysis*. DMV Seminar, Band 2, Birkhäuser, 1981

Diederich, K., and T. Ohsawa

[1] An estimate for the Bergman distance on pseudoconvex domains. Ann. of Math. **141** (1955), 181–190

Diederich, K., and N. Sibony

[1] Strange complex structures on Euclidean space. J. Reine u. Angew. Math. **311/312** (1979), 397–407

Dineen, S.

[1] *The Schwarz Lemma*. Oxford Math. Monographs, 1989

Dineen, S., R. M. Timoney and J-P. Vigué

[1] Pseudodistances invariantes sur les domaines d'un espace localement convexe. Ann. Scuola Norm. Sup. Pisa **12** (1985), 515–529

Dinghas, A.

[1] Ein n-dimensionales Analogon des Schwarz-Pickschen Flächensatzes für holomorphe Abbildungen der komplexen Einheitskugel in eine Kähler-Mannigfaltigkeit. Arbeitsgemeinschaft für Forschung des Landes Nordrhein-Westfalen **33** (1965), 477–494
[2] Über das Schwarzsche Lemma und verwandte Sätze. Israel J. Math. **5** (1967), 157–169

Do Duc Thai

[1] Royden-Kobayashi pseudometric and tautness of normalizations of complex spaces. Bolletino U. M. I. **(7) 5-A** (1991), 147–156

[2] Remark on hyperbolic embeddability of relatively compact subspaces of complex spaces. Ann. Polon. Math. **54** (1991), 9–11

[3] The fixed points of holomorphic maps on a convex domain. Ann. Polon. Math. **56** (1992), 141–148

[4] A note on the Kobayashi pseudo-distance and the tautness of holomorphic fiber bundles. Ann. Polon. Math. **58** (1993), 1–5

[5] On the hyperbolicity and the Schottky property of complex spaces. Proc. Amer. Math. Soc. **122** (1994), 1025–1027

Docquier, F., and H. Grauert

[1] Levisches Problem und Rungescher Satz für Teilgebiete Steinscher Mannigfaltigkeiten. Math. Ann. **140** (1960), 94–123

Dolbeault, P., and J. Ławrynowicz

[1] Holomorphic chains and extendability of holomorphic mappings. *Deformations of Mathematical Structures* (Łódź/Lublin 1985/87), 191–204, Kluwer Acad. Publ., 1989

Dolgachev, I., and M. Kaparanov

[1] Arrangements of hyperplanes and vector bundles on P^n. Duke Math. J. **71** (1993), 633–664

Douady, A.

[1] Le problème des modules pour les sous-espaces analytiques compacts d'un espace analytique donné. Ann. Inst. Fourier **16** (1964), 1–95

Drouilhot, S. J.

[1] A unicity theorem for meromorphic mappings between algebraic varieties. Trans. Amer. Math. **265** (1981), 349–358

Duchamp, T., and M. Kalka

[1] Holomorphic foliations and the Kobayashi metric. Proc. Amer. Math. Soc. **67** (1977), 117–122

Dufresnoy, H.

[1] Théorie nouvelle des familles complexes normales. Applications à l'étude des fonctions algébroïdes. Ann. Sci. École Norm. Sup. **61** (1944), 1–44

Duren, P. L.

[1] *Theory of H^p Spaces*. Academic Press, 1970

Earle, C. J.

[1] On the Carthéodory metric in Teichmüller spaces. Ann. Math. Studies No. 79, Princeton U. Press, 1974, pp. 99–104

Eastwood, A.

[1] A propos des variétés hyperboliques complètes. C. R. Acad. Sci. Paris **280** (1975), 1071–1074

Eisenman, D.

[1] *Intrinsic Measures on Complex Manifolds and Holomorphic Mappings*. Mem. Amer. Math. Soc. No. 96 (1970)

[2] Holomorphic mappings into tight manifolds. Bull. Amer. Math. Soc. **76** (1970), 46–48

[3] Proper holomorphic self-maps of the unit ball. Math. Ann. **190** (1971), 298–305; correction in **202** (1973), 135–136

El Goul, J.

[1] Algebraic families of smooth hyperbolic surfaces of low degree in $\mathbb{P}^3_{\mathbb{C}}$. Manuscripta Math. **90** (1996), 521–532

Erëmenko, A.

[1] Holomorphic curves omitting five planes in projective spaces. Amer. J. Math. **118** (1996), 1141–1151

[2] A counter example to Cartan's conjecture on holomorphic curves omitting hyperplanes. Proc. Amer. Math. Soc. **124** (1996), 3097–3100

Erëmenko, A., and M. L. Sodin

[1] Distribution of values of meromorphic functions and meromorphic curves from the standpoint of potential theory. St. Petersburg. Math. J. **3** (1992), 109–136

Faran, J. J.

[1] Hermitian Finsler metrics and the Kobayashi metric. J. Diff. Geometry **31** (1990), 601–625

Fefferman, C.

[1] The Bergman kernel and biholomorphic mappings of pseudoconvex domains. Invent. Math. **26** (1974), 1–65

Fischer, G.

[1] Fibrés holomorphes au-dessus d'un espace de Stein. *Espaces Analytiques*, pp. 57–69, Bucarest, Acad. Rép. Soc. Roumanie, 1971

Fischer, G., and H. Wu

[1] Developable complex analytic submanifolds. Intern'l J. Math. **6** (1995), 229–272

Fornaess, J. E.

[1] *Dynamics in Several Complex Variables*. CBMS Regional Conf. Ser. in Math. 87, Amer. Math. Soc., 1996

Fornaess, J. E., and N. Sibony

[1] Increasing sequence of complex manifolds. Math. Ann. **255** (1981), 351–360

Fornaess, J. E., and E. L. Stout

[1] Polydisks in complex manifolds. Math. Ann. **227** (1977), 145–153

Forstnerič, F.

[1] Proper holomorphic mappings, a survey. *Several Complex Variables* (Stockholm 1987/ 88), 297–363, Math. Notes 38, Princeton Univ. Press, 1993

[2] An elementary proof of Fefferman's theorem. Exposition. Math. **10** (1992), 135–149

Forstnerič, F., and J-P. Rosay

[1] Localization of the Kobayashi metric and the boundary continuity of proper holomorphic mappings. Math. Ann. **279** (1987), 239–252

Franchis, M. de

[1] Un teorema sulle involuzioni irrazionali. Rend. Circ. Mat. Palermo **36** (1913), 368

Franzoni, T., and E. Vesentini

[1] *Holomorphic Maps and Invariant Distances*. Notas de Mat. 69 North-Holland, 1980

Fridman, B. L.

[1] Biholomorphic invariants of a hyperbolic manifold and some applications. Trans. Amer. Math. Soc. **276** (1983), 685–698

Fu, S.

[1] Some estimates of Kobayashi metrics in the normal directions. Proc. Amer. Math. Soc. **122** (1994), 1163–1169

[2] Asymptotic expansions of invariant metrics of strictly pseudoconvex domains. Canadian Math. Bull. **38** (1995), 196–206

440   References

Fujiki, A.

[1] Closedness of the Douady spaces of compact Kähler spaces. Publ. Res. Inst. Math. Sci. Kyoto Univ. **14** (1978), 118–138

Fujimoto, H.

[1] On the holomorphic automorphism groups of complex spaces. Nagoya Math. J. **33** (1968), 85–106
[2] On holomorphic maps into a real Lie group of holomorphic transformations. Nagoya Math. J. **40** (1970), 139–146
[3] On holomorphic maps into a taut complex space. Nagoya Math. J. **46** (1972), 49–61
[4] Extensions of the big Picard theorem. Tohoku Math. J. **24** (1972), 415–422
[5] Families of holomorphic maps into the projective space omitting some hyperplanes. J. Math. Soc. Japan **25** (1973), 235–249
[6] On meromorphic maps into the complex projective space. J. Math. Soc. Japan **26** (1974), 272–288
[7] On families of meromorphic maps into the complex projective space. Nagoya Math. J. **54** (1974), 21–51
[8] The uniquness problem of meromorphic maps into the complex projective space. Nagoya Math. J. **58** (1975), 1–23
[9] A uniqueness theorem of algebraically non-degenerate meromorphic maps into $P^N(C)$. Nagoya math. J. **64** (1976), 117–147
[10] Remarks to the uniquness problem of meromorphic maps into $P^N(C)$, I. Nagoya Math. J. **71** (1978), 13–24; II, ibid. **71** (1978), 25–41; III, ibid. **75** (1979), 71–85; IV, ibid. **83** (1981), 153–181
[11] The defect relations for the derived curves of a holomorphic curves in $P^n(C)$. Tohoku Math. J. **34** (1982), 141–160
[12] A finiteness theorem of meromorphic maps into a compact normal complex space. Sci. Reports Kanazawa Univ. **30** (1985), 15–25
[13] Finiteness of some families of meromorphic maps. Kodai Math. J. **11** (1988), 47–63
[14] *Value Distribution Theory of the Gauss Map of Minimal Surfaces in* **R**^{m}. Aspects of Math. E 21, 1993

Funahashi, K.

[1] Normal holomorphic mappings and classical theorems of function theory. Nagoya Math. J. **94** (1984), 89–104

Gardiner, F. P.

[1] *Teichmüller Theory and Quadratic Differentials*. Wiley, 1987

Gentili, G.

[1] Regular complex geodesics for the domain $D_n = \{(z_1, \cdots, z_n) \in \mathbf{C}^n : |z_1|+\cdots+|z_n| < 1\}$. *Complex Analysis*, III (College park, Md. 1985–86), 35–45, Lecture Notes in Math. 1277, Springer, 1987
[2] Distances on convex cones. *Geometry Seminar "Luigi Bianchi"*, Lecture Notes in Math. 1022, 1–31, Springer 1983

Gigante, G.

[1] On normal mappings. Atti. Sem. Mat. Fis. Univ. Modena **39** (1991), 235–253

Gigante, G., and G. Tomassini

[1] M-hyperbolicity, evenness, and normality. *Partial Diff. Eq. Appl.* 179–185, Lecture Notes in Pure Appl. Math. 177, Dekker

Gigante, G., G. Tomassini and S. Venturini

[1] M-hyperbolic real subsets of complex spaces. Pacific. J. Math. **172** (1996), 101–115

Gilligan, B.

[1] Equivariant fibrations of homogeneous complex manifolds. *Complex Analysis and Applications* '85 (Varna 1985), 249–258, Bulgar. Acad. Sci. Sofia, 1986

Goloff, D., and W-K. To

[1] Holomorphic mappings, the Schwarz-Pick lemma, and curvature. Michigan Math. J. **42** (1995), 3–15

Graham, I.

[1] Boundary behavior of the Carathéodory and Kobayashi metrics on strongly pseudo-convex domains in $\mathbf{C}^n$. Trans. Amer. Math. Soc. **207** (1975), 219–240
[2] Intrinsic measures on complex manifolds. Mat. Vesnik **38** (1986), 475–488
[3] Intrinsic measures and holomorphic retracts. Pacific J. Math. **130** (1987), 299–311
[4] Holomorphic mappings into strictly convex domains which are Kobayashi isometries at one point. Proc. Amer. Math. Soc. **105** (1989), 917–921
[5] Holomorphic maps which preserve intrinsic metrics or measures. Trans. Amer. Math. Soc. **319** (1990), 787–803
[6] Distortion theorems for holomorphic maps between convex domains in $\mathbf{C}^n$. Complex Variables, Theory & Appl. **15** (1990), 37–42
[7] Sharp constants for the Koebe theorem and for estimates of intrinsic metrics on convex domains. *Several Complex Variables and Complex Geometry*, Part 2 (Santa Cruz 1989), 233–238, Proc. Symp. Pure Math. **52**, Amer. Math. Soc., 1991
[8] Holomorphic mappings into complex domains. *Complex Analysis* (Wuppertal 1991), 127–133, Aspects of Math. E 17, 1991

Graham, I., and H. Wu

[1] Some remarks on the intrinsic measures of Eisenman. Trans. Amer. Math. Soc. **288** (1985), 625–660
[2] Characterizations of the unit ball B^n in complex Euclidean space. Math. Z. **189** (1985), 449–456

Grant, C. G.

[1] Entire holomorphic curves in surfaces. Duke Math. J. **53** (1986), 345–358
[2] Hyperbolicity of surfaces modulo rational and elliptic curves. Pacific J. Math. **139** (1989), 241–249

Grauert, H.

[1] Characterisierung der Holomorphiegebiete durch die vollständige Kählersche Metrik. Math. Ann. **131** (1956), 38–75
[2] Über Modifikationen und exzeptionelle analytische Mengen. Math. Ann. **146** (1962), 331–368
[3] Mordells Vermutung über rationale Punkte auf algebraischen Kurven und Funktionenkörper. Publ. Math. I.H.E.S. 25 (1965), 131–149
[4] Set theoretic complex equivalence relations. Math. Ann. **265** (1983), 137–148
[5] On meromorphic equivalence relations. *Contributions to Several Complex Variables*, Aspects Math. E9, 115–147, Vieweg, 1986
[6] Meormorphe Äquivalenzrelationen. Math. Ann. **278** (1987), 175–183
[7] Jetmetriken und hyperbolische Geometrie. Math. Z. **200** (1989), 149–168

Grauert, H., and U. Peternell

[1] Hyperbolicity of the complement of plane curves. Manuscripta Math. **50** (1985), 429–441

Grauert, H., and H. Reckziegel

[1] Hermitesche Metriken und normale Familien holomorpher Abbildungen. Math. Z. **89** (1965), 108–125

Grauert, H., and R. Remmert

[1] Plurisubharmonische Funktionen in komplexen Räume. Math. Z. **65** (1956), 175–194
[2] *Theory of Stein Spaces*. Grundl. 236, Springer-Verlag, 1979
[3] *Coherent Analtic Sheaves*. Grundl. 265, Springer-Verlag, 1984

Green, M. L.

[1] Holomorphic maps into complex projective space omitting hyperplanes. Trans. Amer. Math. Soc. **169** (1972), 89–103
[2] Some Picard theorems for holomorphic maps to algebraic varieties. Amer. J. Math. **97** (1975), 43–75
[3] On the functional equation $f^2 = e^{2\varphi_1} + e^{2\varphi_2} + e^{2\varphi_3}$ and a new Picard theorem. Trans. Amer. Math. Soc. **195** (1974), 223–230
[4] The complement of the dual of a plane curve and some new hyperbolic manifolds. *Value Distribution Theory*, Vol A, pp. 119–131, Marcel Dekker (1974)
[5] Holomorphic mappings to Grassmannians of lines. Proc. Symp. Pure Math. Amer. Math. Soc. Vol. **27** (1975), 27–31
[6] Some examples and counter-examples in value distribution theory for several variables. Compositio Math. **30** (1975), 317–322
[7] The hyperbolicity of the complement of $2n + 1$ hyperplanes in general position in P_n, and related results. Proc. Amer. Math. Soc. **66** (1977), 109–113
[8] Holomorphic maps to complex tori. Amer. J. Math. **100** (1978), 615–620

Green, M. L., and P. A. Griffiths

[1] Two applications of algebraic geometry to entire holomorphic mappings. *The Chern Symposium 1979*, Springer-Verlag 1980, pp. 41–74

Greene, R. E., and S. Krantz

[1] Stability properties of the Bergman kernel and curvature properties of bounded domains. *Recent Developments in Several Complex Variables*, Ann. of Math. Studies, **100** (1981), 179–198
[2] Stability of the Carathéodory and Kobayashi metrics and applications to biholomorphic mappings. Proc. Symp. Pure Math. **41** (1984), 77–93
[2] Characterizations of certain weakly pseudoconvex domains with non-compact automorphism groups. Lecture Notes in Math. **1276** (1987), 121–157

Greene, R. E., and H. Wu

[1] Curvature and complex analysis. Bull. Amer. Math. Soc. **77** (1971), 1045–1049
[2] *Function Theory on Manifolds Which Possess a Pole*. Lecture Notes Math. 699, Springer-Verlag, 1979

Griffiths, P. A.

[1] Holomorphic mappings into canonical algebraic varieties. Ann. of Math. **93** (1971), 439–458
[2] Two theorems on extensions of holomorphic mappings. Invent. Math. **14** (1971), 27–62
[3] Some remarks on Nevanlinna theory. *Value Distribution Theory*, part A, Dekker, 1974, pp. 1–11
[4] Holomorphic mappings: Survey of some results and discussion of open problems. Bull. Amer. Math. Soc. **78** (1972), 374–382
[5] A Shottky-Landau theorem for holomorphic mappings in several complex variables. Symp. Math. vol. 10 (Convegno Geom. Diff. INDAM, Rome), Academic Press, 1972, pp. 229–243
[6] Two results in the global theory of holomorphic mappings. *Contributions to Analysis, dedicated to L. Bers*, Academic Press, 1974, pp. 169–183
[7] Differential geometry and complex analysis. Proc. Symp. Pure Math., vol. 27, part II, Amer. Math. Soc. 1975, pp. 43–64

[8] *Entire Holomorphic Mappings in One and Several Complex Variables.* Ann. Math. Studies No. 85, Princeton U. Press, 1976

Griffiths, P. A., and J. Harris

[1] Algebraic geometry and local differential geometry. Ann. Sci. École Norm. Sup. **12** (1979), 355–432

Griffiths, P. A., and J. King

[1] Nevanlinna theory and holomorphic mappings between algebraic varieties. Acta Math. 130 (1973), 145–220

Gunning, R. C.

[1] On Vitali's theorem for complex spaces with singularities. J. Math. Mech. **8** (1959), 133–141

Gunning, R. C., and H. Rossi

[1] *Analytic Functions of Several Complex Variables.* Prentice Hall, 1965

Hall, P.

[1] Landau and Schottky theorems for holomorphic curves. Michigan Math. J. **38** (1991), 207–223

Hahn, K. T.

[1] On completeness of the Bergman metric and its subordinate metrics. Proc. Nat. Acad. Sci. USA 73 (1976), 4294; same title II, Pacific J. Math. **68** (1977), 437–446
[2] Inequality between the Bergman metric and Carathéodory differential metric. Proc. Amer. Math. Soc. **68** (1978), 193–194
[3] Some remarks on a new pseudo-differential metric. Annales Polonici Math. **39** (1981), 71–81
[4] Equivalence of the classical theorems of Schottky, Landau, Picard and hyperbolicity. Proc. Amer. Math. Soc. **89** (1983), 628–632
[5] Asymptotic behavior of normal mappings of several complex variables. Canad. J. Math. **36** (1984), 718–746
[6] Asymptotic properties of normal and nonnormal holomorphic functions. Bull. Amer. Math. Soc. **11** (1984), 151–154
[7] Higher dimensional generalizations of some classical theorems on normal meromorphic functions. Complex Variables **6** (1986), 109–121
[8] Non-tangential limit theorems for normal mappings. Pacific J. Math. **135** (1988), 57–64
[9] Boundary behavior of normal and nonnormal holomorphic mappings. *Analysis and Geometry*, 1987 (Taejon), 123–147, Korea Inst. Tech. Taejon, 1987

Hahn, K. T., and K. T. Kim

[1] Hyperbolicity of a complex manifold and other equivalent properties. Proc. Amer. Math. Soc. **91** (1984), 49–53; see Do Duc Thai, ibid. **112** (1994), 1025–1027

Hahn, K. T., and J. Mitchell

[1] Generalization of Schwarz-Pick lemma to invariant volume in Kähler manifolds. Trans. Amer. Math. Soc. **128** (1967), 221–231; the same title II, Canadian J. Math. **21** (1969), 669–674

Hahn, K. T., and P. Pflug

[1] The Kobayashi and Bergman metrics on generalized Thullen domains. Proc. Amer. Math. Soc. **104** (1988), 207–214

Hall, P.

[1] Landau and Schottky theorems for holomorphic curves. Michigan Math. J. **38** (1991), 207–223

Harris, J.

[1] *Algebraic Geometry*. Springer-Verlag, 1992

Harris, L. A.

[1] Schwarz-Pick systems of pseudometrics for domains in normed linear spaces. *"Advances in Holomorphy"*, ed. J. A. Barroso, North-Holland Math. Studies 34, 1979, 345–406

Harris, L. A., and J. P. Vigué

[1] A metric condition for equivalence of domains. Atti Accad. Naz. Lincei **67** (1979), 402–403

Hartshorne, R.

[1] *Ample Subvarieties of Algebraic Varieties*. Lecture Notes in Math. 156, Springer-Verlag, 1970
[2] Amples vector bundles on curves. Nagoya Math. J. **43** (1971), 73–89

Hayashi, M.

[1] The maximal ideal space of the bounded analytic functions on the Riemann surface. J. Math. Soc. Japan **39** (1989), 337–344

Heins, M.

[1] On a class of conformal metrics. Nagoya Math. J. **21** (1962), 1–60

Henkin, G. M.

[1] Integral representations of functions holomorphic in strictly pseudoconvex domains, and some applications. Math. USSR. Sb. **7** (1969), 597–616
[2] An analytic polyhedron is not holomorphically equivalent to a strictly pseudoconvex domain. Soviet Math. Dokl. **14** (1973), 858–862

Herbort, G.

[1] On invariant differential metrics near pseudoconvex boundary points where the Levi form has corank one. Nagoya Math. J. **130** (1993), 25–54; erratum **135** (1994), 149–152

Hirschowitz, A.

[1] Domaines de Stein et fonctions holomorphes bornés. Math. Ann. **213** (1975), 185–193

Hirzebruch, F.

[1] *Topological Methods in Algebraic Geometry*. Springer-Verlag, 1966

Hörmander, L.

[1] *An Introduction to Complex Analysis in Several Variables*. Van Nostrand, 1966

Horst, C.

[1] Compact varieties of surjective holomorphic endomorphisms. Math. Z. **190** (1985), 499–504
[2] Compact varieties of surjective holomorphic mappings. Math. Z. **196** (1987), 259–269
[3] Two examples concerning hyperbolic quotients. Arch. Math. **49** (1987), 456–458
[4] A finiteness crieterion for compact varieties of surjective holomorphic mappings. Kodai Math. J. **13** (1990), 373–376

Horstmann, H.

[1] Carathéodorysche Metrik und Regularitätshüllen. Math. Ann. **108** (1933), 208–217

Howard, A., and Y. Matsushima

[1] Weakly ample vector bundles and submanifolds of complex tori. *Rencontre sur l'Analyse Complexe à Plusieurs Variables et les Systèmes Surdeterminés*, Presse Univ. de Montréal, 1975, 65–104

Howard, A., and A. J. Sommese

[1] On the orders of the automorphism groups of certain projective manifolds. *Manifolds and Lie Groups*, pp. 145–158, Progr. Math. 14, Birkhäuser, 1981

[2] On the theorem of de Franchis. Ann. Scuola Norm. Sup. Pisa **10** (1983), 429–436

Hristov, V. Z.

[1] The Carathéodory pseudometric under factorization of complex manifolds. C. R. Acad. Bulgare Sci. **30** (1977), 643–644

[2] On the topology induced by the Carathéodory metric. C. R. Acad. Bulgare Sci. **33** (1980), 1599–1602

[3] Convergence of sequences of Carathéodory or Kobayashi pseudometrics. PLISKA. Studia. Math. Bulgarica **4** (1981), 62–77

[4] Some topological properties of a natural equivalence relation defined by the Kobayashi and Carathéodory pseudodistances. C. R. Acad. Bulgar. Sci. **37** (1984), 445–448

[5] A new example of typical Carathéodory and Kobayashi pseudodistances. C. R. Acad. Bulgar. Sci. **38** (1985), 973–976

[6] Examples of typical Carathéodory and Kobayashi pseudodistances. C. R. Acad. Bulgar. Sci. **39** (1986), 23–25

[7] A property of typical Kobayashi pseudodistance on compact complex spaces. C. R. Acad. Bulgar. Sci. **40** (1987), 17–19

Hu, P. C., and C-C. Yang

[1] Pseudo volume forms and their applications to holomorphic mappings. Proc. Japan Acad. **69** (1993), 149–153

Huang, X-J.

[1] A non-degeneracy property of extremal mappings and iterates of holomorphic self-mappings. Ann. Scuola Nor. Sup. Pisa **21** (1994), 339–419

Huckleberry, A. T., and M. Sauer

[1] On the order of the automorphim group of a surface of general type. Math. Z. **205** (1990), 321–329

Huckleberry, A. T., and J. Winkelmann

[1] Subvarieties of parallelizable manifolds. Math. Ann. **295** (1993), 469–483

Iannuzzi, A.

[1] Balls for the Kobayashi distance and extension of the automorphisms of strictly convex domains in $\mathbf{C}^n$ with real analytic boundary. Atti. Accad. Naz. Lincei, Cl. Sci. Fis. Mat. Natur. Rend. Lincei Mat. Appl. **5** (1994), 193–196

Iitaka, S.

[1] On D-dimensions of algebraic varieties. J. Math. Soc. Japan **23** (1971), 356–373

[2] Logarithmic forms of algebraic varieties. J. Fac. Sci. Univ. Tokyo **23** (1976), 525–544

[3] On logarithmic Kodaira dimension of algebraic varieties. *Complex Analysis and Algebraic Geometry*, Iwanami, Tokyo, 1977, pp. 175–189

[4] Geometry on complements of lines in P^2, Tokyo J. Math. **1** (1978), 1–19

[5] *Algebraic Geometry*. Graduate Texts in Math. 76, Springer-Verlag, 1981

Imayoshi, Y.

[1] Generalization of de Franchis theorem. Duke Math. J. **50** (1983), 393–408

[2] An analytic proof of Severi's theorem. Complex Variables **2** (1983), 151–155

[3] Holomorphic mappings of compact Riemann surfaces into 2-dimensional compact C-hyperbolic manifolds. Math. Ann. **270** (1985), 403–416

[4] Holomorphic maps of projective algebraic manifolds into compact C-hyperbolic manifolds. J. Math. Soc. Japan **46** (1994), 289–307

Imayoshi, Y., and H. Shiga

[1] A finitness theorem for holomorphic families of Riemann surfaces. *Holomorphic Functions and Moduli II*, pp. 207–219, Springer-Verlag, 1988

Jarnicki, M., and P. Pflug

[1] Three remarks about the Carathéodory distance. Seminar on Deformations, Lódź-Lublin 1985/86, Part II: *Complex Analytic Geometry*, D. Reidel Publ. Co.
[2] Effective formulas for the Carathéodory distance. Manuscripta Math. **62** (1988), 1–20
[3] Bergman completeness of complete circular domains. Ann. Polonici Math. **L** (1989), 219–222
[4] The Carathéodory pseudo-distance has the product property. Math. Ann. **285** (1989), 161–164
[5] The simplest example for the non-innerness of the Carathéodory distance. Result. Math. **18** (1990), 57–59
[6] The inner Carathéodory distance for the annulus. Math. Ann. **289** (1991), 335–339; same title II, Michigan Math. J. **40** (1993), 393–398
[7] Invariant pseudodistances – completeness and product property. Ann. Polonici Math. **55** (1991), 169–189
[8] Some remarks on the product property for invariant pseudometrics. Proc. Symp. Pure Math., Amer. Math. Soc. **52** (1991), Part 2, 263–272
[9] A counter example for the Kobayashi completeness of balanced domains. Proc. Amer. Math. Soc. **112** (1991), 973–978
[10] *Invariant Distances and Metrics in Complex Analysis*. Walter de Gruyter, 1993
[11] Geodesics for convex complex ellipsoids I. Ann. Sculoa Norm. Sup. Pisa **20** (1993), 535–543; II Arch. Math. **65** (1995), 138–140

Jarnicki, M., P. Pflug and J-P. Vigué

[1] The Carathéodory distance does not define the topology – the case of domains. C. R. Acad. Sci. Paris **312** (1991), 77–79
[2] A remark on Carathéodory balls. Archiv Math. **58** (1992), 595–598
[3] An example of a Carathéodory complete but not finitely compact analytic space. Proc. Amer. Math. Soc. **118** (1993), 537–539

Järvi, P.

[1] An extension theorem for normal functions in several variables. Proc. Amer. Math. Soc. **103** (1988), 1171–1174

Joseph, J. E., and M. H. Kwack

[1] Hyperbolic imbedding and spaces of continuous extensions of holomorphic maps. J. Geom. Analysis **4** (1994), 361–378
[2] The topological nature of two Noguchi theorems on sequences of holomorphic mappings between complex spaces. Canadian J. Math. **47** (1995), 1240–1252
[3] Some classical theorems and families of normal maps in several complex variables. Complex Variables **29** (1996), 343–362
[4] Hyperbolic points and imbeddness modulo closed sets. Preprint
[5] Extension and convergence theorems for families of normal maps in several variables. Proc. Amer. Math. Soc. **125** (1997), 1675–1684
[6] A note on hyperbolically imbedded spaces. Preprint

Kaliman, S.

[1] The limiting behavior of the Kobayashi-Royden pseudometric. Trans. Amer. Math. Soc. **339** (1993), 361–371
[2] Some facts about Eisenman intrinsic measures. Complex Variables Theory Appl. **27** (1995), 163–173; II Proc. Amer. Math. Soc. **124** (1996), 3805–3811

[3] Exotic analytic structures and Eisenman intrinsic measures. Israel J. Math. **88** (1994), 411–423

Kaliman, S., and M. Zaidenberg

[1] Non-hyperbolic complex spaces with hyperbolic normalization. Preprint (1990), (ftp://iu-math.math.indiana.edu/pub/scv)
[2] A transversality theorem for holomorphic mappings and stability of Eisenman-Kobayashi measures. Trans. Amer. Math. Soc. **348** (1996), 661–672

Kalka, M.

[1] A deformation theorem for the Kobayashi metric. Proc. Amer. Math. Soc. **59** (1976), 245–251

Kalka, M., B. Shiffman and B. Wong

[1] Finiteness and rigidity theorems for holomorphic mappings. Michigan Math. J. **28** (1981), 289–295

Kani, E.

[1] Bounds on the number of non-rational subfields of a function field. Invent. Math. **85** (1986), 185–198

Kaup, W.

[1] Infinitesimale Transformationsgruppen komplexer Räume. Math. Ann. **160** (1965), 72–92
[2] Reelle Transformationsgruppen und invariante Metriken auf komplexen Räumen. Invent. Math. **3** (1967), 43–70
[3] Holomorphe Abbildungen in Hyperbolische Räume. Geometry of Bounded Homogeneous Domains, C.I.M.E. (1967), 111–127
[4] Hyperbolische komplexe Räume. Ann. Inst. Fourier **18** (1968), 303–330
[5] Holomorphic mappings of complex spaces. Symposia Math. **2** (1968), 333–340
[6] Some remarks on the automorphism groups of complex spaces. Rice Univ. Studies **56** (1970), 181–186

Kay, L. D.

[1] On the Kobayashi-Royden metric for ellipsoids. Math. Ann. **289** (1991), 55–72
[2] On the Kobayashi metric for perturbations of the ball. Math. Z. **212** (1993), 201–209

Kawamata, Y.

[1] On Bloch's conjecture. Invent. Math. **57** (1980), 97–100

Kelley, J. L.

[1] *General Topology*. Van Nostrand, 1955

Kerner, H.

[1] Über die Automorphismengruppen kompakter komplexer Räume. Arch. Math. **11** (1960), 282–288

Kerzman, N.

[1] Taut manifolds and domains of holomorphy in $\mathbf{C}^n$. Notices Amer. Math. Soc. **16** (1969), 675–676

Kerzman, N., and J-P. Rosay

[1] Fonctions plurisousharmoniques d'exhaustion bornées et domaines tauts. Math. Ann. **257** (1981), 171–184

Khenkin, G. M.

[1] An analytic polyhedron holomorphically non-equivalent to a strictly pseudoconvex domain. Soviet Math., Dokl. **14** (1973), 858–862

Kiernan, P. J.

[1] Hyperbolic submanifolds of complex projective space. Proc. Amer. Math. Soc. **22** (1969), 603–606

[2] On the relation between taut, tight and hyperbolic manifolds. Bull. Amer. Math. Soc. **76** (1970), 49–51

[3] Quasiconformal mappings and Schwarz's lemma. Trans. Amer. Math. Soc. **148** (1970), 185–197

[4] Some results concerning hyperbolic manifolds. Proc. Amer. Math. Soc. **25** (1970), 588–592

[5] Extension of holomorphic maps. Trans. Amer. Math. Soc. **172** (1972), 347–355

[6] Hyperbolically imbedded spaces and the big Picard theorem. Math. Ann. **204** (1973), 203–209

[7] Holomorphic extension theorems. *Value Distribution Theory*, Marcel Dekker (1973), 97–107

[8] On the compactifications of arithmetic quotients of symmetric spaces. Bull. Amer. Math. Soc. **80** (1974), 109–110

[9] Meromorphic mappings into compact compact complex spaces of general type. Proc. Symp. Pure Math., Amer. Math. Soc. **30** (1977), 239–244

Kiernan, P. J., and S. Kobayashi

[1] Satake compactification and extension of holomorphic mappings. Invent. Math. **16** (1972), 237–248

[2] Holomorphic mappings into projective space with lacunary hyerplanes. Nagoya Math. J. **50** (1973), 199–216

[3] Comments on Satake compactification and the great Picart theorem. J. Math. Soc. Japan **28** (1976), 577–580

Kim, D. S.

[1] Boundedly holomorphic convex domains. Pacific J. Math. **46** (1973), 441–449

[2] Complete domains with respect to the Carathéodory distance. II. Proc. Amer. Math. Soc. **553** (1975), 141–142

Kim, K-T.

[1] Complete localization of domains with noncompact automorphism groups. Trans. Amer. Math. Soc. **319** (1990), 139–153

[2] Domains in $\mathbf{C}^n$ with a piecewise Levi flat boundaries which possess a noncompact automorphism group. Math. Ann. **292** (1992), 575–586

[3] Asymptotic behavior of the curvature of the Bergman metric of the thin domains in $\mathbf{C}^n$. Pacific J. Math. **155** (1992), 99–110

[4] *Geometry of Bounded Domains and the Scaling Technique in Several Complex Variables*. GARC Lecture Notes Ser. 13, Seoul National University, 1993

Kim, K. T., and J. Yu

[1] Boundary behavior of the Bergman curvature in strictly polyhedral domains. Pacific J. Math. **176** (1996), 141–163

Klembeck, P.

[1] Kähler metrics of negative curvature, the Bergman metric near the boundary and the Kobayashi metric on smooth bounded strictly pseudoconvex sets. Indiana Univ. Math. J. **27** (1978), 275–282

Klimek, M.

[1] Extremal plurisubharmonic functions and invariant pseudodistances. Bull. Soc. Math. France **113** (1985), 231–240

[2] Infinitesimal pseudodistances and the Schwarz Lemma. Proc. Amer. Math. Soc. **105** (1989), 134–140

[3] Metrics associated with extremal plurisubharmonic functions. Proc. Amer. Math. Soc. **123** (1995), 2763–2770

[4] Invariant pluricomplex Green functions. *Topics in Complex Analysis*, 207–226, Banach Center Publ. 31, 1995

Kobayashi, R.

[1] Holomorphic curves into algebraic subvarieties of an abelian variety. Intern'l J. Math. **2** (1991), 711–724

Kobayashi, S.

[1] Geometry of bounded domains. Trans. Amer. Math. Soc. **92** (1959), 267–290

[2] On the automorphism group of a certain class of algebraic manifolds. Tohoku Math. J. **11** (1959), 184–190

[3] On complete Bergman metrics. Proc. Amer. Math. Soc. **13** (1962), 511–513

[4] Invariant distances on complex manifolds and holomorphic mappings. J. Math. Soc. Japan **19** (1967), 460–480

[5] Distance, holomorphic mappings and the Schwarz lemma. J. Math. Soc. Japan **19** (1967), 481–485

[6] Volume elements, holomorphic mappings and Schwarz's lemma. Proc. Symp. Pure Math. vol. **11** (1968), 253–260

[7] *Hyperbolic Manifolds and Holomorphic Mappings*. Marcel Dekker, New York, 1970

[8] *Transformation Groups in Differential Geometry*. Ergebnisse 70, Springer, 1972

[9] Schwarz lemma. *Symmetric Spaces*, Washington Univ. 1969, Marcel Dekker, (1972), pp. 247–254

[10] Some remarks and questions concerning the intrinsic distance. Tohoku Math. J. **25** (1973), 481–486

[11] Some problems on intrinsic distances and measures. Carathéodory Symposium, Sept. 3–7, 1973, Greek Math. Soc., pp. 306–317

[12] On hyperbolic complex spaces and extension problems. Proc. Intern'l Conf. on Manifolds and Related Topics in Topology, Tokyo, 1973, pp. 333–341

[13] Negative vector bundles and complex Finsler structures. Nagoya Math. J. **57** (1975), 153–166

[14] Intrinsic distances, measures, and geometric function theory. Bull. Amer. Math. Soc. **82** (1976), 357–416

[15] Complex manifolds with nonpositive holomorphic sectional curvature and hyperbolicity. Tohoku Math. J. **30** (1978), 487–489

[16] Projective structures of hyperbolic type. Minimal Submanifolds and Geodesics, Kaigai Publ. Tokyo (1978), pp. 85–92

[17] First Chern class and holomorphic tensor fields. Nagoya Math. J. **77** (1980), 5–11

[18] The first Chern class and holomorphic symmetric tensor fields. J. Math. Soc. Japan **32** (1980), 325–329

[19] Holomorphic projective structures and invariant distances. Proc. 1981 Hangzhou Conf. Several Complex Variables, Birkhäuser (1984), pp. 67–71

[20] Projectively invariant distances for affine and projective structures. Differential Geometry, Banach Center Publ., vol. 12, PWN – Polish Sci. Publ., Warsaw, 1984, pp. 127–152

[21] *Differential Geometry of Complex Vector Bundles*. Iwanami Shoten/Princeton U. Press, 1987

[22] A new invariant infinitesimal metric. Intern'l J. Math. **1** (1990), 83–90

[23] Theorem of Busemann-Mayer on Finsler metrics. Hokkaido Math. J. **20** (1991), 205–211

[24] Relative intrinsic distance and hyperbolic imbedding. Intern'l Symp. *Holomorphic Mappings, Diophantine Geometry and Related Topics*, RIMS. Kyoto, 1992, 239–242
[25] A new relative invariant distance and applications. Symposia Math. 36, *Manifolds and Geometry*, (Pisa 1993), 261–266. Cambridge Univ. Press, 1996
[26] On the intrinsic relative distance. *Geometric Complex Analysis*, Proc. 3rd Intern'l Research Inst. Math. Soc. Japan (1995), 1996, 355–361

Kobayashi, S., and K. Nomizu

[1] *Foundations of Differential Geometry*. I, 1963; II, 1969, Interscience, John Wiley & Sons

Kobayashi, S., and T. Ochiai

[1] Mappings into compact complex manifolds with negative first Chern class. J. Math. Soc. Japan **23** (1971), 137–148
[2] Satake compactification and the great Picard theorem. J. Math. Soc. Japan **23** (1971), 340–350
[3] Meromorphic mappings into compact complex spaces of gneral type. Invent. Math. **31** (1975), 7–16

Kobayashi, S., and T. Sasaki

[1] Projective structures with trivial intrinsic pseudo-distance. *Minimal Submanifolds and Geodesics*, Kaigai Publ. Tokyo 1978, 93–99

Kobayashi, S., and H. Wu

[1] On holomorphic sections of certain hermitian vector bundles. Math. Ann. **189** (1970), 1–4

Kodaira, K.

[1] Holomorphic mappings of polydiscs into compact complex manifolds. J. Diff. Geometry **6** (1971), 33–46
[2] *Complex Manifolds and Deformation of Complex Structures*. Grundl. Math. Wiss. 283, Springer, 1985

Kodama, A.

[1] On homogeneous Carathéodory-hyperbolic complex manifolds. Memoirs Fac. Educ. Akita Univ., Natur. Sci. **29** (1979), 41–45
[2] On bimeromorphic automorphisms of hyperbolic complex spaces. Nagoya Math. J. **73** (1979), 1–5
[3] On boundedness of circular domains. Proc. Japan Acad. Sci. **58** (1982), 227–230
[4] Remarks on homogeneous hyperbolic complex manifolds. Tohoku Math. J. **35** (1983), 181–186
[5] On the structure of a bounded domain with a special boundary point. Osaka J. Math. 23 (1986), 271–298; same title II, Osaka J. Math. **24** (1987), 499–519
[6] A simple proof of Kubota's theorem on the Kobayashi and Carathéodory distances. Math. Rep. Toyama Univ. **12** (1989), 201–205

Kodama, A., and Y. Sakane

[1] Remarks on compact complex hypersurfaces on compact complex parallelisable nilmanifolds. Tohoku Math. J. **29** (1977), 57–60

Kodama, A., and H. Shima

[1] Characterizations of homogeneous bounded domains. Tsukuba J. Math. **7** (1983), 79–86

Kollár, J., and T. Matsusaka

[1] Riemann-Roch type inequalities. Amer. J. Math. **105** (1983), 229–252

Korányi, A.

[1] A Schwarz lemma for bounded symmetric domains. Proc. Amer. Math. Soc. **17** (1966), 210–213

Krantz, S. G.

[1] *Function Theory of Several Complex Variables.* Wiley, 1982
[2] *Complex Analysis: the Geometric Viewpoint.* Carus Math. Monograph 23, MAA, 1990
[3] Convexity in complex analysis. Proc. Symp. Pure Math. 52, Part 1, *Several Complex Variables and Complex Geometry*, pp. 119–137, Amer. Math. Soc. 1991
[4] The boundary behavior of the Kobayashi metric. Rocky Mountain J. Math. **22** (1992), 227–233
[5] Invariant metrics and the boundary behavior of holomorphic functions on domains in $\mathbf{C}^n$. J. Geometric Analysis **1** (1991), 71–97
[6] *Geometric Analysis and Function Spaces.* CBMS Amer. Math. Soc. 81, 1993

Krasiński, T.

[1] On biholomorphic invariants related to homology groups. *Analytic Functions* (Blażejewko 1982), 276–284, Lecture Notes in Math. 1039, Springer, 1983

Królikowski, W.

[1] Biholomorphic invariants on relative homology groups. *Analytic Functions* (Blażejewko 1982), 285–319, Lecture Notes in Math. 1039, Springer, 1983

Królikowski, W., and L. M. Tovar

[1] Hyperbolic-type manifolds, homology groups and holomorphic mappings. Intern'l Seminar Algebra and its Appl. (Mexico City 1991), 109–125, Soc. Mat. Mexicana, 1992

Krushkal', S. L.

[1] Invariant metrics on Teichmüller spaces and quasiconformal extendability of analytic functions. Ann. Acad. Sci. Fenn. Ser. A I Math. **10** (1985), 299–303
[2] Hyperbolic metrics on finite-dimensional Teichmüller spaces. Ann. Acad. Sci. Fenn. Ser. A I Math. **15** (1990), 125–132
[3] Strengthening pseudoconvexity of finite dimensional Teichmüller spaces. Math. Ann. **290** (1991), 681–687

Kubota, Y.

[1] On the Kobayashi and Carathéodory distances of bounded symmetric domains. Kodai Math. J. **12** (1989), 41–48
[2] A generalization of a theorem of Landau. Kodai Math. J. **9** (1986), 241–244

Kuranishi, M.

[1] On the locally complete families of complex analytic structures. Ann. of Math. **75** (1962), 536–577
[2] New proof for the existence of locally complete families of complex structures. Proc. Conf. Complex Analysis, pp. 142–154, Minneapolis 1964, Springer, 1965

Kurke, H.

[1] An algebraic proof of a theorem of S. Kobayashi and T. Ochiai. (ETH preprint, 1978)

Kwack, M.

[1] Generalizations of the big Picard theorem. Ann. of Math. **90** (1969), 9–22
[2] A Schwarz lemma for canonical algebraic manifolds. Proc. Amer. Math. Soc. **41** (1973), 219–222
[3] Mappings into hyperbolic space. Bull. Amer. Math. Soc. **79** (1973), 695–697
[4] Some classical theorems for holomorphic mappings into hyperbolic manifolds. Proc. Symp. Pure Math., vol. 27, Amer. Math. Soc. 1975 pp. 99–104

[5] Meromorphic mappings into compact complex manifolds with a Grauert positive bundle of q-forms. Proc. Amer. Math. Soc. **87** (1983), 699–703

[6] *Families of Normals Maps in Several Variables and Classical Theorems in Complex Analysis.* Lecture Notes 33, Res. Inst. Math., Global Analysis Res. Center, Seoul, Korea, 1996

Landau, E.

[1] Über die Blochsche Konstante und zwei verwandte Weltkonstanten. Math. Z. **30** (1929), 608–634

Lang, S.

[1] Higher dimensional diophantine problems. Bull. Amer. Math. Soc. **80** (1974), 779–787

[2] Hyperbolic and diophantine analysis. Bull. Amer. Math. Soc. **14** (1986), 159–205

[3] *Introduction to Complex Hyperbolic Spaces.* Springer-Verlag, 1987

[4] Diophantine problems in complex hyperbolic analysis. *Current Trends in Arithmetic Algebraic Geometry* (Arcata, Cailf. 1985), 229–246, Contemp. Math. 67, Amer. Math. Soc. 1987

[5] The error terms in Nevanlinna theory. Duke Math. J. **56** (1988), 193–218

[6] *Number Theory* III. *Diophantine Geometry*, Encyclopaedia Math. Sci. vol. 60, Springer, 1991

Lang, S., and W. Cherry

[1] *Topics in Nevanlinna Theory.* Lecture Notes in Math. 1433, Springer, 1990

Langmann, K.

[1] Abschätzungen für die Anzahl der holomorphen Abbildungen zwischen algebraischen Räumen. Math. Z. **190** (1985), 411–418

[2] Picard-Borel-Eigenschaft and Anwendungen. Math. Z. **192** (1986), 587–602

[3] Endlichkeits- und Picard-Sätze für quasiprojektive Räume. Math. Z. **197** (1988), 483–504

[4] Picard-Borel Räume. Math. Ann. **284** (1989), 137–163

[5] Gruppennahe Räume. J. Reine Angew. Math. **452** (1994), 1–37

[6] Wertverhalten holomorpher Funktionen auf Überlagerungen und zahlentheoretische Analogien. Math. Ann. **299** (1994), 127–153

[7] Endlichkeitssätze für die Anzahl der holomorphen Abbildungen. Math. Nachr. **173** (1995), 237–257

Lehto, O., and K. I. Virtanen

[1] Boundary behavior and normal meromorphic functions. Acta. Math. **97** (1957), 47–65

Lelong, P.

[1] *Fonctions Plurisousharmoniques et Formes Différentielles Positives.* Gordon and Breach, 1968

Lempert, L.

[1] La métrique de Kobayashi et la représentation des domaines sur la boule. Bull. Soc. Math. France **109** (1981), 427–474

[2] Holomorphic retracts and intrinsic metrics in convex domains. Analysis Math. **8** (1982), 247–261

[3] Intrinsic distances and holomorphic retracts. *Complex Analysis and Applications* '81 (Varna 1981), 341–364, Bulgar. Acad. Sci., Sofia, 1984

[4] A precise result on the boundary regularity of biholomorphic mappings. Math. Z. **193** (1986), 559–579; erratum **206** (1991), 501–504

[5] Complex geometry in convex domains. Proc. Intern'l Congress Math. Berkeley (1986), 759–765

[6] Holomorphic invariants, normal forms, and moduli space of convex domains. Ann. Math. **128** (1988), 43–78

[7] Metamorphoses of the Kobayashi metric. Proc. GARC Workshop on Geometry and Topology '93 (Seoul), 177–210, Lecture Notes Ser. 18 Seoul Nat. Univ., 1993

[8] Elliptic and hyperbolic tubes. *Several Complex Variables*, (Stockholm, 1987/88), 440–456, Math. Notes 38, Princeton Univ. Press, 1993

Leung, K.-W., G. Patrizio, and P.-M. Wang

[1] Isometries of intrinsic metrics on strictly convex domains. Math. Z. **196** (1987), 343–353

Liebermann, D., and D. Mumford

[1] Matsusaka's big theorem. Proc. Symp. Pure Math., Amer. Math. Soc. **29** (1975), 513–530

Lin, E. B., and B. Wong

[1] Boundary localization of the normal family of holomorphic mappings and remarks on existence of bounded holomorphic functions on complex manifolds. Illinois J. Math. **34** (1990), 656–664

Lin, V. Ya., and M. G. Zaidenberg

[1] Finiteness theorems for holomorphic mappings. Encyclopaedia Math. Sci. Vol. 9, *Several Complex Variables* III, 113–172, Springer, 1989

Look, K. H.

[1] Schwarz Lemma and analytic invariants. Sci. Sinica **7** (1958), 435–504

Ma, D. W.

[1] Boundary behavior of invariant metrics and volume forms on strongly pseudoconvex domains. Duke Math. J. **63** (1991), 673–697

[2] On iterates of holomorphic maps. Math. Z. **207** (1991), 417–428

[3] Sharp estimates of the Kobayashi metric near strongly pseudoconvex points. *The Madison Symposium on Complex Analysis* (1991), Contemp. Math. 137, Amer. Math. Soc. 1992, 329–338

[4] Smoothness of Kobayashi metric of ellipsoids. Complex Variables Theory Appl. **26** (1995), 291–298

Maarouf, M. A.

[1] Hyperbolicity problems on the Douady space and its variants. Inst. Fourier preprint no. 338 (1996)

Maehara, K.

[1] Family of varieties dominated by a variety. Proc. Japan Acad. **55** (1979), 146–151

[2] A finiteness property of varieties of general type. Math. Ann. **262** (1983), 101–123

[3] The Mordell-Bombieri-Noguchi conjecture over function fields. Kodai. Math. J. **11** (1988), 1–4

[4] On the higher dimensional Mordell conjecture over function fields. Osaka J. Math. **28** (1991), 255–261

[5] Diophantine problem of algebraic varieties and Hodge theory. Intern'l Symp. *Holomorphic Mappings, Diophantine Geometry and Related Topics*, RIMS, Kyoto, 1992, 167–187

Manin, Y.

[1] Rational Points of algebraic curves over function fields. Izvestia Akad. Nauk **27** (1963), 1395–1440

Markowitz, M.

[1] Holomorphic affine and projective connections of hyperbolic type. Math. Ann. **245** (1979), 55–62

Martens, H. H.

[1] Remarks on de Franchis' theorem. *Complex Analysis*, Springer Lecture Notes in Math. 1013 (1983), 160–163

Martin-Deschamps, M.

[1] Propriétés de descente des variétés à fibre cotangent ample. Ann. Inst. Fourier, Grenoble **33** (1984), 39–64
[2] La construction de Kodaira-Parshin. Séminaire sur les pinceaux arithmétiques: la conjecture de Mordell. Soc. Math. France, Astérisque **127** (1985), 261–271

Martin-Deschamps, M., and R. Lewin-Menegaux

[1] Applications rationelles séparables dominantes sur une variété de type général. Bull. Soc. Math. France **106** (1978), 279–287
[2] Surface de type général dominnées par une variété fixe. Bull. Soc. Math. France **110** (1982), 127–146

Masuda, K., and J. Noguchi

[1] A construction of hyperbolic hypersurfaces of $\mathbf{P}^n(\mathbf{C})$. Math. Ann. **304** (1996), 339–362

Matsuki, K.

[1] An approach to the abundance conjecture for 3-folds. Duke Math. J. **61** (1990), 207–220

Matsumura, H.

[1] On algebraic groups of birational transformations. Rend. Accad. Naz. Lincei **34** (1963), 151–155

Matsumura, H., and P. Monsky

[1] On the automorphisms of hypersurfaces. J. Math. Kyoto Univ. **3** (1964), 347–361

Matsusaka, T.

[1] On canonical polarized varieties, II. Amer. J. Math. **92** (1970), 283–292

Matsushima, Y.

[1] Holomorphic immersions of a compact Kähler manifold into complex tori. J. Diff. Geometry **9** (1974), 309–328

Matsushima, Y., and W. Stoll

[1] Ample vector bundles on compact complex spaces. Rice Univ. Studies **59** (1973), 71–107

Matsuura, S.

[1] The generalized Martin's minimum problem and its applications. Trans. Amer. Math. Soc. **208** (1975), 273–307
[2] Fundamental inequalities on higher order metrics in several complex variables. Bull. Nagoya Inst. Tech. **38** (1986), 93–98

Mazet, P., and J-P. Vigué

[1] Points fixes d'une application holomorphe d'un domaine borné dans lui-même. Acta Math. **166** (1991), 1–26
[2] Convexité de la distance de Carathéodory et points fixes d'applications holomorphes. Bull. Sci. Math. **116** (1992), 285–305

Mazur, T., P. Pflug and M. Skwarczyński

[1] Invariant distances related to the Bergman function. Proc. Amer. Math. Soc. **94** (1985), 72–76

McQuillan, M.

[1] A new proof of the Bloch conjecture. J. Alg. Geometry **5** (1996), 107–117

Mercer, P. R.

[1] Proper maps, complex geodesics and iterates of holomorphic maps on convex domains in $\mathbf{C}^n$. *The Madison Symposium on Complex Analysis*, (1991), Contemp. Math. **137**, Amer. Math. Soc. 1992, 339–342

[2] Complex geodesics and iterates of holomorphic maps on convex domains in $\mathbf{C}^n$. Trans. Amer. Math. Soc. **338** (1993), 201–211

Meyer, R.

[1] The Carathéodory pseudo-distance and positive linear operators. Intern'l J. Math. **8** (1997), 809–824

Milnor, J.

[1] On deciding whether a surface is parabolic or hyperbolic. Amer. Math. Monthly **84** (1977), 43–46

Minda, C. D.

[1] The Hahn metric on Riemann surfaces. Kodai Math. J. **6** (1983), 57–69

Miyano, T., and J. Noguchi

[1] Moduli spaces of harmonic and holomorphic mappings and diophantine geometry. Lecture Notes in Math. No. 1468, 227–253, Springer-Verlag, 1991

Miyaoka, Y., and S. Mori

[1] A numerical criterion for uniruledness. Ann. Math. **124** (1986), 65–69

Moishezon, B. G.

[1] On n-dimensional compact complex varieties with n algebraically independent meromorphic functions. Amer. Math. Soc. Transl. (2) **63** (1967), 51–177

Mok, N-G.

[1] The Serre problem on Riemann surfaces. Math. Ann. **258** (1981/82), 145–168

[2] Rigidity of holomorphic self-mappings and the automorphism group of hyperbolic Stein spaces. Math. Ann. **266** (1983), 433–447

Mori, S.

[1] Threefolds whose canonical bundles are not numerically effective. Ann. Math. **116** (1982), 133–176

Mori, S., and S. Mukai

[1] The uniruledness of the moduli space of curves of geneus 11. *Algebraic Geometry*, Proc. Japan-France Conf. 1982, pp. 334–358, Lecture Notes Math. 1016, Springer

Murata, T.

[1] Holomorphic mappings from the unit disk to algebraic varieties. Osaka J. Math. **28** (1991), 623–637

Nadel, A. M.

[1] Hyperbolic surfaces in $\mathbf{P}^3$. Duke Math. J. **58** (1989), 749–771

Nag, S.

[1] Hyperbolic manifolds admitting holomorphic fiberings. Bull. Australian Math. Soc. **26** (1982), 181–184

Nagano, T., and B. Smyth

[1] Minimal varieties and harmonic maps in tori. Comment. Math. Helv. **50** (1975), 249–265

Nakajima, K.

[1] Homogeneous hyperbolic manifolds and homogeneous Siegel domains. J. Math. Kyoto Univ. **25** (1985), 269–291

Namba, M.

[1] *Families of Meromorphic Functions on Compact Riemann Surfaces.* Lecture Notes Math. 767, Springer, 1979

Narasimhan, R.

[1] *Introduction to the Theory of Analytic Spaces.* Lecture Notes Math. no. 26, Springer-Verlag, 1966
[2] *Analysis on Real and Complex Manifolds.* North Holland, 1968
[3] *Several Complex Variables.* The University of Chicago Press, 1971

Narasimhan, M. S., and R. R. Simha

[1] Manifolds with ample canonical class. Invent. Math. **5** (1968), 120–128

Nevanlinna, R.

[1] *Le Théorème de Picard-Borel et la Théorie des Fonctions Méromorphes.* 1939, Gauthier-Villars

Nishimura, Y.

[1] Automorphismes analytiques admettant des sous-variétés de points fixés attractives dans la direction transversale. J. Math. Kyoto Univ. **23** (1983), 289–299
[2] Applications holomorphes injectives de $\mathbf{C}^n$ dans lui-même qui exceptent une droite complexe. J. Math. Kyoto Univ. **24** (1984), 755–761

Niino, K.

[1] Deficiencies of the associated curves of a holomorphic curve in the projective plane. Proc. Amer. Math. Soc. **59** (1976), 81–88

Nishino, T.

[1] Le théorème de Borel et le théorème de Picard. C. R. Acad. Sci. Paris **299** (1984), 667–668

Nishino, T., and M. Suzuki

[1] Sur les singularités essentielles et isolées des applications holomorphes à valeurs dans une surface complexe. Publ. RIMS **16** (1980), 461–497

Nochka, E. I.

[1] On the theory of meromorphic functions. Soviet Math. Dokl. **27** (1983), 377–381

Noguchi, J.

[1] Meromorphic mappings into a compact complex space. Hiroshima Math. J. **7** (1977), 411–425
[2] Holomorphic curves in algebraic varieties. Hiroshima Math. J. **7** (1977), 833–853
[3] Supplement to "Holomorphic curves in algebraic varieties". Hiroshima Math. J. **10** (1980), 229–231
[4] Lemma on logarithmic derivatives and holomorphic curves in algebraic varieties. Nagoya Math. J. **83** (1981), 213–233
[5] A higher dimensional analogue of Mordell's conjecture over function fields. Math. Ann. **258** (1981), 207–212

[6] A higher dimensional analogue of Mordell's conjecture over function fields and related problems. Proc. 1981 Hangzhou Conf. Several Complex Variables, pp. 237–244, Birkhäuser, 1984

[7] Hyperbolic fibre spaces and Mordell's conjecture over function fields. Publ. RIMS, Kyoto Univ. **21** (1985), 27–46

[8] On the value distribution of meromorphic mappings of covering spaces over $\mathbf{C}^m$ into algebraic varieties. J. Math. Soc. Japan **37** (1985), 295–313

[9] Logarithmic jet spaces and extensions of de Franchis' theorem. *Contributions to Several Complex Variables*, pp. 227–249, Aspects Math. No. 9, Vieweg, 1986

[10] Moduli spaces of holomorphic mappings into hyperbolically imbedded complex spaces and locally symmetric spaces. Invent. Math. **93** (1988), 15–34

[11] Hyperbolic manifolds and diophantine geometry, Sugaku Expositions. Amer. Math. Soc. **4** (1991), 63–81

[12] Moduli space of abelian varieties with level structure over function fields. Intern'l J. Math. **2** (1991), 183–194

[13] Meromorphic mappings into compact hyperbolic complex spaces and geometric diophantine problems. Intern'l J. Math. **3** (1992), 277–289; corrections, ibid. p. 677

[14] An example of a hyperbolic fiber space without hyperbolic imbedding into compactification. *Complex Geometry*, (Osaka, 1990), 157–160, Lecture Notes in Pure and Appl. Math. 143, Dekker, 1993

[15] Some problems in value distribution and hyperbolic manifolds. Intern'l Symp. *Holomorphic Mappings, Diophantine Geometry and Related Topics*, RIMS, Kyoto, 1992, 66–79

[16] Some topics in Nevanlinna theory, hyperbolic manifolds and Diophantine geometry. *Geometry and Analysis on Complex Manifolds*, (Festschrift for S. Kobayashi), ed. by Mabuchi et al., pp. 140–156, World Scientific, 1994

[17] On Nevanlinna's second main theorem. *Geometric Complex Analysis*, Proc. 3rd Intern'l Res. Inst. Math. Soc. Japan, (1995), 489–503, World Scientific, 1996

[18] On holomorphic curves in semi-abelian varieties. Math. Z., to appear

Noguchi, J., and T. Ochiai

[1] *Geometric Function Theory in Several Complex Variables*. Transl. Math. Monogr., vol. 80, Amer. Math. Soc. 1990. (Translation of "Kikagakuteki Kansuron", Iwanami, 1984)

Noguchi, J., and T. Sunada

[1] Finiteness of the family of rational and meromorphic mappings into algebraic varieties. Amer. J. Math. **104** (1982), 887–900

O'Byrne, B.

[1] On Finsler geometry and applications to Teichmüller spaces. Ann. Math. Studies **66** (1971), 317–328

Ochiai, T.

[1] Some remarks on the defect relation of holomorphic curves. Osaka J. Math. **11** (1974), 483–501

[2] On holomorphic curves in algebraic varieties with ample irregularity. Invent. Math. **43** (1977), 83–96

Ohgai, S.

[1] On the relative hyperbolicity of complex analytic spaces. Kumamoto J. Sci. (Math.) **15** (1982), 47–58

Ohsawa, T.

[1] On complete Kähler domains with C^1-boundary. Publ. Res. Inst. Math. Sci. **16** (1980), 929–940

[2] A remark on the completeness of the Bergman metric. Proc. Japan Acad. **57** (1981), 238–240

[3] On the Bergman kernel of hyperconvex domains. Nagoya Math. J. **129** (1993), 43–52; Addendum, ibid. **137** (1995), 145–148

[4] Applications of L^2 estimates and some complex geometry. *Geometric Complex Analysis*, 5-5-523, World Sci. Publ., 1996

Overholt, M.

[1] Injective hyperbolicity of domains. Ann. Polon. Math. **62** (1995), 79–82

Øvrelid, N.

[1] Integral representation formulas and L^p-estimates for the $\bar{\partial}$-equations. Math. Scand. **29** (1971), 137–160

Pang, M.-Y.

[1] Finsler metrics with the properties of the Kobayashi metric on convex domains. Publ. Math. **36** (1992), 131–151

[2] Smoothness of the Kobayashi metric of non-convex domains. Intern'l J. Math. **4** (1993), 953–987

[3] On infinitesimal behavior of the Kobayashi distance. Pacific J. Math. **162** (1994), 121–141

Parshin, A. N.

[1] Algebraic curves over function fields I. Akad. Nauk. SSSR Ser. Mat. **32** (1968), 1145–1170

[2] Finiteness theorems and hyperbolic manifolds. Progress in Math. v. 86, *The Grothendieck Festschrift*, vol. 3, 163–178, Birkhäuser, 1990

Patrizio, G.

[1] On holomorphic maps between domains in $\mathbf{C}^n$. Ann. Sc. Norm. Sup. Pisa **13** (1986), 267–279

[2] The Kobayashi metric and homogeneous complex Monge-Ampère equation. *Complex Analysis and Applications* '85 (Varna 1985), 515–523, Bulgar. Acad. Sci. Sofia, 1986

[3] Disques extrémaux de Kobayashi et équation de Monge-Ampère complexe. C. R. Acad. Sci. Paris **305** (1987), 721–724

[4] On the convexity of the Kobayashi indicatrix. *Deformations of Mathematical Structures*, (Lódź/Lublin, 1985/87), 171–176, Kluwer Acad. Publ. 1989

Patrizio, G., K-W. Leung and P-M. Wong

[1] Isometries of intrinsic metrics on strictly convex domains. Math. Z. **196** (1987), 343–353

Pelles, D. A. (= D. A. Eisenman)

[1] Holomorphic maps which preserve intrinsic measures. Amer. J. Math. **97** (1975), 1–15

Perrone, D.

[1] Remarks on intrinsic distances associated with flat affine structures. Istit. Lombardo Accad. Sci. Lett. Rend. **115** (1981), 279–292

Peternell, T.

[1] Calabi-Yau manifolds and conjecture of Kobayashi. Math. Z. **207** (1991), 305–318

Peters, Klaus

[1] Über holomorphe und meromorphe Abbildungen gewisser kompakter Mannigfaltigkeiten. Arch. Math. **15** (1964), 222–231

Peters, Konrad

[1] Starrheitssätze für Produkte normierter Vektorräume endlicher Dimension und für Produkte hyperbolischer komplexer Räume. Math. Ann. **208** (1974), 343–354

Pflug, P.

[1] Quadratintegrable holomorphe Funktionen und die Serre-Vermutung. Math. Ann. **216** (1975), 285–288
[2] Various applications of the existence of well growing holomorphic functions. *Functional Analysis, Holomorphy and Approximation Theory*, ed. J. A. Barroso, pp. 391–412, North-Holland Math. Studies 71, 1982
[3] About the Carathéodory completeness of all Reinhardt domains. *Function Analysis, Holomorphy and Approximation Theory* II, pp. 331–337, North-Holland Math. Studies 86, 1984

Pinchuk, S. I.

[1] On proper holomorphic mappings of strictly pseudoconvex domains. Siberian Math. J. **15** (1975), 644–649
[2] Holomorphic nonequivalence of some classes of domains in $\mathbf{C}^n$. Math. USSR. Sb. **39** (1981), 61–86

Poletskiĭ, E. A., and B. V. Shabat

[1] Invariant Metrics. *Several Complex Variables* III, G. M. Khenkin (ed.), Springer-Verlag, 1989, pp. 63–112

Pyatezkii-Shapiro, I. I.

[1] Arithmetic groups in complex domains. Russian Math. Survey **19** (1964), 83–109
[2] *Géométrie des Domaines Classiques et Théorie des Fonctions Automorphes*. Dunod, Paris 1966

Rabotin, V. V.

[1] A counter example to problems of S. Kobayashi. *Multidimensional Complex Analysis*, 256–258, Akad. Nauk. SSSR Sibirsk Otdel Inst. Fiz. Krasnoyarsk, 1985

Range, R. M.

[1] The Carthéodory metric and holomorphic maps on a class of weakly pseudoconvex domains. Pacific J. Math. **78** (1978), 173–189

Raynaud, M.

[1] Around the Mordell conjecture for function fields and a conjecture of Serge Lang. *Algebriac Geometry*, Proc. Japan-France Conf. 1982, Lecture Notes Math. 1016, pp. 1–19, Springer

Reckziegel, H.

[1] Hyperbolische Räume und normale Familien holomorpher Abbildungen. Göttingen, Dissertation, 1967

Reiffen, H.-J.

[1] Die differentialgeometrischen Eigenschaften der Invarianten Distanzfunktion von Carathéodory. Schriftenr. Math. Inst. Univ. Münster No. 26, 1963
[2] Die Carathéodorysche Distanz und ihre zugehörige Differentialmetrik. Math. Ann. **161** (1965), 315–324
[3] Invariante Metriken in der Funktionentheorie mehrerer Veränderlichen. Exposition. Math. **12** (1994), 227–241

Remmert, R.

[1] Holomorphe und meromorphe Abbildungen komplexer Räume. Math. Ann. **133** (1957), 328–370

Riebesehl, D.

[1] Hyperbolische komplexe Räume und die Vermutung von Mordell. Math. Ann. **257** (1981), 99–110

Rinow, W.

[1] *Die innere Geometrie der metrischen Räume.* Grundlehren Math. Wiss. vol. 105, Springer-Verlag, 1961

Robinson, R.

[1] A generalization of Picard's and related theorems. Duke Math. J. **5** (1939), 118–132

Rosay, J-P.

[1] Sur une caractérisation de la boule parmi les domaines de $\mathbf{C}^n$ par son groupe d'automorphismes. Ann. Inst. Fourier, Grenoble **29**(4) (1979), 91–97

[2] Un exemple d'ouvert borné de $\mathbf{C}^3$ "taut" mais non hyperbolique complet. Pacific J. Math. **98** (1982), 153–156

Royden, H.

[1] Report on the Teichmüller metric. Proc. Nat. Acad. Sci. **65** (1970), 497–499

[2] Remarks on the Kobayashi metric. Proc. Maryland Conference on Several Complex Variables, Lecture Notes Math. 185 (1971), 369–383

[3] Automorphisms and isometries of Teichmüller space. Ann. Math. Stusies No. 66, *Advances in the Theory of Riemann Surfaces*, 369–383, Princeton Univ. Press, 1971

[4] The extension of regular holomprhic maps. Proc. Amer. Math. Soc. **43** (1974), 306–310

[5] Holomorphic fiber bundles with hyperbolic fiber. Proc. Amer. Math. Soc. **43** (1974), 311–312

[6] Invariant metrics on Teichmüller space. *Contributions to Analysis* (dedicated to Lipman Bers), 393–399, Academic Press, 1974

[7] The Ahlfors-Schwarz lemma in several complex variables. Comment. Math. Helv. **55** (1980), 547–558

[8] The Ahlfors-Schwarz lemma: the case of equality. J. d'Analyse Math. **46** (1986), 261–270

[9] Complex Finsler metrics. *Complex Differential Geometry and Nonlinear Differential Equations* (Brunswick, Maine 1984), 119–124, Contemp. Math. 49, Amer. Math. Soc., 1986

[10] Hyperbolicity in complex analysis. Ann. Acad. Sci. Fennicae, Ser. A.1. Math. **13** (1988), 387–400

Royden, H., and P-M. Wang

[1] Carathéodory and Kobayashi metrics on convex domains. Preprint 1983

Ru, M.

[1] Integral points and the hyperbolicity of the complement of hypersurfaces. J. Reine Angew. Math. **442** (1993), 163–176

[2] Geometric and arithmetic aspects of P^n minus hyperplanes. Amer. J. Math. **117** (1995), 307–321

Ru, M., and P.-M. Wong

[1] Integral points of $P^n(K) - \{2n + 1$ hyperplanes in general position$\}$. Invent. Math. **106** (1991), 195–216

Rudin, W.

[1] Essential boundary points. Bull. Amer. Math. Soc. **70** (1964), 321–324

Sadullaev, A.

[1] Schwarz lemma for circular domains and its applications. Math. Notes **27** (1980), 120–125

Saito, K.

[1] Theory of logarithmic differential forms and logarithmic vector fields. J. Fac. Sci. Univ. Tokyo **27** (1980), 265–291

Saito, M.-H., and S. Zucker

[1] Classification of non-rigid families of K3 surfaces and a finiteness theorem of Arakelov-type. Math. Ann. **289** (1991), 1–31

Sakai, F.

[1] Degeneracy of holomorphic maps with ramification. Invent. Math. **26** (1974), 212–229
[2] Defect relations and ramification. Proc. Japan Acad. **50** (1974), 723–728
[3] Defect relations for equidimensional holomorphic maps. J. Fac. Sci. Univ. Tokyo **23** (1976), 561–580
[4] Kodaira dimensions of complements of divisors. *Complex Analysis and Algebraic Geometry*, pp. 239–257, Iwanami, 1977

Salinas, N.

[1] The Carathéodory and Kobayashi infinitesimal metrics and completely bounded homomorphisms. J. Operator Theory **26** (1991), 433–443

Samuel, P.

[1] *Lectures on old and new results on algebraic curves*. Lectures on Math., no. 36, Tata Inst. 1966

Satake, I.

[1] On compactifications of the quotient spaces for arithmetically defined discontinuous groups. Ann. of Math. **72** (1960), 555–580
[2] Holomorphic imbeddings of symmetric domains into a Siegel space. Amer. J. Math. **87** (1965), 425–461

Schneider, M.

[1] Tubenumgebungen Steinscher Räume. Manuscripta Math. **18** (1976), 391–397

Schumacher, G., and K. Takegoshi

[1] Hyperbolicity and branched coverings. Math. Ann. **286** (1990), 537–548

Seabury, C.

[1] On extending regular holomorphic maps from Stein manifolds. Pacific J. Math. **65** (1976), 499–515

Shabat, B. V.

[1] *Distribution of Values of Holomorphic Mappings*. Transl. Math. Monographs, vol. 61, Amer. Math. Soc. 1985
[2] *Introduction to Complex Analysis* Part II. *Functions of Several Variables*, Transl. Math. Monograph 110, Amer. Math. Soc., 1992

Shiffman, B.

[1] On the removal of singularities of analytic sets. Michigan Math. J. **15** (1968), 111–120
[2] Extension of holomorphic maps into Hermitian manifolds. Math. Ann. **194** (1971), 249–258
[3] Extensions of positive line bundles and meromorphic maps. Invent. Math. **15** (1972), 332–347
[4] Applications of geometric measure theory to value distribution theory for meromorphic maps. *Value Distribution Theory*, part A, Dekker, 1974, pp. 63–95
[5] Nevanlinna defect relations for singular divisors. Invent. Math. **31** (1975), 155–182
[6] Holomorphic curves in algebraic manifolds. Bull. Amer. Math. Soc. **83** (1977), 553–568

[7] Introduction to Carlson-Griffiths equidistribution theory. Springer Lecture Notes in Math. **981** (1983), 44–89

[8] A general second main theorem for meromorphic functions on $\mathbf{C}^n$. Amer. J. Math. **106** (1984), 509–531

Shiga, K.

[1] On holomorphic extension from the boundary. Nagoya Math. J. **42** (1971), 55–66

Sibony, N.

[1] Prolongement analytique des fonctions holomorphes bornées. C. R. Acad. Sci. Paris **275** (1972), 973–976

[2] Fibrés holomorphes et métrique de Carathéodory. C. R. Acad. Sci. Paris **279** (1974), 261–264

[3] Prolongement des fonctions holomorphes bornées et métrique de Carathéodory. Invent. Math. **29** (1975), 205–230

[4] A class of hyperbolic manifolds. Ann. Math. Studies No. **100** (1981), 357–352

[5] Une classe de domaines pseudoconvexes. Duke Math. J. **55** (1987), 299–319

[6] Some aspects of weakly pseudoconvex domains. Proc. Symp. Pure Math. **52** (1991), 199–233

Simha, R. R.

[1] Holomorphic maps into compact complex spaces. Arch. Math. **39** (1982), 262–263

Siu, Y-T.

[1] All plane domains are Banach-Stein. Manuscripta Math. **14** (1974), 101–105

[2] Extension of meromorphic maps into Kähler manifolds. Ann. of Math. **102** (1975), 421–462

[3] Every Stein subvariety admits a Stein neighborhood. Invent. Math. **38** (1976), 89–100

[4] Defect relations for holomorphic maps between spaces of different dimensions. Duke Math. J. **55** (1987), 213–251

[5] Nonequidimensional value distribution theory and subvariety extension. *Complex Analysis and Algebraic Geometry* (Göttingen 1985), Lecture Notes Math. 1194, 158–174, Springer, 1986

[6] Defect relations for holomorphic maps between spaces of different dimensions. Duke Math. J. **55** (1987), 213–251

[7] Problems and recent results in several complex variables. *Complex Analysis and Its Applications*, (Hong Kong, 1993), 38–49, Pittman Res. Notes Math. Ser. 305, 1994

[8] Hyperbolicity problems in function theory. *Five Decades as a Mathematician and Educator*, in honor of Y-C. Wong, 409–513, World Sci. Publ. Co., 1995

Siu, Y-T., and S-K. Yeung

[1] A generalized Bloch's theorem and hyperbolicity of the complement of an ample divisors in an Abelian variety. Math. Ann. **306** (1996), 743–758

[2] Hyperbolicity of the complement of a generic smooth curve of high degree in the complex projective plane. Invent. Math. **124** (1996), 573–618

Skoda, H.

[1] Fibrés holomorphes à base et à fibre de Stein. Invent. Math. **43** (1977), 97–107

Smyth, B.

[1] Weakly ample Kähler manifolds and Euler number. Math. Ann. **224** (1976), 269–279

Sommer, F., and J. Mehring

[1] Kernfunktion und Hüllenbildung in der Funktionentheorie mehrerer Veränderlichen. Math. Ann. **131** (1956), 1–16

Sommese, A. J.

[1] Submanifolds of Abelian varieties. Math. Ann. **233** (1978), 229–256

Stanton, C. M.

[1] A characterization of the polydisc. Math. Ann. **253** (1980), 129–135
[2] A characterization of the ball by its intrinsic metrics. Math. Ann. **264** (1983), 271–275

Stehlé, J-L.

[1] Fonctions plurisousharmoniques et convexité holomorphe de certains fibrés analytiques. C. R. Acad. Sci. Paris **279** (1974), 235–238; also Séminaire P. Lelong, Année 1973–74, pp. 155–179

Steinmetz, N.

[1] Eine Verallgemeinerung des zweiten Nevanlinnaschen Hauptsatzes. J. reine angew. Math. **368** (1986), 134–141

Stoll, W.

[1] Die beiden Hauptsätze der Wertverteilungstheorie bei Funktionen mehrerer komplexer Veränlichen, I. Acta Math. **90** (1953), 1–115; II, ibid. **92** (1954), 55–169
[2] Normal families of non-negative divisors. Math. Z. **84** (1964), 154–218
[3] Aspects of value distribution theory in several complex variables. Bull. Amer. Math. Soc. **83** (1977), 166–183
[4] *Value Distribution on Parabolic Spaces.* Lecture Notes in Math. 600 (1977), Springer-Verlag

Stolzenberg, G.

[1] *Volumes, Limits, and Extensions of Analytic Varieties.* Lecture Notes in Math. vol. 19, Springer-Verlag, 1976

Sunada, T.

[1] Holomorphic mappings into a compact quotient of symmetric bounded domains. Nagoya Math. J. **64** (1976), 159–175
[2] Rigidity of certain harmonic mappings. Invent. Math. **51** (1979), 297–307

Suzuki, Makoto

[1] Moduli spaces of holomorphic mappings into hyperbolically imbedded complex spaces and hyperbolic fibre spaces. Intern'l Symp. *Holomorphic Mappings, Diophantine Geometry and Related Topics*, RIMS, Kyoto (1992), pp. 157–166
[2] Moduli spaces of holomorphic mappings into hyperbolically imbedded complex spaces and hyperbolic fibre spaces. J. Math. Soc. Japan **46** (1994), 681–698
[3] The modell property of hyperbolic fiber spaces with noncompact fibers. Tohoku Math. J. (to appear)

Suzuki, Masaaki

[1] The holomorphic curvature of intrinsic metrics. Math. Rep. Toyama Univ. **4** (1981), 107–114
[2] The intrinsic metrics on the domains in $\mathbf{C}^n$. Math. Rep. Toyama Univ. **6** (1983), 147–177
[3] The intrinsic metrics on the circular domains in $\mathbf{C}^n$. Pacific J. Math. **112** (1984), 249–256
[4] The generalized Schwarz Lemma for the Bergman metric. Pacific J. Math. **117** (1985), 429–442
[5] Complex geodesics on convex domains. Math. Rep. Toyama Univ. **9** (986), 137–147

Tanigawa, H.

[1] Rigidity and boundary behavior of holomorphic mappings. J. Math. Soc. Japan **44** (1992), 131–143

[2] Holomorphic mappings into Teichmüller spaces. Proc. Amer. Math. Soc. **117** (1993), 71–78

Tishabaev, D. K.

[1] Invariant metrics and indicatrices of bounded domains in $\mathbf{C}^n$. Siberian Math. J. **30** (1989), 166–168

Toda, N.

[1] On the functional equation $\sum_{i=0}^{p} a_i f_i^{n_i} = 1$. Tohoku Math. J. **23** (1971), 289–299

Tsai, I-H.

[1] Dominating the varieties of general type. J. Reine Angew. Math. **438** (1997), 197–219

Tsuji, H.

[1] A generalization of Schwarz lemma. Math. Ann. **256** (1981), 387–390

Tsushima, R.

[1] Rational maps to varieties of hyperbolic type. Proc. Japan Acad. Sci. **55** (1979), 95–100

Ueno, K.

[1] *Classification Theory of Algebraic Varieties and Compact Complex Spaces*. Lecture Notes in Math. 439, Springer, 1975

Urata, T.

[1] On meromorphic mappings into taut complex analytic spaces. Nagoya Math. J. **50** (1973), 49–65

[2] Holomorphic mappings into taut complex analytic spaces. Tohoku Math. J. **31** (1979), 349–353

[3] Holomorphic mappings onto a certain compact complex analytic space. Tohoku Math. J. **33** (1981), 573–585

[4] Holomorphic automorphisms and cancellation theorems. Nagoya Math. J. **81** (1981), 91–103

[5] The hyperbolicity of complex analytic spaces. Bull. Aichi Univ. Educ. **31** (Natural Sci.) (1982), 65–75

Vâjâitu, V.

[1] Fiber bundles with hyperconvex fiber and base are hyperconvex. Preprint

Valiron, G.

[1] Sur la dérivée des fonctions algébroïdes. Bull. Soc. Math. France **59** (1931), 17–39

Venturini, S.

[1] Eisenman's measure on complex manifolds. Boll. Un. Mat. Ital. **1** (1987), 49–57

[2] Comparison between the Kobayashi and Carathéodory distances on strongly pseudo-convex bounded domains in $\mathbf{C}^n$. Proc. Amer. Math. Soc. **107** (1989), 725–730

[3] Pseudodistances and pseudometrics on real and complex manifolds. Ann. Mat. Pura Appl. **154** (1989), 385–402

[4] On holomorphic isometries for the Kobayashi and Carathéodory distances on complex manifolds. Atti. Accad. Nz. Lincei Rend. Cl. Sci. Fis. Mat. Natur. **83** (1989), 139–145

[5] Intrinsic metrics in complete circular domains. Math. Ann. **288** (1990), 473–481

[6] The Kobayashi metric on complex spaces. Math. Ann. **305** (1996), 25–44

Vesentini, E.

[1] Variations on a theme of Carathéodory. Ann. Scuola Norm. Sup. Pisa **6** (1979), 39–68

[2] Complex geodesics. Comp. Math. **44** (1981), 375–394

[3] Complex geodesics and holomorphic maps. Symp. Math. **26** (1982), 211–230

[4] Invariant distances and invariant differential metrics in locally convex spaces. Proc. Banach Math. Center, **8** (1982), 493–512

[5] Su un teorema di Wolff e Denjoy. Rend. Sem. Mat. Fis. Milano **53** (1983), 17–25
[6] Fixed points of holomorphic maps. Riv. Mat. Univ. Parma **10** (1984), 33–39
[6] Iteration of holomorphic maps. Russian Math. Survey **40** (1985), 7–11
[7] Semigroups of holomorphic isometries. Adv. Math. **65** (1987), 272–306
[8] Injective hyperbolicity. Ricerche di Mat. suppl. vol. **36** (1987), 99–109

Vigué, J-P.

[1] Sur les applications holomorphes isométriques pour la distance de Carathéodory. Annali Scuola Norm. Sup. Pisa **9** (1982), 255–261
[2] La distance de Carathéodory n'est pas intérieure. Resultate Math. **6** (1983), 100–104
[3] Géodésiques complexes et points fixes d'applications holomorphes. Advances Math. **52** (1984), 241–247
[4] Caractérisation des automorphismes analytiques d'un domaine convexe borné. C. R. Acad. Sci. Paris **299** (1984), 101–104
[5] The Carathéodory distance does not define the topology. Proc. Amer. Math. Soc. **91** (1984), 223–224
[6] Points fixes d'applications holomorphes dans un domaine borné convexe de $\mathbf{C}^n$. Trans. Amer. Math. Soc. **289** (1985), 345–353
[7] Sur la caractérisation des automorphismes analytiques d'un domaine borné. Portugallae Math. **43** (1985–86), 439–453
[8] Points fixes d'une limite d'applications holomorphes. Bull. Sci. Math. **110** (1986), 411–424
[9] Sur les points fixes d'applications holomorphes. C. R. Acad. Sci. Paris **303** (1986), 927–930
[10] Une remarque sur l'hyperbolicité injective. Atti Accad. Nz. Lincei Rend. Cl. Sci. Fis. Mat. Natur. **83** (1989), 57–61
[11] Fixed points of holomorphic mappings in bounded convex domains in $\mathbf{C}^n$. Proc. Symp. Pure Math., Amer. Math. Soc. **52** (1991), Part 2, 579–582
[12] Un lemme de Schwarz pour les domaines bornés symétriques irréductibles et certains domaines bornés strictement convexes. Indiana Univ. Math. J. **40** (1991), 293–304
[13] Les métriques invariantes et la caractérisations des isomorphismes analytiques. *Topics in Complex Analysis* (Warsaw 1992), 373–382, Banach Center Publ. 31, 1995
[14] Sur les domaines hyperboliques pour la distance intégrée de Carathéodory. Ann. Inst. Fourier (Grenoble), **46** (1996), 743–753

Vitter, A.

[1] The lemma of the logarithmic derivatives in several complex variables. Duke Math. J. **44** (1977), 89–104

Vojta, P.

[1] *Diophantine Approximations and Value Distribution Theory*. Lecture Notes Math. 1239, Springer, 1987

Vormoor, N.

[1] Topologische Fortsetzung biholomorpher Funktionen auf dem Rande bei beschränkten streng-pseudokonvexen Gebieten im $\mathbf{C}^n$ mit C^∞-Rand. Math. Ann. **204** (1973), 239–261

Wang, J. T-Y.

[1] The truncated second main theorem of function fields. J. Number Theory **58** (1996), 137–159
[2] An effective Roth's theorem of function fields. Rocky Mt. J. Math. **26** (1996), 1225–1234
[3] Integral points of $\mathbf{P}^n - \{2n + 1$ hyperplanes in general position$\}$ over number fields and function fields. Trans. Amer. Math. Soc., to appear

Wavre, R.

[1] Sur la réduction des domaines par une substitution à *m* variables complexes et l'existence d'un seul point invariant. L'Enseignement Math. **25** (1926), 218–234

Webster, S. M.

[1] Biholomorphic mappings and the Bergman kernel off the diagonal. Invent. Math. **51** (1979), 155–169

Weil, A.

[1] *Variétés Abéliennes et Courbes Algébriques*. Actualités Sci. Ind., 1948, Hermann, Paris

Weng, L.

[1] From subvarities of abelian varieties to Kobayashi hyperbolic manifolds: after P. Vojta and G. Faltings. Intern'l Symp. *Holomorphic Mappings, Diophantine Geometry and Related Topics* (Kyoto 1992), RIMS Reports **819** (1993), 80–95

Weyl, H., and J. Weyl

[1] *Meromorphic Functions and Analytic Curves*. Ann. Math. Studies **12** (1943), Princeton Univ. Press

Whitney, H.

[1] *Complex Analytic Varieties*. 1973, Addison-Wesley

Winkelmann, J.

[1] The Kobayashi-pseudodistance on homogeneous manifolds. Manuscripta Math. **68** (1990), 117–134
[2] On automorphisms of complements of analytic subsets in $\mathbf{C}^n$. Math. Z. **204** (1990), 117–127
[3] Semicontinuity results for the topology of taut manifolds. Intern'l J. Math. **8** (1997), 149–168

Wolf, J. A.

[1] Fine structures of Hermitian symmetric spaces. *Symmetric Spaces*, Washington Univ., 271–357, Marcel Dekker, 1972

Wolf, J. A., and A. Korányi

[1] Generalized Cayley transformations of bounded symmetric domains. Amer. J. Math. **87** (1965), 899–939

Wong, B.

[1] On the holomorphic sectional curvature of some intrinsic metrics. Proc. Amer. Math. Soc. **65** (1977), 57–61
[2] Characterization of the unit ball in $\mathbf{C}^n$ by its automorphism group. Invent. Math. **41** (1977), 253–257

Wong, P.-M.

[1] Holomorphic mappings into Abelian varieties. Amer. J. Math. **102** (1980), 493–501
[2] On the second main theorem of Nevanlinna theory. Amer. J. Math. **111** (1989), 549–583
[3] Diophantine approximation and theory of holomorphic curves. Proc. Symp. Value Distribution Theory in Several Complex Variables, (1990, Notre Dame), 115–156, Notre Dame Lectures No. 12, 1992
[4] Recent results in hyperbolic geometry and Diophantine geometry. Intern'l Symp. *Holomorphic Mappings, Diophantine Geometry and Related Topics*, RIMS, Kyoto, 1922, 120–135

Wright, M.

[1] The Kobayashi pseudometric on algebraic manifolds of general type and in deformations of complex manifolds. Trans. Amer. Math. Soc. **232** (1977), 357–370

[2] The behavior of the infinitesimal Kobayashi pseudometric in deformations and on algebraic manifolds of general type. Proc. Symp. Math. Amer. Math. Soc. **30** (1977), 129–134

Wu, H.

[1] Normal families of holomorphic mappings. Acta Math. **119** (1967), 193–233

[2] Mappings of Riemann surfaces (Nevanlinna theory). Proc. Symp. Pure Math., vol. 11, Amer. Math. Soc. 1968, pp. 48–532

[3] An n-dimensional extension of Picard's theorem. Bull. Amer. Math. Soc. **75** (1969), 1357–1361

[4] *The Equidistribution Theory of Holomorphic Curves.* Ann. Math. Studies, no. 64, Princeton U. Press, 1970

[5] A remark on holomorphic sectional curvature. Indiana Univ. Math. J. **22** (1973), 1103–1108

[6] Some theorems on projective hyperbolicity. J. Math. Soc. Japan **33** (1981), 79–104

[7] Old and new invariant metrics on complex manifolds. *Several Complex Variables*, (Stockholm 1987/88), 640–682, Math. Notes 38, Princeton Univ. Press 1993

Xiao, G.

[1] Bound of automorphisms of surfaces of general type, I. Ann. Math. **139** (1994), 51–77; II, J. Alg. Geom. **4** (1995), 701–793

Xu, G.

[1] Subvarieties of general hypersurfaces in projective space. J. Diff. Geom. **39** (1994), 139–172

Yang, H. C., and Z. H. Chen

[1] On the Schwarz lemma for complete Hermitian manifolds. *Several Complex Variables* (Hangzhou 1981), 91–116, Birkhäuser, 1984

Yano, K., and S. Bochner

[1] *Curvature and Betti Numbers.* Annals of Math. Studies, No. 32, Princeton Univ. Press, 1953

Yau, S. T.

[1] On the curvature of compact Hermitian manifolds. Invent. Math. **25** (1974), 213–239

[2] Intrinsic measures of compact complex manifolds. Math. Ann. **212** (1975), 317–329

[3] A general Schwarz lemma for Kähler manifolds. Amer. J. Math. **100** (1978), 197–203

[4] On the Ricci curvature of a compact Kähler manifold and the complex Monge-Ampère equation, I. Comm. Pure & Appl. Math. **31** (1978), 339–411

Yu, J. Y.

[1] Weighted boundary limits of the generalized Kobayashi-Royden metrics on weakly pseudoconvex domains. Trans. Amer. Math. Soc. **347** (1995), 587–614

[2] Singular Kobayashi metrics and finite type conditions. Proc. Amer. Math. Soc. **123** (1995), 121–130

Yu, Q. H.

[1] Holomorphic maps and conformal transformations on hermitian manifolds. *Several Complex Variables* (Hangzhou 1981), 117–126, Birkhäuser, 1984

Zaidenberg, M. G.

[1] Theorems of Brody and Green on conditions for complete hyperbolicity of complex manifolds. Funct. Analysis Appl. **14** (1980), 311–313

[2] Picard's theorem and hyperbolicity. Siberian Math. J. **24** (1983), 858–867

[3] Deformations and degenerations of complex hyperbolic pseudometrics and pseudovolumes. Func. Analysis and its Appl. **19** (1985), 215–216

[4] On hyperbolic embedding of complements of divisors and the limiting behavior of the Kobayashi-Royden metric. Math. USSR Sbornik **55** (1986), 55–70

[5] Criteria for hyperbolic embedding of complements to hypersurfaces. Russian Math. Surveys **41** (1986), 249–250

[6] The complement of a generic hypersurface of degree $2n$ in $\mathbf{CP}^n$ is not hyperbolic. Siberian Math. J. **28** (1987), 425–432

[7] Stability of hyperbolic embeddedness and construction of examples. Math. USSR Sbornik **63** (1989), 251–261

[8] Transcendental solutions of diophantine equations. Func. Analysis and its Appl. **22** (1988), 322–324

[9] A function-field analog of the Mordell conjecture: A noncompact version. Math. USSR Izvestiya **35** (1990), 61–81

[10] Ramanujam surfaces and exotic algebraic structures in $\mathbf{C}^n$. Soviet Math. Doklady **42** (1991), 636–640

[11] Shottky-Laundau growth estimates for s-normal families of holomorphic mappings. Math. Ann. **293** (1992), 123–141

[12] Hyperbolicity in projective spaces. Proc. Intern'l Conf. on Hyperbolic and Diophantine Analysis, RIMS. Kyoto, 1992, 136–156

Zaidenberg, M. G., and V. Ja. Lin

[1] On bounded domains of holomorphy that are not holomorphically contractible. Soviet Math. Dokl. **20** (1979), 1262–1266

Zalcman, L.

[1] A heuristic principle in complex function theory. Amer. Math. Monthly **82** (1975), 813–817

Zarkhin, Yu. G., and A. N. Parshin

[1] Finiteness problems in Diophantine geometry. Amer. Math. Soc. Transl. (2) Vol. **143** (1989), 35–102

Zhang, J. H.

[1] Chern-Levine-Nirenberg norms on homology class. Dongbei Shuxue **3** (1987), 429–438

[2] Metric S on holomorphy domain. Kexue Tongbae **33** (1988), 353–356

Zwonek, W.

[1] A note on Carathéodory isometries. Arch. Math. **60** (1993), 167–176

[2] Effective formulas for complex geodesics in generalized pseudoellipsoids with applications. Ann. Polon. Math. **61** (1995), 261–294

[3] On an example concerning the Kobayashi pseudo-distance. Proc. Amer. Math. Soc., to appear

Index

Printing: Mercedesdruck, Berlin
Binding: Buchbinderei Lüderitz & Bauer, Berlin